Rivers in the Landscape

Rivers in the Landscape

Ellen Wohl
Colorado State University
Fort Collins, CO, USA

Second Edition

This second edition first published 2020

Edition History
Wiley-Blackwell (1e, 2014)

Registered Office(s)
John Wiley & Sons, Inc., 111 River Street, Hoboken, NJ 07030, USA
John Wiley & Sons Ltd, The Atrium, Southern Gate, Chichester, West Sussex, PO19 8SQ, UK

Editorial Office
9600 Garsington Road, Oxford, OX4 2DQ, UK

For details of our global editorial offices, customer services, and more information about Wiley products visit us at www.wiley.com.

Wiley also publishes its books in a variety of electronic formats and by print-on-demand. Some content that appears in standard print versions of this book may not be available in other formats.

Library of Congress Cataloging-in-Publication Data

Names: Wohl, Ellen E., 1962- author.
Title: Rivers in the landscape / Ellen Wohl.
Description: Second edition. | Hoboken, NJ : John Wiley & Sons, Inc., 2020. | Includes bibliographical references and index.
Identifiers: LCCN 2019032034 (print) | LCCN 2019032035 (ebook) | ISBN 9781119535416 (paperback) | ISBN 9781119535478 (adobe pdf) | ISBN 9781119535430 (epub)
Subjects: LCSH: Rivers–Research. | Fluvial geomorphology–Research.
Classification: LCC GB1201.7 .W65 2020 (print) | LCC GB1201.7 (ebook) | DDC 551.48/3–dc23
LC record available at https://lccn.loc.gov/2019032034
LC ebook record available at https://lccn.loc.gov/2019032035

Cover Design: Wiley
Cover Image: Courtesy of Ellen Wohl

Set in 10/12pt WarnockPro by SPi Global, Chennai, India

Contents

Acknowledgements

The contents of this book reflect the contributions of many individuals. Mike Church helped to explain the evolution of thinking about bedload transport. Bill Dietrich identified pioneering discussions of bedload dynamics in G.K. Gilbert's flume experiments. Angela Gurnell helped me to more fully integrate plants into my thinking about river process and form. Theodore Endreny and Grant Meyer helped me to identify and correct errors in the first edition of the book. Natalie Kramer, Katherine Lininger, Lina Polvi, and Dan Scott provided stimulating discussions about various aspects of river science. I thank each of them and many other colleagues who have made the study of rivers such an enjoyable challenge.

1

Introduction

Rivers are the shapers of terrestrial landscapes. Very few points on Earth above sea level do not lie within a drainage basin. Even points distant from the nearest channel are likely to be influenced by that channel. Tectonic uplift raises rock thousands of meters above sea level. Precipitation falling on the uplifted terrain concentrates into channels that carry sediment downward to the oceans and influence the steepness of adjacent hill slopes by governing the rate at which the landscape incises. Rivers migrate laterally across lowlands, creating a complex topography of terraces, floodplain wetlands, and channels. Subtle differences in elevation, grain size, and soil moisture across this topography control the movement of ground water and the distribution of plants and animals.

Investigators have begun to quantify the extent to which rivers influence the surrounding landscape. Stream ecologists ask, "How wide is a stream?" and address the question by using isotopic signatures to analyze food web data indicating exchanges of matter and energy between aquatic and terrestrial biotic communities (Muehlbauer et al. 2014). Geomorphologists ask, "How large is a river?" and address the question by defining signatures – emergent properties of sets of processes acting on a river landscape – and envelopes – the dynamic penetration of a signature across the landscape (Gurnell et al. 2016b). In each case, the answer is, "Wider and larger than surface appearances might suggest."

Throughout human history, people have settled disproportionately along rivers, relying on them for water supply, transport, fertile agricultural soils, waste disposal, and food from aquatic and riparian organisms. People have also devoted a tremendous amount of time and energy to altering river process and form. We are not unique in this respect: ecologists refer to various organisms, from beaver to some species of riparian trees, as *ecosystem engineers* in recognition of their ability to alter their environment. Humans are unique, however, in the extent to and intensity with which we alter rivers. In many cases, river engineering has unintended consequences, and effectively mitigating these consequences requires that we understand rivers in the broadest sense, as shapers and integrators of landscape.

Geomorphologist Luna Leopold once described rivers as the gutters down which flow the ruins of continents (Leopold et al. 1964). His father, Aldo Leopold, described the functioning of an ecosystem as a "round river," to emphasize the cycling of nutrients and energy. Rivers can be thought of as having a strong unidirectional and linear movement of water, sediment, and other materials. Rivers can also be thought of as more broadly connected systems with bidirectional fluxes of energy and matter between the channels of the river network and the greater environment. This volume emphasizes the latter viewpoint.

Rivers in the Landscape, Second Edition. Ellen Wohl.

Rivers are not simply channels. Various phrases have been used to describe the integrated system of channels, floodplain, and underlying hyporheic zone, including "the river system," "the fluvial system," "the river ecosystem," and "the river corridor." Regardless of the exact words used, the intent is to recognize that the active channel is integrally connected to adjacent surface and subsurface areas by fluxes of material and organisms. The three legs of the tripod of physical inputs that support a river corridor are inputs of water, sediment, and large wood from adjacent uplands. Although large wood has received less attention than water and sediment inputs, the historical abundance of large wood in regions with forested uplands or floodplains, along with observations of the geomorphic effects of large wood in the few remaining natural river corridors, indicates that large wood significantly influences river process and form. The material inputs of water, sediment, and wood are redistributed within the river corridor, stored for varying lengths of time, and eventually transported to the ocean, to another long-term depositional environment (e.g. alluvial fan or delta), or – for water – back to the atmosphere or ground water.

Each of the primary inputs to a river corridor can be described in terms of natural regimes that occur in the absence of human alterations in land cover, river form, flow regulation, and the water table, and in terms of altered regimes associated with human activities. The *natural flow regime* can be characterized with respect to magnitude, frequency, duration, timing, and rate of rise and fall of water discharge (Poff et al. 1997). Human alterations of the flow regime can be quantified using indicators of hydrologic alteration (Richter et al. 1996; Poff et al. 2010). The *natural sediment regime* can be characterized with respect to inputs, outputs, and storage of sediment (Wohl et al. 2015b). Because records of sediment flux analogous to those of gaged stream discharge do not exist, human alterations of the sediment regime can be inferred from the occurrence of sustained changes in river process and form that result from altered sediment dynamics. The *natural wood regime* can be characterized with respect to magnitude, frequency, duration, timing, rate, and mode of wood recruitment, transport, and storage within river corridors (Wohl et al., 2019). As with sediment, insufficient systematic records exist of wood flux in the absence of human influences to quantify changes in the natural wood regime, but the effect of human influences can be inferred from sustained changes in river process and form (e.g. Collins et al. 2012).

The details of how materials from uplands enter a river corridor and move through it are partly governed by the spatial context of the corridor (Figure 1.1). *Context* here includes valley geometry (downstream gradient, valley-bottom width relative to active channel width), position in the network, base-level stability, and substrate erosional resistance (Wohl 2018a). Valley geometry influences the energy available for changes in river form and the space available to accommodate change. Steep river reaches typically correspond to relatively narrow valleys and coarser sediment or bedrock (Livers and Wohl 2015). Lower-gradient reaches are more likely to have wide valley bottoms relative to channel width, as well as floodplains or secondary channels. Position in the network can influence the sensitivity of a river corridor to fluctuations in relative base level: commonly, the lower portions of a river network are more likely to incise in response to relative base-level fall or aggrade in response to relative base-level rise. Base-level stability influences river corridor configuration in that a river reach may be incising or aggrading irrespective of inputs of water, sediment, and large wood because of base-level instability (e.g. Schumm 1993). Substrate erosional resistance describes the ability of the channel and floodplain substrate to resist erosional changes. Resistance derives from substrate composition (grain size, stratigraphy, bedrock lithology; e.g. Finnegan et al. 2005) and from the presence of riparian vegetation (e.g. Gurnell 2014).

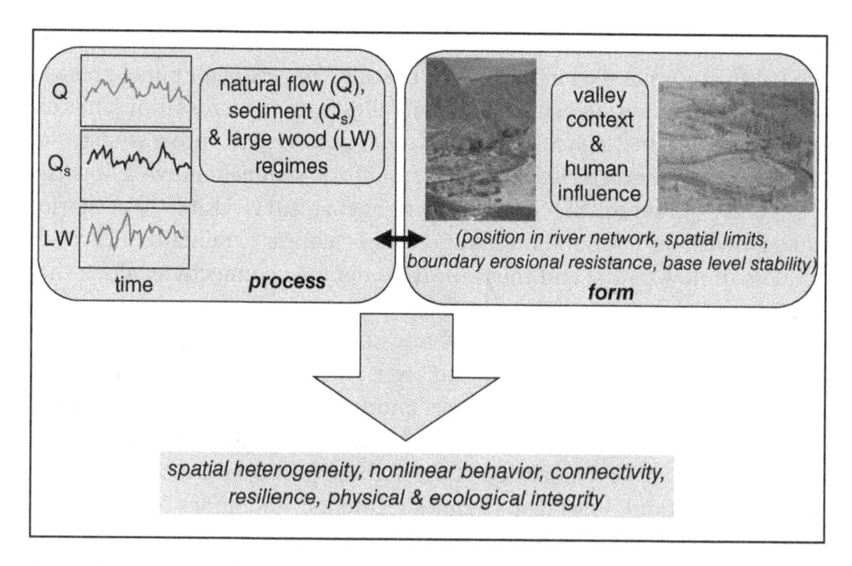

Figure 1.1 Schematic illustration of the primary inputs to river corridors (water, sediment, large wood) and the context in which they interact with one another and with the river form to create the integrative river corridor characteristics listed in the lower portion of the figure. (*See color plate section for color representation of this figure*).

Human activities can modify inputs and context. Although people typically do not alter the actual valley geometry, they do commonly alter the effective valley geometry by building levees, regulating flow and reducing flood peaks, or stabilizing the banks, each of which limits the interactions between channel and floodplain. Analogously, construction of grade controls or dams affects local base-level stability, and land drainage or bank stabilization modifies substrate erosional resistance.

Interactions between inputs and valley context create the characteristics of the river corridor listed in the lower row of Figure 1.1: spatial heterogeneity, nonlinear behavior, connectivity, resilience, and integrity. Connectivity and nonlinear behavior are introduced in this first chapter. The other concepts are covered in subsequent ones.

1.1 Connectivity and Inequality

Contemporary research and conceptual models of river form and process increasingly explicitly recognize the important of connectivity. Connectivity, sometimes referred to as coupling (e.g. Brunsden and Thornes 1979), is multifaceted. *Hydrologic connectivity* can refer to the movement of water down a hillslope in the surface or subsurface, from hillslopes into channels, or along a channel network (Pringle 2001; Bracken and Croke 2007). *River connectivity* refers to water-mediated fluxes within the channel network (Ward 1997). *Sediment connectivity* can refer to the movement, or storage, of sediment down hillslopes, into channels, or along channel networks (Harvey 1997; Fryirs et al. 2007a,b; Kuo and Brierley 2013; Bracken et al. 2015). *Biological connectivity* refers to the ability of organisms or plant propagules to disperse between suitable habitats or between isolated populations for breeding. *Landscape connectivity* can refer to the movement of water, sediment, or other materials between individual landforms such as hillslopes and channels (Brierley et al. 2006). *Structural connectivity* describes the extent to which landscape units – which can range in scale from <1 m for bunchgrasses dispersed across exposed soil to the configuration of hillslopes and valley bottoms across thousands of meters – are physically linked to one another. *Functional connectivity* describes the process-specific interactions between multiple structural characteristics, such as runoff and sediment moving downslope between the bunchgrasses and exposed soil patches (Wainwright et al. 2011). Using the scenario of runoff and sediment moving downslope, temporal variability (connectedness of rainfall) can create spatial variability (connectedness of flow paths) and thus control functional connectivity along the slope (Wainwright et al. 2011).

In general, connectivity describes the efficiency of material transfer between geomorphic system components such as floodplains and channels, hillslopes and river corridors, or longitudinal segments within a river network (Wohl et al. 2019a). Landscapes and individual landforms such as a delta are increasingly conceptualized as networks using the framework of graph and network theory (e.g. Kupfer et al. 2014; Heckmann et al. 2015; Passalacqua 2017). These networks are composed of compartments (e.g. hillslope, valley bottom), links (e.g. channel segments), and nodes (e.g. channel junctions), each of which exhibits connectivity at differing temporal and spatial scales.

Whatever form of connectivity is under discussion, its magnitude, duration, and extent are each important. Magnitude can be thought of as the volume of flux: Is only a trickle of water moving down a channel network, or a flood? Duration describes the time span of the connectivity: Can fish disperse along a river network throughout an average flow year, or only during certain seasons of high flow? Closely associated with duration is the idea of storage. If sediment stops moving downstream during periods of lower discharge, then it is at least temporarily stored in the streambed and banks. Large wood can be stored on a floodplain until overbank flows or bank erosion transport it back into the active channel. Extent is the spatial characteristic of connectivity: Does sediment move readily from the crest to the toe of a hillslope, but not into the adjacent channel because it is trapped and stored in alluvial fans perched on stream terraces? Research focuses on quantifying connectivity or developing indices of connectivity using tools such as high-resolution digital terrain models derived from aerial LiDAR (Cavalli et al. 2013) or direct measurements of fluxes (Jaeger and Olden 2012).

These dimensions of connectivity are important for adequately characterizing fluxes within a landscape, and for understanding how human activities alter those fluxes (Kondolf et al. 2006). Many human actions substantially reduce connectivity within a river network. Dams alter hydrologic connectivity and may effectively interrupt or eliminate connectivity of sediment and some organisms along a river (Magilligan et al. 2016). Levees and bank stabilization interrupt or prevent connectivity between the channel and the adjacent floodplain (Florsheim and Dettinger 2015). Flow diversions, in contrast, may increase connectivity between drainage networks, allowing exotic organisms to migrate with the diverted water and colonize a river network (Zhan et al. 2015). Dredging, channelization, straightening, and other activities that reduce geomorphic complexity and the storage of fine sediment and nutrients typically increase the longitudinal connectivity of rivers and associated downstream fluxes of sediment and solutes. By limiting overbank flows, however, these alterations reduce lateral connectivity between the channel and floodplain. Effective mitigation of undesirable human alterations of rivers requires understanding the details of connectivity.

Connectivity implies an inverse characteristic of disconnectivity. Disconnectivity can result from features that limit movement of material, typically by creating obstructions such as beaver dams (Burchsted et al. 2010) or by enhancing storage such as floodplains storing water during overbank flow (Lininger and Latrubesse 2016) or sediment (Wohl 2015a,b). Disconnectivity can also result from lack of sufficient energy or discharge to transport material in a temporally (Jaeger and Olden 2012) or spatially (Mould and Fryirs 2017) continuous manner.

Although connectivity is commonly regarded as a desirable characteristic, naturally occurring disconnectivity can be critically important. Natural disconnectivity can attenuate peak flows (Lane 2017), for example. It can also enhance retention of sediment and particulate and dissolved nutrients. This retention facilitates biological processing of these nutrients and improves water quality (Battin et al. 2008), as well as enhancing habitat abundance and diversity (Venarsky et al. 2018). A wide variety of metrics have been proposed to quantify the degree of (dis)connectivity for diverse materials (Table 1.1) (Wohl 2017b).

Table 1.1 Selected examples of quantitative metrics of connectivity.

Description	Metric	References
	Primarily hydrologic metrics	
Integral connectivity scale lengths (ICSLs)	Average distance over which wet locations are connected using either Euclidean distances or topographically defined hydrologic distances; 1 of 15 indices of hillslope hydrologic connectivity in Bracken et al. (2013: Table 4)	Western et al. (2001)
Attenuated imperviousness (I) $I = \left(\frac{\sum_j (A_j W_j)}{A_c}\right)$	Weighted impervious area as a percentage of catchment area; *Aj* is the area of the *j*th impervious surface; *Wj* is the weighting applied to *Aj*; *Ac* is catchment area	Walsh and Kunapo (2009)
River connectivity index (RCI) $DCI_p = \sum_{i=1}^{n} \frac{v_i^2}{V^2} * 100$	The size of disconnected river fragments between dams in relation to the total size of the original river network, based on Cote et al.'s (2009) directional connectivity index (DCI) model; size can be described in terms of volume (example at left), length, or other variables	Grill et al. (2014)
	Primarily sediment metrics	
Sediment delivery ratio (SDR) $SDR = \frac{net\ erosion}{total\ erosion}$	Measure of sediment connectivity	Brierley et al. (2006)
Connectivity index (IC) $IC = \log_{10}\left(\frac{D_{up}}{D_{dn}}\right)$ $D_{up} = \overline{WS}\sqrt{A}$ $D_{dn} = \sum_i \frac{d_i}{W_i S_i}$ $W = 1 - \left(\frac{RI}{RI_{MAX}}\right)$ Roughness index (RI) $RI = \sqrt{\frac{\sum_{i=1}^{25}(x_i - x_m)^2}{25}}$	D_{up} and D_{dn} are the upslope and downslope components of connectivity, respectively, with connectivity increasing as IC increases; $\overline{W}$ is the average weighting factor of the upslope contributing area, $\overline{S}$ is the average slope gradient of the upslope contributing area; A is the upslope contributing area; d_i is the length of the flow path along the ith cell according to the steepest downslope direction; W_i and S_i are the weighting factor and the slope gradient of the ith cell, respectively; RI_{MAX} is the maximum value of RI in the study area; 25 is the number of processing cells within a 5 × 5 moving window; x_i is the value of one specific cell of the residual topography within the moving window; x_m is the mean of the 25 cell values	Cavalli et al. (2013)

(Continued)

Table 1.1 (Continued)

Description	Metric	References
Complexity index based on overall relief Dh_{max} $Dh_{max} = E_{max} - E_{min}$ and slope variability SV $SV = S_{max} - S_{min}$	E_{max} and E_{min} are the maximum and minimum elevations, respectively, in the catchment; S_{max} and S_{min} are the maximum and minimum % slope, respectively, within the area of analysis (moving window)	Baartman et al. (2013)
Cluster persistence index (CPI) $CPI_i = \int_{\substack{over\ all \\ times\ t}} M_j^{(i)}(t)dt$	Defines clusters within a river network where mass (sediment) coalesces into a connected extent of the network; the superscript (i) denotes all clusters $M_j^{(i)}$ that occupy link i at time t	Czuba and Foufoula-Georgiou (2015)
	Metrics for diverse fluxes	
$C(t) = \sum_{i=1}^{m(t)} \sum_{j=1}^{n_i(t)} p_{ij}(t)S_{ij}(t)$	Patch connectivity, along with line, vertex, and network connectivity, can be used to characterize landscape connectivity; patch connectivity is the average movement efficiency between patches; C is patch connectivity; $p_{ij}(t)$ is the area proportion of the jth patch in the ith land cover type to the total area under investigation at time t; S is movement efficiency; $0 \leq C(t) \leq 1.1$	Yue et al. (2004)
Directional connectivity index (DCI) $DCI = \frac{\sum_{i=1}^{v} \sum_{j=r+1}^{R} w_{ij} \frac{dx(j-r)}{d_{ij}}}{\sum_{i=1}^{v} \sum_{j=r+1}^{R} w_{ij}}$	i is a node index; j is a row index; r is the row containing the node i; R is the total number of rows in the direction of interest; dx is the relative pixel length along that direction; d_{ij} is the shortest connected structural or functional distance between node i and any node in row j; w_{ij} is a weighting function	Larsen et al. (2012)
Adjacency matrix	Applies a connectivity analysis to a delta by identifying a set of objects (e.g. locations or variables) arranged in a network such that objects are nodes and connections or physical dependencies are links; connections between nodes can be evaluated using the mathematical technique of an adjacency matrix, which captures whether two nodes are connected, as well as link directionality and the strength of the connection	Newman et al. (2006); Heckmann et al. (2015); Passalacqua (2017)

Source: After Wohl (2017a,b,c), Table 2.

Inextricable from connectivity is the idea of reservoirs, sinks, or storage: components of a river channel, river network, or other landscape feature in which connectivity is at least temporarily limited. Being able to quantify the magnitude and average storage time of material in flux is critical to understanding connectivity, as is being able to predict the thresholds that define the upper and lower limits of storage. Sediment moving downslope from a weathered bedrock outcrop toward a stream channel might remain in storage on a debris-flow fan for 2000 years before reaching the stream channel, for example, so that the fan limits connectivity between the slope and channel at time spans of 10^0–10^3 years (Fryirs et al. 2007a,b). Or, the sediment might progressively accumulate on the hillslope until a precipitation or seismic trigger causes the slope to cross a threshold of stability and fail in a mass movement that instantaneously introduces much of the sediment into the stream. Or, the sediment might move quickly downslope and into the channel as soon as it is physically detached from the bedrock outcrop, because the slope angle is too high for sediment storage.

Focusing on coarse sediment transport in streams, Hooke (2003) distinguishes five scenarios. (i) Unconnected channel reaches have local sinks for sediment and lack of transport between reaches. (ii) Partially connected reaches have sediment transfer only during large floods. (iii) Connected reaches have coarse sediment transfer during frequent floods. (iv) Potentially connected reaches are competent to transfer sediment but lack a sediment supply. (v) Disconnected reaches were formerly connected but are now obstructed by a feature such as a dam. Differentiating these scenarios facilitates recognition that most natural and engineered river systems have some degree of retention of water, sediment, solutes, and organisms, and understanding net and long-term fluxes of these quantities involves quantifying both movement and storage.

Connectivity, storage, and fluxes are thus a central component of river process and form. Connectivity does not imply that all aspects of a connected valley segment, river network, or landscape are of equal importance to fluxes of matter and energy. Biogeochemists coined the phrases "hot moment" and "hot spot." A *hot moment* describes a short period of time with disproportionately high reaction rates relative to longer intervening time periods. A *hot spot* describes a small area with disproportionately high reaction rates relative to the surroundings (McClain et al. 2003). A channel-spanning logjam provides an example of a river hot spot (Figure 1.2). The logjam can effectively trap finer sediment and organic matter that might otherwise remain in transport. By storing organic matter for even a few hours, the logjam facilitates access for microbes and macroinvertebrates, which can ingest the organic matter (Bisson et al. 1987; Beckman and Wohl 2014a; Livers et al. 2018). The logjam also creates pressure gradients that facilitate downwelling of water and solutes into the streambed, where subsurface microbial communities enhance processes such as uptake of nitrate (Fanelli and Lautz 2008; Hester and Doyle 2008). The logjam thus creates a biochemical hot spot along the river.

The concepts of hot moments and hot spots are useful because any aspect of river process or form reflects inequalities in time and space. Czuba and Foufoula-Georgiou (2017), for example, identify hot spots of geomorphic change at the scale of river networks. These hot spots result from sediment accumulation or high rates of bed shear stress. Approximately 50% of the suspended sediment discharged by rivers of the Western Transverse Ranges of California, USA comes from the 10% of the basin underlain by weakly consolidated bedrock (Warrick and Mertes 2009). Somewhere between 17 and 35% of the total particulate organic carbon flux to the world's oceans comes from high-standing islands in the southwest Pacific, which constitute only about 3% of Earth's landmass (Lyons et al. 2002). Along bedrock channels with large knickpoints, the great majority of channel incision occurs at the knickpoint.

Temporal inequalities in river networks illustrate hot moments. More than 75% of the long-term sediment flux from mountain rivers in Taiwan occurs in the span of <1% of the year, during typhoon-generated floods (Kao and Milliman 2008). One-third of the total amount of stream energy generated by the Tapi River of India during the monsoon season is expended on the day of the peak flood (Kale and Hire 2007).

Figure 1.2 Channel-spanning logjam in the Rocky Mountains of Colorado, USA. Where logjams are not present, the stream has cobble- to boulder-size substrate, high transport capacity, and minimal storage of fine sediment and organic matter. Each logjam, in contrast, creates a backwater of lower-velocity flow that traps fine gravel, sand, and silt, as well as small logs, branches, and pine cones and needles. In the photograph, flow is from right to left. (*See color plate section for color representation of this figure*).

These are but a few of many examples mentioned in the remainder of this volume. Because not all moments in time or spots on a landscape are of equal importance in shaping rivers, effective understanding and management of rivers requires knowledge of how, when, and where fluxes occur.

1.2 Six Degrees of Connection

Any river network or segment of a single river exists in a rich and complicated context that reflects fluxes of matter and energy between the river and the greater environment, as well as the history of these fluxes. At any given moment in time, the only fluxes that are likely to be obvious are longitudinal fluxes as water and sediment move downstream. Longitudinal fluxes, however, are only one of six degrees of connection between a river and the environment (Figure 1.3) (Wohl 2010b).

(1) The longitudinal connection is the most obvious and intuitive. Water, sediment, and solutes move downstream. Globally, rivers transport an estimated 7819 million tons of sediment to the oceans (Milliman and Syvitski 1992) and approximately 0.9 Pg (1 Pg = 10^{15} g) of carbon per year (Aufdenkampe et al. 2011). Organisms move actively up- and downstream to new habitat and passively drift downstream with the current. Both European (*Anguilla anguilla*) and American (*Anguilla rostrata*) eels migrate from rivers to the Sargasso Sea off Bermuda for spawning, covering a

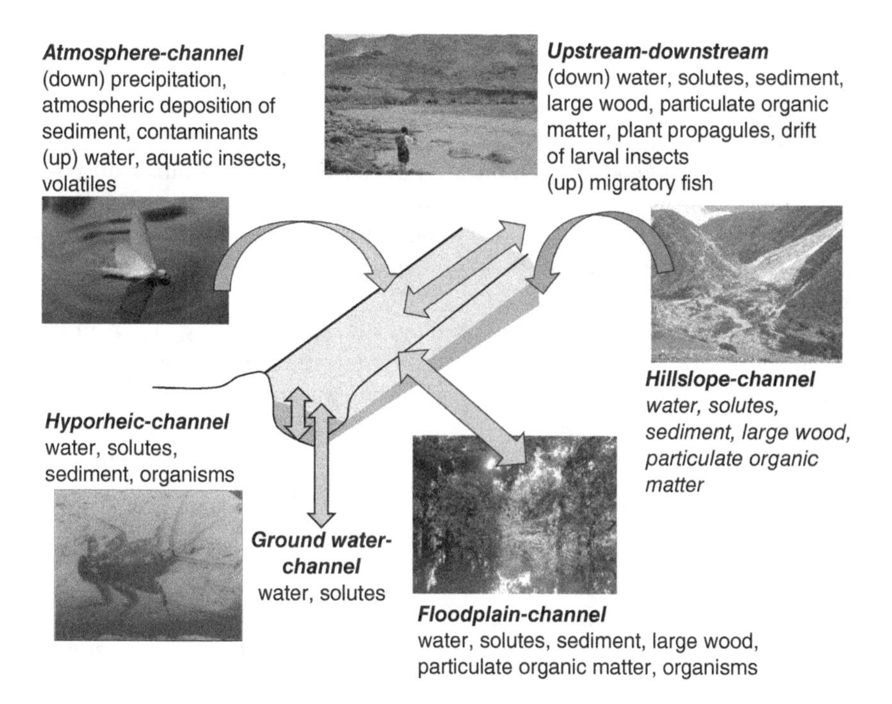

Figure 1.3 Schematic illustration of the six degrees of connection between rivers and the greater landscape. The segment of channel (lighter gray) shown here is connected to: upstream and downstream portions of the river network; adjacent uplands; the floodplain; ground water; the hyporheic zone (darker gray); and the atmosphere. The photograph representing upstream–downstream connection was taken during a flood on the Paria River, a tributary of the Colorado River that enters just downstream from Glen Canyon Dam in Arizona, USA. In this view, the Paria is turbid with suspended sediment whereas the Colorado, which is released from the base of the dam, is clear. The photograph representing hillslope–channel connection shows a large landslide entering the Dudh Khosi River in Nepal. The photograph respresenting floodplain–channel connection was taken along the Rio Jutai, a blackwater tributary of the Amazon River, during the annual flood in early June. In this view, the "flooded forest" is submerged by several meters of water. The photograph representing hyporheic–channel connection shows a larval aquatic insect (macroinvertebrate) as an example of the organisms that can move between the channel and the hyporheic environment. The photograph respresenting atmosphere–channel connection shows a mayfly emerging from the river prior to entering the atmosphere as a winged adult. Source: Image courtesy of Jeremy Monroe, Freshwaters Illustrated. *(See color plate section for color representation of this figure)*.

distance of as much as 5600 km. Numerous species of salmon (*Salmo* and *Oncorhynchus* spp.) typically travel tens to hundreds of kilometers upstream from the ocean to spawn.

(2) The lateral connection between the river channel and adjacent floodplain can operate over time spans including multiple high flows as channels migrate laterally into the floodplain via bank erosion and the floodplain migrates laterally as channel bars and islands accrete to it. The lateral connection is most obvious, however, during periods of flow with sufficient volume to overtop the banks and spread across the unchanneled valley bottom. Water, sediment, solutes, and organisms disperse from the channel onto the floodplain during the rising and peak stages of a flood, and some of these materials concentrate once more in the channel during the falling stage.
High rates of primary production by photosynthetic organisms occur during the rising limb of the flood, providing food for the consumer organisms that follow the flood pulse onto the floodplain.

High rates of decomposition occur during the flood peak, and the resulting nutrients concentrate back in the channel during the descending limb. Sediment moves onto the floodplain during the rising limb, typically remaining in storage within the floodplain until bank erosion returns it to the channel (Dunne et al. 1998). Tropical river ecologists refer to the regular annual fluxes between the channel and the floodplain as the *flood pulse*, a phrase now used to refer to fluxes during floods of any recurrence interval or magnitude sufficient to create overbank flow (Junk et al. 1989; Bayley 1991). *Flow pulses* – fluctuations in surface water below the bankfull level – create similar processes within secondary channels or areas of flow separation along a single, confined channel (Tockner et al. 2000).

Levees, channelization, and flow regulation have so restricted overbank flooding along most of the world's large and medium rivers that it is now easy to underestimate the spatial extent and duration of flooding once present along lowland rivers. The Amazon, by far the world's largest river and still one of the least engineered, can extend across 50 km of floodplain during the seasonal flood, which can last more than 3 months. Along smaller rivers, historic removal of instream large wood and, in the northern hemisphere, beavers has substantially reduced channel–floodplain connectivity (Jeffries et al. 2003; Burchsted et al. 2010).

(3) The lateral connection between adjacent uplands and the river channel is more likely to be a one-way flux, with water, sediment, and solutes moving downslope at the surface and subsurface into the channel. The pathways, rates, and magnitudes of flux from the uplands typically exhibit substantial spatial and temporal variability. During an individual rainstorm, for example, water flowing across saturated ground may become a progressively more important source of runoff as infiltration capacity declines (Dykes and Thornes 2000). During the dry season, soils in the seasonal tropics can develop water repellency, which, along with an extensive network of macropores and pipes, facilitates rapid downslope transmission of runoff early in the wet season. Water repellency declines as the wet season continues, allowing infiltration to increase and runoff to decrease. By the peak of the wet season, however, saturated soils can promote rapid, abundant surface runoff (Niedzialek and Ogden 2005). Another example of temporal variability in lateral connectivity comes from rivers fed by snowmelt, which typically exhibit an ionic pulse when the release of solutes from the snowpack and the flushing of weathering products from the soil create the highest solute concentrations in the stream water at the initiation of snowmelt (Williams and Melack 1991). Mineral sediment and organic matter coming from the uplands can originate in episodic, point sources such as landslides (Hilton et al. 2008a,b) or via more diffuse, gradual erosion.

(4) Vertical fluxes link the channel to the zone of subsurface flow immediately below the channel, with flowpaths that originate and terminate at the stream. This subsurface region is known as the *hyporheic zone*, from the Greek roots *hypo* for under or beneath and *rheo* for flow or current. Water, sediment, solutes, and small organisms such as microbes and macroinvertebrates moving between the surface and subsurface can strongly influence the volume, temperature, and chemistry of flow in the river channel, and hyporheic habitat can account for a fifth of the invertebrate production in a river ecosystem (Smock et al. 1992). The hyporheic zone can extend more than 2 km laterally from the channel in wide valleys and to depths of 10 m (Stanford and Ward 1988).

(5) Deeper vertical fluxes between the river and the saturated zone of the ground water can also occur in both directions, with water and solutes moving into the channel in a *gaining stream* or into the ground water in a *losing stream*. Human activities can create gaining and losing streams. Ground-water withdrawal that lowers the water table sufficiently to prevent ground-water

flow into the channel, for example, can substantially reduce stream flow in dryland rivers (Falke et al. 2011).

As in exchanges between the hyporheic zone and surface flow, exchanges between ground and surface water can influence the temperature and chemistry of river water. Solute concentrations typically increase toward saturation as ground water moves relatively slowly through sediment or bedrock (Constantz and Stonestrom 2003), so ground-water inputs can strongly influence river solute concentrations. The flow of rivers originating from large springs in carbonate terrains or landscapes with layered basalt flows, for example, can come almost entirely from ground water (Gannett et al. 2003).

Hydraulic conductivity, a measure of permeability and ground-water flow rate, can range over 12 orders of magnitude (Domenico and Schwartz 1998). Consequently, the travel times of ground water from areas of recharge to areas of discharge in springs or rivers can range from less than a day to more than a million years (Alley et al. 2002). This means that vertical connectivity between ground water and channels typically influences river dynamics over long time scales relative to hyporheic flow.

(6) The vertical connection between the river and the atmosphere can be obvious when precipitation falls directly on the river or an aquatic insect emerges from the river for the winged, terrestrial, adult phase of its life. Other fluxes involved in this connection are likely to be much less visible. Water evaporates into the atmosphere, especially from the oceans, and moves long distances before falling onto landscapes that drain into rivers. En route, the water vapor acquires very fine particulates. These particulates include dust, which may have traveled from a different hemisphere (Prospero 1999), and nitrates from vehicles, industrial emissions, and agricultural sources. The nitrates are deposited with rain and snow – and as particles and gases – in rivers hundreds of kilometers away (Heuer et al. 2000). Fine particulates also include highly toxic mercury released by vehicles and coal-burning power plants (Grahame and Schlesinger 2007). Volatile organic compounds – solvents such as tetrachloroethylene, chlorinated compounds such as chloroform, and others – volatilize from polluted river water into the air. Although essentially invisible, these fluxes are widespread and important.

Conceptualizing a river as having six degrees of connection with the greater environment emphasizes how diverse aspects of connectivity influence river process and form. This conceptualization also emphasizes the diversity of temporal and spatial scales across which connectivity occurs. River corridor science exemplifies explicit attention to areas outside of the active channel. In hydrology, for example, the *river corridor* – the active channel(s), floodplain, riparian zone, and hyporheic zone – is an increasingly common unit of study, gradually replacing a limited focus on the wetted channel (Harvey and Gooseff 2015). In a river corridor perspective, three-dimensional exchanges and the resulting biogeochemical processing and creation and maintenance of habitat are integral to supporting healthy levels of biomass, biodiversity, water quality, and other ecosystem services associated with rivers.

1.3 Rivers as Integrators

Thanks to the extensive and sometimes subtle fluxes between a river and the greater environment, the river's forms and processes integrate the physical, chemical, and biotic processes – contemporary and historical – within the environment. This may seem obvious when considering Figure 1.3, but it represents the most profound summation possible regarding rivers, because of the implications.

If a river integrates diverse and seemingly unrelated processes within the greater environment, for example, then attempting to manage the river or some segment of the river in isolation from those processes is absurd.

If a river integrates ... then human activities far from the physical boundaries of the channel may strongly influence the river, as when increasing atmospheric dust transport from the deserts of the southwestern United States alters snowpack melting and the resulting spring snowmelt hydrograph and water chemistry in rivers of the Rocky Mountains (Clow et al. 2002). Another example comes from the Mississippi River, where concentrations of nitrate have increased by two to five times since the early 1900s as farmers have applied increasing quantities of nitrogen fertilizers to upland crop fields across the Mississippi's huge drainage basin. The resulting flux of nitrate down the river to the Gulf of Mexico tripled during the last 30 years of the twentieth century, resulting in massive algal blooms that cover a swath of the Gulf as big as New Jersey (~20 000 km^2) each year, and in some years move out of the Gulf and up the eastern coast of the United States (Goolsby et al. 1999).

If a river integrates ... then historical resource uses of which most people are now unaware may continue to strongly influence contemporary river process and form (Macklin and Lewin 2008). Meandering gravel-bedded streams in the eastern United States are typically bordered by fine-grained deposits that were formerly interpreted as self-formed floodplains. Prior to European settlement, however, these river networks consisted of small anabranching channels within extensive vegetated wetlands. These pre-colonial valley bottoms were buried by up to 5 m of slackwater sedimentation behind tens of thousands of seventeenth- to nineteenth-century milldams (Walter and Merritts 2008). The ubiquitous fine sediments are thus fill terraces that reflect ongoing adjustment as the milldams breached and the channels incised. Another example of historical human influences comes from rivers in the Carpathian Mountains of Poland. Agriculture began in the region during the thirteenth and fourteenth centuries, and the increased sediment yield resulted in overbank aggradation along meandering rivers draining the mountains (Klimek 1987). When the proportion of crop lands that remained bare for some portion of the year increased with more widespread cultivation of potatoes during the second half of the nineteenth century, the further increases in sediment yield caused some of the meandering rivers to assume a braided planform that persists today.

If a river integrates ... then altering river process and form at one point in the river network may affect other portions of the network in unforeseen ways. The two Djerdap dams on the Danube River where it flows through Romania were built in 1970 and 1984. These massive dams, along with dozens of smaller upstream dams, have reduced sediment yields to the river's delta by 70% and silica export to the Black Sea by two-thirds relative to fluxes of these materials prior to the last third of the twentieth century. The reduced fluxes have caused erosion of the delta and a shift in the Black Sea's phytoplankton communities from siliceous diatoms to nonsiliceous coccolithophores and flagellates. These changes have stimulated algal blooms and destabilized the Black Sea ecosystem (Humborg et al. 1997). Globally, humans have increased sediment supplied to and transported by rivers as a result of soil erosion, yet reduced sediment yield to the world's oceans by 1.4 billion metric tons per year because of retention behind dams (Syvitski et al. 2005). The result of this reduced coastal sediment yield has been widespread delta and near-shore erosion (Crossland et al. 2005; Yang et al. 2011).

In summary, a river integrates fluxes across a much larger and more diverse environment than the channel itself. Consequently, understanding and effectively managing river process and form is much more challenging than is likely to be recognized if a river segment is manipulated as though it were spatially and temporally isolated.

1.4 Organization of this Volume

The title of this book, *Rivers in the Landscape,* reflects the inherent connections between a river and the landscape. Landscape is defined here as the physical, chemical, and biotic environment of the *critical zone* – Earth's outer layer, from the top of the vegetation canopy to the base of the soil and ground water, which supports life. The critical zone represents the intersection of atmosphere, water, soil, and ecosystems. Recent research increasingly reminds us of what perhaps should always have been obvious: rivers do not merely flow through a landscape in isolation, but rather interact with the landscape in complex and fascinating ways. Riverine vegetation, for example, does not just increase the hydraulic resistance of overbank flow – vegetation can alter the default river planform from braiding to meandering (Tal and Paola 2007). Rivers do not flow passively down steep topography created by tectonic uplift – removal of mass through riverine erosion can increase the upward flux of molten rock and tectonic uplift (Zeitler et al. 2001).

Recognition of the connections between rivers and landscapes implies that the topics traditionally covered in a fluvial geomorphology text – hydraulics, sediment transport, river geometry – should be treated in a manner that explicitly recognizes the influences exerted on river process and form by entities beyond the channel boundaries. Consequently, this book builds from traditional understanding of rivers toward the larger, more comprehensive viewpoint.

Chapter 2 covers the development of channels and channel networks, including how water, sediment, and solutes are produced; how they move from uplands into channels; how channel heads form; and how channel networks extend across the landscape. This chapter addresses the processes by which water moves across and through unchannelized hillslopes and concentrates sufficiently to create channels.

Chapter 3 covers channel processes, with a focus on energy (hydraulics) and quantities (hydrology). Knowledge of the basic mechanics of channelized flow is integral to understanding sediment erosion, transport and deposition, and adjustment of channel form.

Chapter 4 covers the movement of sediment in channels. The discussion begins with the sediment texture of channel beds and the processes that initiate motion of noncohesive and cohesive sediment. Once sediment is mobilized from the streambed and banks, it can be transported in solution, in suspension, or in contact with the bed, and can be organized into bedforms.

Chapter 5 discusses the movement and storage of large wood in river corridors. Starting with how wood is mobilized, transported, and deposited, the discussion explores how it influences river process and form, and the effects on rivers of human alterations of wood dynamics.

Chapter 6 addresses channel form, exploring how movement and storage of water, sediment, and large wood shape channel geometry through time and space. Interactions between process and form are implicit throughout Chapters 3–5, but Chapter 6 explicitly examines feedbacks between process and form at increasingly larger spatial scales, from cross-sectional geometry, through channel planform and longitudinal gradient, to downstream trends along a river and across a river basin.

Chapter 7 summarizes the process and form of fluvially created and maintained features outside of the active channel – floodplains, terraces, alluvial fans, deltas, and estuaries. These river landforms both reflect and influence channel process and form.

Chapter 8 metaphorically steps back to use the knowledge of process and form developed in the preceding chapters as a means to understand rivers in a landscape context. This chapter starts with a discussion of how topography influences the spatial distribution of river networks and energy expenditure within rivers, how rivers influence rates of landscape denudation, and the indicators used to infer relations between rivers and landscape evolution. Spatial differentiation of geomorphic process and form within river basins is discussed, and connectivity is reexamined. Distinctive river characteristics associated with high and low latitudes and arid regions provide examples of the importance of landscape context.

One of the challenges in writing a reasonably concise fluvial geomorphology text is the tremendous volume of research conducted on rivers within the past century. Scientists from diverse backgrounds in geology, geography, civil engineering, and other disciplines study river process and form via:

- direct measurements and experimental manipulations of real rivers;
- indirect measurements using remote sensing imagery from space-based (e.g. aerial photographs, satellite imagery, airborne LiDAR) and ground (e.g. ground-penetrating radar) platforms;
- physical experiments in a laboratory;
- numerical models; and
- integrations of these approaches.

Another fundamental challenge is the diversity of rivers. Water flows downslope under the influence of gravity. The basic physics are the same in any environment, but the ability to generalize beyond the most basic level is typically obscured by the local, place-specific details and history of a particular river. As fluvial geomorphology continues to develop as a discipline, there remains an underlying tension within the community between investigators who emphasize quantification as a means of identifying physical principles and mechanisms acting across a range of specific landscapes (e.g. Dietrich et al. 2003) and investigators who emphasize the use of historical and sedimentary records as a means of identifying the role of contingency and site-specific characteristics in river process and form (e.g. Phillips and Van Dyke 2016).

Until perhaps the 1960s or '70s, the great majority of river research focused on medium-sized, low- to medium-gradient, sand-bed rivers. These were the most accessible rivers for scientists living primarily in the temperate latitudes, and the foundational research conducted on these rivers gave rise to widely used conceptual models and equations for hydraulics, sediment transport, and channel geometry. As investigators have subsequently spent more time quantitatively examining rivers with steeper gradients and more resistant boundaries (gravel-bed rivers, bedrock rivers, mountain rivers) and greater hydrologic variability (seasonal tropics, drylands), as well as rivers at higher (boreal, arctic) and lower (tropical) latitudes, the ability of the foundational models and equations to adequately describe process and form across the known spectrum of rivers has become weaker. Throughout this volume, I explicitly address some of the unique characteristics of rivers beyond temperate-zone sand-bed channels.

My intent in this text is to maintain conciseness while reflecting the diversity of natural rivers and the methods of studying rivers. The references cited are not an exhaustive list, but rather a starting point that combines some foundational studies and particularly integrative or insightful recent studies.

1.5 Understanding Rivers

Recent emphasis on connectivity in landscapes and river networks illustrates the importance of conceptual models and methods of inquiry in governing the questions that scientists ask. If we view rivers as complex systems with multiple interactions between different components, we are more likely to focus on the factors that control those interactions and on ways to quantify and predict them. If we view rivers as predominantly physical systems, we are more likely to neglect the interactions among hydraulics, sediment dynamics, and aquatic and riparian organisms. Even when not explicitly recognized, our conceptual models of rivers tend to constrain the questions that we consider interesting and important and the methods we use to examine them (Grant et al. 2013). Studies of sediment transport, for example, that employ an *Eulerian* framework focus on the flux of sediment within a spatially bounded area. This is a very useful approach for developing a sediment budget, but a *Lagrangian* framework in which specific objects are tracked through time can provide more insight into actual mechanisms of sediment movement (Doyle and Ensign 2009).

A conceptual model results from assumptions about how a river functions. The conceptual model can be qualitative or quantitative. A quantitative model can be more precise than a qualitative model, but is not necessarily more accurate. Drawing on the second chapter of Leopold et al.'s (1964) fluvial geomorphology text for inspiration, the remainder of this section uses a landscape with which I am very familiar to explore the different conceptual models and approaches that investigators employ to understand river segments, river networks, and the greater landscape.

1.5.1 The Colorado Front Range

Atop the Precambrian-age crystalline rocks that form the continental divide in Colorado, you can stand shivering in the cold wind even at the height of summer. Here, 4000 m above sea level, bedrock topography crests in a series of ridges and peaks that divide water flowing west to the Pacific Ocean and water flowing east to the Atlantic (Figure 1.4). In some places, the divide is a sharp-edged ridge of bedrock and periglacial boulders with talus chutes and waterfalls. In other places, small alpine streams meander across broad, gently undulating surfaces.

Sharp or broad, the heights drop precipitously down to glacial cirques and troughs. Rivers alternate between paternoster lakes and steep cascades as they flow through subalpine conifer forests. Beyond the terminal glacial moraine, each valley continues downward, alternating between steep, narrow gorges in which the river flows turbulent and aerated and relatively wide canyons with gentler gradients along which the river flows through pools and riffles. These longitudinal alternations in valley and channel geometry reflect spatial heterogeneity in joint density associated with shear zones and differential weathering of the crystalline rocks. Wide, low-gradient valley segments correspond to zones with relatively densely spaced joints, whereas more widely spaced joints correspond to gorges and waterfalls (Ehlen and Wohl 2002; Wohl 2008; Ortega et al. 2013).

Climate grows progressively warmer and drier at lower elevations, and subalpine forest gives way to more open montane forest with more frequent wildfires and associated debris flows (Veblen and Donnegan 2005). Warm, moist masses of air moving inland from the southeast during summer are forced upward as they near the Colorado Rockies, and the water vapor being transported with the air masses cools, condenses, and falls as rain. Most of this moisture is wrung from the clouds at the lower to middle elevations of the mountains, which can experience flash floods from convective storms, as

Figure 1.4 Landscapes and river corridors in and adjacent to the Colorado Front Range. Upper left: View east from the summit surfaces at the continental divide. The coarse blocks in the foreground are periglacially weathered boulders and bedrock. The surface drops steeply into a glaciated valley that transitions downstream (out of sight) into a fluvial valley. Upper right: View northwest from a hogback, an asymmetrical hill of sandstone and limestone strata dipping steeply to the right in this view, with an intervening valley formed in shales. Lower right: The South Platte River near Fort Morgan, Colorado, in the low-relief environment of the Great Plain. This sand-bed channel was historically much wider and had a braided planform, but flow regulation has resulted in encroachment of riparian vegetation and transformation to a single relatively narrow channel. This river heads high in the mountains. Lower left: View of smaller drainages that head on the Great Plains, here at Pawnee National Grassland. These channels have downcut within the past few decades, largely via piping erosion. (*See color plate section for color representation of this figure*).

well as the late-spring snowmelt floods that flow down from the highest portions of the river network each year.

At the base of the mountains on the eastern side, the rivers gradually change from boulder- to cobble-bed channels as they flow through a series of steeply tilted sedimentary rocks forming asymmetrical hills. Beyond the hills lies the gently undulating topography and steppe vegetation of the semiarid Great Plains, where sand-bed channels shrink back to a trickle after the annual snowmelt peak flow.

The dramatic topography and strong elevational contrasts in climate and vegetation dominate initial impressions of the Colorado Front Range. This leads to questions about how river process and form change moving downstream, and what factors influence this change. At a basic level, we can address these questions using empirical or theoretical approaches. *Empirical* approaches are largely inductive. In logic, to induce is to conclude or infer general principles from particular examples. In an empirical approach, data are collected and analyzed in order to establish relationships between variables. A fundamental challenge to empirical understanding of rivers lies in generalizing from empirical results defined by using a restricted database. If I measure bedload transport along a cobble-bed mountain river segment for a year and demonstrate that the majority of transport occurs when flow equals or exceeds half of the bankfull depth, can I extrapolate from this site to other rivers? What if I repeat the measurements on a sand-bed river of the plains and find that bedload transport begins at a much lower level of flow?

Theoretical approaches formulate and test specific statements based on established principles. To deduce is to reason from the general to the particular. Theoretical approaches are more deductive, but are typically hampered by a relative lack of established geomorphic theory. Consequently, theoretical approaches to river process and form commonly draw heavily on related fields such as hydraulic engineering in which the theory represents a system much more simple than most natural river channels.

Theoretical approaches to bedload transport developed by hydraulic engineers, for example, assume that bedload transport (i) begins once flow energy exceeds a critical level defined by the average grain size of the sediment and (ii) is proportional to the level of excess energy beyond the critical energy. The second assumption is illustrated by a generic equation for bedload transport rate q_b

$$q_b = k(\tau - \tau_c)^n \tag{1.1}$$

where k is an empirical constant, τ is boundary shear stress, τ_c is critical boundary shear stress for entrainment, and n is an empirically derived exponent. This equation implies that bedload transport is proportional to the amount of shear stress above the critical level for moving sediment. Eq. (1.1) is an example of a *flux equation*. For rivers, flux equations usually refer to flow–sediment interactions and processes such as sediment flux within a channel.

The relatively narrow grain-size range of sand-bed channels makes it easy to specify average grain size, and the relative ease of mobility of sand grains makes assumption (ii) reasonable. In a cobble- or boulder-bed channel, however, the wider range of grain sizes means that larger grains can shield smaller grains from the force of the flow and limit the movement of the smaller grains. Consequently, average grain size may not be a particularly useful parameter for specifying the start of bedload transport. Larger grains at the streambed surface can prevent the movement of underlying smaller grains and create turbulence, so that bedload transport is not likely to have a linear relationship with flow energy.

The problem of characterizing bedload transport in mountains and plains rivers can also be described using the dichotomy of deterministic versus probabilistic. *Deterministic* approaches assume that physical laws control river process and form. Once these laws are known, river behavior can be predicted for a given set of conditions.

Deterministic modeling of river processes relies on five basic equations: one continuity equation each for water and for sediment; the flow momentum equation; a flow resistance equation; and a sediment transport equation.

Conservation equations or *continuity equations* are based on the fact that mass, momentum, and energy cannot be created or destroyed in any process. The continuity equation for flow is simply

$$Q = w\,d\,v \tag{1.2}$$

where Q is discharge, w is flow width, d is flow depth, and v is mean velocity. An example of a sediment version is the Exner equation for sediment continuity,

$$(1 - \lambda_p)\frac{\partial \eta}{\partial t} = -\frac{\partial q}{\partial x} \tag{1.3}$$

where λ_p is bed porosity, η is bed elevation, t is time, q is volume transport rate of bed material load per unit width, and x is direction of flow (Parker et al. 2000). Another example of a continuity equation is a sediment budget that equates sediment storage to sediment input minus output.

The flow momentum equation is based on Newton's second law of motion and is well defined theoretically. Momentum is a vector defined by the product of mass and velocity. Momentum per unit time of water in a channel is ρQv, where ρ is water density, Q is discharge, and v is average velocity (Robert 2014).

The flow resistance and sediment transport equations used in deterministic modeling of river processes will include empirically derived coefficients. Deterministic modeling can thus use both empirical and theoretical understanding of a system, but assumes that river process and form can be directly predicted based on knowledge of existing parameters.

As the particular component of a river system being modeled increases in complexity, the interactions are increasingly difficult to represent using a set of closed equations, and predictions become less reliable (Knighton 1998). *Probabilistic* approaches reflect an assumption that natural systems are so complex that complete deterministic explanations are unrealistic because natural systems include inherent randomness. The ability to specify appropriate empirical flow resistance and sediment transport coefficients in boulder-bed mountain streams, for example, is limited by the extreme spatial variability in bed grain size, as well as irregularities in cross-sectional geometry caused by pieces of wood and lateral constrictions from bedrock outcrops or very large boulders. Under these circumstances, it is more effective to acknowledge a substantial level of uncertainty in predicting bedload transport: bedload movement may be described as occurring when discharge falls within upper and lower bounds, rather than as a direct relationship between discharge and bedload transport (Buffington and Montgomery 1997).

Another approach to predicting bedload transport is to use a force equation. *Force equations*, typically the balance of forces involved in erosional and depositional processes, describe a critical level beyond which a process such as movement of sediment on the streambed or erosion of the stream bank begins. An example of a simple force equation for entrainment of a sediment particle in a river is

$$\tau = \gamma_f DS \tag{1.4}$$

where τ is shear stress acting on the sediment γ_f is unit weight of the fluid, D is flow depth, and S is water-surface slope (Andrews 1980). Again, the less spatial and temporal variation there is in a system – think sand-bed (relatively uniform grain sizes), rather than boulder-bed – the simpler it is to specify the forces at work and to accurately assign average values to parameters such as flow depth and water-surface slope in Eq. (1.4).

Because natural rivers are commonly quite spatially and temporally variable, geomorphologists try to simplify process and form using *physical experiments* in which one or more variables are directly and systematically manipulated in order to observe the effect on the whole system. Such manipulations are typically conducted in a laboratory setting (Schumm et al. 1987) or, more rarely, in the field.

Bedload transport in the boulder-bed mountain rivers of the Colorado Front Range occurs 24 hours a day during the snowmelt peak. Much of the transport actually occurs in the early hours of the morning when the previous afternoon's snowmelt runoff comes down the river. Instead of attempting to directly measure bedload movement, and perhaps missing some of the sediment movement by not sampling the entire channel width or sampling continuously, useful insights into sediment dynamics can be gained by creating a scaled-down river in a flume and then measuring changes in bedload transport as discharge is varied. Physical experiments present challenges of scaling forces (can you effectively simulate the turbulence and associated hydraulic forces of a flow that is several meters deep in the real channel?) and of including all relevant variables (can you effectively simulate fluctuations in upland or tributary sediment supply to the main channel?). Experiments can nonetheless provide useful insights into process and form in real channels.

Rivers can also be investigated by developing *numerical simulations* in which those variables and interactions considered to be relevant are quantified (Coulthard and Van de Wiel 2013). Simulation outcomes are then compared to real rivers in order to evaluate the accuracy of parameterization and, once such accuracy is established, to test scenarios such as the effect of altering water or sediment yields to a river. Numerical simulations can be based on some combination of theoretical and empirically derived equations, which can be deterministic or probabilistic. A numerical simulation of bedload transport, for example, might specify channel geometry, streambed grain-size distribution, discharge, and sediment input from upstream, and then use an equation such as Eq. (1.3) to predict bedload flux. Among the challenges of numerical simulation are identifying the relevant variables and processes, and parameterizing them.

In addition to downstream differences in streambed substrate and bedload dynamics, some of the more obvious changes along river networks in the Colorado Front Range are the transitions from alpine meadows to relatively dense subalpine forest of spruce and fir, to more open montane pine forest, and finally to semiarid steppe. Along the forested portions of the river networks, wood recruited from riparian forests can strongly influence channel process and form. These interactions illustrate another commonly used conceptual model of rivers as complex and nonlinear systems.

A *complex system* is composed of interconnected parts that as a whole exhibit one or more properties – including behavior – not obvious from the properties of the individual parts (Bar-Yam 1997). A complex system displays self-organization over time and emergence over scale. *Self-organization* describes the formation of patterns attributable to the internal dynamics of a system, independent of external inputs or controls (Phillips 2003). *Emergence* is defined as patterns that arise from a multiplicity of relatively simple interactions (Goldstein 1999). A tree topples into a river, for example, with a portion of the roots still attached to the bank. The downed tree extends into the river, trapping smaller pieces of wood in transport and forming a logjam. The logjam blocks flow, creating a backwater of lower velocity where sediment in transport settles out. As the elevation of the streambed increases, flow begins to spill over the channel banks and erodes a secondary channel that branches away before rejoining the main channel downstream. Bank erosion during formation of the secondary channel undermines more trees, which fall into the river, forming additional logjams that facilitate further channel branching. Eventually, a network of branching and rejoining channels that enhance wood recruitment and storage is present. The tree fall and its consequent effects resulting in a multithread channel segment are an example of a complex system (Wohl 2011b).

In a *nonlinear system*, output is not directly proportional to input, such that, mathematically, the variable to be solved for cannot be written as a linear combination of independent components because of interactions among those components (Phillips 2003). Pieces of wood floating downstream in a river are influenced by the hydraulic force of the flowing water, for example, but also by the movement of adjacent pieces of wood or the trapping effect of stationary instream wood (Braudrick et al. 1997; Kramer and Wohl 2017). Because the movement of wood down the channel does not depend only on hydraulic force, this movement is an example of a nonlinear system.

Although phrases such as "nonlinear" and "complex systems" were not commonly used until the 1990s, the behavior described by these phrases was recognized decades earlier in descriptions of river process and form within the work of G.K. Gilbert (1877) and, in the mid-twentieth century, Luna Leopold, Stanley Schumm, and others.

Rivers are also viewed as *open systems*, characterized by a continual exchange of matter and energy with the surrounding environment (Chorley 1962). Such exchanges might be obvious at the scale of a channel segment with fluxes of water and sediment *from* upstream and upland sources and *to* downstream portions of the river. As emphasized in the opening discussion of connectivity, even the largest river networks also experience continual inputs of matter and energy from the atmosphere and the lithosphere, sometimes accomplished by the activities of organisms. Snow falling in the Colorado Front Range reflects the dynamics of cold, dry Arctic air masses moving southward and interacting with slightly warmer air carrying much more moisture and moving inland from the Pacific Ocean. The melting of the resulting snowpack, and consequently the timing and volume of snowmelt runoff in the rivers, are influenced not only by air temperature, but also by deposition of wind-blown dust that can come from nearby sources such as the deserts of the southwestern United States, as well as from very distant sources in Asia (Painter et al. 2010).

Viewing a river as a complex, open system implies that, at whatever scale the river is considered, it contains multiple, interacting components. Interactions between components include feedbacks, thresholds, and lag times, and can create equifinality.

Feedback refers to interactions among variables. Self-enhancing feedbacks promote continuing change, as when sand grains saltating across bedrock are preferentially trapped on accumulations of sand that increase with time. The fallen log that initiates a logjam, and eventually a network of secondary channels that promote additional wood recruitment and logjams, is another example of a self-enhancing feedback. Self-arresting feedbacks limit change. For example, a lateral channel constriction causes an increase in flow velocity, which results in erosion of the constriction until the velocity drops below a magnitude capable of causing erosion of the channel boundaries.

Thresholds involve abrupt changes in process or form. *External or extrinsic thresholds* are crossed as a result of changes in external controls. *Internal or intrinsic thresholds* can be crossed in the absence of changes in external variables (Schumm 1979). An example of an external threshold comes from hillslope hydrology, when the early stages of precipitation cause shallow infiltration and relatively slow downslope movement of water via subsurface diffusion. When sufficient water infiltrates to reach deeper layers with preferential flow paths in the form of soil pipes, downslope water delivery to channels abruptly becomes much more rapid. In this example, the abrupt change in downslope water delivery is externally forced by increasing volumes of precipitation.

A hillslope can also be forced across a threshold of stability by intense or prolonged precipitation that saturates regolith and triggers debris flows. During widespread and sustained rainfall in September 2013, more than 1100 debris flows and landslides across the Colorado Front Range moved the equivalent of hundreds to thousands of years of hillslope weathering products (Anderson et al. 2015). Substantial amounts of large wood and organic carbon were also delivered to streams as a result of crossing the threshold of hillslope stability (Rathburn et al. 2017).

The ephemeral tributaries that head on the dry eastern steppe of Colorado provide an example of internal thresholds. Over hundreds to thousands of years, these channels alternately incised to form steep-sided gullies or arroyos and aggraded to form relatively shallow swales (Figure 1.4). These alternating episodes of cut and fill represent crossing of an external threshold in response to fluctuations in precipitation, vegetation, and runoff (Tucker et al. 2006).

Alternating cut and fill can also occur in response to the crossing of an internal threshold of sediment transport within the channel. Stream flow in such channels is brief and infrequent, and sediment can be deposited midway down as discharge declines because of evaporation and infiltration into the streambed. Repeated deposition of sediment partway along the stream's longitudinal profile can develop a steeper section of the bed, at which a headcut eventually forms, triggering a wave of upstream-migrating incision. All of this can occur in the absence of any change in external variables such as precipitation, runoff, or sediment inputs (Schumm and Hadley 1957).

Lag time typically refers to the delay between a change in an external variable, such as an increase in water yield, and the response of the river, such as bank erosion. The cobble-bed streams of the subalpine forest in the Front Range provide an example. Commercial ski resorts in this region divert river water to make snow for their ski runs. When this artificially created snow melts in spring, the runoff commonly goes into a different channel than the source of the water. These receiving channels can have peak flows more than 200% larger than would result from natural runoff. Channels along which dense riparian vegetation and cohesive silt and clay increase bank resistance take longer to respond to increased peak flows than channels with less erosionally resistant banks (David et al. 2009).

Where an external disturbance is very intense or widespread, lag times can be minimal. An intense wildfire in the montane zone of the Colorado Front Range during summer 2012 completely consumed hillslope vegetation over hundreds of hectares of pine forest underlain by weathered granite. The first rainstorms following the fire resulted in widespread erosion of hillslopes and headwater channels, and aggradation of larger channels (Wohl 2013a; Kampf et al. 2016; Rathburn et al. 2018).

Fire-induced sediment accumulation in the larger channels is of particular concern because these rivers supply municipal drinking water to communities along the base of the Front Range. Water managers trying to maintain storage capacity at intake structures and limit turbidity associated with suspended sediment and organic matter could use the force equation for sediment entrainment, the continuity equations for flow and sediment, and the flux equation for bedload transport mentioned earlier to quantify sediment transport. They could also use *diffusion equations*, which describe the movement of matter, momentum, or energy in a medium in response to some gradient, such as the turbulent mixing of suspended sediment driven by gradients in flow energy (Robert 2014). An example particularly relevant to the deposition of hillslope sediment mobilized after wildfire comes from unit sediment flux q in a river depositional system

$$q = \nu \frac{\partial h}{\partial x} \tag{1.5}$$

where ν is diffusivity, h is elevation, and x is distance downstream (Voller and Paola 2010).

Equifinality, also known as convergence, refers to the fact that different processes and causes can produce similar effects. This condition makes it difficult to infer process from form (Chorley 1962). Channel incision leading to terrace formation along the primary rivers of the Front Range may have resulted from lowering of the base level, or from fluctuations in the water and sediment supply to the river associated with the advance and retreat of Pleistocene valley glaciers, or from widespread deforestation and mining during the nineteenth century (Schumm 1991). Data on the age, spatial extent, and stratigraphy of the terraces, as well as independent information on the timing and nature of base-level change, glaciations, and historical land use, are necessary to explain terrace formation.

Underlying conceptualizations such as feedbacks and thresholds is one of the most widely used geomorphic conceptual models: the idea that a river can exhibit various forms of stability, or equilibrium (Gilbert 1877). *Equilibrium* typically refers to a condition with no net change, and is thus very

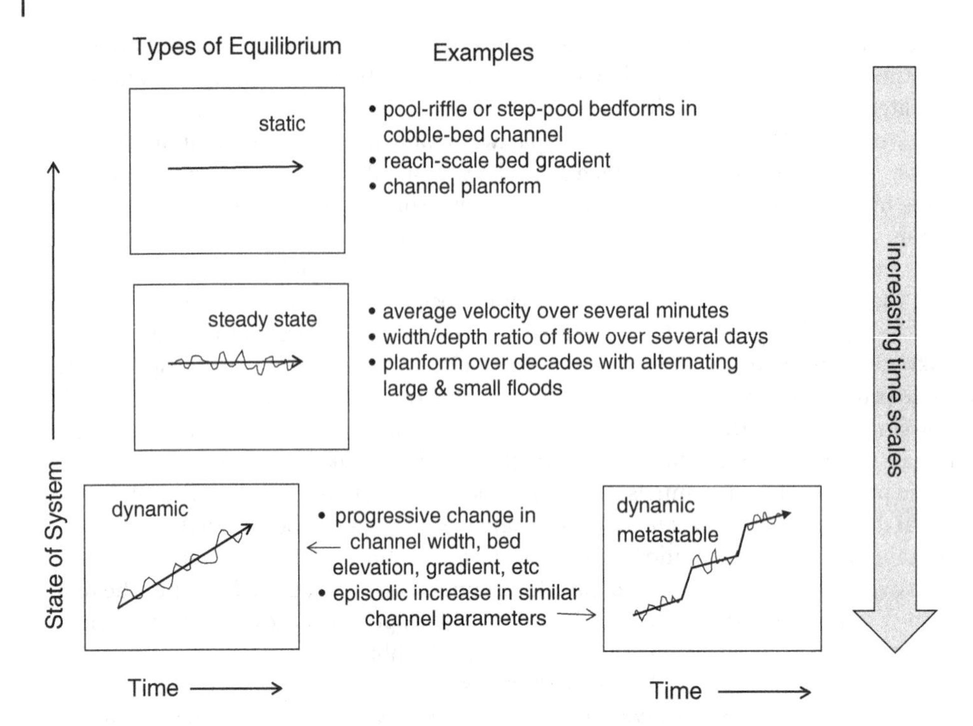

Figure 1.5 Schematic illustration of different types of equilibrium.

dependent on the time span being considered. A river that undergoes substantial channel change during a short-duration flood can nonetheless be in equilibrium when considered over a decade because subsequent smaller flows rework the erosional and depositional features created by the flood. Consequently, different forms of equilibrium can be distinguished with respect to time span (Figure 1.5).

Over the shortest time intervals, any particular variable representing the state of the river system (e.g. channel planform or gradient) is *static* and unchanging. At progressively longer time intervals, the variable may be in *steady state*, with fluctuations about a consistent mean. At the longest time intervals, the mean value of the variable is likely to change, either progressively through time in *dynamic equilibrium* or in a stepped manner that reflects the crossing of thresholds, as in *dynamic metastable equilibrium*. The latter two cases are not, strictly speaking, equilibrium, because the system exhibits net change over the time span being considered. These phrases are, however, widely used.

Equilibrium implies that multiple interacting variables within a river can reach a state of stability. This is reflected in the widely used definition of a *graded river* as a channel in which streambed slope is adjusted to prevailing water and sediment discharges, such that the channel neither aggrades nor degrades and the slope remains constant over the time interval of interest (Mackin 1948).

Equilibrium also implies that a river will change in response to changes in the supply of energy or material. Pleistocene valley glaciation in the upper portion of the Front Range changed water and sediment yields to downstream portions of the river network. Thinking of these river networks within a framework of equilibrium raises questions regarding how, and how rapidly or over what time span, the rivers responded to altered water and sediment supplies during glacial advance and retreat. One way to asses this is to evaluate downstream hydraulic geometry (Leopold and Maddock 1953) relations for rivers in the Front Range. *Downstream hydraulic geometry* relations are empirical equations in the form of *power functions* derived from linear regressions of log-transformed data. These equations relate dependent variables of channel geometry to the independent variable of discharge. For example,

$$w = aQ^b \tag{1.6}$$

where w is channel width, Q is discharge, and the coefficient a and exponent b are determined from the linear regression.

Eq. (1.6) implies that discharge is the primary influence on channel width. One implication is that values of channel width in the Front Range have fluctuated through time as the advance and retreat of valley glaciers has altered discharge downstream.

A river in equilibrium is expected to have well-developed downstream hydraulic geometry such that variations in discharge explain most of the observed downstream pattern of variation in width (Wohl 2004b). Headwater rivers within the glaciated portion of the Front Range exhibit less well-developed downstream hydraulic geometry relations, as indicated by lower values of the regression coefficient for w–Q regressions, than headwater rivers at lower elevations beyond the extent of Pleistocene glaciations. This suggests that rivers in the glaciated zone are still adjusting, more than 10 000 years after glacial retreat, to local variations in gradient, substrate resistance, sediment supply, and other factors that are affected by glaciation and can influence channel width. These rivers may be further from equilibrium than otherwise analogous channels at lower elevations in the mountain range. Downstream hydraulic geometry relations in glacial valleys, for example, include much greater variability than relations in fluvial valleys (Livers and Wohl 2015).

Equilibrium, or its absence, can also be described in terms of steady-state versus transient landscapes. A *steady-state landscape* can be defined with respect to denudation and topography as a landscape in which erosion and rock uplift are balanced such that a statistically invariant topography and constant denudation rate are maintained over a specified time interval (Whipple 2001). A steady-state landscape thus exhibits equilibrium between uplift and erosion. *Transient landscapes* are those experiencing relatively brief (on a geological time scale) increases in erosion rate in response to, for example, active tectonic uplift (Attal et al. 2008). Ongoing change indicates that a transient landscape has not yet reached equilibrium following an external perturbation.

Exhumation of the Denver Basin at the eastern margin of the Colorado Front Range within the last few million years caused relative base-level fall for the major rivers of the Front Range. Base-level fall triggered a wave of incision that has been migrating upstream at an estimated rate of 0.15 mm/yr (Anderson et al. 2006c). The location of contemporary active response to base-level fall appears as a steepening – either a waterfall or a steep section of rapids – in the longitudinal profile of each river. Portions of the river network upstream and downstream from this steeper zone are presently in steady state with respect to the base-level fall, whereas the gradient of the steeper portion of the longitudinal profile is transient.

Contrasting river process and form between different portions of a region such as the Front Range underlies another approach to understanding rivers. Data for understanding rivers can be obtained from direct measurements in a field setting or from remote sensing imagery (Oguchi et al. 2013). Because of the long temporal scales over which river processes such as development of drainage networks or longitudinal profiles act, ergodic reasoning is also commonly used. *Ergodic reasoning* substitutes space for time by comparing features in different stages of development, under the assumption that variables other than time remain relatively constant. For example, drainage networks developed on otherwise comparable basalt flows of widely differing ages within a limited region can be compared to examine network development through time. The challenge of ergodic approaches is that variables other than time likely differ between sites being compared. Even if the basalt flows are identical in composition, for example, fluctuations in climate through time might cause the older networks to represent at least preliminary development under a climate different than the climate present during development of networks on younger basalt flows.

Returning to the example of instream wood in forested streams of the Front Range, one way to investigate the importance of forest stand age to river–wood dynamics is to compare otherwise analogous stream segments flowing through forests of diverse age. Study design can be challenging: ideally, all other important parameters – drainage area, stream flow, sediment supply, valley geometry – are similar between the stream segments, and only the age of the riparian forest varies. Comparisons using this ergodic approach suggest that old-growth forests with average tree age greater than 200 years have more instream wood, larger wood pieces, more closely spaced channel-spanning logjams, and consequently a greater abundance of secondary channels and greater channel–floodplain connectivity (Livers and Wohl 2016).

Stepping back to consider the river networks of the Front Range at a regional scale, many of the questions posed by Leopold et al. (1964) remain highly relevant today:

- What factors control hillslope process and form, and the initiation of channelized flow?
  - What is the rate of bedrock weathering and regolith production?
  - How and how rapidly does regolith move downslope into channel networks?
  - What variables influence the location of channel heads?
  - What processes result in the formation of channel heads?
- What factors govern the longitudinal profile of the rivers?
  - In particular, what is the relative importance of landscape-scale denudation in response to continuing adjustment to uplift, Pleistocene glaciations, and Quaternary relative base-level fall?
  - What is the relative importance of longitudinal variations in bedrock erosional resistance, sediment supply, and flow regime?

Examining rivers in the context of the greater landscape, we can also add a series of new questions. Examples include:

- How do diverse types of connectivity vary throughout these river networks?
  - Are the alpine summit surfaces storing periglacial sediment, for example, or are they strongly coupled to adjacent glaciated valleys (Anderson et al. 2006a,b,c)?
  - Channel–floodplain and channel–hyporheic–ground water connectivity increase within lower-gradient, wider valley segments, and then decrease in steep, narrow segments. What are the specific processes governing these downstream variations in connectivity (e.g. Wegener et al. 2017)?
- What are the magnitude and extent of human alteration of river networks?
  - When people of European descent settled the Front Range during the nineteenth century, they initiated lode and placer mining, extensive deforestation, and widespread flow regulation in the form of dams and diversions (Wohl 2001). Some of these activities ceased a century ago. Do river process and form still differ between networks in which these historical activities occurred and networks that were not altered in this way?
  - How does contemporary flow regulation alter the physical and ecological functions of Front Range rivers (Ryan 1997; McCarthy 2008; Wohl and Dust 2012)?
  - Warming climate is resulting in changes in precipitation, soil moisture, wildfire regime, outbreaks of native insects that kill trees, and forest blowdowns. How do these pervasive alterations of forest dynamics and precipitation–runoff–stream flow influence channel process and form?

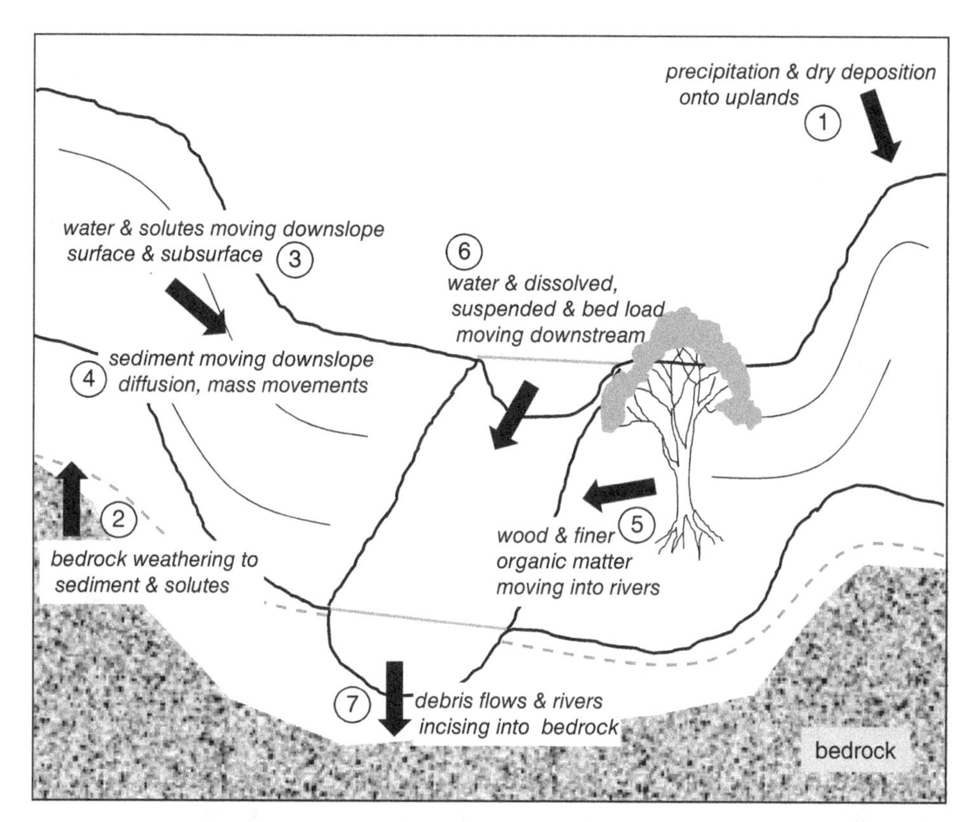

Figure 1.6 Schematic illustration of fundamental fluxes in any landscape. Among those for which some form of geomorphic transport law (GTL) has been proposed are 2, 4, and 7, although these GTLs require additional testing and parameterization for specific field settings. Empirical or theoretical equations have also been proposed for 1, 3, and 6. Again, these equations require testing and parameterization. All of the equations proposed for these fluxes assume that average values can be quantified based on prevailing conditions. Using bedrock weathering as an example, chemical reactions are a function of factors such as temperature and the amplitude of temperature oscillations. Explicitly incorporating connectivity requires quantifying variations in prevailing conditions through time that limit or enhance fluxes, such as short-term variations in weather and longer-term variations in climate that influence temperature and thus chemical weathering rate.

- What river-related geomorphic processes can be quantified in a manner applicable to diverse landscapes?
  - In an influential paper, Dietrich et al. (2003) highlighted the importance of developing geomorphic transport laws (GTLs) in the form of mathematical statements derived from a physical principle or mechanism, which express the mass flux or erosion caused by one or more processes. In order to be useful, it is important that such laws can be parameterized from field measurements, can be tested in physical models, and can be applied over relevant spatial and temporal scales. Existing GTLs include those for soil production from bedrock, linear slope-dependent transport of colluvium, and debris flow and river incision into bedrock.
  - What fundamental processes in the Colorado Front Range can usefully be expressed via GTLs (Figure 1.6)?
  - What processes are not yet adequately described by such laws?
  - How can we integrate GTLs with quantitative measures of connectivity?

- What components of river process and form are significantly influenced by biota?
  - Beaver were much more abundant in the Colorado Front Range prior to the nineteenth century. Have historical reductions in beaver populations and beaver dams influenced rivers regionally, or only local segments of rivers (Laurel and Wohl 2018)?
  - How do instream wood volume and associated geomorphic effects differ between subalpine and montane forests, or between steep, narrow valley segments and wide, lower-gradient valley segments (Livers et al. 2018)?
  - The extent and species diversity of riparian vegetation differ markedly between steep, narrow valley segments and wide, lower-gradient segments (Polvi et al. 2011). How do these differences influence valley-bottom sediment storage, hyporheic and ground water exchange, and water chemistry along Front Range rivers?

To quote Leopold et al. (1964, p. 18), "Partial explanations of these problems can be offered, but more complete explanations require much more knowledge of processes than is presently available."

1.6 Only Connect

E.M. Forster took "only connect" as the epigraph for his novel, *Howards End*. Forster was referring to connections between individual people and different classes within a society, but this phrase is also particularly apt for understanding rivers. If we can extend our understanding sufficiently and

- connect rivers to landscapes
- connect contemporary river configuration to human and geological history
- connect site-specific river characteristics to universal river process and form
- connect field observations to physical experiments and numerical simulations, and
- connect geomorphic knowledge of river process and form to
  - ecological knowledge of aquatic and riparian organisms,
  - geological knowledge of rock substrates and tectonics,
  - social science knowledge of human history and decisions regarding resource use, and
  - biogeochemical knowledge of aqueous chemistry

then we will be making progress in understanding the complex and fascinating world of rivers.

2

Creating Channels and Channel Networks

2.1 Generating Water, Solutes, and Sediment

2.1.1 Generating Water

Rain is the most common form of precipitation (Barry and Chorley 1987). Precipitation includes all liquid and frozen forms of water – rain, snow, hail, dew, hoar frost, fog drip, and rime – moving from the atmosphere toward the ground surface, but rain and snow constitute by far the greatest contributors to precipitation worldwide.

The major types of precipitation can be differentiated into convective, cyclonic, and orographic, based on the primary mode of uplift of the air that causes water vapor to condense into water droplets or ice crystals. *Convective precipitation* can occur (Barry and Chorley 1987): (i) across small (20–50 km^2) areas for short periods of time (generally <1 hour) as a result of scattered convective cells that develop through strong heating of the land surface, typically in summer; (ii) as showers that form from cold, moist, unstable air passing over a warmer surface, or parallel to a cold or warm front; and (iii) in tropical cyclones, where cells become organized about the center in spiraling bands. Convective precipitation is generally of high intensity and includes thunderstorms.

Cyclonic precipitation results from the ascent of air through horizontal convergence of airstreams in an area of low pressure (Barry and Chorley 1987). Cyclonic precipitation is typically of lower intensity, greater spatial extent, and longer duration than convective precipitation. It includes what is sometimes referred to as frontal precipitation, associated with the movement of a warm or cold front, as well as tropical and subtropical cyclones, hurricanes, and monsoonal depressions.

Orographic precipitation occurs because regions topographically higher than the surrounding terrain can: (i) trigger instability by giving an initial upward motion to an air mass or through differential heating of land-surface slopes; (ii) increase cyclonic precipitation by slowing the rate at which an atmospheric depression is moving; or (iii) cause convergence and uplift through the funneling effects of topographic valleys on airstreams (Barry and Chorley 1987). The increase of precipitation with height occurs on mountains throughout the world, although the details vary regionally and seasonally (Barry 2008; Wohl 2010b). Individual mountain ranges can also have substantial areal variation at the same elevation, depending on the orientation (windward versus leeward) of the slopes.

Precipitation measurements indicate consistent global and seasonal patterns of moisture. An equatorial belt of maximum precipitation is slightly displaced into the northern hemisphere. Precipitation totals tend to be very low in high latitudes, with secondary minima in subtropical latitudes. Superimposed on these average patterns are multi-year and multi-decadal, global- to regional-scale

Rivers in the Landscape, Second Edition. Ellen Wohl.

fluctuations in oceanic and atmospheric circulation that strongly influence precipitation. These include the El Niño-Southern Oscillation (ENSO) and the Pacific Decadal Oscillation (PDO) (Neal et al. 2002). ENSO involves an episodic warming of the sea surface in the equatorial Pacific, with associated alterations in precipitation patterns from southern Africa to northeastern North America. One phase of the oscillation is known as El Niño, and the opposite phase is La Niña. PDO involves ocean-temperature anomalies in the northeast and tropical Pacific Ocean that strongly influence precipitation in the North Pacific and North America.

The characteristics of precipitation strongly influence the amount of water reaching channel networks and the pathways that water follows downslope into channels. Most river networks are influenced by persistent types of precipitation, such as predominantly convective storms, although moderate- to large-sized networks commonly experience multiple forms of precipitation, such as winter snowfall and summer convective storms.

Precipitation reaching Earth's surface can be stored for periods of hours to months in a solid form as snow, or for decades to thousands of years as glacial ice. Melting of this solid, stored water strongly influences rivers at high altitudes and latitudes. The relative importance of glacier melt, snowmelt, and rainfall typically varies by latitude and location with respect to atmospheric circulation patterns, and by elevation within a region. Precipitation and runoff patterns within the Himalayan massif exemplify the effect of elevation. Up to 80% of river discharge originates as monsoon rainfall at lower elevations on the southern side of the massif, whereas glacier and snowmelt contribute 50–70% of river discharge at higher elevations and on the northern side of the massif (Gerrard 1990; Wohl and Cenderelli 1998).

Worldwide, glacier and snowmelt are progressively more important at higher latitudes and higher elevations, and the seasonal melt contribution is delayed later into the summer with increasing latitude and elevation. As the proportion of the basin area covered by ice and snow increases, progressively more of the total runoff occurs during summer (Chen and Ohmura 1990; Collins and Taylor 1990). Interannual variation in runoff also declines with greater snow and ice coverage, although the relation is not linear (Collins 2006).

2.1.2 Generating Sediment and Solutes

The bedrock underlying a drainage basin is the starting point for much of the sediment and solutes that eventually move downslope and into rivers, although particulate and dissolved materials can also enter rivers via eolian transport and wet and dry atmospheric deposition (e.g. Schuster et al. 2002; Galloway et al. 2004). An idealized vertical profile through the weathering zone has unweathered bedrock at the base, overlain by *regolith*. Regolith is subdivided into *weathered rock*, which is fractured or chemically weathered but has not been mobilized by hillslope processes or bioturbation, *saprolite* overlying it, which retains the original rock structure yet has been sufficiently altered that it can be augured through or dug with a shovel, and *mobile regolith*, which includes *soil*, organized into horizons by soil-forming processes (Anderson and Anderson 2010). In practice, "mobile regolith" and "soil" are commonly used interchangeably, as they are in this chapter.

At the regional scale, lithology, tectonics, and climate influence processes and rates of bedrock weathering. Minerals that crystallize from molten material at high temperatures are typically less resistant to weathering than minerals such as quartz that crystallize at lower temperatures, so the mineralogical composition of bedrock influences weathering (Anderson and Anderson 2010; Ritter et al. 2011). Tectonic stresses and regional-scale deformation can fracture bedrock at various depths in the crust, increasing surface area and making the rock more susceptible to chemical alteration and to other physical weathering processes such as freeze–thaw cycles (Anderson and Anderson 2010). Chemical weathering results in chemical alteration of the regolith and in dissolution of more reactive minerals, which release ions that are transported into ground and surface waters as solutes. Chemical weathering is facilitated by warm, wet conditions, whereas physical weathering is facilitated by moderately wet climates with low temperatures that promote frost action (Ritter et al. 2011).

At smaller spatial scales, factors such as soil erosional flux, hillslope morphology, and biota influence rates of weathering. Bedrock must weather at a rate equal to or greater than the rate of erosion if soil is to persist, and many soil profiles appear to reach a steady-state thickness such that soil production is balanced by removal (Lebedeva et al. 2010). Fundamentally, soil depth reflects the rate of soil production versus the rate of soil erosion (Heimsath et al. 1999).

Soil production is sometimes described as having a "humped" functional relationship, such that bedrock erosion is maximized under an intermediate soil thickness and decreases as soils become thicker or thinner (Gilbert 1877). Chemical weathering rates, soil production rates, and hillslope curvature decrease with increasing soil depth (Heimsath et al. 1997; Burke et al. 2007). Consequently, any process that alters soil depth by moving soil downslope – mass movements, creep, bioturbation – also alters rates of bedrock weathering and soil production. Humans, in particular, move tremendous amounts of sediment globally (Hooke 2000), resulting in long-term and spatially extensive changes in bedrock weathering and soil distribution and development (Montgomery 2007).

Hillslope morphology is closely connected to erosional fluxes. Upper, convex portions of a hillslope are likely to have steady removal of weathered products, for example, whereas lower, concave portions can accumulate weathered materials. At larger scales, hillslope curvature (and soil production) varies inversely with soil depth (Heimsath et al. 1997).

Plants and animals influence rates of weathering by excreting organic acids and physically disrupting weathered materials, as when plant roots expand into bedrock joints or burrowing animals churn the regolith. In forested regions, treefall can result in pit-mounds, which are microtopographical forms caused by a single uprooted tree (Šamonil et al. 2010). Pit-mounds create an uneven distribution of soil thicknesses across a hillslope (Gabet and Mudd 2010), which can influence downslope movement of water and sediment, as well as rates of bedrock weathering and erosion.

In landscapes such as the Oregon Coast Range, a region of humid temperate rainforest, trees dominate sediment production via processes of

- root wedging as roots grow and increase in diameter,
- biological alteration of the chemical weathering environment via release of organic acids,
- tree throw (Gabet and Mudd 2010; Roering et al. 2010),
- root redistribution of water, which fluctuates diurnally as uptake of water by the tree fluctuates (Brantley et al. 2017), and
- wind-driven tree sway, which translates into the subsurface and helps to propagate fractures in the bedrock (Brantley et al. 2017).

Although bedrock properties in the Oregon Coast Range limit the damage that tree roots can inflict on the underlying bedrock (Marshall and Roering 2014), the rates and processes of tree growth and death strongly influence hillslope process and form.

Heimsath et al. (1997) derives an empirical function for soil production, $-(\delta e/\delta t)$

$$\frac{\delta e}{\delta t} = -\left(\frac{\rho_s}{\rho_t} K \nabla^2 z\right) \tag{2.1}$$

where ρ_s and ρ_t are soil and rock bulk densities, K is a diffusion coefficient with dimensions length^2/time, z is ground surface elevation, $\nabla^2 z$ represents hillslope curvature, e is the elevation of the bedrock-soil interface, and t is time. This empirical function is based on observations that hillslope curvature and soil production vary inversely with soil depth – in other words, deeper soils have slower rates of production.

The analogous function for chemical reaction in a soil can be written in terms of the average reaction rate $\overline{k}$ over the period P (a year)

$$\overline{k} = \frac{1}{P} \int_0^P A \exp\left(\frac{-E_a}{R(\overline{T} + \Delta T \sin(2\pi t/P))} \right) dt \tag{2.2}$$

where A is surface area, E_a is activation energy, R is the gas constant, T is temperature in degrees kelvin, and ΔT describes the amplitude of temperature oscillation over daily or annual cycles (Anderson and Anderson 2010). Eq. (2.2) reflects the fact that mineral reaction and dissolution rates are strongly dependent on temperature, which varies below the ground surface and on daily and annual cycles. The exponential increase in reaction rate with temperature means that periods of high temperature are disproportionately important in mineral weathering.

The use of cosmogenic isotopes such as ^{10}Be has substantially enhanced understanding of soil dynamics. *In situ* ^{10}Be is produced in quartz grains within the soil by cosmic ray bombardment. Meteoric ^{10}Be is produced by cosmic ray interactions in the atmosphere and rained out onto the landscape. Measurement of the concentration of such isotopes can be used to infer rates of soil production and movement downslope (Heimsath et al. 1997; Jungers et al. 2009; Egli et al. 2010).

As noted in Chapter 1, sediment production is rarely uniform across even a small catchment because of differences in lithology, climate, tectonics, biotic communities, hillslope morphology, and land use (e.g. Warrick and Mertes 2009). Disproportionate sediment generation has been demonstrated for the Amazon and Mississippi River basins in relation to spatial variation in topography, climate, and lithology within them (Meade et al. 1990; Meade 2007). Mountain and high-mountain (>1000 m elevation) drainages collectively cover 70% of the global land area but contribute 96% of total river sediment yields (Milliman and Syvitski 1992). Analogous patterns also appear to govern solute production (Lyons et al. 2002).

2.2 Getting Water, Solutes, and Sediment Downslope to Channels

2.2.1 Downslope Pathways of Water

The great majority of water flowing in a river passes over or through an adjacent upland and its regolith before reaching a channel (Kirkby 1988). The pathways followed by the water entering a river exert a strong influence on the volume and timing of flow in the channel. Precipitation falling toward the ground does not necessarily reach a river, however.

Precipitation can be intercepted by plants and either directly *evaporate* from the plants or be taken up by the plant tissues and then released back to the atmosphere through *transpiration*. Evaporation and transpiration are strongly influenced by energy availability at the surface and water availability in the subsurface. Transpiration also reflects plant physiology (Kramer and Boyer 1995) and the ability of plant roots to take up, redistribute, and even selectively extract water from different subsurface depths (Lai and Katul 2000). Recent estimates suggest that transpiration represents the largest water flux from continents (i.e. larger than rivers), composing 80–90% of combined evaporation and transpiration – commonly known as evapotranspiration (Jasechko et al. 2013).

A traditional assumption has been that trees transpire water that would eventually have entered a stream from subsurface storage. However, Brooks et al. (2010) have demonstrated that tightly bound water in the soil, which had been retained through a long summer dry season and used by trees in a seasonally dry temperate-latitude watershed, did not participate in the runoff process during the subsequent wet season. This indicates that the conceptual model of translatory flow (Hewlett and Hibbert 1967), in which infiltrating precipitation displaces water held in the soil prior to the precipitation, forcing the older water deeper into the subsurface and eventually into a stream, does not adequately describe downslope water movement in all scenarios. Measurements from additional field sites now support the *two water worlds hypothesis*, in which vegetation and streams return different pools of water to the hydrosphere (McDonnell 2014), although the mechanisms by which this occurs remain unclear.

Interception losses from evaporation and transpiration can reach 10–20% beneath grasses and crops and up to 50% beneath forests (Selby 1982). Interception can vary substantially during the course of the year in regions with seasons during which plants go dormant or during which soil moisture declines and plants store more water (Link et al. 2005). Plants can also directly intercept cloud water. In cloud forests, this interception can reach 35% of mean annual precipitation (Bruijnzeel 2005). Plant stems can concentrate the movement of precipitation toward the ground via *stemflow* (Levia and Germer 2015), which has been described for environments as diverse as tropical montane forest, semiarid loess terrain, and a seasonal cloud forest, in which stemflow accounted for 30% of total precipitation (Hildebrandt et al. 2007).

Precipitation that does reach the ground surface can be evaporated from bodies of standing or flowing surface water, or from the ground surface. The pathways taken by precipitation reaching the ground surface are strongly dependent on the soil cover. Soil cover changes continuously in response to water movement and processes of weathering, bioturbation, sediment transport, and land use or other changes in land cover (Brooks 2003). Development of crusts, compaction, and sealing (e.g. with clay particles) decrease surface permeability (Slattery and Bryan 1994; Kampf and Mirus 2013), as does development of soil water repellency, or *hydrophobicity*, from burnt plant litter (Martin and Moody 2001). Precipitation is equally dynamic, and the intensity and volume of precipitation reaching the land surface vary substantially through time and space. Consequently, downslope movement of water on the surface and in the subsurface is rarely in a steady-state condition (Wainwright and Parsons 2002).

Water can flow downslope at the surface as *overland flow*, which is used to describe surface flow outside the confines of a channel. Overland flow includes *Hortonian overland flow*, also known as *infiltration excess overland flow* (Horton 1945), if the infiltration capacity is low relative to precipitation intensity (Figure 2.1). Hortonian overland flow is most common where vegetation is sparse, slope gradients are steep, regolith is thin or of low permeability (e.g. clay), and precipitation intensities are high: characteristics of arid and semiarid regions. The great majority of most natural watersheds does not produce Hortonian overland flow, but human land uses that compact the soil or cause erosion of permeable, near-surface soil layers promote it. The most extreme example of this effect is paved, impermeable urban surfaces. Under natural (nonpaved) conditions, the initiation of overland flow is complex and begins at different times in different locations on a hillslope (e.g. Moody and Martin 2015).

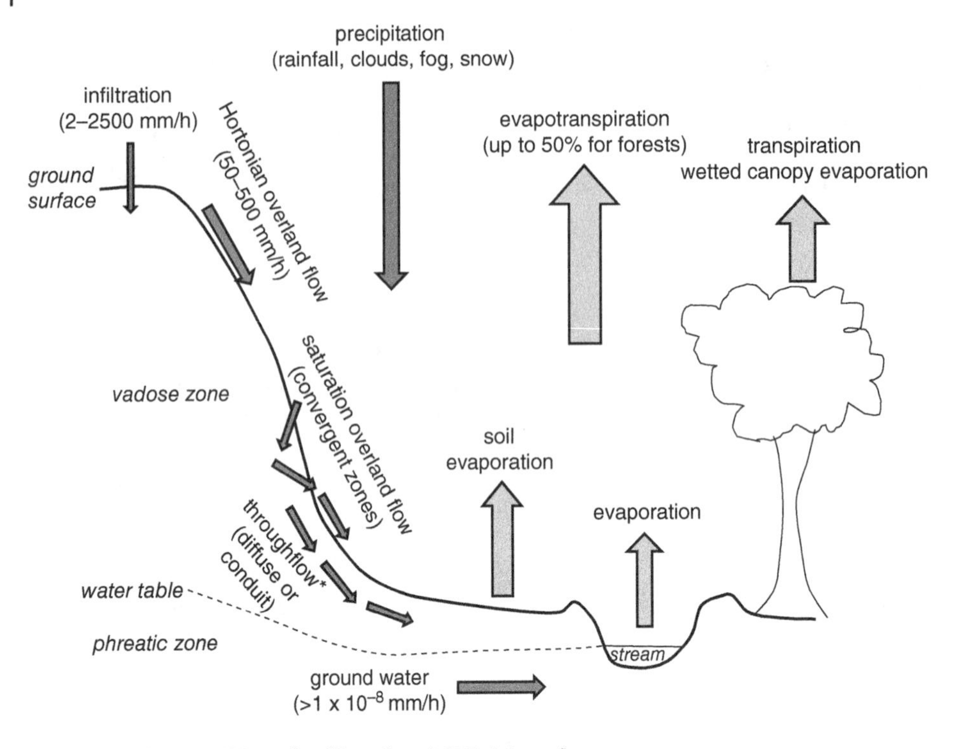

Figure 2.1 Schematic illustration of different types and rates of downslope water movement. Downward arrows reflect movement of water toward (precipitation), into (infiltration), across (overland flow), or beneath (throughflow, ground water) Earth's surface. Upward arrows reflect water movement back into the atmosphere. Ranges of measured rates of movements listed in millimeters per hour.

Overland flow can also move downslope as *saturation overland flow*, also known as *saturation excess overland flow* (Dunne and Black 1970a,b). This is a combination of direct precipitation onto saturated areas and return flow from the subsurface as saturation occurs. The moisture content of the regolith before, during, and after precipitation exerts a particularly important control on saturation overland flow, which is influenced more by antecedent soil moisture and subsurface transmissivity than by slope steepness (Montgomery and Dietrich 2002). As prolonged precipitation allows deeper and less permeable regolith layers to become saturated, subsurface flow expands to include areas progressively closer to the surface (Knighton 1984). Conditions that favor saturation overland flow include high permeability near the surface; a humid climate with high cumulative water input; and gentler slopes with shallow soils that cannot drain as easily as steep slopes in which hydraulic gradients approximately parallel the steep surface topography (Kampf and Mirus 2013).

Saturation overland flow is rare outside of convergent flow zones such as hillslope concavities (Dietrich et al. 1992). Several scenarios can cause saturation flow at other points along a hillslope, however, including: topographic breaks; permeability contrasts (e.g. roads or pavement); areas of low subsurface storage capacity; geologic structures that promote zones that saturate readily (e.g. layered basalt flows) (Mirus et al. 2007); exclusion from frozen soil; and snowmelt over saturated soil (Kampf and Mirus 2013). Saturation overland flow can also occur in tropical rainforests, where a sharp drop in permeability with depth and high rain inputs lead to perched water tables (Bonnell and Gilmour 1978; Elsenbeer and Vertessy 2000).

Overland flow commences when water ponds on the land surface to sufficient depth to begin flowing downslope. Surface roughness from microtopography and vegetation generates flow resistance, which influences overland flow pathways (Bergkamp 1998), particularly in humid environments with dense vegetation and organic litter; however, the flow also modifies the surface roughness. Vegetation influences overland flow by altering infiltration. Shrub mounds in semiarid regions facilitate infiltration and generate less overland flow than intervening spaces, creating downslope pathways strongly coupled to vegetation patterns (Dunne et al. 1991).

Reported infiltration capacities from diverse locations range from 0 to 2500 mm/h (Selby 1982). Infiltration can be extremely variable through space and time within a single small catchment because of the numerous factors that influence it (Figure 2.2). The *variable source area* concept (Hewlett and Hibbert 1967) reflects the fact that the area of a catchment actually contributing water to a channel extends during precipitation and contracts after it ends, varying between 5 and 80% of the catchment (Dunne and Black, 1970b; Selby 1982). Much of this variability likely reflects thresholds of activation for lateral subsurface flow (McDonnell 2003).

Saturation overland flow typically occurs first in downslope portions of a catchment and then expands upslope (Dunne 1978). In cold regions with low topographic relief, however, upslope areas can be the first sources of runoff as a result of varied local topography and water-table position in relation to frozen ground (Spence and Woo 2003). The *element threshold concept* describes a catchment in which runoff is governed by the function of spatially and hydrologically distinct areas (elements), such as bedrock uplands, soil-filled valleys, and lakes, and the hydrologic linkages between elements (Spence and Woo 2006). Each element can store, contribute, or transmit water, and the occurrence of these functions reflects the water balance of the element as well as connections to adjacent elements.

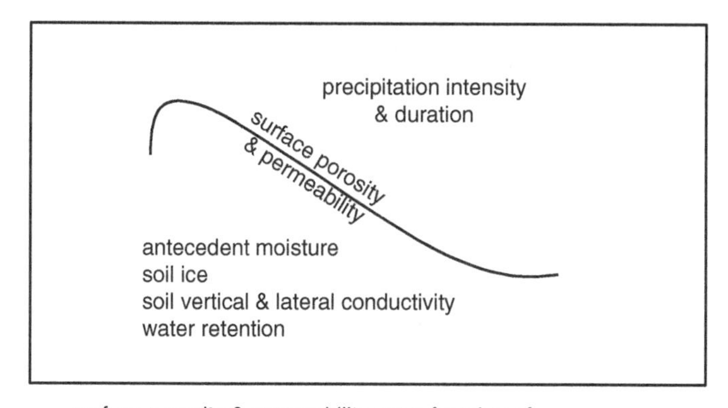

surface porosity & permeability are a function of
hillslope gradient
vegetation
regolith grain size, compaction, depth & areal extent

Figure 2.2 Schematic illustration of the variables influencing infiltration capacity. Line represents a side view of a sloping ground surface. Precipitation characteristics listed at upper right, surface characteristics listed along the line, and subsurface characteristics listed at lower left all help to create changes in infiltration capacity over small spatial and temporal scales.

The primary difference between the variable source area and element threshold concepts involves the assumed greater hydrologic connectivity within catchments operating under the variable source area concept.

Infiltrating water that remains in the subsurface can flow downslope in the vadose (unsaturated) zone above the water table as *throughflow*, or in the phreatic (saturated) zone below the water table as *ground water*. In either case, subsurface water flows through interconnected voids between solid materials. Where the interconnected pores are small, flow through the porous medium is laminar and of low velocity (Kampf and Mirus 2013) and is commonly described using Darcy's law (Darcy 1856). When the void space in a porous medium is filled with water under saturated conditions, the *hydraulic conductivity*, or ease of water flow through the medium, is at its maximum value (Kampf and Mirus 2013). When the void space is not completely filled with water, the connectivity of pore space and the hydraulic conductivity decrease.

Hillslopes are typically unsaturated at the ground surface and water movement is predominantly vertical during infiltration. Moisture is redistributed in the subsurface through vertical and lateral movements that reflect the interaction of infiltration and percolation (wetting) with evaporation and transpiration (drying) (Kampf and Mirus 2013). Water moving into the soil can propagate in the form of diffuse wetting fronts, fingered flow paths that result from wetting-front instabilities, or preferential flow along conduits (Wang et al. 2003). Although lateral unsaturated flow can occur, lateral flow is most common under saturated conditions (Kampf and Mirus 2013).

Diffuse throughflow depends on the general porosity and permeability of the unsaturated zone, whereas concentrated throughflow moves in preferential flow paths such as pipes or macropores (Dunne 1980; Jones 1981). *Macropores* are openings sufficiently large that capillary forces have an insignificant effect on the water running through them (Germann 1990). Preferential flow within the macropore does not have time to equilibrate with slower flow through the surrounding matrix (Šimůnek et al. 2003) and is likely to be turbulent, and thus not adequately described by Darcy's law. Lateral flow through macropores requires pore connectivity and a rate of water supply that exceeds the loss rate to the surrounding soil, which suggests that macropore flow is triggered at a threshold wetness level (Beven and Germann 1982). Once this threshold is exceeded, flow is much faster than diffuse matrix flow. Rapid subsurface stormflow is particularly widespread on densely vegetated hillslopes in steep terrain where dense biological activity creates high concentrations of macropores, although the phenomenon has also been documented in semiarid climates (Kampf and Mirus 2013).

Substantial preferential flow through macropores can facilitate the formation of soil pipes. *Soil pipes* are larger than macropores and are typically formed by subsurface erosion that enlarges animal burrows, root channels, or cracks from desiccation or unloading (Bryan and Jones 1997). Pipes can be only a few centimeters in length and diameter, or they can be 2 m in diameter and hundreds of meters long (Selby 1982). They tend to form just above a zone of lower porosity and permeability or along a cavity created by a burrowing animal or the decay of plant roots. Piping can occur in any region, but is particularly associated with drylands. In some catchments, piping can contribute nearly 50% of stormwater flow (Jones 2010).

Pipe networks can exist at multiple levels in the regolith, with each level being activated by precipitation of different magnitude (Gilman and Newson 1980; Kim et al. 2004). Rapid lateral flow via soil pipes or via discontinuities at the soil–bedrock interface appears to depend on thresholds, such that hillslopes "turn on" when sufficient water infiltrates (Uchida et al. 2001; McDonnell 2003). This may help to explain the *old water paradox*, in which pre-event water largely dominates storm runoff, suggesting that catchments store water for considerable periods of time but then promptly release it during storms (Kirchner, 2003). The existence of thresholds also helps to explain abrupt changes in hydrologic response and water delivery to channels with different hillslope wetness states (Kampf and Mirus 2013).

Downslope water movement below the water table can also be quite complex as a result of spatial variations in the depth and rate of movement of ground water. The water table responds separately in riparian and hillslope zones, for example (Seibert et al. 2003), so that upslope-area ground water can be falling during the early portion of runoff while the riparian water table is rising. The hillslope ground water can be slowly falling as part of the recession from rainfall several days earlier. Ground water close to the stream is more likely to be in phase with runoff.

Other sources of complexity in ground-water dynamics include regional ground-water flow between watersheds (Genereux and Jordan 2006). Small ground-water reservoirs such as those in mountainous regions can respond to seasonal processes such as snowmelt runoff (McDonnell et al. 1998; Clow et al. 2003). Deep infiltration into bedrock can be important where bedrock is close to the surface or where highly porous and permeable or highly fractured bedrock takes up substantial volumes of water (Flint et al. 2008). This type of deep subsurface flow can be reflected in slow or doubly peaked runoff response (Onda et al. 2001), in which the two peaks reflect unsaturated and saturated zone flow.

Subsurface flow typically dominates slopes with full vegetative cover and thick regolith (Dunne and Black 1970a). Thicker soils increase the mean residence time of water on slopes and damp the temporal fluctuations of water movement in response to precipitation inputs (Sayama and McDonnell 2009). Water moving downslope via Hortonian overland flow typically has the most rapid rate of movement (50–500 m/h), with progressively slower rates during saturation overland flow, throughflow, and ground-water flow, which can move as slowly as 1×10^{-8} m/h (Selby 1982). The exception to this generally slower movement from the surface to progressively deeper paths is concentrated throughflow in pipes or macropores, which can move quite rapidly (Figure 2.3).

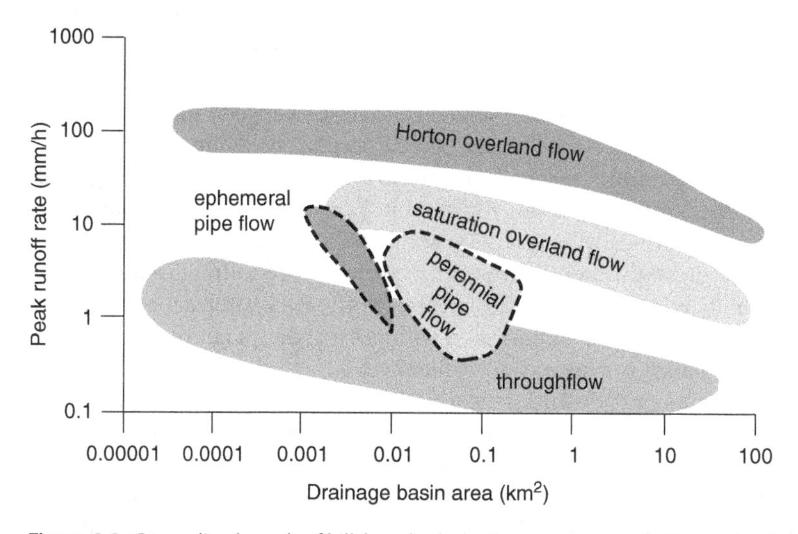

Figure 2.3 Generalized graph of hillslope hydrologic processes in relation to the size of the contributing area (after Jones 2010, Figure 7b). The drainage basin area in this figure reflects the total contributing area of the channel upstream from the point at which runoff enters a channel, but does not necessarily reflect the continuous extent of the area generating runoff. A 100 km^2 drainage basin area, for example, does not mean that Horton overland flow has to persist for tens of kilometers before entering a channel.

The distribution of water among different downslope pathways can alter in relation to precipitation magnitude, intensity, or duration during a storm, or on a regular annual basis in strongly seasonal climatic regimes. The dominant runoff processes also change with spatial scale (McDonnell et al. 2005). The manner in which moisture is released from pores within soil might condition runoff within a soil column, whereas partitioning between preferential and nonpreferential flow governed by soil structure and rain intensity becomes more important at the plot scale. Spatial variation in soil depth strongly influences lateral water movement in transient subsurface saturated areas at the hillslope scale, and the proportion of water derived from different catchment geomorphic units such as hillslopes, riparian zones, and bedrock outcrops influences the river hydrograph at the catchment scale (McDonnell et al. 2005).

Hillslope flow paths can be conceptualized as occurring along a spectrum from predominantly vertical to predominantly lateral (Elsenbeer 2001). Lateral flow, in particular, is highly nonlinear and exhibits threshold behavior that influences downslope connectivity as small depressions in the underlying bedrock topography fill and spill or as laterally discontinuous soil pipes and macropores self-organize into larger preferential flow systems as sites become wetter (Sidle et al. 2001; Hopp and McDonnell 2009).

Karst terrains, cold regions underlain by permanently frozen ground, and the humid tropics form three distinctive subsets in terms of regional patterns of downslope pathways of water. *Karst* terrains have distinctive landforms and drainage as a result of increased rock solubility in the presence of natural water; typically, these terrains are associated with carbonate rocks and evaporites. Water flowing at the surface in karst terrains over rocks of low solubility can move very abruptly downward to the ground water via *swallow holes*, which are open cavities on the channel floor where the channel flows onto carbonate rocks. Swallow holes can divert a portion or all of a surface river's discharge. Precipitation falling on karst terrains can percolate down to the ground water via diffuse infiltration in the zone of aeration, a process known as *vadose seepage*. Precipitation can also move downward via a highly permeable zone – typically formed by the intersection of vertical joints in the bedrock or cylindrical solution openings – at the base of a closed depression, a process known as *vadose flow* or *internal runoff* (Ritter et al. 2011).

The *epikarstic zone* within the vadose zone is a heterogeneous interface between unconsolidated material and solutionally altered carbonate rock (Jones et al. 2003). The epikarstic zone is partially saturated and can delay or store and locally reroute vertically infiltrating water into the deeper, regional phreatic zone of the underlying *karstic aquifer*. Aquifers formed in karst terrains differ from other types of aquifers in that karstic aquifers typically have larger variations in porosity and permeability. Ground water in these aquifers can move through intergranular pores within unfractured bedrock (*matrix permeability*), through joints and bedding planes created in the rock after deposition and lithification (*fracture permeability*), or through conduits with widths exceeding 1 cm that have been enlarged by solutional weathering (*conduit permeability*) (White 1999).

The extent of frozen soil strongly influences downslope pathways of water in cold regions. In very cold regions, an *active layer* that thaws seasonally and ranges from 15 cm to 5 m thickness overlies permanently frozen ground, or *permafrost*. Frozen soil impedes infiltration and limits percolation, with the result that a large portion of snowmelt moves downslope as overland flow and is quickly delivered to channels (Vandenberghe and Woo 2002). As the active layer thaws, the depth and importance of infiltration can change, but the presence of permafrost fundamentally limits deep infiltration and ground-water flow. Retreat of permafrost in response to global warming is creating major changes in downslope pathways of flow in cold regions.

Many catchments in the humid tropics have high runoff coefficients and quick hydrologic responses to precipitation. Infiltration rates in the humid tropics range from 0 to over 200 mm/h (Harden and Scruggs 2003), however, and runoff generation reflects numerous influences, as in temperate environments (Scatena and Gupta 2013). A variety of shallow subsurface flow paths contribute to the typically flashy response and high discharge per unit drainage area common in humid tropical catchments. Among these shallow subsurface flow paths are tension cracks that develop during the dry season and abundant macropores (Niedzialek and Ogden 2005).

The downslope pathways of meltwater through snow and glacial ice are as complex as those through slopes of sediment and bedrock. Outflow from a glacier includes a base flow component supplied by ground-water discharge, runoff from storage zones within the ice, runoff from the firn water aquifer at the glacier's surface, and regular drainage from lakes (Gerrard 1990). The melting of seasonal snow cover on the glacier and on nonglacial surfaces can produce an initial runoff peak, with subsequent melting of the glacial ice producing another runoff peak (Aizen et al. 1995). Runoff from the glacier can also vary during the melt season as channels develop on top of, within, and beneath the glacier (Fenn 1987; Nienow et al. 1998) and as air temperatures fluctuate (Hodgkins et al. 2009).

Runoff from snowpack depends on percolation times from the melt at the surface of the snowpack to the ground and on the distance downslope from the melt source. Travel times through the snowpack dominate runoff in watersheds $<30\,km^2$ in the Sierra Nevada of California, USA, for example, whereas snowpack heterogeneity results in more consistent timing of peak runoff in watersheds $>200\,km^2$ (Lundquist et al. 2005).

Ice- and snow-covered portions of a catchment can sometimes retain meltwater and rainfall to a greater extent than snow-free portions. The ability of the snowpack to retain or transmit rainfall partly depends on its structure, because the presence of ice layers can substantially increase water retention compared to a homogeneous snowpack (Singh et al. 1998). The proportion of a basin covered by ice and snow typically varies through the seasons as the transient snow line shifts, resulting in seasonal variations in the magnitude of rainfall-induced runoff (Collins 1998).

The ability to explain or even conceive of mechanisms of downslope movement of water fundamentally rests on the ability to measure water movement on and within hillslopes (Kirchner 2003, 2006; Robinson et al. 2008). This might sound obvious, but in practice it is extraordinarily difficult to obtain field measurements at relevant spatial and temporal scales, given that such scales span broad ranges. We need to know how individual water droplets diffuse through a porous medium or flow down an irregular surface at very small spatial and short temporal scales, as well as how much larger masses of water move downslope at larger spatial and longer temporal scales. Most importantly, we need to know whether there are thresholds differentiating specific types and rates of downslope water movement, such as those inferred to cause hillslopes to "turn on" when lateral connectivity is established between otherwise distinct subsurface reservoirs or when discrete pipe networks become active.

Because the unconsolidated material overlying bedrock exerts such an important influence on the downslope pathways of water, a great deal of effort is devoted to estimating the spatial distribution of soil depth (Pelletier and Rasmussen 2009; Tesfa et al. 2009) and soil moisture (Ritsema et al. 2009; Rajib et al. 2016). Measurement techniques include those at a point using various types of *in situ* sensors such as a wireless embedded network consisting of a series of probes and infrastructure nodes (Ritsema et al. 2009). Measurements can be obtained at intermediate scales using noninvasive methods such as ground-penetrating radar to map the depth to bedrock (Collins et al. 1989; Neal 2004). Measurements of soil moisture can also be made at basin ($2500–25\,000\,km^2$) and continental scales using remote sensing (McColl et al. 2017). Measurements are most common at point and basin/continental scales, leaving a gap at the intermediate spatial scales of individual hillslopes and small catchments, and limiting the ability to identify the emergent behavior of small watersheds (Kirchner 2006; Robinson et al. 2008). Geographic Information Systems (GIS) provide a mechanism for combining various types of soil moisture data (e.g. high-resolution multispectral images of soil moisture) with additional layers (e.g. topography and aspect) in order to conduct a detailed investigation of the spatial pattern of moisture distribution (Mattikalli and Engman 1997; Das and Mohanty 2006).

Much effort is also devoted to understanding how best to represent the distribution of soil depth and moisture in models (Chappell et al. 2005). Ongoing development of physically based models capable of including greater detail in soil profiles illustrates the linkages between soil hydrology and slope development (Brooks 2003) and drives further field characterization of soils, but representation of subsurface drainage remains one of the core challenges of hydrologic modeling (McDonnell et al. 2007; Ghasemizade and Schirmer 2013).

Indirect evidence of hillslope flowpaths comes from different rates of runoff response and discharge to channels (Onda et al. 2001). This can be taken to a higher level by using *hydrograph separation* based on various isotopes. The basic assumption underlying hydrograph separation is that different water sources (e.g. old hillslope water versus new snowmelt) have different isotopic signatures with respect to, for example, $\delta^{18}O$. Fluctuations in the $\delta^{18}O$ component of stream flow through time can thus be used to infer the relative importance of each water source through time (Mast et al. 1995). This type of approach is most accurate when the chemical signatures from diverse sources are different, are constant over time, and can be measured with high precision relative to this difference, and when the tracer is conservative and does not change in the stream channel but the mixing of water from different sources is complete within the channel (Caine 1989). Other naturally occurring tracers include sodium, chloride, dissolved silica, and dissolved organic carbon (Mast et al. 1995; Miller et al. 1995).

The simplest models of hillslope hydrology use a "black box" approach, in which the amounts or composition of precipitation, along with base flow inputs and stream outputs, are used to quantify catchment response to storms without revealing much about the processes of water movement downslope (Kendall et al. 2001). The first statistical and conceptual hydrologic models treated input parameters as lumped over the entire study catchment by ignoring the spatial variability of the physical system and its processes. Subsequent semidistributed and distributed models require data on the spatial and temporal variations in the hydrologic system, including topography, soils, land cover, land use, and precipitation (Kang and Merwade 2011).

Most models of hillslope hydrology are now semidistributed or distributed, even though these require better input data and can be more difficult to interpret because their output can be strongly sensitive to multiple parameters (Geza et al. 2009). Satellite data on vegetation, soil moisture, precipitation, snow cover, water level, and other hydrologic variables can provide input to distributed hydrologic models (Brown et al. 2008; Chaponniere et al. 2008).

Some hydrologic models include the entire catchment and the basic stages of precipitation, runoff, and stream flow routing during storm-flow and base-flow conditions (Downs and Priestnall 2003; Downs and de Asua 2016). Top-down approaches are based on data and derive a model structure directly from the available data (Post et al. 2005). Bottom-up approaches develop complex models to describe small-scale processes operating in a catchment without considering hydrological processes operating at the catchment scale. The small-scale processes are then upscaled to reproduce the hydrological response of the entire catchment (Post et al. 2005). The Representative Elementary Watershed (REW) approach provides a middle ground in which representative subwatersheds organized around the river network are coupled (Zehe et al. 2005). Thorough and useful reviews of measuring and modeling downslope pathways of water include Beven (2012) on rainfall–runoff modeling and uncertainties in ungaged basins, Brooks (2003) on models of hillslope hydrology, Barnes and Bonell (2005) on how to choose an appropriate model of catchment hydrology, and Götzinger et al. (2008) on models that couple surface and ground water systems. McDonnell et al. (2005) summarize the need to move from models that rely heavily on calibration to models that are based on extraction of first-order process controls and better understanding of flow sources and pathways at the basin scale.

Although knowledge of downslope pathways of water has increased substantially in the past few decades, the fundamental questions remain: Where does water go when it rains? What flow path does it take to streams? And, how long does water reside in the catchment? Knowledge of downslope water pathways is critical to fluvial geomorphology, because the resulting stream flow is a primary driving force in river networks.

2.2.2 Downslope Movement of Sediment

Sediment moves downslope through mass movements as slides, flows, or heave, and through gradual diffusive processes of rainsplash and overland flow. Diffusive processes typically involve individual grains rather than aggregates of grains. Whatever the transport mechanism, the portion of the soil profile actively transported downslope in steeper, forested terrains can reflect the rooting depth and consequent root-wad thickness of fallen trees, which in turn reflects depth to the soil/saprolite boundary (Jungers et al. 2009).

Mass movements involve downslope transport of aggregates rather than individual particles and are typically strongly seasonal as a function of moisture availability and freeze–thaw processes (Hales and Roering 2009). Mass movements are categorized based on the characteristics of the moving mass into slides, flows, and heave, each of which results from a decrease in the shear strength of the material or an increase in the shear stress acting on the material (Carson and Kirkby 1972).

A *slide* occurs when a mass of unconsolidated material moves without internal deformation along a discrete failure plane, which can be curved and produce the rotational movement of a slump or can be relatively straight. Slides typically result from a decrease in the shear strength of the soil as a result of weathering, increased water content, seismic vibrations, freezing and thawing, or human alterations such as deforestation and road construction. An increase in shear stress caused by additions of mass or removal of lateral or underlying support can also trigger a slide. Slides typically transition downslope into *flows*, which occur when the moving mass is sufficiently liquefied or vibrated to create substantial internal deformation during downslope movement.

Mass movements can also begin as flows initiated by runoff-dominated processes that cause progressive sediment entrainment or by infiltration-dominated processes that trigger discrete failures (Cannon et al. 2001). Mass movements resulting from runoff occur when sheetwash and rills entrain progressively more sediment downslope until the runoff concentrates in gullies and channels, eroding additional sediment from these conduits. This succession of events can occur during intense rainfall in arid environments, or after a wildfire that removes surface vegetation and plant litter (e.g. Wohl and Pearthree 1991; Meyer and Wells 1997). Infiltration can initiate local slope failures in the form of a slide that entrains more material as it moves downslope. This can occur during intense precipitation or rapid melting of a snowpack, or following a wildfire. Mass movements can also result from the "firehose effect," in which concentrated peak discharge at the outlet of a fan or the base of a steep, bedrock chute mobilizes accumulated sediment (Johnson and Rodine 1984).

Abrupt mass movements recur frequently – almost annually – in many high-relief terrains, and transport the majority of sediment to or along low-order stream channels (Jacobson et al. 1993; Guthrie and Evans 2007). Mass movements can exert diverse influences on valley geometry and the spatial arrangement of channels in drainage networks (Korup et al. 2010; Korup 2013). Hovius et al. (1998), for example, describe three phases of valley development in Papua New Guinea: initial incision of isolated gorges; lateral expansion and branching by landsliding in patterns influenced by ground-water seepage; and entrenchment by river incision of landslide scars and deposits. Large mass movements can cause substantial sediment accumulation in valley floors, interrupting progressive channel incision into bedrock, as documented along the Indus River in Pakistan by Burbank et al. (1996). The downstream end of mass-movement deposition in a valley bottom can create a knick-point, or steeper segment, along a river's course. Landslide dams can pond river water upstream, but are typically breached within a few days (Costa and Schuster 1988), resulting in an outburst flood with a peak discharge and sediment transport capacity that greatly exceed precipitation-generated floods along the river (Cenderelli and Wohl 2001, 2003). Small catchments can form in the detachment area of a mass movement. Large landslides can move so much mass that they reduce local elevation, cause drainage divides to shift location abruptly, or truncate headwater streams and cause stream piracy. Landslides can even cause drainage reversal. Mass movements also strongly influence spatial and temporal variations in sediment delivery to rivers in steep terrains (Korup et al. 2004, 2010). Mass movements can also occur in low-relief areas, particularly where an impermeable layer such as frozen ground (Kokelj et al. 2015) or shale promotes saturation of overlying materials.

Creep occurs when particles displaced by bioturbation and in wetting–drying or freeze–thaw cycles move downslope under the influence of gravity (Kirkby 1967). Bioturbation via tree throw and burrowing by soil-dwelling rodents can displace substantial amounts of sediment (Heimsath et al. 1997; Roering et al. 2002; Gallaway et al. 2009). Creep in cold climates is strongly influenced by freeze–thaw processes (Hales and Roering 2007) and, with increasing water or ice content, grades into solifluction or gelifluction, respectively, which involve the very slow downslope flow of partially saturated regolith. Creep is greatest in the upper meter of the soil and is proportional to surface gradient (Selby 1982; McKean et al. 1993). This results in soil flux proportional to the depth-slope product (Furbish et al. 2009b).

Gradual diffusive processes, in which individual grains move downslope, typically occur via rainsplash or overland flow. *Rainsplash* occurs when rain falling on a surface loosens or detaches individual particles, making the particles more susceptible to entrainment by overland flow (Furbish et al. 2009a; Dunne et al. 2010). Overland flow can be capable of eroding measurable quantities of sediment where unvegetated, unfrozen slope surfaces are exposed (Dingwall 1972; Rustomji and Prosser 2001). *Thread flow* occurs when overland flow goes around individual roughness elements. Sediment can be stripped evenly from a slope crest and upper zone during *sheet flow,* which submerges individual roughness elements and forms a fairly continuous sheet of water across the slope. Erosion by sheet flow is particularly effective where inter-particle cohesion has been reduced by needle ice, trampling, or disturbance of vegetation (Selby 1982). Microbiotic soil crusts, very coarse particles at the surface, or vegetation cover can substantially reduce sediment detachment and erosion by rainsplash or sheet flow (Uchida et al. 2000). The presence of seasonally or permanently frozen soil can enhance overland flow and soil erosion during the melt season by impeding infiltration (Ollesch et al. 2006).

Horton (1945) proposes the existence of a *critical length of slope,* X_c, above which no erosion occurs via sheet flow. The critical length is a function of surface resistance and slope gradient. As a simple approximation, sheet flow erodes particles as a function of the shear stress τ exerted on the surface by the flowing water ($\tau = \gamma H S$, where γ is the specific weight of water, H is flow depth, and S is hillslope gradient) relative to the resistance of the surface over which the water is flowing. For a given surface resistance and flow depth, X_c will be shorter on steeper slopes, up to about 40°. Sheet flow becomes competent to transport sediment within a few meters of the drainage divide on long hillslopes subject to Hortonian overland flow, but microtopographic mounds generated mostly by biotic processes can force the sheet wash to converge and diverge, increasing depth, velocity, and transport capacity in the converging zones. Sediment can be released from microtopographic mounds into the sheet flow, however, so that the sediment supply is sufficient to prevent rill incision on the upper portion of the hillslope (Dunne et al. 1995).

At some point downslope, surface irregularities concentrate overland flow into slight depressions, which then enlarge as the increasing water depth increases the shear stress acting on the substrate at the base of the flow. This can give rise to *rills* and *gullies*, typically described as parallel channels with few or no tributaries. Sometimes, these names are used interchangeably, but rills can be distinguished as channels sufficiently small to be smoothed by ordinary farm tillage, whereas gullies are sufficiently deep not to be destroyed by ordinary tillage. Rills and gullies can form effective conduits for sediment erosion down slopes and into river networks (Sutherland 1991), and rill-channel networks can exhibit equilibrium scaling characteristics for bifurcation and channel-length ratios similar to those of river networks (Raff et al. 2004).

Sklar et al. (2017) conceptualize the size distribution of sediment delivered from hillslopes to channels as reflecting five key boundary conditions: lithology, climate, life, erosion rate, and topography. The initial size of rock fragments detaching from bedrock is set by the spacing of bedrock joints. It is then modified by weathering, which depends both on chemical weathering as influenced by climate and mineralogy and on particle residence time on a hillslope as influenced by erosion rate and soil depth.

Downslope movement of sediment tends to be extremely spatially and temporally variable as a result of local changes in slope gradient, ground cover, vegetation, and microtopography (Saynor et al. 1994). Most hillslopes are shaped through time by some combination of processes (Jimenez Sanchez 2002). Just as subsurface water in a hillslope can be "old water" stored for relatively long periods of time between precipitation inputs, so sediment that begins to move downslope and into a channel network can be stored for 10^3 years even in mountainous uplands. In the Oregon Coast Range, for example, sediment transit times along colluvial valley segments dominated by debris flows average 440 years, whereas within the headwater fluvial portion of the channel network they average 1220 years (Lancaster and Casebeer 2007). In both debris-flow and fluvial channels, significant volumes of sediment remain in storage for thousands of years. Consequently, terraces, fans, and some floodplain deposits near tributary junctions are effectively sediment reservoirs at time spans of 10^2–10^3 years (Lancaster et al. 2010).

The relatively slow rates of soil production and downslope creep make direct measurement of these processes difficult. An exception occurs when a surface has been disturbed, such as following a wildfire. After such a disturbance, direct measurements can be made over limited areas using techniques such as erosion pins and sediment fences. For erosion pins, lowering of the ground surface relative to the top of rebar pounded into the slope, or some other benchmark, is measured over time scales of days to a few years. Sediment fences are fabric fences installed across small, ephemeral channels that allow water to pass but retain sediment; the accumulated sediment can be measured every few days or weeks. These local measurements are then extrapolated to larger areas based on considerations such as grain size, slope steepness, and severity of disturbance (Wagenbrenner et al. 2006; Larsen and MacDonald 2007).

In most cases, rates of downslope sediment movement are indirectly estimated using techniques such as cosmogenic ^{10}Be accumulation (McKean et al. 1993; Bierman and Nichols 2004). The cosmic neutrons that cause nuclide production are rapidly attenuated with depth, so cosmogenic isotope dating focuses on near-surface processes over time intervals in which a meter or two of rock or soil is eroded (Bierman and Nichols 2004). Such intervals can be less than 1000 years in rapidly eroding landscapes, or millions of years in very stable ones. Bierman and Nichols (2004) provide an excellent overview of cosmogenic isotope geochronologic techniques.

Processes acting more rapidly can be measured directly or simulated using physical experiments. Dunne et al. (2010) used data from lab experiments and simulated rainfall in a field setting, for example, to develop an equation capable of calculating instantaneous rainsplash transport as a function of the detachability of the soil, drop diameter, cover density, and fraction of sediment mass splashed downslope by each raindrop impact. Similarly, Roering (2004), used data from physical experiments to test a model for granular creep. Iverson et al. (2010) summarize numerous physical experiments on debris flows.

At the hillslope scale, feedbacks among erosion, sediment transport, weathering, and deposition are modeled using simulations such as TOPOG (O'Loughlin 1986) or SHETRAN (Burton and Bathurst 1998; Burton et al. 1998; Ewen et al. 2000). These can also be used to model shallow landslide initiation by coupling digital terrain data with near-surface throughflow and slope stability models (Dietrich et al. 1993; Montgomery and Dietrich 1994). Reid et al. (2007) investigated the generation of sediment from hillslopes and channel banks and its delivery to the channel network by coupling SHALSTAB (Montgomery and Dietrich 1994) and TOPMODEL (Beven and Kirkby 1979).

Landscape evolution models such as SIBERIA and CAESAR include hillslope processes as part of their simulation of catchment-wide erosion and deposition over time scales from tens to thousands of years (Hancock et al. 2010). Models of this level of spatial, temporal, and process integration include numerous assumptions that may be more applicable to some catchments than to others. With respect to downslope movement of sediment, for example, nonlinear slope- and depth-dependent models better reproduce actual topography and soil thickness from at least some field sites than do models in which flux varies proportionally with hillslope gradient (Roering 2008).

2.2.3 Processes and Patterns of Water Chemistry Entering Channels

The chemistry of precipitation falling over a drainage basin varies with distance from the ocean, with pollution inputs, and through time. The precipitation then reacts with plants, soil, regolith, and bedrock. The resulting chemistry of water entering a river is usually more influenced by the hillslope flow paths followed by the water than by the chemistry of the original precipitation. And, just as runoff during a storm can be dominated by old water that has been stored in the subsurface for some period prior, so can runoff chemistry be dominated by old water (Anderson and Dietrich 2001).

Waters entering a river via direct precipitation, overland flow, soil water, and ground water typically have distinctly different chemistries (McDonnell et al. 1991). The primary constituents of river chemistry are (Berner and Berner 1987; Allan 1995): the dissolved ions HCO_3^-, Ca^{2+}, SO_4^{2-}, H_4SiO_4, Cl^-, Na^+, Mg^{2+}, and K^+; dissolved nutrients N and P; dissolved organic matter; dissolved gases N_2, CO_2, and O_2; and trace metals. The sum of the concentrations of the dissolved major ions is known as the *total dissolved solids* (TDS), and is typically highly temporally and spatially variable in response to factors such as precipitation input and downslope flow path, discharge, lithology, the growth cycles of terrestrial vegetation, and any factor that influences rates and processes of bedrock weathering and soil development. (Berner and Berner 1987). These factors include: topographic relief; climate; glaciation; snow cover; land use; and episodic events such as mass movements and volcanic eruptions.

The specific downslope flow paths followed by water exert an important seasonal or event-based influence on water chemistry. Seasonal fluctuations in snow-covered catchments can reflect changes in the primary flow path from the snowpack early in the melt season, to the shallow soil zone, and finally to the bedrock aquifer (Finley et al. 1995). Changes in primary flow path in turn correspond to strong temporal variations in solute chemistry because new snow, the snowpack, and snowmelt each have different solute chemistries as a result of isotopic redistribution during snow metamorphism and melting (Taylor et al. 2001). Rainfalls of differing intensity and duration that access different flow paths with distinct weathering environments as a function of depth below the surface can also produce different water chemistries (White et al. 1998; Mul et al. 2007).

Differentiating the effects of individual potential control variables on water chemistry is typically difficult. As stated earlier, high-standing oceanic islands in the southwest Pacific, for example, contribute a disproportionately large amount of the carbon entering the world's oceans relative to their surface area as a percentage of the global whole (Lyons et al. 2002). This disproportionate contribution likely reflects the effects of very high sediment yields per unit drainage area (which in turn reflect short, steep rivers with little sediment storage and high variability of rainfall), as well as human activities including deforestation and agriculture (Lyons et al. 2002).

Although water chemistry has traditionally been neglected by fluvial geomorphologists, understanding the influences on the chemistry of water before the water enters the channel network and once it is there is important for at least one reason: A significant proportion of the management of river process and form is now undertaken in order to meet legislated water quality standards. River characteristics such as flow depth, hyporheic exchange, substrate grain-size distribution, and stability, as well as the characteristics of aquatic and riparian communities, can strongly influence the chemistry of water in a river. These characteristics can potentially be managed to reduce or remove contaminants or other undesirable traits of water chemistry that result from the downslope pathways that water follows before entering a channel. Understanding of the chemistry of water entering the river, however, remains important because of the influence these hydrologic inputs can exert on stream-water chemistry.

2.2.4 Influence of the Riparian Zone on Fluxes into Channels

"Riparian" is a Latin word designating something "of or belonging to the bank of a river" (Naiman et al. 2005). The *riparian zone* is the interface between terrestrial and aquatic ecosystems, and includes sharp gradients of environmental factors, ecological processes, and biotic communities (Gregory et al. 1991). Riparian zones can be difficult to delineate because they include components as diverse as depressions that create floodplain wetlands and higher-elevation natural levees (Figure 2.4).

The presence of a riparian zone can strongly influence the characteristics of water, solutes, mineral sediment, and particulate organic matter entering a river. Hillslope waters tend to be chemically and isotopically distinct from riparian-zone waters, and the degree of expression of hillslope water in the river can be minimal because of chemical reactions as it passes through the riparian zone (Burns et al. 2001; McDonnell 2003). Subsurface water coming from the riparian zone commonly leads the river

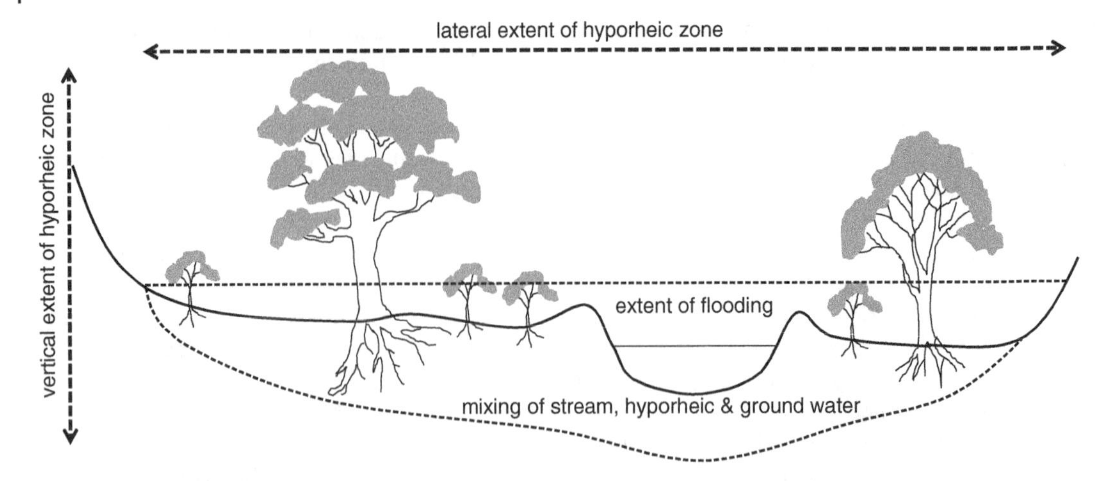

Figure 2.4 Schematic illustration of the spatial boundaries of a riparian zone. The boundaries extend: outward to the limits of flooding – typically defined for a relatively frequent recurrence interval of flood, such as 10 years; upward into the canopy of streamside vegetation; and downward into the zone where ground water, stream water, and hyporheic water mix.

hydrograph, with hillslope input either negligible or dominating the recession limb of the hydrograph after the threshold for activating hillslope pathways is exceeded (McGlynn and McDonnell 2003). These observations support the conceptualization of a catchment as a series of reservoirs with coupled unsaturated and saturated zones and explicit dimensions and porosities, which connect laterally and vertically through time and space in linear and nonlinear ways (McDonnell 2003).

Vegetated riparian zones can effectively trap and store particulate organic matter and mineral sediment of sand size or finer that is being transported by overland flow or by overbank flows from a channel (Naiman et al. 2005). Sediment and organic matter deposition reflects the greater hydraulic roughness, and consequently lower velocities and transport capacities, of flow passing over and through the stems and leaves of the vegetation (Hickin 1984; Griffin et al. 2005). Once the material is deposited, plant roots facilitate stabilization and continued storage of the material (Allmendinger et al. 2005; Tal and Paola 2007). Many riparian species are able to grow successive sets of near-surface roots as the plant's base is progressively buried by sedimentation.

Riparian zones also exert an important influence on the dynamics of carbon, nitrogen, and phosphorus in river corridors. Carbon is of concern because of its potential to serve as a greenhouse gas, and excess levels of nitrogen and phosphorus can cause eutrophication in freshwaters and nearshore areas. Riparian soils can receive substantial inputs of organic matter from riparian vegetation and overbank flooding, and, combined with a high water table that promotes reducing conditions in the soil, these inputs can result in large organic carbon stocks in the riparian zone (Hoffmann et al. 2009; Sutfin et al. 2016). Numerous studies indicate that riparian zones can serve as sinks or buffers for nitrates in agricultural watersheds (Mitsch et al. 2001). The capacity of the riparian zone to retain dissolved and particulate nutrients such as nitrogen, phosphorus, calcium, and magnesium is controlled by two factors. The first is the hydrologic characteristics of the riparian environment, including water table depth, water residence time, and degree of contact between soil and ground water.

The second level of control involves biotic processes, including plant uptake and denitrification. Denitrification is microbially facilitated nitrate reduction that ultimately produces molecular nitrogen, N_2 (Naiman et al. 2005). As an important nutrient and component of greenhouse gases, nitrogen has received particular attention, and the complexities of nitrogen dynamics illustrate the importance of riparian hydrology and biology. Ammonium-nitrogen (NH_4^+) is the primary form of mineralized nitrogen in most saturated soils (Mitsch et al. 2001). Ammonium can be absorbed by plants (through their roots) or anaerobic microbes and converted back to organic matter. Ammonium can also be immobilized through ion exchange onto negatively charged soil particles. The presence of a thin oxidized soil layer can create a gradient between high ammonium concentrations in the underlying, reduced soils and low concentrations in the oxidized layer. This gradient allows the ammonium to diffuse upward and be oxidized by bacteria through a process of nitrification to nitrate-nitrogen (NO_3^-). NO_3^- is mobile in solution and therefore readily enters streams if it is not assimilated immediately by plants or microbes. Denitrification is carried out by microbes in anaerobic conditions, resulting in loss of NO_3^-, which is converted to gaseous nitrous oxide (N_2O) and molecular nitrogen (N_2) (Mitsch et al. 2001).

As a result of these chemical reactions, riparian zones and floodplain wetlands can enhance nitrogen uptake by promoting denitrification of nitrate. Riparian zones can thus serve as buffers that limit nitrate entry from upland runoff (particularly agricultural lands or wastewater point sources) into rivers, if the velocity of the runoff entering the riparian zone is sufficiently reduced to allow sediment to settle from suspension (Mitsch et al. 2001).

Analogously, the balance between release and storage of phosphorus entering river corridors is strongly influenced by factors such as the moisture, pH, temperature, and biotic communities present in riparian soils (Records et al. 2016).

Biogeochemical hot spots for dissolved nutrients include anoxic zones beneath riparian environments, because of the microbial communities present in these zones (Lowrance et al. 1984). Riparian communities in wet temperate and tropical environments can mediate nitrate fluxes into rivers because denitrification and plant uptake remove nitrate that originated elsewhere, such as from ground water within a few meters of travel along shallow riparian flow paths (McClain et al. 1999). The effectiveness of this nitrogen removal can vary seasonally from very high values in late summer to much lower values in winter in temperate regions (Maitre et al. 2003). In contrast, little nitrogen processing may occur during transport from the uplands through riparian areas in arid regions if precipitation moves rapidly across the riparian zone as surface runoff (McClain et al. 1999).

In addition to uptake of nutrients by vascular plants and microbial communities, alluvial surfaces of different ages within the riparian zone can have different soil and water chemistries and infiltrations because vegetation influences evaporation and transpiration, which control soil water potential and the concentration of salts. Vegetation also creates litter: a surface layer of dead plant parts such as leaves or conifer needles that influences infiltration and evaporation from the soil, as well as contributing organic matter to the soil. Some plants also directly fix nitrogen in the soil (Van Cleve et al. 1993).

People have substantially reduced the extent of riparian vegetation and floodplain wetlands worldwide, causing associated changes in fluxes of water, sediment, and dissolved and particulate materials from these environments into channels. Although no global estimates of the total loss or degradation of riparian ecosystems have been published, numerous regional and basin-wide estimates suggest losses of much more than half of the riparian zones in industrialized countries, with consequent loss of flood attenuation, sediment storage, and biological processing of nutrients.

2.3 Human Influences on Fluxes from Uplands to Channels

2.3.1 Climate Change

Systematic sampling of atmospheric gases during the past half century, as well as air bubbles trapped in polar ice sheets, records increasing atmospheric concentrations of CO_2 and other compounds that permit a given volume of air to absorb infrared radiation. Increased CO_2 levels have caused measurable increases in air temperature since the start of systematic measurements, and particularly since the 1950s. International scientific panels estimate that continued increases in CO_2 will cause a rise in average global temperature during the twenty-first century of anywhere from 1 to 5 °C (IPCC 2008). Increases in atmospheric temperature cascade through atmospheric, oceanic, freshwater, and terrestrial systems as changes in numerous characteristics of these systems (Table 2.1). Even a brief

Table 2.1 Examples of how changing global climate will impact rivers.

Description	References
Because temperature and precipitation thresholds influence the magnitude and frequency of mass movements, climate change is increasing sediment yields to mountain rivers in the Himalaya and Tibetan Plateau	Lu et al. (2010)
In the semiarid/arid western United States, changes in air temperature, moisture, and thunderstorms alter the frequency and severity of wildfires, which causes changes in water, sediment, and nutrient yield to streams, as well as changes in forest structure and wood recruitment to streams; predictions include loss of floodplain forests	Hauer et al. (1997) Westerling et al. (2006) Rood et al. (2008)
In Alaska, USA, warming increases glacial runoff, which changes flow regime, water temperature, and sediment discharge, causing changes in channel substrate, bedforms, channel stability, leaf litter quantity and quality, and habitat complexity	Oswood et al. (1992)
Along arid-region rivers, changes in flood magnitude and frequency can cause channel incision, which removes the deep hyporheic sediments that support microbial communities; changes in precipitation and runoff can also alter the availability of nitrogen, which is a limiting nutrient in these rivers	Grimm and Fisher (1992)
As the proportion of precipitation coming in the form of rain (rather than snow) increases at high elevations, winter rains increase flood hazards and decrease ground water and summer stream flow by up to 50% in the mountains of central Europe	Eckhardt and Ulbrich, (2003)
Droughts in the United Kingdom during the late twentieth century highlighted the sensitivity of water resources to climatic fluctuations; predictions suggest that rainfall is likely to become more intense and less frequent in future	Wilby (1995)
In Australia, changes in rainfall will create the greatest runoff changes in arid catchments; some regions will receive more runoff (more intense and frequent rainfall in eastern Australia), others will receive less; frequency and severity of droughts will also increase	Chiew and McMahon (2002)
Changes in temperature and hydrology that promote disturbances to vegetation such as wildfire, insect outbreaks, and drought-related die off are predicted to increase sediment yield to rivers in the northern US Rocky Mountains; increased sediment yields will affect aquatic habitat and downstream water-storage reservoirs	Goode et al. (2012)
Increasing air temperatures in the monsoonal Mahanadi River basin of India during the twentieth century correlate with declining river flows	Rao (1995)

consideration of the extent and complexity of local to regional changes in river process and form associated with global changes in air temperature reveals the difficulty of predicting and adapting to the effects.

For river networks, changes in precipitation associated with changing atmospheric temperature are the most immediate manifestation of climate change. Most studies of the effects of climate change on rivers start with an examination of how altered precipitation characteristics will likely influence the river's flow regime. Predictions of future precipitation commonly rely on General Circulation Models (GCMs). At present, these models typically have a spatial resolution based on grid cells as small as 25 km (Molod et al. 2015). They can thus be used in examining large river basins, but smaller spatial scales require either statistical down-scaling to relate local climate variables to large-scale meteorological predictions (Gyalistras et al. 1998; Andreasson et al. 2003; Wang et al. 2016) or a more detailed regional model nested within the GCM (Giorgi et al. 1994; Marinucci et al. 1995). Inferred precipitation characteristics must be coupled with simulations of soil moisture, land cover, infiltration, and runoff (e.g. Azari et al. 2016).

Examples of the complexity of such modeling efforts come from studies in mountainous regions, where even a small increase in air temperature can significantly influence the distribution, volume, snow water equivalent, and snowmelt timing of mountain snowpacks (López-Moreno et al. 2009; Gillan et al. 2010). Nonlinear responses complicate attempts at modeling and prediction. Snowpack on a glacier, for example, reduces absorbed solar radiation and melt rate (Oerlemans and Klok 2004). Removal of the snowpack can increase daily discharge amplitude by more than 1000% (Willis et al. 2002). Upward retreat of rainfall–snow elevation limits may thus substantially increase rates of glacial melting. Changes in glacial mass balance are particularly important in mountainous regions because even relatively small alpine glaciers provide large reserves of freshwater, which sustain summer peak flows and autumn base flows. In populous, relatively dry lowlands such as parts of South America, India, and Pakistan, the disappearance of mountain glaciers and the associated runoff from higher elevations will substantially reduce water supply and likely cause social upheaval (Hasnain 2002; Barnett et al. 2005; Singh et al. 2006; Carey et al. 2017).

Changes measured during the past few decades can provide insight into such multifaceted alterations. Shifts in large-scale atmospheric circulation have resulted in altered flood frequency across Switzerland (Schmocker-Fackel and Naef 2010). Earlier snowmelt, reduced snow accumulation, and altered stream flow are particularly well documented for the European Alps, Himalaya, and western North America (Kundzewicz et al. 2007; Wagner et al. 2017). Regional-scale studies document how differences in characteristics such as subsurface drainage and the seasonal distribution of precipitation influence between-watershed differences in the sensitivity of stream flow to climate warming. In the Oregon Cascade Range of the western United States, for example, differences in bedrock geology between watersheds correspond to differences in the volume and seasonal flux of subsurface water (Tague et al. 2008; Tague 2009). Watersheds dominated by extensive, low-relief basaltic lava flows have deeper ground-water flow and are predicted to show greater absolute reduction in summer stream flow under expected temperature increases than are watersheds dominated by shallow subsurface flow, where stream flow is already flashy and highly seasonal.

At the broadest scale, modeling suggests that precipitation totals will increase in high latitudes (Bintanja and Selten 2014) and in mid-latitudes during winter (Osborne et al. 2015). Precipitation variability and the intensity of extreme precipitation will increase in the tropics (e.g. Knutson et al. 2010). Snow-cover duration will decrease and mid-latitude soil moisture may decrease in summer (Kattenberg et al. 1996; Mote et al. 2005). The largest hydrological changes are predicted for snow-dominated basins of mid to higher latitudes (Nijssen et al. 2001; Berghuijs et al. 2014). Glacial retreat, earlier snowmelt, and reduced snow accumulation already affect most mountainous regions in the mid-latitudes (Kundzewicz et al. 2007), as illustrated by warming-induced changes in the snowmelt runoff of the Rocky Mountains in western North America (Sridhar and Nayak 2010). Spring and summer snowmelt in rivers of the Rocky Mountains typically makes up 50–80% of the total flow (Hauer et al. 1997). The peak of snowmelt runoff shifted toward earlier in the year by more than 20 days at some measurement sites during the period 1948–2000 (Stewart et al. 2004, 2005) as a result of warmer springtime air temperatures. These changes will continue, and snowmelt-dominated basins at lower elevations will likely shift to rain-dominated flow regimes. Sites with multidecadal climate and stream-flow records indicate a decrease in the proportion of snow to rain, the snow-water equivalent, and the length of the snow season, with corresponding shifts in stream flow (Mote et al. 2005; Nayak et al. 2010). Rivers in the central and southern Rocky Mountains will decline to half of their historical spring runoff volumes, whereas spring runoff may substantially increase in rivers of the northern Rockies (Byrne et al. 1999). Rapidly retreating glaciers and snow cover in other mountainous regions of the world will cause decreased summer peak flows and autumn base flows (Barnett et al. 2005; Hagg and Braun 2005; Zierl and Bugmann 2005; Koboltschnig et al. 2007). At least one modeling study of how climate changes will alter stream flow has now been published for most regions of the world and for most major river basins.

Understanding and preparing for the implications of these changes at the scale of individual drainage basins requires a great deal more effort, but it is worth emphasizing that hydroclimatic changes that influence river ecosystems are not hypothetical. Many regions of the world already exhibit statistically significant differences between the latter half of the twentieth century and earlier periods (Marchenko et al. 2007; Van Der Schrier et al. 2007; Rood et al. 2008). As of 2012, successive records for the smallest June snow cover extent had been set each year in Eurasia since 2008, and in three of the past five years in North America (Derksen and Brown 2012). These trends have continued since 2012 (Hori et al. 2017).

2.3.2 Altered Land Cover

Alterations in land cover began thousands of years ago when people started to grow crops, domesticate grazing animals, and even use fire to alter vegetation communities in favor of plants preferred by the animals that they hunted. Primary types of land-cover alterations include deforestation, afforestation, grazing, crop growth, urbanization, upland mining, land drainage, and, more locally, commercial recreational property development. Alterations in land cover can change downslope pathways and quantities of water, sediment, solutes, and organic matter entering river corridors.

2.3.2.1 Deforestation

People have reduced global forest cover to about half of its maximum extent during the Holocene and have eliminated most old-growth forests (Montgomery et al. 2003a). Forestry (wood harvest, with anticipated regrowth of the forest) is now practiced in many industrialized countries, but historical deforestation was seldom conducted in a manner conducive to forest regrowth, and minimal attention is currently paid to regrowth in many areas experiencing deforestation in developing countries. Consequently, the discussion that follows applies primarily to clearcutting, or complete removal of forest cover over areas of varying spatial extent, rather than to selective cutting of a limited number of trees within a forest.

An extensive literature documents the effects of deforestation on rivers from mountains to lowlands and from the boreal regions to the tropics. Thorough reviews are provided in Foley et al. (2005), Scanlon et al. (2007), Douglas (2009), and Wohl (2010). Cutting of trees and the associated building of roads typically greatly increase hillslope sediment yield over a period of a decade or less, and water yield over periods of multiple decades, until vegetation recovers. Deforestation in the tropics, however, can decrease water yield, because of reduced transpiration and precipitation (Costa 2005; Wohl et al. 2012a). At present, the three main humid tropical forest regions of South America, Africa, and southeast Asia have particularly accelerated and widespread deforestation (Drigo 2005; Curtis et al. 2018).

Sediment yields increase as mineral soils exposed and compacted during deforestation become more susceptible to erosion via overland flow, rilling, and landslides and debris flows. Bank erosion, headward expansion of channels, and windthrow of remaining trees can also increase sediment yields. Roads decrease hillslope stability by redistributing weight, changing surface angles, reducing infiltration, altering conveyance via culverts under roads, and initiating gullies. Unpaved roads continue to contribute increased fine sediment yields to channels long after harvested vegetation has regrown (Jones et al. 2000). Water yields increase as removal of vegetation reduces interception and transpiration, and compaction decreases infiltration. Changes in sediment and water yield cause changes in stream flow, stream chemistry, and channel morphology. Depending on the magnitude and timing of alterations in sediment and water yields, numerous channel changes can result indirectly from timber harvest and road building. Changes in channel morphology can persist for at least several decades (Madej and Ozaki 2009). Associated processes such as recruitment of instream wood may require two centuries to return to pre-cutting levels, because trees must grow to maturity and then be recruited into and retained within channels (Bragg et al. 2000).

The degree to which water and sediment yields, and thus rivers, are altered depends on:

- the portion and spatial distribution of deforested areas within a drainage basin (Whitaker et al. 2002);
- the methods used to harvest trees – clearcutting typically has substantially larger effects on water and sediment yields than selective logging (Cristan et al. 2016);
- site characteristics including climate, topography, and soils (Liébault et al. 2002); and
- the connectivity between deforested areas and channels – this can be strongly influenced by the extent and placement of roads, which are major sources of sediment and can provide effective conduits for water and sediment from deforested areas to adjacent channels, as shown by gully development from road outlets to streams in southeastern Australia (Croke and Mockler 2001).

Clearcutting an entire drainage causes major changes in water and sediment yield and rivers. Selective cutting, with riparian buffer strips of unaltered vegetation left in place, minimizes change.

2.3.2.2 Afforestation

Afforestation, or the regrowth of forest cover, whether natural or human-induced, can reverse some of the trends created by deforestation. Documented effects of afforestation include increased infiltration and base flow, decreased runoff and sediment yield, and channel narrowing and deepening (e.g. López-Vicente et al. 2017). The magnitude and timing of these changes can be complicated by continuing remobilization of sediment eroded during deforestation and stored in colluvial and alluvial features within the drainage basin (Larsen and Román 2001). If afforestation occurs as a result of declines in population in regions that have been extensively deforested for centuries, as is now occurring in portions of the European Alps, channels that had become stable under former land use patterns can become unstable under newly changing water and sediment yields (Latocha and Migoń 2006).

2.3.2.3 Grazing

Upland grazing can have many of the same effects as timber harvest. These include reduced and altered vegetation cover, soil compaction, reduced infiltration, increased runoff and sediment yield, and associated changes in stream flow, channel morphology, and channel stability (Trimble and Mendel 1995). The more intense and widespread the grazing, the more severe these effects. Upland grazing can be practiced with minimal effect if the density of grazing animals is kept below a level that negatively impacts vegetation type and density, and if it does not compact soils. Overgrazing characterizes mountain livestock husbandry worldwide (Hamilton and Bruijnzeel 1997), however, leading to land degradation and altered river process and form.

2.3.2.4 Crop Growth

As with deforestation and upland grazing, the planting of crops alters soil infiltration capacity, soil exposure and erosion, and thus water and sediment yields to channels (e.g. Kuhnle et al. 1996). Compacted footpaths and roads around crop lands exacerbate these effects (Ziegler et al. 2000). Land drainage associated with crops can increase or decrease runoff (Blann et al. 2009).

The magnitude of alterations in water and sediment yield depends on the type and extent of the crops, as well as topographic and soil characteristics at the site. All crops do not create equal changes. Replacement of grains with potatoes in nineteenth-century Poland, for example, increased sediment yields and flood peaks to the point that meandering channels became braided (Klimek 1987).

The most common effect of planting crops is increased sediment yield and associated changes in channel pattern and stability. This effect can be documented in sedimentary records of prehistoric land use (Mei-e and Xianmo 1994). Compilation of numerous case studies indicates that soil erosion under conventional agriculture exceeds rates of soil production and geological erosion by up to several orders of magnitude – a situation that is clearly unsustainable with respect to soil fertility (Montgomery 2007). Crops can also increase nutrient fluxes to channels as soil nutrients are lost and excess fertilizers are mobilized from uplands (CENR 2000; Wang et al. 2004; Boyer et al. 2006).

2.3.2.5 Urbanization

Wolman (1967a), in a classic study of the effects of urbanization on sediment yield and channel response, delineates a sequence of changes within a small watershed in the Piedmont region of Maryland, USA (Figure 2.5). Subsequent studies from the tropics to the high latitudes document similar sequences that differ only in the details of magnitude and timing (Chin 2006; Gurnell et al. 2007; Chin et al. 2013).

Removal of vegetation and leveling or artificial contouring of the land surface during the initial phase of construction dramatically elevates sediment yields and causes aggradation and planform changes in receiving channels. Completion of construction stabilizes ground surfaces beneath roads, buildings, and lawns, causing sediment yield to decline to a negligible value and triggering bed coarsening, incision, and bank erosion in receiving channels. Water yield increases as impervious surface area within the contributing basin increases. Storm sewers rapidly drain urban areas, further decreasing the infiltration and conveyance time of water, creating flood peaks of shorter duration and higher magnitude for at least small to moderately sized precipitation inputs, and exacerbating channel erosion. (Where storm sewer drainage is fed into the sanitary sewer system, or into buffer

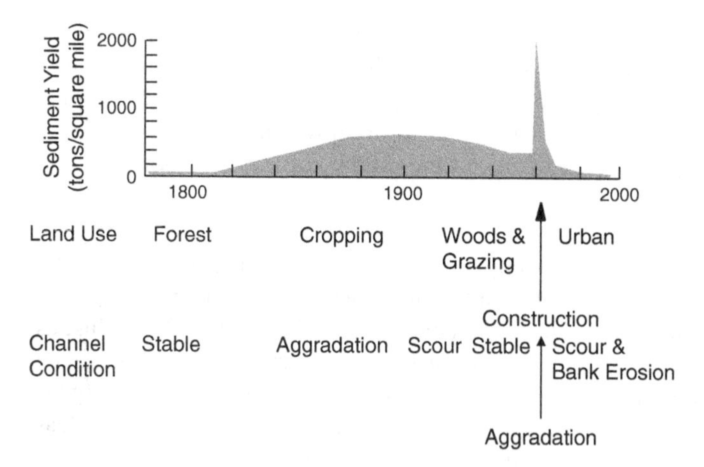

Figure 2.5 Schematic illustration of changes in sediment yield per unit drainage area through time in relation to changes in land use and channel condition. Source: After Wolman (1967a), Figure 1, p. 386.

areas planted with vegetation, these effects can be minimized.) The magnitude, extent, and speed of change in channel networks reflect factors such as the percentage of the catchment that becomes impervious, the connectivity and conveyance of impervious surfaces, and the characteristics of the receiving channels (Bledsoe and Watson 2001).

Urbanization typically also involves direct alteration of channels via channelization, bank protection, and other engineering, as well as substantial changes in water chemistry because of contaminants moving with runoff and sediment (Paul and Meyer 2001; Booth et al. 2016). Headwater channels are commonly paved or otherwise filled, or else diverted into sewer systems, effectively eliminating this portion of the river network (Elmore and Kaushal 2008).

Urban wastewater treatment plants can introduce substantial nutrients such as nitrogen and phosphorus to rivers, particularly given that effluent from these plants can at times constitute the entire discharge of dryland streams (Carey and Migliaccio 2009). Rivers in or downstream from urban settings typically have elevated concentrations of a variety of chemical contaminants, from trace metals such as copper, mercury, lead, and zinc, through road salts that can corrode infrastructure and mobilize lead and copper in drinking-water pipes (Kaushal et al. 2018), to synthetic compounds such as pesticides that have varying levels of environmental persistence and toxicity (Wohl 2004a).

2.3.2.6 Upland Mining

Upland mining that occurs within a watershed but outside of the channel and floodplain can take the form of hard-rock mining of metals or of surface or subsurface excavation of coal, construction aggregate, or building stone. Extensive surface disruption or the creation of large tailings piles can increase sediment yield to channels (Harden 2006). Alteration of runoff pathways can increase peak runoff (McCormick et al. 2009). Treatment of ores or the concentration of naturally occurring toxic contaminants can substantially alter the chemistry of surface and subsurface waters and fluvial sediments (e.g. Stoughton and Marcus 2000). Acid mine drainage of sulfide minerals (Sullivan and Drever 2001), for example, can acidify surface and subsurface waters, although additional factors such as the presence of bacteria that enhance reactions between the water and sulfide minerals (Gough 1993) also become important. Extreme examples of upland mining impacts come from mountain-top removal in coal-bearing regions of the eastern United States (Palmer et al. 2010a), in which material overlying coal seams is removed to vertical thicknesses up to 300 m and dumped into adjacent headwater valleys, obliterating surface flow and valley topography. Contemporary surface mining commonly increases peak and total runoff in watersheds with mountain-top removal, but individual studies reveal significant variability in hydrologic response through time (Miller and Zegre 2014).

2.3.2.7 Land Drainage

Land drainage has been undertaken for centuries in many parts of the world in order to decrease the extent of standing water at the surface, lower the water table, and improve access to low-lying lands for agriculture and settlement (Davis 2006; Stephens and Stephens 2006; Simco et al. 2010). Just as most people today cannot really imagine how much wood was historically in rivers and how many rivers had a multi-thread planform, so we have trouble understanding how extensive various types of wetlands – marshes, swamps, fens, and mires – once were, and how much their drainage and loss has altered water and sediment dynamics across large regions (Vileisis 1997).

Field drainage systems are designed to control surface water and the water table via surface features such as bedded systems used on flat lands growing rice and graded systems used on sloping lands for other crops. Field drainage can also use subsurface features such as horizontal or slightly sloping channels or trenches, wells with pumps, and buried pipe drains. In many cases, an extensive network of collector and main drains exists to transfer water to a gravity outlet structure or pumping station. Drainage can result in soil compaction and increased runoff, soil erosion, and suspended sediment transport, although the changes are not necessarily substantial (Walling et al. 2003; Gramlich et al. 2018).

2.3.2.8 Commercial Recreational Property Development

Commercial ski resorts and other spatially extensive changes in land cover undertaken for recreational purposes, such as golf courses, can involve deforestation, road construction, alteration of topography, and the transfer and application of large volumes of water or pesticides. Such activities can alter water, sediment, and nutrient and contaminant yields to nearby channels, resulting in channel change (Keller et al. 2004; David et al. 2009). Very few studies have systematically evaluated the effects of commercial recreational property development on river networks (De Jong 2015).

An important consideration in reviewing the effects of changes in land cover on river process and form is that very few locations have had only one change in land cover through time. Most river networks have experienced multiple changes that overlapped in time and space. This complicated history, combined with factors such as equifinality, thresholds, lag times, and complex response, can make it very difficult to decipher cause and effect in relation to any particular past land cover alteration, or to predict responses to ongoing or likely future alterations in land cover.

A common research technique is to compare otherwise similar rivers with and without a particular land-cover change in order to detect the influence of the land-cover change on river process and form (e.g. Pizzuto et al. 2000). The river without human impacts reflects reference conditions (Section 8.5.1). This *paired watershed approach* is effective to the extent that other potential control variables for river process and form can be held constant. Where the absence of unaltered drainage basins makes such comparisons infeasible, paleoenvironmental records of past changes or numerical modeling of river response to altered water and sediment yield can be used to infer cause and effect. Understanding of process domain or river style (Section 8.2) can also be used to infer reference conditions and the evolutionary trajectory of a river (Fryirs et al. 2012) in the absence of a single, highly similar, reference watershed.

2.4 Channel Initiation

Channel initiation is a threshold phenomenon in which surface or subsurface flow concentrates and persists sufficiently to produce a discrete channel. The upstream boundary of concentrated water flow and sediment transport between definable banks is the *channel head* (Montgomery and Dietrich 1988, 1989), which separates the process domains of hillslopes or unchanneled hollows from channel networks (Dietrich and Dunne 1993). Banks can be defined in the field based on the presence of sediment transport (wash marks, small bedforms, armored surfaces) and an observable sharp break in slope. The channel head does not necessarily coincide with the location where perennial flow occurs, which is the *stream head.* Channel segments of ephemeral and intermittent flow can be present downslope from the channel head but upslope from the stream head even in wet regions.

Channel heads can occur on planar segments of a hillslope, but are most commonly found within hillslope concavities that concentrate surface and subsurface flow sufficiently to create channel erosion. A concavity that accumulates sediment upslope from a channel head is known as a *colluvial hollow* (Dietrich and Dunne 1978). Unchannelized hollows with convergent contour lines are known as zero-order basins (Sidle et al. 2018).

The locations of individual channel heads in even a small channel network can have substantially different drainage areas. This is not surprising, given the multiple factors that influence the location of a channel head. These factors include gradient, drainage area, infiltration capacity, porosity and permeability, and cohesion, each of which influences surface and subsurface flow paths and the erodibility of near-surface materials. The location of a channel head can also vary through time as a result of changes in climate or land cover that affect runoff, surface erodibility, and sediment supply (Montgomery and Dietrich 1992). Following a wildfire that killed vegetation and burned the surface layer of litter and duff on forested hillslopes in the semiarid Colorado Front Range, USA, surface runoff became more common, causing channel heads to migrate upslope and form at minimum drainage areas two orders of magnitude smaller than pre-fire minimum drainage areas (Wohl 2013a).

The distribution of channel heads across a drainage basin can reflect primarily surface or subsurface flow, some combination of the two, or mass movements (Figure 2.6). Regardless of the mechanisms, flow convergence facilitated by topography and stratigraphy promotes the concentration of flow that initiates channels (Dunne 1990; Dietrich and Dunne 1993).

Channel heads that reflect primarily surface processes can form via Hortonian or saturation overland flow that leads to rilling. The greatest surface irregularities and erodibility result in flow concentrated in a *master rill* to which adjacent slopes are cross-graded (Horton 1945). Steep slopes, high rainfall intensity, low infiltration rates, and high erodibility favor rilling (Dietrich and Dunne 1993). These conditions are characteristic of arid, semiarid, or disturbed landscapes dominated by Hortonian overland flow (Kampf and Mirus 2013).

Rills can develop nearly simultaneously across a terrain and then integrate into a network (Dunne 1980), although drylands typically have discontinuous headwater networks of short, actively eroding channel reaches separated by unchanneled or weakly channeled, vegetated, stable reaches (Tucker et al. 2006). Alternatively, channels can extend downstream during slow warping or intermittent

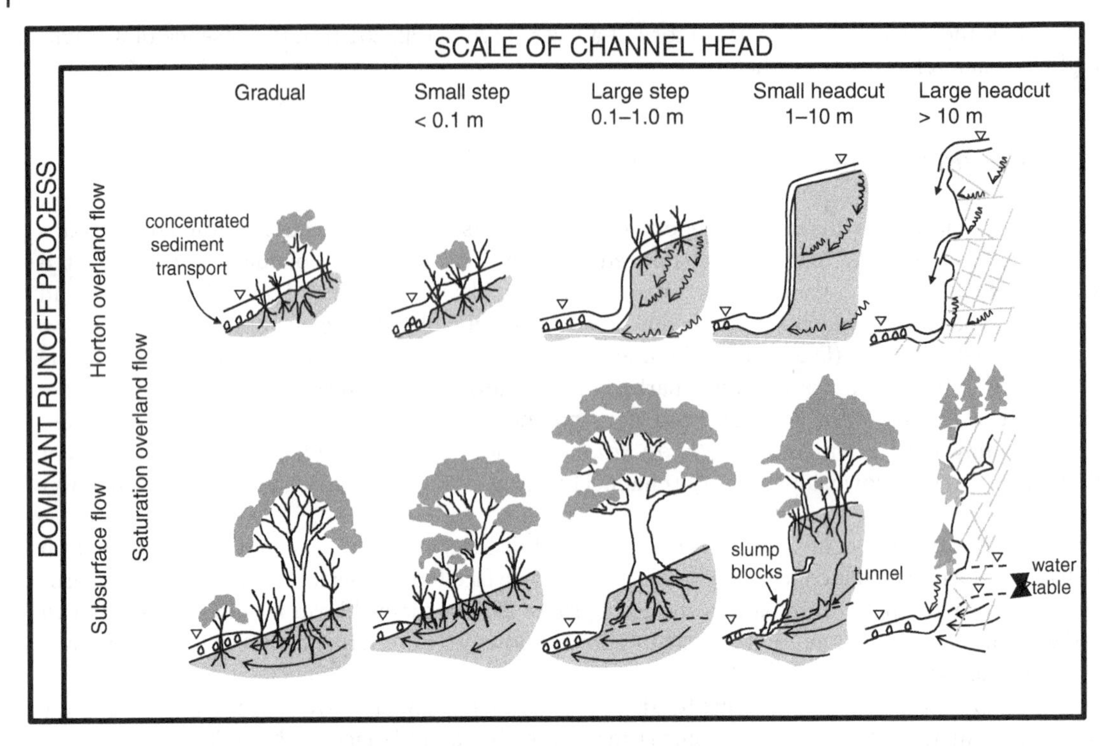

Figure 2.6 Classification of channel heads based on incision depth and dominant runoff process. Sketches indicate flow paths for Hortonian overland flow and subsurface flow. Smooth arrows indicate saturated flow; wiggly arrows indicate unsaturated percolation, including flow through macropores. Even at sites with substantial Hortonian overland flow, the face of a large headcut can allow the emergence of erosive seepage. Saturation overland flow drives erosion that includes features from both of the other runoff types. Source: After Dietrich and Dunne (1993), Figure 7.6.

exposure of new land on a rising land surface, or upstream in response to an increase in slope or the lowering of base level (Dunne 1980).

Erosional hot spots occur where topographic constrictions or locally steep gradients amplify hydraulic forces sufficiently to overcome surface resistance and initiate *headcuts* (Tucker et al. 2006), which may coincide with the channel head or occur downstream. Headcuts that coincide with the channel head are vertical faces that separate upslope unchanneled environments from downslope channels. Headcuts downstream from the channel head separate upslope presently stable channel segments from downslope recently incised or incising channel segments (Figure 2.7).

Channel heads dominated by subsurface processes can form via piping or sapping, or from shallow landsliding on steep slopes that creates a topographic low where subsurface flow can begin to exfiltrate (Montgomery et al. 2002; Kampf and Mirus 2013). *Piping* occurs in the unsaturated zone when flow is sufficiently concentrated to erode or dissolve subsurface materials and create physical conduits for preferential flow. If a pipe enlarges sufficiently to cause collapse of overlying materials, it can have a surface expression as a channel head. Piping typically occurs in unconsolidated material.

Sapping occurs in the saturated zone, which can intersect the surface to form a spring. The return of subsurface flow to the surface enhances physical and chemical weathering and creates a pore-pressure gradient that exerts a drag on the weathered material (Dunne 1980). Sapping can occur in unconsolidated material or bedrock, but is not likely to be an effective mechanism of

Figure 2.7 View upstream to a headcut along an ephemeral channel in the grasslands of eastern Colorado, USA. The headcut is just over 2 m tall. Circled, approximately 1 m to the left of the survey rod, is a buried tree stump, indicating a past cycle of incision followed by aggradation, prior to the present phase of incision. The channel upstream of this headcut is stable now, but has incised in the past.

erosion in the latter (Lamb et al. 2006; Jefferson et al. 2010) unless it is accompanied by chemical dissolution in carbonate rocks.

Early field-based studies of channel initiation indicated that the source area above the channel head decreases with increasing local valley gradient in steep, humid landscapes with soil cover (Montgomery and Dietrich 1988). For slopes of equal gradient, source area can vary in relation to total precipitation or precipitation intensity, as these characteristics influence the concentration of runoff (Henkle et al. 2011). Empirical, site-specific relations between topographic parameters typically use bounding equations to quantify the range in channel head locations (Montgomery and Dietrich 1989, 1992). An example comes from work by Prosser and Abernethy (1996) in Gungoandra Creek, Australia

$$A = 30 \tan \theta^{-16} \tag{2.3}$$

where A is specific catchment area (the ratio of upslope catchment area to lower contour width) and $tan\ \theta$ is hillslope gradient.

The inverse slope–area relationship is not always present, however, because of differences in runoff processes. Low-gradient hollows with convergent topography and seepage erosion can differ from steeper topography where channel initiation is more likely to reflect Hortonian or saturation overland flow, or even landsliding (Montgomery and Dietrich 1989; Montgomery and Foufoula-Georgiou 1993). In terrains with substantial flow through fractured bedrock, bedrock topography is likely to exert a greater influence on channel head locations than does surface topography (McDonnell 2003; Adams and Spotila 2005; Jaeger et al. 2007). Heterogeneities in the bedrock, such as spatial variation in joint density, can influence both the location of channel heads and the spatial distribution of channels within a river network, with channels tending to follow more densely jointed bedrock (Loye et al. 2012).

If channel heads form where saturation overland flow exerts a boundary shear stress that exceeds the critical value for substrate erosion, the channel initiation threshold, C, can be expressed as the product of contributing catchment area, A, and hillslope gradient, S (Dietrich et al. 1992, 1993)

$$AS^{\alpha} \geq C \tag{2.4}$$

Substantial variability in values of A and S reflects the influence of factors such as vegetation, slope aspect, surface versus subsurface flow paths, and substrate grain size (Montgomery and Foufoula-Georgiou 1993; Prosser et al. 1995; Istanbulluoglu et al. 2002; Yetemen et al. 2010) (Figure 2.8). In the absence of field-based data, many investigators assume that channel heads lie near reversals or inflections in averaged hillslope profiles (Ijjász-Vásquez and Bras 1995), although this can result in significant over- or underestimations of contributing area (Tarolli and Dalla Fontana 2009; Henkle et al. 2011).

Syntheses of area–slope relations for diverse regions suggest that channel head locations can be predicted with reasonable accuracy where channel heads form predominantly via surface erosion, but area–slope relations are much less consistent and predictable where subsurface processes strongly influence channel heads (Wohl 2018b).

Human-induced changes in land cover can alter the location of channel and stream heads by changing infiltration and the balance of water conveyed downslope in surface and subsurface flowpaths. The most typical scenario involves decreased infiltration and smaller contributing areas for channel and stream heads (e.g. Montgomery 1994). The location of a channel head over time spans of years to decades is best characterized as an average location that can and does change through time in a manner that reflects the disturbance regime of the hillslope and the speed with which hillslope-channel processes recover following disturbance (Montgomery and Dietrich 1994; Wohl and Scott 2017a; Wohl 2018b). Over longer time spans of centuries to millennia, the locations of channel heads reflect the interactions between processes of channel head extension and processes of diffusional infilling of valleys, as influenced by events such as glacial retreat, volcanic eruption and deposition of new surfaces, and deformation along faults that changes hillslope gradients or exposes new surfaces (Montgomery and Dietrich 1994).

The location of channel heads on the landscape, the processes that create channel heads, and the locational stability of channel heads are all important for several reasons (Wohl 2018b). First, channel heads are the start of the channel network and thus reflect the locations of thresholds that differentiate geomorphic process domains for water, solute, and sediment fluxes. Because distinct process domains are most effectively understood and modeled with different governing equations, being able to locate channel heads improves the accuracy of physically based models for hillslope hydrology, sediment routing, and landscape evolution (Montgomery and Foufoula-Georgiou 1993; Tucker and Hancock 2010). Second, the locations of channel heads reflect the degree of dissection of the landscape and thus provide insight into landscape evolution. Third, stream order (Strahler 1952) is commonly used in classifying streams for diverse characteristics, from habitat potential and nutrient uptake to stream assessment and monitoring. Stream order derives from mapped channel networks, and the accuracy of stream ordering depends on the accuracy of identifying channel heads. In the absence of mapped channel heads, remote methods that under- or overestimate the contributing area necessary to form a channel head can significantly over- and underestimate, respectively, the extent of first-order streams

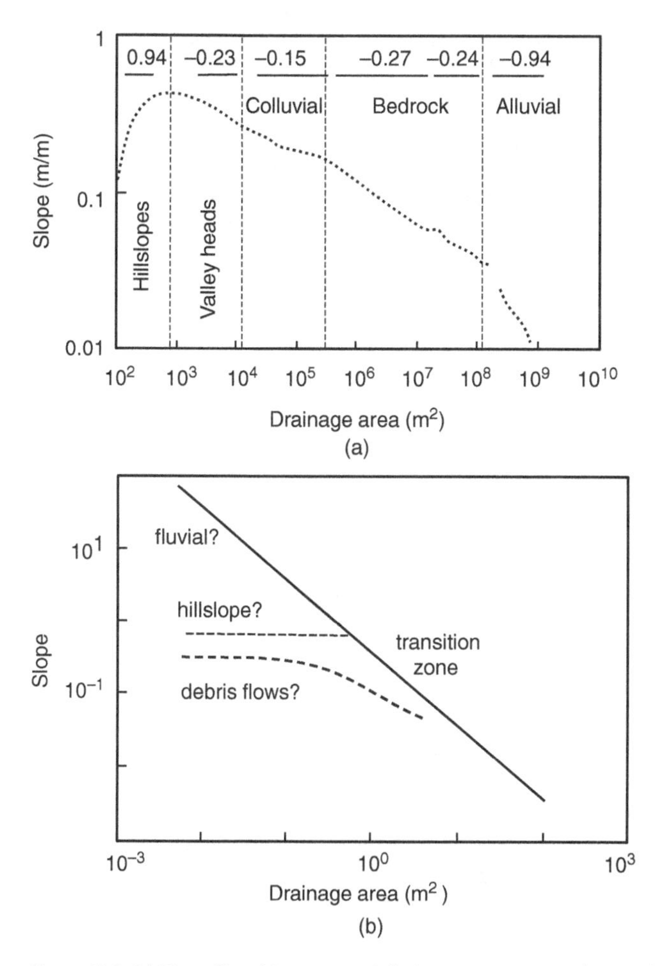

Figure 2.8 (a) Plot of log-bin averaged drainage area versus slope relationship for the Olympic Mountains of Washington, USA. Plot shows the mean slope of individual 10-m grid cells for each 0.1 log interval in drainage area. Numbers at the top are the exponent for a power function regression of values in the segments of the plot indicated by horizontal lines below. Dashed vertical lines divide the plot into areas considered to reflect different geomorphic zones of the landscape, or process domains (Source: After Montgomery 2001, Figure 5A). (b) Hypothetical topographic signatures for hillslope and valley processes. Area and slope are measured incrementally up valley mainstem to the valley head (Source: After Stock and Dietrich 2003, Figure 1a). Source: After Wohl (2010b), Figure 2.4.

and therefore result in erroneous ordering of larger streams. Finally, channel heads may be important in a regulatory context where the designation of a landscape feature as a river governs regulatory jurisdiction, as in the United States (Leibowitz et al. 2008).

2.5 Extension and Development of the Drainage Network

Glock (1931) proposes that networks go through five stages of development. Channel heads form and rills integrate through cross-grading during the stage of initiation. Stage two, extension by elongation,

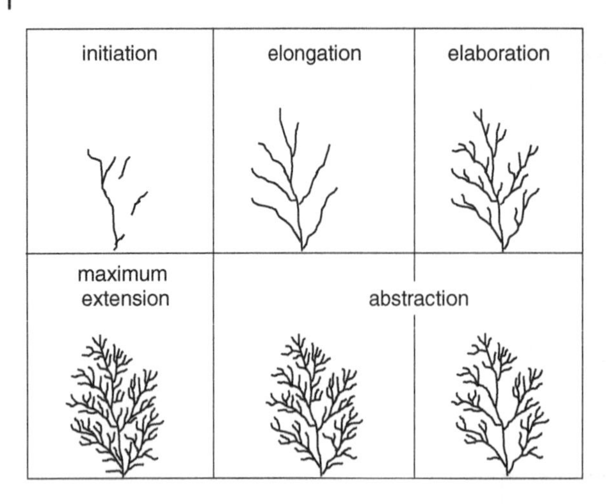

Figure 2.9 Illustration of Glock's stages of drainage network development. During initiation, a few, potentially longitudinally discontinuous channels are present. With elongation, these channels grow upslope and integrate into a continuous network. Additional, shorter, first-order streams are added during elaboration and maximum extension. As secondary slopes and small-scale relief are reduced by continued channel erosion, some first-order streams are lost along the downstream margins of the main channel during abstraction.

features headward growth of channels. During stage three, extension by elaboration, tributaries are added. This results in a stage of maximum extension, followed by the final stage of abstraction as local relief is reduced and tributaries are lost (Figure 2.9). Subsequent field-based studies that substitute surfaces of different ages for time (e.g. successive basalt flows or glacial deposits in the same region) tend to support this general model of network development, as do physical experiments (Schumm et al. 1987). Contemporary research is more likely to emphasize the role of geological variability through time and space (e.g. tectonic activity, lithology, rock structure) in creating spatial differences in the configuration and rate of evolution of drainage networks (e.g. Delcaillau et al. 2011; Prince et al. 2011).

Computer simulations and landscape evolution models based on statistical models, cellular automaton approaches, and stochastic rules are used to study drainage network development (e.g. Willgoose et al. 1990; Coulthard et al. 1997; Collins and Bras 2010). Models can be used to examine trends through time, as well as sensitivity to physical inputs and to assumptions about channel initiation and extension (Willgoose et al. 1991). Coulthard and Van de Wiel (2013) review types and applications of numerical simulations in fluvial geomorphology.

2.5.1 Morphometric Indices and Scaling Laws

Initial descriptions of drainage networks focused on their relationship to topography. An example is J.W. Powell's nineteenth-century characterization of antecedent and superimposed drainage networks. Powell inferred that a river cuts across a mountain range rather than flowing down and away from both sides of a mountain drainage divide either because the river maintained its location and cut downward as the surface was deformed around it (*antecedent*) or because the river eroded down to a buried structure (*superimposed*). Early descriptions also emphasized the categorization of networks based on their planform appearance (Zernitz 1932; Parvis 1950). Networks with a dendritic appearance typically indicate relatively low relief – homogeneous substrate, for example – whereas a rectangular network could indicate strong regional joint control that facilitates right-angle channel junctions where joints intersect.

Starting in the mid-twentieth century with work by Horton (1932, 1945), descriptions of drainage networks became increasingly quantitative and nondimensional. Horton, Schumm (1956), and

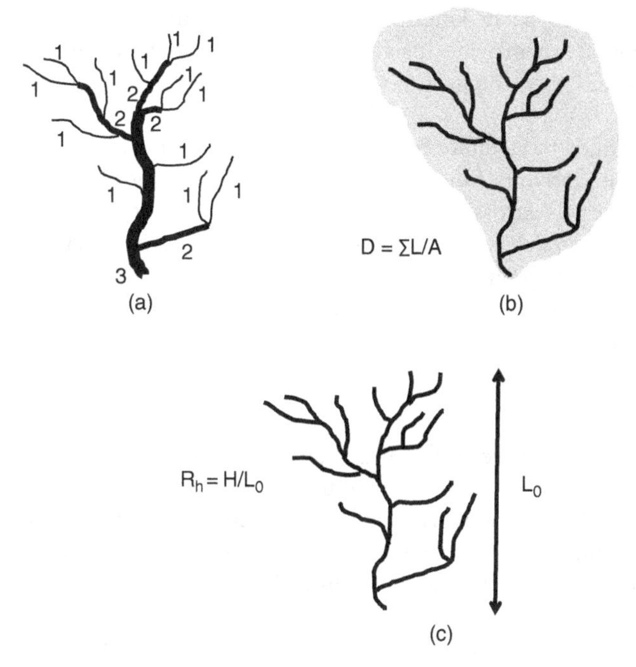

Figure 2.10 Commonly used stream morphometric indices. (a) Stream order, after Strahler (1952). This stream network is third-order. (b) Drainage density: the ratio of the sum of stream lengths per unit drainage area (area shaded gray). (c) Relief ratio, R_h: the ratio of the maximum topographic relief (elevation difference) H to the longest horizontal distance of the basin measured parallel to the main stream L_0.

Strahler (1957) modified earlier morphometric indices and developed new ones, helping to disseminate the idea of characterizing river networks using stream order, drainage density, relief ratio, and other indices (Figure 2.10).

Stream order is a number assigned to a stream segment. A segment with no tributaries is a first-order stream. Where two first-order segments join, they form a second-order segment; two second-order segments form a third-order segment, and so on (Strahler 1952). Stream order provides a convenient means of indicating the relative size of a channel segment or of a drainage network. Stream order is sometimes replaced by a link at the basic unit of network composition. A *link* is an unbroken section of channel between successive nodes (sources, junction, or outlet) (Shreve 1966). An *exterior link* corresponds to a first-order stream, in that it extends from a source to the first junction downstream. An *interior link* connects two successive junctions, or the last junction and the outlet (Knighton 1998).

Drainage density is the ratio of total length of streams to basin area. Drainage density reflects the degree to which a basin is dissected by channelized flow. Basin dissection reflects the relative influences of substrate resistance (higher resistance typically results in lower density), rainfall–runoff–land cover (high drainage densities typically occur in semiarid regions that receive more rainfall than arid regions but lack the continuous vegetation cover and associated higher infiltration of humid regions), and the age of the network (as reflected in Glock's stages of drainage development).

Relief ratio is the ratio of total basin relief to basin length. Relief ratio indicates the steepness of topography within a basin. Steepness can provide insight into tectonic uplift and rock resistance, as well as downslope pathways and the storage time of water, sediment, and solutes (e.g. Figueroa and Knott 2010).

The various morphometric indices are designed to facilitate comparisons across channel networks of diverse sizes, and to provide insight into underlying, fundamental characteristics of energy distribution, erosion, and the distribution of channels across a landscape. Horton (1945), for example, proposes a law of stream numbers and a law of stream lengths, and interprets their consistency across networks as suggesting independence of the specific geomorphic processes in a network. Hack (1957) proposes that consistent scaling relations can be used to describe network characteristics such as the relation of stream length L to drainage area A

$$L = 1.4\,A^{0.6} \tag{2.5}$$

when the units are miles (L) and square miles (A). Shreve (1966) and Smart (1968) compare statistical properties among populations of natural channels not strongly influenced by geologic controls, interpreting statistical similarity as indicating that the networks are topologically random and thus a consequence of random development according to the laws of chance. Milton (1966) and Kirchner (1993), however, propose that, as all branching phenomena could equally satisfy the Horton criteria, the laws of drainage network configuration are geomorphically irrelevant.

Similar controversy has arisen over more recent work describing drainage network patterns using fractal approaches (e.g. Rodríguez-Iturbe and Rinaldo 1997). *Fractals* are geometrical structures with irregular shapes that retain the same degree of irregularity at all scales and are thus self-similar. Investigators who have observed fractal properties in networks argue that this reflects a scale independence to landscape dissection. Beauvais and Montgomery (1997), however, note that a minimum contributing area is needed to concentrate surface or subsurface flow sufficiently to form channels, indicating a scale dependence to landscape dissection. Drainage networks in the western United States, for example, do not exhibit the scaling properties required for fractal geometry (Beauvais and Montgomery 1997).

Regardless of the insights that may or may not result from statistical self-similarity or consistent power functions between discrete variables, individual morphometric indices remain useful for inter-drainage comparisons and for understanding drainage network history.

A fundamental limitation on the ability to characterize any drainage network is how to accurately map actual channel locations. This is not particularly difficult for large channels, but can be very problematic for the smallest headwater tributaries. The default approach is typically to use the extent of "blue lines" indicating channels on topographic maps or digital elevation models (DEMs). The correspondence between these lines and real channels varies with the spatial resolution and type of data used to generate the map or DEM, and the consistency of channel extent through time (e.g. channel extent can vary substantially over a period of years in drylands or in areas with rapidly eroding headwaters). Another approach is to use contour crenulations (the degree to which contour lines are distorted when crossing a valley) as indicators that there is sufficient valley incision to create channelized flow (e.g. Pelletier 2013; Hooshyar et al. 2016), but the threshold value chosen for crenulation varies between studies and the correspondence between crenulation and channelized flow varies between regions. As noted in Section 2.4, reversals or inflections in averaged hillslope profiles can also be used to indicate the location at which channelized flow begins across a drainage basin. Each of these methods is subject to errors, and it is important to be aware of these errors as a source of uncertainty in network analysis.

2.5.2 Optimality

Various aspects of river form and process have been described using *extremal hypotheses*, which characterize the tendencies toward which rivers evolve based on the balance between the energy available to move water and sediment and to shape channels, and the resistance of the channel boundaries. The extremal hypothesis of *optimal channel networks* has been applied to understanding the spatial arrangement of channels in a drainage network. Optimal channel networks display three principles of optimal energy expenditure: (i) minimum energy expenditure in any link of the network; (ii) equal energy expenditure per unit area of channel anywhere in the network; and (iii) minimum total energy expenditure in the network as a whole (Rodríguez-Iturbe et al. 1992). The effect of the tendencies underlying these principles can be considered in terms of local and global optimal energy expenditure. In *local optimal energy expenditure*, the channel properties (e.g. channel cross-sectional geometry and gradient) of river networks adjust toward a constant rate of energy dissipation per unit channel area. In *global optimal energy expenditure*, the topological structure of the networks (e.g. drainage density, bifurcation ratio) adjusts to minimize total energy dissipation rate (Molnár and Ramírez 1998; Molnár 2013). Energy expenditure can be quantified using variables such as shear stress or stream power.

These principles largely grew out of work summarized in Leopold and Langbein (1962), Langbein (1964), and Langbein and Leopold (1964). Leopold and Langbein propose that the mean form of river channels represents a *quasi-equilibrium state*: the state most likely to occur, because it balances the opposing tendencies of minimum total rate of work in the whole river system and uniform distribution of energy expenditure throughout the system.

The tendency toward uniform distribution of energy expenditure can be understood by considering a portion of channel with higher rates of energy expended against its boundaries because of a lateral constriction or larger gradient that results in locally higher velocity. Assuming the channel boundary is erodible, the greater rate of energy directed against it should enlarge it until the rate of energy expenditure declines. An example comes from the Colorado River in the Grand Canyon, USA. Flash floods and debris flows along steep, ephemeral tributaries to the river create tributary debris fans that constrict the main channel. During large floods, these constrictions force the flow to become critical or supercritical (Section 3.1.1). The higher energy directed against the channel boundaries through the constriction during floods erodes the toe of each debris fan until the main channel is sufficiently widened to permit subcritical flow even during large floods (Kieffer 1989). Erosion then decreases and the channel cross-sectional geometry becomes stable, until the next tributary input again creates a constriction. In other words, sites of very high energy expenditure are transient in an erodible channel. Channel geometry adjusts to reduce energy expenditure at the site and create more uniform distribution of energy expenditure relative to upstream and downstream sites. Numerous adjustments of the type described for the Grand Canyon along the course of a river or throughout a drainage network can result in uniformity and minimization of energy expenditure, given sufficient time and energy relative to boundary resistance.

One of the points of discussion regarding extremal hypotheses is whether most natural channels ever actually attain an optimal state, given the potential for highly resistant boundaries or continual perturbations such as tectonic uplift or sediment inputs. An assumption underlying hypotheses of optimal channel networks is that place-specific details (e.g. lithology, structure, tectonics) that influence substrate resistance are absent or negligible. This is often not the case. Consequently, extremal hypotheses may describe the conditions toward which a system is evolving, rather than its actual state. Many natural river networks, however, do approximate the structure of optimal networks (Rodríguez-Iturbe and Rinaldo 1997; Molnár and Ramírez 1998; Rinaldo et al. 2014).

2.6 Spatial Differentiation Within Drainage Basins

Schumm (1977) conceptualizes a drainage basin as being longitudinally zoned with respect to sediment dynamics, differentiating a headwater production zone, mid-basin transfer zone, and downstream depositional zone. These zones represent predominant processes: the headwater zone does include transfer and deposition, for example, but is most characterized by sediment production. Subsequent research has elaborated on and quantified these distinctions. Benda et al. (2005), for example, describe headwater streams as sediment reservoirs at time scales of 10^1–10^2 years, along which sediment is episodically evacuated by debris flows or gully erosion. Sediment yield per unit area, or *sediment delivery ratio*, decreases and sediment residence time increases as stream order and drainage area increase as a result of increasing storage on hillslopes or valley bottoms (Schumm and Hadley 1961; Dietrich and Dunne 1978). These changes may also be a step function, with relatively large rates of change in sediment storage volume and time at transitions in geomorphic processes, such as where debris flows give way to predominantly fluvial processes (Lancaster and Casebeer 2007).

Rice (1994) describes a continuum of sediment transfer to channels, which represents another way of describing the sediment connectivity and storage discussed in Chapter 1. At one end of the continuum are *strongly coupled links*, in which sediment is transferred from hillslopes to channels relatively rapidly and continuously. At the other are *completely buffered links*, along which floodplains or valley-fill deposits protect hillslopes from basal erosion and limit direct sediment supply from hillslopes to channels. Schumm's headwater production zone would be best characterized as comprising strongly coupled links, whereas the proportion of completely buffered links would increase progressively downstream in the transfer and depositional zones. Again, these are general patterns: headwater channels can be buffered and downstream channel segments can be strongly coupled. Most importantly, the degree of coupling for any channel segment can vary through time.

Spatial differentiation of geomorphic process domains can be used to distinguish six basic types (Montgomery et al. 1996; Sklar and Dietrich 1998; Montgomery 1999) (Figures 2.8 and 2.11):

- *Hillslopes*: Hillslope in this context refers to the unchannelized, largely straight or convex portion of a slope, typically dominated by diffusive transport or slopewash.
- *Unchanneled hollows or zero-order basins*: Hillslope concavities that do not have channelized flow, but serve as sediment storage sites and are important points for the initiation of mass movements (Dietrich and Dunne 1978; Montgomery et al. 2009).
- *Debris-flow channels (or colluvial)*: These can be influenced by fluvial processes, but nonfluvial processes dominate sediment dynamics and channel geometry.
- *Bedrock-fluvial channels*: These have bedrock exposed along the channel boundaries or at sufficiently shallow depth that overlying alluvium is readily mobilized during higher flows, so that the underlying bedrock limits channel-boundary erosion. The distinctions among bedrock fluvial channels and coarse- and fine-bed alluvial channels reflect differences in the balance between

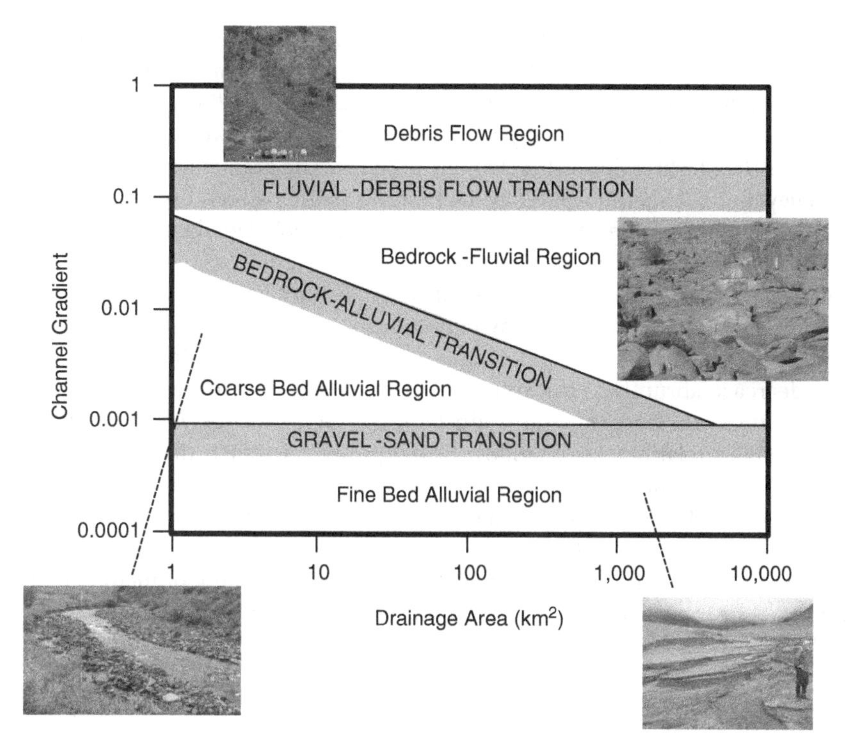

Figure 2.11 Hypothesized distribution of channelized erosional process domains (fluvial versus debris flow) and channel substrate types in relation to drainage area and slope. Source: After Sklar and Dietrich (1998), Figure 1, p. 241. (*See color plate section for color representation of this figure*).

sediment supply and river transport capacity. Where transport capacity exceeds sediment supply, bedrock is exposed in the channel bed.

- *Coarse-bed alluvial channels*: Channels with unconsolidated sediment coarser than sand size. Sediment mobility in these channels is likely to be limited by flow energy.
- *Fine-bed alluvial channels*: Channels with noncohesive sand-sized sediment forming the bed. Sediment mobility is more likely to be limited by sediment supply than by flow energy.

The transition between debris flow and completely fluvial channels is marked by changes in channel process and form resulting from differences in flow mechanics. Flow conditions can vary among debris flow, hyperconcentrated flow, and water flow downstream and with time during a flow as a result of changes in sediment concentration (O'Connor et al. 2001). Water and sediment move together in a *debris flow* as a single visco-plastic body that can be up to 90% sediment by weight, with a bulk density of 1.8–2.6 g/cm^3 (Costa 1984; Iverson 2005). A *hyperconcentrated flow* is a water flow with 40–70% sediment by weight and a bulk density of 1.3–1.8 g/cm^3 (Costa 1984; Pierson 2005). A *water flow* carries only 1–40% sediment by weight and has a bulk density in the range of 1.0–1.3 g/cm^3 (Costa 1984).

Differences in bulk density are important because bedload transport rate and maximum clast size can increase with increasing fluid density if the flow around the grains is not laminar (Rickenmann 1991). Debris flows can be highly erosive in steep or confined channels, and can create substantial aggradation in lower-gradient, less confined channel segments or at channel junctions (Benda 1990; Wohl and Pearthree 1991). Because they are capable of mobilizing clast sizes and volumes of sediment greater than those mobilized by fluvial processes, debris flows drive cycles of aggradation and degradation in the mountainous portions of many river networks (Benda and Dunne 1987; Anderson et al. 2015). Debris flows also strongly influence the gradient of side slopes, disturbance regime, and valley and channel morphology, including aquatic habitat, organic matter, and the structure and composition of riparian vegetation (Swanson et al. 1987; Florsheim et al. 1991; Hewitt 1998; Benda et al. 2003b; Rathburn et al. 2017).

Valleys consistently subject to different flow processes can have distinctly different geometries (De Scally et al. 2001). Valley-side slopes dominated by debris flows and fluvial erosion are more incised and contain more closely spaced channels, for example, relative to valleys dominated by landslides. Valley gradient can decrease abruptly at the transition from debris flow to fluvial channels (Stock and Dietrich 2006). Downstream from this transition, strath terraces (Section 7.2.1) can form and drainage area–stream gradient relations (A–S) follow fluvial power laws such that S varies as an inverse power law of A (Stock and Dietrich 2003). Debris flows in forested regions can entrain substantial volumes of wood, which they subsequently deposit as a dam, giving rise to aggradation, terraces, and outburst floods (Lancaster et al. 2003; Comiti et al. 2008; Rigon et al. 2008). Mountainous headwater channels subject to debris flows can display downstream coarsening of median bed surface grain size (Brummer and Montgomery 2003), in contrast to the more typical downstream fining in fluvial channels (Section 4.5).

2.7 Summary

Every process discussed in this chapter is characterized by diverse levels of spatial and temporal variability. Beginning with inputs and downslope movement of water, the details of how precipitation is produced and how water moves downslope via surface and subsurface pathways vary significantly across even a small catchment and through relatively brief intervals (hours to days) of time. Water movement, in particular, is characterized by thresholds. Once precipitation or glacier or snowmelt begins, the downslope pathways and rates of water movement change substantially as different components of a hillslope turn on, such as when a subsurface pipe network becomes active or infiltration gives way to surface flow. Similarly, sediment produced through bedrock weathering does not move uniformly downslope into rivers, but instead moves abruptly during mass movements or diffusive creep, which occurs primarily during precipitation, with sediment coming disproportionately from steep or otherwise more readily erodible portions of a catchment. Solutes entering the river come disproportionately from more readily weathered minerals or lithologies and are strongly influenced by movement through biochemically active portions of the catchment, such as riparian zones.

Channel and stream heads that mark the start of a river network reflect the processes that govern downslope movement of water and sediment, as these processes cause water to concentrate sufficiently to create definable, persistent channels. Threshold conditions of drainage area and hillslope gradient at which channels form vary among different catchments and within a catchment through time in response to differences in surface and subsurface downslope pathways of water and sediment.

A variety of morphometric indices have been used to characterize the spatial distribution of channel networks and topography within catchments as a means of facilitating comparison among catchments. Stream order, drainage density, and relief ratio are particularly widely used. Investigators have also used mathematical frameworks such as optimality and fractals to search for universal physical laws that underlie and can provide insight into observed patterns in river networks. This approach remains controversial at least in part because it is unclear to what degree actual river networks, with their departures from ideal mathematical forms as a result of spatial variation in substrate resistance, sediment inputs, and so forth, can be effectively described by optimality or fractals. At a more general level, most catchments can be usefully viewed as being spatially differentiated into:

- headwaters, characterized by relatively efficient movement of water, solutes, and sediment from uplands into first- and second-order channels;
- mid-basin zones, characterized by downstream transfer of water, solutes, and sediment, but also by some storage of these materials in valley-bottom environments such as alluvial fans and floodplains; and
- lower-basin depositional zones, characterized by more spatially extensive and longer-term storage of water, solutes, and sediment in well-developed floodplains and deltas.

Part I

Channel Processes I

3

Water Dynamics

This chapter introduces the basic physical properties of water flow within a channel. It begins with a discussion of hydraulics, which starts by explaining how flow is classified based on velocity, the ratio of viscous to inertial forces, and the ratio of inertial to gravity forces. It then describes measures of the energy of water flowing in a channel and the implications of energy level for stability of the flow and adjustments of the channel boundaries. Energy and stability are closely connected to sources of flow resistance and equations used to quantify resistance, as well as velocity and turbulence resulting from the interactions between available flow energy and resistance. Finally, the section on hydraulics returns to the idea of energy and introduces variables used to express the energy exerted against the channel boundaries. Understanding the physical properties of flowing water is necessary for everything that follows in this volume, because the quantity of energy available and the resistance created by the channel boundaries govern the movement of sediment and wood and adjustment of channel geometry.

The second part of the chapter first addresses methods used to estimate the volume of flow over differing time periods. This is followed by a discussion of the effects on stream discharge of surface–subsurface water exchanges, and the effects on discharge of flow regulation and channel engineering.

3.1 Hydraulics

Water flowing down a channel converts potential energy (PE) to kinetic energy (KE), some of which is dissipated. The rate and manner in which energy is expended depend on the configuration of the channel, including the frictional resistance of the channel boundaries and the amount of sediment being transported. Velocity is one of the most commonly measured hydraulic variables and is particularly variable in time and space because of its sensitivity to frictional resistance. The basic flow continuity equation introduced in Chapter 1, $Q = w\ d\ v$, is quite mathematically simple. Any of the dependent variables can be very difficult to predict in natural channels, however, even if Q is known, because each is influenced by other properties of the flow and channel boundaries. Width, for example, reflects Q, but also the ability of flow to erode the channel banks. Depth reflects Q, but also the erodibility of the bed. Velocity reflects Q, but also the frictional resistance of the channel boundaries. Width, depth, and velocity also influence one another. Increasing flow depth changes the resistance to flow, and hence the velocity associated with a particular bed grain-size distribution, for example, as the grains protrude into a progressively smaller proportion of the total flow depth.

Rivers in the Landscape, Second Edition. Ellen Wohl.

Knowledge of basic hydraulic properties is necessary to understand the complex interactions among flowing water, channel geometry, and sediment transport, and to understand the assumptions that underlie many of the equations applied to processes in natural channels.

3.1.1 Flow Classification

Much of the following discussion of basic hydraulics in open-channel flow comes from Chow (1959), a classic hydraulics textbook, from Fox and McDonald (1978), a classic fluid mechanics textbook, and from Robert (2014), which contains a thorough and comprehensive treatment of water and sediment dynamics in rivers.

Water flowing in a conduit can have either open-channel flow or pipe flow. *Open-channel flow* has a free surface at the boundary between the water and the atmosphere, and this free surface is subject to atmospheric pressure (Chow 1959). Figure 3.1 illustrates several important parameters used to characterize open-channel flow. For simplicity, individual flow lines are parallel and moving

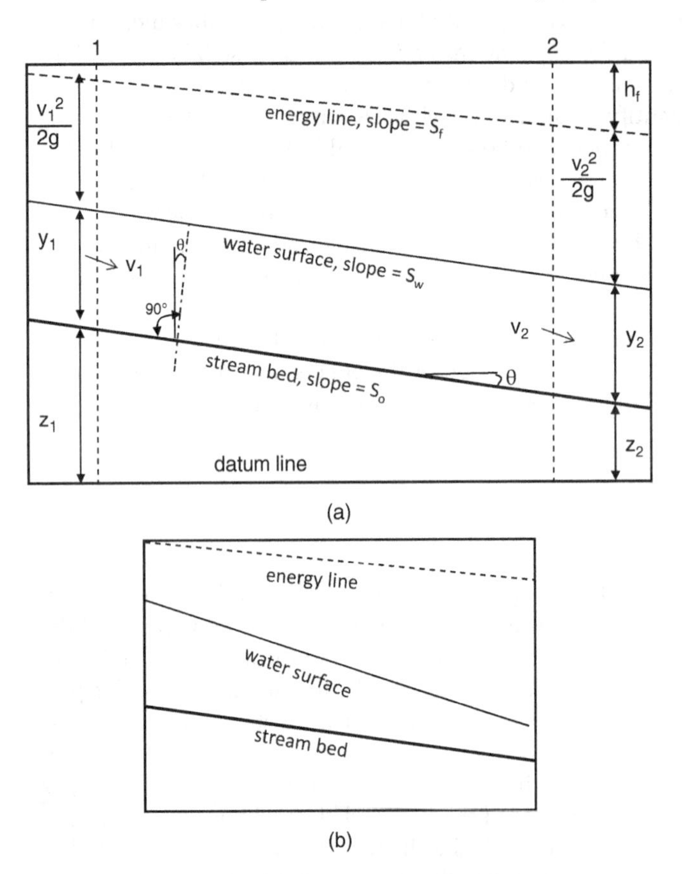

Figure 3.1 Parameters of open-channel flow; variables are defined in the text. Longitudinal view of a short channel segment, with flow from left to right. (a) Uniform flow; assumed parallel flow with a uniform velocity distribution and small channel slope. (b) Gradually varied flow, with velocity increasing and depth decreasing downstream. Source: After Chow (1959), Figure 1-1.

at the same velocity and the slope of the channel is small. The total energy in the flow of the section with reference to a datum line is the sum of the elevation z, the flow depth y, and the velocity head $v^2/2g$, where v is the mean velocity of flow and g is gravitational acceleration (9.8 m/s^2). The energy is represented by the *energy grade line* or *energy line*, and the loss of energy resulting from water flowing from section 1 to section 2 is h_f (Chow 1959). The depth of flow, the discharge, and the slopes of the channel bottom and the water surface are interdependent. The flow parameters in Figure 3.1 form the basis for procedures such as step-backwater calculations, used to estimate discharge from paleostage indicators (Section 3.2.2).

Water flowing in an open channel is subject to gravity and friction. External friction results from the channel boundaries. The portion of the flow closest to the boundaries typically moves more slowly than portions farther away from the boundaries as a result of external friction. External friction can vary widely in natural channels because of the presence of an irregular surface over individual grains and of bedforms, bends, changes in channel width, and other factors, as discussed in more detail in Section 3.1.3.

Internal friction results from eddy viscosity (Chow 1959). *Viscosity* represents the resistance of a fluid to deformation. The *molecular* or *dynamic viscosity*, μ, is the internal friction of a fluid that resists forces tending to cause flow (Robert 2014). The greater the dynamic viscosity, the smaller the deformation within the fluid for a given applied force and the lower the degree of mixing and turbulence. *Kinematic viscosity*, υ, is the ratio of molecular viscosity to fluid density ($\upsilon = \mu/\rho$; typically $\upsilon \sim 1 \times 10^{-6}$ m^2/s). Both types of viscosity decrease significantly as water temperature increases (Fox and McDonald 1978). The effect of changing density on kinematic viscosity is such that very high suspended sediment concentrations can increase kinematic viscosity (Colby 1964).

Eddy viscosity is friction within the flow that results from the vertical and horizontal circulation of turbulent eddies. Eddy viscosity expresses the vertical and horizontal transfer of momentum, or exchange between slower- and faster-moving parcels of water, and varies with position above the bed. The coefficient of eddy viscosity, ε, represents momentum exchange or turbulent mixing (Robert 2014). Considering only the vertical dimension within a channel, eddy viscosity can be expressed as

$$\epsilon = \iota^2\, dv/dz \tag{3.1}$$

where ι is the *mixing length* – the characteristic distance traveled by a particle of fluid before its momentum is altered by the new environment (Chanson 1999), dv is change in velocity, and dz is change in height above the bed (Robert 2014). The mixing length represents the degree of penetration of vortices within the flow and depends on distance from the boundary (a vortex is the rotating motion of water particles around a common center; Lugt 1983). The mixing length is assumed to be

$$\iota = \kappa z \tag{3.2}$$

where κ is the von Karman constant, which is 0.41 in clear water flowing over a static bed (Robert 2014). Understanding of eddy viscosity is important because eddies allow dissolved or particulate material carried in the water to spread throughout the flow field.

Open-channel flow is classified into types based on four criteria. The classifications segregate based on properties important to natural processes (Table 3.1). *Uniform flow* occurs when the depth of flow is the same at every section of the channel (Figures 3.1a and 3.2), and is very rare in natural channels. Flow is varied if the depth changes along the length of the channel, and can be either *rapidly varied* (Figure 3.2), where the depth changes abruptly over a comparatively short distance, or *gradually varied* (Figures 3.1b and 3.2).

Table 3.1 Types of open-channel flow.

Type of flow	Criterion
Uniform/varied	Velocity is constant with position/Velocity is variable with position
Steady/unsteady	Velocity is constant with time/Velocity is varied with time
Laminar/turbulent	$R_e < 500/R_e > 2500$ (transitional flow between)
Subcritical/critical/supercritical	$F < 1/F = 1/F > 1$

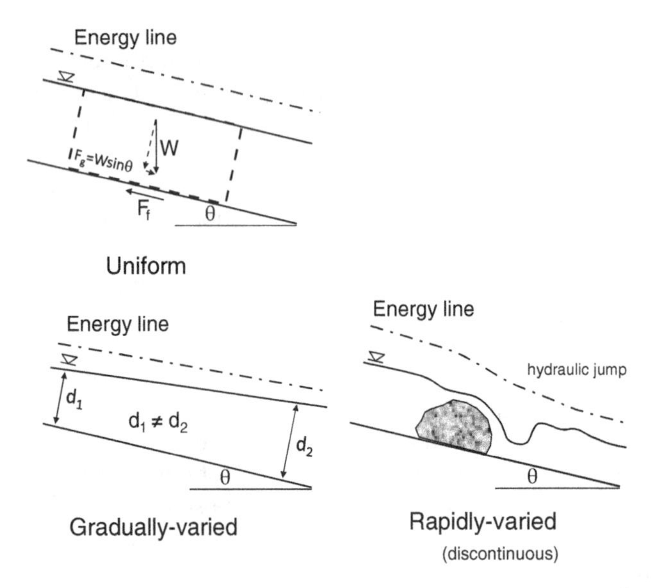

Figure 3.2 Illustration of the differences between uniform and gradually and rapidly varied flow. Longitudinal view of a short channel segment, with flow from left to right. Source: Courtesy of David Dust.

Uniform flow can be steady or unsteady. In *steady flow*, the local depth does not change, or can at least be assumed to be constant during the time interval under consideration. Like uniform flow, steady flow is rare in natural channels. *Unsteady flow* can occur as gradual variation among a series of steady flows, or as rapid variation (Figure 3.3).

Most natural flows are varied and unsteady, although the variations in time and space can be gradual. Although uniform flow was traditionally assumed for simplicity when calculating hydraulic parameters in natural channels (Chow 1959), many applications now assume gradually varied flow.

The state of open-channel flow is governed by the effects of viscosity and gravity relative to the inertial forces of the flow. The effect of viscosity relative to inertia can be represented by the *Reynolds number*

$$R_e = \frac{vR\rho}{\mu} \tag{3.3}$$

where v is velocity, R is hydraulic radius (cross-sectional area divided by the wetted perimeter; wetted perimeter is the wetted length of bed and banks at the cross section), ρ is mass density, and μ is dynamic viscosity.

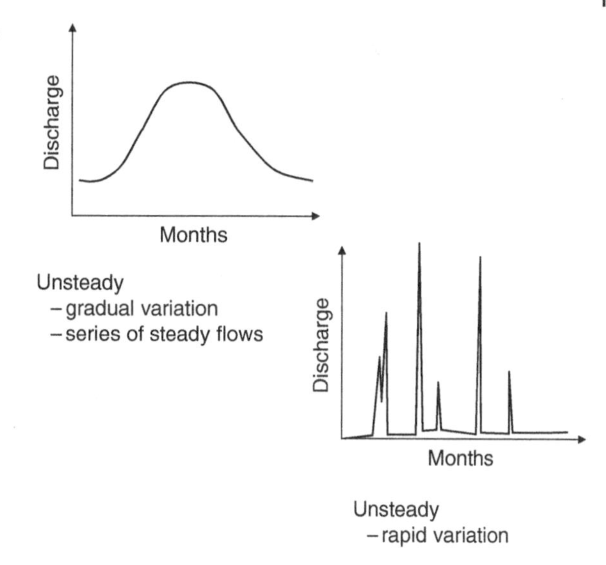

Figure 3.3 Idealized hydrographs for different types of unsteady flow. Source: Courtesy of David Dust.

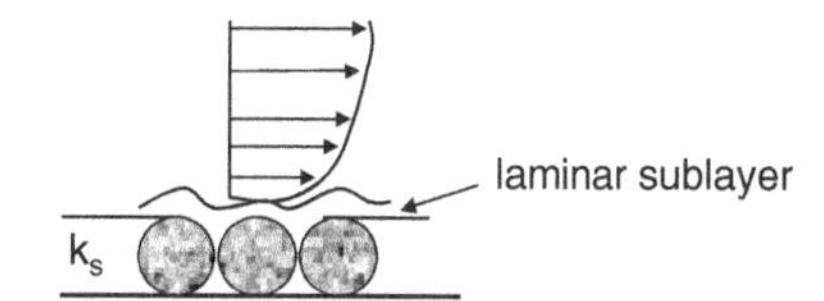

Figure 3.4 Side view of vertical velocity distribution in a channel (flow from left to right). Arrows indicate relative velocity and k_s indicates roughness height. Source: Courtesy of David Dust.

The Reynolds number can be used to differentiate laminar, turbulent, and transitional flow. In *laminar flow*, the viscous forces are sufficiently strong relative to the inertial forces that viscosity exerts a significant influence on flow behavior. Each fluid element moves along a specific path with uniform velocity and no significant mixing between adjacent layers. When the velocity or hydraulic radius exceeds a critical value, the flow becomes *turbulent*. The viscous forces are weak relative to the inertial forces. The fluid elements follow irregular paths and mixing occurs, involving transfer of momentum by large-scale eddies.

The transition from laminar to turbulent depends on the energy dissipation within the fluid: high viscosity equates to a need for more energy to mix the flow. A smooth or glassy water surface in a channel does not mean that the flow is laminar. The flow in most channels is turbulent, but laminar motion can persist in a very thin layer (typically <1 mm thick) next to the channel boundary known as the *laminar sublayer* (Figure 3.4).

The Reynolds number provides a useful indicator of the magnitude of mixing within a natural channel. The flow is typically turbulent when $R_e > 2300$ and laminar when $R_e < 2300$, although there is no single value of R_e at which the flow changes from laminar to turbulent (Fox and McDonald 1978).

The effect of gravity on the state of flow is represented by the ratio of inertial forces to gravity forces, as given by the Froude number, F

$$F = \frac{v}{\sqrt{gd}} \tag{3.4}$$

where d is the hydraulic depth, defined as the cross-sectional area of the water normal to the direction of flow in the channel divided by the width of the free surface (Chow 1959).

When $F = 1$, $v = \sqrt{gd}$ and the flow is in a critical state. When $F < 1$, $v < \sqrt{gd}$ and the flow is *subcritical*. When $F > 1$, $v > \sqrt{gd}$ and the flow is *supercritical*. *Critical flow* can occur when the flow is constricted or subject to a substantial increase in bed slope and passes through a state of minimum specific energy ($v/\sqrt{gd_c} = 1$, where d_c is critical hydraulic depth), as it funnels through or drops over the channel contraction or slope break (Section 3.1.2).

Subcritical flow has lower velocity and greater depth for a given channel configuration and discharge, and is stable and persistent. Subcritical flow is the usual flow state. Supercritical flow is inherently unstable, and changes in channel bed gradient or cross-sectional area can cause it to become subcritical, with implications for erosion and deposition along the channel, as discussed in the next section.

3.1.2 Energy, Flow State, and Hydraulic Jumps

As noted earlier, water flowing downhill within a channel is converting potential energy ($PE = mgh$, for water mass m and height h above a given datum) to kinetic energy ($KE = ½ mv^2$), some of which is dissipated at the microscale of turbulence and the molecular scale of heat. A *streamline* is a visual representation of the flow field in the form of a line drawn in the flow field such that, at a given instant of time, the streamline is tangent to the velocity vector at every point in the flow field (Fox and McDonald 1978). In other words, streamlines indicate the paths of flow, and there can be no flow across a streamline. The total energy in any streamline passing through a channel section can be expressed as the total head H, which is equal to the sum of the elevation above a datum, the product of the depth below the water surface d and the cosine of the bed angle θ, and the velocity head (Figure 3.1)

$$H = z + d\cos\theta + \frac{\alpha v^2}{2g} \tag{3.5}$$

For channels of small slope, $\theta \sim 0$ and $cos\ \theta \sim 1$. Because velocity is not uniformly distributed over a channel section, the velocity head of open-channel flow is typically greater than the value of $v^2/2g$. The true velocity head is expressed as $\alpha v^2/2g$, where α is the energy coefficient and typically varies from about 1 to 2 for natural channels (Chow 1959). The value is higher for small channels and lower for large, deep channels. Every streamline passing through a cross-section will have a different velocity head because of the nonuniform velocity distribution in real channels, but the velocity heads for all points in the cross-section can be assumed to be equal in gradually varied flow (Chow 1959).

The line representing the elevation of the total head of flow is the *energy line*, and the slope of the energy line is known as the *energy gradient*, S_f (Chow 1959). The water-surface slope is S_w and the slope of the channel bed is $S_o = sin\ \theta$, where θ is the slope angle of the channel bed. In uniform flow, $S_f = S_w = S_o = sin\ \theta$ (Chow 1959). Based on the principle of the conservation of energy, the total energy head at upstream section 1 (Figure 3.1) should equal the total energy head at downstream section 2 plus the loss of energy h_f between the two sections in parallel or gradually varied flow

$$z_1 + d_1 + \frac{\alpha_1 v_1^2}{2g} = z_2 + d_2 + \frac{\alpha_2 v_2^2}{2g} + h_f \tag{3.6}$$

This is known as the *energy equation*. When $\alpha_1 = \alpha_2 = 1$ and $h_f = 0$, this becomes the Bernoulli equation, which holds in friction-free flow

$$z_1 + d_1 + \frac{v_1^2}{2g} = z_2 + d_2 + \frac{v_2^2}{2g} = \text{constant} \tag{3.7}$$

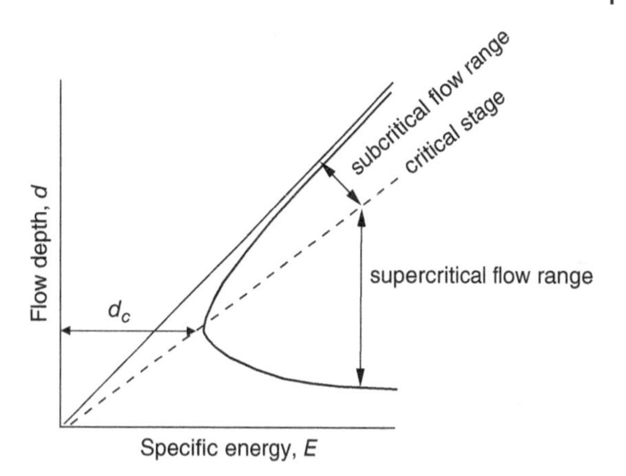

Figure 3.5 Specific energy curve; d_c is critical flow depth. Source: After Chow (1959), Figure 3-2, p. 42.

Specific energy in a channel section is the energy per kilogram of water at any section of a channel measured with respect to the channel bottom (Chow 1959). Using Eq. (3.5), with $z = 0$, specific energy E is

$$E = d\cos\theta + \frac{\alpha v^2}{2g} \tag{3.8}$$

For a channel of small slope and $\alpha = 1$,

$$E = d + \frac{v^2}{2g} \tag{3.9}$$

so that specific energy is equal to the sum of the flow depth and the velocity head. Because $v = Q/A$, $E = d + Q^2/2gA^2$. Therefore, for a given channel section and discharge, E is a function of only the depth of flow (Chow 1959). A *specific energy curve* is a plot of E and d for a given channel section and discharge (Figure 3.5).

For a given specific energy, two possible depths exist: the low stage and the high stage, each of which is an alternate for the other. At point C on the curve, known as *critical depth*, d_c, the specific energy is a minimum, which corresponds to the critical state of flow. At greater depths, the velocity is less than the critical velocity for a given discharge and flow is subcritical. At lower depths, the flow is supercritical. Specific energy changes as discharge changes, as shown in Figure 3.5. The criterion for critical flow is that the velocity head equals half the hydraulic depth ($v^2/2g = d/2$), assuming that flow is parallel or gradually varied, the channel slope is small, and the energy coefficient equals 1 (Chow 1959). Otherwise, $\alpha v^2/2g = (d \cos\theta)/2$.

A flow at or near the critical state is unstable because a minor change in specific energy will cause a major change in depth (Chow 1959). Consequently, extended areas of critical or supercritical flow are relatively uncommon in natural channels. The flow is likely to return to a subcritical state via a hydraulic jump.

When a rapid change in the flow depth from low stage to high stage occurs, the water surface rises abruptly in a *hydraulic jump* (Figure 3.6). This is common where a steep channel slope abruptly decreases. An *undular jump* forms when the change in depth is small and the water passes from low to high stage through a series of undulations that gradually diminish in size (Chow 1959). A *direct jump* occurs at a large change in depth and involves a relatively large amount of energy loss through dissipation in the turbulent flow within the jump.

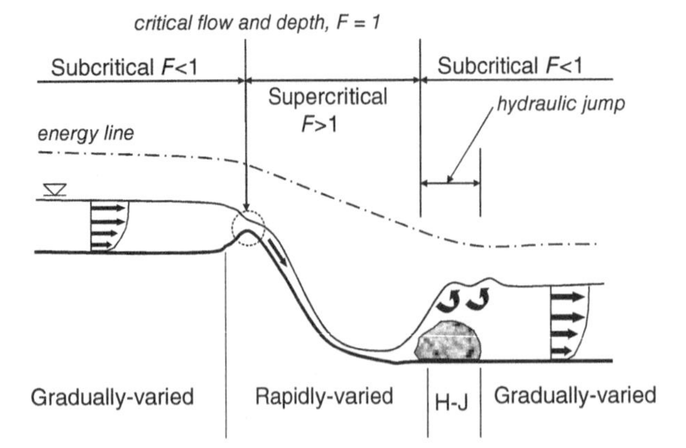

Figure 3.6 Longitudinal view of a hydraulic jump and the associated velocity distributions and flow conditions in channel segments immediately upstream and downstream from it. Flow from left to right. *F* is Froude number. Length of arrows indicates relative velocity. The boulder at the change in slope can localize the jump, but is not necessary for the jump to occur. The abrupt change in slope can be sufficient to create a hydraulic jump. Source: Courtesy of David Dust.

The geomorphic significance of a hydraulic jump is that it creates intense turbulence and large kinetic energy losses. Flow velocity decreases substantially across the jump and sediment can be deposited downstream, stabilizing the jump's position (Carling 1995). If the channel boundaries are adjustable under a given flow, sites of supercritical or critical flow are likely to be preferentially eroded so that the channel cross-section enlarges or the gradient declines until subcritical flow occurs (Kieffer 1989; Grant 1997). Natural channels thus exhibit feedbacks between flow energy and channel geometry that tend to maintain subcritical flow (Grant 1997).

3.1.3 Uniform Flow Equations and Flow Resistance

Both velocity distribution within a channel and average velocity are sensitive to boundary roughness that retards flow. Velocity can be directly measured with a velocity meter and the measured value can be used to calculate flow resistance. Or, flow-resistance coefficients can be used to estimate velocity under conditions in which direct velocity measurements are not feasible.

Water flowing down an open channel encounters resistance that is counteracted by gravity acting on the water in the direction of motion. Uniform flow occurs when the resistance is balanced by gravity forces. Velocity is related to flow resistance or boundary roughness using one of three commonly applied uniform flow equations, the Chezy, Darcy–Weisbach, and Manning equations. Because steady uniform flow is rare in natural channels, the results obtained by applying these equations are approximate and general (Chow 1959). The equations continue to be widely used because they are relatively simple to apply and provide satisfactory approximations for many natural channels.

The French engineer Antoine Chézy developed the first uniform-flow formula

$$v = C\sqrt{RS} \tag{3.10}$$

where v is mean velocity, R is hydraulic radius, S is the slope of the energy line S_f (commonly approximated as water surface slope S_w or bed slope S_o), and C is a factor of flow resistance that represents the ratio between the driving force RS and the velocity sustained in the presence of frictional resistance to flow. Secondary equations developed to calculate the value of C typically rely on an empirically determined coefficient of roughness.

Henry Darcy, Julius Weisbach, and other nineteenth-century engineers developed the approach now known as the Darcy–Weisbach equation

$$v = \left(\frac{8gRS}{f}\right)^{\frac{1}{2}} \tag{3.11}$$

This equation was developed primarily for flow in pipes, but the form stated here can also be applied to uniform and nearly uniform flows in open channels. The variable f is sometimes preferred over other friction factors because it is nondimensional and has a more sound theoretical basis.

Natural channels have irregular boundaries that can be more difficult to characterize in terms of resistance compared to flow in pipes. The Gauckler–Manning empirical formula was developed in the late nineteenth century by French engineer Philippe Gauckler, and subsequently modified by Irish engineer Robert Manning, to describe the relation between velocity and flow resistance in natural channels. The formula was later modified to its present form for metric units (i.e. v in m/s)

$$v = \frac{R^{2/3}S^{1/2}}{n} \tag{3.12}$$

where n is the coefficient of resistance, typically known as Manning's n. This is the most widely used formula for uniform flow.

The flow-resistance factors C and n relate to one another as

$$C = \frac{1}{n}\left(\frac{D_H}{4}\right)^{\frac{1}{6}} \tag{3.13}$$

where D_H is the hydraulic diameter, which is four times the hydraulic radius R (Chanson 1999).

Determining an n value – or a value for C or f – is the most difficult aspect of applying these equations, in part because n incorporates so many forms of roughness and flow resistance. Each of the resistance coefficients listed here includes multiple sources of resistance (Figure 3.7) (Chow 1959). Values of n are commonly quite variable in space and time because the n value tends to decrease as stage and discharge increase, although the rate of change in n value can be nonlinear because of interactions between the flow and the channel boundaries. The most common approach is to visually estimate the n value using standard tables (Chow 1959) or comparisons to field-measured values from diverse sites (Barnes 1967; Yochum et al. 2014b).

Numerous empirical equations also exist for estimating n values for particular types of channels, especially steeper, gravel-bed rivers (Table 3.2). Most methods for estimating a flow-resistance coefficient are not designed specifically for steep channels, for which n values are much greater than in low-gradient streams with similar relative grain submergence (R/D_{50}) values (Jarrett 1987). (In this and other flow-resistance formulas, D_x is the grain size for which x percentage of the cumulative grain size distribution is finer.) As gradient increases, energy losses increase as a result of wake turbulence and the formation of localized hydraulic jumps downstream from boulders (Jarrett 1992). Equations for steeper rivers with a wider distribution of grain sizes and coarse grains than sand-bed channels recognize the importance of relative roughness by including a ratio of hydraulic radius to a representative grain size. Much effort has been devoted to determining the appropriate representative grain size to use in steep channels with a wide range of grain sizes. Values in various studies range from

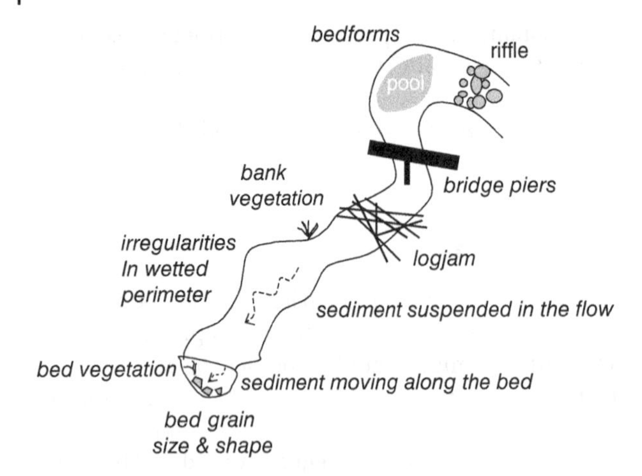

Figure 3.7 Schematic illustration of some of the sources of resistance (named in italicized text) that are lumped together in resistance coefficients such as *C*, *f*, and *n*. Sources of resistance in natural channels include boundary irregularities, vegetation along the channel boundaries, sediment forming the bed and banks, sediment in transport, obstructions, and channel bends.

$1.25D_{35}$ to $3.5D_{90}$ (Millar and Quick 1994), with many using $3.5D_{84}$ (Hey 1988) or $2D_{90}$ (Parker and Peterson 1980).

A key consideration in applying these uniform flow equations is that many natural channels have local accelerations and decelerations associated with protruding grains or bedforms, so that bed, water surface, and energy slopes may not be equal and uniform flow assumptions are not met. Under these circumstances, a uniform-flow resistance equation can be used to relate reach-averaged velocity to reach-averaged depth and bed or water surface slope if slope is measured over multiple obstacles and bedforms, starting and finishing at sections with the same depth and velocity (Ferguson 2012). Otherwise, S must be the energy line. The existence of protruding grains and bedforms also makes estimation of R very difficult. The most common approximation is to use a cross-sectionally or reach-averaged value of R in the flow-resistance equations.

The Chezy, Darcy–Weisbach, and Manning equations are based on the recognition that the velocity of water flowing in a channel is influenced by the channel boundary roughness. Within the *boundary layer*, velocity varies according to distance from the channel surface. Outside the boundary layer, the velocity distribution is practically uniform (Chow 1959). In many natural channels, the boundary layer extends to the water surface. As noted in Section 3.1.1, a laminar sublayer can be present at the base of the boundary layer. The top surface of the laminar sublayer corresponds to the transitional zone from laminar to turbulent flow and gives way to the turbulent boundary layer, in which the vertical velocity distribution is approximately logarithmic. The thickness of the laminar sublayer, δ, is defined by

$$\delta = 11.6\,\upsilon/v^* \quad (3.14)$$

where υ is kinematic viscosity and v^* is shear velocity, which has the dimensions of velocity and is determined from gravity g, hydraulic radius R, and channel gradient S

$$v^* = \sqrt{(gRS)} \quad (3.15)$$

Table 3.2 Empirical equations for Manning and Darcy–Weisbach friction factors developed for gravel-bed rivers.

Equation	Characteristics of data set	References
$n = \dfrac{0.1129R^{0.167}}{1.16 + 2.0\log\left(\dfrac{R}{D_{84}}\right)}$	S 0.00068–0.024; Q 5.6–427 m^3/s; D_{84} 2–75 cm; R/D_{84} 0.9–47.2	Limerinos (1970)
$n = \dfrac{0.113R^{0.167}}{1.09 + 2.2\log\left(\dfrac{R}{D_{84}}\right)}$	S 0.00022–0.015; Q 5.5–8140 m^3/s; R/D_{84} 11–85	Bray (1979)
$n = \dfrac{0.1129R^{0.167}}{2.03\log\left(\dfrac{aR}{3.5D_{84}}\right)}$	S 0.0090–0.031; Q 0.995–190 m^3/s; D_{84} 0.046–0.25 m; R/D_{84} 0.97–17.24; a varies from 11.1 to 13.46 in relation to channel cross-sectional shape	Hey (1979)
$n = \dfrac{0.1129R^{0.167}}{0.76 + 1.98\log\left(\dfrac{R}{D_{50}}\right)}$	S 0.000085–0.011; Q 0.05–1540 m^3/s; D_{50} 0.013–0.301 m; R/D_{50} 3–53	Griffiths (1981)
$\left(\dfrac{8}{f}\right)^{0.5} = 5.75\left[1 - \exp\left(-0.05\left(\dfrac{R}{D_{90}}\right)\left(\dfrac{1}{S^{0.5}}\right)\right)\right]^{0.5}\log\left(8.2\left(\dfrac{R}{D_{90}}\right)\right)$		Smart and Jaeggi (1983)
$n = \dfrac{0.3193R^{0.17}}{56.2\log\left(\dfrac{R}{D_{84}}\right) + 4}$	S 0.004–0.04; Q 0.14–195 m^3/s; D_{84} 0.113–0.74 m; $R/D_{84} < 10$	Bathurst (1985)
$n = 0.32\,S_e^{0.38}R^{-0.16}$	S 0.002–0.052; Q 0.34–127 m^3/s; D_{84} 0.1–0.8 m; R 0.15–2.2 m (S_e is energy gradient)	Jarrett (1987)
$\left(\dfrac{8}{f}\right)^{1/2} = 3.10\left(\dfrac{d}{D_{84}}\right)^{0.93}$		Bathurst (2002)
$\left(\dfrac{8}{f}\right)^{\frac{1}{2}} = \dfrac{V}{u_*} = a_1\left(\dfrac{d}{D}\right)^{\frac{1}{6}}$ deep flows $\left(\dfrac{8}{f}\right)^{\frac{1}{2}} = \dfrac{V}{u_*} = a_2 d/D$ shallow flows $(R/D_{84} < 4)$	d mean flow depth; D representative grain diameter; u_* shear velocity; $a_1 \approx 7-8$, $a_2 \approx 1-4$	Ferguson (2007)

The thickness of the laminar sublayer decreases with an increase in shear stress as turbulence penetrates closer to the bed (Richards 1982; Robert 2014), but, as noted earlier, the sublayer is typically <1 mm thick.

The thicknesses of the laminar and turbulent portions of the boundary layer partly depend on the characteristics of the boundary roughness. The effective height of the irregularities forming the rough boundary is the *roughness height*, k_s (Figure 3.8). The ratio of the roughness height to the hydraulic radius, k_s/R, is the *relative roughness* (Chow 1959). If k_s is only a small fraction of the thickness of the laminar sublayer, the surface is hydraulically smooth. If the effects of the roughness elements

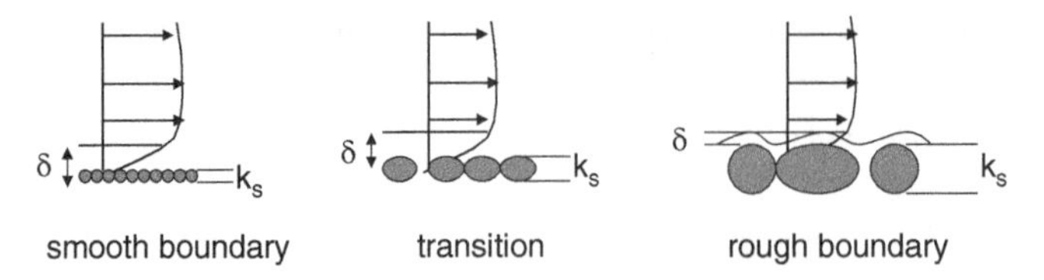

Figure 3.8 Hydraulically smooth and rough boundaries, showing k_s roughness height and laminar sublayer δ. In these beds of uniform grain size, k_s is effectively the grain diameter. In beds with a range of grain sizes, k_s is better approximated by some multiplier of a characteristic grain size (e.g. $3.5D_{84}$), as explained in the text. Source: After Julien (1998), Fig 6.2, p. 95.

extend beyond the laminar sublayer, the surface is hydraulically rough and the velocity distribution depends on the form and size of the roughness projections. Because the laminar sublayer is commonly extremely thin and most natural channels have individual grains and other features that extend beyond this distance into the flow, natural channels have hydraulically rough surfaces and flow resistance increases significantly with increasing relative roughness across a range of channel gradients and Froude numbers (Lamb et al. 2017).

The use of $3.5D_{84}$ for k_s originated with experiments using quasi-homogeneous sand roughness and assuming a logarithmic velocity distribution (Nikora et al. 1998). Wiberg and Smith (1991) demonstrate theoretically that the log formula for hydraulic resistance with $k_s = 3.5D_{84}$ is valid even for flows with large relative roughness despite velocity profile deviations from the log law in these flows. This log formula is

$$f = \left[2.03 \log\left(\frac{12.2R}{k_s}\right)\right]^{-2} \tag{3.16}$$

The value of k_s also depends on the concentration of bed roughness elements (Wiberg and Smith 1991; Nitsche et al. 2012). The same grain-size distribution can offer greater or lesser resistance to flow, depending on packing (Ferguson 2007). Consequently, statistical and spectral analysis of the bed microtopography may better characterize roughness height than a multiplier of some characteristic grain size (Nikora et al. 1998; Aberle et al. 2010; Smith 2014). The standard deviation of bed elevation (σ_z) has been used but is laborious to obtain in field settings (Aberle and Smart 2003; Yochum et al. 2012). The use of terrestrial laser scanning, however, can greatly enhance the ability to derive such measures (e.g. Schneider et al. 2015).

More recent attempts to develop empirical equations for predicting resistance in steep channels employ f rather than n because f is nondimensional and is physically interpretable as a drag coefficient if resistance is equated with the gravitational driving force per unit bed area and assumed proportional to the square of velocity (Ferguson 2007). Ferguson (2007) combines the Manning equation for deep flows with a linear resistance relation for roughness layers in shallow flows to develop a pair of nondimensional hydraulic geometry relations and a variable-power resistance equation

$$\left(\frac{8}{f}\right)^{0.5} = \frac{a_1 a_2 \left(\frac{d}{D}\right)}{\left[a_1^2 + a_2^2\left(\frac{d}{D}\right)^{1.67}\right]^{0.5}} \tag{3.17}$$

where a_1 is a constant, typically 6.5 if D_{50} is used as a roughness scale or 8.2 if D_{84} or D_{90} is used; a_2 is a constant, typically ~1–4; d is mean flow depth; and D is a representative grain diameter. This provides a single resistance equation applicable to shallow and deep flow over coarse beds. Values for n or f tend to be very high in steep streams relative to lower-gradient channels and vary substantially in relation to stage and to channel morphology (Reid and Hickin 2008; David et al. 2010).

The turbulent boundary layer is of most interest for river processes because it is within this layer that velocity is measured, shear stress is estimated, and sediment transport is linked to hydraulic parameters (Robert 2014). The *law of the wall* describes the variation of velocity with height above the bed surface within the turbulent boundary layer and is used to derive shear stress and roughness height from measured velocity profiles (Robert 2014). The law of the wall is

$$v_z = 2.5v^* ln(z/z_0) \tag{3.18}$$

where v_z is the mean one-dimensional velocity at a height z above the bed, z_0 is the projected height above the bed at which velocity is zero, and the constant 2.5 is equal to $1/\kappa$.

The law of the wall can be integrated from $z = z_0$ to $z = d$, where d is flow depth, to calculate mean one-dimensional velocity at a vertical section via the Keulegan equation

$$v/v^* = 2.5\ ln\,(d/k_s) + 6.0 \tag{3.19}$$

where v is the average velocity at a vertical, v^* is shear velocity, and k_s is roughness height (Robert 2014). The primary challenge is the determination of k_s. With densely packed grains of uniform size and shape and a flat, immobile bed, k_s is approximately the median diameter of the bed sediment (Robert 2014). With poorly sorted, heterogeneous bed sediment and bedforms, k_s can vary widely. Commonly used, empirically based approximations include $3.5D_{84}$, $6.8D_{50}$, and $2D_{90}$. Average values of the roughness dimensions can be used to represent surfaces of variable roughness, as in channels with a wide range of grain sizes, although much effort has been devoted to determining how to most accurately characterize an average. In at least some streams, simple D_{84} compares well to bed-roughness measures derived from semivariograms and standard deviation of point clouds from laser-scanned bed surveys (Schneider et al. 2015).

The loss of energy in turbulent flow over a rough boundary results from the formation of wakes behind each roughness element, so the longitudinal spacing λ of the roughness elements is particularly important. *Isolated-roughness flow* occurs where individual roughness elements are sufficiently far apart that the wake and vortex at one are completely developed and dissipated before the flow reaches the next (Figure 3.9). Roughness results from form drag on the roughness elements and friction drag on the wall surface between elements. *Wake-interference flow* occurs where the roughness elements are sufficiently closely spaced for the wake and vortex at one to interfere with the wake and vortex developed at the next. This results in intense and complete vorticity and turbulence mixing, as well as the highest values of flow resistance. The transition from isolated-roughness to wake-interference flow occurs when the ratio of downstream spacing between obstacles L to the height of the obstacles H is approximately 9–10 (Wohl and Ikeda 1998). *Skimming flow* occurs where the roughness elements are so closely spaced that the flow skims the crests of the elements. Large roughness projections are absent and the surface acts hydraulically smooth (Chow 1959). Each depression between elements has extremely low velocity and a stable eddy. Isolated roughness elements can inhibit entrainment of surrounding grains and affect bedload motion. These effects become more pronounced as roughness elements are more closely spaced on the bed.

As noted earlier, resistance to flow along the bed in most natural channels results from diverse features. These features are commonly divided into three categories (Griffiths 1987; Robert 2014). The

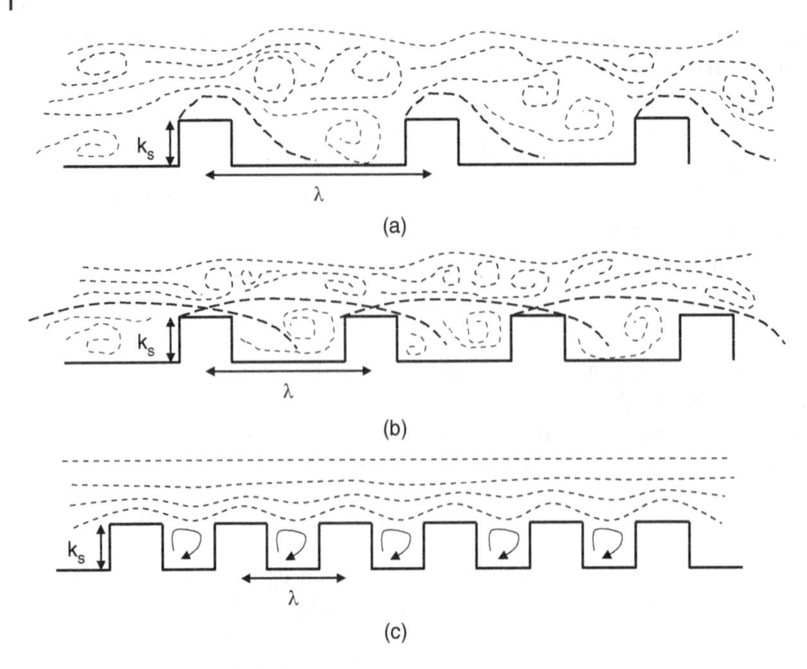

Figure 3.9 Longitudinal view of differing types of bed roughness configurations. (a) Widely spaced roughness elements create isolated roughness flow. (b) More closely spaced roughness elements create wake-interference flow, leading to intense turbulence and complex vorticity. (c) Roughness elements sufficiently close together to create flow that skims the crest of each element and creates quasi-smooth flow. Source: After Chow (1959), Figure 8-4, p. 197.

friction created by individual grains within heterogeneous bed sediments is designated *grain resistance* (or skin or grain friction or grain roughness). Friction from individual grains organized into bedforms such as dunes, pools and riffles, or steps and pools is known as *form resistance* (or form roughness or form drag). Individual pieces or accumulations of large wood in channels can directly contribute to flow resistance by creating grain or form resistance. Large wood can also indirectly contribute to form roughness by altering the dimensions of bedforms (e.g. increasing the magnitude of steps in step-pool sequences; Curran and Wohl 2003; MacFarlane and Wohl 2003). Friction also results from pressure and viscous drag on sediment in transport above the bed surface (Griffiths 1987). Sometimes, a distinction is also drawn between relatively small bedforms and large bed undulations that create *spill resistance* because of a vertical drop in the water surface (Leopold et al. 1960).

When resistance is partitioned into grain and form resistance, grain resistance C_G is typically quantified using a modified form of Keulegan's relation (Parker and Peterson 1980)

$$C_G^{-1/2} = \frac{1}{\kappa}\left[\ln\left(\frac{D}{D_{90}}\right) + A\right] \tag{3.20}$$

where κ is von Karman's constant (0.41 for clear water) and

$$A = \ln\left(\frac{11}{m}\right) \tag{3.21}$$

where $m = k_s/D_{90}$.

Another way to think about resistance is that the forces responsible for generating resistance result from shearing forces and pressure forces (Powell 2014). Shearing forces are generated by transfers of fluid momentum and create grain resistance. Pressure forces result from pressure gradients around roughness elements and create form resistance. The relative importance of grain and form resistance varies with relative submergence. Grain resistance dominates when flow depth is much greater than the size of the bed material. Models for such conditions assume a logarithmic velocity profile and can be approximated by a power equation of the form

$$\frac{1}{\sqrt{f}} \propto \left({}^{R}\!/_{D}\right)^{b} \tag{3.22}$$

where f is the Darcy–Weisbach roughness coefficient and the exponent is 1/6. When flow depth is approximately equal to or less than the bed material diameter, the assumption of a logarithmic velocity profile is inaccurate and flow-resistance models are based on either hydraulic geometry or roughness layer theory, as approximated by

$$\frac{1}{\sqrt{f}} \propto \left({}^{R}\!/_{D}\right) \tag{3.23}$$

This is essentially the form of Eq. (3.22), with an exponent of 1, which implies that a variable power relation provides a single equation that can be used in deep and shallow flows and explicitly accounts for the changing sources of flow resistance as relative submergence changes (Powell 2014). Predictive errors can still be significant, however, suggesting the need for continuing research.

An important consideration is that roughness is related to uniform flow equations. Anything that causes the water surface to shift to gradually or, especially, rapidly varied flow (such as a large bedform or a logjam) is an obstruction that does not formally fall under the category of form roughness as used by hydraulic engineers. In practice, however, any large obstacle to flow is lumped under the category of form roughness in most geomorphic discussions of flow resistance in natural channels, and "roughness" is used generically to refer to various forms of flow resistance. In the discussion that follows here, "resistance" is used to include roughness elements and elements that cause the water surface to shift to varied flow. (In engineering hydraulics, roughness is a coefficient that represents the effects of boundary shear stress in uniform flow equations, resistance is a force with a vector [directional roughness], and energy dissipation is everything that retards flow, including flow transitions such as hydraulic jumps, eddies, and grain and form resistance.)

Einstein and Banks (1950) use flume experiments to demonstrate that grain, f_g, and form, f_b, resistance could be additive

$$f = f_g + f_b \tag{3.24}$$

Grain resistance is negligible in channels formed in sand-sized and finer sediment. Individual particles do not protrude into the flow above the rest of the bed and, except at very shallow flow depths, influence only a small portion of the flow. Grain resistance can become quite substantial in channels with larger grains, particularly where the grains are poorly sorted and are large relative to flow depth. Poorly sorted substrates with a wide range of grain sizes allow large grains to protrude into the flow above the rest of the bed. Individual grains may create the greatest resistance in shallow, steep, bouldery headwaters, whereas bedforms such as step-pool and pool-riffle sequences (Section 4.3.2) formed in gravel dominate flow resistance in the middle segments of drainage networks (Prestegaard 1983). Other sources of flow resistance become more important in the deeper, low-gradient segments of drainages.

The importance of grain resistance can be expressed using R/k_s when k_s is a function of grain size. This ratio is known as the *relative grain submergence*, commonly expressed as R/D_{84}, and is the inverse of relative roughness. Relative grain submergence can be used to distinguish between large-scale roughness ($0 < R/D_{84} < 1$), in which individual grains protrude a substantial distance into the flow, intermediate-scale roughness ($1 < R/D_{84} < 4$), and small-scale roughness ($R/D_{84} > 4$), in which grains protrude above the bed only a small proportion of flow depth (Figure 3.10)

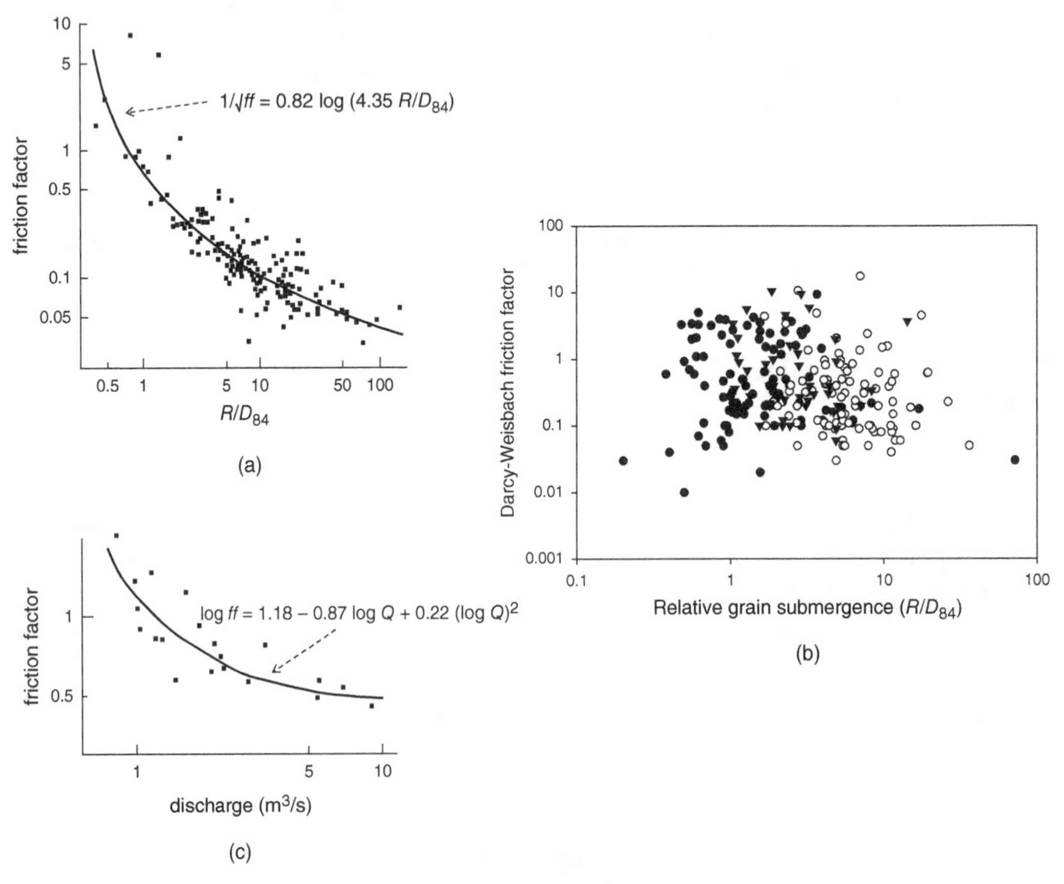

Figure 3.10 The relationship between Darcy–Weisbach friction factor and relative grain submergence (R/D_{84}) differs between (a) low-gradient alluvial channels with sand bedforms in which friction factor decreases as grain roughness is drowned out (Source: After Knighton 1998, Figure 4.2A) and (b) high-gradient alluvial channels with cobble to boulder beds in which the range of values for friction factor is relatively constant (Source: After Wohl and Merritt 2008, Figure 9). Step-pool channels are solid circles, pool-riffle channels are open circles, and plane-bed channels are solid triangles. (c) Friction factor tends to decrease with discharge, as shown in this plot of data from the River Bollin, UK (Source: After Knighton 1998, Figure 2B), although the rate of decrease varies among channels as a function of the sources of roughness.

(Bathurst 1985). Relative form submergence, R/H, where H is bedform amplitude, can be similarly used to express the importance of form roughness (Wohl and Merritt 2008).

The phrases "grain resistance" and "form resistance" can be misleading in that individual large grains actually create form resistance. Grain resistance is the cumulative effect of many individual grains, which retard water flowing over the bed in a uniform manner, without accelerations or decelerations and with parallel streamlines and the same mean depth and velocity (Ferguson 2013). Resistance results from viscous shear over the grain surfaces and local pressure gradients around the grains. Large grains that protrude into the flow create form resistance as larger-scale flow acceleration and deceleration generate large turbulent eddies in which energy is dissipated by viscous forces (Ferguson 2013). Form resistance thus includes the effects of drag on large roughness elements such as protruding boulders or logs, spill losses over steps or riffles, standing waves around protruding obstacles, and transverse accelerations caused by sharp changes in flow direction (Ferguson 2013). Grain and form resistance can be very difficult to distinguish in many natural channels.

A widely used approach (Robert 2014) for partitioning resistance in natural channels has been to select an appropriate value of k_s for grain resistance and use this to calculate f_g via the equation

$$1/\sqrt{f_g} = 2.11 + 2.03\, log_{10}\,(d/k_s) \tag{3.25}$$

Grain resistance is then subtracted from total resistance, f, to estimate form resistance. However, flume studies that combine grain resistance and form resistance with resistance from individual instream wood pieces and jams indicate the existence of substantial interaction effects between resistance components. Consequently, a simple additive approach can be inaccurate (Wilcox et al. 2006).

An alternative to assuming that any portion of f not included in f_g is form resistance f_b involves directly calculating f_b

$$f_b = 0.5 C_d\, h\, \rho\, v^2 \tag{3.26}$$

where C_d is a drag coefficient that varies with obstacle shape and h is the height of bed undulations associated with the bedforms (Robert 2014). Quantifying C_d is difficult, however, because form drag is influenced by the pressure field, and thus the velocity field, above the bedforms. The *pressure field* is the instantaneous water pressure exerted on a given surface, such as the bed, within a control volume of fluid, expressed as a function of three spatial coordinates and time. The *velocity field* is the instantaneous velocity of the center of gravity of a volume of fluid instantaneously surrounding a point in the flow (Fox and McDonald 1978). In other words, C_d for a given obstacle is not constant, but varies with discharge and with the configuration of the channel geometry and bed resistance in the immediate vicinity of the obstacle, as these influence local velocity and pressure distributions. Quantifying and partitioning flow resistance, particularly in coarse-grained channels, continues to be an elusive goal.

Although more attention has been devoted to bed resistance, bank resistance can also exert an important influence on total resistance and the distribution of hydraulic forces (Wohl et al. 1999; Kean and Smith 2006a,b; Powell 2014). Bank resistance results from: individual grains; bank vegetation; irregularities such as those produced by slumping of bank material or the presence of vegetation; and repetitive variations in the channel planform such as bends. In Lost Creek, a small (4 m wide), gravel-bed channel with rough banks (± 20 cm amplitude), additional flow resistance created by drag on bank topographic features substantially reduces near-bank velocity and shear stress (Kean and Smith 2006a). Neglecting these effects in Lost Creek results in a 56% overestimate of discharge. Enhanced bank resistance caused by vegetation can alter the spatial distribution of velocity, shear stress, and sediment deposition, and thus channel geometry (Griffin et al. 2005; Gorrick and Rodríguez 2012; Curran and Hession 2013). Friction associated with woody riparian vegetation on the lateral boundaries of the sand-bed Rio Puerco in New Mexico, USA (average channel width ~11 m) reduced perimeter-averaged boundary shear stress by almost 40% and boundary shear stress in the channel center by 20% (Griffin et al. 2005). The effects of bank resistance may be more important along channels formed in finer sediment, such as sand, than in more coarse-grained cobble- to boulder-bed channels in which bed resistance is very large (Yochum et al. 2012).

Momentum equations define the hydrodynamic forces, such as drag, exerted by flow. Houjou et al. (1990) propose that the velocity field in flows for which lateral shear is an important factor can be calculated using a momentum equation in the form of

$$-\rho g S = \frac{\partial}{\partial z}\left(\kappa \frac{\partial u}{\partial z}\right) + \frac{\partial}{\partial x}\left(\kappa \frac{\partial u}{\partial x}\right) \tag{3.27}$$

where ρ is fluid density, g is acceleration due to gravity, S is the downstream water-surface gradient, κ is kinematic eddy viscosity, z is the coordinate normal to the bed, x is the transverse coordinate, and u is the downstream velocity component. In a channel with a rectangular cross-section, flow structure reflects the width–depth ratio and the ratio of wall and bed resistance (Houjou et al. 1990). Eq. (3.27) permits computation of velocity and stress fields affected by both the channel bed and the banks.

Water flowing through a bend experiences a transverse force proportional to v^2/r, where r is the radius of curvature. This adds to the total flow resistance in sinuous channels (Ferguson 2013). Total head loss along a reach scales with R/r^2 (Chang 1978). Radius of curvature increases more rapidly than v or R as river size increases, so bend losses decrease in importance with river size and are insignificant in large rivers whose banks are readily erodible and in which regular meanders can develop (Ferguson 2013).

3.1.4 Velocity and Turbulence

Velocity is a vector quantity, with magnitude and direction, and is one of the most sensitive and variable flow properties. Because of the presence of a free surface and friction along the channel boundaries, velocity is not uniformly distributed within a channel. It varies with distance from the bed, across the stream, downstream, and with time (Figure 3.11). At a cross-section, the maximum velocity typically occurs just below the free surface. The location of the maximum velocity with respect to distance between the two channel banks depends on the symmetry or asymmetry of the channel. Velocity tends to remain constant or increase slightly downstream as resistance from the channel boundaries affects a smaller proportion of the total flow volume. Velocity varies at short time intervals of a few seconds as a result of flow turbulence and at longer time intervals as a result of roughness of the channel boundaries and fluctuations in discharge during unsteady flow.

As noted earlier, velocity can be characterized in terms of the velocity field. At a given instant in time, the velocity field is a function of the space coordinates x or w (cross-stream), y or u (downstream), and z or v (vertical with respect to the bed), so that a complete representation of the velocity field is given by $V = V(x, y, z)$ (Fox and McDonald 1978). The usual convention is for velocity terms u, v, w to correspond to spatial coordinates x, y, z: hence the use of z for depth or vertical position in some of the equations in this chapter. A flow field described in this manner is *three-dimensional* because the velocity at any point in the flow field depends on the three coordinates required to locate the point in space. A flow can be classified as one-, two-, or three-dimensional, depending on the number of space coordinates required to specify the velocity field.

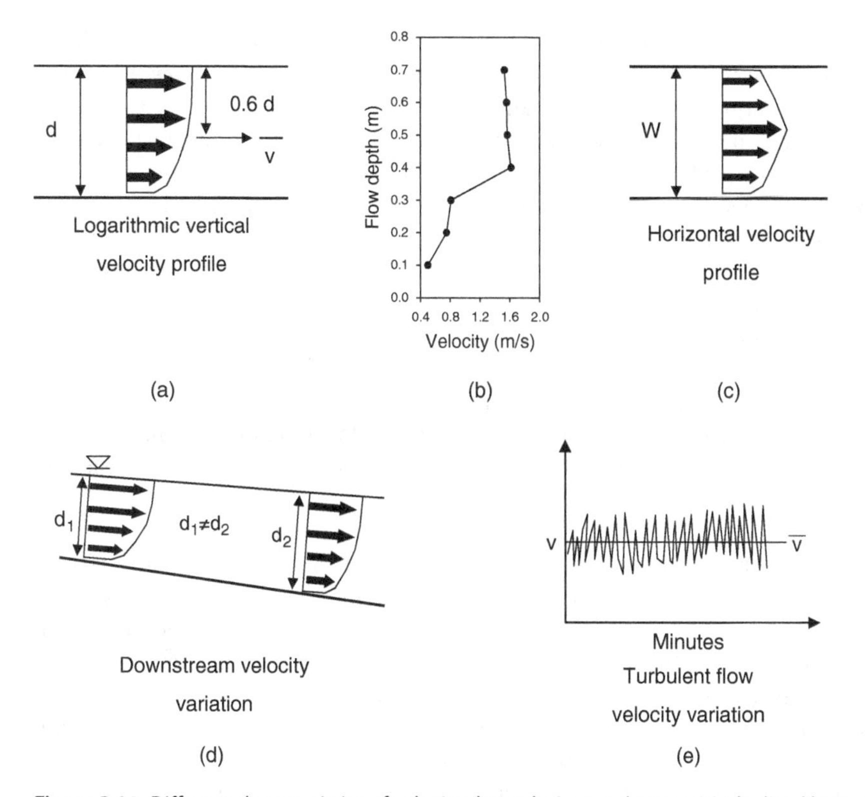

Figure 3.11 Different characteristics of velocity through time and space. (a) Idealized logarithmic vertical velocity profile (flow from left to right, length of arrows indicates relative velocity). (b) Vertical velocity profile from a boulder-bed channel. Abrupt increase in velocity between 0.3 and 0.4 m flow depth reflects top of boulders and transition from highly turbulent flow with low average velocity to rapid flow in upper profile. (c) Horizontal velocity profile looking down on a channel of width *w*. (d) Longitudinal view illustrating downstream variations in velocity, as reflected in differences within successive vertical velocity profiles. (e) Temporal variations in velocity at short time intervals associated with turbulence. Source: Parts (a), (c), (d), and (e) courtesy of David Dust.

One-dimensional flow occurs where velocity is represented by only one dimension, such as along the channel centerline of a long, straight channel with constant cross-section, where velocity can be primarily influenced by distance above the channel bed. One-dimensional representations have cross-sectionally averaged velocity along the hydraulic axis that follows the thalweg. From the continuity equation, mean one-dimensional velocity is discharge divided by cross-sectional area ($v = Q/A$).

Two-dimensional flow occurs when the velocity field is predominantly influenced by two of the space coordinates, such as distance above the bed and from the bank. Two-dimensional representations of flow have vertically integrated velocity with only the cross- and downstream components specified.

The complexity of analysis increases substantially with the number of dimensions of the flow field. Consequently, the simplest case of one-dimensional flow is widely used to provide approximate solutions for evaluating features such as thalweg velocity profile (Fox and McDonald 1978).

Flow along channels with low gradients and relatively uniform bed material is commonly approximated by a semilogarithmic velocity profile in which velocity varies with distance from the bed

(Leopold et al. 1964); this is an example of assuming one-dimensional flow. The profile includes a laminar sublayer, a turbulent boundary layer with logarithmic profile, and an outer layer that deviates slightly from logarithmic (Ferguson 2007). For channels with rough boundaries or irregular bed material, velocity is proportional to roughness-scaled distance from the bed, rather than just distance from the bed. In effect, the velocity profile becomes segmented, with low-velocity flow between larger boulders or below the average surface elevation of the bed and an abrupt transition to higher-velocity flow above the level of the boulders (Figure 3.11b). This two-part velocity profile has been described as s-shaped, although the profile varies substantially in relation to channel-boundary roughness. The law of the wall is used to describe the variation of velocity with height above the bed surface in the bottom 20% of fully turbulent flows (Robert 2014).

Velocity in natural channels can be measured using current meters or dilution tracers. Point measurements use various types of current meters, including mechanical impellors, electromagnetic and ultrasound meters, and laser velocimeters (Clifford and French 1993b). One of the limitations of point measurements is the number and spatial density of measurements needed to characterize mean velocity.

Acoustic Doppler current profilers (ADCPs) can obtain spatially dense point measurements in a very short period of time and are now widely used for measuring velocity in moderate to large rivers, as well as turbulence characteristics, suspended sediment concentration, spatial variations in shear stress, and bedload transport rates (e.g. Parsons et al. 2013). These instruments can be of limited usefulness in very shallow (<0.5 m depth) or highly aerated flow. Doppler current meters use the Doppler effect of sound waves scattered back from particles within the water column. The meter generates and receives sound signals. The traveling time of a sound wave can be used to estimate distance and the change in sound wavelength can be converted to velocity.

Dilution tracer techniques using a fluorescent dye (David et al. 2010; Runkel 2015) or a chemical such as NaCl (Garcia Parra et al. 2016) can also be used to characterize the average velocity for a channel segment. The tracer can be introduced steadily over a finite time period or as a slug injection that is introduced instantaneously (Elder et al. 1990). Mean velocity can be calculated using tracer concentrations measured with a fluorometer or conductivity meter at two points separated by a known distance (Planchon et al. 2005). The travel time of the tracer can be calculated using time between peaks (Wilcox and Wohl 2006), time between centroids (Kratzer and Biagtan 1997), or a spatial harmonic mean travel time (Walden 2004). Dilution tracers are inexpensive and fast but provide no information on the distribution of velocity within the measured channel segment.

Most recently, remote sensing platforms have been used to estimate velocity and other hydraulic parameters based on characteristics such as water-surface topography and the reflectance of the surface (e.g. Legleiter et al. 2017a). Although in the early stages of development, this type of approach holds promise for the rapid measurement of velocity along large portions of rivers.

As water particles move in irregular paths, momentum is exchanged between different portions of the water (Chanson 1999). *Turbulence* occurs when water parcels flow past a solid surface or past an adjacent water parcel with a different velocity (Clifford and French 1993a,b). Turbulence appears as irregular velocity fluctuations. Time series of velocity in natural channels commonly reveal intriguing variations in the magnitude of even one-dimensional measurements, let alone three-dimensional flow, and a great deal of human energy is devoted to characterizing the patterns of velocity fluctuations in order to reveal the underlying processes and thus improve understanding of the relations among resistance, hydraulics, and channel adjustment. Understanding of velocity and turbulence is

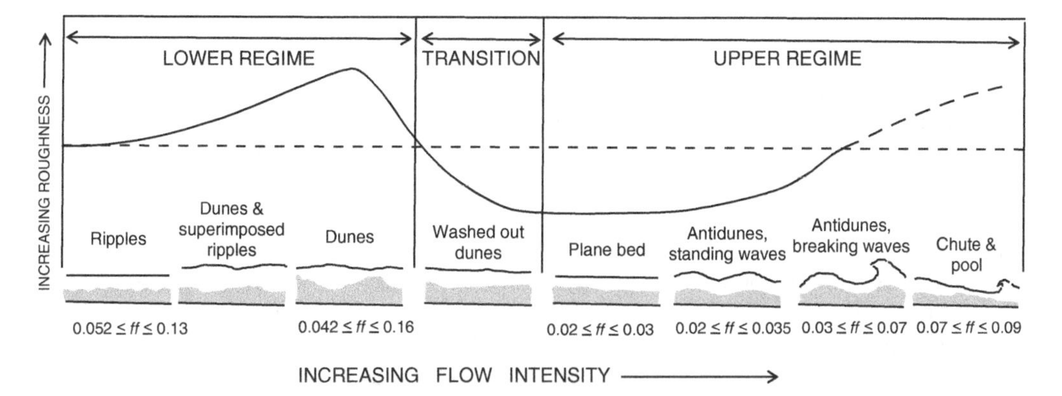

Figure 3.12 Sequence of bedforms that develop with increasing discharge or velocity in mobile-bed channels, and associated flow resistance. Small drawings at the base of the figure are longitudinal views of a segment of bed, indicating the bed profile (gray shading) and the water surface configurations (upper black line). Note that resistance, as indicated by the Darcy–Weisbach friction factor *f*, generally increases going from ripples to dunes in the lower regime, declines dramatically when the dunes wash out to a plane bed, and then increases to a lesser degree in the progression toward chutes and pools within the upper regime, as indicated by the undulating line partway up the y-axis in this figure. Source: After Simons and Richardson (1966) and Knighton (1998), Figure 4.3.

important because of the feedbacks among turbulence, channel morphology, and sediment transport (Figure 3.12) (Leeder 1983; Buffin-Bélanger et al. 2013; Schmeeckle 2014).

Channel morphology creates resistance and contributes to the variability in the intensity of turbulent exchanges with the water column (Legleiter et al. 2007). The structure of turbulent flows consists of the mean structure, as reflected in the one-dimensional vertical velocity profile, and the temporal fluctuations of the structure associated with turbulence. Both of these elements influence sediment mobility. Mobile sediments can alter the flow structure and contribute to bedforms, which then influence flow resistance and turbulence.

Turbulence is quantified based on measurements of velocity fluctuations at a point in the flow. The *turbulence intensity* of any of the three velocity components (*x, y, z*) is the average magnitude of the deviation from the mean for a given velocity series (Robert 2014). Variability around the mean for a normal or near-normal distribution is represented by the standard deviation of the velocity distribution, known as the root-mean-square value, RMS, which is used to represent turbulence intensity (Robert et al. 1996; Robert 2014).

$$RMS_y = \sqrt{\left(\frac{\left(\sum v_y^2\right)}{N}\right)} \tag{3.28}$$

where v_y is the downstream component of velocity and N is the total number of observations in a given series. The RMS value can be computed for any of the three dimensions of velocity, and the three expressions can be combined into an index of total turbulence intensity (Robert 2014).

Flow structures in the vicinity of large obstacles increase the local turbulence intensity in the near-bed region. These obstacles can be created by protruding clasts or large wood (David et al. 2013; Curran and Tan 2014), or by rigid vegetation growing within the active channel (trees sometimes grow within an active ephemeral channel) or in overbank areas (e.g. Yager and Schmeeckle 2013). Flexible vegetation, particularly when submerged, also tends to increase flow depth and modify the vertical structure of the flow by creating a zone of slower flow within the vegetation and a zone of faster flow above the vegetation (Luhar and Nepf 2013; Le Bouteiller and Venditti 2015).

The greater the value of RMS, the greater the turbulence and the greater the potential for sediment entrainment and transport. The turbulent kinetic energy of the flow, TKE, is

$$TKE = 0.5\rho\,(RMS_x^{\,2} + RMS_y^{\,2} + RMS_z^{\,2}) \tag{3.29}$$

where ρ is water density (Clifford and French 1993a; Robert 2014). TKE is the mean kinetic energy per unit mass associated with eddies in turbulent flow and can be conceptualized as the energy extracted from the mean flow by turbulent eddies (Bradshaw 1985; Robert 2014). The downstream component of TKE is typically dominant in natural channels (Robert 2014).

Turbulence can also be studied by following the trajectories of tracers traveling with the flow. This *flow visualization* approach uses hydrogen bubbles, colored fluids, or fine particles of neutral submerged density (Buffin-Bélanger et al. 2013). Flow visualization does not provide the quantitative information of Eqs. (3.28) or (3.29), but does provide insight into the location, size, and structure of turbulence.

Reynolds stresses quantify the degree of momentum exchange at a point in the flow (Robert 2014). Six Reynolds stresses can be calculated as the product of the negative value of water density ($-\rho$) and the product of the average fluctuations of either a single velocity component (e.g. $v_x^{\,2}$) or two different velocity components (e.g. $v_x v_y$) (Buffin-Bélanger et al. 2013). Like TKE, Reynolds stresses are important descriptors of the intensity of turbulent exchanges and are linked to sediment transport (Buffin-Bélanger et al. 2013). Values of RMS, TKE, and Reynolds stresses are each widely used to quantify the turbulence characteristics of flow in natural channels.

Quadrant analysis is commonly used to understand and describe the structure of turbulence in two dimensions. The instantaneous horizontal and vertical velocities are divided into four quadrants based on their deviation from the mean (Robert 2014). Quadrant I represents positive deviation in the horizontal and vertical. Quadrant II represents negative deviation in the horizontal and positive deviation in the vertical. Quadrant III represents negative deviation in the horizontal and vertical. Quadrant IV represents positive deviation in the horizontal and negative deviation in the vertical.

Quadrant II events have slower than average downstream flow velocity and positive vertical flows away from the channel boundary, and are known as ejection events or *bursts* because water is ejected from the bed upward into the outer flow (Robert 2014). Quadrant IV events have greater than average downstream flow velocity and negative vertical flows toward the bed and are known as *sweeps*. Alternate zones of low- and high-velocity water near the bed create *low-speed streaks*: relatively narrow zones of low-velocity water near the bed (Robert 2014). The spacing of low-speed streaks reflects the shear velocity and fluid kinematic viscosity. Low-speed streaks culminate in ejections upward from the bed (bursts). To preserve continuity, bursts are followed by high-velocity outer-layer flow penetrating the near-bed flow (sweeps). Sweeps lose momentum as they impact the bed and diffuse laterally (Best 1993; Robert 2014). This sequence of bursts and sweeps exerts an important control on sand-bed bedforms such as ripples and dunes by moving sand grains and influencing bedform size (Best 1992, 2005) (Section 4.3.1).

Bursts and sweeps occur only a small fraction of the time but cause most of the momentum exchange in a channel and can strongly influence sediment transport (Robert 2014). This is the key point regarding turbulence in natural channels, and one of the reasons so much effort is focused on quantifying turbulence. Numerous studies of sediment mobility and associated channel stability indicate that turbulent fluctuations exert a stronger influence on sediment dynamics, particularly

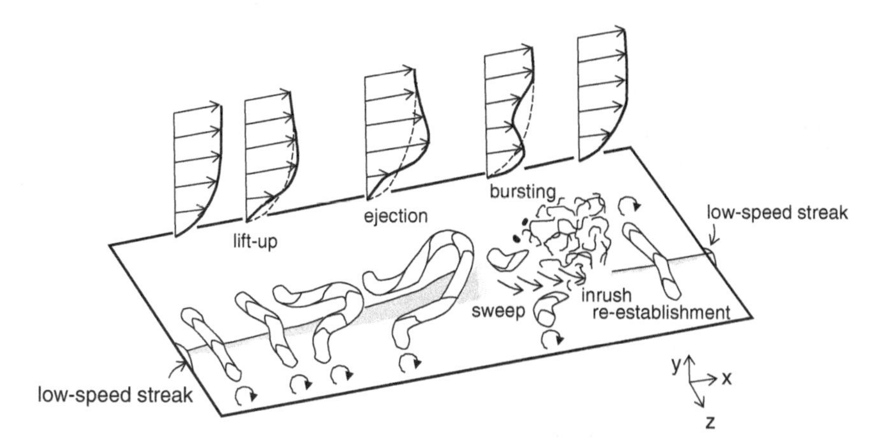

Figure 3.13 Schematic representation of turbulent bursting, according to Allen (1985). Flow from left to right, with length of arrows representing relative velocity. A developing horseshoe vortex lifts, stretches, and bursts. Associated velocity profiles shown at rear. Source: After Bridge (2003), Figure 2.12a, p. 29.

initiation of motion, than does mean velocity, especially in channels with beds composed of sediment coarser than sand size (e.g. Diplas et al. 2008; Valyrakis et al. 2011).

Turbulent boundary layers include various types of vortices and irregular but spatially and temporally repetitive flow patterns known as *coherent flow structures.* Coherent flow structures are self-perpetuating (Smith 1996; Robert 2014). Low-speed streaks alternate with zones of high-velocity water and give rise to coherent flow structures such as horseshoe vortices. *Horseshoe vortices,* which are named for their shape (Figure 3.13), are advected upward from channel beds.

A second major group of vortices and coherent flow structures are present over coarse-grained surfaces in which vortices are induced by individual large particles or bedforms (Figure 3.14) (Roy et al. 1999; Robert 2014). Eddies are shed from the downstream end of obstacles and from the shear layer along the flow separation zone downstream from obstacles. Standing vortices form upstream of individual obstacles (Robert 2014). Horseshoe vortices form from the flow separation zone downstream from isolated obstacles.

The frequency of eddy shedding downstream from isolated obstacles is expressed by the dimensionless *Strouhal number, Str*

$$Str = \frac{(feD)}{v} \tag{3.30}$$

where fe is the frequency at which eddies are shed from obstacles, D is obstacle size, and v is mean flow velocity (Robert 2014). Large Strouhal numbers ($\sim$1) indicate that viscosity dominates fluid flow. Low Strouhal numbers ($\leq 10^{-4}$) indicate that the high-speed, quasi-steady-state portion of the fluid flow dominates. Intermediate Strouhal numbers reflect the formation and rapid shedding of vortices (Sobey 1982).

Eddy shedding may be the dominant mechanism of energy dissipation in coarse-grained channels (Robert 2014) and may cause the greater-than-expected flow resistance of coarse-grained channels (Clifford et al. 1992). Pseudo-periodic oscillations in time series of velocity fluctuations may reflect the periodicity at which eddies are shed (Robert 2014).

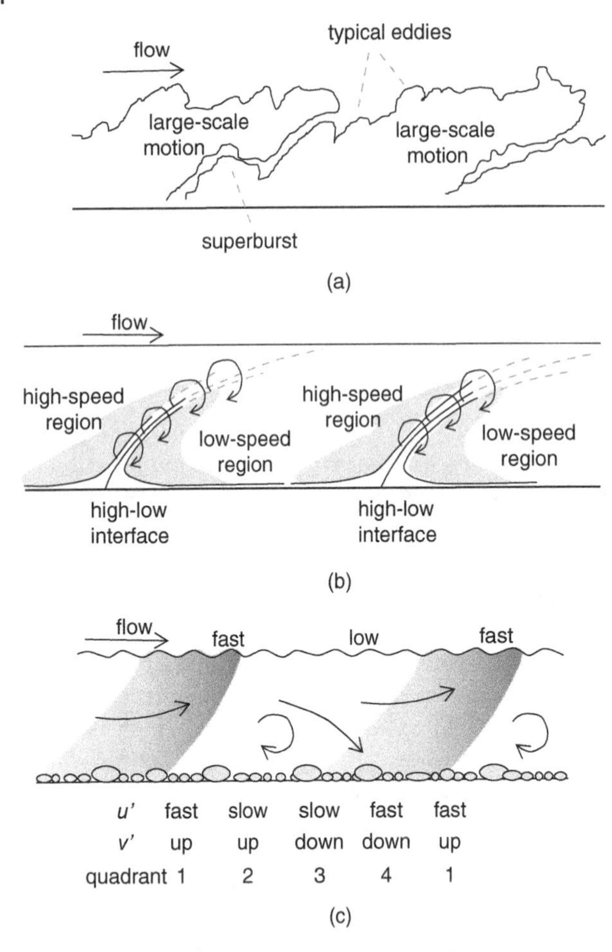

Figure 3.14 Longitudinal views of turbulence. (a) Large-scale flow motions with the presence of typical eddies at their boundaries. (b) High- and low-speed regions within the flow. (c) A sequence of turbulent flows associated with the passage of high- and low-speed wedges. Source: After Buffin-Bélanger et al. (2000), Figure 1 and Robert (2014), Figure 2.12.

Very large roughness elements or an abrupt change in channel boundary or orientation creates *separated flows*, or portions of the channel in which there is no downstream flow (Robert 2014). Examples of sites inducing flow separation include channel bends, channel expansions, pools, bedforms, and very large grains or pieces of wood. The boundary between the lower-velocity separation zone and the higher-velocity main flow is a zone of rapid change in velocity, intense mixing, and high turbulence intensity known as a *shear layer* (Figure 3.15). The acceleration of flow around the obstacle can allow the turbulent boundary layer to detach or separate from the bed or bank. The *reattachment point* occurs where the turbulent boundary layer reattaches to the bed or bank downstream from the obstacle. The area between the separation and reattachment points is the zone of *recirculating flow* (Buffin-Bélanger et al. 2013). Zones of recirculating flow are typically sites where finer sediment is deposited as a result of lower velocity. Recirculating flow can also form an important aquatic habitat

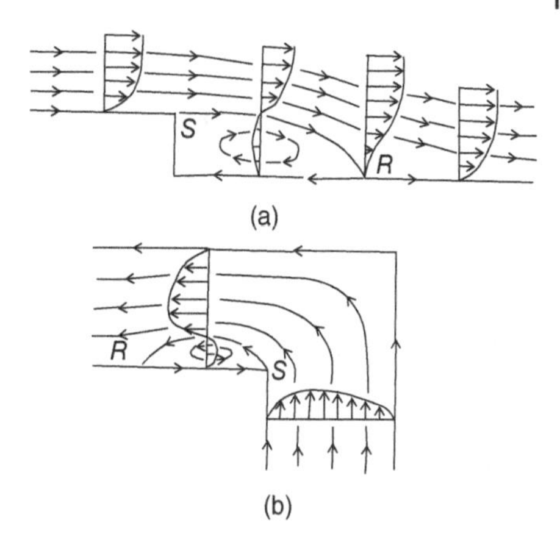

Figure 3.15 Examples of flow separation (*S* is separation point, *R* is reattachment point) as indicated by time-averaged streamline patterns at (a) a downward transverse step, as seen in side view, and (b) a sharp bend in an open channel, as seen in planform. Source: After Robert (2014), Figure 2.1.

by concentrating organic matter in transport and providing a low-velocity resting area for organisms such as fish. Strong gradients in water properties such as velocity or density where two rivers join can also promote lateral momentum exchanges and large and intense shear layers between the flows (Buffin-Bélanger et al. 2013). The energy dissipated at these boundaries can facilitate sediment deposition.

In summary, the spatial distribution and magnitude of velocity and turbulence reflect both the magnitude of flow and the characteristics of the channel boundaries. Individual roughness elements such as protruding clasts can create local perturbations in velocity and turbulence, and, in shallow flows, these can affect the entire water column. As flow depth increases, flow structure is more likely to represent features at the cross-sectional to reach scale, such as pool-riffle sequences and associated variations in bankfull width and depth. Consequently, flow depth relative to the sources of flow resistance along the bed can be conxsidered a first-order control on flow structure.

3.1.5 Measures of Energy Exerted Against the Channel Boundaries

The force driving a flow, F_d, along a unit length of channel can be expressed as

$$F_d = W \sin \theta \tag{3.31}$$

where W is the weight of the water and θ is the channel bed slope. Weight equals mass times gravitational acceleration, g, and $W \sin \theta$ represents the downslope components of mass acted on by g. Mass equals mass density, ρ, and the volume of water, V (i.e. mass $= \rho V$), so

$$F_d = \rho g V \sin \theta \tag{3.32}$$

Stress is force per unit area (A), so the average shear stress exerted by the flow on the bed, τ_0, under conditions of steady uniform flow is

$$\tau_0 = \rho g R \sin \theta \tag{3.33}$$

where R is hydraulic radius. For small slopes, sin θ is approximately equal to tan θ, which is approximated by slope, S, so that

$$\tau_0 = \rho g R S \tag{3.34}$$

This is the basic equation for average *bed shear stress* in natural channels (Robert 2014). Although steep channels violate the assumption that bed slope approximates sin θ, Eq. (3.28) is commonly applied to steep channels. As noted in Section 3.1.3, shear stress can also be derived from a measured velocity profile.

The shear stress actually available to do work such as entraining or transporting sediment is commonly less than average bed shear stress over coarse-grained and poorly sorted beds with relatively immobile large particles. The large particles bear a significant fraction (as much as 50%) of the total shear stress and reduce the stress available to transport finer, more mobile sediment (Yager et al. 2007). Consequently, calculating average shear stress does not necessarily provide a simple index of flow energy available for work against the channel boundaries.

Stream power is also commonly used to express energy exerted against the channel boundaries. Stream power is the rate of doing work in transporting water and sediment. Stream power can be expressed as *total power*, Ω

$$\Omega = \gamma Q S \tag{3.35}$$

where γ is the specific weight of water (assumed to be 9800 N/m^3), Q is discharge, and S is channel gradient.

Stream power can also be expressed as *stream power per unit area*, ω

$$\omega = \tau_0 v \tag{3.36}$$

where τ_0 is bed shear stress and v is velocity.

Finally, stream power can be expressed as *specific stream power*, ω

$$\omega = \Omega / w \tag{3.37}$$

where w is channel width.

Each measure of stream power quantifies the energy available to perform geomorphic work against the channel boundaries in the form of entrainment and erosion.

Spatial variations in stream power along a river are addressed in a model developed by Knighton (1999). Depending on the downstream rates of change for Q and S, the model predicts that total power peaks at an intermediate distance between the drainage divide and the river outlet. Specific stream power is more sensitive to the rate of change in S and peaks closer to the headwaters. The model accurately predicts observed conditions along lowland alluvial rivers (Knighton 1999), but limited tests for mountain rivers reveal substantial deviations from expected patterns because of local factors that influence Q, S, and w (Brummer and Montgomery 2003; Fonstad 2003; Wohl and Wilcox 2005).

All of the measures of stream power noted here are essentially instantaneous values at a point or cross-section and do not account for variations in flow – and thus stream power – through time. Temporal variations in stream power relate mainly to variations in discharge. In one study along the Tapi River, India, the annual peak flood contributed up to 34% of the total energy expended during a monsoon season (Kale and Hire 2007). Integrating variations in stream power through time and along a channel is necessary to effectively understand the influence of energy expenditure on channel form and process (Costa and O'Connor 1995).

3.1.6 Numerical Models of Hydraulics

Computational fluid dynamics (CFD) is the use of applied mathematics, physics, and computational software to examine how water flows in a channel and affects the channel boundaries. For rivers,

CFD models can be one-dimensional models that focus on the downstream component of flow and estimate cross-sectionally averaged velocity distribution and water-surface elevation based on discrete channel cross sections. CFD can also be the basis for two-dimensional models that include downstream and cross-stream or vertical flow components, or for three-dimensional models that include all primary components of flow. Two- and three-dimensional models use spatially continuous data on channel geometry. The progression from one to three dimensions represents progressively greater computational running time, but all CFD approaches solve momentum (or energy) and conservation-of-mass equations to estimate flow characteristics (Tonina and Jorde 2013).

The solution of CFD equations is difficult in natural streams because of limitations in resolving spatial and temporal fluctuations of turbulence. Most applications in natural channels average the equations to estimate mean flow properties. The integral or partial differential form of the momentum and conservation of mass equations is transformed into a set of algebraic equations that provide results at a finite number of points, in a process known as *discretization*. Discretization divides the channel into an arbitrary number of control volumes, and CFD then provides numerical results at a finite number of points known as nodes. The spatial resolution of survey data for the channel boundaries influences the node spacing, which in turn influences the spatial resolution and accuracy of the CFD output (Lane and Richards 1998; Tonina and Jorde 2013). Consequently, the number of dimensions included in the model, the discretization approach used, and the spatial resolution and accuracy of the model results depend on available input data, available computational running time, and the intended use of the model outputs (e.g. estimating sediment entrainment or quantifying fish habitat). Model output is typically validated by comparison to a set of data not used during the model calibration process, although validation does not necessarily guarantee that a model is correct (Lane et al. 2005). Numerous CFD models are now available online for free public use or to purchase. Tonina and Jorde (2013) provide a comprehensive introduction to CFD modeling in natural channels.

3.2 Hydrology

3.2.1 Measuring Discharge

The basic data on river discharge come from gaging stations, typically operated by either government agencies or private companies that require knowledge of water supply. In the United States, for example, the majority of gages are operated by the US Geological Survey, although state agencies also operate many gages. The Environment Agency operates most stream gages in England and Wales. In these countries, discharge data are publicly accessible online at no charge. Government agencies in other countries, such as India, consider stream records sensitive information and will not readily release the data.

Although the continuity equation ($Q = w\ d\ v$) suggests a simple means to measure discharge at a gage, w, d, and v are not measured continuously. Instead, they are initially measured over a range of flows and used to establish a *rating curve* for the cross-section. This curve relates Q to *stage* (water-surface elevation), which is in turn related to d, the most readily measured parameter. A rating curve is typically recalibrated periodically to ensure that gradual, progressive changes in cross-sectional geometry do not affect the accuracy of discharge calculations. This method of calculating discharge is most accurate where cross-sectional geometry is relatively simple, without split flow or extensive backwaters, and stable through time. Cross-sections with bedrock exposures or with banks or beds stabilized by concrete or other artificial materials are chosen for gaging sites where possible.

The simplest technique for measuring flow depth is to install a calibrated vertical scale along one bank and use it to read depth. This technique was originally developed for use along the Nile River circa 620 AD. Most gages now use a stilling well, in which the stage equals the height of the water surface in the channel. A pressure sensor in the well is connected to instruments that record fluctuations in the stage through time, using something as simple as a paper strip chart and a stylus, or sending the electronically transduced measurements to a central receiving and recording station via satellite transmission. Discharge can be measured at regular intervals, the length of which typically reflects some trade-off between accuracy and cost. Various statistical measures, such as mean daily, mean monthly, and mean annual discharge, can be computed from these data.

For larger rivers, stream gage information can be combined with remotely sensed imagery, data on image brightness, and Manning-based estimates of stream roughness to calculate water depths from aerial photographs or hyperspectral imagery without ground verification at the time of image acquisition (Fonstad and Marcus 2005; Legleiter et al. 2009). Discharge can also be measured from a helicopter, using a pulsed Doppler radar to measure surface velocities and ground-penetrating radar to measure river depth (Melcher et al. 2002).

Direct discharge measurements are relatively limited. Some large rivers have continuous records extending back more than a century. More commonly, records are of short duration or include time gaps, and provide only partial spatial coverage of a river network, such that the main channel and a limited number of tributaries have one or more gaging stations, but many tributaries have few or no gages. Under these circumstances, a number of techniques can be used to indirectly estimate the discharge of a discrete event, such as a flood, or to estimate statistical characteristics of discharge through time, such as mean annual flow.

3.2.2 Indirectly Estimating Discharge

In regions with reasonably complete spatial and temporal coverage of discharge gages, the existing records can be composited to create *regional discharge–drainage area relations.* These relations can then be applied to estimate flows at ungaged sites based on the drainage area of the site (Sanborn and Bledsoe 2006; https://streamstats.usgs.gov/ss). Related to this is the practice of *storm transposition,* in which a particularly extreme storm for which records exist at one gage site is used to estimate the discharge that would result if it occurred at another site (Changnon 2002). This type of extrapolation can be inaccurate where discharge–drainage area relations are nonlinear as a result of changes in hydroclimatology with elevation or changes in rainfall–runoff relations with geology or land cover (Pitlick 1994).

The most extreme precipitation input physically likely at a given site, the *probable maximum precipitation,* is defined as the greatest depth of precipitation for a given storm duration. This value is used to estimate the resulting *probable maximum flood,* which in the United States is mandated for design of structures such as dams where failure of the structure would result in loss of human life (Felder et al. 2017).

In the absence of direct measurements at discharge gages, flow can be indirectly estimated using several approaches. Most of these are based on a rearrangement of the *Manning equation,* as presented earlier

$$Q = \frac{1}{n} A R^{\frac{2}{3}} S^{\frac{1}{2}} \tag{3.38}$$

where A is cross-sectional area, R is hydraulic radius, S is channel gradient, and n is the Manning roughness coefficient. The Manning equation is based on the assumption of steady, uniform flow such that slope, discharge, and velocity are constant with time and space along a segment of channel. This assumption may not apply very well to flow in natural channels, particularly during high flows. Nonetheless, the commonly used indirect methods of slope-area and step-backwater computations assume steady, uniform flow and use one-dimensional hydraulic theory as their basis (Webb and Jarrett 2002).

Both the slope-area and step-backwater methods use the conservation of mass, conservation of energy, and Manning equations to calculate flow. Slope-area uses known water-surface elevations to compute discharge (Dalrymple and Benson 1967), whereas step-backwater uses discharge to compute stage (O'Connor and Webb 1988). Both of these approaches, and use of the Manning equation, assume that a water-surface profile can be used with surveyed cross-sectional geometry to calculate R as an approximation of flow depth. The accuracy of this calculation depends on the representativeness of the surveyed cross-sectional geometry and water-surface profile. Geometry and water-surface elevation can vary along a river reach and through time, even during a single flood, because of scour and fill of the bed, rapid changes in flow, substantial sediment transport, and flow transitions between subcritical and supercritical (Jarrett 1987; Sieben 1997).

Variations of parameters through time are particularly problematic when the Manning equation is used to calculate flood discharge. The accuracy of indirect flood-discharge estimations based on the Manning equation typically declines in at least four scenarios. The first scenario is rivers with very short duration and high-magnitude discharges that do not have steady, uniform flow. This scenario is characteristic of smaller catchments, particularly in arid regions and the tropics (e.g. Alexander and Cooker 2016). The second scenario is rivers with seasonal ice in which much of the runoff during the season when the ice is melting occurs on top of the ice (Priesnitz and Schunke 2002). The third scenario is small, steep rivers with very rough boundaries (Bathurst 1990). Finally, the fourth scenario is rivers such as sand-bed channels that undergo substantial changes in cross-sectional geometry during a flood as a result of local bed scour and fill and bank erosion throughout the flood hydrograph (Church 2006).

The Manning equation can also be used to estimate the magnitude of discharges that occurred in the distant past, if cross-sectional geometry and water-surface profiles are preserved. Information on water-surface elevation can come from *historical records* created by people, including physical marks on buildings or bedrock channel walls, as well as diaries, journals, and damage and insurance reports (Table 3.3) (Gurnell et al. 2003; Kjeldsen et al. 2014). Historical records are limited by the length of human occupation of a region. Historical photographs can provide useful information on hillslope and channel change caused by floods or more persistent hydrologic conditions (Webb 1996; Webb et al. 2004). Contemporary photographs and remote sensing imagery can also be used to infer flow characteristics in ungaged basins, as for example when remote imagery is used to estimate changes in water storage in small irrigation reservoirs, which are widespread in semiarid Africa (Liebe et al. 2009).

Information on water-surface elevation can also come from botanical indicators (Table 3.3) (Yanosky and Jarrett 2002; Ballesteros-Cánova et al. 2015a,b). *Botanical records* can take the form of lateral zoning of plants along channels, with the age and type of the plants reflecting the magnitude and frequency of flows (Sigafoos 1964; Hupp and Bornette 2003; Ruiz-Villanueva et al. 2010). The lowest elevation of plants intolerant of annual submergence, for example, can be used to estimate mean annual peak flow (Pike and Scatena 2010).

Table 3.3 Types of paleoflood and paleoflow indicators.

Category	Types and information
Historical	Written records of date, extent, and depth of a flood
	Photographs of a flood
	Maps of flood extent
	Physical marks recording peak stage
Botanical	Vegetation structured by age or type
	Impact scars recording flood stage and date
	Adventitious stems or split-base sprouts recording flood date
	Adventitious roots indicating burial by overbank sedimentation
	Exposed roots indicating date of bank erosion
	Variations in tree-ring width and symmetry indicating high and low flows
Geologic	Regime-based: channel dimensions related to mean flow
	Competence: average clast size, bedforms, bedding structure related to mean flow; maximum clast size related to peak flow
	Paleohydraulic: depositional or erosional indicators of hydraulics (e.g. lateral gravel berms, berms at downstream end of plunge pools, lateral potholes)
	Stage indicators: scour lines, lichen limits, truncation of landforms impinging on channel (e.g. alluvial fans), silt lines, organic debris, boulder bars, slackwater sediments

Variations in ring width can also be used to reconstruct annual variations in runoff (Meko et al. 2015; Elshorbagy et al. 2016), as explained in more detail at the TreeFlow website (www.treeflow.info). Because fluctuations in regional climate leave unique sequences of wider growth rings during good years and narrower growth rings during years of limited resources for trees across the region, the oldest rings of living trees can be matched to the youngest rings of dead trees. Using this type of cross-correlation, tree-ring records can be extended back more than 11 000 years (Becker and Kromer 1993). Botanical records can thus provide chronologically precise information for a wide range of discharges, although these records are still limited by the presence and age of trees (Wohl and Enzel 1995; Yanosky and Jarrett 2002).

Where large floods cause lateral channel movement across the floodplain, erosion associated with the lateral movement typically removes existing riparian trees. Deposition during the waning stages of the flood creates new germination sites, however, and a cohort of trees typically germinates soon after the flood. The age and distribution of patchy stands of trees across a floodplain can thus be used to infer the history of channel movement and the chronology of floods sufficiently large to trigger such movements (Sigafoos 1961; Everitt 1968; Miller and Friedman 2009; Bollati et al. 2014).

Riparian trees that are not removed by floods can sustain enough damage to leave scars that appear in their rings. Impact scars occur when floodborne debris traveling at or near the water surface hits a tree with sufficient impact to kill the cambium, which is the growth layer immediately below the bark surface (Figure 3.16) (Yanosky and Jarrett 2002). Even if the bark subsequently grows over this point, the scar remains in the tree's interior and can be detected when coring it. Because trees only grow upward at the tips of branches, the elevation of the impact scar remains constant with time and can

Figure 3.16 Flood-scarred Ponderosa pine (*Pinus ponderosa*) along Rattlesnake Creek, Arizona, USA. Two scars are visible in this photo: a larger scar at center, the base of which has been cut for tree-ring sampling, and a smaller scar at upper right. (*See color plate section for color representation of this figure*).

be used as a stage indicator. Scars on multiple trees are typically needed to define the water-surface profile of a flood (McCord 1996; Stoffel and Corona 2014).

Floodborne debris can also hit a tree with sufficient force to tilt the trunk. When this happens, subsequent annual growth rings are asymmetrical, allowing the time of tilting to be determined. The most severe impacts can shear off the trunk of the tree. This level of damage would kill many upland trees, but riparian trees are more likely to resprout from the remaining trunk, producing adventitious or split-base sprouts that can be dated to determine the time of the damaging flood (Hupp 1988; Stoffel and Corona 2014).

The root characteristics of riparian trees can also record river dynamics. Some species put out adventitious roots just below the ground surface or exhibit a change in ring characteristics when overbank sedimentation buries the lower portion of their trunk (Figure 3.17) (Hupp 1988; Friedman et al. 2005). Exposure of the roots through bank erosion can also trigger anatomical changes in the roots (Sigafoos 1964; Malik and Matyja 2008).

Regime-based reconstruction uses sedimentary features to reconstruct the cross-sectional geometry of relict channels preserved on a floodplain or in cutbanks of the contemporary channel

Figure 3.17 Adventitious roots of a mesquite tree exposed along a cutbank on Cienega Creek, Arizona, USA. Repeated burial during overbank sedimentation has caused the tree to grow at least three sets of roots, as visible here. Hat for scale. (*See color plate section for color representation of this figure*).

(Figure 3.18) (Jacobson et al. 2016; van de Lageweg et al. 2016). Channel geometry is then related to a relatively frequent flow, such as mean annual flood, using an approach such as the Manning equation. Specific aspects of channel geometry, such as bankfull cross-sectional area or meander wavelength, correlate with a relatively frequent flood. Changes in channel geometry through time can thus be used to infer differences in past discharge (Knox 2006; Carson et al. 2007).

Competence estimates use the average clast size, bedforms, or bedding structure to estimate past mean flow (Ryder and Church 1986; Nott and Price 1994), or the largest clasts likely to have been transported by fluvial processes to estimate the shear stress, stream power, or velocity of the associated peak flow (Wohl and Enzel 1995; Alexander and Cooker 2016). In either case, this involves two levels of inference. First, clast size or bedding must be related to a hydraulic parameter, and then the hydraulic parameter – typically, velocity, shear stress, or unit stream power – must be related to discharge using cross-sectional geometry and assumptions regarding flow depth. Competence estimates are usually less precise than other methods of indirectly estimating discharge and hydraulics (Alexander and Cooker 2016). When applied to maximum clast size, competence estimates provide information about the greatest flow magnitude, but not the timing or frequency of flows of this

Figure 3.18 Paleochannel exposed in the cutbank of a contemporary arroyo in eastern Colorado, USA. The fill of the paleochannel displays a fining upward sequence from cobbles to silt and clay at the top of the channel fill. Daypack on ground below paleochannel for scale. (*See color plate section for color representation of this figure*).

magnitude. Most equations relating clast size or bedding to hydraulics are not very precise or readily transferable between sites because of the site-specific effects of grain size and sorting (Komar and Carling 1991; Wilcock 1992). Most equations also focus on entrainment, whereas paleohydrology is based on deposition (O'Connor 1993). However, competence-based estimates can provide some insight into past flow conditions when no other records are available.

Paleohydraulic approaches use erosional or depositional indicators of flow hydraulics, in connection with channel geometry, to infer ungaged flow conditions. Lateral gravel or boulder berms deposited adjacent to hydraulic jumps in bedrock channels, for example, can be used to estimate the Froude number of flow (Carling 1995). Bars at the downstream ends of plunge pools can be used to estimate hydraulic force through the pools (Carling and Grodek 1994; Nott and Price 1994). Lateral potholes scale with flow Reynolds number and can be used to estimate this number (Zen and Prestegaard 1994).

Paleostage indicators in the form of erosional or depositional records of maximum stage can be used to reconstruct a water-surface profile at sites such as bedrock canyons where the cross-sectional geometry changes relatively slowly (Table 3.3) (Wohl and Enzel 1995; Jarrett and England 2002; Benito and O'Connor 2013).

Erosional records include scour lines eroded into regolith along valley walls (Benito and O'Connor 2013), the lower limits of lichen growth (Gregory 1976), and the truncation of landforms such as alluvial fans that impinge on the channel (Shroba et al. 1979). Depositional indicators include silt lines (Figure 3.19) of very fine sediment and organic fragments adhering to valley walls (O'Connor et al. 1986), and coarser organic material such as floodborne wood (Carling and Grodek 1994), boulder bars (Elfström 1987; Cenderelli and Cluer 1998), and fine-grained slackwater sediments deposited from suspension in areas of flow separation and lower velocity (Figure 3.20) (Baker 1987; O'Connor et al. 1994; Benito and O'Connor 2013; Zha et al. 2015). Using each of these types of information, A, R, and S are directly measured from existing channel geometry and high-water marks, and n is

(a)

(b)

Figure 3.19 (a) Silt lines along the back wall of an alcove formed in sandstone along the banks of the Paria River, Utah, USA. The overhanging roof of the alcove protects the silt lines from erosion. The top of a sequence of slackwater deposits is visible at lower right. (b) Close-up view of a silt line, showing the fine organic matter adhering to the bedrock wall in association with the silt. This organic matter can be used to radiocarbon date the age of the flood deposits. (*See color plate section for color representation of this figure*).

Figure 3.20 Slackwater deposits protected by an overhanging alcove in the bedrock walls along Buckskin Gulch in Arizona, USA. (*See color plate section for color representation of this figure*).

estimated, allowing at least an approximate value of Q to be calculated. The assumption is always that the erosional or depositional indicator corresponds to peak flood stage, although the accuracy of this assumption varies with the type of indicator (Baker and Kochel 1988; Jarrett and England 2002; Springer 2002).

The use of any proxy record of hydraulics or stage, including regime, competence, paleohydraulic, and paleostage indicators, is based on the assumption that surveyed channel geometry accurately represents the channel geometry present during the flow that created the feature. This in turn requires that (i) channel geometry has not changed substantially since the flood and (ii) scour and fill were minimal during the flood. Competence, hydraulic, and stage indicators are most commonly used along channels with stable boundaries formed in bedrock or very coarse alluvium. Channels with a relatively deep, narrow cross-section that maximizes the stage change in relation to discharge facilitate differentiation of indicators from different floods. Competence estimates, or the use of lichen limits, scour lines, boulder bars, or erosional landforms, may provide information about only the largest flood that occurred along the channel, but slackwater deposits can record numerous floods. The chronology of depositional units within a slackwater depositional site can be established using numerous techniques. These include radiocarbon dating of organic material in the sediment and luminescence (Lian 2007; Rhodes 2011) or ^{137}Cs dating of the sediment (Ely et al. 1992; Dezileau et al. 2014).

3.2.3 Modeling Discharge

Different measures of discharge, from flood peak flow to average annual discharge, can also be numerically simulated. The use of existing discharge records to estimate flow at ungaged sites is based on statistical analyses and the assumption that represented trends are consistent and continuous and can therefore be extrapolated across space and time, as well as extrapolated to the very large flows that may not be present in the gage records – although statistical techniques to address nonstationarity are evolving (Luke et al. 2017). *Stationarity* is the assumption that any statistical property of the flow record, such as mean annual flow or mean annual flood peak, has a probability density function that does not vary through time. This assumption facilitates extrapolation of trends in hydrologic data forward in time and prediction of the magnitude of future 100-year floods, for example, based on the magnitude of 100-year floods in the past, or in a situation where the record length is less than 100 years. Numerous studies now indicate that stationarity is unlikely to occur at most locations, however (e.g. Milly et al. 2008).

An alternative to the statistical approach of estimating discharge is a deterministic approach based on a mechanistic concept used to simulate precipitation inputs and discharge outputs under imposed boundary conditions (Sieben 1997). Deterministic approaches include a variety of models, such as the US Army Corps of Engineers' HEC-HMS model, which estimates discharge resulting from rainfall (http://www.hec.usace.army.mil/software/hec-hms).

Hydrologic models of rainfall–runoff processes at the watershed scale depend on input data. Input data can include topography of the catchment surface and parameters that are spatially distributed over (land cover, land use) and beneath (soil texture, moisture) the surface or lumped into averages at differing spatial scales (Beven 2012). Input data can also include the spatial characteristics of the river channel network (e.g. drainage density) and channel geometry. Many of these data are acquired using remote sensing and manipulated using GIS software. This scale of hydrologic modeling presents numerous challenges because catchment-scale models integrate data derived from a wide range of sources, at a range of spatial scales and resolution, and over a period of time (Downs and Priestnall 2003; Downs and de Asua 2016). Understanding the sources and magnitude of uncertainty in the diverse input data is particularly important to evaluating the accuracy and representativeness of these models (Taormina and Chau 2015). Data on the spatial distribution of basic hydrological variables such as rainfall and runoff can be combined with digital terrain analysis to estimate discharge at varying points in a channel network and to create spatially explicit water budgets (Montgomery et al. 1998; Marks and Bates 2000).

Another approach to numerically modeling floods involves combining gaged and ungaged stream flows with physical and statistical data to develop probability models for all uncertain parameters (Campbell 2005). Hydrologic model prediction is not deterministic, but rather must explicitly include an estimate of uncertainty. Uncertainty in model predictions reflects measurement errors for input and output parameters; model structural errors from the aggregation of spatially distributed real-world processes into a mathematical model; and errors of parameter estimation. Bayesian model selection uses the rules of probability theory to select among different possibilities, automatically choosing simpler, more constrained models. This approach can be applied to rainfall–runoff models (Sharma et al. 2005) or to models of stream flow. Both types of models can also be run using a Generalized Likelihood Uncertainty Estimation scheme, or GLUE (Beven and Binley 1992). The GLUE framework was one of the first attempts to represent prediction uncertainty. This framework conditions the parameter distributions and generates prediction uncertainty envelopes that incorporate parameter uncertainties (Wyatt and Franks 2006). A great deal of attention is given to spatial variability in parameters and processes and to uncertainty in parameter estimation (Sivapalan et al. 2006; Beven 2012).

Although direct measurements of flow stage at gaging stations include errors and uncertainty, these data remain the standard against which indirect estimates, statistical extrapolations, and numerical simulations are validated. Changing priorities in government expenditures in countries such as the United States, however, continue to result in declining numbers of active gages. Estimation of discharge using satellite- or ground-based remote imagery is helping to fill this widening gap in discharge data, and such approaches are rapidly being developed (e.g. Legleiter et al. 2009; Flener et al. 2012). Examples include thermal imaging of flow velocities using particle image velocimetry, which quantifies advection of thermal features by the flow, and spectrally-based depth retrieval from passive optical image data (Legleiter et al. 2017b).

3.2.4 Flood Frequency Analysis

The frequency with which a flood of a given magnitude recurs is critical to understanding the physical and ecological significance of floods, as well as mitigating hazards associated with flooding. Flood frequency analysis uses different techniques to relate flood magnitude to recurrence interval. The record of floods at a site can be considered in terms of an *annual maximum series*, which includes only the largest discharge in each year of record, or a *partial duration series*, which includes all floods above a specified discharge. The partial duration series is likely to be more geomorphically important, particularly if the specified discharge has some physical relevance to sediment transport or channel morphology.

The simplest way to calculate recurrence interval T (years) for either annual maximum or partial duration series is via the Weibull formula

$$T = (n + 1)/N \tag{3.39}$$

where n is the number of years of record and N is the rank of a particular flood when floods are ranked from largest to smallest (Dalrymple 1960). T can then be plotted against Q to create a *flood-frequency curve*. In any year, the probability of the largest flow exceeding a flood with a recurrence interval of 10 years is 1/10. A flood with an average recurrence interval of 10 years is commonly referred to as a 10-year flood, but this type of abbreviated description can be very misleading. There is no physical reason that a 100-year flood cannot occur during two consecutive years, for example: it is improbable, but not impossible.

There are no strong theoretical justifications for applying any particular statistical distribution to hydrologic data. Log-Pearson Type III is commonly used in the United States because it is a skew distribution bounded on the left and therefore of the general shape of most hydrologic distributions. The Gumbel extreme value distribution is commonly used in the United Kingdom.

Regardless of the distribution used, treating the highest known discharges for a site is particularly problematic because these discharges tend to be outliers that do not follow trends present in the remainder of the data. Historical and paleoflood information can be combined with systematic gage measurements of discharge to extend the length of the flow record at a site. The historical and paleoflood data must be treated differently than the systematic data because they represent what is known as a *censored record*: only floods above a magnitude threshold are recorded, rather than all flows being recorded, as at a stream gage. Early work used threshold-exceedance maximum likelihood estimators, which employed the number of floods that exceeded a known threshold, without differentiating the magnitude of each flood (Stedinger and Baker 1987). Subsequent work relies on moments-based parameter estimation procedures (Cohn et al. 2001; England et al. 2003; Kjeldsen et al. 2014) or simply uses field evidence of paleoflood magnitude to quantify nonexceedance bounds for discharge over a time interval defined using various geochronologic techniques (England et al. 2010, 2014).

Another difficult issue for flood-frequency analysis is the existence of *mixed distributions.* Many sites include floods produced by more than one hydroclimatic mechanism – snowmelt floods and flash floods resulting from convective thunderstorms, for example, or flash floods and longer-duration floods caused by dissipating hurricanes. Analyses that separate populations with different flood-generating mechanisms result in improved parameter estimates of the component distributions and a better understanding of the physical basis for extreme floods (Hirschboeck 1987; Smith et al. 2011).

The presence of nonstationarity is a primary concern for any form of flood-frequency analysis. Assumptions of stationarity can be incorrect if the particular period of record represents some deviation from longer-term averages. Precipitation can exhibit long-period variability such that more zonal or meridional circulation patterns characterize periods up to several decades in length at regional to continental scales (Hirschboeck 1988). More meridional circulation tends to produce more severe flooding, so using 20 years of record from a period of meridional circulation to predict future flood frequency might overestimate the recurrence interval of a particular flood magnitude. Other climatic circulation patterns can produce fluctuations in flooding at time scales of decades (Webb and Betancourt 1990) to centuries and millennia (Ely et al. 1993; Benito et al. 1996; Redmond et al. 2002) across broad regions. Human alterations of rainfall–runoff relations as a result of changing land cover or flow conveyance in river networks can also compromise the assumption of stationarity (Kundzewicz 2011). Finally, ongoing global climate change associated with CO_2-induced warming is rendering the assumption of stationarity inappropriate in many regions. Efforts are underway to develop nonstationarity probabilistic models (Milly et al. 2008; Cheng et al. 2014), but no consensus yet exists on the best approach (e.g. Bayazit 2015).

3.2.5 Hydrographs and Flow Regime

A *hydrograph* is a curve of discharge plotted against time. An *event* or *flood hydrograph* represents a single flood. An *annual hydrograph* represents flow over the course of an average year as derived from averages of some flow interval (typically, minutes to one day) over the period of gage records. A *unit hydrograph* is based on runoff volume adjusted to a unit value (e.g. 100 mm of precipitation spread evenly over the drainage basin) (Rodriguez-Iturbe and Valdes 1979). A *geomorphologic unit hydrograph* relates the unit hydrograph to the morphologic parameters of the river network and defines the travel time distribution of water particles to the outlet of the basin (Rodriguez-Iturbe and Valdes 1979).

Any hydrograph includes base flow and direct runoff, a rising limb, a peak, and a falling limb (Figure 3.21). *Base flow* results primarily from ground-water inputs to a channel and is the stable, low flow to which a river returns following precipitation inputs. *Direct runoff* results from the combined inputs of overland flow and flow in the unsaturated zone (Section 2.2.1). The *rising limb* is the portion of the hydrograph where discharge increases as a result of direct runoff inputs, the *peak* represents the maximum discharge within a particular time span, and the *falling or recession limb* reflects progressive declines in direct runoff with time until discharge returns to base flow.

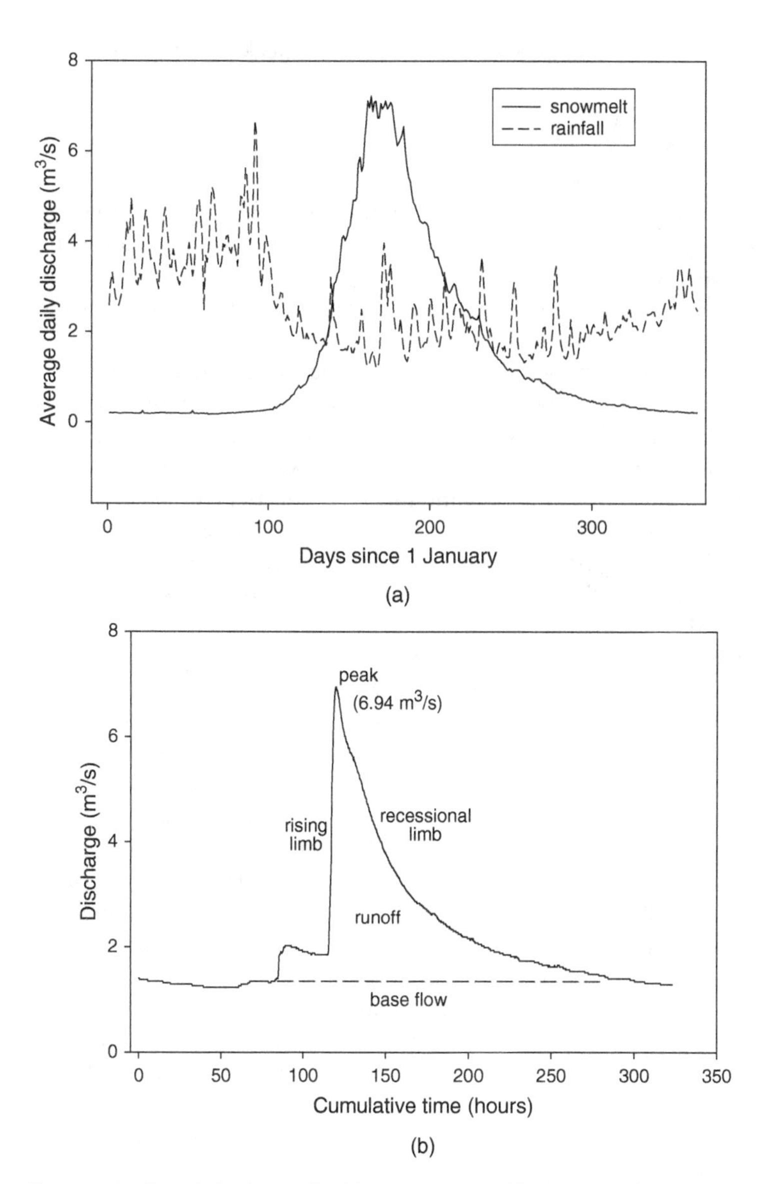

Figure 3.21 Sample hydrographs. (a) Average annual hydrographs based on average daily discharge over the period of record for a snowmelt-dominated stream (North St. Vrain Creek, Colorado, USA; drainage area 91 km^2) and a rainfall-dominated stream (Cedar Creek, South Carolina, USA; drainage area 161 km^2). (b) Flood hydrograph for a single rainfall-generated flood (Gills Creek, South Carolina, USA; drainage area 167 km^2).

The term *flashy* is commonly applied to hydrographs with short-duration peaks and rapid rise and recession. Baker et al. (2004) propose a flashiness index

$$\text{Richards} - \text{Baker Flashiness Index} = \frac{\sum_{i=1}^{n} |q_i - q_{i-1}|}{\sum_{i=1}^{n} q_i} \quad (3.40)$$

where q is mean daily flow, q_i is flow on a particular day, and q_{i-1} is flow on the following day. Although many drainages are described as being flashy, the criteria used for this designation vary widely and are not necessarily comparable between studies.

The *lag time* or *basin lag* indicates the delay between the center of mass of precipitation inputs and the center of mass of streamflow. Lag time reflects the manner in which precipitation is transmitted from hillslopes into the channel, as well as the size of the catchment, spatial arrangement of channels in the river network, and geometry of valleys and channels. Lag time for a given type of precipitation input increases with basin size because direct flow must travel longer distances along hillslope paths before reaching a channel and longer distances within the river network before reaching the basin outlet (McGlynn et al. 2004). Lag time is lower for equant-shaped basins than for linear basins. Equant-shaped basins tend to produce larger, shorter peak discharges because of efficient concentration at the basin outlet (Strahler 1964). Lag time also increases where broad valley bottoms and floodplains allow direct runoff to move relatively slowly downstream in overbank areas or multiple channels (Lininger and Latrubesse 2016).

Hydrograph characteristics can also be strongly influenced by the *hydroclimatology*, or precipitation regime. Ignoring other influences such as drainage basin size, snowmelt runoff tends to be spread over a longer time and to produce a peak flow of longer duration and lower peak than rainfall runoff (Pitlick 1994). Different types of rainfall produce differently shaped hydrographs (Roberts and Klingeman 1970). Convective rainfall, because of its greater intensity and short duration, tends to produce a more peaked hydrograph than cyclonic rainfall. Rain-on-snow, as the name implies, occurs when rain falls on a snowpack. The resulting runoff depends on the magnitude of the rain and the water equivalent and spatial extent of the snowpack, and can be quite large but of shorter duration than snowmelt (McCabe et al. 2007). Because of these characteristic differences in event and annual hydrographs, a river's flow regime is typically described as being *snowmelt-dominated* or *rainfall-dominated*. These categories are not mutually exclusive, because different types of flooding can occur during different seasons or over a period of many years in the same region. A snowmelt-dominated hydrograph can have secondary peaks associated with late-summer convective storms, for example, or a rainfall-dominated hydrograph can have predominantly frontal cyclonic precipitation with occasional hurricane rainfall.

In addition to describing flow regime in terms of the dominant type of runoff, flow regime can also be characterized by its spatial and temporal continuity. *Perennial flow* is continuous through time and space: some level of flow is always present throughout the river network so designated. *Intermittent flow* is spatially discontinuous, such that some portions of a river contain flow while other portions are dry. Intermittent rivers can also flow continuously for only a limited portion of the year, when the water table intersects the ground surface (Wohl et al. 2017d). Intermittent flow can occur where a channel crosses into a different climate, as when perennial river segments flowing from mountainous highlands with large annual precipitation enter drier lowlands and the surface flow evaporates or infiltrates. Intermittent flow can also result from longitudinal contrasts in subsurface permeability, as when flow in a channel crossing a thick, permeable alluvial layer infiltrates and moves downstream in the subsurface, returning to the surface channel at places where an impermeable layer close to the surface forces the water upward once more. Ground-water pumping that lowers alluvial water tables can increase the spatial extent and duration of intermittency (e.g. Falke et al. 2011). *Ephemeral flow* is temporally discontinuous, with periods of surface flow shortly after precipitation inputs to the basin interspersed with periods of no flow. Ephemeral channels do not have ground-water inputs or base flow.

Intermittent and ephemeral flows are most likely to occur in dry environments such as arid and semiarid regions of the polar, temperate, and tropical latitudes, and in very small catchments that may have limited base flow inputs. Intermittent flow can also occur in *karst terrains*, where underground drainage networks developed through chemical dissolution of carbonate rocks capture surface flow in some portions of a network (Ford and Williams 2007). Surface–subsurface exchanges in karst environments can produce *blind valleys* that end suddenly where a stream disappears underground, and *exsurgence* where an underground stream with no surface headwaters reaches the surface (Ford and Williams 2007).

The phrase *temporary rivers* is commonly used to describe rivers that periodically cease flowing (Datry et al. 2011), and thus includes both ephemeral and intermittent rivers. Temporary rivers are present in all climates and are growing in extent as a result of warming climate and land uses that limit infiltration and ground-water recharge. Temporary rivers have become a flashpoint in the context of river and water-resources management because, although river scientists argue that protection of these portions of the channel network is critical to the geomorphic and ecological integrity of river systems (e.g. Wohl et al. 2017d), legislation to protect and regulate temporary rivers is limited and controversial (e.g. Acuña et al. 2014).

The spatial and temporal extent and magnitude of flow also influence the connectivity of river networks. Ecologists distinguish *riverine connectivity*, which indicates spatial linkages within rivers, and *hydrologic connectivity*, which is the water-mediated transport of matter, energy, and organisms within or between elements of the hydrologic cycle (Freeman et al. 2007). Both forms of connectivity are vital to maintaining the *ecological integrity* of riverine ecosystems, where ecological integrity is the undiminished ability of an ecosystem to continue its natural path of evolution, its normal transition over time, and its successional recovery from perturbations (Westra et al. 2000). Graf (2001) defines *physical integrity* as a set of active river processes and landforms such that the channel, near-channel landforms, sediments, and river-corridor configuration maintain a dynamic equilibrium, with adjustments not exceeding limits defined by societal values. In other words, a river corridor has physical integrity to the extent that the physical components of the corridor are able to adjust to changing boundary conditions such as water and sediment inputs and relative base level. As with ecosystems, riverine and hydrologic connectivity are essential to maintaining the physical integrity of river corridors (e.g. Graf 2006).

The typical scenario is that discharge and occurrence of perennial flow increase downstream within a river network as contributing drainage area increases. If the rate of increase in discharge with drainage area is known, the more easily measured drainage area can be used as a surrogate for discharge. This may not be the case, however, in drylands (semiarid and arid regions). Low annual precipitation and high evaporation result in low annual runoff, and the interannual variability of runoff increases in the driest regions. The spatial variability of precipitation and runoff limits the use of drainage area as a surrogate for discharge in drylands (Tooth 2013). Instead, most dryland river networks exhibit downstream decreases in discharge during floods as a result of transmission losses from infiltration into unconsolidated alluvial channel beds and evaporation and transpiration, as well as a common absence of appreciable tributary inflows in the lower parts of many dryland river networks (Tooth 2013). Infiltration losses from channels are a major source of ground-water recharge in arid regions (Shanafield and Cook 2014). Maximum flood peaks and discharge per unit drainage area can be quite large in dryland rivers, particularly where small, steep catchments and abundant low-permeability surfaces associated with bedrock or crusted soils are present. This situation can result in highly skewed flood-frequency distributions and steep flood-frequency curves that reflect a high ratio of large to small flows (Knighton and Nanson 2002; Tooth 2013).

3.2.6 Other Parameters Used to Characterize Discharge

Discharge measurements that are continuous during a year and over several years can be used to establish a *flow-duration curve* in which mean discharges over a specified time interval (e.g. daily mean) are grouped into selected classes of discharge magnitude and plotted against the percentage of time that class is equaled or exceeded (Figure 3.22). The shape of the flow-duration curve reflects the drainage basin's response to precipitation inputs (Coopersmith et al. 2012). The steeper the flow-duration curve, the more quickly storm runoff enters the channel.

At sites with measurements of suspended sediment concentration at varying discharges that allow construction of a *sediment rating curve*, the flow-duration curve can be used to develop a *cumulative sediment transport curve* indicating how flows of differing magnitude and recurrence interval contribute to sediment transport during an average year (e.g. Biedenharn and Thorne 1994).

Ecologists developed the *flood-pulse concept* for large lowland rivers in recognition of the importance of the magnitude, timing, duration, and rates of rise and fall of annual or seasonal floods to aquatic and riparian ecosystems. Natural, predictable floods – the flood pulse – that inundate at least a portion of the floodplain are associated with much higher biological productivity. The flood pulse provides clear, shallow water in overbank areas for primary, photosynthetic production. The flood pulse also creates fish nursery habitat and feeding, as well as more abundant and diverse habitat for a variety of other organisms. Inundation of overbank areas and subsequent recession of stream flow into channels during the flood pulse facilitates the exchange of nutrients, organic matter, and organisms between the channel and the floodplain (Junk et al. 1989; Bayley 1991).

Flow pulses are fluctuations in surface waters below the bankfull level of a channel (Tockner et al. 2000). Although these fluctuations do not cause overbank flow, they change the extent of flow in secondary channels and in areas of flow separation along a single, confined channel and provide some of the same morphological changes, biogeochemical exchanges, and habitat abundance and diversity as overbank flows (Bertoldi et al. 2010).

A third key concept in riverine ecology is the *natural flow regime* (Poff et al. 1997). This describes the magnitude, frequency, duration, timing, and rate of change of flow conditions in the absence of human interference through processes such as flow regulation. Channel geometry and aquatic and riparian communities are typically adjusted to the natural flow regime, so that altering this flow regime is likely to cause corresponding alterations in the physical and ecological characteristics of the river. Management of rivers with flow regulation increasingly emphasizes protecting or restoring at least some approximation of the natural flow regime.

3.2.7 Hyporheic Exchange and Hydrology

The hyporheic zone is the portion of unconfined, near-stream aquifers where stream water is present. Hydrologists define this zone as a flow-through subsurface region containing flow paths that originate

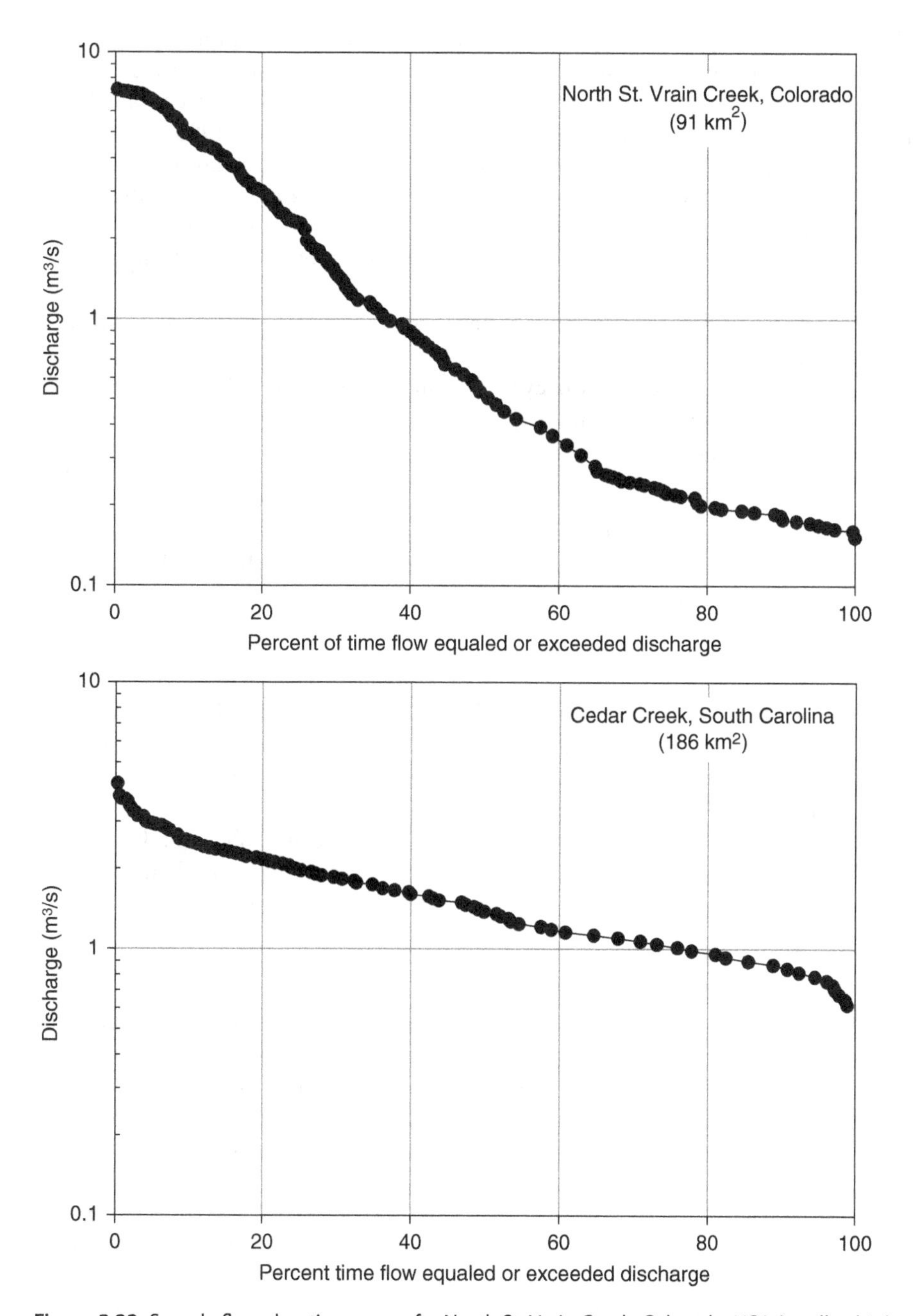

Figure 3.22 Sample flow-duration curves for North St. Vrain Creek, Colorado, USA (smaller, high-relief, snowmelt flow regime) and Cedar Creek, South Carolina, USA (larger, lower relief, rainfall flow regime).

and terminate at the stream. The hyporheic zone can be defined based on the time scale of flow, analogous to definitions of floodplains based on inundation frequency, such that there are 2-hour to 24-hour hyporheic zones (Gooseff 2010). The hyporheic zone can extend up to 2 km laterally from the active channel along rivers with broad, gravel floodplains (Stanford and Ward 1988). Hyporheic flow may constitute only a minor portion of stream discharge (<1%) in steep, headwater channels with limited alluvium (Wondzell and Swanson 1996), but can be 15% or more of surface discharge in larger, lowland alluvial rivers (Laenen and Risley 1997). In the latter case, ignoring hyporheic flow can lead to errors in discharge estimation, particularly given that the characteristics of the hyporheic zone vary widely in space and time.

Water exchange between the channel and adjacent aquifer is controlled by vertical and horizontal hydraulic gradients, although pressure and velocity fluctuations in surface flow decrease exponentially in the subsurface, primarily within the first two layers of grains (Detert et al. 2008). The upwelling or downwelling hyporheic flux, e, can be represented as

$$e = -KA\frac{d^2h}{dl^2} - K\frac{dA}{dl}\frac{dh}{dl} - A\frac{dK}{dl}\frac{dh}{dl} \tag{3.41}$$

where K is the sediment hydraulic conductivity, A is the cross-sectional area of a sediment volume within the riverbed, and dh/dl is the spatial gradient of the energy head h, which is total streambed pressure. Energy head is expressed in meters of water and defined as

$$h = z + h_p + C\frac{U^2}{2g} \tag{3.42}$$

where z is the elevation head (changes in bed elevation), h_p is the static pressure head (changes in flow depth), $C\ (U^2/2g)$ is the dynamic pressure head (changes in flow velocity and momentum), C is a generic loss representing changes in momentum resulting from form drag or channel contraction/expansion, and U is mean flow velocity (Buffington and Tonina 2009a).

The turnover length, L_m, is the length scale for complete mixing between surface and hyporheic waters. L_m depends on river discharge Q_r, wetted channel perimeter P, and rate of exchange q_h (the average downwelling flux per unit streambed area) (Buffington and Tonina 2009a)

$$L_m = \frac{Q_r}{q_h P} \tag{3.43}$$

Downwelling to the hyporheic zone occurs where stream-water slope increases or where permeability or bed depth increases. These changes in slope and depth can occur in association with streambed topography such as at the transition from pools to steeper riffles (Wondzell and Gooseff 2013; Blois et al. 2014); mobile bedforms such as sand ripples or dunes (Packman and Brooks 2001; Gomez-Velez et al. 2015); and structures such as individual pieces of wood and logjams (e.g. Hester and Doyle 2008; Sawyer et al. 2011).

Upwelling to the channel occurs where stream-water slope decreases, such as the transition from riffles to pools, or where permeability or bed depth decreases downstream. Flow exchanges can be so complex and localized, however, that some channel segments concurrently lose and gain water (Payn et al. 2009). The relative importance of mechanisms driving hyporheic exchange likely varies in relation to channel morphology, as do the rate and extent of hyporheic exchange (Buffington and Tonina 2009b; Gomez-Velez and Harvey 2014).

Hyporheic flow paths recharged by stream water are generally short, on the order of 1–10 m (Harvey and Bencala 1993). However, average flow path lengths tend to increase from headwaters to larger rivers (Anderson et al. 2005) because the spacing between upwelling and downwelling is influenced by the spacing of channel units such as pools and riffles (Marion et al. 2002; Gooseff et al. 2006). Spatial variations in ground-water upwelling along the river corridor can also influence hyporheic flow paths (Caruso et al. 2016). Stratified beds favor the development of horizontal flow paths within the bed relative to homogeneous beds (Marion et al. 2008a; Sawyer and Cardenas 2009). Temporal variations in hyporheic exchange can reflect variations in stream discharge (Schmadel et al. 2017) or the indirect effects of changing discharge as bed structure changes (Wondzell and Swanson 1996, 1999). Seasonal ice cover can also influence hyporheic exchange, promoting shallower but larger fluxes (Cardenas and Gooseff 2008).

Conceptual models of surface–hyporheic flow paths and exchanges have existed for decades (Vaux 1968), but automated streambed instrumentation that facilitates the mapping of spatial and temporal patterns of flow and exchange has dramatically enhanced quantification of these processes (Constantz 2008). Tracers used to quantify exchange include water temperature, based on instantaneous temperature differences between instream and hyporheic water (Arrigoni et al. 2008; Lautz 2010). Introduced tracers include dissolved sodium chloride, which is electrically conductive, so that near-surface electrical resistivity imaging can be used to image spatial and temporal hyporheic dynamics *in situ* (Ward et al. 2010a,b; Busato et al. 2019). Tracer-based field and flume data can be used to calibrate numerical models, which vary from one to three dimensions (Storey et al. 2003; Cardenas et al. 2004; Gooseff et al. 2006). The starting point for modeling solute transport in rivers (STIR), including dissolved tracers, is to use a one-dimensional advection-dispersion analysis (Bencala and Walters 1983). Release of a tracer into a stream forms a "cloud" that moves downstream in a process known as *advection. Dispersion* occurs as small-scale mixing processes cause the tracer to spread out, or disperse, as it moves downstream.

The "tail" of a solute tracer pulse – the relatively long period of time over which low concentrations of tracer continue to be present after the main pulse of tracer has passed a measurement point in the stream – is not adequately characterized using a convection-dispersion analysis. Long tails have been simulated by allowing for storage zones or "dead" zones along the channel. Storage zones of relatively stagnant water are present downstream from protruding logs or boulders, around vegetation in the shallows, along the edges of pools with recirculating flow, and so forth (Ensign and Doyle 2005; Gooseff et al. 2007). Solute mass is removed from the main flow into these storage zones during the rising phase of a solute pulse, retained in storage until the pulse passes, and then returned to the stream. The exchange of solute between this *transient storage zone* and the main channel is assumed to be proportional to the difference in concentration between the main channel and the storage zone (Marion et al. 2008b). The residence time distribution of water and solute in storage is assumed to be exponential (Gooseff et al. 2003). Modeling this exchange requires knowledge of the storage zone cross-sectional area and an empirical exchange coefficient. In effect, transient storage models (Bencala and Walters 1983; Hart 1995) modify the advection-dispersion model to include processes such as lateral inflow and outflow, and transient storage (Runkel et al. 2003).

Early approaches included great simplifications. Subsequent models such as STAMMT-L (solute transport and multirate mass transfer-linear coordinates; Haggerty and Reeves 2002) attempt to overcome these simplifications with an additional source/sink term representing mass exchanges with the storage domains through a convolution integral of the instream solute concentration and a residence time distribution (Marion et al. 2008b). Comparative tests reveal that the transient storage model OTIS (one-dimensional transport with inflow and storage; Runkel and Chapra 1993) may be more accurate over short time scales of circa 30 minutes, whereas STAMMT-L does better over periods of circa 24 hours (Gooseff et al. 2003). The STIR model uses a stochastic method to derive a relation between instream solute concentration and the resident time distributions in surface and hyporheic retention zones (Marion et al. 2008b). Numerous models now exist to simulate transient storage in the surface and subsurface portions of a river corridor (McCluskey and Grant 2016).

Distinguishing surface retention zones associated with eddies and pools from hyporheic retention zones is difficult when measuring and modeling hyporheic exchange (Runkel et al. 2003; Gooseff et al. 2005). Approaches to date include using ADCP data and wavelet decomposition to separate stream flow into regions of differing velocity and to estimate the relative sizes of main-channel and storage zones (Phanikumar et al. 2007). Wavelet decomposition allows the original signal – in this case, velocity data – to be split into different components so that each can be studied with a resolution appropriate for its scale. For velocity measurements, the data can be decomposed into slowly changing (low-frequency) and rapidly changing (high-frequency) features using low-pass and high-pass wavelet filters at different scales in a multiscale decomposition (Phanikumar et al. 2007). In essence, instream transient storage zones can be identified objectively using wavelet decomposition of three-dimensional velocity data obtained from ADCP surveys. Tracer breakthrough curves from the main channel, cross-sectional stream velocity distributions, and stream tracer concentration time-series data from multiple locations in the main channel and adjacent surface transient storage zones can be used as input to a transient storage model with surface and subsurface storage zones (Briggs et al. 2009; Johnson et al. 2014). Gonzalez-Pinzon et al. (2015) review the advantages and limitations of diverse techniques that can be used to quantify surface–subsurface exchanges and transient storage.

Enhanced hyporheic exchange is increasingly sought during stream restoration because of its multiple effects on water-quality parameters such as temperature, dissolved oxygen, and dissolved nutrients and organic matter (e.g. Merrill and Tonjes 2014).

3.2.8 River Hydrology in Cold Regions

Hyporheic and ground-water exchange can complicate the relations between precipitation inputs and outputs of river flow, but in most basins, greater precipitation inputs equate to higher river stage and greater discharge. This is not necessarily the case in high-latitude rivers with strongly developed seasonal ice cover. A river-ice season that lasts more than 100 days between autumn freeze-over and spring break-up characterizes many rivers of high latitudes, including some high-elevation or interior rivers as far south as 42 °N in North America and 30 °N in Eurasia (Prowse and Beltaos 2002). The hydraulic resistance added by an ice cover elevates river water levels, particularly when the ice cover is hydraulically rough during freeze-over and break-up (Ettema and Kempema 2012; Shen 2016).

Forms of river ice (Hicks 2009) include *anchor ice* frozen to the bed of the channel, *frazil ice* composed of granular ice transported within the flow, and *aufeis* or icings. Aufeis consists of sheet-like masses of layered ice that form during the winter via freezing of successive flows of water (from ground-water springs) onto the top of the river ice (Morse and Wolfe 2015). Aufeis typically occurs in the same portion of a river each year and can extend along more than 100 km^2 at thicknesses exceeding 10 m (Pavelsky and Zarnetske 2017).

River ice can modify the quantity of flow in a channel via three mechanisms (Prowse and Beltaos 2002; Ettema and Kempema 2012). The first involves reduction of contributing area, when anchor ice frozen to the river bed cuts off ground-water inflow. The second involves direct storage of water in river ice, which typically involves slower abstraction during freeze-over and rapid resupply during break-up. The third mechanism, which can produce the most significant effect, is hydraulic storage in the channel. This involves increased water levels because of the increased hydraulic resistance caused by ice cover. The latter effect can be particularly significant when ice jams form during break-up (Ettema and Kempema 2012).

Ice-jam frequency, and the frequency of the floods that result, is highly variable from year to year (Boucher et al. 2012) as a result of the strong influence of interannual fluctuations in air temperature. The reduction in flow created by the three mechanisms just discussed can be equivalent to nearly 30% of the flow that would otherwise occur at a particular site (Prowse and Carter 2002). Release of this water during break-up can account for nearly 20% of the spring peak flow (Prowse and Carter 2002). Because of these effects, early winter freeze-over can create the lowest discharge of the year, even though runoff is lowest during late winter. Analogously, break-up frequently establishes the annual maximum water levels, even though maximum discharge is more likely to result from spring snowmelt or summer rainfall later in the year (Prowse and Ferrick 2002). One implication of the presence of ice is that drainages with ice cover may not exhibit the direct relationship between flow and stage that exists in most rivers (Prowse and Beltaos 2002).

Flow regimes in high-latitude rivers can be distinguished as nival, proglacial, wetland, and prolacustrine (Woo and Thorne 2003). Snowmelt and river ice break-up generates the largest flows in *nival regimes*. High flows are prolonged into the summer by glacial melt water in *proglacial regimes*; some of the highest flows occur in mid- to late summer, when glacier melt is most pronounced. Rivers below wetlands also have prominent snowmelt peaks because the wetlands have low storage capacity while frozen, but once the ground thaws, they can retain water and retard summer flows in *wetland regimes*. The influence of large lakes creates a fairly even runoff throughout the year in *prolacustrine regimes*, although the outflow channel of an arctic lake is more likely to be blocked during freeze-over and inflow to the lake can be very small, minimizing winter outflow.

Warming climate is reducing the duration, spatial extent, and thickness of river ice (Park et al. 2016), and may thus be predicted to eventually reduce ice-related hydrologic and geomorphic effects on some high-latitude rivers. At present, however, increasing discharge and precipitation are driving increased ice jams in some temperate-latitude rivers (e.g. Carr and Vuyovich 2014) and exacerbating earlier-season (including mid-winter) ice-jam floods on arctic rivers because of earlier ice break-up (Cooley and Pavelsky 2016).

3.2.9 Human Influences on Hydrology

Human activities can alter flow regimes in channels indirectly through changes in climate, topography, and land cover that influence the amount and downslope pathways of water entering channels from uplands, as reviewed in Chapter 2. Direct human alterations of stream flow involve changes in the volume and timing of flow as a result of flow regulation and changes in the ability of channels to convey flow downstream as a result of river corridor engineering (Table 3.4).

3.2.9.1 Flow Regulation

Flow regulation includes dams and diversions. Most of the world's rivers are affected to some degree by flow regulation. Dams have been built since circa 2800 BC (Smith 1971) for diverse purposes, including water supply, flood control, navigation, and hydroelectric power generation. The construction of dams larger than 15 m tall has accelerated substantially since the 1950s (Goldsmith and Hildyard 1984; Nilsson et al. 2005; Grill et al. 2015). Although relatively few large dams are now

Table 3.4 Types of direct human influences on stream flow.

Type of influence	Effects	Sample references
Flow regulation	Dams typically reduce the mean and coefficient of variation of annual peak flow, increase minimum flow, shift seasonal variability, and increase diurnal fluctuations; diversions typically reduce base flows and flood peaks in the source stream and increase base and peak flows in the receiving stream	Graf (2006) Ryan (1997); Wohl and Dust (2012)
Channel engineering	Levees prevent or limit overbank flow and thus typically increase the magnitude and decrease the duration of peak flows	Kondolf (2001)
	Bank stabilization can decrease channel complexity and increase conveyance	
	Channelization (straightening and dredging) increases channel conveyance and typically increases the flashiness of stream flow	Shankman and Pugh (1992)
Alterations of riverine biota	Removal of riparian vegetation decreases near-bank and overbank roughness, can decrease bank stability, and decreases transpiration	Anderson et al. (2006a)
	Introduction of exotic riparian vegetation can change vegetation density and function, typically with increased density and roughness that facilitates channel narrowing	Dean and Schmidt (2011)
	Removal of beaver decreases channel–floodplain connectivity, increases flow velocity and peak flow magnitude, and decreases the elevation of the alluvial water table	Wohl (2001)

built in Europe and North America, a great many are being built or have been proposed for rivers in Africa, Asia, and South America (e.g. Kondolf et al. 2014; Rubin et al. 2015). Multiple dams are under consideration in the few large river basins not yet extensively regulated, including the Amazon (Latrubesse et al. 2017) and the Congo (Wohl 2011a). Although dams are sometimes promoted as an environmentally benign, clean source of hydroelectric power, flow regulation nearly always substantially disrupts physical process and form, water chemistry, and ecological communities along rivers. In addition, research indicates that although hydroelectricity generates lower volumes of greenhouse gases than fossil-fuel combustion (hence, the "clean" designation for hydroelectric power), reservoirs can in some cases emit significant amounts of greenhouse gases, particularly in tropical regions (e.g. Galy-Lacaux et al. 1999; Fearnside 2015).

As reviewed in Petts and Gurnell (2005), geomorphologists began to pay increasing attention to the effects of dams in the late 1960s and into the '70s (e.g. Wolman 1967b; Gregory and Park 1974; Petts 1979). Coupled with advances in measurement techniques and process-based studies, this led to a series of influential papers during the 1980s on the geomorphic effects of dams on river networks (e.g. Petts 1984; Williams and Wolman 1984; Carling 1988).

Dams can be small barrages or weirs that are frequently overtopped during moderate to high discharges but serve to pond water upstream during low discharges. At the opposite end of the size scale, large dams are those taller than 15 m and major dams are taller than 150 m. More than 45 000 large dams and more than 300 major dams currently exist. Over half (172 of 292) of the world's large river systems are affected by dams (Nilsson et al. 2005; Lehner et al. 2011). Only about 2% of the total river kilometers in the United States are not affected by dams (Graf 2001), and other industrialized countries have equally high rates of flow alteration by dams. Global manipulations of river flow have resulted in a doubling to tripling of the residence time of continental runoff and a 600–700% increase in fresh water stored in channels within reservoirs (Vörösmarty et al. 2004).

Dams vary in their design and operation. Those that store water for flood control or water supply, for example, release the stored water in large, sustained pulses. Dams designed for hydroelectric power generation typically have substantial daily fluctuations in water release to maximize power production during periods of high demand. Not all dams substantially change flow hydrology. Some are operated as run-of-river facilities that do not significantly alter downstream hydrographs. Dams that store large amounts of water, however, do substantially alter the timing, magnitude, frequency, and duration of flow (Magilligan et al. 2013). The most common effect of dams in general is to homogenize regional river flow regimes (Poff et al. 2007) by reducing peak flows and the variability of lower flows.

The specific effects of flow regulation depend on the character of the alterations in water and sediment discharge and on the characteristics of the channel (Grant et al. 2003; Salant et al. 2006; Do Carmo 2007; Magilligan et al. 2013). The presence of a dam typically: reduces the mean and the coefficient of variation of annual peak flow (Williams and Wolman 1984; Poff et al. 2007); increases minimum flows (Hirsch et al. 1990); shifts the seasonal flow variability (Hirsch et al. 1990); and greatly increases diurnal flow fluctuations, if the dam is operated for hydroelectric power generation (Magilligan and Nislow 2001). Although an individual dam may have only local effects, the cumulative effects of numerous dams (estimated at one dam per 48 km of river in third- through seventh-order watersheds in the United States) is to regionally homogenize stream flow (Poff et al. 2007).

Numerous studies document the response of the downstream channel to changes in hydrology and to the changes in sediment supply as the majority of bedload and, in some cases, suspended load is trapped upstream from the dam. Geomorphic response reflects the ratio of sediment supply below the dam to that above, as well as the ratio of pre-dam to post-dam flood discharge and the fractional change in frequency of flows transporting sediment (Figure 3.23) (Grant et al. 2003; Schmidt and Wilcock 2008). Where sediment supply decreases substantially and flow competence remains sufficiently high, bed coarsening, channel incision, and bank erosion occur. These effects can extend hundreds of kilometers downstream from a large dam (Galay 1983; Xu 1996). Alternatively, the reduction in peak flows and flood scouring can result in bed fining, aggradation, and narrowing, especially if large sediment sources such as tributary inputs are present just downstream from the dam or encroaching riparian vegetation helps to stabilize sediment (Curtis et al. 2010). Channel adjustment varies with distance downstream from a dam, but many rivers have sequential dams that cause substantial cumulative disruption of water and sediment fluxes (Skalak et al. 2013).

Where dams have been present for many decades and discharge measurements are sparse, quantifying the changes in flow regime associated with a dam can be difficult. Several approaches have been used to quantify changes in flow regime caused by flow regulation. *Indices of hydrologic alteration* (IHA) can be used to assess the interannual variability of 67 streamflow parameters that reflect the

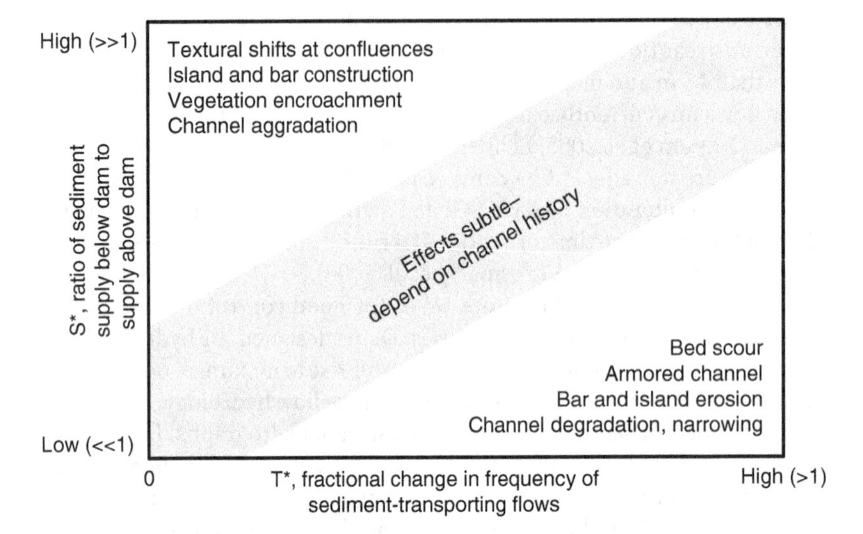

Figure 3.23 Response domain for channel adjustments predicted in response to the presence of a dam. Adjustments are conceptualized in relation to the fractional change in frequency of sediment-transporting flows (T*) and the ratio of sediment supply downstream from the dam to that upstream from the dam (S*). End-member textural and morphologic adjustments are shown. Response of rivers plotting within the shaded diagonal region is likely to be strongly influenced by geological factors, including the history of floods and landslides that leave legacies of large volumes of coarse material and bedrock incision in the valley and channel bottom. Source: After Grant et al. (2003), Figure 3, p. 209.

magnitude, timing, frequency, duration, and rate of change of discharge before and after regulation (Richter et al. 1996, 1997, 1998; Peñas et al. 2016). The Ecological Limits of Hydrologic Alteration (ELOHA) is a variation of this approach that focuses on ecologically relevant hydrologic variables (Poff et al. 2010). Another approach is to use the mathematical tool wavelet transform to extract the dominant modes of variability from statistically nonstationary signals (Zolezzi et al. 2009; Wu et al. 2015). Finally, the seasonal probability density function of natural and regulated stream flows can be compared to assess changes (Botter et al. 2010). Of these techniques, IHA is the most commonly used.

In addition to altering water and sediment flux and resultant changes in channel form, dams alter water temperature and chemistry and the movement of nutrients, plant propagules, instream wood, and organisms. Numerous studies highlight the effects of dams on fish populations (Brooker 1981; Ligon et al. 1995; Bunn and Arthington 2002; Oliveira et al. 2018), including physical blockage of migration, loss of floodplain spawning and nursery habitat when reduced peak flows limit overbank flooding, and restricted access to spawning areas. Fine sediment deposition on gravels used for spawning limits the survival of eggs and embryos. Rapid submergence and exposure through dewatering of the *varial zone* – the shallow borders of channels – can expose and kill fish eggs and embryos. And, disruption of seasonal thermal cues that regulate the timing of lifecycles can stress or eliminate fish populations.

Riparian communities also experience numerous disruptions where dams are present, including decreases or elimination of surfaces for seedling establishment, prolonged submersion by increased base flows, and disrupted downstream transport of seeds to germination sites (Nilsson et al. 1997;

Nilsson and Berggren 2000; Katz et al. 2005; Merritt and Wohl 2006). The integrated effects of these diverse changes are illustrated by comparison of free-flowing and regulated rivers, which indicates decreases in plant diversity and changes in species composition along the latter (Jansson et al. 2000).

Dams began to be removed in some portions of the United States at the end of the twentieth century (O'Connor et al. 2015). These were mostly relatively small, nineteenth-century dams built for a purpose that no longer existed, such as powering a long-gone mill. An estimated 750 dams were decommissioned in the country by 2008, with 20–50 added to the list each year of the first decade of the twenty-first century (O'Connor et al. 2008). A primary concern with dam removal is remobilization of sediment stored behind the dam, particularly if this sediment contains toxic contaminants or high nutrient concentrations (Hart et al. 2002; Pizzuto 2002). Issues of increased mobility for aquatic organisms and loss of lentic (still-water) habitat can also be important (Stanley and Doyle 2003). Removal of a dam can help restore the natural flow regime, with attendant changes in channel process and form over a period of years.

Decommissioning of two large dams on Elwha River in Washington, USA involved some of the largest structures removed to date (East et al. 2015). Removal of large dams is typically not economically or politically feasible, but disruptions to river ecosystems can be reduced by modifying their operation (Graf 2001). The most widely studied example involves experimental flood releases from Glen Canyon Dam on the Colorado River just upstream from Grand Canyon National Park in Arizona, USA (Collier et al. 1997; Wright et al. 2008; Melis et al. 2015). These releases are timed to achieve mobilization and downstream redistribution of sediment inputs from tributaries just downstream from Glen Canyon Dam. The objective behind redistributing sediment is to enhance channel-margin sand deposits and associated camping sites and channel-margin fish habitat. The experimental releases also provide opportunities to test conceptual and numerical models of river dynamics and response to altered flow regime (Collier et al. 1997; Wright et al. 2008; Melis et al. 2012, 2016).

The complexity of most river ecosystems, and the complications introduced by other factors such as exotic species, makes it extraordinarily difficult to predict and create targeted effects on these ecosystems by modifying dam operations, particularly when each experimental flow release is expensive and comes under intense public scrutiny. Nonetheless, the adaptive management being used on the Colorado River, in which hypotheses regarding river response to experimental flow releases are tested and modified hypotheses are used to design the next experimental release (Wohl et al. 2008; Melis et al. 2015), is now being applied to other regulated rivers (Bednarek and Hart 2005; Jorde et al. 2008; Olden et al. 2014). Target flow regimes are sometimes known as *environmental flows* (Rathburn et al. 2009; Richter 2010), as discussed in more detail in Section 8.5.3.

Flow diversion can either reduce flow in the source stream or augment flow in the receiving stream. In extreme cases, all water is removed from the source stream for at least a portion of the year (e.g. Baker et al. 2011) or the peak flow is more than doubled in the receiving stream (e.g. David et al. 2009). Although flow diversion has historically been most common on small to medium-size rivers in primarily dry regions, diversions are present in diverse environments (e.g. Kellerhals et al. 1979), and massive diversion projects on very large rivers are now underway or being planned in China and parts of Africa and the Middle East (e.g. Wohl 2011a; He et al. 2014).

Diversion has most commonly been undertaken for water supply or flood control. It tends to reduce base flow and flood peaks on the source channel and increase them in the receiving channel, although the specifics depend on the proportion of total flow diverted and the frequency and duration of diversion. The downstream extent of the altered flow partly depends on tributary and ground-water inputs below the point of diversion (McCarthy 2008). As with dams, changes in flow regime associated with

Figure 3.24 La Poudre Pass Creek in Colorado, USA. Flow in this creek is augmented by diversions from the headwaters of another river. (a) One portion of the creek at (1) 0.15 m^3/s, (2) 0.70 m^3/s, and (3) 4.52 m^3/s. During low flows, eroded banks are exposed and flow is exceptionally shallow. Water in the channel freezes to the bed during the winter, effectively precluding overwinter survival of fish in the creek. Prior to flow regulation, winter base flow was slightly higher and fish were able to survive through the winter. (b) Average annual hydrographs for the creek under natural flow conditions and augmented flow. (c) Plot of reach-averaged bankfull width against drainage area (used here as a surrogate for discharge in ungaged sites), indicating that the sites along La Poudre Pass Creek that have augmented flow have substantially greater width than would be expected under natural flow conditions. The additional sites on the plot are unregulated streams in the region. Source: After Wohl and Dust (2012), Figure 4A. (*See color plate section for color representation of this figure*).

diversions disrupt sediment dynamics, channel morphology, and aquatic riparian communities in both source and receiving streams. The magnitude of these disruptions depends on the magnitude of hydrologic changes and the characteristics of the affected channels (Figure 3.24) (Ryan 1997; Parker et al. 2003; Wohl and Dust 2012; Hillman et al. 2016).

Among the most insidious ecological effects of flow diversions is the introduction of new species from one river network to another. Millions of dollars are now being spent in the United States, for example, to construct barriers using concrete, wire mesh, and electrical fields in order to prevent four

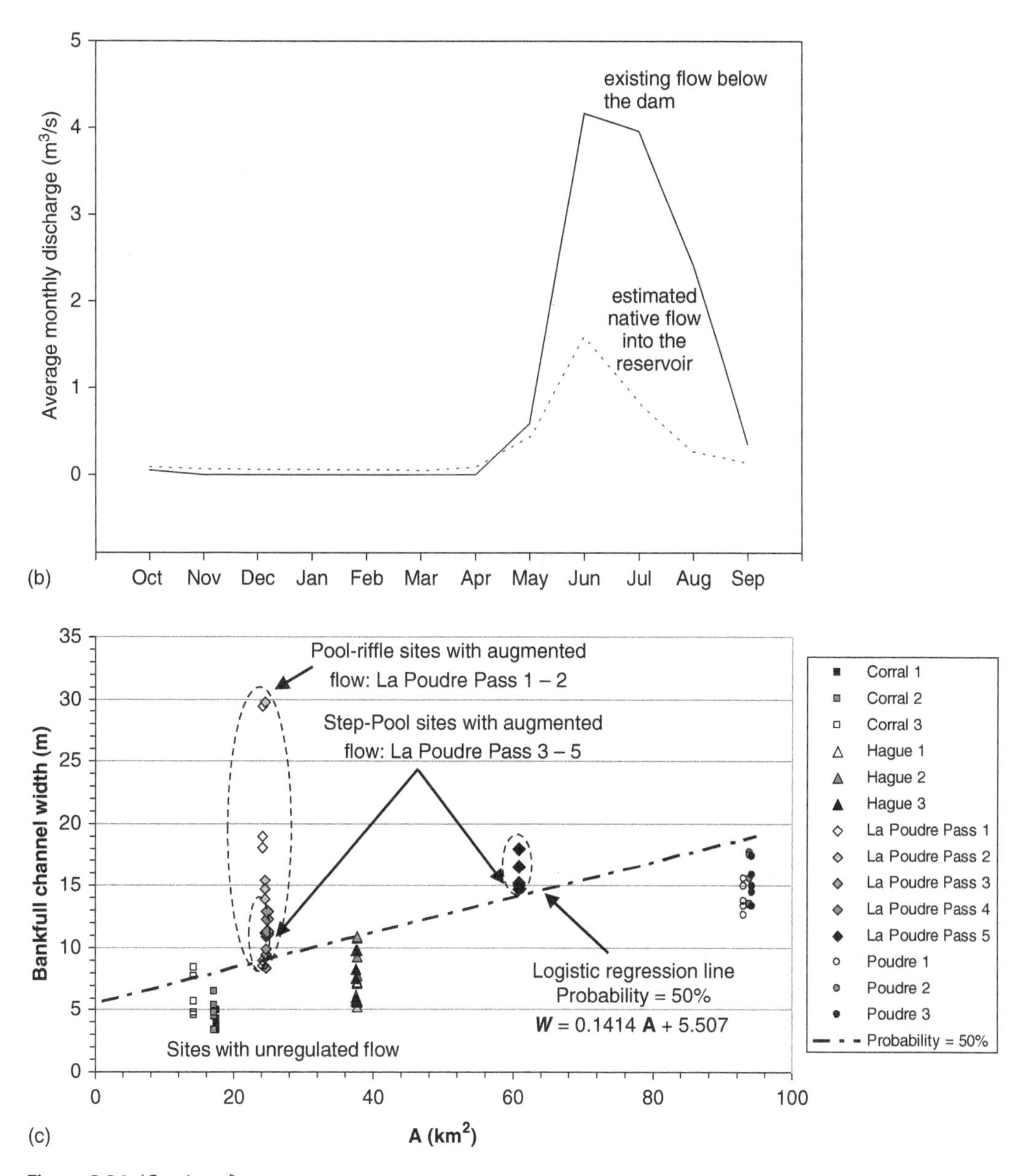

Figure 3.24 *(Continued)*

introduced species of Asian carp from migrating through a canal connecting the headwaters of the Illinois River system, where the fish are established, into the Great Lakes ecosystem at Lake Michigan (Sandiford 2009; Jerde et al. 2013). In southeastern Australia, water impounded and diverted from the headwaters of the eastward-draining Snowy River into the headwaters of the westward-draining Murray River catchment since 1967 has been accompanied by pioneering fish species, including climbing galaxis (*Galaxias brevipinnis*), which may compete with many of the already-endangered native species of the Murray–Darling basin (Waters et al. 2002).

3.2.9.2 River Corridor Engineering

Direct alterations of channel and floodplain form that change flow conveyance and at least some characteristics of the hydrograph take several forms. Construction of artificial levees prevents or limits overbank flows, increasing the magnitude and velocity of water conveyed within the channel during higher discharges (Heine and Pinter 2012). This can result in channel instability and reduced boundary complexity. Levees are constructed to protect land use or structures in the floodplain (Blanton and Marcus 2009).

Channel engineering can take the form of bank stabilization, dredging, confining braided or anastomosing channels to a single-thread planform, or straightening sinuous channels. All of these modifications are sometimes lumped together as *channelization*. Channelization has been undertaken across the full spectrum of channels, from small headwater streams in mountainous environments (Wohl 2000a; Habersack and Piégay 2008) to large rivers such as the Danube, the Rhine, and the Mississippi (Wohl 2011a). The intent behind these activities typically includes flood control, increased human use of floodplains, and navigation (Schoof 1980). Channelization usually increases cross-sectional areas, reduces the geometric complexity and hydraulic roughness of channel boundaries, decreases access to overbank areas during peak flows, and limits exchanges between sediment in transport and sediment in the channel bed and banks (e.g. Habersack et al. 2013). These alterations typically result in larger-magnitude but shorter-duration flood peaks (e.g. Sholtes and Doyle 2011; Koebel and Bousquin 2014).

Beaver are herbivorous rodents that build dams out of wood and sediment along small to medium-sized channels in North America (*Castor canadensis*) and Europe (*Castor fiber*) (Pollock et al. 2003). These dams serve to pond water, along the margins of which the beaver build a lodge of wood (although beaver are present along very large rivers, they dig dens into the streambanks rather than damming the flow). Beaver were historically widespread and abundant in forested environments of the northern hemisphere, stretching from Alaska to northern Mexico (Naiman et al. 1986, 1988), from Britain east across the whole Eurasian continent, and from the Mediterranean Sea north to the tundra (Hartman 1996). They were completely or almost completely removed from much of their historical range by fur trapping during the sixteenth to nineteenth centuries, with a consequent loss of beaver dams, flood attenuation, and channel–floodplain connectivity.

Channels inhabited by beaver typically have a stepped appearance because of the numerous dams along their length. Beaver dams create a backwater that reduces velocity and promotes sediment accumulation. This can result in loss of conveyance and promote overbank flows. Overbank flows can enhance infiltration and raise riparian water tables, as well as attenuating flood peaks (Burns and McDonnell 1998; Pollock et al. 2003). Beaver dams increase the extent, depth, and duration of overbank floods, as well as greatly elevating the floodplain water table during periods of high and low flow (Westbrook et al. 2006). These effects extend up- and downstream from the dam, helping to create and maintain hydrologic regimes that support the formation and maintenance of floodplain wetlands (Hood and Bayley 2008). Such wetlands are sometimes known as beaver meadows (Ives 1942) because of the feedbacks among beaver dams, overbank flows, high water tables, and the deciduous riparian vegetation (*Salix* spp., *Populus* spp.) that beaver prefer to eat. Where multiple beaver dams and associated ponds and secondary channels enhance surface and subsurface water storage, beaver meadows can help to enhance base flow to downstream portions of the river network (Wegener et al.

2017). Removal of beaver and their dams typically reduces overbank flooding and associated floodplain wetlands and increases flood peak magnitude and velocity by confining flood discharges to a main channel (Green and Westbrook 2009; Pollock et al. 2014).

Alteration of riparian vegetation occurs either through removal of vegetation or through introduction of exotic species with substantially different characteristics than the native species. Riparian vegetation can influence the characteristics of stream flow by taking up soil water or stream water through roots and releasing it into the atmosphere via transpiration or evapotranspiration (Doody and Benyon 2011; Humphries et al. 2011). Evapotranspiration rates for diverse plant communities and river corridors can be estimated using remote sensing approaches or ground-based measurements (e.g. Nagler et al. 2005).

Vegetation can also alter bank stability and hydraulic resistance along the channel margins, influencing velocity and stage of flow (Griffin et al. 2005; Tal and Paola 2007; Vincent et al. 2009). The degree to which vegetation alters bank stability depends both on the root characteristics (tensile strength, spatial density, depth, and lateral extent) of individual types of plants and on the sediment texture, stratigraphy, and moisture of the stream-bank substrate (Polvi et al. 2014). Analogously, the effect of vegetation on hydraulic resistance along the channel margins depends on factors such as the flexibility, stem density, and spatial density of the plants (Le Bouteiller and Venditti 2015).

Alteration of vegetation, either by removal or by replacement with exotic riparian species that have different rates of transpiration or different morphologies and stand densities, can thus alter the effects of vegetation on diverse aspects of stream flow (Tabbachi et al. 2000; Doody and Benyon 2011; Hultine and Bush 2011).

3.2.10 The Natural Flow Regime

As mentioned earlier, the natural flow regime has become a key conceptual model in river ecology. The concept of a natural flow regime was first articulated by Poff et al. (1997), who review how the ecological integrity of rivers depends on their natural dynamic character and how this dynamic character in turn depends on the quantity and timing of flow. They then explain how the five components magnitude, frequency, duration, timing, and rate of change can be used to describe a river's flow quantity, timing, and variability.

Prior to the 1997 paper, efforts to protect rivers as ecosystems typically focused on water quality and minimum flows. The natural flow regime emphasizes that aspects of the annual hydrograph beyond base flow are critical to sustaining river process, form, and biotic communities. The existence of systematic records of river flow in both relatively natural and human-altered watersheds facilitates quantification of both the natural flow regime and alterations of that regime (e.g. Richter et al. 1996; Poff et al. 2010). The natural flow regime has given rise to environmental flows, which quantify the flow regimes necessary to sustain desired river processes and forms (e.g. minimum water quality standards, pool volume, or floodplain inundation) (e.g. Tharme 2003).

A vital consideration in quantifying human-induced deviations is the recognition that a natural flow regime varies over diverse time scales. Floods and droughts occur regardless of human activities, although human activities can exacerbate or ameliorate the effects of these fluctuations. The existence of temporal variability implies that multiple years of record are needed to adequately characterize the natural flow regime. The length of record necessary partly depends on what aspect of the regime is of most interest and on the characteristics of the river. Two decades of records may be adequate to quantify the mean annual flow in a river with relatively low hydrologic variability, for example, whereas well over a century of records are required to effectively quantify the extreme flood in a river with greater hydrologic variability. Under the latter scenario, indirect flow records derived from historical, botanical, and geological information are required to supplement directly gaged discharge. In general, understanding of the *natural range of variability* (Morgan et al. 1994; Richter et al. 1997) is needed to characterize the natural flow regime.

The concept of the natural flow regime has been enormously influential in river management through: (i) emphasizing the importance of the full range of flows, rather than just a minimum flow, in sustaining river ecosystems; (ii) providing quantitative metrics to characterize the natural flow regime and human-induced deviations from natural conditions; and (iii) providing a conceptual basis for tying specific flow characteristics to desired outcomes in river process and form.

3.3 Summary

Equations used in basic engineering hydraulics start from the assumptions that flow in a channel is uniform and steady and that the channel has relatively smooth boundaries and limited turbulence. Natural channels seldom meet these characteristics, and the degree to which flow is varied and unsteady determines how far actual hydraulic parameters deviate from expected conditions. The instabilities associated with hydraulically rough boundaries, flow transitions, and spatial and temporal variations in velocity and turbulence are inherent in natural channels, as are the interactions between flowing water and the channel boundaries that result in sediment movement and channel adjustment. A substantial amount of research is focused on quantifying and predicting hydraulic variables such as resistance (n or f), velocity, turbulence intensity, shear stress, and stream power, because these variables can be used to predict sediment movement and channel geometry. Our ability to accurately characterize hydraulics in natural channels is greatest in those channels that best approximate the assumptions underlying engineering hydraulics: sand-bed channels with readily deformable boundaries and simple geometry. Natural channels with more hydraulically rough boundaries, more complex and spatially variable geometry, and greater erosional thresholds – gravel-bed, boulder-bed, and bedrock channels – are typically poorly characterized using standard assumptions from engineering hydraulics.

The energy of flow quantified in hydraulic equations inherently reflects the volume of water moving down a channel, so a substantial amount of research is focused on quantifying and predicting discharge. Systematic, gage-based measurements of discharge seldom provide sufficient length and spatial density of record to infer the characteristics of very large, infrequent flows. A variety of indirect methods are used to estimate the magnitude and recurrence interval of a wide range of flows, from rare floods to the shape of an average annual hydrograph. Human activities have directly and indirectly altered river discharge for centuries across much of the world, making it challenging to infer a natural flow regime in the absence of human manipulation.

Part II

Channel Processes II

4

Fluvial Sediment Dynamics

This chapter examines the movement of particulate and dissolved sediment within river channels. The first section describes the channel bed and initiation of sediment motion from the bed. Within this part of the chapter, the first subsection addresses a stationary streambed and the methods used to characterize bed sediment. The ability to characterize the bed substrate is important to understanding one source of sediment supply when sediment is in transport, as well as sources of flow resistance that influence the distribution of hydraulic forces moving sediment. The next two subsections discuss the initiation of motion of individual particles on a noncohesive bed and the removal of aggregates of particles from a cohesive bed.

The second section focuses on the processes of sediment transport and deposition. Sediment can be transported downstream in solution, as particles suspended in the flow, or as particles that remain in contact with the bed. Each mode of transport is influenced by similar hydraulic forces but is described by a distinct suite of processes and equations.

In the third section, sediment moving in contact with the bed is differentiated into readily mobile and infrequently mobile bedforms. This division recognizes a fundamental distinction among rivers. Some rivers, particularly sand-bed channels, have relatively *mobile beds* in which sediment moves frequently and sometimes at relatively high rates, at least in terms of grain numbers. These channels are also referred to as live-bed or regime channels. In contrast, rivers with a bed dominated by gravel and larger particles are *threshold channels* in which bed sediment moves relatively rarely and at low rates.

The fourth section reviews processes and forms of sediment deposition within the channel boundaries. The fifth looks at downstream trends in grain size.

The sixth section discusses factors that control bank stability and processes of bank erosion. Bank erosion receives substantial attention because of the potential for property loss and damage to infrastructure as a result of bank failure. Erosion of banks is distinct from that of streambeds for at least two reasons. First, bank erosion commonly occurs via mass failures, as well as grain-by-grain detachment. Second, bank stability and erosion can be strongly influenced by vegetation.

The seventh section examines sediment budgets, which use information on sediment input and storage to quantitatively estimate fluxes of sediment past a given point in a watershed. Sediment yield is characterized by great spatial and temporal variability, with much of the sediment coming from a limited portion of a drainage basin during a relatively short period of time. Globally, a disproportionate amount of sediment originates in small, mountainous drainages.

Rivers in the Landscape, Second Edition. Ellen Wohl.

The final sections of the chapter integrate knowledge of sediment dynamics into the conceptual model of natural and altered sediment regimes and review how human activities have altered sediment, inputs, outputs, and storage in many drainage basins.

4.1 The Channel Bed and Initiation of Motion

4.1.1 Bed Sediment Characterization

Knowledge of the size distribution, packing, sorting, particle shape, and stratigraphy of bed sediments is important for understanding the initiation of particle motion on the bed, sediment transport, and channel change. Accurately measuring characteristics of the bed sediments becomes more difficult as these sediments become more coarse-grained and heterogeneous, not least because progressively larger samples are necessary to accurately characterize the grain-size distribution. The default method remains the 100 clasts originally proposed by Wolman (1954), which is usually sufficient for measures of the central tendency such as mean grain size. Fripp and Diplas (1993), Rice and Church (1996), and Bunte and Abt (2001) describe how to determine the minimum sample size based on the size fraction of interest and the level of precision required (e.g. Lisle and Eads 1991).

The grain-size distribution of bed sediments can be measured by taking a bulk sample and measuring the sediments in a laboratory. For sediment with an intermediate diameter < 0.0625 mm (silt and clay), a wet separation technique with flow-through particle size analyzers is commonly used. For sediment in the size range 0.0625–16.0 mm, sieving is typically used to determine grain-size distribution. For sediment coarser than approximately 16 mm, the intermediate diameter of each clast is typically measured, either by hand or by digitizing from an image.

Laboratory measurements work well for pebble-size or finer sediments, but quickly become impractical with coarser sediments because of the large sample size needed. As the intermediate axis of the largest clast approaches 64 mm, the sample size required to accurately characterize the distribution exceeds 400 kg (Church et al. 1987). Samples in which the weight of the largest clast is <1% of the total weight can provide unbiased estimates of mean grain size (Mosley and Tinsdale, 1985), but this makes sampling coarse sediments almost entirely impractical. *In situ* measurements of clast size can be more practical in terms of avoiding sediment collection, but such methods may not consistently sample clasts <15 mm in diameter (Fripp and Diplas 1993). Volumetric samples of finer sediment can also be combined with *in situ* measurements of coarser sediment in a bed with mixed grain sizes (Bunte and Abt 2001).

In situ methods for characterizing bed sediment can employ direct measurements of clast size along a grid or a random walk (Wolman 1954) or systematic subsampling in which all particles exposed at the surface within a defined area are measured (e.g. Diplas and Sutherland 1988). The accuracy of *in situ* measurements can be limited because different methods of surface sampling and different operators may not produce the same results (Wolcott and Church 1991; Marcus et al. 1995; Wohl et al. 1996). Spatial variations in sediment size on the bed associated with bedforms such as pools and riffles can also make bulk or *in situ* sediment measurements challenging. Subsamples of equal volumes can be obtained from each population (Wolcott and Church 1991) to create a representative composite sample. Another approach is to create a *facies map* of the streambed by classifying textural patches based on the relative abundance of major size classes and subcategories of the dominant size (Buffington and Montgomery 1999a).

Increasingly, photographs (Warrick et al. 2009; Dugdale et al. 2010), laser-scanning images (Hodge et al. 2009), and structure-from-motion photogrammetry (e.g. Westoby et al. 2015) are used to characterize bed grain-size distributions in combination with either software that digitizes the imagery or spectral decomposition that provides statistical estimates of grain size (Buscombe et al. 2010; Turley et al. 2017). Photographs can be ground-based or can be obtained from airplanes, helicopters, balloons, drones, and other platforms (Carbonneau et al. 2005; Warrick et al. 2009; Vazquez-Tarrio 2017).

In addition to size distribution, bed sediment can be characterized in terms of density, shape, sorting, and packing. *Grain density* is the mass per unit volume of the sediment, which is usually assumed to be 2.65 g/cm^3 – the density of quartz. (Bulk density of aggregates of grains is more typically in the range of 1.4–1.8 g/cm^3). Grain density affects the fall or settling velocity, which is the velocity at which a grain moves down a water column. *Specific gravity* is the density of the sediment relative to the density of water (2.65 g/cm^3/1 g/cm^3 = 2.65).

Grain shape includes the form and surface texture of a sediment grain. Shape affects fall velocity: flatter particles fall more slowly than spherical particles of similar weight and density. Shapes can be described as spheres, blades, discs, or rods, and with respect to roundness and sphericity (Folk 1980). Shape factors such as grain elongation (the ratio of the long axis to the intermediate axis) can also be quantified (Robert 2014). A grain's shape influences the surface area exposed to the flow and the ease with which the grain can be rolled along the bed.

Sorting describes the range of particle sizes present. Well-sorted sediments have a narrow range of sizes. Various measures of "sorting" exist, including the widely used inclusive graphic standard deviation (Folk 1980)

$$\sigma_I = \frac{\phi 84 - \phi 16}{4} + \frac{\phi 95 - \phi 5}{6.6} \tag{4.1}$$

where ϕx is the phi size for which x percent of the distribution is finer-grained: $\phi = -\log_2 x$, where x is grain size in mm. Values of $\sigma_I < 0.35\phi$ are very well sorted, $0.35–0.5\phi$ are well sorted, $0.5–0.71\phi$ moderately well sorted, $0.71–1\phi$ moderately sorted, $1–2\phi$ poorly sorted, $2–4\phi$ very poorly sorted, and $> 4\phi$ extremely poorly sorted.

Packing describes the organization of particles on the bed. Packing can be qualitatively categorized as imbricated, interlocked, or open. *Imbricated* clasts lie with their long axes parallel to flow and dipping slightly up-current. Imbrication makes it more difficult to entrain clasts. Imbricated clasts can provide useful paleocurrent indicators when preserved in stratigraphy. Individual coarse particles touch one another in an *interlocked bed*, which is *framework-supported*, whereas coarse particles are surrounded by finer sediment in a *matrix-supported bed*. *Open-framework gravels* lack finer sediment in the pore spaces between individual gravel particles. The distinction between framework- and matrix-supported can be important in contexts such as aquatic habitat. A framework-supported or open-framework bed, for example, allows water, dissolved oxygen, and nutrients to circulate more readily around fish eggs laid in the streambed than does a matrix-supported bed.

4.1.2 Entrainment of Noncohesive Sediment

Most grains of sand size and coarser (Table 4.1) on a streambed are noncohesive. *Entrainment* refers to the initiation of motion of sediment. Similarly, *incipient motion* describes the threshold condition when the hydrodynamic moment of forces acting on a particle balances the resisting moment of

Table 4.1 Grain size categories.

Category	Size range (mm)	Phi units[a] (ϕ)
Boulder	≥ 256	−8 to −12
Cobble	64–256	−6 to −8
Gravel[b]	2–64	−1 to −6
Sand[c]	0.062–2	4 to −1
Silt	0.004–0.062	8 to 4
Clay	≤ 0.004	≥ 8

a) phi units are calculated from $\phi = -\log_2 x$, where x is grain size in mm
b) Gravel is sometimes divided into pebble (4–64 mm, −2 to −6 phi) and granule (2–4 mm, −1 to −2 phi)
c) Sand is sometimes divided into very coarse sand (1–2 mm, 0 to −1 phi), coarse sand (0.5–1 mm, 1 to 0 phi), medium sand (0.25–0.50 mm, 2 to 1 phi), fine sand (0.125–0.25 mm, 3 to 2 phi), and very fine sand (0.0625–0.125 mm, 4 to 3 phi)
Source: After Wentworth (1922) and Folk (1980).

force (Shields 1936; Julien 1998). Being able to predict when sediment will be entrained is critical to understanding bed erosion, movement of bedforms, maintenance and stability of aquatic habitat, sediment transport, and hydraulic roughness.

Studies of entrainment initially focused on well-sorted, sand-bed channels, and these remain the channels for which entrainment can be predicted most accurately. As cohesion between particles increases in grains finer than sand, or in bedrock, bed sediment is less likely to detach as individual particles. Under these conditions, substrate characteristics such as porosity and permeability or bedrock jointing influence entrainment. As the grain-size distribution becomes wider in coarse-grained channels, effects such as packing, sorting, shielding of finer particles, protrusion of coarser particles, and particle shape influence entrainment.

G.K. Gilbert's pioneering quantitative flume experiments examined the effect of potential control variables such as discharge and channel gradient on sediment entrainment (Gilbert 1914, 1917). Many subsequent investigations of entrainment and transport used these flume data (e.g. Wiberg and Smith 1989; Shih and Diplas 2018). Modern studies of entrainment in sand-bed channels, however, start with Shields (1936), who conducted flume experiments on the initiation of motion in channels with relatively uniformly sized sand grains. Shields quantifies entrainment using *dimensionless critical shear stress*, τ_c^*, calculated by dividing the critical shear stress by an approximation of the weight of an immersed grain

$$\tau_c^* = \frac{\tau_c}{(\rho_s - \rho)gD} \tag{4.2}$$

where τ_c^* is the dimensionless critical shear stress, ρ_s is the sediment density, ρ is the water density, g is the acceleration due to gravity, and D is the grain size. D_{50} is typically used for D when the grain-size distribution is relatively narrow. This approach, although developed for sand-bed channels, is now also the starting point for studying entrainment across a much wider range of bed grain sizes.

Dimensionless critical shear stress is commonly known as the *Shields number*. This empirically derived number typically varies between 0.03 and 0.08 for channels with $S < 0.03$, but can include values between 0.01 and 0.2 (Buffington and Montgomery 1997). The wide variation in Shields number results from at least three factors (Ferguson 2013; Yager and Schott 2013): (i) diverse studies use different criteria to define the initiation of grain motion, from vibration in place or dilation of the bed prior to downstream movement to the initiation of downstream movement, and different methods of measuring the initiation of motion; (ii) differences in the grain-size distribution and flow roughness in diverse channels produce a range of values for the Shields number; and (iii) the use of reach-averaged conditions results in a greater range of values.

It is also important to distinguish between the apparent Shields number and the actual Shields number. The *apparent Shields number* is calculated using Eq. (4.2). It is widely used because it is derived from relatively simple measurements of the reach-averaged shear stress and grain-size distribution. It is relatively simplistic, however, because entrainment is typically not spatially or temporally uniform, and thus is not accurately represented by a single shear stress for a particular grain or a single grain size for a streambed (Yager and Schott 2013). The actual Shields number of a particular grain reflects the force balance acting on the grain, which depends on the local particle arrangement and flow turbulence. Any given grain size thus has a range of possible critical shear stress values (Figure 4.1). Flume and field studies suggest that a grain may have a higher critical Shields stress for entrainment on steeper slopes, for example, as a result of grain protrusion and local fluctuations in velocity and turbulence, which correlate with bed gradient (Lamb et al. 2008; Scheingross et al. 2013). Research on entrainment in beds of mixed grain sizes focuses on the force balances acting on the grains, the grain properties, and the turbulence parameters that influence motion.

4.1.2.1 Forces Acting on a Grain

The balance of forces acting on a grain is schematically illustrated in Figure 4.2. The downstream and upward driving forces are the downstream component of particle weight F_g, buoyancy F_B, lift F_L, and drag F_D. The frictional resisting forces are F_r, the bed-perpendicular component of particle weight, and the friction angle. Entrainment occurs when the driving forces exceed the resisting forces. Each of these forces reflects the local turbulence and particle characteristics, including diameter, drag coefficient, friction angle (resisting pocket angle of the grain), and protrusion of the grain into the flow (Yager and Schott 2013).

For spatially and temporally averaged flow conditions, the forces acting on a grain can be derived as follows (Wiberg and Smith 1987; Yager and Schott 2013). Drag force results from flow of water

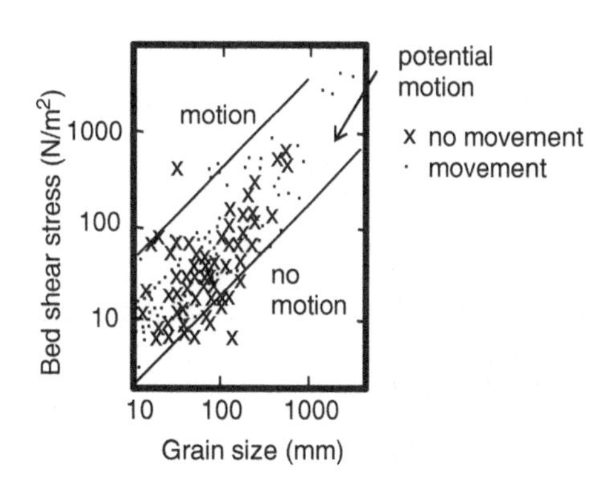

Figure 4.1 Bed shear stress versus grain size, showing range of entrainment. Source: After Knighton (1998), Figure 4.5C, p 110.

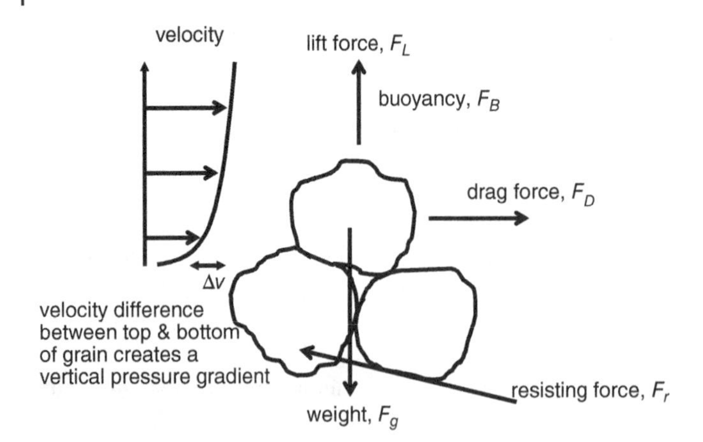

Figure 4.2 Schematic illustration of the forces acting on a noncohesive grain under steady uniform flow on a nearly horizontal surface Source: After Knighton (1998), Figure 4.4, p. 108 and Julien (1998), Figure 7.4, p. 115.

past a grain in the downstream direction and is

$$F_D = \frac{1}{2}\rho A C_D \langle v^2(z)\rangle = \frac{1}{2} A C_D \tau_b \left\langle f^2\left(\frac{z}{z_o}\right)\right\rangle \tag{4.3}$$

where C_D is the drag coefficient and is a function of the particle Reynolds number (the particle Reynolds number describes the ratio of kinetic to viscous forces applied on a moving particle), shape, size, orientation, and relative submergence; ρ is the water density; $<v^2(z)>$ is the square of the fluid velocity averaged over the grain; τ_b is the reach-averaged boundary shear stress; $f(z/z_o)$ is a function that determines the form of the velocity profile around the grain; and A is the cross-sectional grain area, which varies with relative submergence and protrusion.

Lift force results from differences in velocity and pressure at the top and bottom of a grain and is

$$F_L = \frac{1}{2}\rho A C_L (v_T^2 - v_B^2)^2 = \frac{1}{2} A C_L \tau_b \left\langle f^2\left(\frac{z_T}{z_o}\right) - f^2\left(\frac{z_B}{z_o}\right)\right\rangle \tag{4.4}$$

where z_T and z_B are the heights of the top and bottom of the grain, respectively; C_L is the lift coefficient; and v_T and v_B are the flow velocities of the top and bottom of the grain, respectively. The larger the velocity and pressure gradients between the top and the bottom of a grain, the more important lift force becomes.

The force exerted by particle weight is

$$F_g = (\rho_s - \rho) g V \tag{4.5}$$

where g is the gravitational acceleration and V is the particle volume. This force is the difference between the grain weight ($\rho_s gV$) and buoyancy ($\rho_w gV$), which is the amount of fluid weight displaced by the grain of interest. The net gravitational force is divided between the downslope (driving force) and bed-perpendicular (resisting force) components.

The driving forces can then be written as

$$F_D + F_L tan\phi + F_g sin\beta \tag{4.6}$$

where ϕ is the friction angle and β is the channel slope. The friction angle is the angle through which a grain will pivot when it is dislodged in the downstream direction.

The resisting force is given by

$$F_g cos\beta tan\phi \tag{4.7}$$

For entrainment, Eqs. (4.6) and (4.7) can be equated, Eqs. (4.3) and (4.4) can be substituted for F_D and F_L, respectively, and the equation can be solved for the critical shear stress to yield

$$\frac{\tau_c}{(\rho_s - \rho)gD} = \frac{2V(tan\phi cos\beta - sin\beta)}{C_D AD \left\langle f^2\left(\frac{z}{z_o}\right)\right\rangle \left(1 + \left[\frac{F_L}{F_D}\right] tan\phi\right)} \tag{4.8}$$

This equation requires an assumed velocity profile. The logarithmic profile is commonly used. Other representations of the force balance use different forces and alter the parameterization of each (Yager and Schott 2013). Bagnold (1977), for example, proposes using stream power per unit area rather than shear stress to evaluate entrainment and transport.

Although Eq. (4.8) addresses the basic forces acting on a grain, it is in practice cumbersome to apply to real channels with a range of grain sizes. This reflects the substantial scatter in actual, site-specific values of individual forces associated with grain properties and flow turbulence.

4.1.2.2 Grain Properties

The grain properties that influence entrainment include grain size, shape, orientation, packing, and imbrication. As already noted, the original Shields curve of dimensionless shear stress and sediment transport rate was empirically derived for relatively uniformly sized sediment. Larger grains have a higher critical shear stress because of their greater weight, and weight-driven transport represents one end-member for entrainment from a bed of mixed grain sizes. Equal-mobility transport represents the other end-member.

Equal mobility can be defined as the same critical shear stress (τ_c) for all grain sizes (Parker et al. 1982). Under this condition, τ_c^* decreases with increasing grain size because coarser particles protrude into the flow and are easier to move (relative to a bed of uniform grain size), whereas finer particles are hidden and therefore harder to move. Equal mobility can also be defined as occurring when the grain-size distribution of the bed load equals that of the channel bed (Parker and Toro-Escobar 2002).

Under conditions of intermediate mobility between the two end-members of weight-driven and equal mobility, fine sediment is likely to be more mobile than coarse sediment, but less mobile than it would be on a bed of uniform fine particles. This is known as *selective entrainment* (Parker 2008).

The competing effects of grain weight and hiding can be represented by hiding functions that estimate the influence of mixed sediment sizes and flow conditions on entrainment, as in this example from Parker et al. (1982)

$$\frac{\tau_{ci}^*}{\tau_{c50}^*} = \left(\frac{D_i}{D_{50}}\right)^b \tag{4.9}$$

where τ_{ci}^* and τ_{c50}^* are the dimensionless critical shear stresses for the ith grain size (D_i) and the median grain size (D_{50}), respectively, and b indicates the relative importance of grain weight and hiding effects. When grain weight dominates, $b = 0$. When hiding effects dominate and create equal mobility, $b = -1$ (Yager and Schott 2013).

Because hiding functions are derived from site-specific empirical measurements, they are not transferable between locations (McEwan et al. 2004). Accounting for hiding effects when trying to quantitatively predict entrainment becomes progressively more important, however, as the grain-size distribution grows wider and entrainment of the smallest grains is more influenced by hiding.

In addition to grain size, grain protrusion, p, grain exposure, e, and friction angle, ϕ, influence entrainment. *Protrusion* is the distance a particle protrudes above the mean bed elevation. *Exposure* is the distance a grain extends above or below the neighboring grains. *Friction angle*, also known as pivoting angle, is about 30–32° for uniform sediments (Robert 2014). In heterogeneous sediments, the angle through which a grain must pivot to be entrained varies significantly depending on the size distribution of the surrounding grains. Protrusion, exposure, and friction angle can vary with relative grain size, shape, and packing, and are typically empirically estimated.

Beds with increasingly wide distributions of grain size experience increasingly larger effects from protrusion, exposure, and hiding. Large, protruding, immobile grains can be particularly important in creating higher values of dimensionless critical shear stress in steep, shallow flows. The additional resistance created by these grains and by immobile bedforms reduces the shear stress available for grain entrainment, so that the Shields number increases with slope (Ferguson 2013). Yager et al. (2007) accounts for this by scaling the predicted sediment transport rate by the proportion of bed area that is occupied by the mobile fraction of grains

$$q_{sm}^* = 5.7(\tau_m^* - \tau_{cm}^*)^{1.5} \frac{A_m}{A_t} \tag{4.10}$$

where q_{sm}^* is the dimensionless transport rate of the mobile sediment, τ_m^* is the dimensionless stress borne by the mobile sediment, τ_{cm}^* is the dimensionless critical shear stress of the mobile sediment, A_m is the bed-parallel area of mobile sediment, and A_t is the total bed area.

4.1.2.3 Turbulence

Near-bed turbulent bursts, sweeps, and inward and outward interactions can influence entrainment by inducing fluctuations in the pressure and velocity fields around a particle and thus influencing drag and lift forces (Nelson et al. 1995). The relationship between local turbulent forces and grain entrainment is quite complex, and sediment motion can be dominated by drag, lift, or some combination of both (Schmeeckle et al. 2007). Both the magnitude and the duration of turbulent fluctuations influence entrainment (Diplas et al. 2008).

Spatial and temporal variability in bed texture and hydraulic forces make it effectively impossible to predict exactly when a particular grain will be entrained. Entrainment of a particular grain or a population of mixed grain sizes is instead commonly described in terms of a range of critical values for velocity, shear stress, or some other parameter.

4.1.2.4 Biotic Processes

Biotic processes can also influence entrainment (Riggsbee et al. 2013). Instream wood creates large-scale hiding effects and alters the distribution of pressure and velocity near the bed in ways that can either increase or decrease entrainment immediately around the wood (Wohl and Scott 2017b). Microbial mats and macroinvertebrates such as net-spinning caddisfly larvae can increase the cohesion of sand-size and finer sediments and thus increase the shear stress necessary for entrainment (Statzner et al. 1996; Gerbersdorf and Wieprecht 2015). Algae and microorganisms can facilitate deposition of calcium carbonate, which limits sediment mobility and enhances the formation of quasi-permanent steps and pools (Marks et al. 2006). Bioturbation by invertebrates and the building of spawning mounds or depressions by fish can alter surface grain-size distribution and topography (Statzner and Sagnes 2008; Rice et al. 2010).

The magnitude of these effects varies in part with the abundance of the biota. Instream wood was historically widespread and abundant along rivers in forested regions and undoubtedly strongly influenced sediment dynamics, but centuries of removing wood from channels has diminished these effects. Bed sediment dynamics are substantially influenced by spawning activities in rivers that still have abundant salmon populations (Hassan et al. 2008; Buxton et al. 2015), although such rivers are now rare.

4.1.3 Erosion of Cohesive Beds

Channels formed in cohesive material can have a thin, continuous or discontinuous veneer of unconsolidated sediment along the bed. The cohesive material underlying this alluvium limits the rate and manner of bed and bank erosion and exerts a strong influence on cross-sectional geometry and hydraulics during large discharges because the boundary is not readily erodible. Cohesive material in this context includes channel boundaries with bedrock and boundaries with sufficient silt and clay to create strong interparticle cohesion and limit the detachment of individual silt or clay particles. Cohesive boundaries are not necessarily more resistant to erosion than alluvial boundaries. A very soft siltstone or sandstone bedrock may have a lower erosional threshold than an alluvial channel formed in boulders. In general, however, cohesive material has a higher erosional threshold than noncohesive alluvium.

4.1.3.1 Erosion of Bedrock

Because individual particles are not readily detached from cohesive material, sediment entrainment is not the only process by which channels in cohesive beds erode, although sediment entrainment can be an important component of other erosive processes. The primary erosive processes in indurated, cohesive material (bedrock) are corrosion, cavitation, abrasion, and quarrying (Whipple et al. 2013).

Corrosion is the chemical dissolution of bedrock. Most chemical weathering occurs outside the channel, where water moves more slowly through or past the rock matrix. Chemical weathering can be effective in eroding carbonate rocks along channels, however, and in weakening other rock types, particularly those with carbonate cement (Springer et al. 2003). Chemical weathering can thus make the bedrock more susceptible to processes of physical erosion (Hancock et al. 2011; Murphy et al. 2016). Degree of rock weathering increases with height above the thalweg at low rates of incision (Shobe et al. 2016), and vertical incision rates can be faster in seasonally exposed, weathering-dominated portions of the high-flow channel than in perennially wet, abrasion-dominated portions of the low-flow channel (Collins et al. 2016).

Cavitation involves shock waves generated by the collapse of vapor bubbles in a flow with rapidly fluctuating velocity and pressure. The minute irregularities along channel boundaries caused by joints or crystal or grain boundaries in the bedrock can facilitate the formation and collapse of vapor bubbles. Although this effect may sound insignificant, millions of bubbles imploding and sending out shock waves during a flood can very effectively weaken and erode bedrock – a phenomenon common on the concrete spillways of dams (Barnes 1956; Eckley and Hinchliff 1986; Wohl 1998).

The importance of cavitation relative to the processes of quarrying and abrasion remains largely unquantified for natural channels, but is considered to be high in pothole erosion (Beer et al. 2015).

Quarrying refers to the entrainment of blocks at least partially detached from the surrounding bedrock, typically along joints. Lift force is particularly important in quarrying, and the large lift forces generated in shallow, swift flow can effectively quarry blocks much greater than 1 m across (Wohl 1998; Whipple et al. 2000a; Chatanantavet and Parker 2009). Quarrying can be the dominant mechanism of bedrock channel erosion in strongly jointed bedrock, but is likely to be limited by transport of the blocks (Lamb and Fonstad 2010). Quarrying can be particularly effective in detaching jointed blocks through sliding at a knickpoint face (Dubinski and Wohl 2013). Lamb et al. (2015) propose a threshold equation for block quarrying

$$\tau^*_{pc} = \frac{\tau_b}{(\rho_r - \rho_w)gH} \tag{4.11}$$

in which τ^*_{pc} is the critical Shields stress for quarrying, τ_b is bed shear stress, ρ_r is the density of rock, ρ_w is the density of water, g is gravity, and H is block height. This equation reflects observations that block height is the most important length scale in the resistive force per unit area because of particle weight (Hancock et al. 1998). Observations also suggest that the protrusion height relative to block length strongly influences the threshold of entrainment (Coleman et al. 2003). Two years of measurements of bedrock channel erosion along a stream in Switzerland indicated that a single quarrying event created erosion rates two orders of magnitude greater than those associated with abrasion (Beer et al. 2017).

Abrasion is abrasive erosion of bedrock by clasts carried in the flow. *Macroabrasion* is sometimes used to refer to the combined effects of particle impacts that fracture bedrock into fragments which can be quarried (Chatanantavet and Parker 2009). The effectiveness of abrasion depends on (i) the relative hardness of the clasts in transport and the cohesive boundaries, (ii) the amount of sediment in transport, (iii) the frequency and duration of flows containing clasts capable of creating abrasion, and (iv) position in the channel. The efficiency of abrasion is inversely proportional to rock tensile strength (Sklar and Dietrich 2001), which has been measured using the Brazilian splitting tensile strength test (ISRM 1978). Abrasion has a nonlinear relation with the amount of sediment available. Very low sediment supply limits the tools for abrasion, but high sediment supply can cover and protect the bed from abrasion (Figure 4.3) (Sklar and Dietrich 2004). This so-called tools-and-cover effect has been verified in numerous studies (e.g. Johnson and Whipple 2010; Beer et al. 2015). Abrasion rates can be greatest at flow obstacles, with progressively lower rates on the channel bed and walls (Beer et al. 2017). Although the majority of research focuses on bed-load abrasion, suspended sediment is also capable of effectively eroding bedrock channel boundaries (e.g. Scheingross et al. 2014).

The relative importance of any of the erosive processes for cohesive beds varies with flow and substrate properties (Springer et al. 2003; Wohl and Springer 2005). Quarrying tends to dominate strongly jointed rocks and create faster erosion than the abrasion that erodes massive bedrock (Hancock et al. 1998; Whipple et al. 2000a; Hartshorn et al. 2002). Along the Ukak River in Alaska, USA, well-weathered and fractured bedrock typically coincides with broad channels, which incise relatively rapidly, whereas solid, cohesive bedrock corresponds to deep, narrow inner channels and slower incision (Whipple et al. 2000b). Most channels have some combination of erosive processes, with interactions among them. Erosion of potholes in large jointed blocks along the Orange River in South Africa, for example, reduces the mass of the blocks and facilitates their removal via quarrying (Springer et al. 2006).

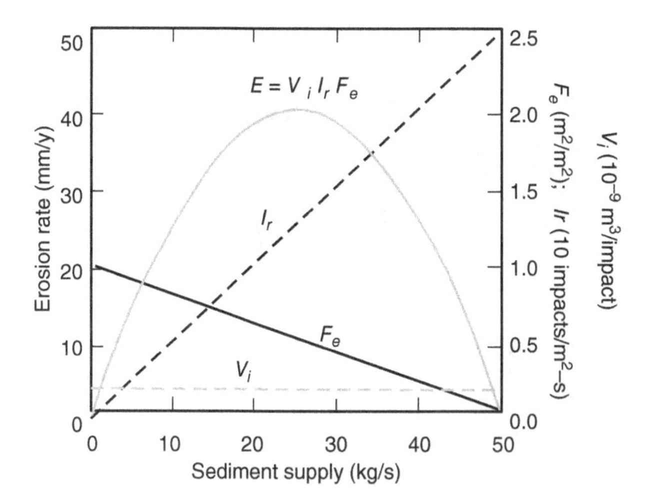

Figure 4.3 Erosion rate, E, as a function of sediment supply, as predicted by the saltation-abrasion model of Sklar and Dietrich (2004). This plot includes the model component terms. Volume eroded per unit impact, V_i, remains constant as sediment supply changes. Impact rate per unit area, I_r, increases steadily with sediment supply. The fraction of the streambed exposed, F_e, declines with sediment supply. The resulting erosion rate peaks at intermediate levels of sediment supply. Source: After Sklar and Dietrich (2004), Figure 10.

Bed erosion in cohesive materials typically does not occur evenly across a cross-section or throughout a reach, and bedrock channels seldom have a planar bed. Instead, erosion tends to occur most actively over a fraction of the bed width that varies with bed load supply and transport capacity (Finnegan et al. 2007). This creates features such as inner channels and variable bed gradients (Baker 1978; Wohl 1998; Inoue et al. 2016), as well as sculpted forms such as potholes and plunge pools.

Flume experiments illustrate the progressive interactions between sculpted forms and mobile sediment through time. Sediment concentrated in topographic lows in the streambed enhances local bed erosion and creates sculpted forms and tortuous flow paths until sufficient energy is dissipated by the rough bed to limit sediment mobility and inhibit further erosion (Johnson and Whipple 2007). Numerous flume and field studies indicate analogous interactions among bedrock erosion, channel morphology, and hydraulics (Wohl and Ikeda 1997; Wohl et al. 1999; Wohl and Merritt 2001; Johnson and Whipple 2010; Goode and Wohl 2010a,b; Scheingross and Lamb 2016).

Just as alluvial bedforms such as ripples and dunes reflect and influence boundary roughness, velocity, and turbulence, so feedbacks among boundary roughness, rate of erosion, velocity, turbulence, and sediment mobility influence the location and style of erosion in bedrock channels (Goode and Wohl 2010b; Lamb et al. 2015; Collins et al. 2016) and can interact to maintain patches of alluvial cover in bedrock channels (Hodge and Hoey 2016). Johnson (2014) proposes a one-dimensional model in which bedrock surface roughness controls partial alluvial cover. This model predicts that when bedrock roughness is much greater than grain size, alluvial streambed cover increases approximately linearly with the ratio of sediment supply to transport capacity. When bedrock roughness is much less than the sediment diameter, the model predicts abrupt shifts between full bedrock exposure and alluvial cover. A distinctive result of localized erosion is the creation of *sculpted forms* such as potholes (Figure 4.4), grooves, and undulating walls (Richardson and Carling 2005). The locations and dimensions of some sculpted forms can be used to infer hydraulics during formative flows (Wohl 2010b).

Figure 4.4 Numerous potholes along the Ocoee River, Tennessee, USA. Hand tape resting on bedrock at left-center of photo for scale. Flow is from left to right. (*See color plate section for color representation of this figure*).

Equations have been proposed for specific erosive processes in bedrock channels. Building on work in Sklar and Dietrich (2004) that models bedrock abrasion, E, in relation to bed load supply, Lamb et al. (2008) relate incision, E^*, to the three dimensionless quantities of normalized sediment supply, normalized effective impact velocity cubed, and relative supply of bed load

$$E^* = \frac{E\sigma_T^2}{\rho_s Y (gD)^{\frac{3}{2}}} = \frac{A_1}{k_v}\left[\frac{q}{(UH\chi + U_b H_b)}\right]\left[\frac{w_{i,eff}}{(gD)^{\frac{1}{2}}}\right]^3\left[1 - \frac{q_b}{q_{bc}}\right] \tag{4.12}$$

where σ_T is rock tensile strength, ρ_s is sediment density, Y is Young's modulus of elasticity, D is sediment diameter, A_1 is the cross-sectional area of a sediment particle, k_v is an empirical rock erodibility coefficient, q is volumetric sediment supply per unit channel width, U is depth-averaged streamwise flow velocity, U_b is depth-averaged streamwise bed load velocity, H is depth of flow, χ is a dimensionless integral relating the flux of suspended sediment to near-bed volumetric sediment concentration, flow depth, and flow velocity, H_b is the thickness of the bed load layer, $w_{i,eff}$ is effective impact velocity, q_b is volumetric bed load flux per unit channel width, and q_{bc} is volumetric bed load transport capacity per unit channel width. Eq. (4.12) predicts that bedrock erosion rate equates to the product of the impact rate of mobile particles, the mass loss per particle impact, and a bed coverage term. It also predicts that impact rate scales linearly with the product of the near-bed sediment concentration and the impact velocity (Lamb et al. 2008).

Chatanantavet and Parker (2009) model bedrock incision via abrasion, quarrying, and macroabrasion as

$$E = E_a + E_p = \left\{ \left[\beta\alpha \left(\frac{m_b + 1}{\chi^{m_b}} \int_0^{\chi} E\chi_1^{m_b} d\chi_1 \right) \right] + q_p p_{pb} \right\} * \left[1 - \frac{\left(\frac{m_b+1}{\chi^{m_b}} \int_0^{\chi} E\chi_1^{m_b} d\chi_1 \right)}{q_{ac}} \right]^{n_0} \tag{4.13}$$

where E is total bedrock erosion rate, E_a is vertical rate of bed incision via abrasion, E_p is rate of bedrock incision via quarrying and macroabrasion, β is the abrasion coefficient, α is the fraction of the load that consists of particles coarse enough to accomplish wear, m_b is an exponent defined as $n_b/(1 - n_b)$ (where n_b is an exponent in width-area relation), χ is a surrogate for distance x defined as drainage area/channel width, q_p is volume rate of entrainment of quarried chunks into bed-load transport per unit bed area per unit time, p_{pb} is the volume fraction of material in the battering layer that consists of chunks which can be quarried, and q_{ac} is the capacity transport rate of wear particles. Eq. (4.13) is a physically based empirical model developed from flume experiments. This equation implies that quarrying-macroabrasion processes can operate more efficiently than abrasion, although lower values of rock tensile strength facilitate the latter.

As with alluvial channels, bedrock channel erosion ultimately depends on how often and for how long thresholds for removal of material are exceeded (Johnson et al. 2010). Such thresholds could represent cavitation, quarrying, or the entrainment of coarse sediment that creates abrasion (Stock et al. 2005). Most investigations suggest that relatively large flows are needed to overcome the higher erosional thresholds of cohesive material (e.g. Baker and Pickup 1987), although field studies from some bedrock rivers suggest that relatively frequent flows of moderate intensity can dominate bedrock erosion (Hartshorn et al. 2002).

4.1.3.2 Erosion of Cohesive Sediment

Erosive processes in cohesive but nonindurated materials, such as silt and clay, are similar to those in bedrock channels, but are also influenced by upward-directed seepage and matric suction. *Upward-directed seepage* is seepage directed vertically upward. This leads to static liquefaction, which creates an additional driving force. *Matric suction* occurs when negative pore water pressure above the water table increases the apparent cohesion of a soil; this is also called matrix suction. Matrix suction creates an additional force of resistance to bed erosion (Simon and Collison 2001). Much of the research on erosion of cohesive, nonindurated materials focuses on stream banks rather than the streambed, as discussed in Section 4.6.

4.2 Sediment Transport

4.2.1 Dissolved Load

The sediment carried in solution is sometimes ignored as a component of sediment transport, but it can be substantial in some rivers: a survey of the world's large rivers indicates anywhere from 2% of the total load is carried in solution in the Huang He River, China to 93% in the St. Lawrence River, Canada (Knighton 1998).

Solute concentration is highest in water entering a channel through subsurface pathways, in which generally slower rates of flow provide longer reaction times in relation to the surrounding matrix. Solute concentration typically declines with increasing discharge as water moving more rapidly and along surface flow paths enters a river. Total dissolved load continues to increase with discharge, but much of the dissolved load is carried by relatively frequent flows (Webb and Walling 1982).

Solute concentration also displays *hysteresis effects*, with higher solute concentrations on the rising limb than on the falling as a result of mobilization of soluble material that accumulates prior to the flood (Walling and Webb 1986). Rivers with strongly seasonal flow regimes typically have an annual flush of high dissolved loads, as in the rising limb of a snowmelt-dominated river (e.g. Williams et al. 2009).

A global survey of 370 rivers indicates that solute concentration, C, varies with discharge, Q, as

$$C = aQ^b \tag{4.14}$$

where b is mostly 0 to −0.4, with a mean value of −0.17 (Walling and Webb 1986).

As noted in Chapter 2, the primary constituents of dissolved load are the dissolved ions HCO_3^-, Ca^{2+}, SO_4^{2-}, H_4SiO_4, Cl^-, Na^+, Mg^{2+}, and K^+; dissolved nutrients N and P; dissolved organic matter; dissolved gases N_2, CO_2, and O_2; and trace metals (Berner and Berner 1987; Allan 1995). The sum of the concentrations of the dissolved major ions is known as *total dissolved solids* (TDS). An average natural value for rivers is 100 mg/L, and pollution adds on average another 10 mg/L. Ca^{2+} and HCO_3^- from limestone weathering tend to dominate TDS (Berner and Berner 1987). Because dissolved fluxes in rivers partly reflect rates of continental weathering, which in turn reflect rates of tectonic uplift, the largest contemporary fluxes occur in rivers draining the Himalayan and Andean mountains and the Tibetan Plateau (Raymo et al. 1988).

Reach-scale dissolved load can also strongly reflect interactions between surface and hyporheic water. Downwelling into the hyporheic zone transfers solutes and surface water rich in dissolved oxygen, whereas upwelling flow can transfer nutrients to streams (Tonina and Buffington 2009). Hyporheic exchange exposes solutes in stream water, including nutrients, to alternating anoxic and oxic zones in the bed that are composed of geochemically reactive sediments and microbial communities (Lautz and Siegel 2007). Anoxic zones that indicate the presence of sulfate, iron, and manganese reduction typically occur upstream of streambed structures, as in low-velocity pools. Oxic zones that reflect the production of nitrate are typically found downstream of bed structures, as in turbulent riffles (Lautz and Fanelli 2008). These zones can enhance biogeochemical reactions and increase nutrient utilization, and may be critical to maintaining stream water quality (Hancock et al. 2005).

Patrick (1995) reviews the importance of river chemistry to aquatic organisms. Dissolved organic matter provides an important energy and nutrient source. Ca, Mg, oxidized S, N, and phosphates, along with small amounts of Si, Mg, and Fe, are desirable for many species. The pH of the water affects the solubility and availability of elements.

Nitrogen and carbon are important nutrients in river water. Rivers are major conduits for nitrogen and carbon transport, but they can also remove and transform dissolved nitrogen and carbon in transport (Hall and Tank 2003; Wohl et al. 2017a). The details of nitrogen and carbon transport versus removal provide a nice example of how physical, chemical, and biochemical processes interact in rivers to govern transport of solutes.

4.2.1.1 Nitrogen

Of the total nitrogen lost from the land, only about 18–20% is carried to the oceans by rivers (Van Breemen et al. 2002), because of removal and transformation en route. Rates of dissolved nitrogen uptake from river water are especially high in shallow streams with algae and microbes in attached biofilms (Hall and Tank 2003), but riverine processing of nitrogen is heterogeneous in time and space. Biogeochemical hot spots in which accelerated chemical reactions occur (McClain et al. 2003) include anoxic zones beneath riparian environments (Lowrance et al. 1984) and the convergence of ground and surface waters in hyporheic zones (Harvey and Fuller 1998). Examples of hot moments include rare high flows that occupy secondary channels in arid environments (McGinness et al. 2002) and snowmelt that enhances leaching of dissolved nutrients in high-elevation catchments (Boyer et al. 2000).

Much of the nitrogen removal within rivers occurs in stream sediments (Nihlgard et al. 1994) and riparian zones (McClain et al. 1999) through the activities of benthic (bottom-dwelling) organisms such as bacteria, algae, and insects, as well as riparian soil organisms and plants that take up nitrogen. Any source of fluvial complexity – logjams, beaver dams, channel-margin irregularities, hyporheic zones, and shallow overbank flow across floodplains – that promotes flow separation and retention of organic matter can substantially increase nitrogen uptake (Battin et al. 2008). By retaining organic matter, these sites provide biota an opportunity to access and ingest the organic matter (Naiman et al. 1986; Fanelli and Lautz 2008). Partly for this reason, river management increasingly emphasizes protection or restoration of physical channel complexity, as well as protection or restoration of riparian corridors that can "buffer" channels from excess nitrogen (McMillan et al. 2014). Small streams can be very important in nitrogen removal, but reduction of drainage density and the cumulative length and complexity of small streams in a river network likely shifts the location of nitrogen removal downstream (Helton et al. 2018).

Human activities, including the use of inorganic fertilizer and emissions from fossil-fuel combustion, have dramatically increased nitrogen inputs to watersheds (Boyer et al. 2006). Simultaneously, river engineering that limits channel boundary complexity and retention, as well as channel–floodplain connectivity, has reduced the ability of many rivers to process nitrogen inputs (Tuttle et al. 2014). The result is excess nitrogen loads and eutrophication in coastal areas (Boyer et al. 2006). Many large river systems are reaching the limit of their ability to process excess nitrogen. Pre-industrial nitrogen fluxes were greatest from the largest rivers, such as the Amazon. Post-industrial fluxes are greatest from rivers in the industrialized zones of North America, Europe, and southern Asia (Green et al. 2004). Although river restoration commonly seeks to limit downstream fluxes of nitrogen by restoring in-channel retention and increasing channel–floodplain connectivity, the spatial extent and functionality of restoration projects are not keeping pace with increasing watershed-scale nitrogen inputs (Bernhardt and Palmer 2011). Steffen et al. (2015) identify fluxes of nitrogen as exceeding planetary boundaries for sustainability and resilience.

4.2.1.2 Carbon

Organic matter, which is predominantly carbon, can be present in river water in particulate and dissolved forms (Berner and Berner 1987). Particulate organic matter enters rivers as fossil carbon from sedimentary rocks (Galy et al. 2015) and as carbon from contemporary soils and vegetation (Schuur et al. 2015) (Figure 4.5). Carbon inputs can be gradual and continuous. They can also be episodic, as in active mountain belts where sediment yield to rivers is dominated by landslides triggered by

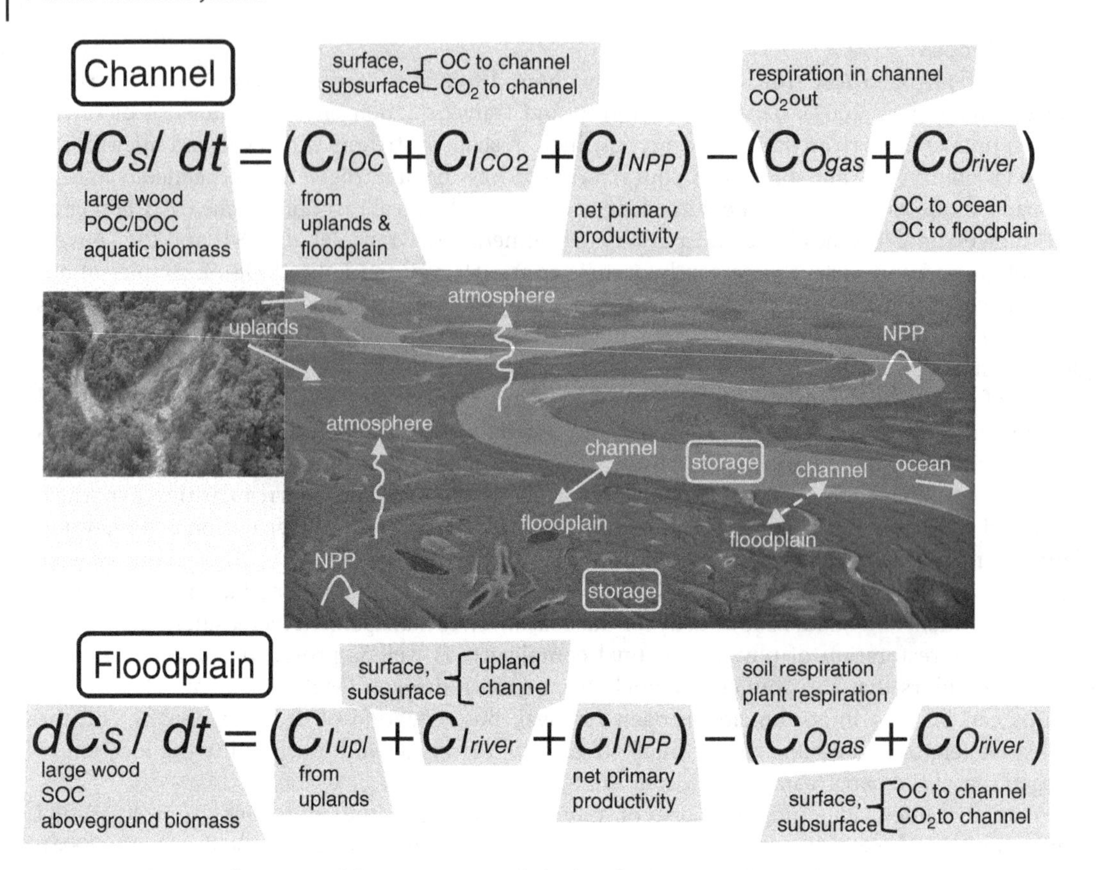

Figure 4.5 Schematic illustration of the components of a budget for organic carbon within a river corridor at the reach scale. C_S indicates storage, C_I indicates inputs, C_O indicates outputs. NPP is net primary productivity, POC is particulate organic carbon, DOC is dissolved organic carbon, SOC is soil organic carbon. Gray shading is intended to make it easier to see which explanatory terms (smaller font) correspond to variables in the budget (larger, italicized font). Within the central photo, yellow arrows indicate fluxes (dashed arrows are subsurface fluxes, squiggly arrows are fluxes to the atmosphere) and yellow boxes indicate storage. (*See color plate section for color representation of this figure*).

tropical cyclones, which also generate floods that result in large riverine fluxes of carbon to the ocean (Hilton et al. 2008a,b, 2011a,b).

Dissolved organic matter includes different classes of organic compounds that differ in reactivity and ecological role, as well as varying in quantity with time and space in response to seasonal variation and precipitation inputs. Dissolved organic matter is usually expressed as dissolved organic carbon (DOC). Values in river waters are typically 2–15 mg/L, but can reach 60 mg/L in rivers draining wetlands (Drever 1988). DOC varies with the size of the river, the climate, and vegetation (Thurman 1985). Values are typically high for temperate and tropical rainforests and taiga, for example, and low for arid and semiarid environments.

DOC is highly reactive and influences riverine ecosystems by controlling microbial food webs (Kaiser et al. 2004). The efficiency with which rivers retain and oxidize organic carbon depends on the presence of microbial communities in response to riverine features that increase the residence time of organic molecules in transport, as explained earlier for nitrogen (Battin et al. 2008, 2009). Simply put, storage of organic matter depends on channel and floodplain morphology and connectivity. More spatially heterogeneous morphology enhances at least temporary storage of dissolved and particulate organic matter, and this enhances the opportunities for biological (microbial and macroinvertebrate) processing and uptake of organic matter, as well as for burial of organic matter in sediment. Vertical connectivity between the surface and hyporheic zone enhances biological processing and uptake. Lateral connectivity between the channel and floodplain also enhances biological processing and uptake, along with burial of organic matter in the floodplain.

Substantial differences in carbon dynamics are present between rivers with well-developed floodplains, which exchange more than half of the recent biogenic carbon in transport with floodplain carbon, and small mountainous rivers that transport carbon directly to the ocean (Galy et al. 2008; Leithold et al. 2016). Even small mountainous rivers can include segments with substantial carbon storage, however, if physical channel complexity creates flow obstructions and enhanced channel–floodplain connectivity (Wohl et al. 2012c; Sutfin et al. 2016). Consequently, river management that facilitates maintenance or restoration of river-corridor complexity has numerous benefits, including carbon and nitrogen retention and increased diversity of aquatic habitat (Wohl 2016; Wohl et al. 2018b).

DOC is an important component of the global carbon cycle. Estimates of the amount of carbon delivered from terrestrial sources to freshwater environments each year range from 2.7 (Aufdenkampe et al. 2011) to 5.1 (Drake et al. 2018) petagrams (Pg). Of this, somewhere between 0.2 and 1.6 Pg are buried in sedimentary reservoirs such as floodplains or deltas. An estimated 1.2 Pg are released to the atmosphere as a result of metabolism and respiration of riverine organisms, and approximately 0.9 Pg are carried by rivers to the ocean. River dynamics strongly influence the partitioning between sedimentary reservoirs, release to the atmosphere, and transport to the ocean (Raymond et al. 2013; Hotchkiss et al. 2015; Wohl et al. 2017a). The range of estimated values reflects uncertainties as measurements continue to be refined, as well as differences in what is included (e.g. the range for burial in sedimentary reservoirs reflects different studies that do and do not attempt to estimate burial in artificial reservoirs).

The relative importance of DOC versus particulate organic carbon exports varies among rivers. In small mountainous rivers, for example, coarse particulate organic carbon export exceeds DOC export (Turowski et al. 2016).

4.2.1.3 Trace Metals

Trace metals typically occur at concentrations less than 1 mg/L in river waters. Trace metals can be derived from rock weathering or from human activities that include mining, burning of fuels, smelting of ores, and disposal of waste products (Drever 1988). Sediments contaminated with trace metals associated with human activities are commonly referred to as *legacy sediments* (as are other types of deposits that reflect past human activities) (James 2013). These contaminated sediments create a long-term toxic legacy. Trace metals receive most attention as contaminants occurring at concentrations above background levels or hazards occurring at levels potentially harmful to organisms. Examples of trace metals that commonly create pollution are arsenic, cadmium, chromium, copper, lead, mercury, nickel, selenium, and zinc, each of which is very toxic and biochemically accessible to living organisms. Because many trace metals are adsorbed to fine sediment, the mobility and storage of the fine sediment strongly influence the spread of trace metals through a river network and the bioavailability of the metals (Wohl 2015a). Metals adsorbed to silt and clay traveling in suspension or resting on the surface of the streambed, for example, are more readily ingested by aquatic organisms than are metals adsorbed to fine sediment deeply buried in a floodplain.

4.2.1.4 Other Environments

As with other aspects of hydrology, cold-region rivers and rivers of the humid tropics have some distinctive chemical characteristics. Cold-region rivers in alpine or permafrost terrains with high levels of impermeable surface area may be more influenced by precipitation chemistry than other types of rivers with greater infiltration and deeper, slower downslope pathways of water (Meixner et al. 2000). Selective weathering in glaciated catchments can allow more reactive minerals to contribute disproportionately to dissolved load relative to their abundance in the local rock. Chemical denudation rates can be higher for glaciated areas than for adjacent nonglaciated catchments. Both seasonal snowmelt and glacial melt can create large temporal and spatial variations in stream-water chemistry as the source and contact time of meltwater with sediment and rock change through the melt season (Taylor et al. 2001; Anderson et al. 2003). Total organic carbon export from cold-region rivers is increasing as climate warms because of accelerated decomposition of soil organic matter as the seasonal duration of frozen soil decreases (Trumbore and Czimczik 2008; Holmes et al. 2012).

Rivers of the humid tropics have warmer water, higher annual exports of dissolved constituents (Lyons et al. 2002), and lower seasonal variability in water temperature and chemistry than do rivers at higher latitudes (Scatena and Gupta 2013). Carbon exports occur primarily as DOC, and both DOC and particulate carbon exports can increase significantly when tropical storms create widespread defoliation or landsliding (Hilton et al. 2008a,b). Consequently, humid tropical rivers can create important global sinks of carbon by transporting carbon to oceanic burial (Alin et al. 2008; Goldsmith et al. 2008; Kao et al. 2014).

4.2.2 Suspended Load

The particulate sediment transported by rivers can be subdivided in different ways. One distinction is between wash load and bed-material load, another is between suspended load and bed load. The former reflects different mechanics of transport, but the coarsest grain size moving as wash load is not easy to measure directly, blurring this distinction. The latter distinction primarily reflects measurement technology, with some forms of bed load transport (saltation) being transitional between the two end-members. A third distinction is between mineral sediment and particulate organic matter.

Wash load is the finest size fraction of the total sediment load, typically grains with intermediate diameter ≤ 0.062 mm. Wash load consists of particles typically not found in large quantities on the bed surface. These particles have settling velocities so small that they move at approximately the same velocity as the flow and only settle from suspension when velocity declines substantially. Wash load is vertically mixed by turbulent flows, so that concentration varies little with flow depth. Because relatively little energy is needed to transport wash load, transport rates typically reflect sediment supply rather than flow energy. Much of the wash load comes from bank erosion and surface erosion across the drainage basin (Skalak and Pizzuto 2010).

Interparticle forces are greater than the gravity force among portions of the wash load finer than 0.004 mm in diameter, and these very fine sediments are cohesive. Cohesive sediments typically travel as aggregated or flocculated material, rather than as single grains (Kuhnle 2013). Unlike coarser sediments, for which transport may be limited by stream energy (and which are thus known as *transport limited*), fine cohesive sediments are more likely to be *supply limited*, such that the amount in transport is limited by the amount supplied to the flow. During supply-limited conditions, the actual *transport rate* is likely to be less than the *transport capacity*, which reflects what the flow is capable of transporting.

Bed-material load includes grains typically coarser than 0.062 mm. These grains move either in contact with the bed by rolling, sliding, or saltating as *bed load*, or in suspension just above the bed, with concentration declining upward from it. Bed-material load can also be subdivided into suspended load and bed load. If sediment is divided in this way only, then the suspended load includes wash load.

Noncohesive sediment will remain in suspension if the strongest vertical velocity fluctuations exceed the particle fall velocity. Close to the bed, the root mean square of the vertical velocity fluctuations reaches a maximum that is approximately equal to the magnitude of the *shear velocity*, v_*, which provides an approximate criterion for suspension (Kuhnle 2013)

$$v_* = \sqrt{\frac{\tau_0}{\rho}} \tag{4.15}$$

where τ_0 is the bed shear stress and ρ is water density. Noncohesive suspended sediment concentration typically decreases with distance away from the bed in a manner described using the Rouse equation

$$\frac{C}{C_a} = \left(\frac{(a)(d-\gamma)}{(\gamma)(d-a)}\right)^Z \tag{4.16}$$

where C is the concentration of sediment at a distance γ above the mean bed elevation, C_a is the concentration of sediment at the reference level a above the bed, d is flow depth, and the exponent, Z, known as the *Rouse number*, is defined as

$$Z = \frac{\omega_s}{\kappa u_*} \tag{4.17}$$

where ω_s is the particle fall velocity, κ is the von Karman constant, and u_* is the shear velocity (Kuhnle 2013).

The Rouse number is basically the ratio of the grain fall velocity to the strength of the flow. The relation between the Rouse number and the distribution of suspended sediment through the water column (Figure 4.6) reflects the fact that, as the value of the flow strength (u_*) increases relative to the fall velocity of the sediment grains, the resulting gradient of sediment concentration through the flow depth grows less steep (Kuhnle 2013).

Large suspended sediment concentrations commonly occur during high discharge. Using an equation similar to Eq. (4.14), the exponent b is generally in the range of 1 to 2 (Walling and Webb 1986). Wolman and Miller (1960) were first to suggest that most suspended sediment is carried by relatively frequent flows with a recurrence interval of 1–2 years, although the magnitude of flow carrying the greatest proportion of annual suspended sediment load – also known as the *effective discharge* – varies widely among catchments. Wolman and Miller (1960) assume that the rate of suspended sediment transport can be represented by a power law (curve a in Figure 4.7) and the frequency of occurrence of transport can be represented by a log-normal law (curve b in Figure 4.7). These assumptions greatly simplify the scatter seen in actual suspended sediment data, so sites with abundant actual data that can constrain the true shapes of the curves in Figure 4.7 can better

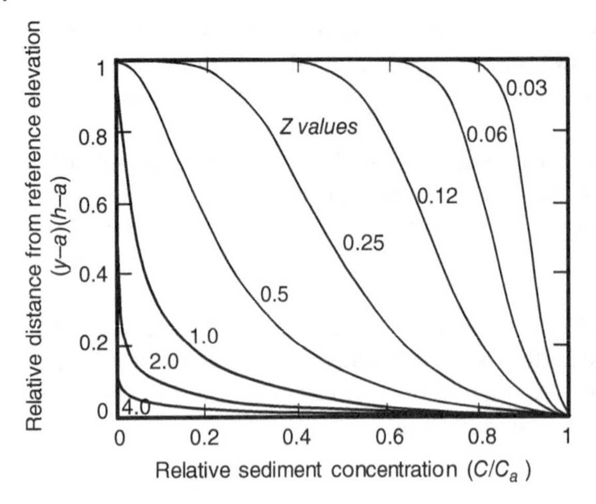

Figure 4.6 Diagram illustrating the effect of a range of Rouse numbers (Z values) on the prediction of suspended sediment concentration profiles. The variables on the x and y axes are as follows: a is a reference level above the bed, h is depth of flow, C is the concentration of sediment at a distance y above the mean bed elevation, and C_a is the concentration of sediment at the reference level a. Z values reflect the ratio of grain-fall velocity to the strength of the flow. Greater values of flow strength relative to the fall velocity (smaller Z values) of the grains result in a less steep gradient of sediment concentration with flow depth; that is, greater concentrations of suspended sediment high in the flow than at large Z values. Source: After Kuhnle (2013), Figure 2.

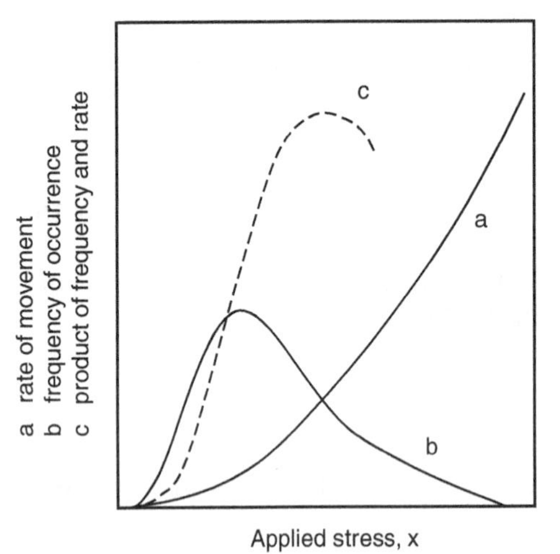

Figure 4.7 Effective discharge can be defined in terms of the magnitude and recurrence interval of flow that transports the most suspended sediment, originally defined by Wolman and Miller (1960) as the flow that accomplishes the most geomorphic work. Source: After Wolman and Miller (1960), Figure 1.

constrain estimates of effective discharge (Ferro and Porto 2012). Most studies indicate that effective discharge has a recurrence interval in the range of <1 to 5 years (Wolman and Miller 1960; Simon et al. 2004; Crowder and Knapp 2005; Ferro and Porto 2012), although small basins can have effective discharge during extreme flows with annual flow duration much less than 1% (Ashmore and Day 1988). Effective discharge is also sometimes calculated relative to bed load transport.

Gaging suspended sediment requires measuring sediment concentration and water velocity. The suspended sediment flux, q_s, past a single vertical in a river cross-section is

$$q_s = \int c_s(z) v_s(z) dz \tag{4.18}$$

where c_s is concentration of suspended sediment, v_s is the downstream velocity of the suspended sediment, and z is depth (Hicks and Gomez 2016). The concentration of suspended sediment varies with depth as a function of turbulence and the fall velocity of the sediment. Suspended sediment is usually assumed to move at the same velocity as the water.

The integral in Eq. (4.18) can be determined either (i) by collecting point samples of water and point velocity measurements at intervals over the flow depth, with six to eight samples usually needed to define a profile, or (ii) by using a depth-integrating sampler such as the DH-48 to collect water and sediment samples continuously along a vertical profile (Hicks and Gomez 2016). Spatial variations in suspended sediment across the channel can be addressed by sampling at multiple verticals across a cross-section. The total suspended sediment discharge for a section is the product of the discharge-weighted mean concentration at each sampling vertical and the water discharge within each vertical.

Details about the numerous samplers and methods available can be found in technical manuals (Edwards and Glysson 1999) and online (http://water.usgs.gov/fisp/). Most equipment samples only to within 75–100 mm of the bed, to avoid sediment moving as bed load (Hicks and Gomez 2016).

Samples can be collected manually, using either handheld instruments or instruments lowered on a cable from a boat, bridge, or other measuring platform. Automated samplers programmed to collect samples based on time interval, stage change, or flow volume can pump and store a limited number of samples.

Suspended sediment data can also be collected using various types of sensors that do not actually collect samples, but record some type of signal calibrated to suspended sediment concentration. Examples include optical sensors that measure transmissivity or back-scattering (Hicks and Gomez 2016), diffraction of a laser beam, or back-scattering of a sound source (acoustic sensor). Other approaches use remote-sensing imagery for which relations between suspended sediment concentration and visible/near-infrared reflectance have been established (Kuhnle 2013; Umar et al. 2018).

The long-term average suspended sediment yield can be estimated using a sediment rating curve of mean suspended sediment concentration versus water discharge. This approach may only result in a very approximate estimation because the relation between suspended sediment concentration and discharge can vary substantially through time in response to different runoff sources or changes in land cover or sediment supply. Suspended sediment concentration can also be modeled using an empirical multivariate relation for suspended sediment concentration, water discharge, and other influential variables such as season or event hysteresis (Hicks and Gomez 2016).

Values of suspended sediment concentration, typically reported in mg/L, vary widely with time and among rivers. Data from 93 streams in the United States include values from <1 mg/L up to a high value of 29 100 mg/L on the North Fork Toutle River in Washington after the 1980 volcanic eruption of Mt. St. Helens (Williams and Rosgen 1989). Values can also be relatively high for rivers draining active glaciers (e.g. up to 3130 mg/L for Alaska's Talkeetna River) and for rivers draining erodible sedimentary rocks in a semiarid climate (e.g. up to 3020 mg/L for Colorado's Yampa River). Catchments with dense vegetation or crystalline rocks typically have much lower values, as reflected in a range of 1–110 mg/L for Oak Creek in Oregon.

Most rivers display some form of hysteresis for suspended sediment, but the details can vary widely. Williams (1989) identifies three forms of hysteresis (Figure 4.8). One is a clockwise loop in which the peak sediment concentration precedes the peak discharge because available sediment is depleted before runoff peaks, a situation more common in small basins. The second is an anticlockwise loop in which peak sediment concentration lags the peak discharge, a situation more common in large basins. The third is a figure-of-eight loop in which peak sediment concentration precedes peak discharge, but the shape of the sediment output is skewed relative to the flood peak. This can occur where suspended sediment increases more rapidly than discharge and thus peaks first, but

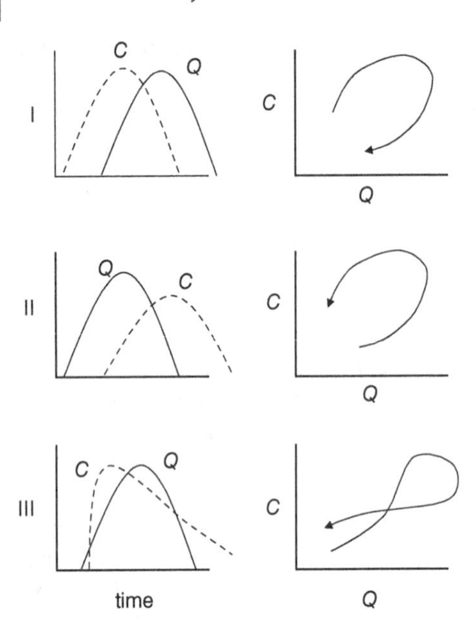

Figure 4.8 Three types of hysteresis in relations between suspended sediment and discharge when *C* (concentration) and *Q* (discharge) are not synchronous. Source: after Knighton (1998), Figure 4.8D, p. 123 and Williams (1989).

post-peak sediment availability and transport are sufficiently high to create sediment concentrations that decrease more slowly than water discharge.

Sources of suspended sediment are typically not uniformly distributed across a drainage basin, particularly in larger basins. This is dramatically illustrated by Meade's (1996) "sedigraphs" for drainage basins such as the Mississippi River in North America (Figure 4.9). The majority of discharge in this drainage comes from the humid–temperate Ohio River basin and the headwaters of the Mississippi, whereas the majority of suspended sediment comes from the semiarid Missouri River basin. The relative importance of different sub-basins as sources of suspended sediment reflects the influence on sediment yields of such factors as rock type, topography, climate, and land use. Sediment budgets for the Amazon River basin show similar spatial disparities as those for the Mississippi, with tributary basins draining highlands contributing substantial suspended sediment, and other tributaries contributing negligible amounts (Meade 2007).

Suspended sediment dynamics along the Amazon also indicate the importance of the floodplain and channel banks. Much of the suspended sediment in transport during the rising and peak stages of the annual flood is deposited in the floodplain, where the sediment remains until it is returned to fluvial transport via bank erosion (Dunne et al. 1998).

Suspended sediment can also be strongly influenced by land use that disrupts land cover, by flow regulation and channel engineering, and by spatial and temporal variations in precipitation and discharge. Upland clearing for agriculture increased sediment yield in the Lago Loíza basin of Puerto Rico, for example. Sediment from this period was stored along channels and floodplains for several decades and then remobilized when second-growth forest took over croplands (Gellis et al. 2006). In the Upper Mississippi River, the dominant source of sediment has shifted from agricultural soil erosion to accelerated erosion of stream banks during the past 150 years as a result of increasing precipitation and river discharge, as well as modification of the channel network by agricultural ditches and tile drains that reduce surface runoff (Belmont et al. 2011).

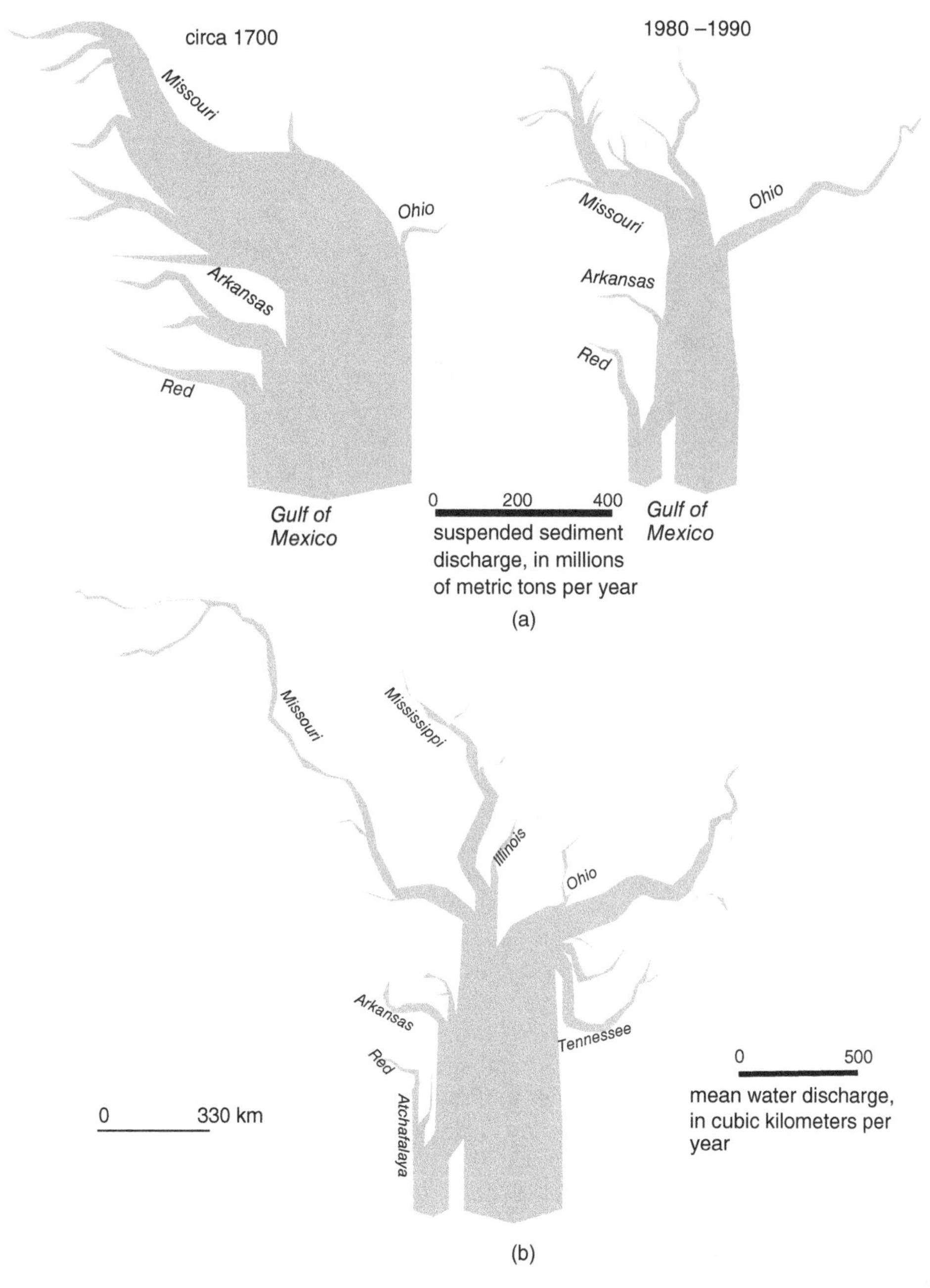

Figure 4.9 (a) Schematic illustration of historical and contemporary contributions to suspended sediment (top) and water (bottom) discharge of the Mississippi River (Source: After Meade 1996, Figures 5 and 6). The majority of the suspended sediment carried in the Mississippi continues to come from the Missouri River, although this value has declined substantially through time as successively more large dams have been built along the Missouri, resulting in substantial sediment storage in reservoirs. The Ohio River is the single largest contributor of discharge to the Mississippi. (b) Schematic representation of downstream changes in suspended sediment and discharge in the Amazon River (Source: After Meade 2007, Figure 4.6, p. 53). The Rio Negro and Rio Madeira contribute similar magnitudes of discharge to the Amazon, but the Rio Madeira contributes substantially more suspended sediment.

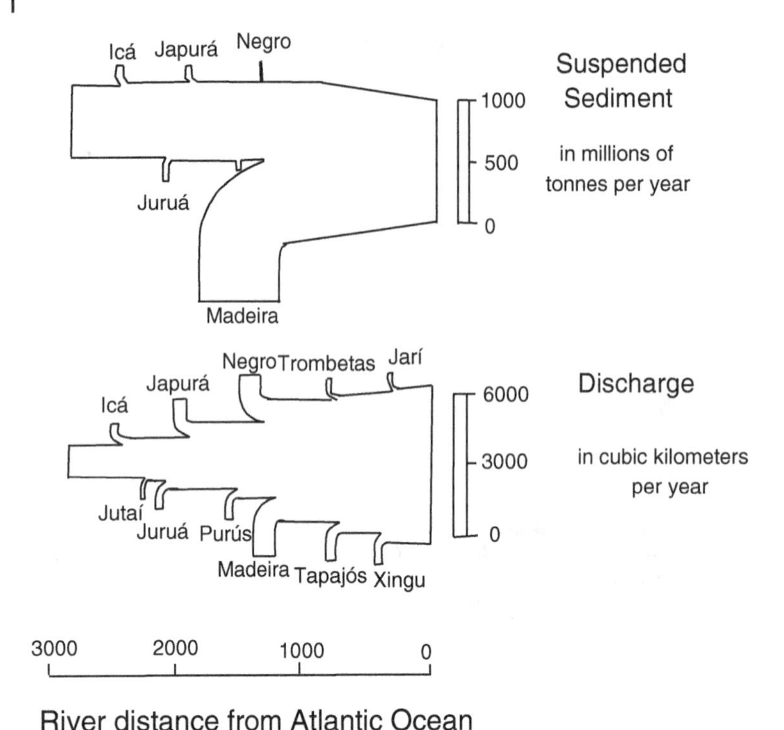

Figure 4.9 *(Continued)*

Suspended load is the dominant component of sediment transport on many rivers, particularly those draining areas larger than about 5000 km^2 (Kuhnle 2013). Suspended load is also the most commonly measured component of total sediment load. Globally, suspended sediment likely accounts for more than 70% of total fluvial sediment transport to the oceans (Milliman and Meade 1983). Small, mountainous basins produce much greater yield per unit area (tons/km^2) and account for a disproportionate amount of erosion and sediment yield (Milliman and Syvitski 1992; Larsen et al. 2014).

Suspended sediment can influence hydraulics, total sediment transport, and channel form in diverse ways. Large concentrations of suspended sediment can increase flow viscosity, reduce settling velocities, and thus decrease turbulence and facilitate the transport of coarser grains than are found in clear water (Simons et al. 1963). Suspended sediment can also create turbidity that limits photosynthesis for aquatic plants (Kemp et al. 2011). Infiltration of fine suspended sediment into an alluvial streambed can increase bed cohesion and reduce the porosity and permeability of the sediment matrix (Packman and Mackay 2003). Reduced porosity and permeability can in turn reduce hyporheic exchange and the ability of stream organisms from aquatic insects to fish embryos to survive within the bed matrix. High suspended sediment concentrations can correlate with high rates of overbank deposition and lateral and vertical accretion of floodplains. High suspended loads can also correlate with stream banks with a greater percentage of silt and clay, greater bank cohesion, and relatively narrow, deep cross-sections (Schumm 1960).

A variety of pollutants, including heavy metals, pesticides, and excess nutrients, can travel adsorbed to cohesive fine sediments, so that the transport and deposition of fine sediments can strongly influence the distribution of contaminants within the river corridor. Nineteenth-century silver and gold mining using a mercury-amalgamation procedure led to the release of thousands of kilograms of mercury-contaminated tailings into the Carson River in Nevada, USA. Mercury, which is associated with fine-grained sediment fractions ($< 63\,\mu m$), has subsequently been dispersed downstream for tens of kilometers via overbank deposition during floods (Miller et al. 1999).

Particulate organic matter (POM) can also travel in suspension. Rates of POM transport increase with discharge, and data from multiple channels in Switzerland show relatively little scatter about a regression line for POM transport and discharge (Turowski et al. 2013). The increase in POM transport with rising flow likely reflects the combined effects of mobilizing large wood that effectively traps and stores POM during lower flows and greater fluxes of POM from adjacent hillslopes (Jochner et al. 2015).

4.2.3 Bed Load

Although the contribution of bed load to total sediment transport is relatively small except in mountainous river networks (Turowski et al. 2010), bed-load dynamics strongly influence river process and form. The movement of bed load influences the stability of the bed and the disturbance regime of aquatic habitat. Moving bed material influences flow resistance and river morphology, as well as lateral channel movement and floodplain formation. In tectonically active regions, the rates of channel incision and landscape evolution fundamentally reflect the rate at which bed load is moved out of the channel network.

Grain motion within bed load occurs via rolling and sliding at lower flows, saltation at intermediate flows, and sheetflow during high flows, although the three types of motion can coexist (Haschenburger 2013). Grains rolling or sliding along the bed can make microleaps across distances smaller than the grain diameter, or can begin to saltate. *Saltation*, from the Latin for "leap," describes grains launched into the water column that follow a predictable, ballistic trajectory before once more contacting the bed. The maximum saltation height, which can be up to 10 times the grain diameter, defines the upper extent of the bed load layer. *Sheetflow* occurs when grains move in multiple granular sheets with a grain concentration approaching that of the underlying stationary bed. ("Sheetflow" is also used to refer to sediment movement under very shallow flows on hillslopes.)

The ratio of dimensionless shear stress to critical dimensionless shear stress (τ_*/τ_{*c}) correlates with the type of sediment transport. Bed load typically dominates where τ_*/τ_{*c} is ~ 1 to 3, bed load and suspended load occur where τ_*/τ_{*c} is between 3 and 33, and suspended load dominates where τ_*/τ_{*c} exceeds ~ 33 (Church 2006).

Most bed load comes from the streambed and lower banks. Bed-load flux depends on grain velocity and concentration. At low concentrations, bed-load particles collide primarily with stationary bed grains. Collisions between mobile grains become more frequent as bed-load concentrations increase. Bed-load grains are supported by grain collisions in a granular-fluid flow at high concentrations. Consequently, the mechanics of bed-load motion vary because fluid–grain and grain–grain interactions change with concentration and viscosity (Haschenburger 2013; Houssais et al. 2015; Houssais and Jerolmack 2017).

As in the discussion of bed-sediment characterization, it is useful to distinguish bed load in predominantly sand-bed channels from bed load in channels with coarser substrate and a wider range of grain sizes. For both types of channels, bed-load movement is typically episodic and spatially heterogeneous across the channel and downstream.

In sand-bed channels, much of the bed load moves as migrating bedforms such as ripples, dunes, and antidunes, which are treated in the next section. Transport rate reflects the energy of the flow, because the individual sand grains are relatively mobile and transport is not typically limited by sediment supply. Sand-bed channels commonly exhibit bed scour during the rising stages of flow, but then fill to approximately pre-flood levels on the falling stage (Leopold et al. 1964). Scour and fill is commonly localized, however, even during large floods (e.g. Church 2006).

In channels with substrate coarser than sand, bed-load movement is more episodic in time and space and is typically limited by sediment supply (e.g. Kleinhans et al. 2002). Supply limitations on transport, as well as large spatial and temporal variations in local flow field and boundary configuration, make it difficult to predict bed-load transport using foundational equations of bulk sediment transport developed for uniformly sized grains (typically sand sizes) in flumes. Consequently, research has shifted toward describing two or three phases of bed-load transport (e.g. Ryan et al. 2002; Lenzi et al. 2004) and applying equations that differentiate phases based on primary sediment size mobilized (e.g. sand versus gravel) and rates of transport (e.g. limited transport over a stable bed versus bed mobilization) by incorporating a nonlinear effect of sand content on gravel transport (Wilcock and Crowe 2003). Gravel-bed channels can also exhibit scour during the rising limb of a flood and fill during the falling limb (Trayler and Wohl 2000). Laboratory experiments can be used to overcome some of the difficulties of characterizing processes in natural rivers (Yager et al. 2015).

4.2.3.1 Bed Load in Channels with Coarse-Grained Substrate: Coarse Surface Layer

In streams with coarser bed material, flow energy is important, but the finer portion of the grain-size distribution can also be supply-limited by the existence of a coarse surface layer. Many channels with relatively coarse, poorly sorted grain-size distributions have at least slightly finer-grained sediments immediately below the coarse surface layer (Figure 4.10). The coarser grain sizes at the surface increase the threshold of hydraulic force that must be exceeded before the bed is mobilized, as well as the hydraulic roughness of the bed. The coarse surface layer is characteristically one grain diameter in thickness and is both more coarse-grained on average and better sorted than the underlying sediment (Powell 1998). The underlying sediment has a grain-size distribution similar to that of the bed load in many gravel-bed rivers (Dietrich et al. 1989).

The coarse surface layer can be called pavement, armor, or a censored layer. Each of these terms is used differently by different investigators. *Pavement* is most commonly used for a coarse surface layer that is rarely disrupted. *Armor* most commonly refers to a coarse lag layer developed at waning flows that is regularly disrupted. *Censored layer* commonly refers to a layer that forms as finer matrix material is removed from around the surface framework particles as flow increases (Carling and Reader 1982).

The mechanism by which the coarse surface layer forms, and the stability and frequency of mobilization of the layer, vary among channels and with time for a given channel. Coarse surface layers typically are well developed when local bed-load supply from upstream is less than the ability of the flow to transport the bed load. This situation is common downstream from dams (Dietrich et al. 1989) and may develop with time following a pulse sediment input such as that from a volcanic eruption (Gran and Montgomery 2005). Ephemeral channels typically have less developed coarse surface layers than perennial channels (Hassan et al. 2006). Coarse surface layers can also be spatially variable at the reach scale as a result of variations in shear stress and sediment dynamics associated

Figure 4.10 Coarse surface layer (a) and underlying finer sediment (b) on an alternate bar along the Poudre River, Colorado, USA. Scale with inches (right) and centimeters (left) at upper left in (b). (*See color plate section for color representation of this figure*).

with alternate bar topography, hydraulic resistance caused by instream wood, or variations in sediment supply (Buffington and Montgomery 1999b).

Several studies suggest that the coarse portion of a surface layer moves down a river at nearly the same rate as the finer sediment in the layer. Coarse grains are intrinsically less mobile than finer grains, but the coarse surface layer may provide an equalizing mechanism by exposing proportionally more coarse grains to the flow. The coarse surface layer can thus be a mobile-bed phenomenon present under a range of flows, with only a few clasts of diverse sizes being entrained at any instant by even the peak flows (Andrews and Erman 1986; Clayton and Pitlick 2008; Powell et al. 2016).

4.2.3.2 Bed Load in Channels with Coarse-Grained Substrate: Characteristics of Grain Movements

In coarse-bed streams, grains move as individual particles or in discontinuous pulses. Pulses have been described as *sheets* in which migrating slugs of bed load one to two grain-diameters thick alternate between fine and coarse particles (Kuhnle and Southard 1988). Pulses of bed load can also take the form of *traction carpets* (Reid et al. 1985), which are highly concentrated bed-load layers, or *streets*: longitudinally continuous tongues of bed load that do not span the full width of the channel. Pulses can also occur as *waves* (Ashmore 1991a), which are groups of low-amplitude bars migrating downstream in braided rivers, with plane bed transport across the upper bar surface and avalanching on downstream faces. Each of these forms of transport reflects the fact that coarse bed load is unlikely to be equally distributed in time and space along a reach of channel.

In channels with mixed grain sizes, rolling and sliding grains move episodically. Field observations suggest that rolling and sliding grains move less than 20% of the time during which flow conditions are sufficient for entrainment (Haschenburger 2013), although periods of transport increase as rates of grain entrainment increase. Turbulent bursts and sweeps cause much of the grain motion (Nelson et al. 1995; Wu and Jiang 2007).

Step length is the travel distance of a particle during continuous motion. The mean step length of grain motion appears to exceed 100 grain diameters, although step length increases with flow and decreases with increasing grain size and decreasing sphericity (Schmidt and Ergenzinger 1992). In particular, step length varies only moderately for grains smaller than D_{50}. Step length of larger grains correlates inversely with grain size because the step length is limited by the energy required for transport of these larger grains (Church and Hassan 1992). Grain collisions or areas of lower hydraulic force or rougher surface topography initiate periods of rest. The distributions of step lengths and rest periods have been described using gamma and exponential density functions, as well as a Poisson distribution (Olinde and Johnson 2015) and a Weibull distribution (Furbish et al. 2016).

The virtual velocity of grains quantifies the rate of downstream movement, including both step and rest periods during flows when entrainment can occur. This is known as *grain velocity*. Grain velocity, which is not particularly sensitive to flow velocity but does reflect grain characteristics such as density, can also be described with a gamma density function (Hassan and Bradley 2017). Mean virtual velocity increases with flow and has been documented at up to 100 m per hour (Hassan et al. 1992).

The *path length* of moving bed load is the total downstream displacement of particles, typically through multiple steps, during a period of competent flow (Haschenburger 2013). Local flow and bed conditions strongly influence the starts and ends of step lengths and path lengths. Longer-duration flow can create longer path lengths when grains move over relatively smooth beds. Bed morphology such as bars or pools can limit path lengths for even longer-duration flows, such that the modal path length correlates to spacing of pool–riffle–bar or step–pool units (Pyrce and Ashmore 2003).

The number and size of grains entrained from the surface increase with flow, and when the entire surface is mobile, grains entrain from the subsurface. As the bed transitions from being immobile to partial mobility and then to full mobility, the upper limit to partial mobility is *equal transport mobility*, where all grain sizes are transported in proportion to their presence on the bed (Parker and Toro-Escobar 2002). This condition, which is relatively uncommon in coarse-grained channels, occurs when all surface grains are mobilized at flows exceeding entrainment thresholds by about four times (Wilcock and McArdell 1997).

The distinction between partial and full mobility is also referred to as phase I versus phase II transport (Andrews and Smith 1992). *Phase I transport* starts when flow energy is just sufficient to rotate some of the gravel-sized particles out of their resting place, sending them rolling or bouncing downstream. As shear stress increases, more particles become mobile. Where a well-developed coarse surface layer is present, phase I involves transport of a finer fraction over a stable coarse bed. Sometimes, three phases of bed transport are distinguished. In this case, phase I is transport of a finer fraction over a stable coarse bed and phase II is partial and usually size-selective local entrainment (Ashworth and Ferguson 1989; Warburton 1992). Most of the bed is mobile during *phase II transport* (or phase III in a three-phase conceptualization). The two phases can be quantified using dimensionless shear stress or a critical discharge (e.g. Schneider et al. 2016).

Houssais et al. (2015) describe the onset of sediment transport in terrestrial and aquatic environments as a continuous transition from creeping to granular flow. A three-phase diagram for sediment transport distinguishes bed-load transport as a dense granular flow bounded by creep below and suspension above. Each of the three phases spans a range of τ^*/τ_c^*.

The nonlinear effects of sand on the transport of coarse particles in a bed with mixed grain sizes illustrate the complexities of bed-load movement (Wilcock and Kenworthy 2002). Sand can be preferentially transported at low discharges that do not move coarse sediment. As the amount of sand present increases, local sandy areas decrease the protrusion of large grains. Moderate amounts of sand can thus decrease the entrainment of large grains. Once a large grain is entrained, though, it can move faster over the relatively smooth sand bed and may move farther because sand fills the pore spaces between large grains and reduces resting places. G.K. Gilbert's (1914, 1917) pioneering flume experiments, undertaken in connection with his studies of nineteenth-century mining sediment in the Sierra Nevada of California, USA, first demonstrated that more fine sediment can result in greater mobility of coarse clasts.

In addition to moving downstream, grains on the bed surface are also vertically exchanged with those in the subsurface through local scour and fill. Under some conditions, the bed is inactive and mobile grains pass over it. Finer grains can preferentially infiltrate, particularly if framework gravels on the bed dilate prior to movement (Allan and Frostick 1999). (Dilation refers to a phenomenon in which grains lift and move apart from one another just before entrainment; Allan and Frostick 1999; Marquis and Roy 2012.) More active beds facilitate grain exchange and sorting, although finer particles are still exchanged into the subsurface more quickly than coarser particles. Increased flow magnitude or duration increases the frequency and depth of vertical exchange (Wong et al. 2007), with exchange depths reaching up to 10 times the diameter of D_{90}, but more typically $2D_{90}$ (Haschenburger 2013; Hassan and Bradley 2017).

Entrainment and transport of alluvium along bedrock rivers likely vary along a continuum from an end-member with continuous alluvial cover, in which processes operate as described in Section 4.1.2, to a bedrock end-member in which sediment entrainment and transport are independent of size and more likely to reflect the location of local roughness and hydraulics (Goode and Wohl 2010a) that stabilize alluvial sediment patches (Hodge et al. 2011, 2016).

Cold-region rivers have unique forms of sediment transport related to the formation and movement of frazil and anchor ice. *Frazil ice* is a collection of loose, randomly oriented, needle-shaped ice crystals that resembles slush. It forms sporadically in open, turbulent, supercooled water, typically on clear nights when the air temperature drops to −6 °C or lower. *Anchor ice* is submerged ice attached to the streambed. It can form when large-scale turbulence mixes suspended ice crystals and frazil ice across the full depth of flow, allowing some of the ice to adhere to the bed or individual boulders. This is particularly common in riffles, and anchor ice can form in flows as deep as 20 m (Ettema and Daly 2004). Under sufficiently cold temperatures and substantial flow turbulence, extensive areas of the streambed can become covered by it. Larger amounts of anchor ice typically form on coarser beds, which facilitate heat flux from a sub-bed zone that is 1–2 °C above freezing to supercooled flow over the full depth above this zone. Anchor ice is less likely to form on fine, noncohesive sediment, which is readily lifted by ice and therefore cannot hold a large accumulation (Ettema and Daly 2004).

Both frazil and anchor ice can be at least briefly attached to the bed and then float up with sediment frozen into the ice. Diurnal formation of these types of ice can result in repeated ice-rafting of sediment along a river during the cold season. Although observations indicate that sediment is entrained and rafted downstream by these forms of ice, there is little quantitative information on exactly how important this process is relative to other forms of sediment transport (Ettema and Daly 2004). Much of the sediment entrained by the ice is stored in a seasonal ice cover until the cover breaks up during warmer weather. Large blocks of river ice that form ice jams can enhance bed and bank erosion (Prowse 2001; Beltaos et al. 2018).

4.2.3.3 Bed Load in Channels with Coarse-Grained Substrate: Controls on Bed-Load Dynamics

Spatial and temporal discontinuities in bed-load movement have been explained in relation to several potential controls, each of which may dominate in different channels or as conditions within a channel change. Multiple controls can also influence bed-load movement at a particular time and place.

Temporal variations in bed load can be distinguished as occurring at three scales. (i) Long-term variations reflect the rate of sediment supply to the channel (Gilbert 1914, 1917; Vericat et al. 2006). Increased sediment can enter the channel in prolonged, widespread inputs that produce responses such as an increase in gravel bars with a broad-scale wave-like form and move downstream over periods of tens to hundreds of years (Jacobson and Gran 1999). Long-term variations in bed-load movement can also reflect interactions between decadal- to century-long variations in sediment input and how that sediment moves through the river network. Landform impediments such as tributary alluvial fans that limit longitudinal sediment connectivity can influence downstream sediment delivery (Fryirs et al. 2007a).

Altered sediment supply from outside the channel can occur as discrete sediment inputs of large volume, such as a landslide. The resulting sediment pulse can translate downstream as a wave or disperse. Dispersal is more common and typically occurs when: the pulse volume is large relative to the dimensions of the receiving channel; the added sediment has a wide grain-size distribution and includes sediment coarser than the pre-existing streambed; the Froude number is close to 1; and selective transport and particle abrasion create downstream fining (Lisle et al. 2001; Lisle 2008). Greater downstream variations in channel width and associated pool–riffle topography also enhance pulse dispersion (Nelson et al. 2015). Natural sediment pulses have been described in field studies (Madej 2001), flume experiments (Lisle et al. 1997), and numerical modeling (Cui and Parker 2005). The human-induced analog of a natural sediment pulse is a gravel augmentation pulse, during which bed load-sized sediment is added to the channel, typically with the intent of improving spawning habitat, increasing bed mobility to facilitate flushing of interstitial fine sediment, and rebuilding pool–riffle topography to increase habitat diversity (Bunte 2004). Laboratory experiments indicate that addition of finer sediment pulses significantly increases bed-load flux and can mobilize substantial portions of the bed surface (Venditti et al. 2010). Pulse translation is more likely to occur during gravel augmentation because the added sediment is finer than the bed material and the volume of added sediment is typically not as large as a natural sediment pulse (Sklar et al. 2009).

(ii) Short-term variations in bed load reflect temporary changes in sediment supply and the movement of bedforms, such as dunes (Gomez et al. 1989). Bedforms may not span the entire channel cross-section, and time gaps can occur between passage of successive bedforms. During a single flood lasting hours to weeks, bed load can display hysteresis analogous to that of suspended sediment, with greater bed-load transport rates during either the rising or the falling limb of a flood (Trayler and Wohl 2000; Mao et al. 2014). Discontinuities in bed-load movement have also been attributed to longitudinal sediment sorting, which results in pulses of coarser or finer material moving downstream (Whiting et al. 1988). Lateral shifting of bed-load streets, which typically do not span the entire channel width, can create spatial discontinuities in bed load and apparent temporal discontinuities at a point (Ergenzinger and de Jong 2002). Sediment storage in secondary channels can fluctuate nonlinearly with changes in stage, resulting in discontinuities in supply to the main channel. Bed-load discontinuities can reflect cross-sectional to reach-scale heterogeneity of bed structure and relief that influences both sediment supply and transport. In a channel with mixed grain sizes, for example, protruding clasts can create lee storage of finer sediment until fluctuating discharge alters the local hydraulics and the sediment is entrained (Thompson 2008). Studies of very steep, coarse-grained channels indicate that both clast transport distance and the volume of bed load transported exhibit approximate power law scaling with peak stream power and the cumulative energy of individual floods (Schneider et al. 2014). Bed-load volume scales more steeply than transport distance with peak stream power and cumulative flow energy.

(iii) Instantaneous variations in bed load reflect the inherently stochastic and probabilistic nature of the processes governing entrainment (e.g. Nelson et al. 1995), as examined in Section 4.1.2. Typically, instantaneous variations increase with channel boundary roughness and increasing spread of bed grain sizes (Rickenmann 2001; Redolfi et al. 2018).

The frequency of the effective discharge for bed load decreases as the size of the bed sediment increases. In other words, coarser beds are mobile less frequently than finer beds. Rates of bed-load transport increase rapidly and nonlinearly within a channel once sand becomes mobile and can span up to seven orders of magnitude (Milhous 1973; Hoey 1992).

The migration of readily mobilized bedforms can account for the majority of bed-load movement in coarse-grained channels, as in sand-bed channels. Bedforms can also strongly influence the hydraulics that control sediment transport. In pool–riffle sequences, the pools have higher sediment transport during high flows, whereas riffles and alternate bars are sites of sediment deposition during high flows (Thompson et al. 1996). The steepened water-surface gradient over riffles and bars during low flows can promote dissection of the riffles and bars and removal of finer particles, which are then stored in pools until the next high flow. Clasts in the pools of step–pool sequences can be preferentially entrained during the rising limb of a flood and deposited during the falling limb (Schmidt and Ergenzinger 1992), although the degree to which preferential entrainment and deposition correlate with bedforms depends on the magnitude and duration of the flood relative to mean annual flood (Lamarre and Roy 2008), with preferential movement stronger during the largest floods.

Instream wood can be a major source of cross-sectional to reach-scale heterogeneity of bed structure and relief, and can exert a particularly strong influence on bed-load transport in channels with mixed grain sizes (Wohl and Scott 2017b). By creating form roughness and obstructions, wood increases flow resistance and flow separation, leading to enhanced entrainment of material from portions of the bed and banks subject to scour because of current deflected by wood. Wood appears to be most effective, however, in promoting localized sediment storage. Many studies document preferential storage of bed material near individual wood pieces or logjams and greater overall storage – particularly of finer sediment – in channel segments with abundant wood (Piégay and Gurnell 1997; Buffington and Montgomery 1999b; Montgomery et al. 1996, 2003b; Skalak and

Pizzuto 2010). Wood that breaks or is mobilized releases a pulse of bed load. Long-term removal of wood can change a depositional channel reach to a net sediment source (Brooks et al. 2003).

In addition to the physical processes acting to move bed load, aquatic organisms can displace or sort bed material and alter the susceptibility of bed sediment to entrainment and transport (e.g. Buxton et al. 2015). Various species of fish create redds or nests for their eggs or larvae. Where large numbers of fish such as salmon concentrate in a channel during the spawning season, their disruption of the bed can cause nearly half of the annual bed load yield (Hassan et al. 2008, 2011). Crayfish also disrupt bed sediments in the course of moving about and burrowing into the bed. These activities can significantly influence clast entrainment and transport during low flows (Statzner and Peltret 2006; Albertson and Daniels 2018).

The magnitude, frequency, and duration of bed-load transport also influence aquatic biota by limiting the stability of the channel substrate and physically disturbing plants and animals via abrasion, dislodging, and displacement or burial. Even if the coarse surface layer of a mixed-grain-size bed remains mostly stable, sand and finer gravel moving across the coarse surface can disturb benthic or bottom-dwelling organisms. Other factors being equal, sand-bed channels typically have lower abundance and diversity of benthic organisms than less mobile substrates, and less mobile features such as instream wood attached to the bank have a disproportionately large concentration of organisms in sand-bed channels (Minshall 1984; Benke and Wallace 2003; Pilotto et al. 2014).

4.2.3.4 Estimating Bed-Load Flux

Bed load can be examined using a Lagrangian approach, which tracks grains along a river network, or an Eulerian approach, which describes grain passage at a fixed location such as a limited segment of channel (Doyle and Ensign 2009). Using an Eulerian perspective, bed-load flux can be estimated using three methods (Haschenburger 2013).

(i) Flux can be estimated from grain kinematics – the movement of grains as described by their displacement and velocity. This approach relies on coupling grain entrainment rate from a given bed area with the associated displacement length of the mobilized grains. Entrainment rate and displacement length are then extended to explicitly account for grain-size fractions (e.g. Wilcock 1997; McEwan et al. 2004). Bed-load transport rates can also be derived from the spatial concentration of bed load traveling at a given grain velocity. Or, transport rates based on the displacement of grains over some fixed time, together with the portion of the bed that is regularly mobilized, can be used to estimate flux (Monsalve et al. 2016).

(ii) Flux can be estimated from changes in river morphology that reflect bedform migration or bed aggradation and degradation (Trayler and Wohl 2000; Haschenburger 2006; Williams et al. 2015). This approach depends on being able to directly access the bed using survey instruments, or indirectly image the bed with fathometers or other remote-sensing methods, at time intervals and spatial resolutions relevant to bed-load flux.

(iii) Bed-load transport rates can be derived from (semi)empirical equations, typically based on laboratory experiments. Starting with Shields' flume experiments in sand-bed channels (Section 4.1.2), investigators initially developed *bulk models of bed-load flux* with equations based on one representative grain size. Most of these models were variations on Shields' formula and assumed that transport is proportional to a bulk flow parameter such as shear stress or stream power. An example of a bed-load transport equation that follows this approach and remains widely used is the Meyer-Peter and Mueller (1948) equation, which can be written in dimensionless form as

$$\frac{g_b}{\left[\left(\frac{\rho_s-\rho}{\rho}\right)gD^3\right]^{\frac{1}{2}}} = 8(\theta' - \theta_c)^{\frac{3}{2}} \tag{4.19}$$

where ρ is water density, ρ_s is sediment density, g is acceleration due to gravity, D is the size of the bed material, g_b is the transport rate (volumetric transport rate per unit width, in m^2/s), θ' is the dimensionless grain stress, and θ_c is the critical dimensionless shear stress (McLean et al. 1994). In the original form of the equation, θ_c was set equal to 0.047 (Robert 2014). Actual transport rate is assumed to equal theoretical transport capacity, although rate is likely to be lower than capacity in coarse-grained channels because of limited sediment supply.

Starting in the 1980s, size-specific formulae were developed for bed-load flux. This approach assumes that different grain sizes move independently of one another, although movement or stationarity by grains of a given size can influence the movement of other grains via effects such as shielding. An example is the *two-fraction transport model* proposed by Wilcock and Kenworthy (2002). The two fractions of sand and gravel are described by the dimensionless transport rate W_i^* via

$$W_i^* = \frac{(s-1)gq_{bi}}{F_i u_*^3} \tag{4.20}$$

where s is the ratio of sediment to water density, g is gravity, q_{bi} is volumetric transport rate per unit width of size i, u_* is shear velocity [$u_* = (\tau/\rho)^{0.5}$, where τ is bed shear stress], F_i is proportion of fraction i on the bed surface, and the subscript i represents either the sand or gravel fraction.

More recent research focuses on *grain–grain interactions* during transport (Frey and Church 2012; Clark et al. 2017). Experiments with sand and gravel tracers reveal their diffusion, statistically defined as the spreading of particles downstream. Diffusion reflects the fact that individual grains move at different rates as a result of the complexity of grain–grain and grain–fluid interactions (Martin et al. 2012).

The progression of differing approaches through time reflects the application of increasingly sophisticated physics to the problem of understanding and predicting bed-load flux, particularly in channels of mixed grain size. Changes in flow and bed-load transport rate typically do not correspond well in space or time in these channels, for the many reasons already outlined, making prediction of bed-load transport extremely difficult. Table 4.2 lists additional, commonly used bed-load transport equations.

Whatever their form, most bed-load transport equations do not accurately predict fluxes in channels with coarse or mixed grain sizes. The discrepancy reflects at least three factors (Haschenburger 2013). First, the equations require a specific channel response based on defined hydraulic conditions. In other words, they require equilibrium transport conditions, which do not always occur in coarse-grained channels. Second, the equations do not integrate the spatial and temporal complexities that influence entrainment and transport, such as spatial and temporal heterogeneities in bed roughness, which can be substantial in coarse-grained channels. Finally, the equations rely on input data that are typically difficult to characterize, such as spatial heterogeneity of the bed. Because of these shortcomings, predicting bed-load transport in coarse-grained channels remains an area of active research.

Table 4.2 Examples of widely used bed-load transport equations.

Equation	Characteristics of dataset and equation	References
$q_v = \sqrt{g(s-1)D^3}\left[\sqrt{\frac{2}{3}+\frac{36\nu^2}{g(s-1)D^3}}-\sqrt{\frac{36\nu^2}{g(s-1)D^3}}\right]$	Flume, sand, and gravels	Brown (1950) and Einstein (1950)
$q_v = \frac{2.5}{\rho_s/\rho}S^{3/2}(q-q_c)\ q_c = 0.26(s-1)^{5/3}\frac{D_{40}^{3/2}}{S^{7/6}}$	Flume and field data, $S<1\%$ (D_{40} refers to the subsurface)	Schoklitsch (1962)
$q_v = \frac{0.1}{f}\sqrt{g(s-1)D_{50}^3}\theta^{5/2}$	Total (bed + suspended) transport	Engelund and Hansen (1967)
$q_{st} = 0.025\frac{qD_{35}}{d}\left[\frac{F_{gr}}{0.17}-1\right]^{1.15}$	Total transport, flume, $\mathrm{Fr}<0.8$, $0.4<D<14$ mm	Ackers and White (1973)
$F_{gr} = \frac{1}{\sqrt{g(s-1)D_{35}}}\left[\frac{u}{\sqrt{32}\log\left(\frac{10d}{D_{35}}\right)}\right]$		
$q_v = 11.2\sqrt{g(s-1)D_{50}^3}\frac{(\theta-0.03)^{4.5}}{\theta^3}$	Flume and field	Parker (1979)
$q_v = 4\sqrt{g(s-1)D_{50}^3}(D_{90}/D_{30})^{0.2}S^{0.6}\frac{U}{u*}\theta^{0.5}(\theta-\theta_c)$	Flume, $3<S<20\%$, $2<D<10.5$ mm	Smart and Jaeggi (1983)
$q_v = 0.053\sqrt{g(s-1)D_{50}^3}\frac{T^{2.1}}{D_*^{0.3}}$	Semiempirical, sand	Van Rijn (1984)
$q_v = 1.5(q-q_c)S^{1.5}$ *for* $0.0004<S<0.2$ $q_v = \frac{12.6}{(s-1)^{1.6}}\left(\frac{D_{90}}{D_{30}}\right)^{0.2}(q-q_c)S^2$ *for* $0.03<S<0.2$	Flume, $0.0004<S<0.2$, $0.4<D<10$ mm	Rickenmann (1991)
$W_i^* = \frac{(s-1)gq_{vi}}{f_i u_*^3} = \begin{cases} 0.002\phi^{7.5} & \text{for } \phi<1.35 \\ 14\left(1-\frac{0.894}{\phi^{0.5}}\right)^{4.5} & \text{for } \phi\geq 1.35 \end{cases}$		Wilcock and Crowe (2003)
$q_v = 3.97\sqrt{g(s-1)D_m^3}(\theta-0.0495)^{1.5}$	Flume, reanalysis of Meyer-Peter and Mueller (1948)	Wong and Parker (2006)
$q_v = \frac{14\sqrt{g(s-1)D_{84}^3}\theta^{2.5}}{[1+(\theta_m/\theta)^4]}$		Recking (2010)

For all equations, d is flow depth (m), D_x is sediment diameter for which x% of the distribution is finer (m), f is flow resistance coefficient, q_s is bed-load transport rate per unit width (kg/s/m), q_v is volumetric solid discharge per unit width (m^3/s/m), s is relative density ($s = \rho_s/\rho$), S is energy slope (m/m), U is vertically averaged flow velocity (m/s), $u*$ is shear velocity (m/s) ($u* = \tau/\rho$), θ is Shields dimensionless parameter, ρ is flow density (kg/m^3), ρ_s is sediment density (kg/m^3), and W is channel width (m)

4.2.3.5 Field Measurements of Bed Load

Bed load is typically difficult to measure because of temporal and spatial heterogeneities in bed-load transport and because direct measuring devices can perturb the flow field and alter bed-load transport. Direct measurements involve either pit traps that capture everything moving across a portion of the bed or sampling devices placed on the bed. Indirect measurements of bed-load transport use tracer clasts, repeated channel surveys, or various types of sensors on the bed.

Pit traps involve some type of sampler embedded in the streambed, typically in channels less than 15 m wide. Designs, which have changed through time, include a concrete trench (Leopold and Emmett 1976), slot samplers that use a water-filled pressure pillow with pressure transmitters to measure filling rates (Reid et al. 1980; Laronne et al. 2003), and a vortex-tube in which slots perpendicular to the flow trap sediment and vortices in the slot propel the sediment to a weighing device at the river edge (Klingeman and Milhous 1970). Comparisons of pit traps and other types of samplers suggest the former have nearly 100% efficiency for material larger than 2.8 mm (Sterling and Church 2002).

A sediment trap can also be built in the form of a storage area excavated at the downstream end of a study reach, which can be monitored for volume of sediment fill following a single flood or an entire flow season. Examples include the 5 km^2 Rio Cordon catchment in Italy (Lenzi et al. 1990) (Figure 4.11), the 0.7 km^2 Erlenbach in Switzerland, the 284 ha Arnás catchment in the Spanish Pyrenees (Lana-Renault et al. 2006), the 8 km^2 East St. Louis Creek catchment in Colorado, USA, and the 0.6 km^2 Nahal Yael catchment in Israel (Schick et al. 1987). As with pit traps, these sediment traps are restricted to relatively small catchments.

The most commonly employed sampling device that is placed directly on the bed is the *Helley–Smith sampler*, which can be hand-held on a rod or lowered from a boat or bridge. Helley–Smith samplers typically have either 76- or 152-mm intakes. Using a sampler with an intake size at least five times larger than bed grain size increases the accuracy of sampling (Vericat et al. 2006). Sampler efficiency drops off after the sampler is 40% full (Emmett 1980), so the duration of sampling times must balance instantaneous transport variations and progressive filling. As a result of the spatial and temporal heterogeneities in bed-load movement, sampling errors decrease with increasing number of samples collected and with increasing number of traverses of the channel over which samples are collected (Gomez and Troutman 1997). For smaller rivers, an alternative to the Helley–Smith sampler is a *portable bed-load trap* with a 0.3×0.2 m opening that is positioned on a ground plate anchored in the streambed to minimize disturbance during sampling and facilitate longer sampling times in channels with minimal fine sediment (Bunte et al. 2004, 2007).

A subset of marked *tracer clasts* that are assumed to represent some proportion of total movement can also be used to estimate bed-load transport (Milan et al. 2002; Hassan and Ergenzinger 2003; Hassan and Roy 2016). Markers include miniature radios (Bradley and Tucker 2012; Olinde and Johnson 2015), naturally occurring magnetic minerals (Carling et al. 2006), artificially emplaced magnets (Schmidt and Gintz 1995; Ferguson et al. 2017a), paint (Goode and Wohl 2010a; Erwin et al. 2011), and radioactive injections. Recovery rates are typically in the range of 33–69%, so tens to hundreds of tracers are generally used.

Bed sensors can be used to estimate bed-load transport by relating some type of signal to the rate of clast movement. Examples include hydrophones or geophones installed in the bed that measure acoustic signals caused by the impacts of bed-load grains transported over the measuring section

Figure 4.11 Upstream view of the discharge and sediment measuring station on the Rio Cordon in the Dolomite Mountains of Italy. Drainage area above station is 5 km^2. The measuring station includes an inlet flume; an inclined grid for the separation of coarse particles (> 20 mm diameter), highlighted in this photograph by the irregularly edged ice present along the left side of the grid; a storage area for coarse sediment deposition downstream from the grid; and an outlet flume to return water and fine sediment to the stream, at the far left of the photograph. Bed-load volume is measured at 5-minute intervals by 24 ultrasonic sensors fitted on a fixed frame over the storage area and via topographical surveys after each flood. Suspended sediment is measured by turbidimeters. Flow samples are gathered automatically using a pumping sampler installed at a fixed position in the inlet channel. Measurement details from Mao and Lenzi (2007). Source: After Wohl (2010b), Figure 3.38. (*See color plate section for color representation of this figure*).

(Rickenmann 1994, 2017). Impact sensors measure the impacts associated with grain movement over a small portion of the bed (Richardson et al. 2003). Pressure pillows can be used to calculate the thickness of overlying sediment (Kurashige 1999). Magnetic induction systems use a sensor that produces a signal when passage of a magnetically sensitive mineral through the potential field of the sensor generates an electromotive force that is recorded as a potential excursion (Hassan et al. 2009). Seismic signals can also be used to infer bed-load transport (e.g. Hsu et al. 2011; Roth et al. 2017; Gimbert et al. 2019). As the latter approach is refined, it may overcome some of the upper limits on channel width imposed by the use of sediment traps, samplers placed on the bed, and bed sensors.

4.3 Bedforms

Bedforms are bed undulations that result from sediment transport. Bedforms in sand-bed channels are smaller than channel-scale bar forms. Sand-bedforms are sometimes distinguished as:

- *microforms* such as ripples whose occurrence and geometry are controlled by boundary-layer characteristics including bed grain size;
- *mesoforms* such as dunes that are controlled by boundary layer thickness or flow depth; and
- *macroforms* such as bars that have lengths of the same order as channel width and heights comparable to the flow producing the form (Bridge 2003).

Bridge (2003) also distinguishes between a bedform – a single geometrical element such as a ripple – and a *bed configuration* – the assemblage of bedforms of a given type that occurs in a particular bed area at a given time. "Bedform" will be used here in a more general sense that includes both individual bedforms and the bed configuration.

Explaining bedforms involves considering both sediment transport dynamics and channel geometry. Migration of bedforms creates bed-load flux and alters the conditions for mobility of sediment not within a bedform. In the presence of bedforms, a significant proportion of the total stress is caused by form drag on the bedforms, and this component is not effective in moving sediment. Bedforms therefore reduce bed-load transport rates relative to a flow with the same mean velocity.

The presence and characteristics of bedforms are interconnected with channel geometry to the extent that the occurrence of certain types of bedforms reflects channel-geometry variables such as flow width and depth or bed gradient, but also influences these variables. Step–pool sequences, for example, form within a restricted range of bed gradient. Steps and pools also alter flow-path length by creating a stepped rather than a uniform longitudinal profile.

As noted in the introduction, the subdivisions within this section recognize an important difference in bed mobility. At one extreme are sand-bed channels in which readily mobile bed substrate produces relatively high bed-load fluxes and quickly changing bedforms. At the other are cohesive-bed channels in which the bed deforms only by erosion. Between these end-members are coarse-grained channels in which a larger threshold of flow energy (relative to sand-bed channels) must be exceeded before the bed material begins to move and bedforms are created or mobilized.

4.3.1 Readily Mobile Bedforms

Readily mobile bedforms are primarily those in sand-bed channels, which occur in the broad classes of small-scale ripples and larger-scale dunes, both of which have their long axes perpendicular to downstream flow. Other types of bedforms, which develop under specific hydraulic conditions and are rarer, include upper-stage plane beds and antidunes (see Figure 3.12).

Understanding of bedforms in sand-bed channels is particularly important, for at least three reasons (Venditti 2013). First, bedforms are the most important source of flow resistance at the local scale, creating two to three times the resistance caused by grains. Second, bedform initiation, growth, and migration dominate sediment dynamics in sand-bed channels, both by serving as a basic transport process and by affecting the near-bed flow field and distribution of hydraulic force and turbulence. Finally, bedforms leave characteristic depositional records, the interpretation of which is fundamental to understanding past river environments.

A typical sequence of bedforms evolves as velocity increases (Figure 3.12) (Simons and Richardson 1966). At very low velocities, sand grains on a bed will not move. As velocity increases, the lift and drag forces exceed the submerged grain weight and other resisting forces. The bed begins to develop millimeter-scale grooves parallel to the flow and spaced at the scale of the high- and low-speed streaks that are ubiquitous in shear flow. This is a *lower-regime plane bed*, which can persist for long periods of time or develop into bedforms as a result of either defects or instantaneous initiation (Venditti et al. 2005). Defects such as a small depression or protrusion into the flow create flow separation and local shear stresses capable of moving sediment while the average shear stress remains below the threshold of motion. At higher shear stress and with general sediment transport, bedforms appear over the entire bed nearly instantaneously. This presumably reflects instability in the form of turbulence and velocity fluctuations at the water–sediment interface that rapidly increase the extent of entrainment and grain–grain interactions, allowing grains to start collecting into bedforms.

The stages in Figure 3.12 up through dunes are *lower-flow-regime bedforms*. These typically exist in subcritical flow and have relatively small bed-load transport rates. The size of lower-regime bedforms varies widely between diverse rivers, from a few centimeters to several meters in height and a few tens of centimeters to as much as a kilometer in length. Bedform height, H, and length, L, are empirically related (Flemming 1988) as

$$H = 0.677L^{0.8098} \tag{4.21}$$

The existence of two subpopulations in a plot of H versus L is used as evidence that ripples and dunes are distinct forms (Ashley 1990). Ripples do not affect the water surface and form only in sediment where $D < \sim 0.7$ mm. When dunes form, waves develop on the water surface that are out of phase with the dune troughs and crests.

Dunes form in sediment from fine sand to gravel size. The ability to form dunes thus reflects flow magnitude, rather than grain size, as evidenced by dunes formed of boulder- to gravel-sized particles during late-Pleistocene outburst floods (Baker 1978). Dunes are typically less steep than ripples, although slope angles for the two types of bedforms overlap.

Lower-regime bedforms can be low-angle symmetric (downstream slope ~10°) or angle-of-repose asymmetric (average upstream slope ~2–6° and downstream slope ~30°). Symmetric forms are common in large rivers and estuaries, whereas asymmetric forms seem to be more common in small channels, although it remains unclear how mechanics of formation differ between the two types (Venditti 2013)

The planimetric morphology of lower-regime bedforms can be distinguished as being two- or three-dimensional. Two-dimensional bedforms have fairly regular spacing, height, and length, and straight or slightly sinuous crestlines oriented perpendicular to the mean flow lines. Three-dimensional bedforms have irregular spacing, height, and length, with highly sinuous or discontinuous crestlines (Figure 4.12) (Venditti 2013). As with symmetry, differences in the mechanics of formation of two- versus three-dimensional bedforms remain under discussion, although three-dimensional forms may reflect persistent, unidirectional flows, whereas two-dimensional bedforms reflect variable flows. Lower-regime bedforms can also be superimposed on one another, as when ripples and dunes are superimposed on bars or ripples are superimposed on dunes (Venditti 2013).

As the Froude number approaches 1, dunes are increasingly likely to wash out to a plane bed, which is the start of the *upper-flow regime bedforms* (plane beds can be present at Froude numbers well below 1). These bedforms exist in supercritical flow and have relatively large bed-load fluxes and small flow resistance. Upper-stage plane beds have low-relief bed waves only a few millimeters in

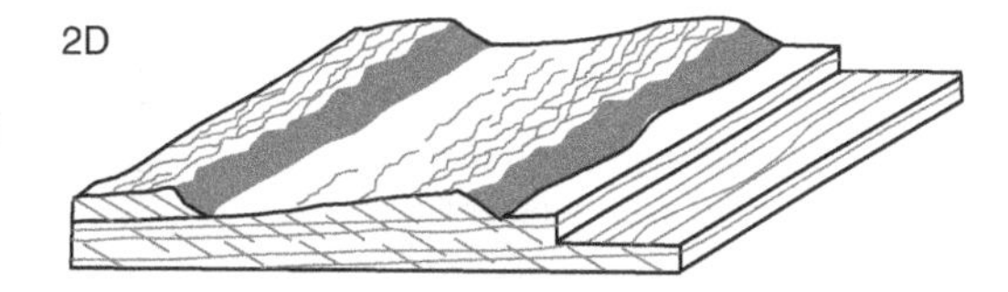

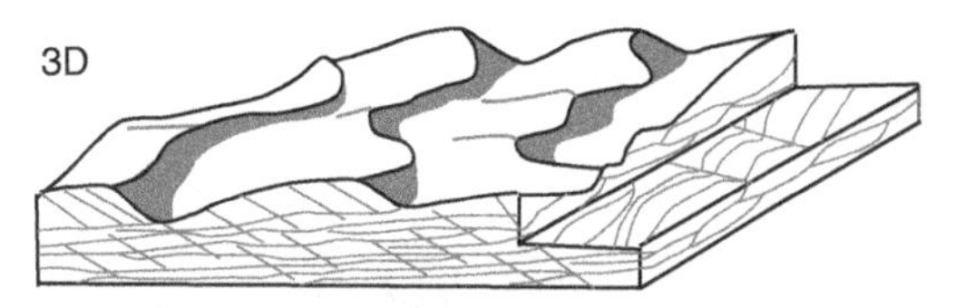

Figure 4.12 Schematic illustration of the planimetric morphology of two-dimensional (2D) sand bedforms with planar cross-stratification and three-dimensional (3D) sand bedforms with trough cross-stratification. Source: After Venditti (2013), Figure 7.

height. These bed waves migrate downstream beneath a water surface with waves that are out of phase with the bed waves. At the high-velocity end of upper-flow-regime bedforms, antidunes occur in all grain sizes and the water-surface waves are in phase with the bed-sediment waves. Antidunes are upstream-moving undulations that typically grow to an unstable steepness, break, and then reform (Southard 1991). Antidunes can form from dunes or upper-stage plane beds as velocity increases (Venditti 2013).

Bedforms can also be described using phase diagrams. These diagrams illustrate the distribution of bedform types in relation to various hydraulic and sediment variables, including fall velocity, stream power, Froude number, and dimensionless excess shear stress (Venditti 2013). Examining bedforms in the context of a phase diagram emphasizes the relation of individual types of bedforms to a continuum defined by hydraulic and sedimentary variables (Southard and Boguchwal 1990; Southard 1991).

Most examinations of flow over sand bedforms have focused on asymmetric, two-dimensional dunes. Flow accelerates and converges over the upstream side of these dunes, then separates at the dune crest. The zone of flow separation is associated with a turbulent wake and shear layer originating at the dune crest and extending and expanding downstream. Flow reattaches at a distance downstream about four to six times the bedform height. An internal boundary layer grows from the reattachment point downstream beneath the wake towards the crest. An outer, overlying wake region is present above the level of flow separation and reattachment (Figure 4.13a) (Venditti 2013).

Flow over symmetric dunes includes a well-defined region of decelerated flow in the lee of the bedform, a shear layer between the main flow and the decelerated flow, and turbulent eddies generated along this shear layer that dominate the macroturbulent flow structure (Figure 4.13b) (Venditti 2013). Symmetric dunes have less intense mixing along the shear layer in the dune lee and only intermittent flow separation.

All of the forms of coherent flow structure and turbulence over asymmetric and symmetric dunes effectively increase the resistance of a channel with such bedforms and influence sediment entrainment and mobility. Symmetric dunes have lower flow resistance, however, than asymmetric dunes.

The structure of turbulence over ripples and dunes controls the flow resistance. The flow structure over dunes is dominated by depth-scale turbulent eddies, known as *macroturbulence* or *kolks* when they are in the water column and *boils* when they disturb the water surface (Venditti 2013). Kolks and boils likely originate from the combined effects of boundary-layer bursts, instabilities along the shear layer, and vortex shedding.

Individual bedforms continue to change their geometry under constant flow conditions, as well as under changing flow. Bedform changes typically lag changes in flow and thus display *bedform*

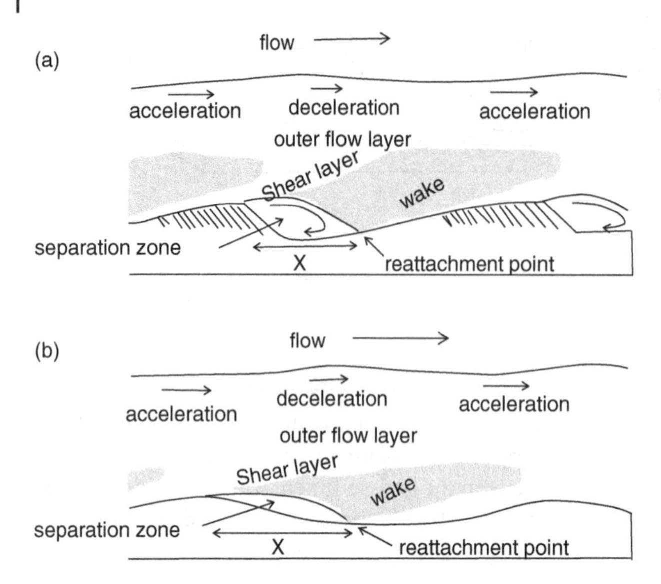

Figure 4.13 Hydraulics over sand bedforms. (a) Two-dimensional asymmetric dunes at the angle of repose. (b) Low-angle asymmetric dunes. *X* is the reattachment length. Source: After Venditti (2013), Figure 9a and b.

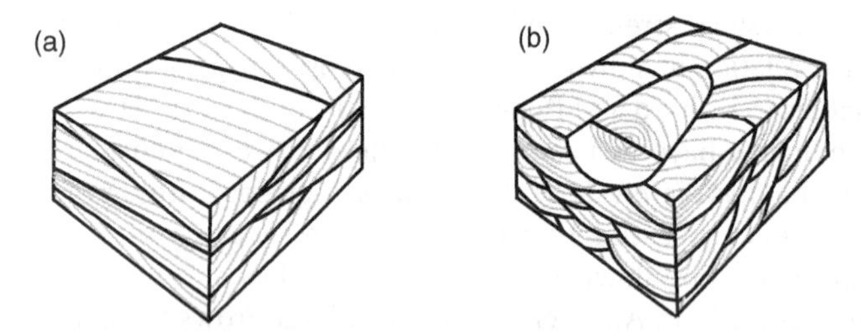

Figure 4.14 Block diagram showing (a) planar cross-bedding and (b) trough cross-bedding, as seen in horizontal, transverse, and longitudinal sections. Units in planar cross-beds are tabular to wedge-shaped and bedding surfaces are planar. Units in trough cross-beds are festoon-shaped and bedding surfaces are curved. Source: After Reineck and Singh (1980), Figures 154 (A) and 155 (B).

hysteresis that can be minutes to hours for ripples and as much as days to months for dunes (Martin and Jerolmack 2013; Venditti 2013).

In addition to changing their geometry, bedforms migrate with time, thus contributing to sediment transport. Asymmetric bedforms migrate by erosion on the upstream side and deposition on the downstream side via avalanching down a steep slipface. Over symmetric dunes, migration can also involve large amounts of sediment entrained into suspension on the upstream side and deposited from suspension on the downstream side.

Migration creates inclined packets of sediment that can be preserved as cross-strata in the sedimentary or rock record. Ripples typically create cross-strata millimeters thick, and dunes can create cross-strata millimeters to centimeters thick. Planar cross-strata formed by two-dimensional bedforms (Figure 4.14) are found parallel to the bedform slipface. These beds form over a planar erosional surface scoured by a laterally continuous recirculation cell as the bedform trough passes downstream. Trough cross-strata formed by three-dimensional bedforms (Figure 4.14) accumulate over a curved erosional surface resulting from flow separation in the lee of a saddle-shaped bedform crest (Bridge 2003).

4.3.2 Infrequently Mobile Bedforms

Infrequently mobile bedforms are primarily those occurring in gravel-bed channels. The cobbles and coarser sediment in these channels have a higher entrainment threshold that may be exceeded much less than half of the time during an average flow year, although minor adjustments associated with individual grain movements or the mobility of the finer fraction of the bed sediment can occur over a wide range of flows. Lower-regime plane beds and dunes, as well as upper-stage plane beds and antidunes, can occur in gravel-bed channels (Bridge 2003).

As with the more readily mobilized sand-bed bedforms, infrequently mobilized bedforms in coarser-grained channels both respond to and strongly influence flow resistance, the distribution of velocity, turbulence, and other hydraulic variables, and sediment transport. The interactions among bedforms, flow, and sediment in transport within gravel-bed channels typically occur over larger spatial scales and longer time scales than analogous interactions in sand-bed channels, but exhibit similar levels of complexity. The primary bedforms treated here are particle clusters, transverse ribs, steep alluvial channel bedforms, pool–riffle channels, step–pool channels, and bars.

4.3.2.1 Particle Clusters

Particle clusters are closely nested groups of clasts aligned parallel to flow, typically 0.1–0.2 m in length in the downstream direction and about twice as long as they are wide (Brayshaw 1984). These features are also known as pebble clusters, imbricate clusters, cluster bedforms, and microforms (Venditti et al. 2017). Particle clusters can take a variety of shapes, including comet-shaped, triangular, rhomboid, and diamond. Typically, an obstacle clast anchors an upstream-side accumulation of imbricated particles with a tail on the downstream side (Papanicolaou et al. 2003).

Particle clusters are most common in gravel-bed channels with low rates of bed-load transport and a stable coarse surface layer, and usually occur in riffle, alternate bar, or plane-bed sections of the channel (Hendrick et al. 2010). Although more than one mechanism of formation may exist, clusters appear to form during the recessional limb of a flood as bed load is deposited around large, protruding clasts (Brayshaw 1984). Clusters create abrupt roughness transitions along the bed and increase average boundary roughness (Hassan and Reid 1990). This leads to localized flow separation and turbulence (Lacey et al. 2007). By creating relatively stable points on the streambed, clusters also delay incipient motion and limit the availability of bed material for transport (Brayshaw 1984).

4.3.2.2 Transverse Ribs

Transverse ribs, like particle clusters, are microtopographic features that can be difficult to see when looking at a streambed. Ribs are formed by a series of regularly spaced pebble, cobble, or boulder ridges perpendicular to the flow (Dell'Agnese et al. 2015). The spacing between ribs is proportional to the size of the largest clasts in the rib crest (Robert 2014). Spacing is typically on the order of decimeters to meters, and heights are one to two clast diameters (Bridge 2003). Ribs appear to be most common on bars in gravel-bed rivers.

Transverse ribs likely represent the crests of antidunes (Koster 1978). Water-surface waves commonly break upstream of antidune crests, creating temporary hydraulic jumps that can promote coarse sediment deposition. As with other bedforms, the presence of transverse ribs influences boundary resistance and sediment entrainment and transport.

4.3.2.3 Steep Alluvial Channel Bedforms

The classification of reach-scale channel morphology proposed by Montgomery and Buffington (1997) for steep, relatively laterally confined gravel-bed channels is largely based on dominant bedforms in these channels (Figure 4.15). The channels with the highest bed gradients are typically *cascade channels* with disorganized, very coarse bed material and small pools that partially span the channel and are spaced less than a channel width apart. In these channel segments, flow may not be competent to organize the very coarse-grained bed material into bedforms.

At gradients of ~0.03–0.10 m/m, *step–pool channels* become particularly common. Although substrate mobility typically remains limited in these channel segments, flow is competent to create regularly spaced bedforms and bed variability in elevation and grain size, primarily in the downstream direction.

Plane-bed channels lack well-defined, rhythmically occurring bedforms and are most common at gradients of 0.01–0.03 m/m. The relatively planar bed formed in coarse clasts or bedrock can persist through conditions of bed stability and mobility. Despite the similar name, coarse-grained plane-bed channels are much less mobile than plane-bed sand channels.

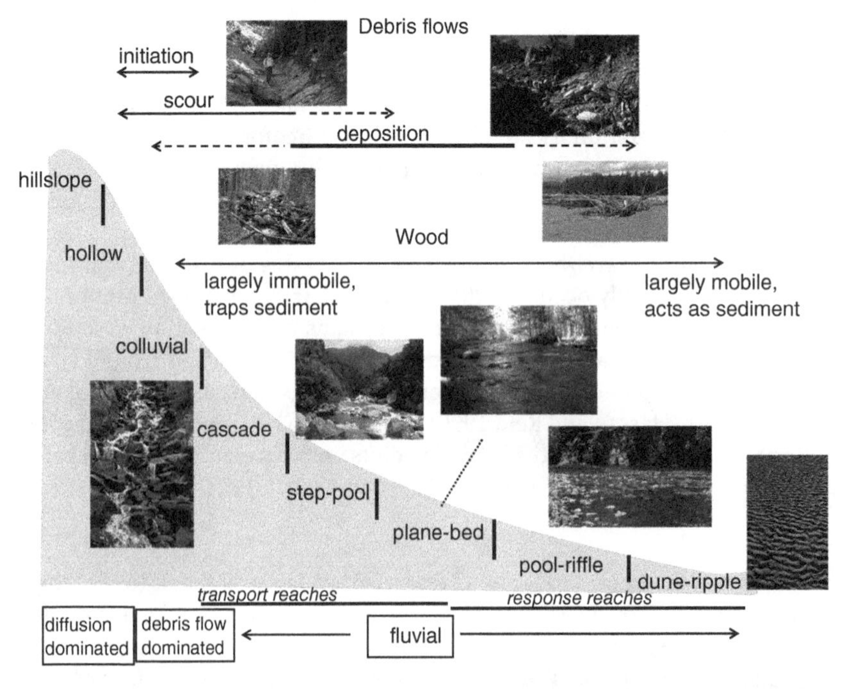

Figure 4.15 The Montgomery and Buffington (1997) channel classification based on dominant bedforms. This schematic illustration of downstream trends includes changes in reach-scale gradient (light gray shading), channel type (cascade/step–pool/plane–bed/pool–riffle/dune–ripple), indication of transport versus response reaches, dominant sediment transport process (hillslope diffusion/debris flow/fluvial), and the behavior of debris flows and instream wood. Source: After Montgomery and Buffington (1997), Figure 4. (*See color plate section for color representation of this figure*).

Alternating pools and riffles characterize channels with gradients less than 0.015 m/m, which exhibit consistent patterns of bed variability both downstream and across the channel. *Pool–riffle sequences* transition at lower gradients to *dune–ripple channels* with sand beds and readily mobile bedforms.

The progression with decreasing reach-scale gradient from cascade to dune–ripple channel corresponds to decreasing average bed grain size and increasing bed mobility. This represents a general downstream trend that can be interrupted or reversed. Mountain rivers are commonly longitudinally segmented and alternate repeatedly downstream between high- and low-gradient reaches (Wohl 2010b), so that cascade segments can be downstream from pool–riffle segments and vice versa. Comparison of portions of mountainous catchments with and without Pleistocene glaciation also suggests that the characteristic channel gradients and other potential control variables at which distinct bedform types occur can vary between glaciated and unglaciated portions of a river network (Livers and Wohl 2015). Step–pool and pool–riffle channels are particularly common and well-studied channel types within the progression of steep alluvial channel bedforms.

4.3.2.4 Step–Pool Channels

The dominant bedforms in step–pool channel segments are channel-spanning steps of clast, wood, or bedrock that alternate downstream with a plunge pool at the base of each step (Figure 4.16) (Chin and Wohl 2005; Zimmermann 2013). Step–pool sequences are most common where relatively immobile large clasts or wood trap sediment in a wedge tapering upstream, with flow plunging over the immobile obstacle to scour a plunge pool in the bed downstream. Although steps likely form via more than one mechanism, those in mobile sediments (rather than bedrock) appear to result from the interlocking of large, *keystone* clasts (Zimmermann and Church 2001) under conditions of limited sediment supply, leading to a *jammed state* (Zimmermann et al. 2010).

Individual steps can be destroyed by erosion of the keystones and subsequent mobilization of the step (Lenzi 2001, 2004). Step–pool sequences can give way to braided or plane-bed channels during extremely large discharges that mobilize the entire bed and greatly enhance the sediment supply. Such mobilization occurs anywhere from annually in some channels to every 50 years or more in others

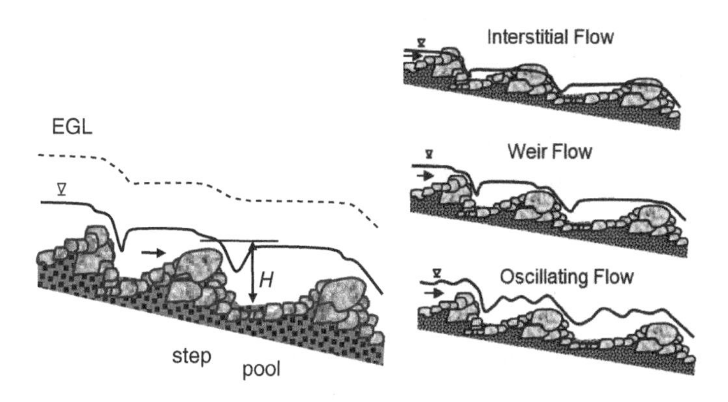

Figure 4.16 Longitudinal view of the components and hydraulics associated with step–pool bedforms. *H* is bedform amplitude: here, the step height. EGL is energy grade line. Interstitial, weir, and oscillating flow represent flow regimes under progressively increasing discharge. Flow is from left to right. Source: Courtesy of David Dust.

(Wohl 2010b). As the mobilizing flood recedes and sediment supply once again declines, steps and pools gradually re-form during subsequent, smaller flows. These smaller flows can mobilize the finer sediments exposed at the surface, but not necessarily the keystone clasts.

Flume experiments provide insights into the mechanisms of step destabilization and reformation. As fine-grained bed load is added to a channel with stable steps and pools, bed roughness decreases and step-forming clasts become more mobile (Hohermuth and Weitbrecht 2018). Bed-load transport significantly decreases flow resistance and increases near-bed velocity, facilitating step mobilization at discharges up to 30% lower than under clear-water conditions (Hohermuth and Weitbrecht 2018). Flume experiments also indicate that interactions between surface roughness and grain size promote clustering of coarse grains during bed stabilization, which then strongly influence flow hydraulics (Johnson 2017).

Step–pool channels have little hyporheic exchange flow relative to pool–riffle channels, with exchanges mainly limited to a shallow depth. Flume experiments and numerical modeling suggest that downwelling fluxes occur at the lip of steps and are deepest during moderate discharges, with steeper slopes causing deeper surface–subsurface exchanges (Hassan et al. 2015). Each step causes downwelling of surface waters starting at a distance upstream that is approximately 1.5 times the step height, with upwelling of hyporheic waters at the step base.

Steps and pools create extremely large values of flow resistance relative to lower-gradient channels (Curran and Wohl 2003). Step geometry is typically defined by height, H, and downstream spacing, L, and many step–pool sequences have a ratio of $H/L/S$ (S is bed gradient) between 1 and 2. Bedforms with this ratio have been interpreted to maximize flow resistance and channel stability (Abrahams et al. 1995). Wood incorporated into clast steps can increase step height, backwater effects, and dissipation of flow energy by making the steps taller and trapping interstitial fine sediment so that the backwaters better retain water (Wilcox et al. 2011). The large grain and form resistance of step–pool channels also creates high turbulence intensities, particularly in pools and at high discharges (Wilcox and Wohl 2007).

Lower discharges in step–pool sequences can create interstitial flow, with water primarily passing through gaps between the step-forming clasts. With increasing discharge, free falls over steps create nappe flow or weir flow, with a hydraulic jump below each step. At the highest discharges, the hydraulic jump disappears and an oscillating or skimming flow regime develops, in which the water flows as a coherent stream and recirculating vortices occur at the base of each step (Figure 4.16) (Chanson 1996; Dust and Wohl 2012a). Flow resistance decreases significantly when flow transitions from nappe to skimming conditions (Comiti et al. 2009a,b; D'Agostino and Michelini 2015).

Values of resistance coefficients such as Manning's n are very difficult to estimate in step–pool channels (Yochum et al. 2014b), which have nonuniform flow. An alternative approach to estimating discharge is to apply the equation developed for broad-crested weirs, under the assumption that a step approximates such a weir

$$Q = C^* g^{0.5} W h^{3/2} \tag{4.22}$$

where C^* is a dimensionless discharge coefficient, W is the crest width, g is the acceleration of gravity, and h is the upstream flow depth above the step crest (Dust and Wohl 2012a). Several studies also suggest that the standard deviation of bed elevation, which captures both grain and bedform roughness, effectively predicts flow resistance (David et al. 2010; Yochum et al. 2014a; D'Agostino and Michelini 2015).

Bed-load transport is extremely spatially and temporally variable and very difficult to quantify or predict in step–pool channels. Yager et al. (2012) propose a modified version of the Parker (1990) bed-load equation that includes the resistance associated with steps and selective transport of relatively mobile sediment using a range of hiding functions. Particles in pools are preferentially entrained and transported for longer distances (Schmidt and Ergenzinger 1992). Particles finer than or equal to D_{40} of the bed surface in step–pool channels do not exhibit substantial differences in travel distance during floods, whereas particles larger than the bed surface D_{84} have very limited mobility (Lenzi 2004).

Step–pool channels are known as *transport reaches* (Montgomery and Buffington 1997) because they are relatively insensitive to changes in water and sediment supply. Field studies support the idea that, when water and sediment supplied to a river network change, step–pool bedforms are less likely to alter their dimensions than are bedforms present at lower channel gradients (Ryan 1997; Wohl and Dust 2012). The resistance to change likely reflects the combined effects of high values of boundary irregularity and flow resistance (which effectively dissipate flow energy) and very coarse, relatively interlocked clasts in steps.

4.3.2.5 Pool–Riffle Channels

Pool–riffle bedform sequences are analogous to meandering in the vertical dimension (Keller and Melhorn 1978) in that regularly spaced deeps with typically finer sediment (pools) and coarser-grained shallows (riffles) create an undulating longitudinal profile at the reach scale (Figure 4.17). Riffles are usually wider and shallower at all stages of flow than are pools.

Pool and riffle bedforms occur in alluvium and in bedrock, and are likely formed and maintained by diverse mechanisms. The amplitude and wavelength of these bedforms are typically described in terms of the average downstream spacing of pools, which has been related to channel width (Keller and Melhorn 1978). Pools are commonly described as being spaced at around five to seven times the average channel width. Despite this commonly cited rule, the relation between pool spacing and channel width is highly variable. Pool spacing varies with substrate, such that pools tend to be more widely spaced in more resistant substrate such as bedrock (Roy and Abrahams 1980; Wohl and Legleiter 2003). Pool spacing also varies with valley width and associated flow convergence (Nelson et al. 2015). Riffles tend to persist in locally wide areas of a valley, and pools are commonly associated with long valley constrictions, despite variations such as channel incision and planform change in response to frequent floods (White et al. 2010). The details of how differing flow magnitudes interact with topographic heterogeneities occurring at various spatial scales, and how these interactions influence channel morphology (Pasternack et al. 2018a,b), likely partly depend on channel boundary resistance. As boundary resistance increases, larger-scale variations in valley geometry that influence the distribution of flow energy during high-magnitude floods are likely to become progressively more

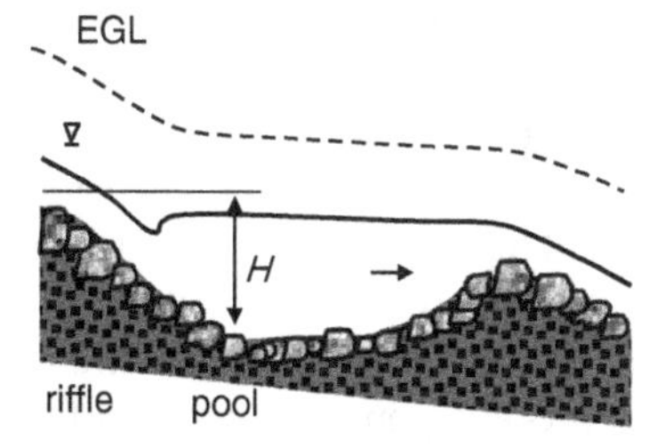

Figure 4.17 Longitudinal view of the components and water surface associated with pool–riffle bedforms. *H* is bedform amplitude: here, the elevation difference between the riffle crest and the next pool thalweg. EGL is energy grade line, shown here for relatively low flow conditions. During high flow, the EGL over riffles flattens, whereas that over pools grows steeper. Flow is from left to right. Source: Courtesy of David Dust.

important in creating a persistent template of channel morphology, such as the spacing of pools and riffles.

Pool spacing also varies with the longitudinal spacing of obstructions such as wood (Montgomery et al. 1995; Buffington et al. 2002; Thompson and Fixler 2017) and large, immobile boulders (Lisle 1986; Thompson 2001). Obstructions pond water upstream and create flow convergence, higher water-surface slopes, and higher flow velocity through the constricted zone. This creates localized scour, which produces pools. As the flow passes downstream beyond the obstruction, flow divergence and deceleration along the downstream end of the pool center facilitate sediment deposition, leading to formation of a riffle (Lisle 1986; Thompson et al. 1998; Thompson 2013). Pool–riffle sequences associated with local scour and deposition around an obstruction are sometimes referred to as *forced pool–riffle sequences* (Montgomery et al. 1995).

Pools and riffles formed in relatively well-sorted alluvium with few obstructions are dominated by feedbacks between hydraulics and sediment transport, and are more likely to exhibit pool spacing at five to seven times channel width. Pools and riffles formed in poorly sorted, coarse-grained alluvium with abundant obstructions are also strongly influenced by hydraulic–sediment interactions, but local variations in channel-boundary configuration exert a much stronger influence on bedform characteristics.

The locations of freely formed (rather than forced) pools and riffles have been explained in terms of at least two mechanisms of hydraulics and bed-load movement. In some channels, kinematic waves of bed load appear to create bars and riffles where the wave stops moving during the falling limb or where transport capacity locally declines (Langbein and Leopold 1968). Accelerating flow converging downstream from the bar then scours a pool. Another scenario involves persistent point sources of coarse sediment, such as tributary junctions, that create riffles (Webb et al. 1989). The riffles create accelerated flow that leads to pool scour downstream (Dolan et al. 1978). In each of these explanations, creation of an initial bed irregularity in the form of a riffle perturbs hydraulics and bed-load transport sufficiently to initiate downstream bed undulations that lead to successive pools and riffles (e.g. Clifford 1993). Average longitudinal pool spacing likely reflects a minimum distance related to the backwater and turbulent conditions needed for pool formation (Thompson 2012).

The *velocity-reversal hypothesis* has been used to explain the maintenance of pools and riffles (Gilbert 1914; Keller 1971). Field observations indicate that the near-bed velocity increases more rapidly with discharge in a pool than in a riffle, so that flow in pools is more competent at high stage than is flow over riffles. This explains the common observation that pools scour at high flow and fill at low flow, whereas riffles are depositional sites at high flow. The velocity reversal has been difficult to explain, however, given the tendency of pools to have greater cross-sectional area (and therefore presumably lower velocity) than riffles. Hydraulic modeling of the River Severn in England indicates that channels in which pools are hydraulically rougher than riffles during high flow and riffles are substantially wider than pools are more likely to exhibit velocity reversal (Carling and Wood 1994).

Numerous field observations and flume experiments illustrate the difficulty of demonstrating a velocity reversal, at least in part because of the difficulty of measuring bed velocity during high discharges. Cross-sectionally averaged velocity does not exhibit velocity reversal, but the presence of strong eddy flow along the margins of pools during high discharges permits the formation of a central jet of high-velocity flow that does exhibit velocity reversal with respect to riffle velocity (Figure 4.18) (Thompson et al. 1998, 1999). Two-dimensional numerical models in which bed topography is altered to reflect different pool–riffle configurations suggest that the presence and strength of a velocity reversal depends strongly on factors such as relative channel width and the depth of the pools and

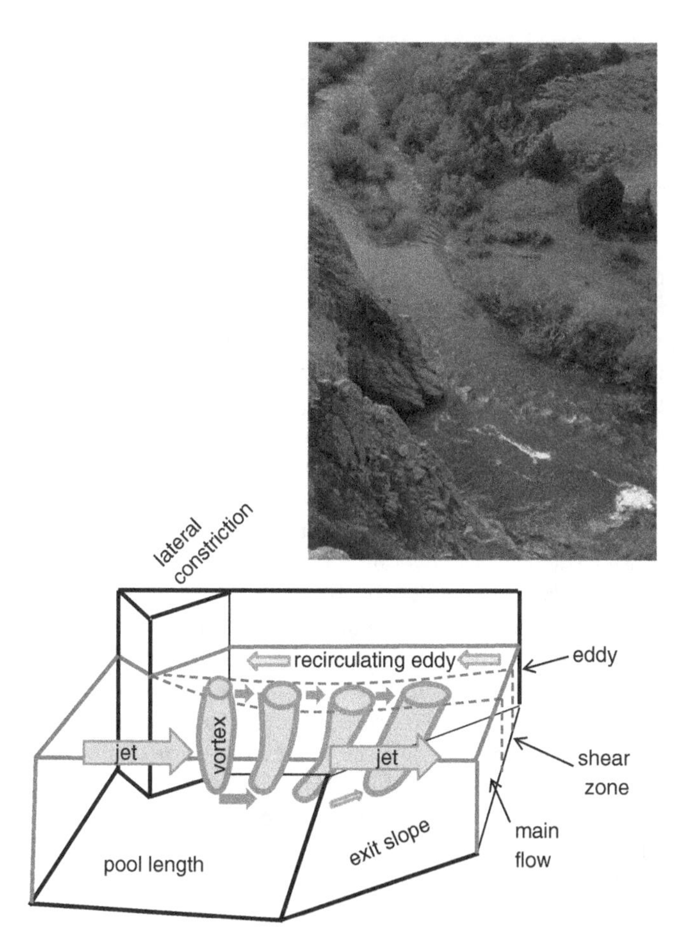

Figure 4.18 Photograph shows a pool with a lateral bedrock constriction, at river left. In this view, during high flow, naturally occurring foam at the water surface highlights the line of flow separation between the eddy at the base of the photo and the jet of high-velocity flow in the pool center, which has a series of standing waves. The deepest portion of the pool is just downstream of the bedrock constriction that appears at the middle left of the photo. As flow diverges in exiting the pool, transport capacity drops and a bar and riffle form. The bar has some vegetation, largely because this is a regulated river and flood peaks have declined, facilitating the encroachment of riparian vegetation. The channel is approximately 20 m wide downstream from the lateral bedrock constriction. Diagram shows an idealized version of flow patterns in a laterally constricted pool. Source: Diagram after Thompson (2004), Figure 1. (*See color plate section for color representation of this figure*).

riffles (Jackson et al. 2015). This is expressed by a criterion developed by Caamaño et al. (2009)

$$\left(\frac{B_r}{B_p}\right) - 1 = D_z/h_{Rt} \tag{4.23}$$

in which B_r and B_p are riffle and pool water-surface widths, respectively; D_z is residual pool depth; and h_{Rt} is riffle thalweg depth. This simplified, one-dimensional criterion may not effectively predict weak, local, or transient velocity reversals (Jackson et al. 2015). It also may not predict the near-bed or peak velocity reversals characteristic of forced pools (MacVicar et al. 2010).

Bed load in pool–riffle channels tends to move during competent flows in spatially discrete steps that are strongly influenced by pool and riffle spacing (Thompson et al. 1996; Pyrce and Ashmore 2003). Clasts entrained from a pool during high flow are deposited on the next riffle downstream. Clasts on a riffle can be moved into the next pool downstream during waning or low flows as progressively steeper water-surface profiles over riffles at lower stages result in dissection of the riffles (Harvey et al. 1993). Studies in urban pool–riffle streams indicate that particulate pollutants such as heavy metals and excess nutrients accumulate preferentially in the hyporheic zones beneath riffles, with increased pollutant concentration associated with decreased hydraulic conductivity at the transition between riffles and pools (Namour et al. 2015).

In contrast to step–pool sequences, pool–riffle channel segments are *response reaches* (Montgomery and Buffington 1997). Because pool–riffle channels are more likely to be transport-limited with respect to sediment, the bedform dimensions and bed grain-size distributions of pool–riffle channels are more likely to change in response to altered water and sediment discharge than are those in supply-limited step–pool channels. Increased sediment load typically causes preferential pool filling. Filling pools reduces form roughness and creates a more uniform reach-scale bed gradient and flow depth, which enhance the ability of moderate flows to transport bed load (Lisle 1982). Under moderate sediment loading, adjustment of the overall bed slope may occur (Nelson et al. 2015). Sand can be preferentially stored on bars and riffle margins in gravel-bed pool–riffle channels, especially if sand is supply-limited during the falling limb of the hydrograph (Milan and Large 2014). Under extreme sediment loading, a pool–riffle channel may lose all or most of its bed topography (Wohl and Cenderelli 2000) or assume a braided morphology as pools preferentially fill (East et al. 2015). Bed relief can also exert an important control on responses to increased sediment supply, with moderate increases in sediment causing bar deposition and increased bed relief in initially low-relief beds, as opposed to pool deposition and decreased bed relief in initially high-relief beds (Zunka et al. 2015). Decreased sediment load or increased discharge can lead to enhanced pool scour and bank erosion (Wohl and Dust 2012). The bedform amplitude and wavelength of pool–riffle sequences can thus adjust to variations in flow and sediment supply.

An important implication of the distinction between transport and response reaches is that diverse types of infrequently mobile bedforms are not likely to respond uniformly to changes in water and sediment yield associated with changes in climate, land cover, or other external controls. If timber harvest within a catchment increases sediment yield, for example, bedform dimensions in the lower-gradient response reaches will change disproportionately, while those in the higher-gradient transport reaches may remain relatively unaffected.

River reaches with different bedforms exhibit differences in *sensitivity* (the ability to react to a stimulus) and *resilience* (the ability to return to initial conditions following a disturbance such as a flood or debris flow) (Brunsden and Thornes 1979). These reach-scale differences within a river network have important implications for channel stability, aquatic and riparian habitat, water quality, and river management (Montgomery and Buffington 1997; Wohl et al. 2007). Sensitive channel segments may be less stable, for example, whereas resilient segments may be more so. Fryirs (2017) discusses different spatial scales of sensitivity, from landscape-scale sensitivity through sensitivity at the reach scale and the scale of individual morphological units within a river. Each of these scales reflects differing controls, although the details of water and sediment fluxes influence sensitivity at all scales.

By influencing flow velocity and depth, substrate size and stability, dissolved oxygen, hyporheic exchange, and solute fluxes, infrequently mobile bedforms also influence habitat and the distribution of aquatic biota. In mountainous headwater channels, for example, pool–riffle channel segments typically have greater pool volume and fish habitat than step–pool or cascade channels (Moir et al. 2004).

4.3.2.6 Bars

Diverse types of bars occur in rivers, including berms, alternate bars in straight channels, point bars in sinuous channels, transverse bars that form riffles or rapids, braid bars or mid-channel bars in braided channels, and bars at tributary–main channel junctions. *Bars* have lengths comparable to channel width. They form when flow energy is sufficient to transport bed material and simultaneously scour other portions of the channel bed, such as pools. As with other bedforms, bars are both influenced by boundary resistance, hydraulic forces, and sediment dynamics and themselves influence roughness, hydraulics, and sediment dynamics.

Berms are accretionary features that form at sites of energy loss associated with hydraulic jumps or the shear zone of flow separation (Carling 1995). They can form downstream from lateral channel constrictions or at the downstream end of a plunge pool below a vertical step such as a waterfall. They can be relatively small features the height and lateral thickness of a single cobble or boulder, or more than a meter tall and wide. Berms tend to be persistent features that form during high discharges and remain during subsequent lower discharges (Longfield et al. 2018).

Alternate bars occur on alternating sides of the channel in a downstream progression. They can form in mixed grain-size channels when concentrations of coarse clasts at the downstream end of bed-load pulses aggrade along portions of the channel with high flow roughness (Lisle et al. 1991). Alternate bars are typically asymmetrical longitudinally and generally migrate in the downstream direction (Bridge 2003) as particles being transported along the top of the bar reach the downstream edge and cascade down the bar front. These bars are discussed more in Section 6.2 as part of the treatment of meandering and braided channels.

Point bars form along the inside edge of meander bends, where fine grains are swept inward over the point bar and coarse grains are routed outward toward the pool (Clayton and Pitlick 2007). Depending on the channel substrate and sediment supply, they can be formed of sand- to boulder-size grains and are scaled to channel width. They are essentially alternate bars in sinuous channels. Point bars migrate and adjust in size and geometry as sediment supply and the distribution of hydraulic forces fluctuate with discharge and as meander geometry changes through time (e.g. Anthony and Harvey 1991). The central portion of a point bar can be armored by coarsening-upward gravels during extreme floods, whereas helical flow creates the more commonly observed pattern of finer grains being swept onto the point bar during annual floods (Ghinassi et al. 2018; Hagstrom et al. 2018). This suggests that the area affected by secondary helical circulation shifts downstream and upstream of the bend apex during extreme and annual floods, respectively, creating persistent changes in bar stratigraphy (Ghinassi et al. 2018). Point bars are also discussed more in Section 6.2.

Studies of bar grain-size texture indicate that rate of flow recession exerts an important control on bars. Ephemeral channels commonly have bars that are coarser-grained than the adjacent thalweg, a pattern that holds across diverse planforms, whereas bars in perennial rivers may be finer-grained than the adjacent thalweg (Storz-Peretz and Laronne 2018).

4.3.3 Bedforms in Cohesive Sediments

Characteristic repetitive bed undulations occur in streambeds with sufficient silt- and clay-sized particles to act as cohesive materials (Wohl and Ikeda 1997; Johnson and Whipple 2007), or in bedrock channels (Wohl 1993; Wohl and Ikeda 1998). These bedforms are strictly erosional and display a sequence with increasing flow strength analogous to that of lower- and upper-regime bedforms for sand-bed channels.

In cohesive streambeds, straight longitudinal grooves and ridges form initially (Allen 1982). These features are parallel to mean flow direction and reflect streaks in the viscous sublayer. Flute marks appear as velocity increases. Flute marks are shaped like the depression in a spoon and reflect a site where locally enhanced near-bed turbulence results in differential bed erosion and flow separation (Bridge 2003). Further velocity increase creates transverse ridge marks shaped like ripples, which reflect supercritical flow (Bridge 2003). Potholes form a distinctive and particularly well-studied subset of erosional features, but they are not typically evenly spaced downstream.

As with depositional bedforms in noncohesive sediments, hydraulics can be inferred from the type and dimensions of erosional bedforms in cohesive sediments. Richardson and Carling (2005) presents a catalog of cohesive bedforms, sometimes known as *sculpted features*, characteristic of bedrock channels.

4.4 In-Channel Depositional Processes

Each of the bedforms discussed in Sections 4.3.1 and 4.3.2 is also a depositional feature, particularly when it stops moving. Deposition occurs when the flow or shear velocity falls below the settling velocity of a particle. This is a lower threshold than that required for entrainment (Figure 4.19). Deposition can be highly localized and limited to a zone of lower velocity and flow energy, or more widespread across and down a length of channel.

Local deposition occurs in association with isolated roughness features such as protruding boulders, instream wood, and aquatic or riparian vegetation, or with irregularities in the channel margins such as tributary junctions and bends. Local deposition alters boundary roughness, grain-size distributions at the channel surface, and subsequent particle entrainment. Local deposition that occurs

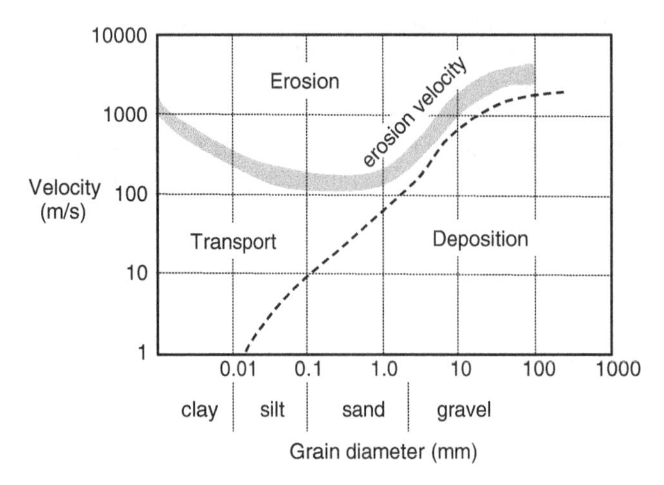

Figure 4.19 Hjulström's diagram of erosion, transport, and deposition thresholds for varying grain sizes as a function of velocity. Source: After Hjulström (1935).

in isolated but numerous locations across the channel produces features such as particle clusters, ripples and dunes, and various types of bars. The size and spacing of bars, in particular, can change with sediment supply and flow energy (e.g. Wathen and Hoey 1998; Surian 1999; Kondolf et al. 2002; Ritchie et al. 2018).

Some portions of the channel that serve as depositional sites at lower discharge can be scoured as discharge increases. Pools typically accumulate finer sediment during low flow, but converging flow off a riffle or a central jet developing in a laterally constricted pool results in removal of sediment during high flows. Similarly, lee deposits accumulating around protruding clasts during lower discharges can be mobilized as hydraulic forces around the clasts change with increasing discharge (Thompson 2008).

Deposition across the entire channel is facilitated by channel-spanning obstructions such as logjams (Nakamura and Swanson 1993; Wohl and Scott 2017b) and beaver dams (Butler and Malanson 2005; Pollock et al. 2007). These obstructions can create sufficient backwater to accumulate a wedge of sediment as thick at the downstream end as the obstruction is tall, with the wedge tapering upstream in a manner dictated by channel gradient and sediment supply. Sediment accumulating upstream from the obstruction is likely to be finer-grained than that elsewhere on the streambed, enhancing the diversity of aquatic habitats and the storage of fine organic material (Bilby 1981; Beckman and Wohl 2014). The pressure gradient associated with backwater at the obstacle also enhances hyporheic exchange within the deposited sediment (Hester and Doyle 2008; Briggs et al. 2013). Sufficiently large and persistent obstructions can create an alluvial channel at the reach scale in a setting that would otherwise be associated with a bedrock channel (Massong and Montgomery 2000).

A very large obstruction such as a sediment dam associated with a rockfall, landslide, or debris flow creates a correspondingly large deposit, although natural dams caused by slope instability typically breach within a year (Costa and Schuster 1988). Some types of large-scale natural obstructions, such as lava dams, can persist for 10^2–10^3 years.

Riparian vegetation can be especially effective in facilitating deposition within the channel and along the channel margins. Increases in the density of vegetation can facilitate sediment accumulation and channel narrowing (Tal et al. 2004). Deposition of large pieces of driftwood or entire trees that are capable of regrowth enhances sediment deposition (Gurnell and Petts 2006) and, where sufficient sand and finer sediment is available, facilitates island and floodplain development by trapping sediment and allowing more vegetation to become established. Establishment of riparian vegetation can initiate a self-enhancing feedback as plants trap and stabilize sediments and organic matter capable of providing germination sites for other plants, which in turn reinforce the development of floodplains, islands, and other landforms (Gurnell and Petts 2006). Gurnell et al. (2012) refer to this process as *vegetation-mediated landform development*. The floodplain large-wood cycle hypothesis (Collins et al. 2012), in which logjams force channel avulsion and create stable alluvial patches that increase floodplain spatial heterogeneity, is another example of feedbacks among living and dead riparian vegetation, water and sediment fluxes, and channel form.

The balance between sediment mobility and the ability of vegetation to germinate, persist, and stabilize sediment can change through time (Gran et al. 2015). Changes in vegetation density can reflect natural fluctuations in flood magnitude and frequency, with large floods removing vegetation, which then regrows during periods of lesser flow (e.g. Friedman and Lee 2002). Vegetation density can also change as a result of invasive exotic species (Graf 1978; Dean and Schmidt 2011) or a reduction in flood peaks associated with flow regulation that allows germinating seedlings to successfully colonize lower portions of channel margins (Nadler and Schumm 1981; Shafroth et al. 2002).

Cross-channel, longitudinally continuous deposition typically represents a reduction in transport capacity or an increase in sediment supply. Transport capacity can decline as a result of decreased discharge or decreased channel gradient. Sediment supply can increase following climate change (e.g. Fuller et al. 2009), hillslope instability (e.g. Dadson et al. 2004), volcanic eruptions (Gran et al. 2006; Gran 2012), or changes in land cover or land use. Cross-channel, longitudinally continuous deposition can occur as a relatively minor or temporary effect that results in finer sediment being deposited on the surface of a coarse-grained channel. Although the effects on cross-sectional geometry can be minor, fining of the bed sediment can alter hyporheic exchange and habitat suitability for benthic organisms such as algae, aquatic insects, and larval fish.

More substantial and sustained deposition throughout a channel is known as *aggradation* and can result in sufficient loss of channel cross-sectional area to cause enhanced overbank flooding and lateral channel movement or the formation of secondary channels. A common cause of aggradation leading to loss of cross-sectional area is the clearance of native vegetation and start of agriculture within a drainage basin, as discussed in more detail in Section 4.8.

4.5 Downstream Trends in Grain Size

Average bed grain size decreases downstream in most rivers at scales of tens to hundreds of kilometers – a pattern known as *downstream fining*. Bed material also becomes better sorted and individual particles become more rounded with distance from sediment source areas (Knighton 1998; Miller et al. 2014). Downstream fining has been attributed to selective sorting (entrainment, transport, and deposition), abrasion in place or during transport, or some combination of these processes (Powell 1998; Miller et al. 2014). The relative importance of selective sorting versus abrasion reflects factors such as clast erodibility; abrasion is more important where clasts abrade readily (Parker 1991).

Abrasion is commonly quantified using some form of Sternberg's (1875) law

$$D = D_0 e^{-\alpha x} \tag{4.24}$$

where initial grain size D_0 wears down to D at distance x from the origin at a rate given by the rock erodibility parameter α, which sets the abrasion length scale, $1/\alpha$ (Sklar et al. 2006). Values of α can vary by at least two orders of magnitude, from 10^{-3}/m to 10^{-5}/m, but tend to decrease as the bed material grows finer. In this general form of Sternberg's law, α incorporates both selective sorting and abrasion (Knighton 1998). Fracturing of clasts is typically lumped under abrasion, although Chatanantavet et al. (2010) suggest distinguishing these processes.

In some rivers, gravel-to-sand transitions occur over distances too short for substantial abrasion. Ferguson (2003) attributes these transitions to nonlinearities in bed-load transport in which small increases in sand content lead to large increases in the mobility of sand and gravel. As the effective bed roughness decreases due to sand deposition, near-bed velocities increase, shear stress and turbulent kinetic energy decrease, burst and sweep events become less frequent, and the sand fraction remains more mobile than coarser particles. All of these factors contribute to creating a sand-bed channel downstream (Sambrook Smith and Nicholas 2005), as do decreases in transport capacity (Ohmori 1991; Parker et al. 2008).

Downstream fining can be interrupted by coarse sediment from hillslope or tributary inputs (Attal and Lavé 2006) or from knickzone erosion (Deroanne and Petit 1999). Mountainous catchments with close coupling of hillslopes and channels may have downstream grain-size distributions that closely mirror the size distribution of sediment supplied from hillslopes, because local resupply from hillslopes offsets the influence of channel processes that create downstream fining (Sklar et al. 2006). Slopes that are steeper and have predominantly physical weathering can produce coarse sediment that rapidly enters the channel network and influences abrasional processes in the channel (Riebe et al. 2015). Sklar et al. (2017) have developed a modeling framework for predicting the size distribution of sediment produced on hillslopes and supplied to channels, starting with the fracture spacing in unweathered rock and continuing through processes of weathering and erosion.

Mountain rivers can also have downstream coarsening. Grain size coarsens to a maximum value at the location along the channel where transport dominated by debris flows gives way to transport dominated by fluvial processes. Downstream fining occurs below this transition (Brummer and Montgomery 2003).

Significant lateral sediment sources that influence downstream fining can be identified by using drainage basin area, network magnitude, and the basin area–slope product to define individual channel links within which downstream fining occurs (Rice 1998). The travel distance required to abrade coarser tributary sediments to the size of mainstem inputs from upstream can be predicted as

$$L^*_{\Delta D} = \frac{1}{\alpha} \ln\left(\frac{D_t}{D_m}\right) \tag{4.25}$$

where $L^*_{\Delta D}$ is the distance over which the grain size perturbation decays, D_t is mean grain size from the tributary, D_m is mean grain size in the mainstem, and α is the rock erodibility parameter (Sklar et al. 2006).

Sedimentary links can be designated as lengths of channel between tributaries that supply particularly coarse and abundant bed load (Rice 1999). These links structure longitudinal changes in slope, morphology, and bed grain-size distribution along the main channel. The *link discontinuity concept* (Rice et al. 2001) describes how response variables such as bed particle size and elevation, channel geometry, and ecosystem structure change abruptly downstream at tributary junctions or other point sources of water and sediment to the main channel. The *network dynamics hypothesis* (Benda et al. 2004b) takes this idea to the network scale with theories of how basin size, shape, drainage density, and network geometry interact to govern the spatial distribution of heterogeneity in river habitats. Rice (2017) builds on this framework and ties tributary confluence effects to network biodiversity.

Changes in cross-sectional geometry and transport capacity (Constantine et al. 2003; Rengers and Wohl 2007) and river engineering (Surian 2002) can also create substantial variation in downstream grain-size trends. Studies in diverse environments indicate minimal downstream fining, or disruption or reversal of system-wide fining trends, along reaches in which the channel is laterally confined (and, typically, steeper).

4.6 Bank Stability and Erosion

The ability of flow to erode the banks of a channel, and the processes and rates of bank erosion, strongly influence sediment supply, channel geometry, riparian vegetation, and rates of lateral channel migration. Bank erosion thus represents, like bedforms, a means of adjusting channel geometry, external resistance to flow, and sediment supply.

Bank erosion can result in property loss and damage to infrastructure such as buildings and bridges. Human alterations of water and sediment supply to the channel commonly result in altered rates of bank erosion. Even unaltered, "natural" bank erosion is in many cases regarded as an indicator of channel instability and a sign that mitigation is needed to limit bank erosion. This represents a fundamental misunderstanding of river processes. A stable channel transporting bed material must entrain an amount of sediment equivalent to that deposited. Most in-channel deposition occurs in bars, and most compensating erosion comes from banks as the channel expands to maintain conveyance. Consequently, local lateral instability in the form of local bank erosion is a natural feature of stable rivers transporting bed material, but is commonly misinterpreted as a sign of general channel instability. Given the tendency to build immediately adjacent to channels and then attempt to prevent bank erosion, much effort is devoted to analyzing and modeling bank stability.

As with bed erosion, the forces driving sediment removal from banks must exceed the resisting forces for banks to erode. This becomes complicated to assess in part because stream banks are commonly layered, with coarser sediment at the bottom as a result of deposition within the channel and finer sediment toward the top as a result of overbank deposition. Bank sediment typically grows progressively finer-grained downstream or along channel segments of lower gradient, but banks can be very spatially heterogeneous. The finer sediments in the upper bank add cohesion, and the roots of riparian vegetation can further enhance the resistance of the bank sediment to erosion (Figure 4.20).

The strength of stream banks reflects the frictional properties of the bank sediment, effective normal stress, and effective cohesion (Simon et al. 1999). *Effective normal stress*, σ', is the difference

Figure 4.20 The lower, weakly cohesive bank along this river in northern Alaska, USA erodes rapidly, but the rate of bank retreat is limited by the overlying finer sediment layer, which is effectively stabilized by the roots of various types of plants. Bank face is approximately 1.5 m tall. (*See color plate section for color representation of this figure*).

between normal stress (the perpendicular component of total stress, σ) and pore pressure, u: in other words, $\sigma' = \sigma - u$. Effective normal stress results from static friction that keeps particles together in a stream bank, minus the effect of pore pressure that keeps particles separate. *Effective cohesion* results from true cohesive forces, from matric suction within the unsaturated portion of the bank, or from root reinforcement via vegetation (Eaton 2006). Acting against these resisting forces are the driving forces of gravity and hydraulic force, expressed as shear stress.

Pore-water content and pressure is one of the more important influences on bank stability (Rinaldi and Darby 2008), and involves multiple effects. Pore water reduces shear strength by increasing lubrication between sediment particles. Pore water increases the unit weight of the bank material, making unsupported material more susceptible to failure. Pore water destabilizes banks by facilitating the presence of water in tension cracks. Pore water also creates seepage forces that can stabilize or destabilize banks.

Pore-water pressure results from the pressure of water filling the voids between particles. Negative pore-water pressures above the water table reflect the surface tension of pore water in voids, which creates a suction effect on surrounding particles and stabilizes the banks. Positive pore-water pressures below the water table, in contrast, help to force particles apart and destabilize banks. Pore-water pressures are extremely transient in response to changes in precipitation and stream flow, but they typically promote bank failure during the falling limb of the hydrograph when the bank sediment is at or near saturation and the confining pressure of the river water is removed.

Vegetation also influences stream banks in diverse ways (Merritt 2013). Plants increase the mass of stream banks and can thus facilitate bank failure. However, plants near the water's edge create flow resistance, which reduces near-bank velocity and shear stress and increases bank stability (Griffin et al. 2005; Gorrick and Rodríguez 2012). Plant roots increase the resistance of the sediment to shearing (Pollen and Simon 2005). And, plants alter bank pore-water pressure by affecting infiltration, evaporation, and transpiration (Griffin et al. 2005; Pollen-Bankhead and Simon 2010). Vegetation exerts the greatest influence on bank stability along low, shallow banks in weakly cohesive sediments and along small channels (Eaton and Millar 2004).

Not all vegetation is created equal with regard to bank stability. Channels with forested banks tend to be wider than those in grasslands, for at least two reasons. First, wood recruited to the channel from the riparian zone promotes channel widening. Second, grass grows readily on point bars, facilitating more rapid deposition of suspended sediment and narrower channels (Allmendinger et al. 2005). Among woody vegetation, densely growing species such as willows (*Salix* spp.) can limit bank erosion and channel widening more effectively than species in which individual plants are more widely dispersed (David et al. 2009). Dividing vegetation into functional groups of trees, shrubs, graminoids, and forbs reveals significant differences in lateral root extent, maximum root diameter, and root tensile strength (Polvi et al. 2014). Woody trees and shrubs have higher values of these three parameters and thus greater potential for stabilizing stream banks.

Cyclic bank erosion can occur over decades when small volumes of sediment are removed from between large trees on the bank, creating a scalloped bank morphology buttressed by large trees (Figure 4.21). The trees are gradually undercut and topple into the channel, resulting in a larger volume of bank erosion and a new round of the cycle (Pizzuto et al. 2010).

The primary processes eroding stream banks involve hydraulic action or mass failure. Fluvial detachment via hydraulic action is likely to be more important in coarse-grained or smaller rivers. Mass failure becomes progressively more important downstream as bank heights increase and bank sediments become finer-grained (Lawler 1992).

Figure 4.21 Scalloped bank along a small creek in Connecticut, USA, caused by locally increased bank resistance as a result of tree roots.

Hydraulic action results from shear stress exerted against a bank and is related to the near-bank velocity. Individual particles can be detached from the bank face in non- or weakly cohesive sediment and the base of the bank can be preferentially eroded (Rinaldi and Darby 2008). Even cohesive sediments can be made more susceptible to hydraulic action by processes that weaken and detach sediment, such as shrink–swell or freeze–thaw cycles (Wynn et al. 2008). Conversely, flow resistance associated with riparian vegetation can reduce bank erosion through hydraulic action. Hydraulic action is typically quantified using an excess-shear-stress equation similar to Eq. (4.15), with fluvial bank erosion rate per unit time substituted for bed-load transport rate (Rinaldi and Darby 2008).

Processes that weaken and detach bank sediment promote mass failure through slumping or toppling of a slab. Seepage also reduces the cohesion of bank sediment by removing clay and, in some cases, promoting piping in the bank. Seepage can be particularly effective where rapid reductions in flow stage leave saturated banks without lateral support, a situation common in rivers regulated for hydropower generation. Trampling by large numbers of grazing animals congregating along streams can further enhance bank erosion by breaking down the banks and creating hydraulic roughness (Trimble and Mendel 1995).

Mass failure of banks takes several forms (Osman and Thorne 1988; Knighton 1998; Rinaldi and Nardi 2013). Shallow slips in which relatively thin segments of bank detach and then disintegrate dominate in noncohesive sediment. Slab-type failures in which a vertical slab detaches from the bank and then topples are most common in weakly cohesive sediment and near-vertical banks. Deep-seated rotational slips dominate in cohesive sediment. In these failures, a mass of sediment slides down the bank along a curved failure surface, rotating so that the toe of the failure protrudes into the river.

All forms of mass failure are enhanced by scour at the base of the bank that over-steepens it. In strongly layered banks, removal of the coarser, noncohesive sediment in the lower bank can create overhangs that eventually collapse as blocks into the channel (Figure 4.20). Such blocks can either quickly break up and disperse, or persist for more than a year, in part because of the dense, fine plant roots within the block. Persistent blocks act similarly to boulders and can protect the lower bank from further erosion or deflect the current toward the bank toe and enhance basal scouring. Bank collapse via overhanging blocks is particularly common in permafrost regions where slow thawing of the soil ice facilitates the persistence of collapsed blocks (Walker et al. 1987) and in wet meadows where bank erosion creates peat blocks (Warburton and Evans 2011). Root reinforcement of bank sediment by riparian vegetation can limit mass failure, particularly where the vegetation grows along the bank toe or at the intersection of the failure plane with the floodplain surface (Van De Wiel and Darby 2007).

Cold-region rivers with a seasonal ice cover can have a distinctive bank morphology in the form of a two-level bank structure that reflects ice scouring (Boucher et al. 2012). Elevation of the ice surface during freeze-over and jamming of ice blocks during break-up lead to abrasion of the banks by ice and the formation of a steep segment of bank above the bankfull stage. Ice gouging and overbank sedimentation during ice-jam flooding can create an elevated ridge or bench along some rivers, which is referred to as a *bechevnik*, from the Russian word for tow rope, because these benches formed convenient paths for towing boats upstream along Siberian rivers (Ettema and Kempema 2012).

Lawler (1993) reviews seven primary techniques for measuring or estimating bank erosion. Over relatively short time spans of a decade or less, repeat cross-sectional surveys (Lane et al. 1994), planimetric surveys, or spatially continuous, three-dimensional surveys (Prosdocimi et al. 2015; Leyland et al. 2017) can be used to measure bank erosion, as can erosion pins placed a set depth into the bank and then measured at intervals of time (Wolman 1959). An automated version of an erosion pin is the photo-electronic erosion pin (PEEP) (Lawler 1991, 2005, 2008; Papanicolaou et al. 2017). A PEEP is a sensor with a row of photovoltaic cells connected in series and enclosed within a transparent, waterproof tube. The cells generate a voltage proportional to incident radiation, which is recorded by a data logger. Erosion of the bank face exposes more cells to light, and accretion buries cells.

Over periods of multiple decades, bank erosion can be estimated from analysis of channel planform changes from aerial photographs (Wolman 1959) or other historical sources such as maps and surveyors' notes (Lewin and Hughes 1976). Over time scales of decades to centuries, rates of bank erosion can also be inferred by using tree-ring dating to determine the timing of changes in tree roots exposed by bank erosion (Malik and Matyja 2008) or rates of bank accretion around trees in which the principal root just below the ground surface is buried more deeply (Hupp and Simon 1991). Regime-based paleohydrological techniques can sometimes provide information on rates of lateral channel movement over centuries to millennia if the ages of stream sediments can be well constrained (Lawler 1993).

Several models quantify and predict bank stability (Pizzuto 2003; Rinaldi and Darby 2008). Models can be differentiated as *mechanistic models* based on the physics of particular erosional processes and *parametric models* that relate bank erosion to potential controlling parameters – typically near-bank velocity or shear stress – using empirical coefficients (Pizzuto 2003). One of the most widely used and comprehensive models available is BSTEM (Bank Stability and Toe Erosion Model; Simon et al. 2000), which is integrated with hydraulic and sediment transport models such as HEC-RAS (Klavon et al. 2017).

Bank erodibility parameters are modeled using methods similar to those for entrainment of bed sediments, with modifications to account for the effect of bank angle on the downslope component of particle weight and for partly packed and cemented sediments (Rinaldi and Darby 2008). Near-bank hydraulic force is either directly measured or estimated using hydraulic models. Models developed for noncohesive banks include the mechanistic model in Kovacs and Parker (1994), which uses a bed-load transport model to compute bed-load transport on steeply sloping banks coupled with the near-bank velocity. Because vegetation, moisture, and even small amounts of fine sediment add cohesion to banks, most models focus on cohesive banks.

Mechanistic models for cohesive banks are generally two-dimensional models that evaluate bank stability in terms of the soil strength and bank geometry (Pizzuto 2003). For example, the mechanistic BSTEM model applies to layered, cohesive stream banks (Simon et al. 2000). This model combines the Coulomb equation for saturated banks with the Fredlund et al. (1978) equation for unsaturated banks. The Coulomb equation is

$$S_t = c + \sigma' \tan \phi \tag{4.26}$$

where S_t is shear strength, c is cohesion, ϕ is the angle of internal friction, and σ' is the effective normal stress. Total normal stress, which tends to hold sediment together, is the sum of effective normal stress and pore pressure. Under saturated conditions, pore pressure is positive and effective normal stress is lower. In partially saturated soils, effective normal stress is increased. Consequently, bank sediment is more susceptible to mass failure under saturated conditions (Robert 2014).

The failure criterion of Fredlund et al. (1978) is

$$\tau = c' + (\sigma - u_a) \tan \phi' + (u_a - u_w) \tan \phi^b \tag{4.27}$$

where τ is shear strength, c' is effective cohesion, σ is normal stress, u_a is pore air pressure, ϕ' is the friction angle in terms of effective stress, u_w is pore water pressure, $(u_a\text{-}u_w)$ is matric suction, and ϕ^b is the angle expressing the rate of increase in strength relative to the matric suction.

Parameter uncertainties in bank stability models are typically so large as a result of natural variability in the parameters that the likelihood of generating unreliable predictions exceeds 80% (Samadi et al. 2009). In addition, processes that weaken and strengthen banks interact in sometimes unpredictable manners. Consequently, rather than use a deterministic model with a single value for each bank material property, a probabilistic representation of effective bank material strength parameters may be the most appropriate approach (Parker et al. 2008).

4.7 Sediment Budgets

A sediment budget quantifies fluxes of sediment past a given point in a watershed. In his work on sediment generated by nineteenth-century hydradulic mining practices in the Sierra Nevada of California, USA, G.K. Gilbert (1917) laid out a framework for understanding sediment inputs, fluxes, and outputs in reference to a specified length of channel. This framework was later formalized as a sediment budget. Sediment budgets are based on the very simple formula of inputs, I, minus the change in mass or volume of sediment, ΔS, stored in the channel reach during a specified time interval

$$I - \Delta S = \varphi \tag{4.28}$$

where φ is the mass or volume of sediment output from the channel reach during the specified time interval (Reid and Dunne 1996, 2016). φ is also known as *sediment yield.* Sediment inputs represent sediment coming from upstream and subsurface sources on the main channel and tributaries, as well as from sources beyond the channel, including hillslopes, glaciers, terraces and other valley-bottom deposits, and eolian inputs. Storage occurs within the channel bed, banks, bedforms, and in overbank areas such as the floodplain. Output is transported within the channel as dissolved, wash, suspended, and bed-load sediment (Figure 4.22).

The mathematical simplicity of Eq. (4.28) is deceptive because each of the three primary variables can fluctuate greatly through time, across a drainage basin, and between basins, making it extremely difficult to accurately quantify input, storage, and output over even relatively short time intervals. Sediment inputs can be gradual, as in slope wash, average tributary sediment transport, and soil creep, or they can be abrupt, as in debris flows and tributary flash floods. Inputs can also be seasonally driven, aperiodic, or variable over time spans of thousands of years because of fluctuations in base level, climate, or land use.

Sediment inputs tend to be temporally and spatially heterogeneous in a wide variety of basins. Research in mountainous basins provides examples: 75% of the long-term sediment flux in Taiwan occurs during typhoon-generated floods occupying <1% of the flow-duration curve (Kao and Milliman 2008). Over a period of 70 years, half of the sediment load from a river in California, USA was delivered in less than 5 weeks (Farnsworth and Milliman 2003). Sedimentation at the inlet of a reservoir during a few days of widespread, high-intensity rainfall in the Colorado Rockies during September 2013 was equivalent to 100 years of average reservoir sedimentation (Rathburn

Figure 4.22 Schematic illustration of the basic components of a sediment budget. Inputs, I, to a river segment can come from uplands and glaciers, eolian deposition, mainstem and tributary transport, or erosion of terraces, alluvial fans, floodplains, and other valley-bottom deposits. Storage, S, of sediment can occur in the floodplain or in various portions of the channel. Sediment outputs, φ, occur via downstream transport as dissolved, suspended, and bed-load sediment. (*See color plate section for color representation of this figure*).

et al. 2017). Sediment yield in drainages with periodic disturbances such as volcanic eruptions and wildfires is likely to be dominated by episodic inputs from both disturbed hillslopes and continuing channel instability (Gran et al. 2011; Pelletier and Orem 2014).

In large river basins as diverse as the Amazon and the Mississippi, one portion of the catchment supplies the great majority of total sediment output (Figure 4.9), as noted earlier. In the case of the Amazon, the Andes of Peru and Bolivia supply more than 80% of the sediment load but constitute only ~10% of the basin area (Meade et al. 1985; Meade 2007).

The relative importance of diverse sediment sources varies greatly among basins. High-relief basins tend to be dominated by hillslope sources (e.g. Dadson et al. 2004). Laterally mobile rivers with extensive floodplains can obtain sediment predominantly from bank erosion (Dunne et al. 1998; Meade 2007). High-latitude basins can be dominated by glacial sources (Gurnell 1995).

The relative importance of diverse sediment sources also varies among solute, wash, suspended, and bed-load sediment. In most catchments, solutes come primarily from ground-water inputs. Bank sediments can contribute nearly 40% of the total suspended sediment, even in relatively low-energy catchments (Walling et al. 1999), with up to 80% of the total suspended sediment yield coming from bank sources in some highly unstable, incised channel networks (Simon and Darby 2002). Bed load can come primarily from mobilization of streambed sediments, inputs from smaller, steeper tributary catchments, or, particularly in high-relief catchments, upland sources such as hillslope mass movements.

The predominant source of sediment can also vary through time. This is illustrated by the Upper Mississippi River, USA, where the dominant source of suspended sediment shifted from agricultural soil erosion in the mid-twentieth century to accelerated erosion of stream banks by 1980 as discharge increased due to climate change and channelization (Engstrom et al. 2009; Belmont et al. 2011).

The existence of sediment storage is reflected in the *sediment delivery ratio*, or SDR. This ratio represents the difference between volume of sediment generated and volume of sediment stored or transported from the basin, typically as a function of drainage area, A

$$\gamma = \alpha A^{\varphi} \tag{4.29}$$

where γ is SDR and α and φ are empirical parameters (note that φ is defined differently in Eqs. (4.28)) and (4.29)). SDR is typically calculated from observed sediment yields, y, and measured or estimated gross erosion rates, e

$$y = \gamma e \tag{4.30}$$

Combining Eqs. (4.29) and (4.30) results in

$$y = bA^{\theta} \tag{4.31}$$

where y is areal average sediment yield and θ is an empirical parameter that varies from −0.52 to 0.12 (Milliman and Meade 1983; Lu et al. 2005).

SDR is typically between 0 and 1, indicating that only a fraction of the total sediment detached from eroding sources actually leaves a catchment. Although SDR is commonly related to drainage area, other factors such as topography, land cover, land use, lithology, connectivity, and cyclic channel processes such as gully erosion and filling can strongly influence SDR (De Vente et al. 2007; Wohl 2010b; Heckmann and Vericat 2018). In general, smaller, steeper catchments have a greater SDR because they have proportionally less sediment storage in features such as floodplains (e.g. Pelletier 2012).

Worrall et al. (2014b) criticizes the use of SDR to interpret suspended sediment fluxes. Although declining sediment yields for larger catchments are explained in terms of greater storage, measured rates of sediment deposition do not produce the magnitude of storage required on annual to decadal time scales. In addition, plotting sediment yield against drainage area on a non-log scale indicates that sediment yields decline most rapidly in the upper parts of a catchment, where floodplain storage is smaller. Lower rates of change in sediment yield occur in the lower portion of a catchment, where floodplain storage is commonly larger. In other words, observed rates of change in sediment yield within different portions of a drainage basin do not correspond to spatial changes in sediment storage within a basin. The SDR approach also assumes that suspended sediment is not reactive within the river corridor. However, because organic matter can form a substantial portion of suspended sediment flux (Worrall et al. 2014a), production and decay of organic matter within the river corridor can change suspended sediment volumes. Examining a large number of catchments in the United Kingdom, Worrall et al. (2014b) found that linear extrapolation of the SDR approach overpredicts source terms and underpredicts fluxes for large catchments. Change in yield with catchment area may reflect a change in sediment supply from channels rather than a change in delivery from hillslope sources. Consequently, change in suspended sediment flux with catchment area might be better modeled as a step function representing catchment-area thresholds at which the rate of suspended sediment flux changes (Worrall et al. 2014b).

All rivers with large sediment loads originate in mountains, and the majority of sediment load within most river basins comes from the mountainous portion of the basin (Milliman and Syvitski 1992), although recent estimates of global denudation rates based on cosmogenic isotopes suggest that the majority of sediment reaching the oceans comes from more slowly eroding, but areally extensive, lower-gradient surfaces (Willenbring et al., 2013). The correlation between sediment load and small, steep catchments reflects the fact that such catchments typically have greater within-hillslope connectivity, hillslope-channel connectivity, and within-channel connectivity for surface water and sediment. Steep hillslopes, numerous first-order channels, and steep, narrow valley bottoms limit sediment storage in colluvial (hillslope) and alluvial sites, creating efficient delivery of sediment into and through river networks. Magnitudes and time scales of sediment storage tend to increase in larger basins where wide valley bottoms buffer channels from hillslope inputs and provide larger storage areas with terraces, floodplains, and alluvial fans, although the relationship between drainage area and sediment yield (Walling 1983; De Vente et al. 2007) or residence time (Dietrich and Dunne 1978; Wohl 2015b) is not particularly strong or linear because of complications associated with factors such as land cover and land use.

The influence of climate on sediment inputs appears in plots of sediment yield versus a variable such as effective precipitation (the precipitation that actually contributes to runoff). Such plots originated with Langbein and Schumm (1958), who used sedimentation records from the United States to demonstrate that sediment yield peaks at ~300 mm effective precipitation. Lesser values of precipitation produce insufficient runoff to mobilize large quantities of sediment. Continuous vegetation cover effectively limits erosion at higher precipitation values. Subsequent work indicates a secondary sedimentation peak in the seasonal tropics (e.g. Douglas 1967). Despite the mostly continuous vegetation cover in the tropics, the great intensity of precipitation may limit the ability of vegetation to retard runoff and surface erosion or, in steep terrains, mass movements such as landslides. Cosmogenic isotopes can be particularly useful in understanding Quaternary-scale changes in sediment yield associated with climatic variability through time (Dosseto and Schaller 2016; Garcin et al. 2017).

The influence of lithology on sediment yield reflects relative rates of bedrock weathering and erosion, as well as the characteristics of weathering products. Other factors being equal, more erodible or clastic sedimentary rocks tend to yield greater sediment volumes than rocks that are more resistant to weathering and erosion. Chemical sedimentary rocks such as limestone also typically have low yields of granular sediment because most weathering products move as dissolved load.

The dominant influence on sediment yield in many catchments is land use. Hooke (2000) estimates that humans move more sediment globally than any other geomorphic agent, and sediment yields from areas where natural vegetation and topography have been altered are among the highest in the world. Section 4.8 discusses in more detail how diverse human activities influence sediment yield. Remobilization of legacy sediments introduced to river corridors as a result of human activities can be a particularly important source of sediment inputs in some river corridors. In the eastern United States, for example, erosion of stream banks formed in historic mill-pond sediments not only produces abundant fine-grained sediment (e.g. Schenk and Hupp 2009) that can degrade downstream depositional areas, but also mobilizes contaminants such as excess nutrients and trace metals adsorbed to the sediment particles (e.g. Niemitz et al. 2013).

The volume and residence time of sediment storage in depositional features can also vary greatly. The residence time of floodplain sediments, for example, reflects the balance between rates of sediment accumulation through vertical and lateral accretion, rates of floodplain erosion through bank migration, and the size of the floodplain (Section 7.1). The mainstem Amazon in Brazil is bordered by $\sim$ 90 000 km^2 of floodplain, and floodplain width increases downstream. Estimated mean residence times for floodplain sediments in this portion of the river are $\sim$1000 to 2000 years, and can reach periods as long as 10 000 years (Mertes et al. 1996). Floodplain storage can attenuate the movement of sediment originating from changes in land cover or other disturbances in uplands (Fryirs and Brierley 2001; Fryirs 2013). Differences in travel time within channels and the magnitude and duration of floodplain storage can create temporal discontinuities in sediment yield between tributaries and the mainstem of a drainage basin subject to upland disturbance (e.g. Trimble 1983, 2013).

The residence time of sediment in floodplains and along river networks typically increases with drainage area (Wohl 2015b). At larger drainage areas, flows of sufficient magnitude to "turn over" (i.e. erode) the entire floodplain and the majority of storage sites along the channel are very rare (Harvey 2002), in part because precipitation capable of creating a high percentage of contributing area within a drainage becomes infrequent at very large drainage areas. A relatively small convective cell can produce heavy rainfall over all of a small drainage basin, increasing the chances that the entire basin will be contributing and that widespread erosion will occur (Hirschboeck 1988). Such basin-wide precipitation is more common in small basins, creating shorter residence times for sediment and higher delivery ratios.

Sediment output varies not only with the occurrence of precipitation and flows capable of mobilizing sediment in colluvial and alluvial storage areas, but also with the duration of flows capable of transporting the sediment. Once the entrainment threshold in a channel is exceeded, sediment transport tends to increase with flow duration if the system does not become supply-limited (Wolman and Miller 1960; Andrews and Nankervis 1995). The magnitude of the entrainment threshold thus exerts an important control on sediment output. The majority of the sediment output in sand-bed channels with relatively low entrainment thresholds can occur during relatively frequent flows (Wolman and Miller 1960). In contrast, the majority of sediment output in very coarse-grained or otherwise resistant-boundary channels may occur during infrequent, high-magnitude flows (e.g. Lenzi et al. 2004; Rathburn et al. 2017).

Sediment inputs, storage, and yield from a catchment can be measured using field data or estimated using empirical equations (Table 4.3). Field measurements should be interpreted with caution because several studies demonstrate that rates based on relatively short-term monitoring typically exceed long-term rates estimated using techniques such as cosmogenic isotope dating (Clapp et al. 2000; Bierman et al. 2005). Disparities between short- and long-term rates can reflect changes in climate, tectonic environment, land cover, and land use that influence sediment production, storage, and delivery (Gellis et al. 2004; Page et al. 2008). Short-term rates can also either over- or underrepresent the influence of large, infrequent sediment inputs and fluxes, depending on whether the time period of direct measurement does or does not include such episodic events. The time-scale dependence of erosion rates is likely to apply to almost every geomorphic system (McElroy et al. 2017).

Net changes in sediment storage over a specified time interval can also be used to constrain inputs and storage. The increasing availability of aerial and terrestrial LiDAR (Heritage and Hetherington 2005) and boat-based sonar and laser scanners (Alho et al. 2009) facilitates these approaches, as do repeat topographic surveys with high spatial resolution, which can be used to produce digital elevation models of difference maps for estimating the net change in sediment storage (Wheaton et al. 2010; Cavalli et al. 2017).

The ability to effectively integrate measured sediment storage into a sediment budget depends on understanding chronology. Directly measured changes in sediment volume cover only short time intervals that, like sediment yield, may or may not represent longer time periods. The age or residence time of sediment in storage can be estimated using several techniques that cover longer time intervals (Table 4.3).

Sediment source areas can be inferred using mineralogic or petrologic characteristics specific to distinct subareas within a catchment (Vezzoli 2004; Hatfield and Maher 2009), an approach known as *sediment fingerprinting* (e.g. Gellis and Walling 2011; Pulley et al. 2015). The spatial characteristics of the sources of sediment and processes of sediment transfer or storage can be modeled using GIS platforms (e.g. Wilkinson et al. 2006).

4.8 Human Influences on Sediment Dynamics

Human effects on almost every aspect of river form and function are now nearly ubiquitous, and sediment dynamics are no exception. Global-scale sediment fluxes are very difficult to quantify accurately as a result of spatial and temporal fluctuations within each drainage basin, as well as incomplete systematic measurements of sediment flux in most basins, but global-scale estimates have been published. People have simultaneously increased the sediment transport by global rivers through soil erosion (Hooke 2000) by more than 2 billion metric tons per year (over a total estimated 14 billion tons per year prior to human alteration), but reduced the flux of sediment reaching the world's coasts by 1.4 billion metric tons per year due to retention within reservoirs behind dams (Syvitski et al. 2005).

At the scale of individual drainage basins, human activities influence the weathering of bedrock and production of sediment in upland areas and the movement of sediment from uplands into river corridors through alterations of land cover and topography, as discussed in Section 2.3. Although deforestation typically increases sediment yield to channels (e.g. Kondolf et al. 2002) and urbanization typically decreases sediment yield, enlargement of first-order streams can cause increases in sediment yield between components of the channel network in mature urban areas (Smith and Wilcock 2015).

Table 4.3 Techniques used to quantitatively estimate parameters in sediment budgets and to constrain the chronology of processes.

Description of technique	Sample references
Sediment inputs	
Field measurements and methods of inferring rates and magnitudes	
• sediment fences capture sediment mobilized upslope from the fence; erosion pins record rates of ground-surface lowering	Robichaud et al. (2008)
• rates of surface lowering can be estimated from exposed tree roots	Carrara and Carroll (1979)
• sediment inputs from bank erosion can be inferred from rates of bank migration, the height of banks, and the longitudinal extent of bank erosion	
• inputs from bed erosion can be inferred from the thickness and extent of bed lowering	
• cosmogenic isotopes can be used to infer rates of bedrock outcrop erosion and sediment generation, as well as the flux of sediment down hillslopes and through river networks over longer time spans than those of direct measurements	Bierman and Nichols (2004), Nichols et al. (2005), and Dosseto and Schaller (2016)
• net changes in sediment storage over a specified time interval can be used to infer sediment inputs based on changes in total volume (e.g. aerial photographs, field measurements) or changes in the mobility of tracer clasts	Wathen et al. (1997) and Wichmann et al. (2009)
Models	
• upland sediment supply can be estimated using modifications of the Universal Soil Loss Equation (USLE), originally developed for croplands	Wischmeier and Smith (1978) and Smith and Dragovich (2008)
• models developed for steeper catchments and drainage basin evolution models such as CAESAR	Coulthard et al. (2005), Martinez-Carreras et al. (2007), and Coulthard and Van De Wiel (2017)
Sediment yield	
Field measurements	
• quantify sediment accumulation in natural or artificial reservoirs based on coring and dating or repeat bathymetric surveys	Lisle and Hilton (1992), Gellis et al. (2006), and Schiefer et al. (2006)
• quantify sediment fluxes (solute, wash, suspended, and bed-load) past a reference point	Ferrier et al. (2005)
Equations and models	
• suspended or bed-load transport equations for in-channel transport	Sear et al. (2003)
• models that include integrated hillslope–channel processes (e.g. SHETRAN/SHESED)	Wicks and Bathurst (1996) and Pandey et al. (2016)
Sediment storage	
Field measurements	
• shallow geophysical methods such as ground-penetrating radar to quantify sediment deposit thickness	Schrott et al. (2003) and Kramer et al. (2012)
• automated routines based on slope-gradient criteria to extract areas of post-glacial fluvial and lacustrine valley fills from digital topography at large spatial scales of entire mountain ranges	Straumann and Korup (2009)

Table 4.3 (Continued)

Description of technique	Sample references
Chronology	
• soil isotopic ratios influenced by nuclear weapons tests since the 1950s	Stokes and Walling (2003), Walling and Foster (2016)
• manufacture dates of garbage buried in alluvial sediments	Kurashige et al. (2003)
• radiocarbon dating of organic material in the sediment, which can cover approximately the past 80 000 years	Bradley (1985)
• luminescence dating, which measures energy accumulated as the result of decay of naturally occurring radioactive uranium isotopes following sediment burial, with exposure of the sediments to sunlight releasing stored energy, typically over time spans less than 100 000 years	Fuchs and Lang (2009)
• naturally occurring cosmogenic radionuclides	Reusser and Bierman (2010)

The net effect of human activities is to increase sediment supply to river corridors, especially sand-size and finer sediment: in 2011, the US Environmental Protection Agency listed excess fine sediment as the most widespread river pollutant.

Within the river corridor, flow regulation influences the energy available to transport sediment, and channel and floodplain engineering influence the balance between transport and storage of sediment. These effects can partially offset the greater yield of sediment to river corridors, as reflected in a study of 10 large river basins in the southeastern United States. ^{10}Be-based estimates of background erosion prior to European settlement of the area and associated changes in land cover indicate that, following peak disturbance of uplands, rates of hillslope erosion exceeded background rates more than 100-fold (Reusser et al. 2015). However, even though basin-scale sediment yields increased 5–10 times above pre-settlement rates, rivers transported only about 6% of the eroded material. The bulk of the anthropogenically disturbed sediment was stored at the base of hillslopes and along valley bottoms, partly because of flow regulation and abundant mill dams and ponds (Walter and Merritts 2008).

Table 4.4 lists the types of human activities occurring within the river network that directly alter sediment dynamics. The net effect on sediment dynamics of each depends on the context. Channel engineering in the form of channelization, for example, commonly exacerbates channel-boundary erosion within the channelized reaches and sediment deposition downstream, whereas levees can exacerbate channel-boundary erosion within the leveed reach.

Likely the greatest alteration of suspended and bed-load transport in many rivers is associated with flow regulation. Dams effectively trap all bed load and much of the suspended load moving down a river, leading to dramatically reduced sediment supplies downstream and a range of channel adjustments, from enhanced bank erosion to bed coarsening and incision (Kondolf et al. 2014). These effects are exacerbated by multiple dams in a watershed (e.g. Rubin et al. 2015; Latrubesse et al. 2017). Globally, over 100 billion metric tons of sediment are now stored in dams (Syvitski et al. 2005).

The effect of a dam on downstream water and sediment fluxes and channel response can be conceptualized using dimensionless ratios of sediment supply upstream from the dam to sediment supply downstream of the dam (S^*) and the fractional change in sediment transporting flows from pre- to post-dam states (T^*) (Grant et al. 2003). When $T^* \gg S^*$, the channel will coarsen, incise, erode bars

Table 4.4 Types of human activities within river networks that alter sediment dynamics.

Category
Altered channel boundary erosion
Flow magnitude altered by changes in land cover or flow regulation
Bank resistance altered by changes in riparian vegetation (including those caused by riparian grazing)
Bed and bank resistance altered by placer and aggregate mining
Distribution of hydraulic forces altered by channel engineering (channelization, bank stabilization, levees)
Distribution of hydraulic forces altered by removal of natural obstructions such as large wood or beaver dams
Increased or decreased flow transport capacity
Flow magnitude, frequency, and duration altered by dams and diversions
Blockage of downstream sediment movement and storage
Dams

and islands, or widen. When $S^* \gg T^*$, the channel will develop a finer bed substrate, aggrade, develop larger bars and islands, or narrow.

Schmidt and Wilcock (2008) quantify this conceptualization by proposing three metrics for assessing the downstream effects of dams based on: (i) sediment mass balance, as reflected in pre- and post-dam ratios of slope, rate of sediment supply, discharge, and grain size; (ii) bed incision potential, as reflected in the Shields number; and (iii) flood reduction, as reflected in pre- and post-dam ratios of the 2-year flood. The metrics are designed to identify conditions of sediment deficit or surplus resulting from the balance between changes in sediment supply and transport capacity.

One mitigation measure for sediment-depleted river segments downstream from dams is to artificially supply sediment to the channel. Such *gravel augmentation* requires decisions regarding volume and grain-size distribution of the sediment supplied, frequency and timing of augmentation, and method of sediment delivery (placement in the channel bed or injection from the channel margins) (Bunte 2004). Effective mitigation also requires some understanding of how the sediment will be stored or mobilized and transported downstream (Sklar et al. 2009) and of how biota are likely to utilize the introduced gravel (Staentzel et al. 2018). Sites along the Mokelumne River, California, USA that received up to 1300 m^3 of gravel, for example, lost up to 50% augmented gravel over 4 years (Merz et al. 2006).

Dams are also being removed in many locations across the United States. At least 121 were removed between 1930 and 1999, and 20–50 were decommissioned each year during the first decade of the twenty-first century (O'Connor et al. 2008). These dams can have substantial accumulations of sediment upstream, and their physical removal can rapidly introduce large volumes of fine sediment into the downstream river. Numerous field and flume studies describe the effects of dam removal (e.g. Stanley et al. 2002; Lorang and Aggett 2005; Major et al. 2008), and several time-lapse videos of sediment erosion from behind a dismantled dam are now available online. One of the largest dam removals to date involved two dams on the Elwha River in Washington, USA. The removal was completed in 2014, and multiple papers chronicle subsequent sediment dynamics and channel response (e.g. East et al. 2015; Magirl et al. 2015; Warrick et al. 2015).

The cumulative, watershed-scale effects of human alterations of sediment dynamics appear along coastlines, with accelerating erosion resulting from depleted riverine sediment transport. Interruptions of watershed-scale sediment cascades can start with artificial hillslope stabilization and trapping of sediment in sediment-detention basins. This decreases sediment supply to rivers, triggering accelerated bed and bank erosion. Channel engineering designed to stabilize beds and banks exacerbates depletion of sediment supply to depositional areas such as deltas and estuaries, resulting in enhanced coastal erosion, as discussed in Section 7.4.4.

4.9 The Natural Sediment Regime

A sediment budget can provide either a conceptual or a quantitative framework for understanding the relative magnitudes of inputs, outputs, and storage, as well as temporal and spatial variations in sediment dynamics. Interactions among sediment inputs, outputs, and storage and the valley context over a specified time span and spatial scale can also be described as the *sediment regime* of a river corridor. Valley context includes valley geometry, position in the network, channel substrate and vegetation, and base-level stability (see Figure 1.1). Water and sediment interact within the valley context to govern river geometry, aquatic and riparian habitat, and the disturbance regime for river biota (Bellmore and Baxter 2014).

Valley context is likely to exert an especially important influence on sediment connectivity, via its influences on sediment production and transport to the river corridor, as well as transport and storage within the corridor (Bracken et al. 2015). Sediment is in storage during the majority of the time that the sediment is present within the river corridor. This storage reflects sediment disconnectivity caused by the combined effects of a lack of sufficient transport energy (e.g. during base flow) and features such as alluvial fans, which interrupt sediment delivery to the channel, and valley constrictions, which limit downstream transport of sediment by creating backwater effects (Hooke 2003; Fryirs et al. 2007a; Fryirs 2013).

The physical configuration of the river corridor and aquatic and riparian communities present within the river corridor reflect the spatial pattern and temporal variability of interacting water and sediment regimes. The sediment regime is much more difficult to quantify than the flow regime of water, however, for at least two reasons. First, there are relatively few sites at which suspended sediment discharge is measured compared to measurements of water discharge, and sites with long-term measurements of bed-load discharge are almost nonexistent (Wohl et al. 2015b). Second, storage of sediment is as important as fluxes of sediment, and sediment can remain in storage for thousands of years.

The *natural sediment regime* describes conditions prior to the construction of dams and intensive human disturbance of topography and land cover (Wohl et al. 2015b). A natural sediment regime is now rare because of the intensity and ubiquity of human alteration of land cover (sediment inputs) and river corridor modification (sediment storage and transport). Consequently, it may be more practical in human-altered watersheds to manage for a balanced sediment regime. A *balanced sediment regime* is present when the energy of flow available to transport sediment is in balance with sediment supply, such that the form of the river corridor remains dynamically stable over a specified time period (Wohl et al. 2015b). A balanced sediment regime can occur in the absence of human alteration, as in a natural sediment regime, or under human-altered conditions in which both altered water and sediment supplies are in balance. Time period is critical in this context. River process and form can change substantially during a single large flood, which can be interpreted as a lack of balance in the sediment regime. Over longer time periods, however, river process and form might show little or no net change. "Dynamically stable" is used in the description of a balanced sediment regime to indicate no net change, rather than static conditions.

If water and sediment supply and other conditions in a watershed have been altered by human activities, the resulting dynamically stable river corridor can be distinctly different than what would be present under natural conditions. Under this scenario, critical management questions are likely to be, What are the supplies of water and sediment? And, What river system structure and function can be achieved under a modified flow regime and balanced sediment regime? (e.g. Wilcock 2012). Managing for a balanced sediment regime may involve either restoring more natural water and sediment inputs to a river or adjusting flow regime to create desired levels of sediment transport given an existing sediment supply (Schmidt and Wilcock 2008). In either scenario, the effective management of river condition requires knowledge of sediment regimes.

4.10 Summary

Natural channels differ from engineered channels in being able to adjust boundary roughness, channel geometry, and sediment transport in response to changes in water, sediment, and large wood supply. Most of these adjustments involve sediment dynamics – the entrainment, transport, and deposition of individual particles and aggregates of particles.

Historically, quantification of sediment dynamics began with the simple scenario of a channel bed with relatively uniform, sand-sized grains that could be readily collected and sieved to quantify the grain-size distribution. Sand-bed channels have a relatively low entrainment threshold and can be adequately represented by bulk sediment transport equations. The entrainment and transport equations developed for sand-bed channels are based on estimation of the magnitude of shear stress above a critical value necessary to mobilize sediment, as formulated by Shields. These equations remain the starting point for characterizing sediment dynamics.

Quantitative studies of sediment dynamics have expanded to include a progressively broader range of natural channels, from fine-grained cohesive beds, to gravel- or boulder-beds, to bedrock beds. Entrainment and transport in these types of channels are not adequately approximated by an excess-shear-stress relation for entrainment or a bulk-transport relation because of effects such as shielding and protrusion and limited sediment supply. Mobilization of particles from beds of mixed grain sizes occurs in phases. Equations using two-fraction transport or grain–grain interactions better capture the processes operating in these channels.

The distinction between readily mobile and infrequently mobile bedforms reflects basic differences in thresholds of entrainment and mobility between sand-bed and other channel types. The readily mobile bedforms of ripples, dunes, and antidunes respond to a range of flow magnitudes. Infrequently mobile bedforms, including particle clusters, transverse ribs, and step–pool and pool–riffle sequences, change most actively under relatively high flow magnitudes.

Sediment can also be eroded from or deposited along stream banks. As with streambeds, bank erosion reflects the frictional properties of the sediment. Bank erosion is also strongly influenced by bank stratigraphy, pore-water pressure, and riparian vegetation, and is more likely than bed erosion to take the form of mass failure. As with bed sediment, efforts to quantitatively model and predict bank erosion are limited by large spatial and temporal variability in sediment properties and in the forces acting on the sediment.

A mass balance of sediment in the form of a sediment budget reflects the inputs, storage, and outputs of sediment across a reference area that can vary from a small subcatchment to the Amazon. Regardless of spatial scale or geographic setting, sediment budgets reveal that sediment comes disproportionately from limited areas of a catchment and moves disproportionately during limited intervals of time. Sediment export includes solute, wash, suspended, and bed-load sediment, but the majority moved from most river catchments is transported in suspension. When combined with the valley context, a sediment budget can be used to characterize the sediment regime. A natural sediment regime reflects sediment dynamics and valley configuration unaltered by humans. A balanced sediment regime occurs when the energy of flow available to transport sediment is in balance with sediment supply and the form of the river corridor remains dynamically stable over a specified time period. This may correspond to a natural or an altered sediment regime.

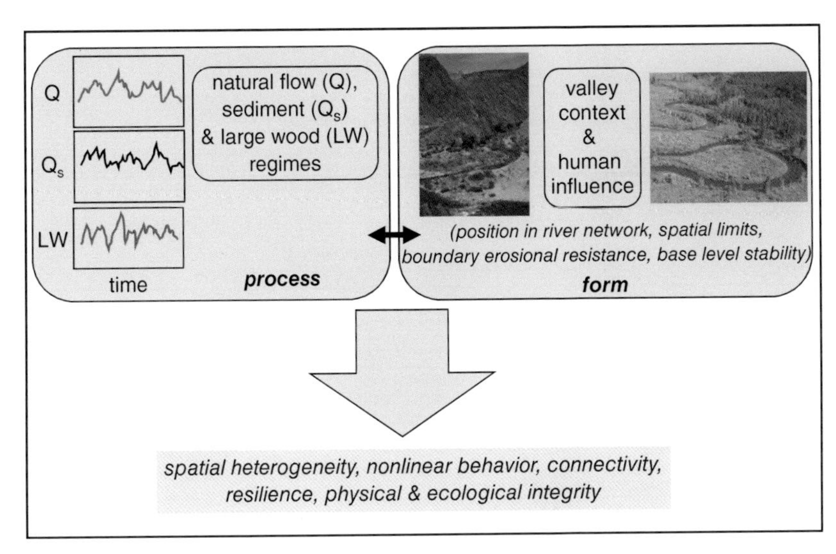

Figure 1.1 Schematic illustration of the primary inputs to river corridors (water, sediment, large wood) and the context in which they interact with one another and with the river form to create the integrative river corridor characteristics listed in the lower portion of the figure.

Figure 1.2 Channel-spanning logjam in the Rocky Mountains of Colorado, USA. Where logjams are not present, the stream has cobble- to boulder-size substrate, high transport capacity, and minimal storage of fine sediment and organic matter. Each logjam, in contrast, creates a backwater of lower-velocity flow that traps fine gravel, sand, and silt, as well as small logs, branches, and pine cones and needles. In the photograph, flow is from right to left.

Rivers in the Landscape, Second Edition. Ellen Wohl.

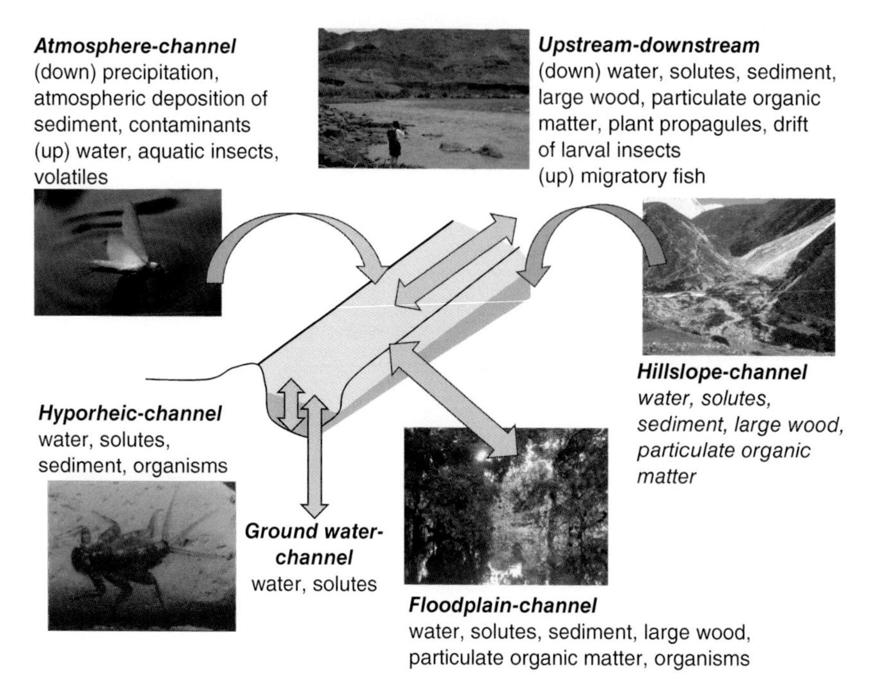

Figure 1.3 Schematic illustration of the six degrees of connection between rivers and the greater landscape. The segment of channel (lighter gray) shown here is connected to: upstream and downstream portions of the river network; adjacent uplands; the floodplain; ground water; the hyporheic zone (darker gray); and the atmosphere. The photograph representing upstream–downstream connection was taken during a flood on the Paria River, a tributary of the Colorado River that enters just downstream from Glen Canyon Dam in Arizona, USA. In this view, the Paria is turbid with suspended sediment whereas the Colorado, which is released from the base of the dam, is clear. The photograph representing hillslope–channel connection shows a large landslide entering the Dudh Khosi River in Nepal. The photograph respresenting floodplain–channel connection was taken along the Rio Jutai, a blackwater tributary of the Amazon River, during the annual flood in early June. In this view, the "flooded forest" is submerged by several meters of water. The photograph respresenting hyporheic–channel connection shows a larval aquatic insect (macroinvertebrate) as an example of the organisms that can move between the channel and the hyporheic environment. The photograph respresenting atmosphere–channel connection shows a mayfly emerging from the river prior to entering the atmosphere as a winged adult. Source: Image courtesy of Jeremy Monroe, Freshwaters Illustrated.

Figure 1.4 Landscapes and river corridors in and adjacent to the Colorado Front Range. Upper left: View east from the summit surfaces at the continental divide. The coarse blocks in the foreground are periglacially weathered boulders and bedrock. The surface drops steeply into a glaciated valley that transitions downstream (out of sight) into a fluvial valley. Upper right: View northwest from a hogback, an asymmetrical hill of sandstone and limestone strata dipping steeply to the right in this view, with an intervening valley formed in shales. Lower right: The South Platte River near Fort Morgan, Colorado, in the low-relief environment of the Great Plain. This sand-bed channel was historically much wider and had a braided planform, but flow regulation has resulted in encroachment of riparian vegetation and transformation to a single relatively narrow channel. This river heads high in the mountains. Lower left: View of smaller drainages that head on the Great Plains, here at Pawnee National Grassland. These channels have downcut within the past few decades, largely via piping erosion.

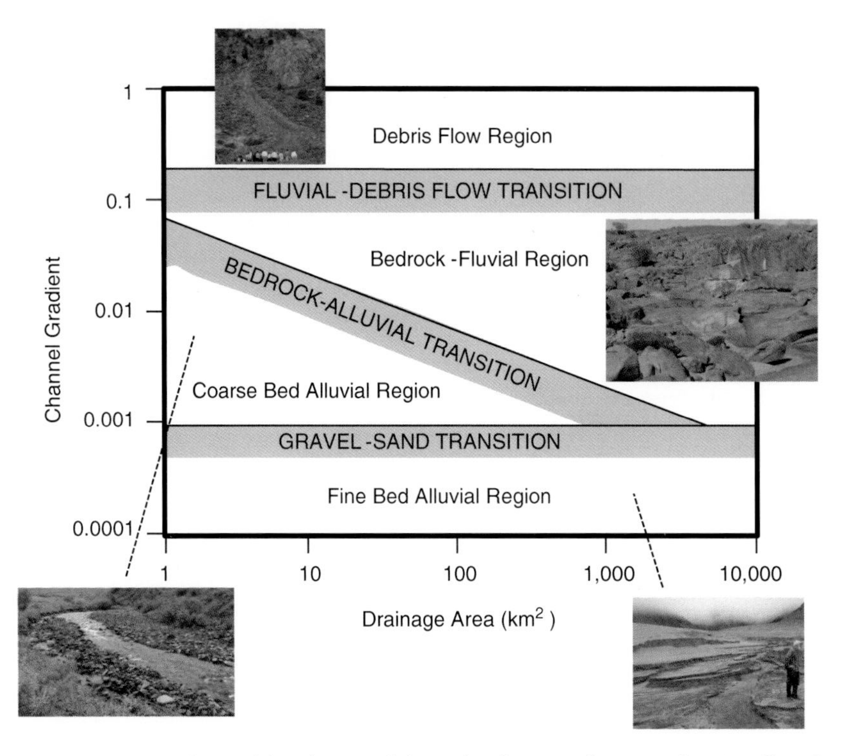

Figure 2.11 Hypothesized distribution of channelized erosional process domains (fluvial versus debris flow) and channel substrate types in relation to drainage area and slope. Source: After Sklar and Dietrich (1998), Figure 1, p. 241.

Figure 3.16 Flood-scarred Ponderosa pine (*Pinus ponderosa*) along Rattlesnake Creek, Arizona, USA. Two scars are visible in this photo: a larger scar at center, the base of which has been cut for tree-ring sampling, and a smaller scar at upper right.

Figure 3.17 Adventitious roots of a mesquite tree exposed along a cutbank on Cienega Creek, Arizona, USA. Repeated burial during overbank sedimentation has caused the tree to grow at least three sets of roots, as visible here. Hat for scale.

Figure 3.18 Paleochannel exposed in the cutbank of a contemporary arroyo in eastern Colorado, USA. The fill of the paleochannel displays a fining upward sequence from cobbles to silt and clay at the top of the channel fill. Daypack on ground below paleochannel for scale.

Figure 3.19 (a) Silt lines along the back wall of an alcove formed in sandstone along the banks of the Paria River, Utah, USA. The overhanging roof of the alcove protects the silt lines from erosion. The top of a sequence of slackwater deposits is visible at lower right. (b) Close-up view of a silt line, showing the fine organic matter adhering to the bedrock wall in association with the silt. This organic matter can be used to radiocarbon date the age of the flood deposits.

Figure 3.20 Slackwater deposits protected by an overhanging alcove in the bedrock walls along Buckskin Gulch in Arizona, USA.

Figure 3.24 La Poudre Pass Creek in Colorado, USA. Flow in this creek is augmented by diversions from the headwaters of another river. (a) One portion of the creek at (1) 0.15 m^3/s, (2) 0.70 m^3/s, and (3) 4.52 m^3/s. During low flows, eroded banks are exposed and flow is exceptionally shallow. Water in the channel freezes to the bed during the winter, effectively precluding overwinter survival of fish in the creek. Prior to flow regulation, winter base flow was slightly higher and fish were able to survive through the winter. Source: After Wohl and Dust (2012), Figure 4A.

Figure 4.4 Numerous potholes along the Ocoee River, Tennessee, USA. Hand tape resting on bedrock at left-center of photo for scale. Flow is from left to right.

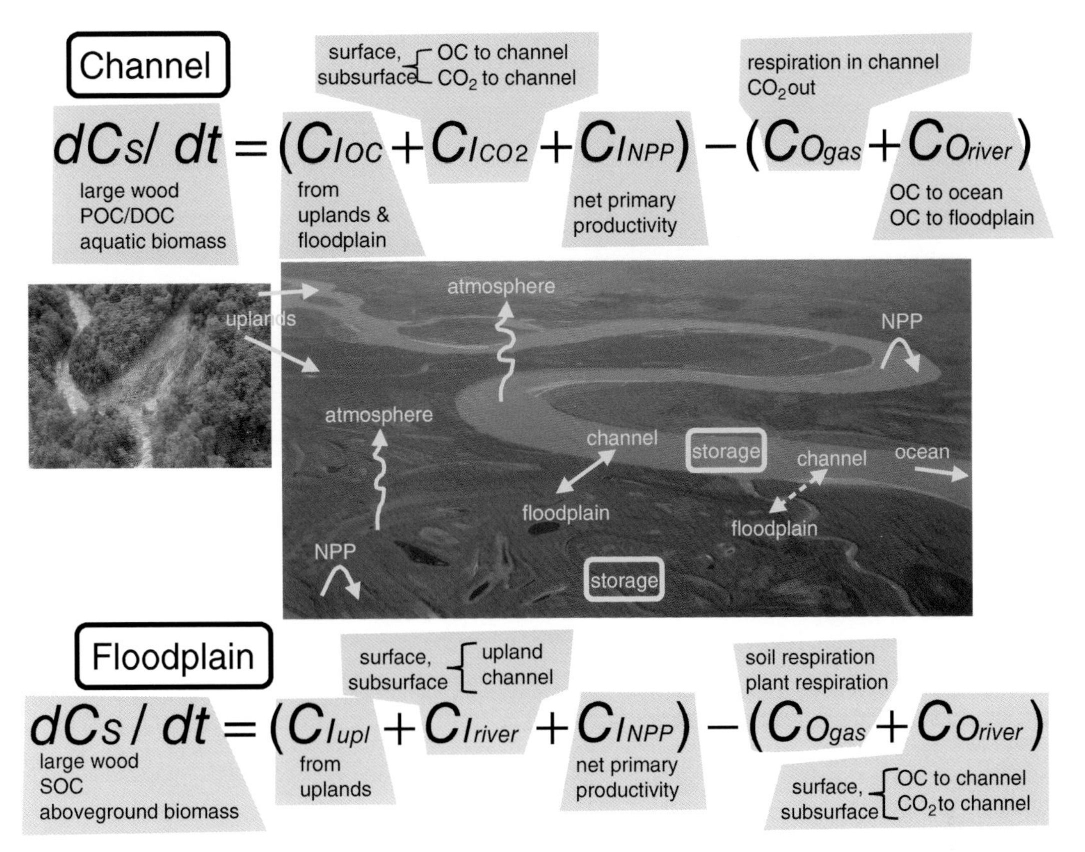

Figure 4.5 Schematic illustration of the components of a budget for organic carbon within a river corridor at the reach scale. C_S indicates storage, C_I indicates inputs, C_O indicates outputs. NPP is net primary productivity, POC is particulate organic carbon, DOC is dissolved organic carbon, SOC is soil organic carbon. Gray shading is intended to make it easier to see which explanatory terms (smaller font) correspond to variables in the budget (larger, italicized font). Within the central photo, yellow arrows indicate fluxes (dashed arrows are subsurface fluxes, squiggly arrows are fluxes to the atmosphere) and yellow boxes indicate storage.

(a)

(b)

Figure 4.10 Coarse surface layer (a) and underlying finer sediment (b) on an alternate bar along the Poudre River, Colorado, USA. Scale with inches (right) and centimeters (left) at upper left in (b).

Figure 4.11 Upstream view of the discharge and sediment measuring station on the Rio Cordon in the Dolomite Mountains of Italy. Drainage area above station is 5 km^2. The measuring station includes an inlet flume; an inclined grid for the separation of coarse particles (> 20 mm diameter), highlighted in this photograph by the irregularly edged ice present along the left side of the grid; a storage area for coarse sediment deposition downstream from the grid; and an outlet flume to return water and fine sediment to the stream, at the far left of the photograph. Bed-load volume is measured at 5-minute intervals by 24 ultrasonic sensors fitted on a fixed frame over the storage area and via topographical surveys after each flood. Suspended sediment is measured by turbidimeters. Flow samples are gathered automatically using a pumping sampler installed at a fixed position in the inlet channel. Measurement details from Mao and Lenzi (2007). Source: After Wohl (2010b), Figure 3.38.

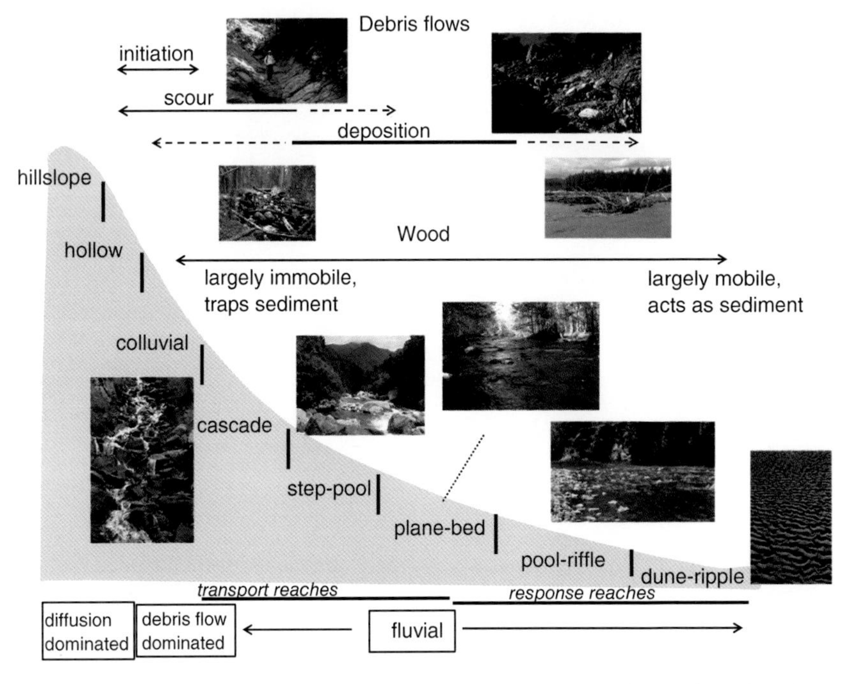

Figure 4.15 The Montgomery and Buffington (1997) channel classification based on dominant bedforms. This schematic illustration of downstream trends includes changes in reach-scale gradient (light gray shading), channel type (cascade/step–pool/plane–bed/pool–riffle/dune–ripple), indication of transport versus response reaches, dominant sediment transport process (hillslope diffusion/debris flow/fluvial), and the behavior of debris flows and instream wood. Source: After Montgomery and Buffington (1997), Figure 4.

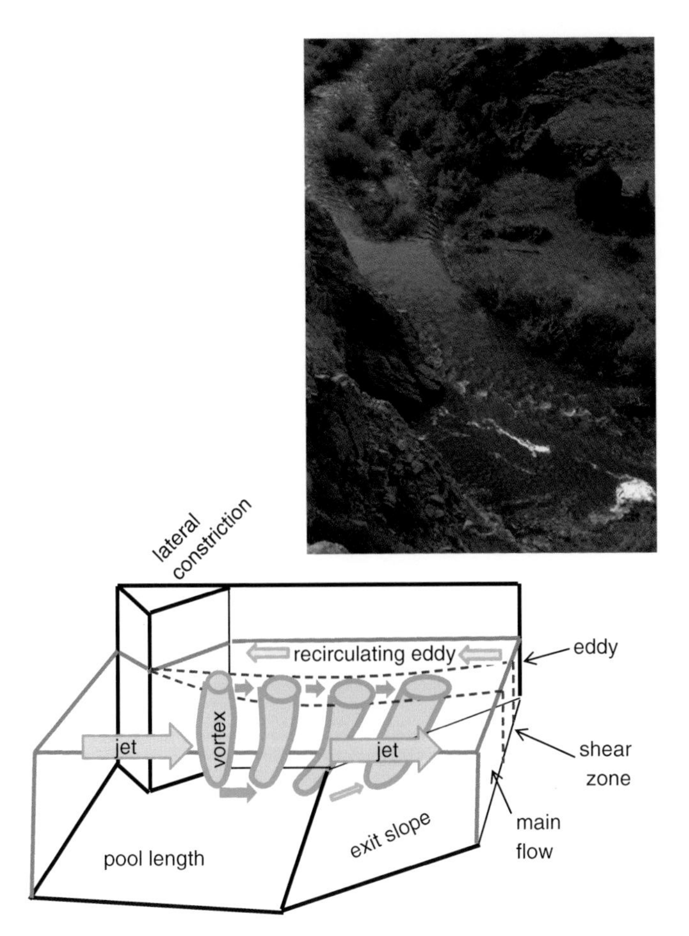

Figure 4.18 Photograph shows a pool with a lateral bedrock constriction, at river left. In this view, during high flow, naturally occurring foam at the water surface highlights the line of flow separation between the eddy at the base of the photo and the jet of high-velocity flow in the pool center, which has a series of standing waves. The deepest portion of the pool is just downstream of the bedrock constriction that appears at the middle left of the photo. As flow diverges in exiting the pool, transport capacity drops and a bar and riffle form. The bar has some vegetation, largely because this is a regulated river and flood peaks have declined, facilitating the encroachment of riparian vegetation. The channel is approximately 20 m wide downstream from the lateral bedrock constriction. Diagram shows an idealized version of flow patterns in a laterally constricted pool. Source: Diagram after Thompson (2004), Figure 1.

Figure 4.20 The lower, weakly cohesive bank along this river in northern Alaska, USA erodes rapidly, but the rate of bank retreat is limited by the overlying finer sediment layer, which is effectively stabilized by the roots of various types of plants. Bank face is approximately 1.5 m tall.

Figure 4.22 Schematic illustration of the basic components of a sediment budget. Inputs, I, to a river segment can come from uplands and glaciers, eolian deposition, mainstem and tributary transport, or erosion of terraces, alluvial fans, floodplains, and other valley-bottom deposits. Storage, S, of sediment can occur in the floodplain or in various portions of the channel. Sediment outputs, ϕ, occur via downstream transport as dissolved, suspended, and bed-load sediment.

Figure 5.2 Illustration of different sources of lateral wood recruitment to a river. Inset photos clockwise from upper left: (i) A floodplain in Costa Rica with buried wood being exposed by bank erosion (1 m-long umbrella for scale within yellow oval). (ii) A beaver dam and pond along the Duke River in the Yukon Territory, Canada. (iii) The toe of a landslide along the Upper Rio Chagres in Panama that introduced large volumes of wood to the river (most of the wood was mobilized by streamflow before this photo was taken; person in yellow oval for scale). (iv) Trees leaning toward the Poudre River in Colorado, USA. (v) Ouzel Creek in Rocky Mountain National Park, Colorado, USA. The forest along this portion of the creek burned in 1978, and standing dead trees have been gradually toppling into the river ever since.

Figure 5.5 *In situ* or autochthonous jams and transport or allochthonous jams. In the case of the *in situ* jams, the smaller pieces have been transported downstream, but have been trapped by a larger, ramped piece that fell into the channel from the adjacent stream bank.

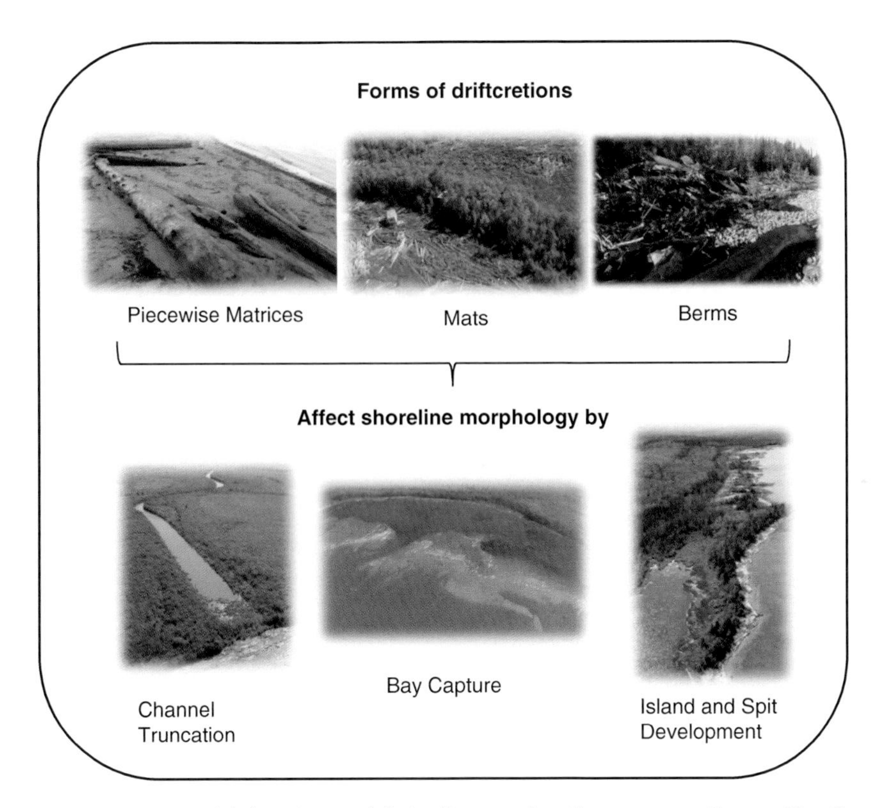

Figure 5.6 Types of driftcretions and their effects on shoreline processes. Source: After Kramer and Wohl (2015), Figure 1.

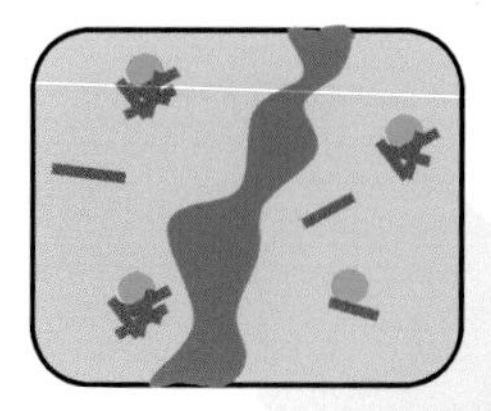

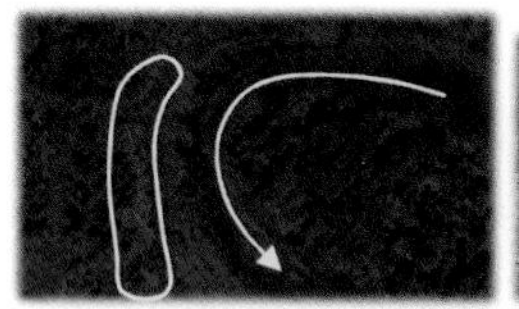

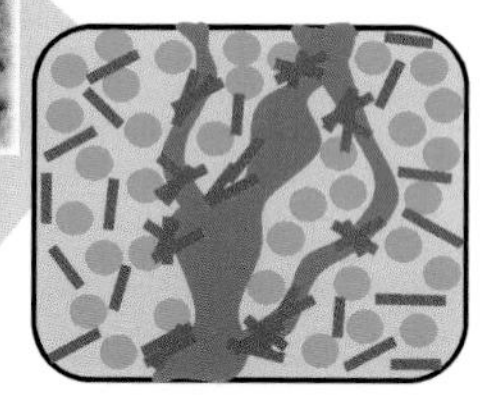

Figure 5.7 End-members for the spatial distribution of large wood on floodplains. At upper left, idealized plan view shows isolated living trees (ovals), with small logjams at the upstream side of each, and a few dispersed pieces of wood. Inset photos show (left to right) jams at the base of widely spaced trees along the Middle Fork Gila River in New Mexico, USA; an aerial view of an extensive floodplain jam (circled) on the inside of a bend along the Middle Fork Gila River; and jams at the base of trees along the Brisbane River in Australia. At lower right, plan view shows densely spaced living trees, with logjams concentrated along the margins of the main channel and secondary channels, and more abundant dispersed wood across the floodplain. The inset photos from the Swan River in Montana, USA show (left to right) an aerial view of a long jam (circled) in an abandoned channel, a ground view of wood in an abandoned channel, and a ground view of wood concentrated along the entrance of a secondary channel. In all photos, arrows indicate flow direction. LW, large wood.

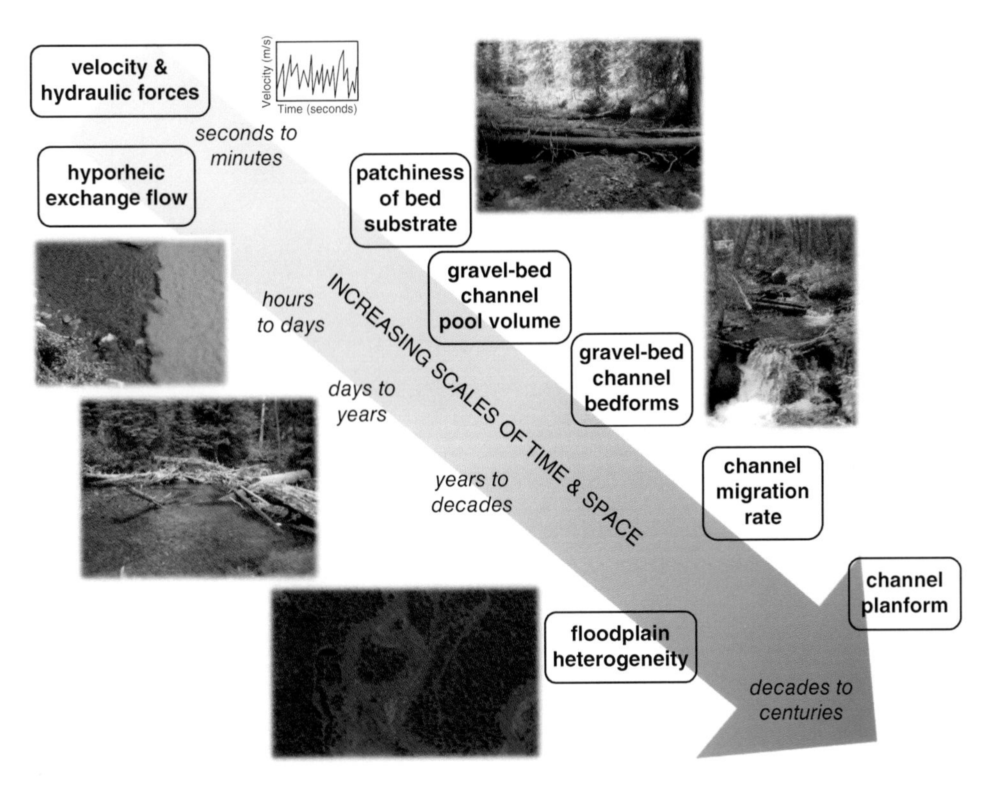

Figure 5.8 Schematic illustration of the diverse spatial and temporal scales at which large wood influences river corridor process and form, from seconds to minutes at the upper left to decades to centuries at the lower right.

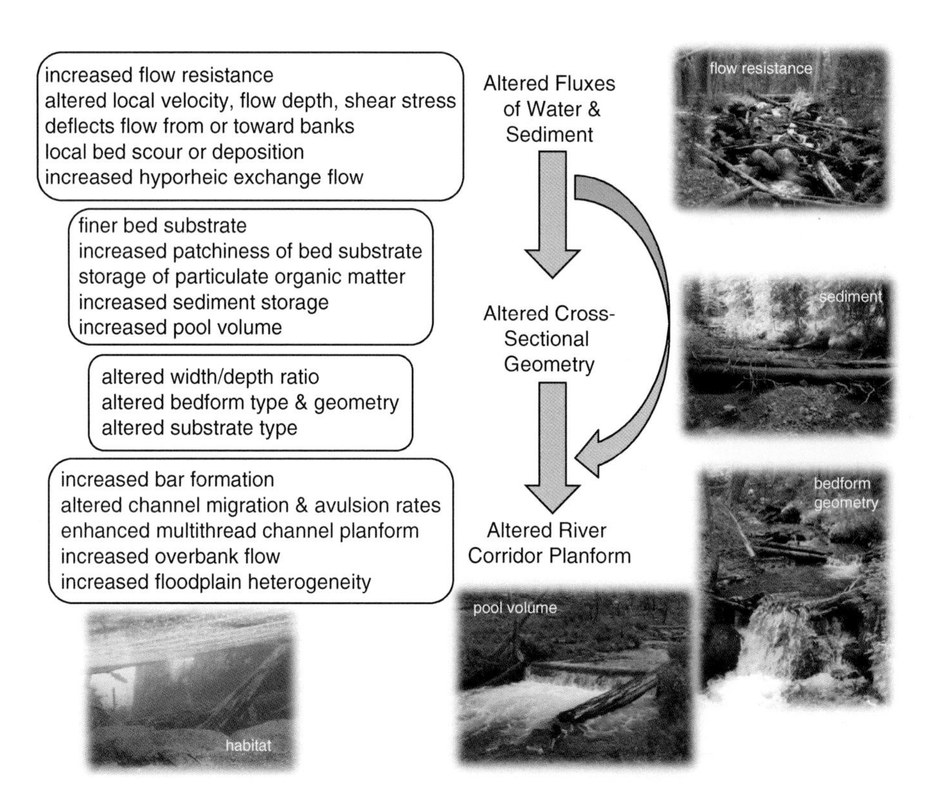

Figure 5.9 The diverse effects of stationary large wood in the river corridor include influences on process (fluxes of water, sediment, mobile large wood, and particulate organic matter) and form from the local scale of a single channel unit (e.g. pool or bar) through cross-sectional geometry to reach-scale planform.

Figure 6.10 (b) Photograph of a wide (~1 km) braided river in northern Alaska, USA during low flow. Flow is from left to right.

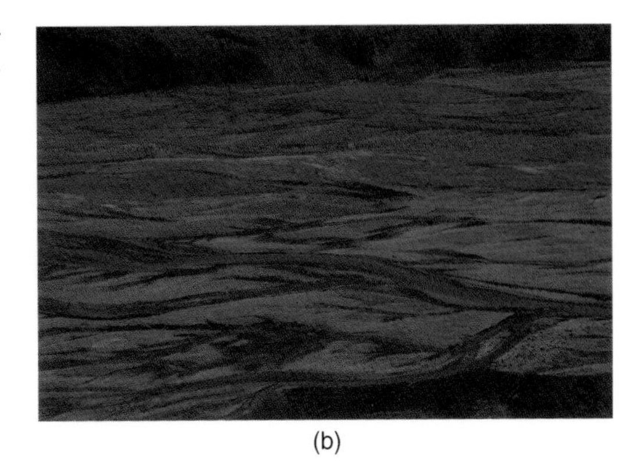

(b)

Figure 6.12 Anabranching planform along a portion of the Yukon River in central Alaska, USA. View is looking upstream.

(a)

(b)

Figure 6.18 Persistent and ephemeral knickpoints along channels. (a) Bedrock knickpoint where a tributary enters the Wulik River in Alaska, USA, representing the inability of the tributary to incise through the bedrock as rapidly as the mainstem river. (The Wulik River flows right to left in the foreground of this view.) (b) Knickpoint along an ephemeral channel tributary to the South Fork Poudre River in Colorado, USA, formed in response to enhanced water yield after a wildfire burned the catchment 4 months before the photograph was taken.

Figure 7.1 The floodplain of the Yukon River in central Alaska, USA. In this view taken from a small airplane, the extent of the floodplain and the complexity of soil moisture, plant communities, small-scale topography, and depositional history are apparent.

Figure 7.3 Examples of scroll bars on the floodplains of sinuous rivers in Alaska, USA. (a) Scroll bars highlighted by differences in vegetation along a river on the coastal plain near Kotzebue in western Alaska. (b) Sandy scroll bars on the inside of a meander bend along a river in central Alaska. As in (a), differences in vegetation highlight the locations of the bars.

(a)

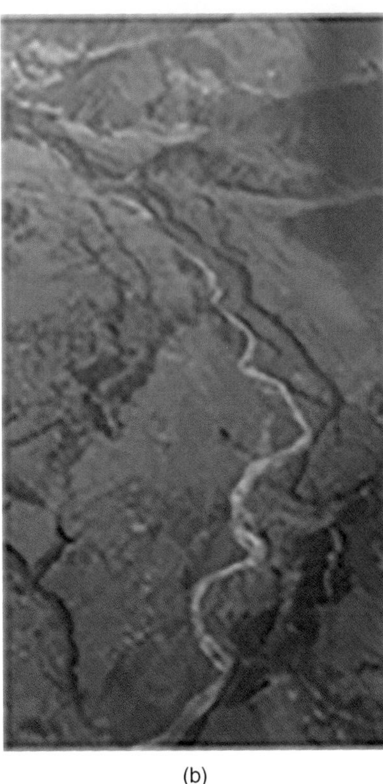

(b)

Figure 7.5 Examples of river terraces. (a) Two small fill terraces (top outer edge of each indicated by a dashed white line) are visible in this view of a small creek in northern California, USA. Successive waves of aggradation and incision along the creek created the higher fill terrace just visible at the base of the forest, then the younger fill terrace about 1 m in elevation below it, and finally lowered and widened the channel, killing the riparian trees and leaving the stumps visible in the photograph. Flow in the channel is from left to right. (b) Aerial view of numerous fill terraces along a river in Nepal. (c) A strath terrace along the Mattole River in California, USA. The white material exposed in the lower portion of the cutbank is bedrock. An upward-fining alluvial sequence of cobbles to loamy soil overlies the bedrock strath. Flow is from the foreground toward the rear in this view. The cutbank is approximately 6 m tall. (d) Distant view of a very tall strath terrace along the Duke River in northwestern Canada. The strath surface is approximately 100 m above the active channel. The lower view shows a close-up of bedrock–fill contact, which is indicated by a dashed line in both photographs. Arrow indicates flow direction on the main channel.

Figure 7.5 (*Continued*)

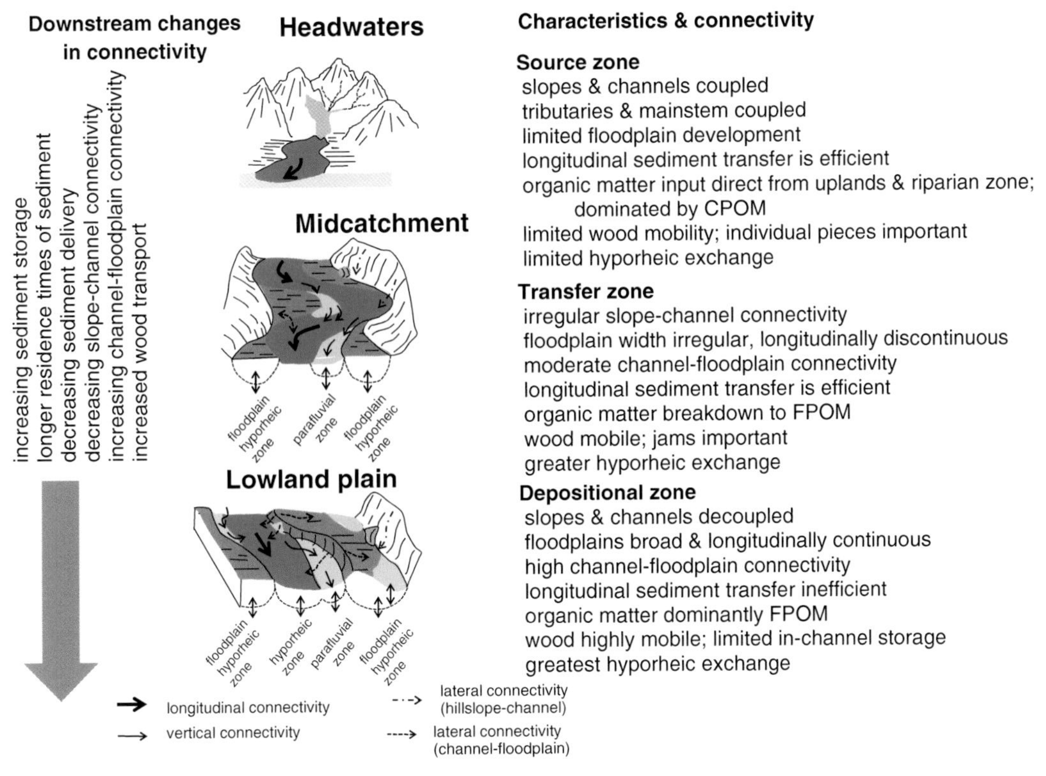

Figure 8.10 Schematic illustration of changes in connectivity with distance downstream along a river with high-relief headwaters. Moving downstream, the river flows through headwater valleys with relatively thin, narrow alluvial veneers over bedrock and then through progressively wider and deeper alluvial valleys with greater floodplain development and hyporheic exchange. The presence of a floodplain buffers the mainstem river from hillslope and tributary inputs by creating depositional zones along its length, and progressively more extensive floodplains typically equate to greater average residence time of sediment, surface flow during overbank floods, and subsurface flow. CPOM is coarse particulate organic matter (>1 mm in diameter), FPOM is fine particulate organic matter (0.45 μm–1 mm). Source: After Brierley and Fryirs (2005), Figure 2.10, p. 44.

Figure 7.6 Examples of alluvial fans. (a) A fan created at the junction of a tributary with the Colorado River in the Grand Canyon, USA. The older fan deposits are outlined with a dashed white line, the most recent deposits with a dotted white line at far right center. Vegetated deposits in the foreground are the floodplain. Mainstem flow is left to right. (b) An event-based alluvial fan created by a damburst flood down the Roaring River in Colorado, USA. The Roaring River enters the Fall River valley; the Fall River flows from lower left to upper right in this view. The outburst flood occurred in 1982, and the resulting alluvial fan appears lighter in color (outlined in dashed white line) in this photograph, taken nearly 30 years later, because of slow growth of vegetation on the fan surface.

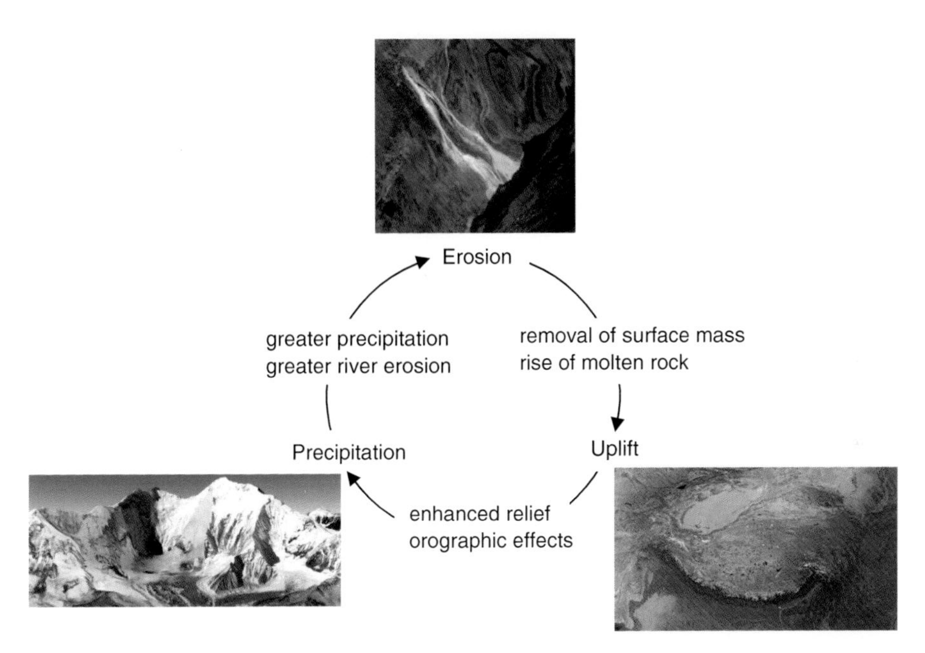

Figure 8.3 Schematic illustration of the interactions among tectonics, topography and climate, as illustrated by research in the Himalaya in southern Asia. The inset photos illustrate (clockwise from top) a landslide along a valley wall in the Nepalese Himalaya, the Tibetan Plateau as seen from space, and snow, ice, and glaciers around Mount Everest. Source: Inset photographs courtesy of Google Earth.

(a)

(b)

Figure 8.5 Aufeis along the Kongakut River, which flows north to the Arctic Ocean in northern Alaska, USA. Aufeis is shelf ice that forms along river margins as ground water continues to flow from adjacent uplands into the river corridor during autumn after air temperatures have dropped below freezing. (a) View of aufeis remaining along the braided channel in late June. Flow is toward the rear in this view, and the valley bottom is approximately 700 m wide. (b) Closer view of the river-side edge of the aufeis. Person in inflatable raft at right center for scale.

Figure 8.6 Permafrost exposed in a cutbank along the Yukon River in the interior of Alaska, USA. White stripes of ice alternate with frozen sediment in the cutbank exposure. Cutbank is approximately 3 m tall.

(a)

(b)

Figure 8.7 The Upper Rio Chagres in Panama. (a) Low-level aerial view of landslides that occurred during widespread intense rainfall shortly before the photo was taken. The landslides introduced substantial sediment and large wood into the channel. White arrow indicates flow direction. Active channel is approximately 35 m wide. (b) View up the mouth of a tributary channel. A logjam approximately 7 m tall formed at the mouth of the tributary and created a thick wedge of sediment and large wood (not visible here) on its upstream side, which extended more than 100 m up the tributary. Within 2 years, this jam was breached and much of the sediment wedge had been eroded. Person at upper left within white oval for scale. White arrow indicates flow direction on main channel.

Figure 8.8 Downstream view along Bumblebee Creek in central Arizona, USA. Although the wetted channel is only about 1.5 m wide under the base flow conditions shown here, the white oval highlights organic material deposited high along the channel margins by a flash flood.

Figure 8.9 A portion of the Rio Grande in Big Bend National Park, Texas, USA. Here, the active channel has narrowed substantially as a result of upstream flow regulation and dense growth of invasive exotic plants (*Tamarix* spp., *Arundo donax*). Wetted channel is approximately 35 m wide.

Part III

Channel Processes III

5

Large Wood Dynamics

Large wood was historically abundant in river corridors with upland or riparian forests. Riparian forests can be present even in otherwise treeless regions, such as deserts and grasslands. Centuries of deforestation, river engineering, and active removal of wood in river corridors have resulted in a wood-poor state for rivers in many parts of the world. However, the influences of large wood on river corridor process and form can be inferred both from historical and geologic records and from the limited number of river corridors that still retain natural levels of wood recruitment, transport, and storage. Large wood can create important physical and ecological effects even in river corridors in which past or continuing human activities have reduced the presence of large wood. River management now increasingly seeks to protect and restore many of these wood-related influences. Increasing recognition of the role of large wood in rivers suggests that wood, water, and sediment form the tripod of physical inputs on which river ecosystems develop.

5.1 The Continuum of Vegetation in River Corridors

Large wood commonly refers to downed, dead wood pieces that are least 10 cm in diameter and 1 m in length (Wohl et al. 2010). However, large wood is part of a continuum from upright, living woody vegetation through standing dead trees to downed wood. This continuum includes living wood that is bent in a manner that causes interactions with channel and floodplain processes (Opperman et al. 2008); living roots that cross small channels or living rootwads that create embayments along the bank of larger channels (Gurnell et al. 2005; Pfeiffer and Wohl 2018); and trees or portions of trees that have fallen or been transported but have resprouted (Gurnell 2014). Dead wood includes not only large pieces at the surface, but also smaller pieces of downed, dead wood, as well as wood buried in the floodplain (Gurnell and Grabowski 2016; Gurnell et al. 2016a).

Each of these categories of woody vegetation can create significant changes in water and sediment fluxes and river corridor form, and there are complex interactions among vegetation on this continuum (Figure 5.1). Large wood deposited on bars, for example, can create sheltered sites in which sediment and plant propagules are deposited and plants preferentially germinate. The plants then help to deflect flow and stabilize the sediment, creating an obstruction that traps additional large wood (Gurnell et al. 2001, 2012). Mobile large wood can damage riparian vegetation, creating opportunities for new germination (e.g. Johnson et al. 2000). Logjams can enhance channel avulsion and then become partly buried in the floodplain, creating more erosionally resistant substrate that can eventually support patches of old-growth forest that are subsequently recruited to the channel as

Rivers in the Landscape, Second Edition. Ellen Wohl.

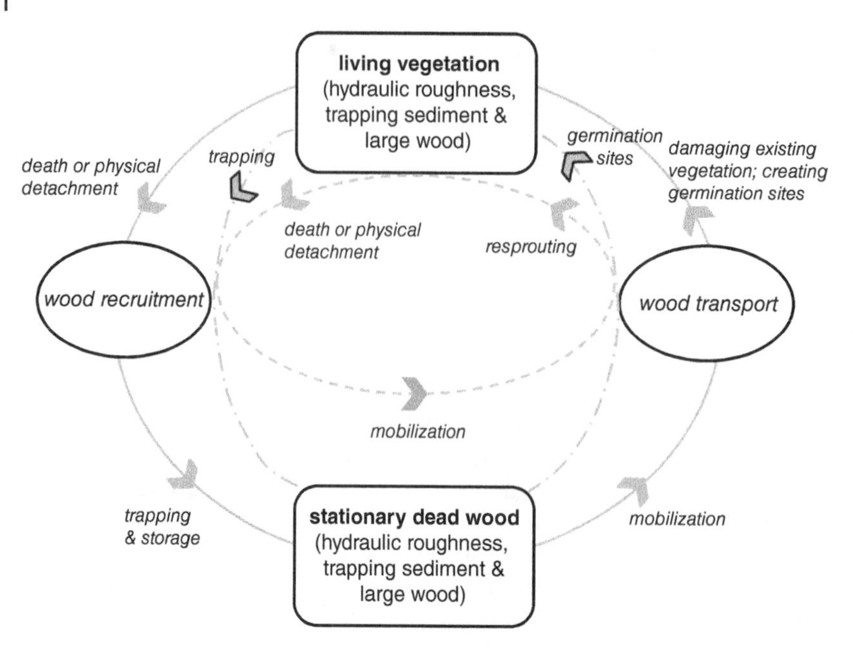

Figure 5.1 Interactions between living vegetation and stationary and mobile large wood. Both living vegetation and stationary dead wood increase hydraulic roughness and alter the distribution of hydraulic forces, as well as trapping and storing mineral sediment, smaller particulate organic matter, and large wood. Living vegetation and large wood are linked through wood recruitment and transport, but these linkages can involve at least three pathways, indicated here by three gray ovals: wood recruitment, storage, and subsequent mobilization (outer oval); living vegetation trapping and storing mobile large wood, which then provides germination sites for new living vegetation (intermediate gray oval); and wood recruitment, transport, and resprouting (inner gray oval). Living vegetation and stationary and mobile large wood also influence river form.

downed, dead wood over a time period of centuries (Collins et al. 2012). Smaller pieces of wood also accumulate and create physical and ecological effects in channels of diverse sizes (e.g. Culp et al. 1996), including channel-spanning jams in headwater channels (Jackson and Sturm 2002). Smaller pieces can fill the interstices among larger wood in logjams, thereby increasing the flow obstruction and secondary geomorphic effects associated with the jam (e.g. Manners et al. 2007).

Geomorphologists began to recognize the effects of living vegetation on river corridor process and form during the 1960s (e.g. Zimmerman et al. 1967), but recognition of the importance of downed, dead wood began about a decade later (e.g. Keller and Swanson 1979). Much of the initial work on large wood occurred in the Pacific Northwest region of the United States. Wood-related research did not expand to include significant numbers of studies from other regions of the world until the 1990s (Gurnell 2013; Le Lay et al. 2013b; Wohl 2017a). The delay in recognizing the importance of large wood in river corridors may reflect the fact that wood has been systematically removed for centuries in many of the river corridors in which the foundational research in fluvial geomorphology was conducted, as discussed in more detail in Section 5.7. Consequently, geomorphologists did not initially recognize that large wood was historically much more abundant and influential in river corridors.

Gurnell et al. (2016a) trace the history of geomorphic thinking about vegetation in river corridors. Early studies tended to focus on how vegetation responded to water and sediment fluxes and the resulting patterns of vegetation in river corridors. Subsequent work gave greater attention to vegetation as an influence on channel-boundary erosional resistance, the distribution of hydraulic forces in a channel, and rates and styles of channel change (e.g. widening, narrowing, avulsion). Since circa 2000, investigators have proposed conceptual models of the interactions between river processes and vegetation. An example is the five zones of vegetation–hydrogeomorphic interactions described in Gurnell et al. (2016a). These zones are spatially and temporally variable and have transitional boundaries, but they are broadly described with respect to frequency of inundation, frequency and grain-size of sediment mobility, and type of plants (e.g. aquatic versus riparian). The inclusion of aquatic vegetation reflects the expansion from a more traditional emphasis on woody, relatively rigid-stemmed riparian vegetation to recognition that flexible, submerged, or emergent aquatic vegetation can also influence hydraulics and fine sediment mobility in low-gradient channels (Luhar and Nepf 2013; Le Bouteiller and Venditti 2015) (see Section 3.1.4).

5.2 Recruitment of Wood to River Corridors

The movements and distribution of large wood within river corridors can be conceptualized in terms of a wood budget that, like a sediment budget, examines how differences between inputs and outputs result in wood storage. As initially proposed in Benda and Sias (2003), the change in wood storage, ΔS, over a river reach of length x and time interval t is

$$\Delta S = \lfloor L_i - L_o + {}^{Q_i}/_{\Delta x} - {}^{Q_o}/_{\Delta x} - D \rfloor \, \Delta t \tag{5.1}$$

where S is commonly expressed as volume of wood per surface area (m^3/ha) or volume per length of channel (m^3/100 m), L_i is lateral inputs, L_o is lateral outputs, Q_i is fluvial transport of wood into the river reach, Q_o is transport out of the reach, and D is decay. Although this equation could be applied to any time scale from a single flood to centuries, the decay term is only likely to be relevant at longer time scales. The processes of wood input are also referred to as wood recruitment.

Lateral inputs can be further described in terms of specific processes (Figure 5.2)

$$L_i = I_m + I_f + I_{be} + I_s + I_e + I_{bv} \tag{5.2}$$

where I_m is individual tree mortality, I_f is mass mortality (e.g. via windstorm, ice storm, wildfire, insect infestation), I_{be} is wood recruitment via bank erosion, I_s is recruitment via hillslope instability (e.g. avalanche, debris flow, landslide), I_e is remobilization of wood buried in or resting on the floodplain, and I_{bv} is beaver-recruited wood (Benda and Sias 2003; Wohl 2017a). Because Eqs. (5.1) and (5.2) focus on the active channel, wood moved from the floodplain into the channel is considered a lateral input and wood moving from the channel onto the floodplain a lateral output. If the wood budget is focused on the entire river corridor, Eqs. (5.1) and (5.2) can be combined and reformulated as

$$\Delta S = \lfloor L_{alloch} + L_{autoch} + {}^{Q_i}/_{\Delta x} - {}^{Q_o}/_{\Delta x} - D \rfloor \, \Delta t \tag{5.3}$$

where L_{alloch} refers to allochthonous wood sourced from outside the river corridor via individual tree mortality, mass mortality, hillslope instability, and beaver activity, L_{autoch} refers to dead wood sourced from living wood within the river corridor (individual or mass mortality among floodplain trees) or wood buried in the floodplain and remobilized via floodplain erosion, and other terms are as in Eq. (5.1).

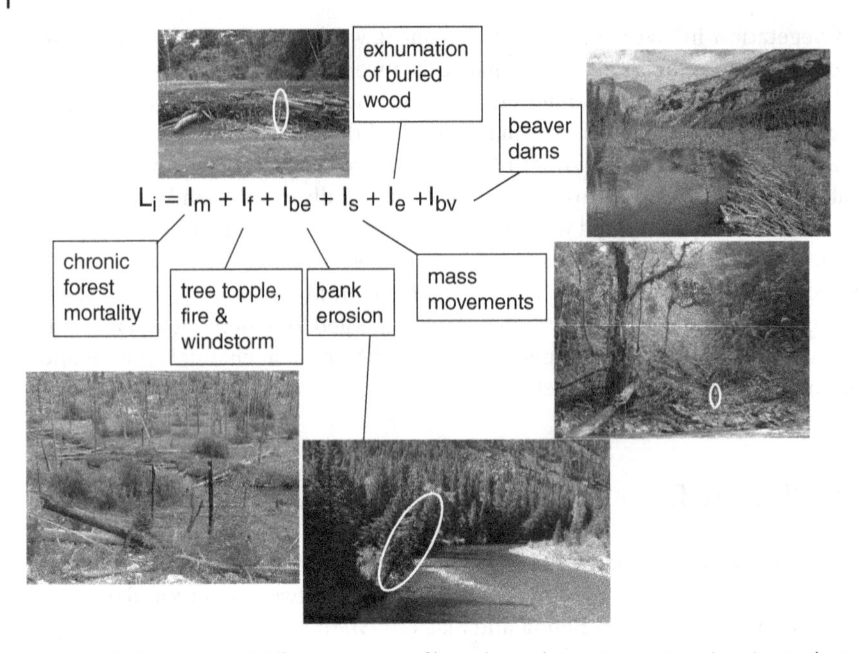

Figure 5.2 Illustration of different sources of lateral wood recruitment to a river. Inset photos clockwise from upper left: (i) A floodplain in Costa Rica with buried wood being exposed by bank erosion (1 m-long umbrella for scale within yellow oval). (ii) A beaver dam and pond along the Duke River in the Yukon Territory, Canada. (iii) The toe of a landslide along the Upper Rio Chagres in Panama that introduced large volumes of wood to the river (most of the wood was mobilized by streamflow before this photo was taken; person in yellow oval for scale). (iv) Trees leaning toward the Poudre River in Colorado, USA. (v) Ouzel Creek in Rocky Mountain National Park, Colorado, USA. The forest along this portion of the creek burned in 1978, and standing dead trees have been gradually toppling into the river ever since. (*See color plate section for color representation of this figure*).

Wood budgets developed for a single flood focus more on volumes of wood recruited and transported. Lucía et al. (2015) and Comiti et al. (2016) propose an adapted wood-budget equation to describe wood dynamics within a river segment in terms of volume

$$V_{o,ds} = V_{i,us} + V_{i,LS} + V_{i,LF} - V_{d,L} \tag{5.4}$$

where downstream exported wood volume ($V_{o,ds}$) during a flood derives from the volume of large wood arriving from the upstream reach ($V_{i,us}$) and wood recruited along the reach by lateral input ($V_{i,L}$), minus the volume of wood deposited within the reach during the flood ($V_{d,L}$). The volume of large wood input laterally during a flood can come from either the river corridor ($V_{i,LF}$) via erosion of forested alluvial surfaces such as banks and floodplains and thus associated with channel widening, or from hillslopes ($V_{i,LS}$) as a result of landslides and debris flows.

The relative importance of different recruitment processes varies greatly through time and across space at scales from an individual reach of river corridor to an entire drainage basin. Using a hypothetical mountainous drainage basin as an example, mass recruitment from hillslope instability is likely to be more important in laterally confined valley segments (e.g. May and Gresswell 2003), whereas recruitment via bank and floodplain erosion becomes increasingly important downstream as valley bottoms widen (e.g. Piégay et al. 2017). In some drainages, recruitment may be a relatively continuous process of individual tree fall (Wohl et al. 2012b), whereas in other drainages or other portions of the same drainage, recruitment can be strongly influenced by episodic mass tree mortality (Phillips and Park 2009).

Recruitment over time spans longer than a single event (flood, storm) or a few years is commonly estimated using numerical models of forest dynamics, as reviewed in Gregory et al. (2003). Early models focused on individual tree mortality and forest characteristics (species, age, stand density) as these determined the trees available to be recruited to the river corridor or active channel. These models were mostly not stochastic and thus did not incorporate variation associated with processes such as mass recruitment during storms. The early models also did not incorporate interactions between processes or user-defined parameters for variables such as land use (Gregory et al. 2003). These simplified models nevertheless provided interesting insights, such as the multidecadal to centuries recovery times (with respect to the size and volume of wood pieces recruited to the river corridor) following a severe disturbance such as a stand-killing wildfire or clearcutting (e.g. Bragg 2000).

The characteristics of wood recruitment can strongly influence the response of the river corridor to large wood. Widespread blowdown of trees during hurricanes, for example, may have been an important process in recruiting wood for the enormous wood raft historically present on the Red River in Louisiana, USA (Phillips and Park 2009). This raft spanned the channel, trapping all wood coming downstream for centuries after the blowdown and facilitating channel avulsions and the formation of secondary channels, which in turn trapped more wood (Triska 1984). On much smaller channels, formation of channel-spanning logjams as a result of wood recruitment from bank erosion or during small-scale blowdowns can also promote formation of secondary channels and associated additional recruitment and storage of wood (e.g. Wohl 2011b). Widespread bank erosion during an extreme flood can rapidly recruit wood and form numerous, channel-spanning logjams that obstruct flow and cause channel aggradation, enhancing overbank flooding and floodplain aggradation that persists even after the jams gradually break up (Oswald and Wohl 2008). Wildfires that burn a river corridor can increase treefall, creating channel-spanning logjams that form local base levels, which facilitate aggradation and raise the riparian water table, enhancing regrowth of the forest (Larsen et al. 2016). In these examples, the details of how wood is recruited and transported or retained within the river corridor create feedbacks that facilitate retention of wood, sediment, and water (as hyporheic exchange and overbank flows), as well as changes in river corridor form.

5.3 Wood Entrainment and Transport

The transport terms in Eq. (5.1) can be as difficult to predict and measure as sediment transport. The forces acting on an individual piece of wood can be specified to predict initiation of motion using a balance-of-force approach analogous to that used for sediment (Figure 5.3). Wood differs from sediment in important ways, however, including the ability of most wood to float (some wood is naturally very dense or becomes saturated and sinks); the tendency of wood pieces to deviate strongly from a spherical shape, so that piece orientation becomes very important; and the irregular shape of branching wood pieces or pieces with a rootwad.

Drag and buoyancy are the main forces mobilizing wood (Alonso 2004). The hydraulic conditions under which wood is entrained are also influenced by factors such as

- piece angle relative to flow direction – pieces parallel to flow tend to be more stable (Braudrick and Grant 2000);

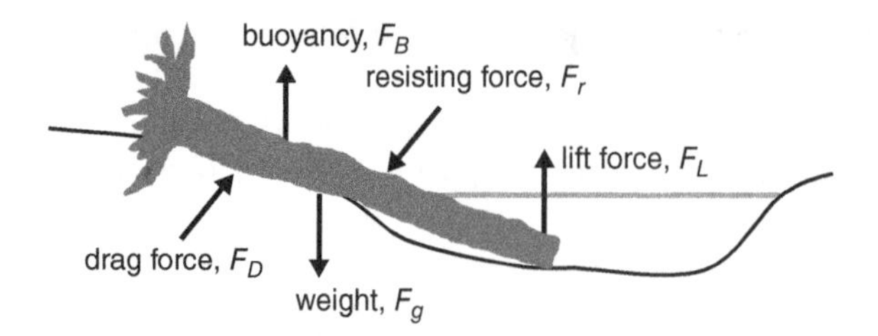

Figure 5.3 Idealized free-body diagram for a single log. The forces acting on the log can be influenced by factors such as its partial burial, its surface friction and that of adjacent sediment or wood surfaces, its shape and density, the angle at which it rests on the bed, and its protrusion into the flow relative to the adjacent bed or channel bank. In this illustration, the end of the log rests on the channel margin above the bankfull level (indicated by a horizontal line).

- the presence of a rootwad, which tends to decrease entrainment;
- wood density, which partly governs whether a log rolls, slides, or floats;
- piece length and diameter (Braudrick and Grant 2000);
- inclusion in a jam, which tends to stabilize individual pieces and decrease mobility (Wohl and Goode 2008); and
- the presence of standing vegetation that can help to trap wood.

One approach to quantifying wood entrainment is to track the movement of large numbers of wood pieces and statistically analyze correlations between piece characteristics and mobility (e.g. Merten et al. 2010). This can be useful for predicting individual piece mobility at a site, although empirical analyses of different sites are likely to select a slightly different combination of most significant predictor variables.

Because wood, like sediment, is strongly influenced by its surroundings and by interactions with the channel boundaries and other wood pieces, another approach to constraining the conditions under which wood is mobilized is to quantify the flow conditions under which a substantial number of wood pieces are mobilized within a reach of river corridor or transported from a reach or an entire watershed. This effectively treats wood as a population and examines the conditions under which that population is mobilized. This approach typically correlates wood flux past a point (using time-lapse or video photography; e.g. MacVicar and Piégay 2012; Kramer and Wohl 2014) or into a reservoir (e.g. Seo et al. 2008; Senter et al. 2017) with hydrologic variables.

The potential for mobilization and removal of individual logjams is also important, especially because wood reintroduction now commonly employs engineered logjams, which are usually expected to remain in place and stable (Bisson et al. 2003; Gallisdorfer et al. 2014; Fryirs et al. 2018). The stability of a logjam can be predicted using a balance-of-forces approach that quantifies the balance between forces acting to mobilize the jam (e.g. drag force) and those tending to keep the jam stable (e.g. frictional resistance) (D'Aoust and Millar 2000; Manners et al. 2007; BR and ERDC 2016). However, this approach does not account for several distinctive characteristics of logjams. The exact dimensions and shape of a jam are difficult to quantify because of individual protruding pieces and porosity within the jam. Individual pieces can move during high flows without the entire jam becoming mobilized, and time-lapse photography of jams during fluctuating flows reveals that they can dilate under high flow, or partially float and expand such that frictional resistance between individual pieces decreases, and then condense again during low flow without the entire jam mobilizing. Scott et al. (2019) describe a procedure for estimating jam stability using an evolving logjam characteristics and dynamics database hosted online.

Once mobilized, wood can move by floating or by rolling or sliding along the channel-bed or floodplain surface. Floating wood can be categorized as undergoing congested transport in which logs move as a single mass occupying more than a third of the channel area; as uncongested transport in which piece-to-piece contacts between logs are rare; or as semicongested transport intermediate between the two end-members (Braudrick et al. 1997). Wood that is mineralized, waterlogged, or very dense can move in contact with the bed by rolling or sliding (Buxton 2010).

The potential for wood to remain in transport is strongly influenced by the dimensions of the wood pieces relative to the flow width and depth. In the context of wood transport, Gurnell et al. (2002) categorize rivers as being small (channel width less than median wood piece length), medium (channel width greater than the length of most wood pieces), or large (channel width greater than the length of all wood pieces). Kramer and Wohl (2017) add the category of "great river" to recognize the potential for mobilization of large masses of wood during higher flows.

Wood transport occurs mainly above a threshold discharge that has been expressed as a proportion of bankfull stage, an absolute value, or a recurrence interval (MacVicar and Piégay 2012; Wohl 2017a). The sequence of flows is also important: lower wood fluxes occur during successive floods of equal magnitude (Moulin and Piégay 2004).

Flow magnitude, duration, and rate of rise and fall also influence wood mobilization and transport (Kramer and Wohl 2017). Wood flux generally increases with discharge on the rising limb of a flood, but the relationship is nonlinear and highly variable (Kramer and Wohl 2014; Ravazzolo et al. 2015). The observed variability likely results from multiple causes, including changing wood inputs during the course of a flood as a result of recruitment processes such as bank erosion or floating of previously stored wood and piece-to-piece interactions among wood in transport (Braudrick et al. 1997; Bertoldi et al. 2014; Kramer and Wohl 2017). Rapid increase in discharge may cause rapid wood recruitment that leads to congested transport and increased wood deposition in jams, for example, and jams may limit the transport of single pieces during the falling limb (Davidson et al. 2015). These interactions create hysteresis, with greater wood flux on the rising limb of a flood hydrograph.

Wood-piece characteristics including anchoring (e.g. burial, presence of a rootwad), length, diameter, orientation, and tree species also influence wood mobilization and transport (Kramer and Wohl 2017). Smaller pieces tend to be more mobile, as do pieces perpendicular to flow when mobilized. Anchored pieces are commonly less mobile. Tree species influences decay rates, wood density, susceptibility to abrasion, and branching complexity. Species more susceptible to abrasion, decay, and breakage tend to be more mobile. Complexly branched pieces are less mobile. The few studies that have examined variations among tree species within a river network suggest that tree type is a strong predictor of downstream travel distance (e.g. Ravazzolo et al. 2015; Ruiz-Villanueva et al. 2016c).

Along with flow and wood-piece characteristics, river-reach characteristics form a third primary category of influences on wood mobilization and transport (Kramer and Wohl 2017). Reach-scale channel and floodplain characteristics can promote transport or retention of wood that is already mobilized. High transport capacity for wood is associated with more uniform, hydraulically smoother boundaries and fewer obstructions such as large boulders or stationary wood. Braudrick and Grant (2001) propose a dimensionless debris roughness index based on the ratios of wood-piece and channel dimensions

$$DR = \left(a_1 \frac{L_{\log}}{w_{av}} + a_2 \frac{L_{\log}}{R_c} + a_3 \frac{d_b}{d_{av}} \right) \tag{5.5}$$

where DR is debris roughness; a_1, a_2, and a_3 are coefficients that vary according to the relative importance of each variable; L_{log}/w_{av} is the ratio of piece length to mean channel width; L_{log}/R_c is the ratio of piece length to mean radius of curvature for channel bends; and d_b/d_{av} is the ratio of buoyant depth to average channel depth. In addition to limits on transport associated with channel dimensions, the presence of stationary wood pieces and logjams (Beckman and Wohl 2014b) or protruding boulders and living vegetation (Jacobson et al. 1999) can limit large wood transport.

Mineral sediment and large wood share several characteristics (Gurnell 2007) in addition to those already mentioned. Wood enters a river corridor and is moved predominantly by high flows and, in some steep mountain streams, debris flows. Wood recruitment and transport are commonly intermittent, with relatively long periods of storage between episodes (Kramer and Wohl 2017). Wood characteristics can change while the wood is stationary and in transport, via processes of decay, abrasion, and breakage that are analogous to the weathering, abrasion, and breakage of mineral sediment, although rates of change in wood are usually faster than those in sediment. In addition, both the movement and the storage of wood influence river process and form, and both are influenced by river process and form (Keller et al. 1995; Wohl et al. 2018c).

Figure 5.4 illustrates the potential for multiple temporary storage sites between the entry of a wood piece into the river corridor and its final export from the river network. The different lateral dimensions of the loops on this spiraling diagram indicate that wood pieces can have long residence times and limited mobility in small rivers, greater mobility and associated shorter residence times in medium rivers, and long residence times in large and great rivers in which wood is effectively trapped within extensive floodplains. At any point along a river, however, a wood piece can have relatively short or long residence times, depending on the trapping efficiency of the location at which it is deposited.

The presence of mobile wood within river corridors is important for at least three reasons. (i) Wood mobility facilitates the presence of wood in varying stages of decay, which in turn enhances habitat and biodiversity within river corridors. Aquatic macroinvertebrates ingest or burrow into wood after

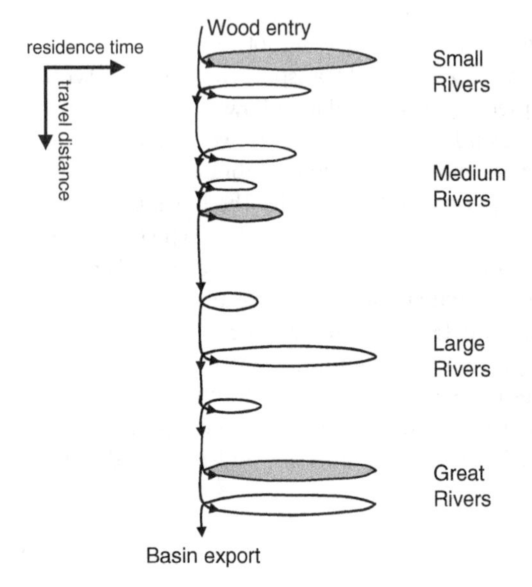

Figure 5.4 Schematic illustration of the spiraling movement of large wood through a river network. The lateral dimension of loops represents the residence time at a single location before a wood piece reenters downstream transport. The cumulative residence time for a piece of wood from its starting location in the river corridor to its final resting place is the sum of all the intermediary residence times and the travel time, which is likely very short relative to the residence times. The distance traveled is the sum of all step lengths. Along its path, the wood piece undergoes vertical, lateral, and downstream exchanges. Within any spiral, a wood piece can decay in place, ending its path, as indicated by filled gray loops. Source: Adapted from Latterell and Naiman (2007) and Kramer and Wohl (2017).

it has been colonized and softened by microbes (Harmon et al. 1986; Bilby 2003). A greater diversity of decay stages among stored wood can increase microbial and invertebrate biodiversity. (ii) Wood in transport can cause mechanical damage to living plants and channel banks (e.g. Johnson et al. 2000), and this can create germination sites for new plants, maintaining the age and species diversity of riparian forests (Merritt 2013). Episodic influxes of new wood that are stored within the river corridor also periodically create germination sites for plants (e.g. Gurnell et al. 2001; Pettit and Naiman 2006; Collins et al. 2012; Osei et al. 2015) and habitat diversity for aquatic and riparian animals (e.g. Benke and Wallace 2003; Matheson et al. 2017). (iii) Fluxes of wood from rivers to lake (Marburg et al. 2006; Kramer and Wohl 2015) and marine (Maser and Sedell 1994) ecosystems represent a significant source of nutrients, as well as creating water-surface (Taquet et al. 2007) and sea-floor habitat. These effects extend from nearshore (Simenstad et al. 2003) to deep-sea (Schwabe et al. 2015) environments and from polar (Rämä et al. 2014) to lower latitudes.

Wood mobility has been studied using (i) direct documentation under field conditions of wood transport (commonly, with time-interval photos); (ii) inferred mobility based on changes in wood load through time (e.g. Wohl and Goode 2008; Boivin et al. 2017); (iii) physical experiments (e.g. Bocchiola et al. 2006; Davidson et al. 2015); and (iv) numerical models. Time-interval or video monitoring can give information on the timing of wood transport relative to discharge, the size of wood pieces and quantity of wood if piece size can be accurately estimated, and, when combined with other sources of information, a wood budget for a reach (e.g. MacVicar and Piégay 2012; Kramer and Wohl 2014; Kramer et al. 2017). Two-dimensional numerical models of wood dynamics now include functions for recruitment, entrainment, transport, and deposition (e.g. Mazzorana et al. 2011; Ruiz-Villanueva et al. 2016a).

5.4 Wood Deposition

During wood-transporting flows, wood is preferentially deposited in hydraulically rougher (Ruiz-Villanueva et al. 2016b) and less laterally confined (Wohl and Cadol 2011; Dixon and Sear 2014; Lucía et al. 2015) reaches. The presence of obstacles such as large boulders; standing trees along the channel margins, within the active channel (Bocchiola et al. 2006), or on the floodplain (Wohl et al. 2018a); or stationary downed wood (Beckman and Wohl 2014b) can be particularly important in facilitating wood deposition. Channel-margin irregularities such as bends or constrictions and expansions can also facilitate deposition of wood (Braudrick and Grant 2001; Wohl and Cadol 2011). Living vegetation is so effective at trapping wood in overbank areas and on mid-channel bars (e.g. Jacobson et al. 1999; Bertoldi et al. 2015) that details such as the frequency and length of inundation during which it obstructs flow may be effective predictors of wood transport distance and may explain wood transport hysteresis: living vegetation retains dead wood on the rising limb of the hydrograph before wood deposition occurs during the falling limb as a result of declining flow depth (Kramer and Wohl 2017).

As discussed earlier in the context of wood recruitment, the details of wood deposition can influence subsequent process and form in river corridors. Abbe and Montgomery (2003) distinguish *in situ* accumulations from transport jams (Figure 5.5). *In situ* accumulations form around a key piece that has not moved downstream, such as a ramped piece of wood still attached to the channel bank at the rootwad. In steep, low-order channels, such accumulations commonly form a backwater pool and fine sediment storage upstream and a scour pool downstream. Transport jams, in contrast,

Figure 5.5 *In situ* or autochthonous jams and transport or allochthonous jams. In the case of the *in situ* jams, the smaller pieces have been transported downstream, but have been trapped by a larger, ramped piece that fell into the channel from the adjacent stream bank. (*See color plate section for color representation of this figure*).

form around wood pieces transported downstream to a site. These jams are common in lower-order channels, where they can influence bar formation, braiding index, and lateral channel mobility (Abbe and Montgomery 2003).

Wood deposition can also influence receiving bodies of water. Wood transported from the Slave River into Canada's Great Slave Lake is redistributed by surface waves in the lake. Wood accumulations along the lake shore can become a *driftcretion* if the accumulation is stabilized and vegetated until the wood is buried or decays in place. Driftcretions can take various forms (Figure 5.6) (Kramer and Wohl 2015), but all increase rates of shoreline progradation, create habitat diversity, protect the shoreline from erosion, and facilitate the storage of organic carbon. Over time spans of decades to hundreds of years, driftcretions also truncate tributary channels on the Slave River delta and create linear islands, spits, and peninsulas.

5.5 Wood Storage

Wood can be stored on the floodplain surface, in contact with the channel bed or banks, buried within the floodplain, or buried in the channel. Wood can also be stored as individual pieces or as aggregates that form logjams, wood rafts, or driftcretions. The details of how and where wood is stored, along with the size of the piece, the portion of the tree from which it is sourced (e.g. branch versus trunk),

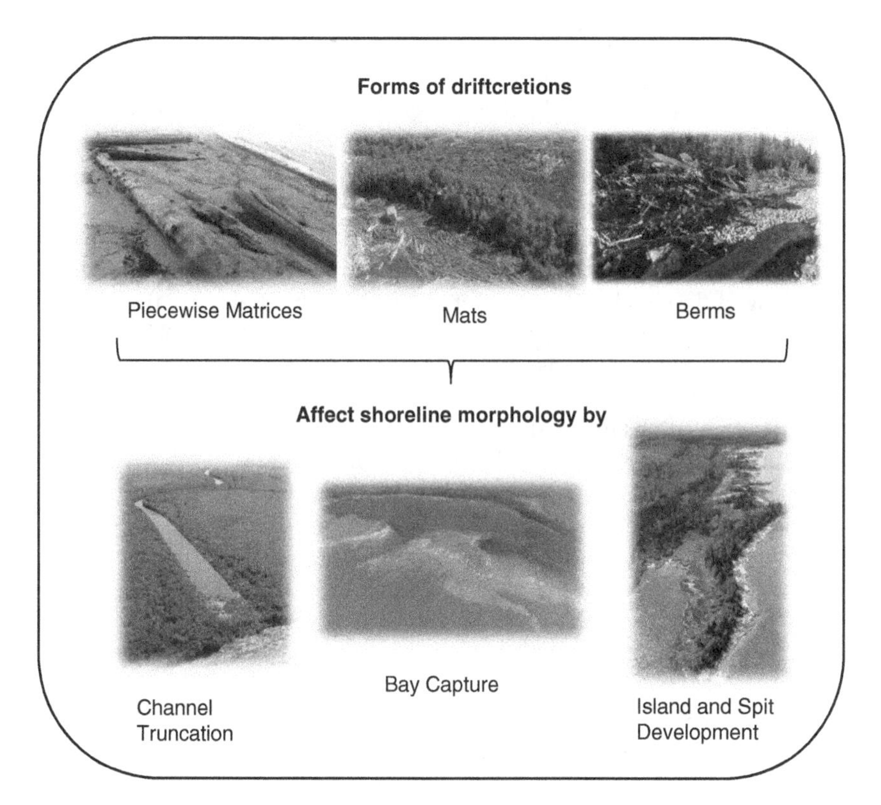

Figure 5.6 Types of driftcretions and their effects on shoreline processes. Source: After Kramer and Wohl (2015), Figure 1. *(See color plate section for color representation of this figure).*

and the species of that tree all influence the extent to which processes such as decay, abrasion, and breakage modify the wood while it remains stationary.

Decay is biogeochemical alteration of wood. Wood decay is influenced by the presence of oxygen, fragmentation of the wood as it is altered, textural complexity of the wood, surface-to-volume ratio, tree species, water temperature and dissolved nutrients, and biotic communities such as microbes, fungi, macroinvertebrates, fish, and plants (Harmon et al. 1986; Bilby 2003; Le Lay et al. 2013a,b). Forest ecology studies of downed wood indicate that decay rate broadly reflects climate, with average rates of <10 years for complete wood decomposition in the tropics (e.g. Clark et al. 2002), 10–100 years in humid temperate climates (Harmon 1982), and 50–100 years in dry climates (Ellis et al. 1999). Complete decay in cold climates can require much longer (e.g. 600 years in Colorado, USA subalpine forests; Kueppers et al. 2004), and individual pieces of very large wood can persist for significant periods. Wood that remains saturated or buried below the depth of most soil microbial activity can also remain relatively intact for very long periods: >1400 years in the Queets River of Washington, USA (Hyatt and Naiman 2001), for example, and >10 000 years when buried in the floodplain of rivers in Tasmania, Australia (Nanson et al. 1995) or Missouri, USA (Guyette et al. 2008).

Abrasion and breakage can enhance decay by increasing the ratio of surface area to volume of wood pieces and providing fresh substrate for bacterial colonization (Bilby 2003). Repeated wetting and drying can also accelerate decay relative to noninundated or continually inundated sites (Noetzli et al. 2008). As might be expected, breakage is likely to exert a stronger control than decay on wood piece size, shape, and mobility over short time periods (e.g. Merten et al. 2013).

Wood in storage is commonly referred to as "wood load" and, as noted earlier, quantified as volume per unit area or per unit length of the channel or river corridor (Van der Nat et al. 2003). Wood loads and the spatial distribution of wood within the active channel are the most commonly described characteristics of large wood dynamics, and several papers synthesize the extensive literature relating to these topics (e.g. Gurnell 2013; Ruiz-Villanueva et al. 2016c; Wohl et al. 2017b). Compilations of published wood loads indicate substantial variability within and between river networks, with ranges of 10 to ~230 m^3/ha for unmanaged floodplains (Lininger et al. 2017; Wohl et al. 2018a) and 0 to ~5000 m^3/ha for unmanaged channels (Ruiz-Villanueva et al. 2016c; Wohl et al. 2017b). Wood loads vary substantially through time and space in response to changes in wood recruitment, transport, and deposition.

Most studies that examine spatial variations in wood load focus on patterns in relation to drainage area or bankfull channel width, both within a river network and between networks (e.g. Gurnell 2003; Fox and Bolton 2007; Ruiz-Villanueva et al. 2016c; Wohl et al. 2017b). Although significant trends exist between channel wood loads and predictor variables such as drainage area or bankfull channel width within a bioclimatic region such as humid temperate rainforest (e.g. Fox and Bolton 2007), relationships are highly variable and do not span data from multiple bioclimatic regions (Gurnell 2013; Wohl et al. 2017b). Insufficient data exist to evaluate patterns of floodplain wood loads within or across regions, but the limited amounts that are available indicate significant differences between regions (Lininger et al. 2017).

In most regions, wood load decreases downstream as transport capacity increases (Lienkaemper and Swanson 1987; Bilby and Ward 1989; Hassan et al. 2005; Gurnell 2013). Network-scale changes in wood load may reflect a transition from transport-limited conditions with respect to wood in headwaters to supply-limited conditions in larger channels (Marcus et al. 2002). The size and number of logjams may be greatest in the middle portions of river networks, where sufficient transport capacity exists to mobilize substantial volumes of wood, but local decreases in transport capacity associated with features such as mid-channel bars, zones of flow expansion, and channel bends can trap large volumes of wood (Wohl and Jaeger 2009). Historical records of very large wood accumulations in large, lowland rivers, such as the Great Raft on the Red River in Louisiana, USA (Triska 1984), however, suggest that the observed downstream decrease in wood load on contemporary rivers may reflect a long history of wood removal from channels, rather than changes in transport capacity (Wohl 2014).

Studies of longitudinal variations in wood load within a single channel or river network document significant longitudinal variability in relation to local factors such as valley-bottom width and gradient, with wider and hydraulically rougher channel and floodplain reaches retaining more wood (e.g. Wohl and Cadol 2011; Ruiz-Villanueva et al. 2016b; Wohl et al. 2018c). These longitudinal variations can be described using Neighbor K statistics, which indicate how much individual pieces and accumulations of wood deviate from random (e.g. Kraft and Warren 2003; Wohl and Cadol 2011). Stout et al. (2017) propose the use of the Weibull statistical distribution to reflect the spatial distribution of wood within a river reach; specifically, the central tendency and the variability of wood loads within the reach. Comparison of multiple reaches suggests that the shape of the Weibull distribution reflects the degree to which wood pieces are aggregated and thus the ability of a river to transport and redeposit wood within a reach (Stout et al. 2017).

Few studies have examined spatial patterns of wood storage on floodplains. The lateral distribution of floodplain wood with respect to the location of the active channel can be conceptualized as

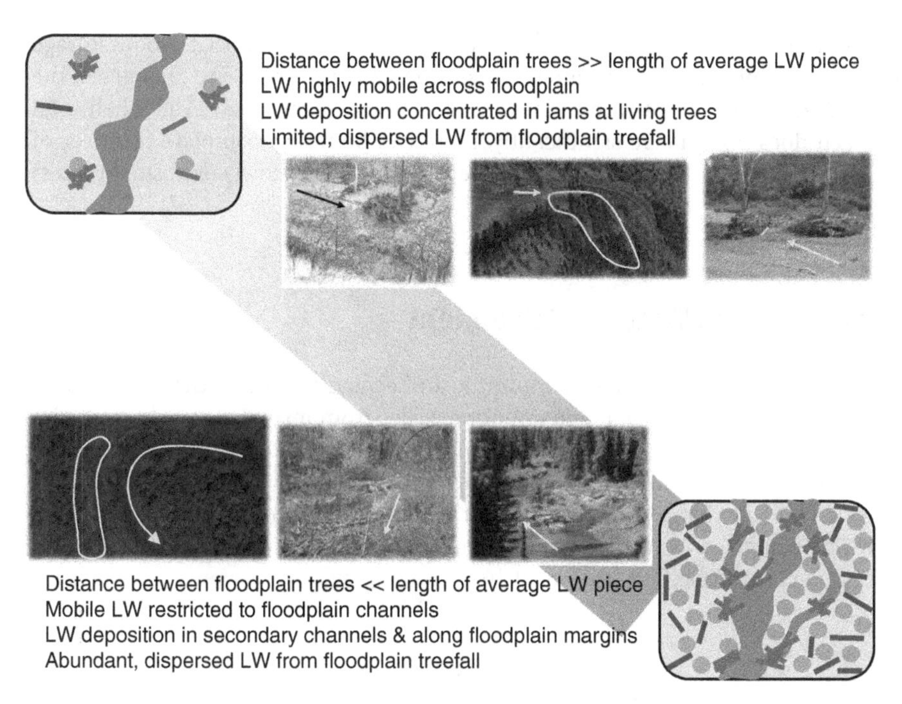

Figure 5.7 End-members for the spatial distribution of large wood on floodplains. At upper left, idealized plan view shows isolated living trees (ovals), with small logjams at the upstream side of each, and a few dispersed pieces of wood. Inset photos show (left to right) jams at the base of widely spaced trees along the Middle Fork Gila River in New Mexico, USA; an aerial view of an extensive floodplain jam (circled) on the inside of a bend along the Middle Fork Gila River; and jams at the base of trees along the Brisbane River in Australia. At lower right, plan view shows densely spaced living trees, with logjams concentrated along the margins of the main channel and secondary channels, and more abundant dispersed wood across the floodplain. The inset photos from the Swan River in Montana, USA show (left to right) an aerial view of a long jam (circled) in an abandoned channel, a ground view of wood in an abandoned channel, and a ground view of wood concentrated along the entrance of a secondary channel. In all photos, arrows indicate flow direction. LW, large wood. (*See color plate section for color representation of this figure*).

occurring along a continuum. The end-members are floodplains where standing trees are widely spaced and those where forests are relatively dense (Figure 5.7). Widely spaced floodplain trees act as trapping points for logjams when overbank flow transports wood across the entire floodplain (e.g. Wohl et al. 2018a). Densely forested floodplains limit overbank transport of wood into the floodplain, so that fluvially transported wood is concentrated along channel margins and in secondary or abandoned channels (e.g. Wohl et al. 2018c).

Studies of temporal variations in wood load within a river reach are less abundant than those examining spatial variations. The majority of work investigating temporal variations uses indirect methods such as correlations between wood load and forest stand age (e.g. Hedman et al. 1996) or numerical modeling (e.g. Ruiz-Villanueva et al. 2016a). Direct studies of variations in wood load through time focus either on a single event such as an extreme storm or large flood that recruits and redistributes wood over a few months to a few years (e.g. Wohl et al. 2009; Phillips and Park 2009) or on multi-year monitoring of a limited length of river using repeat surveys or high-resolution aerial imagery (e.g. Wohl and Goode 2008; Boivin et al. 2017; Kramer et al. 2017). Whether focused on a single event or on monitoring over several years, these studies indicate that wood storage within a reach can fluctuate significantly on an interannual timescale in response to changes in the recruitment, transport, and deposition of wood. These studies also indicate that, even when total wood load within a reach does not change significantly, there can be nearly complete turnover of stored wood (i.e. mobilization of stored wood pieces and replacement with newly deposited pieces) (e.g. Wohl and Goode 2008; Cadol and Wohl 2010).

5.6 Wood Interactions with Water and Sediment

Wood within the river corridor creates diverse geomorphic and ecological effects that span spatial scales from the channel unit to the river reach and temporal scales from instantaneous to centuries (Figure 5.8). At the most basic level, stationary wood in the river corridor alters fluxes of water and sediment, as well as the erosional resistance of the channel and floodplain boundaries, and these effects set up a series of secondary changes in river corridor process and form (Figure 5.9; Table 5.1). Mobile wood can physically damage living vegetation (Johnson et al. 2000) and erode the channel bed and banks and floodplain surfaces (Comiti et al. 2016).

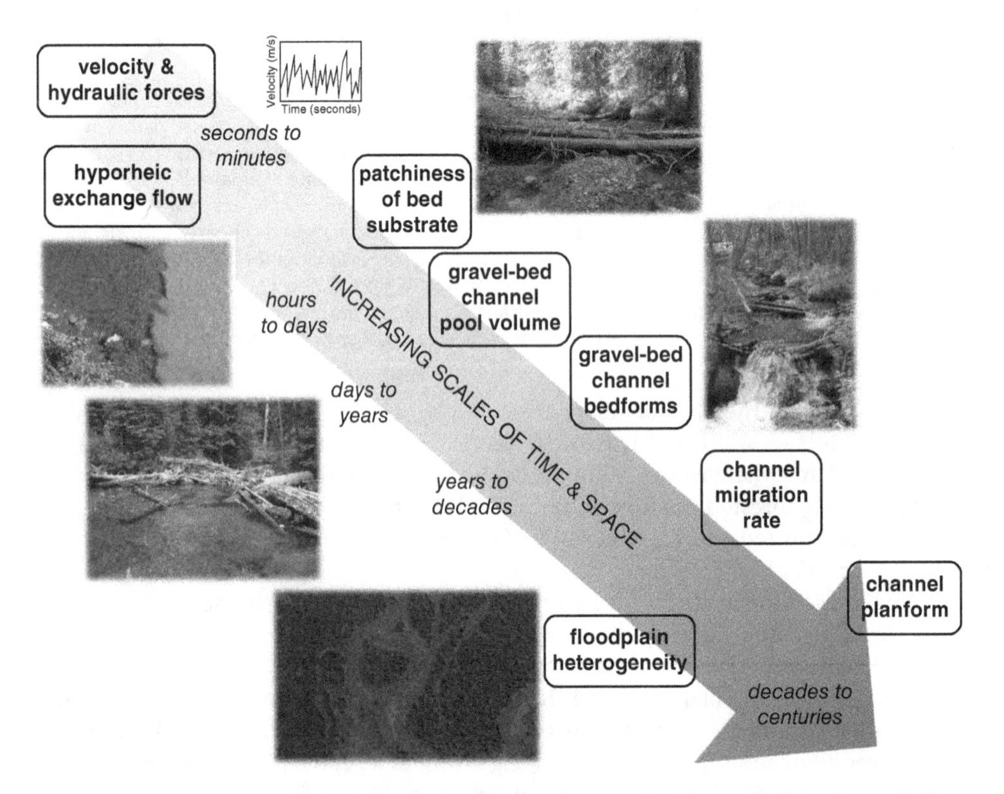

Figure 5.8 Schematic illustration of the diverse spatial and temporal scales at which large wood influences river corridor process and form, from seconds to minutes at the upper left to decades to centuries at the lower right. (*See color plate section for color representation of this figure*).

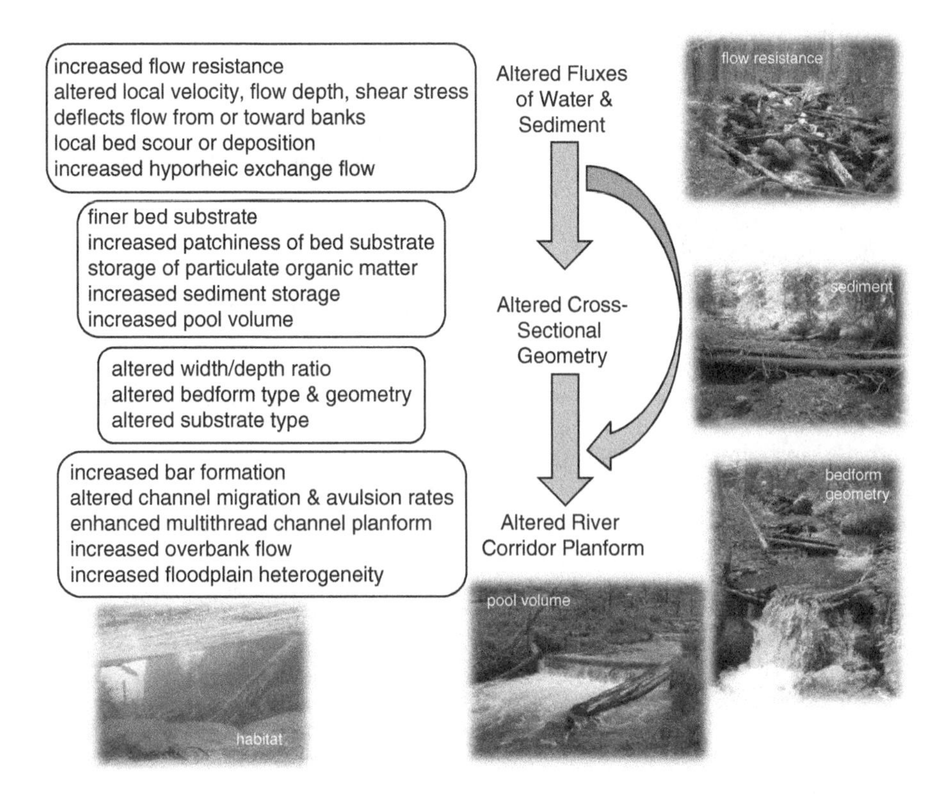

Figure 5.9 The diverse effects of stationary large wood in the river corridor include influences on process (fluxes of water, sediment, mobile large wood, and particulate organic matter) and form from the local scale of a single channel unit (e.g. pool or bar) through cross-sectional geometry to reach-scale planform. (*See color plate section for color representation of this figure*).

Physical changes associated with wood in river corridors are also important in an ecological context. Wood increases habitat abundance and diversity for a variety of aquatic and riparian organisms, including microbial and macroinvertebrate communities, plants, and fish (Benke and Wallace 2003; Chen et al. 2008; Schenk et al. 2015). By increasing spatial heterogeneity, the retention of particulate organic matter, and hyporheic exchange flows, wood facilitates biological uptake and processing of nutrients (e.g. Wallace et al. 1995; Battin et al. 2008). Analyses of biomass and biodiversity in relation to wood load indicate that greater abundance of wood corresponds to greater biological productivity and diversity (Herdrich et al. 2018; Venarsky et al. 2018).

The geomorphic effects of individual wood pieces decrease downstream as each piece interacts with a progressively smaller proportion of the channel boundaries and cross-sectional flow, but aggregate effects can be substantial in large channels. Historical accounts exist from diverse forested environments of enormous volumes of wood concentrated in portions of large alluvial rivers (Triska 1984). These *log rafts* and *congested transport* significantly influenced channel and floodplain dynamics along many forested rivers prior to extensive modification by human activities such as timber harvest and snagging (removal of instream wood) (Wohl 2013b).

The interactions of wood with water and sediment vary across space and through time, as exemplified by wood-induced patterns of sediment storage. In one mountain stream network, for example,

Table 5.1 Physical effects of stationary large wood in river corridors.

Description of effect	Sample references
Increased flow resistance	Shields and Smith (1992); Curran and Wohl (2003); Wilcox et al. (2011)
Increased flow depth and decreased average flow velocity due to greater hydraulic roughness and flow obstruction	Gippel (1995); Davidson and Eaton (2013)
Enhanced storage of fine sediment and organic matter and increased patchiness of bed substrate in channels	Buffington and Montgomery (1999b); Andreoli et al. (2007); Beckman and Wohl (2014a); Wohl and Scott (2017b)
Changed substrate type, as when wood traps sufficient sediment to create forced alluvial reaches in channel segments that would otherwise have bedrock substrate	Massong and Montgomery (2000); Montgomery et al. (2003b)
Flow deflection and local scour of bed and banks	Hassan and Woodsmith (2004)
Enhanced pool volume	Richmond and Fausch (1995); Wohl and Scott (2017b)
Increased channel width on small to medium channels	Nakamura and Swanson (1993)
Altered bedform dimensions (e.g. wood tends to create taller and more widely spaced steps in step–pool sequences)	MacFarlane and Wohl (2003); Mao et al. (2008)
Enhanced bar growth, lateral channel movement, and channel avulsion	O'Connor et al. (2003); Little et al. (2013)
Greater soil moisture and nutrient content beneath floodplain large wood	Zalamea et al. (2007)
Enhanced overbank flooding and floodplain heterogeneity	Jeffries et al. (2003)
Enhanced hyporheic exchange	Sawyer et al. (2011)
Formation of multithread channel planform	Wohl (2011b); Collins et al. (2012)

third- and fourth-order streams store disproportionately large volumes of wood and coarse and fine sediment, whereas first-order streams store much of the coarse particulate organic matter (Pfeiffer and Wohl 2018). Although wood can enhance sediment storage within the river corridor at any point within a river network, wood in steep, laterally confined channels is likely to form channel-spanning logjams that trap a wedge of sediment upstream from the jam (Nakamura and Swanson 1993). Dispersed wood in lower-gradient, less confined channel segments can result in widespread bed aggradation (Brooks et al. 2003). Log rafts in very large, floodplain channels can result primarily in floodplain aggradation and overbank sediment storage (Triska 1984).

Variations in the interactions of wood with water and sediment through time primarily reflect differing inputs and storage of water, sediment, and wood. Unmanaged rivers in forested environments tend to be wood-rich. Although recruitment of wood to the river fluctuates through time in response to disturbances such as forest fires, debris flows, and blowdowns, wood decay is sufficiently slow in most rivers of the temperate zone that at least some wood is always present in the stream. This wood helps to trap and retain newly recruited wood. If instream wood is removed or recruitment is eliminated for a period of time by clearcutting, the river can enter a wood-poor condition in which the absence of instream wood reduces the likelihood that any subsequently recruited wood will be retained in the river rather than transported downstream (Wohl and Beckman 2014). Wood-rich and wood-poor conditions can be conceptualized as alternative states.

Ecologists use *alternative states* to describe a scenario in which an ecosystem can exist under multiple states or sets of unique physical and biological conditions. Alternative states are stable over ecologically relevant time spans, but ecosystems can transition from one stable state to another when perturbed sufficiently to cross a threshold. Collins et al. (2012) and Livers et al. (2018) apply the concept of alternative states to wood influences in forested river corridors. One stable state is present in rivers with abundant wood and associated multichannel planforms, high levels of organic matter retention and processing, diverse habitat, high levels of biomass and biodiversity, and spatially heterogeneous floodplains. An alternative state occurs in rivers without wood that have single-channel planforms, less retention, habitat, and biodiversity, and less diverse floodplains.

5.7 Human Influences on Wood Dynamics

The volume of instream wood within most rivers flowing through forested drainage basins has decreased so substantially because of human activities that our perceptions of natural wood loads and of the geomorphic importance of instream wood are distorted (Chin et al. 2008). Historical records of wood abundance in rivers of regions such as North America (Maser and Sedell 1994; Montgomery et al. 2003a; Wohl 2014) and Australia (Gippel et al. 1992), which were not subject to extensive deforestation and river engineering until circa 200 years ago, suggest that diverse rivers in these regions had orders of magnitude more wood than is currently present. Analogously, the history of deforestation and cut-log rafting to collection points along major rivers in Europe suggests much greater historical wood abundance in European rivers (e.g. Schama 1995). The combined effects of upland and riparian deforestation (reduced wood recruitment); direct removal of wood, channelization, and levee construction (reduced potential for deposition of fluvially transported wood within river corridors); and flow regulation (reduced potential for wood transport) have resulted in significantly altered wood dynamics in rivers throughout much of the world. Although some portions of the United States, such as the Pacific Northwest, now commonly incorporate large wood in river management and restoration (e.g. Cramer 2012), annual or event-based wood removal remains the norm in river management throughout many densely settled regions such as Europe and the United States.

Wood has been actively removed from channels for centuries primarily because of concerns related to navigation and flooding (Wohl 2014). Individual pieces or accumulations of large wood can create obstacles that damage boats or trap materials such as cut logs that are being floated downstream. By creating hydraulic roughness, wood in a channel can also increase flood stage and overbank flooding. Mobile large wood can hit infrastructure along the river corridor with considerable force, and mobile wood can accumulate at constrictions such as bridge piers, where the wood enhances bed scour (e.g. Ruiz-Villanueva et al. 2013). One result of the long history of wood removal from rivers is that people commonly have negative perceptions of wood in river corridors, viewing the wood as hazardous, unsightly, and even unnatural (e.g. Chin et al. 2008). Among river managers, these negative perceptions typically change to appreciation with increasing level of education, training, and field experience (Chin et al. 2014).

Where retaining or reintroducing large wood is an explicit component of river management and restoration, passive and active strategies are employed. Passive strategies focus on restoring conditions that allow for natural wood recruitment (e.g. reforestation) or restoring river corridor form in a manner that facilitates wood retention (e.g. increasing the spatial heterogeneity of the channel and channel–floodplain connectivity). Although likely to be relatively self-sustaining, such strategies can require a long time to show results. Examining recovery of wood loads after removal of instream wood along the King River in southeastern Australia, Stout et al. (2018) used Monte Carlo simulations to estimate a recovery period of ~250 years before wood loads return to natural levels.

Active strategies, which are more widespread, focus on deliberate reintroduction of large wood to the river corridor, commonly in the form of engineered logjams or individual pieces, which are typically fixed in place (e.g. Slaney and Martin 1997; Gallisdorfer et al. 2014). Most of these efforts have been successful with respect to habitat improvement and fish response, although the details of river response to the emplaced wood depend on local factors and watershed-scale processes such as sediment supply and transport (Roni et al. 2015).

Several projects involving active reintroduction of unanchored large wood are being undertaken in the western United States. These are sometimes referred to as stage 0 restoration projects because the objective is to return the valley bottom to what Cluer and Thorne (2014) refer to as a stage 0 channel in a channel evolution model. A stage 0 channel has an anastomosing channel planform, typically because of beaver dams or abundant logjams that facilitate channel avulsion and the formation of secondary channels (Powers et al. 2019). Restoration designed to restore stage 0 conditions in forested channels can involve reintroducing substantial quantities of large wood and allowing floods and channel adjustments to freely redistribute the wood within the river corridor.

Where large wood is present and may create hazards, a stepped decision process can be used to evaluate the relative benefits and harms associated with individual wood pieces and accumulations (Wohl et al. 2015a). The process starts with a simple checklist based on visual assessments and can continue to more detailed numerical modeling.

5.8 The Natural Wood Regime

Analogous to the natural flow and sediment regimes, the natural wood regime can provide an organizing framework for understanding wood dynamics in river corridors. A wood regime consists of recruitment, transport, and storage in river corridors. Each of these three primary components can be characterized in terms of magnitude, frequency, rate, timing, duration, and mode (Wohl et al. 2019b). A *natural wood regime* is present when past and present human activities do not significantly alter the components of the wood regime. A *target wood regime* is present when wood recruitment, transport, and storage balance desired geomorphic and ecological characteristics with mitigation of wood-related hazards over a specified time span.

Both natural and target wood regimes can be classified in terms of wood process domains (Figure 5.10). Different criteria can be used to distinguish these process domains, depending on the components of the wood regime that are of most interest. Process domains focused on recruitment, for example, might differentiate high-relief, laterally confined portions of a river in which hillslope instability dominates large wood inputs, and lower-relief portions in which bank erosion dominates

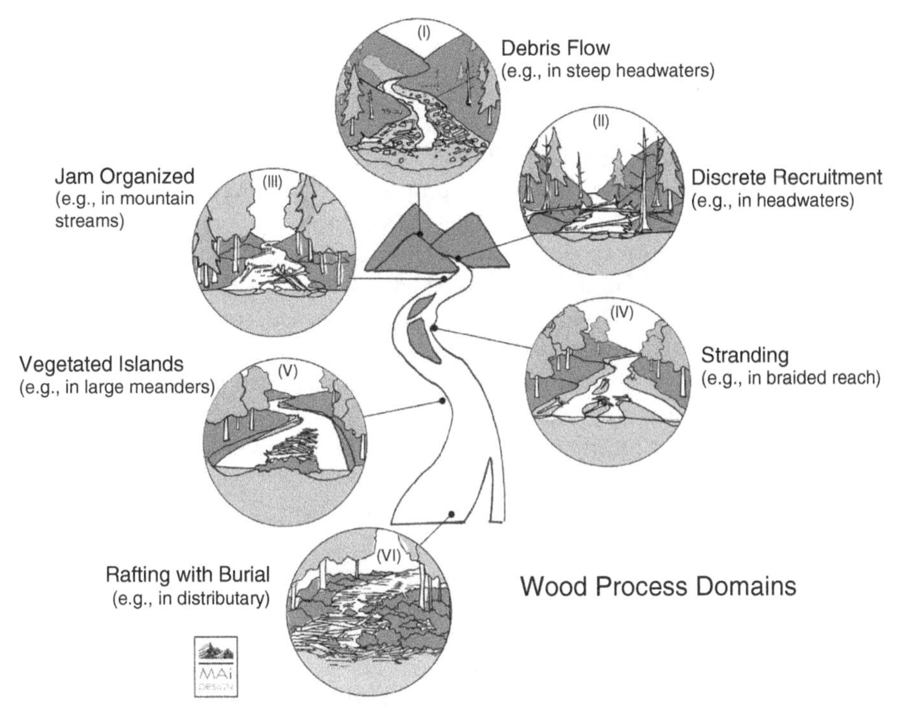

Figure 5.10 Hypothetical wood process domains along a river continuum. Each example domain has defining wood regime characteristics that result in a distinct regime over a specified time. Domains depicted are examples and are not necessarily mutually exclusive. (i) Debris flow: wood is delivered infrequently *en masse* with long to permanent residence times. (ii) Discrete recruitment: recruitment of individual trees left dispersed *in situ* over time leads to abundant storage due to limited transport capacity. (iii) Jam organized: flow is sufficient and frequent enough to mobilize and deflect pieces over short transport durations into concentrated jam features. (iv) Stranding: dispersed wood is stranded on bars and margins as flows recede; transport timing and duration are predictably associated with flow level and frequency. (v) Vegetated islands: wood is frequently floated and transported for long durations until concentrated at deposition sites, such as the heads of islands, facilitating re-vegetation and island expansion. (vi) Rafting with burial: large concentrated rafts obstruct channels; long residence times interact with depositional environments to facilitate abundant accumulation, re-vegetation, and wood burial. Source: From Wohl et al. (2019b), Figure 3. Illustrations by MAi Design llc (www.maisierichards.com).

wood recruitment to partly confined to unconfined channels. Process domains focused on transport might distinguish river reaches in which the physical complexity of the river corridor enhances wood storage from those in which transport is enhanced. Process domains defined in a management context of hazard reduction might differentiate high-risk portions of a river from which wood is actively removed from those with minimal infrastructure and in which wood retention is desired.

The key point to understand is that, analogous to the natural flow regime, the natural wood regime can be used as a benchmark to evaluate human-induced changes in wood recruitment, transport, and storage, in order to develop management strategies that can mitigate the negative effects of these changes, such as loss of habitat abundance and diversity or reduced hyporheic exchange, organic matter retention, and water quality.

5.9 Summary

After initial work in the US Pacific Northwest during the 1970s and '80s, geomorphic and ecological investigations of large wood in river corridors have expanded to include a diverse array of river environments around the world. This substantial body of work clearly demonstrates that, in forested regions, wood historically formed a key component of river process and form. As recognition of the broad array of wood-related geomorphic and ecological effects has increased, river management has gradually changed to emphasize the protection and restoration of processes of wood recruitment, transport, and storage in channels and on floodplains. Stationary wood within channels affects the distribution of hydraulic forces and obstructs flow, thereby altering the transport of solutes, sediment, and particulate organic matter, as well as channel cross-sectional geometry and planform. Stationary wood on or buried within floodplains affects hydraulic roughness and patterns of sediment deposition, floodplain erosional resistance, and soil characteristics. Mobile wood can alter channel beds and banks, floodplain surfaces, and living vegetation, as well as providing a new supply of wood to be deposited within a river corridor. Wood mobilization varies in relation to characteristics of the wood piece or jam, flow, and site from which the wood is being entrained. Wood transport commonly displays hysteresis, with greater wood flux on the rising limb, as well as thresholds in wood mobilization and movement. Wood is disproportionately deposited in river reaches with lower gradient and greater spatial heterogeneity that enhances hydraulic roughness and creates trapping sites. Although wood continues to be removed in many rivers to facilitate navigation and reduce hazards to people and infrastructure, wood reintroduction is becoming increasingly common in some regions as a means of restoring geomorphic and ecological functions. During the past decade, wood has been recognized as a foundational component of river corridor process and form, along with water and sediment.

6

Channel Forms

Channel form can be examined at many levels, as reflected in diverse classifications for rivers. The predominant bedforms present in a river can be used to distinguish step–pool from pool–riffle channels (Section 4.3.2), for example, or the focus on channel form can be at the cross-sectional scale, based on descriptors such as width/depth ratio or exponents of at-a-station hydraulic geometry relations, both of which are introduced in this chapter. Channel planform is commonly used to distinguish straight, meandering, braided, and anabranching channels with differing degrees of sinuosity and single- or multithread channels. Channel form can also include the vertical dimension over the entire length of a river, as when longitudinal profiles are categorized in terms of concavity or stream gradient indices. Just as river classifications attempt to identify consistent thresholds that distinguish channel forms existing within a continuum, so the distinction of cross-sectional, planform, and longitudinal channel forms in this chapter represents an arbitrary division of aspects of river morphology that intergrade with and influence one another. Channel classifications that integrate multiple form factors consistently classify reach types into similar groups based on the dominant geomorphic characteristics of each. Consequently, debates as to which classification system is best may be overstated (Kasprak et al. 2016).

Pasternack et al. (2018a,b) provide an example of an approach designed to incorporate components of channel form across diverse spatial scales. The classification is based on flow convergence routing, which quantifies longitudinally varying convergence and divergence of flow and sediment based on the nonuniform topography inundated by a specified flow magnitude (Pasternack et al. 2018a). Standardized width and detrended, standardized bed elevation are used to differentiate the categories of normal channel, constricted pools (low width, low bed elevation), oversized channel (high width, low bed elevation), nozzles (low width, high bed elevation), and wide bars (high width, high bed elevation). A field-test of the approach indicates that base-flow and bankfull categories are nested within types of valley landforms, with the valley type controlling channel morphodynamics during moderate to large floods (Pasternack et al. 2018b). The field test from a gravel-bed river supports earlier work suggesting that the bankfull channel topography of boulder- to gravel-bed rivers is controlled by flows with a magnitude higher than morphological bankfull discharge (e.g. Wohl 1992; Lenzi et al. 2006; Bertoldi et al. 2010).

Rivers in the Landscape, Second Edition. Ellen Wohl.

width can be top width or weighted average channel width
depth can be maximum depth or weighted average
cross-sectional area is the product of width and depth
wetted perimeter is the length of the channel boundary beneath the water surface (gray shading)
hydraulic radius is cross-sectional area divided by wetted perimeter

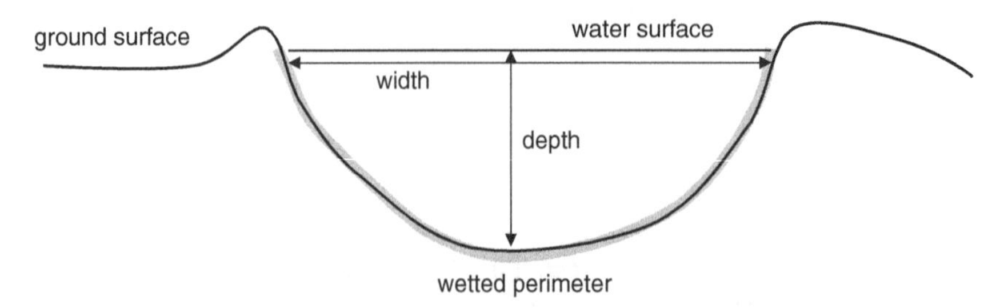

Figure 6.1 Downstream view in a channel, indicating cross-sectional geometry form parameters.

6.1 Cross-Sectional Geometry

Process and form in natural channels typically exhibit large spatial and temporal variability. Much work has been devoted to identifying parameters that can usefully describe the mean state and the variations in process and form. In Chapter 4, channel form was characterized in terms of the dominant bed material (e.g. sand-bed channel, gravel-bed channel) or the predominant bedforms (e.g. pool–riffle channel, step–pool channel). Channel form can also be described with respect to cross-sectional geometry, with a focus on mean cross-sectional characteristics.

Cross-sectional form parameters such as width, depth, cross-sectional area, wetted perimeter, and hydraulic radius (Figure 6.1) are typically measured at bankfull stage, under the assumption that bankfull stage defines a channel area that is filled by flow at least once every 2 years. Bankfull flow is a problematic concept that is used quite differently in different studies, but nonetheless forms the standard for measuring cross-sectional geometry. Cross-sectional parameters can also be measured for specific flow magnitudes, such as the mean annual flood or base flow.

6.1.1 Bankfull, Dominant, and Effective Discharge

Bankfull discharge is one of the most widely used reference discharges in fluvial geomorphology, yet the definition and implications of this flow remain controversial. Wolman and Leopold (1957) define bankfull discharge as the stage just before flow begins to overtop the banks. Bankfull discharge has subsequently assumed various implications because it has been equated to a specific recurrence interval and geomorphic function. Numerous studies indicate that a discharge that nearly overtops the banks recurs approximately every 1–2 years on many channels (Leopold et al. 1964; Castro and Jackson 2001). Consequently, bankfull discharge is sometimes defined based on recurrence interval rather than stage with respect to channel morphology. Because a discharge that recurs every 1–2 years transports the majority of suspended sediment in many rivers (Simon et al. 2004), bankfull discharge has been interpreted as the most important flow magnitude for controlling channel process and form (Wolman and Miller 1960; Dunne and Leopold 1978).

Problems arise in that bankfull discharge can be difficult to define based on channel morphology. Designating bankfull stage is relatively straightforward where a clearly defined, regular top of bank separates the channel from overbank areas. Bankfull stage can be difficult to define morphologically if a river has inset channels. The channel may be so deeply incised, for example, that the top of the bank has little relevance to flow volume. The top of the bank may be at different levels on each side of the channel. The channel can also display multiple convexities along its side slopes that reflect different flow magnitudes. Comparisons of morphological definitions using various criteria proposed in individual studies – top of bank, bank inflection, ratio of channel width to mean depth, level of significant change in the relation between wetted area and top channel width, and first maximum local bank slope – indicate that, on average, discharge estimates vary by a factor of three at a given site (Radecki-Pawlik 2002; Navratil et al. 2006).

Another source of uncertainty is that bankfull discharge defined in relation to a channel morphologic feature can have very different recurrence intervals among different sites. In hydroclimatic regions with extreme annual and interannual flow variability, such as arid and semiarid regions, the morphologically defined bankfull flow may have a recurrence interval much longer than 1–2 years. Recurrence intervals associated with a consistent morphological feature can vary by a factor of two between channel segments within regions as diverse as Puerto Rico (Pike and Scatena 2010) and snowmelt rivers in the Rocky Mountains of the USA (Segura and Pitlick 2010).

Finally, whether defined from channel morphology or recurrence interval, bankfull discharge does not necessarily transport the majority of suspended sediment or exert a greater influence on channel form than other magnitudes of flow (e.g. Roy and Sinha 2014). Analyses of suspended and bedload transport on a variety of rivers indicate that discharge with a recurrence interval of 1–2 years transports much of the sediment moved in many rivers, but there are exceptions to this generalization. Channel morphologic features are likely to reflect multiple recurrent discharge magnitudes (Pickup and Warner 1976; O'Connor et al. 1986; Turowski and Rickenmann 2009).

As a broad generalization, large-magnitude, infrequent flows are more likely to strongly influence sediment transport and channel morphology in rivers with higher thresholds for mobilizing bed and bank materials and greater temporal variation in discharge magnitude. In other words, sediment transport and channel morphology are more likely to reflect larger, infrequent flows in gravel-bed, boulder-bed, and bedrock rivers, and rivers in very dry or seasonal tropical climatic regimes (e.g. Baker 1988; Kochel 1988; Milan 2012).

The concept of a *dominant discharge*, which is sometimes equated with bankfull discharge, reflects the idea of a single flow magnitude that, if maintained, will maintain the same average dimensions and channel morphology as those which result from a stable stream's entire hydrologic regime (Crowder and Knapp 2005). When used in numerical modeling, dominant discharge is the flow that moves the same amount of bed material, with the same size distribution, as the actual flow regime. The assumption is that channel morphology under a dominant discharge will also remain similar to the morphology under the actual flow regime.

Effective discharge is the discharge that transports the largest amount of sediment (Schmidt and Morche 2006). This magnitude of discharge may or may not equate to bankfull stage.

Bankfull discharge, effective discharge, mean annual flood, and the 1.5-year flow have all been proposed as constituting dominant discharge (Rosgen and Silvey 1996; Griffiths and Carson 2000). The idea that a single magnitude of flow strongly dominates river process and form is an oversimplification of the complexity of interactions among flow, sediment, and channel geometry through time. This oversimplification can have important consequences when used in river restoration as an index value for scaling channel dimensions and designing stable channels, particularly if excessive emphasis on dominant discharge results in neglecting the geomorphic and ecological importance of less frequent, larger flows or the importance of base flows (e.g. Poff et al. 1996; Stromberg et al. 2007).

6.1.2 Width-to-Depth Ratio

The ratio of channel top width to mean flow depth (*w*/*d*) is one of the most commonly used parameters of cross-sectional geometry. The *w*/*d* ratio can be used to infer the limit strength and relative erodibility of bed and bank materials, relative base-level stability, and consistency of water and sediment inputs (e.g. Schumm and Khan 1972; Wohl and Ikeda 1997; Finnegan et al. 2005; Yanites et al. 2010).

Channels of diverse size flowing on similar bed materials become wider more rapidly than they become deeper as discharge increases with time at a cross-section or while proceeding downstream in a river network. Widening a channel requires less sediment transport capacity than deepening a channel. Bank erosion simply requires eroding the banks, but the resulting sediment can be stored in the channel. Bed erosion requires both entraining the bed sediment and transporting the sediment downstream. Banks are also more likely to become unstable as they grow taller, unless they are in very cohesive material, so that deepening a channel typically results in associated widening as over-steepened banks collapse. In addition, the channel bed typically consists of coarser sediment than the banks and thus requires more flow energy to entrain and remove.

Relative erodibility of bed and bank materials can reflect differences in properties such as cohesion or grain size between the bed and the banks that influence the ability of stream flow to remove boundary material. As banks become more erodible, *w*/*d* increases (Finnegan et al. 2005): braided channels typically have greater *w*/*d* ratios than single-thread channels (Schumm and Khan 1972). Low bank erodibility can reflect high percentages of bedrock or cohesive sediment in the bank, or effective bank stabilization by vegetation.

If other factors are equal, channels with forested banks tend to be wider and have lower rates of bank erosion and channel migration than channels with grassy banks (Allmendinger et al. 2005). As noted in Section 4.6, the degree to which forest or other vegetation influences bank stability and *w*/*d* ratio varies. Riparian trees with dense root networks and high stem density that increases overbank roughness tend to result in narrower channels than channels in more open woodlands (e.g. David et al. 2009). This is illustrated by plots of bankfull width versus discharge for gravel-bed channels in Colorado, USA (Andrews 1984) and the United Kingdom (Hey and Thorne 1986). In both cases, rivers with banks heavily vegetated with trees and thick brush have narrower channels for a given unit discharge than do rivers with banks primarily covered in grasses and brush.

Grazing animals can also increase channel width and *w*/*d* ratios, both indirectly by selectively grazing on and removing riparian vegetation and directly by trampling stream banks and enhancing bank erosion. Both domesticated animals (e.g. Trimble and Mendel 1995) and wild animals such as bison and elk (Butler 2006; Beschta and Ripple 2012) can create these effects.

Channel *w*/*d* ratio can also reflect base-level constraints. Rivers do not incise below base level, so an increase in discharge when base level remains constant can result in channel widening or some planform change such as formation of secondary channels, rather than in deepening of the river (e.g. Brandt 2000). Conversely, channels develop lower *w*/*d* ratios (i.e. deepen faster than they widen) under relative base-level fall (e.g. Bowman et al. 2010).

Channel cross-sectional geometry can also reflect changes or relative consistency in water and sediment inputs. An increase in sediment yield is likely to cause bed aggradation and channel widening, leading to an increase in w/d ratio (e.g. Simon and Rinaldi 2006; Nelson and Dubé 2016). A decrease in sediment yield can cause bed erosion, but is also likely to result in bank erosion, leading to less predictable changes in w/d ratio (e.g. Draut et al. 2011). Four conceptual models commonly used to describe and predict changes in cross-sectional geometry are discussed in the next four sections.

6.1.3 Hydraulic Geometry

Hydraulic geometry describes changes in channel dimensions and flow velocity in relation to changes in discharge either at a cross-section through time or downstream along a river. Initially proposed by Leopold and Maddock (1953) based on regime theory, the conceptual framework of hydraulic geometry has been applied to understanding natural and engineered channels and underlies much of contemporary river restoration.

6.1.3.1 At-A-Station Hydraulic Geometry

At-a-station hydraulic geometry is used to describe changes in cross-sectional parameters at a site with changes in flow and to compare rates of change in these parameters between sites. At-a-station hydraulic geometry characterizes how changing discharge alters width, depth, and velocity at a cross-section (Figure 6.2). Starting with the continuity equation and assuming that discharge is the primary influence on hydraulic variables, Leopold and Maddock (1953) propose that

$$w = aQ^b \tag{6.1}$$

$$d = cQ^f \tag{6.2}$$

$$v = kQ^m \tag{6.3}$$

where w is width, d is mean flow depth, v is mean flow velocity, Q is discharge, and a, c, k, b, f, and m are numerical constants. Based on the continuity equation, the product of a, c, and k is 1 and the sum of b, f, and m is 1.

The rates of change of w, d, and v reflect:

- the shape and relative erodibility of the channel – channels with nearly vertical, cohesive banks, for example, have a very low rate of change in width (Knighton 1974);
- sediment transport – channels carrying large amounts of bed load are typically wide and shallow, and depth increases little with discharge, whereas velocity increases rapidly (Wilcock 1971);
- the slope of the water surface; and
- the roughness of the wetted perimeter – velocity increases rapidly in a channel in which increasing flow depth quickly makes the protrusion height of roughness elements a small proportion of flow depth (Ferguson 1986; Eaton 2013).

Studies of steep, coarse-grained channels suggest that channel segments in which velocity increases more rapidly than w and d are dominated by grain resistance, because relative grain submergence increases quickly as discharge increases and associated flow resistance declines. Channel segments in which width and depth increase more rapidly are dominated by form resistance (David et al. 2010). Rates of change in depth and velocity in relation to resistance may not be linear, however, if different sources of resistance occur with changing stage. Although velocity may increase more rapidly once

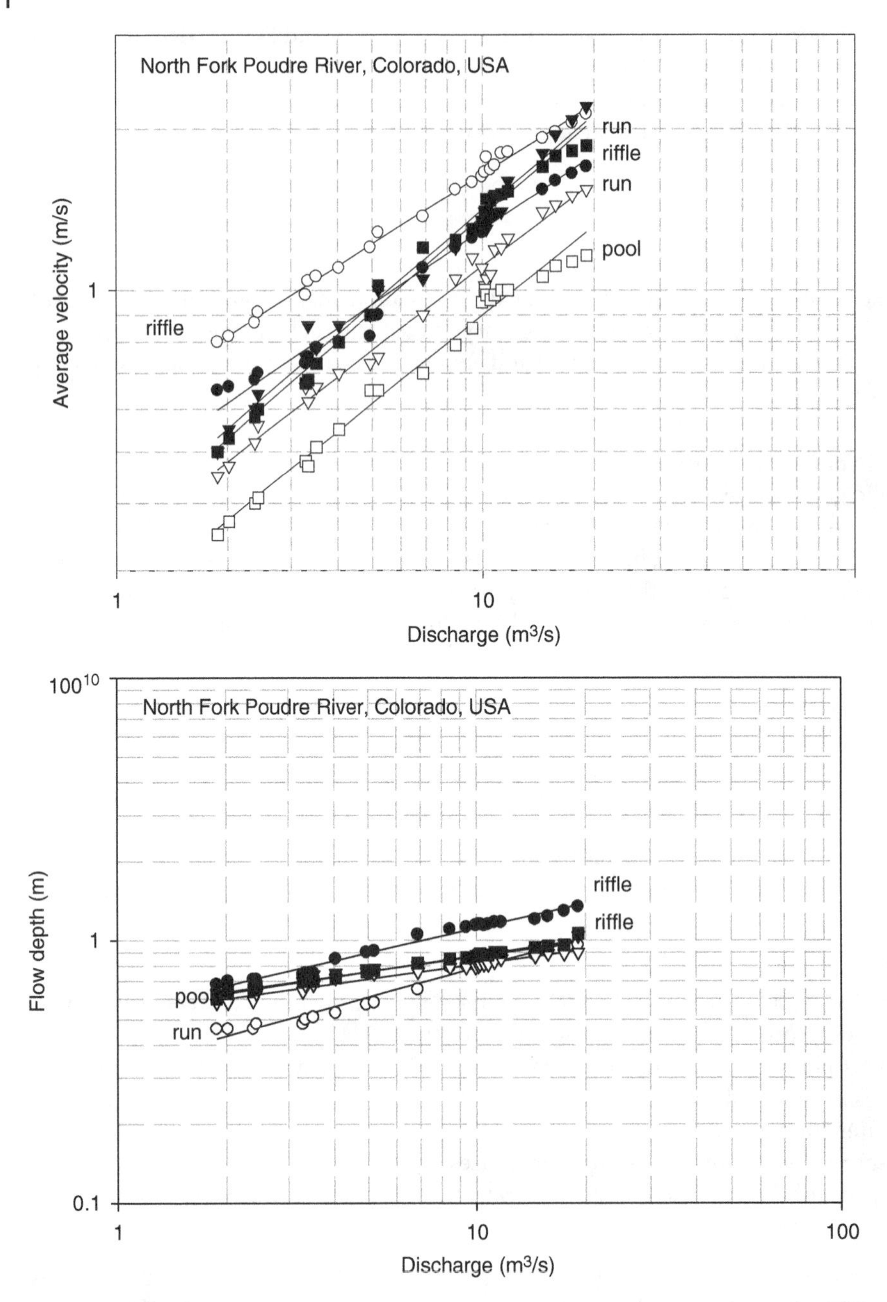

Figure 6.2 At-a-station hydraulic geometry curves for depth and velocity on the North Fork Poudre River in Colorado, USA (drainage area 980 km^2). Individual curves within each plot are for different channel units along this pool–riffle channel.

grain resistance becomes negligible, for example, form resistance associated with a mobile bed can become more important at higher flows and slow the rate of velocity increase at the largest discharges.

Although the complexity of at-a-station hydraulic geometry relations limits generalizations, the width exponent primarily reflects channel geometry and boundary composition. The depth and velocity exponents reflect cross-sectional form as well as hydraulic resistance and sediment transport, which tend to be more variable than form parameters (Knighton 1998).

Average values of at-a-station hydraulic geometry exponents are $b = 0.23, f = 0.42$, and $m = 0.35$ (Park 1977). These values indicate that depth typically increases more rapidly with discharge than does width or velocity, although the exponents vary widely among channels and among different cross-sections on a single channel (Reid et al. 2010). Inflection points that mark variations in the rate of increase in w, d, or v with increasing Q can reflect increasing submergence of grain- or form-resistance elements or the initiation of overbank flow. Hydraulic geometry curves can be divided into three portions (Knighton 1998). The first reflects low discharges below the threshold of sediment movement, when flow characteristics reflect a cross-sectional geometry remaining from earlier high flows. The middle portion indicates conditions when sediment is being entrained from the bed. The upper portion represents overbank flow when flow width expands rapidly but flow depth increases relatively slowly.

An iteration of at-a-station hydraulic geometry uses a characteristic scaling law – a river's at-many-stations hydraulic geometry – with satellite imagery to quantitatively estimate river discharge (Gleason and Smith 2014). The basis for at-many-stations hydraulic geometry is that empirical parameters of at-a-station hydraulic geometry are functionally related along a river (Gleason and Wang 2015). That is, at-a-station hydraulic geometry is dependent on the at-a-station hydraulic geometry of other cross-sections within a specified reach of river. This type of understanding also underlies hydraulic topography measurements (Gonzalez and Pasternack 2015), which are discharge-dependent, spatially averaged river hydraulics. The increasing availability of remote-sensing methods (e.g. Sofia et al. 2015) and hydraulic modeling tools facilitates this type of reach-scale analysis, rather than measurements specific to a cross-section.

Hydraulic geometry is also used to examine downstream changes in the relations between discharge and channel form. At-a-station hydraulic geometry exponents are typically smaller and more variable than those for downstream hydraulic geometry (DHG) relations, indicating lower rates of change with discharge at a cross-section than down a river.

6.1.3.2 Downstream Hydraulic Geometry

DHG relations take the same basic form as at-a-station relations, with power functions relating width, depth, and velocity to discharge. DHG relations describe changes in w, d, and v as a discharge of the same frequency varies in magnitude downstream (Leopold and Maddock 1953; Gleason 2015). Mean annual or bankfull discharge is typically used for DHG analyses (Park 1977; Bieger et al. 2015), with the assumption that channel dimensions primarily reflect the forces exerted by a flow with this recurrence interval. DHG relations are essentially scaling functions for changes in cross-sectional geometry resulting from downstream changes in discharge magnitude. These relations can be applied within a single river or in a physiographic region (e.g. Bieger et al. 2015).

Typical exponent values for the relations between Q and w, d, and v are 0.4–0.5, 0.3–0.4, and 0.1–0.2, respectively (Park 1977). DHG is used in the engineering of stable channels under the assumption that $w \sim Q^{0.5}$ when Q has a recurrence interval of 1–2 years (e.g. Hey and Thorne 1986). As noted in the earlier discussion of bankfull and dominant discharge, this can be a reasonable assumption for channels with limited hydrologic variability, but may be an inappropriate assumption for channels with greater hydrologic variability in which channel width reflects a discharge of longer recurrence interval. DHG is also used to predict the effects of flow regulation and to understand the geomorphic role of floods or infer the magnitude of past floods based on channel dimensions (Ferguson 1986).

Strong correlations between Q and w, d, and v exist where channel boundaries have relatively low erosional thresholds and where substantial variations in sediment supply or imposed gradient are not present. Numerous site-specific case studies, however, document weak or poorly developed DHG relations in mountainous regions with persistent geomorphic effects from glaciation or strong downstream contrasts in substrate erodibility or grain-size distribution of sediment inputs (Wohl 2010b).

Wohl (2004b) proposes an empirical threshold for well-developed DHG relations in high-relief drainage basins as a function of the ratio of stream power (hydraulic driving forces) to sediment size (substrate resistance). Higher values of stream power associated with large discharge per unit drainage area, as in the tropics, can result in well-developed DHG relations even in channels with strong colluvial or bedrock influences (Pike et al. 2010). DHG relations for bedrock channels are within the range of those in alluvial channels, although bedrock channels tend to be narrower and deeper so that depth increases more rapidly than width as discharge increases (Montgomery and Gran 2001; Wohl and David 2008; Ferguson et al. 2017b).

Both forms of hydraulic geometry are scaling relations between discharge and the other variables of the continuity equation. The exponents of the hydraulic geometry relations reflect the relative magnitude of response among w, d, and v to changes in Q and can be used to infer the nature of channel adjustment. The DHG average width exponent of ~0.5, for example, indicates that approximately half of the channel adjustment to downstream increases in discharge occurs via channel widening, whereas the at-a-station average depth exponent of 0.42 indicates that adjustment to greater flows at most cross-sections occurs primarily via increased flow depth rather than increased width.

Precisely predicting channel response to changing discharge is difficult because of two factors not included in hydraulic geometry relations: interactions among w, d, v, and other channel form variables (e.g. bedform amplitude), and the influence of sediment and large wood supplies. At-a-station and DHG are nonetheless useful in understanding the manner in which channel form adjusts to fluctuating discharge (Gleason 2015).

6.1.4 Lane's Balance

Lane's balance refers to a conceptual model of channel adjustment that includes several cross-sectional parameters and accounts for water and sediment discharge. Lane (1955) conceptualizes equilibrium within a channel segment as reflecting a balance among discharge, Q_w, channel gradient, S, sediment load, Q_s, and sediment size, D_s

$$Q_w S \propto Q_s D_s \tag{6.4}$$

This relation is referred to as Lane's balance and has been widely depicted using a drawing of a balance (Figure 6.3), which is intuitively appealing and easy to understand (Grant et al. 2013). An increase in sediment load ($\uparrow Q_s$), for example, will cause aggradation, coarsening ($\uparrow D_s$), and steepening ($\uparrow S$) of the stream.

The ability of Lane's balance to describe river adjustments is inherently limited, however, because the expression does not account for changes in cross-sectional, planform, and bedform geometry that

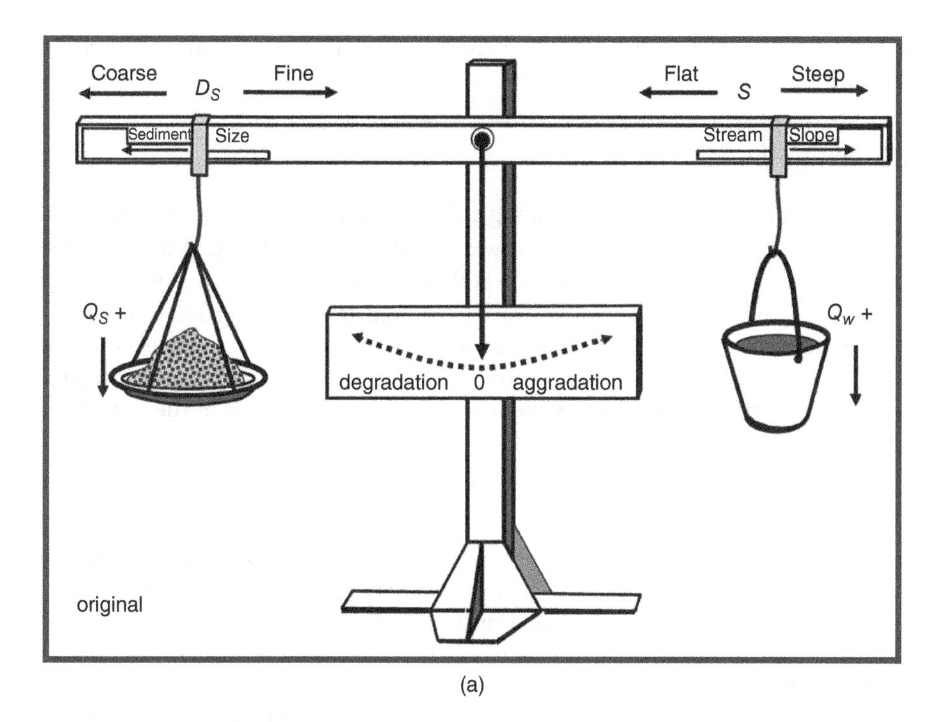

(a)

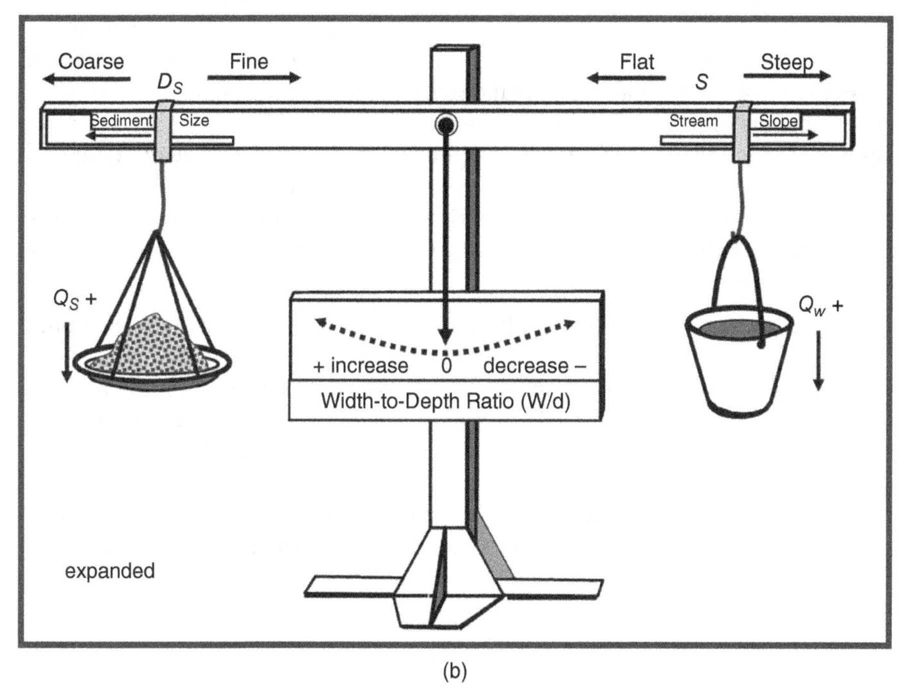

(b)

Figure 6.3 An illustration of Lane's balance, which is widely used to conceptualize channel adjustment in response to changes in water or sediment yield. (a) The original illustration, based on a drawing by Whitney Borland of Colorado State University. (Source: After Dust and Wohl 2012b, Figure 1.) (b) An illustration of an expanded version of Lane's relation, including adjustments to channel width-to-depth ratio. (Source: After Dust and Wohl 2012b, Figure 9.)

are commonly associated with channel adjustments to changes in discharge and sediment load. Dust and Wohl (2012b) expand Eq. (6.4) to include these terms

$$Q_w \left(\frac{\Delta z}{P\overline{H}_a} \right) \propto Q_s D_s \left(\frac{w}{d} \right) \tag{6.5}$$

where S is proportional to total change in elevation along a channel (Δz) and inversely proportional to sinuosity (P) and bedform amplitude ($\overline{H}_a$), and w/d is width-to-depth ratio. Eq. (6.5) suggests that an increase in sediment load ($\uparrow Q_s$) will cause aggradation ($\uparrow z$), decreased sinuosity ($\downarrow P$) and bedform amplitude ($\downarrow \overline{H}_a$), and decreased width-to-depth ratio ($\downarrow w/d$), as well as coarsening and steepening.

Either version of Lane's balance is useful primarily as a conceptualization of the various potential ways in which channel geometry can adjust to the controlling variables of Q_w and Q_s. An analogous method for quantitatively predicting channel adjustment remains elusive because of the numerous interdependent variables.

6.1.5 Complex Response

Schumm (1973) and Schumm and Parker (1973) introduce the phrase *complex response* to describe asynchronous, discontinuous river response to a single external perturbation. During flume experiments, base-level fall initiated a headcut that migrated upstream (Schumm and Parker 1973). This increased sediment load to channel segments downstream from the headcut, which then began to aggrade (Figure 6.4). Sediment supply declined once the headcut reached its farthest upstream point, causing downstream reaches to incise and sometimes creating a second headcut that migrated upstream. Consequently, the upstream portion of a river can be incising while the downstream portion is aggrading. Any given point within the channel alternates between incision and aggradation with time, sometimes going through multiple cycles of incision/aggradation in response to the initial base-level fall before the channel eventually stabilizes once more.

Subsequent studies have applied the idea of complex response to a variety of spatial and temporal scales and types of rivers (Marston et al. 2005; Harvey 2007). In general, although a river can incise or aggrade throughout its length, the more common scenario is that aggradation in some part of the river's length is likely associated with enhanced erosion of the channel boundaries elsewhere. Among the implications of complex response is that:

- multiple terraces along a river may not represent multiple external changes (Womack and Schumm 1977) (Section 7.2);
- incision or aggradation within a segment of channel can be a transient response to upstream or downstream changes in channel geometry, base level, and sediment supply, with the time span of the transience depending on factors such as the size of the channel and the erodibility of the bed and banks (Trimble 2013); and
- restoration or management that seeks to limit incision or aggradation may actually enhance undesirable channel change if implemented just as the direction of channel response (i.e. incision versus aggradation) is changing (Schumm et al. 1984).

6.1.6 Channel Evolution Models

The ideas underlying Lane's balance and complex response, in particular, are incorporated into *channel evolution models*. Alluvial channels can develop very small w/d ratios when undergoing rapid

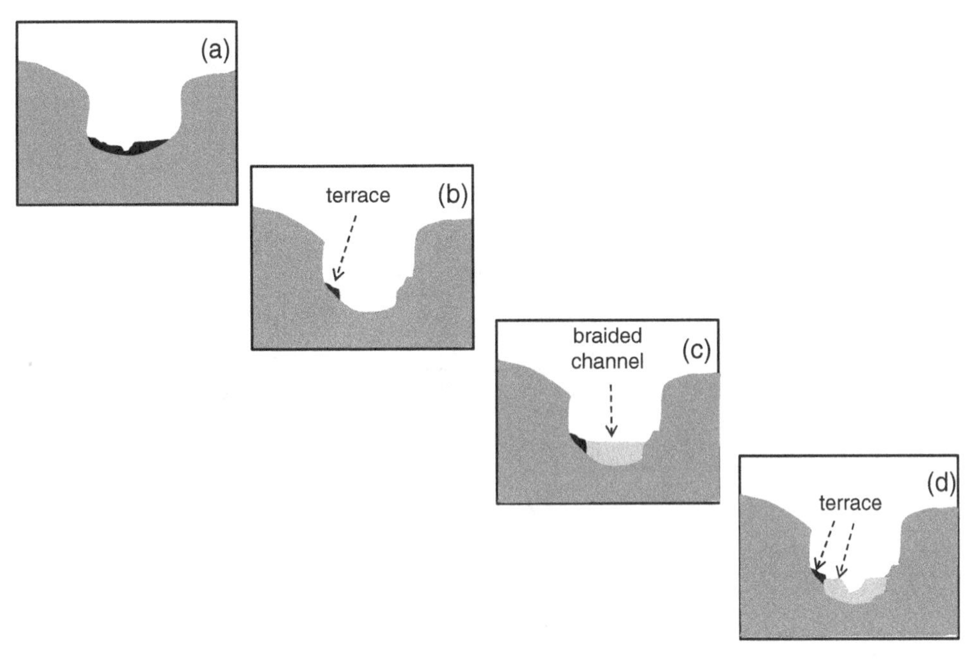

Figure 6.4 Complex response of a channel to a single fall in base level: perspective is upstream within the valley (gray shading) at a point midway along the channel between the mouth and the headwaters. (a) Stream alluvium (black) deposited before base-level lowering. (b) After base-level lowering, a knickpoint migrates upstream, causing the channel to incise into alluvium and bedrock of the valley and leaving the alluvium as a terrace. Following incision, bank erosion widens the channel and partially destroys the terrace. (c) Continued upstream migration of the knickpoint causes increased sediment discharge to the mid-portion of river. Inset alluvial fill is deposited as sediment discharge from upstream increases. A broad, shallow, braided channel (lighter gray shading) develops. (d) Upstream migration of the knickpoint ceases and the sediment supply to downstream channel segments declines. The channel becomes deep and narrow and incises slightly, creating a second terrace. Source: After Schumm and Parker (1973), Figure 1.

incision, but this is typically a transient condition. Alluvial channels in diverse environments that incise in response to base-level fall or changes in water and sediment supplied to the channel pass through a characteristic sequence of channel geometry with time that is summarized in channel evolution models. The channel adjustment described in these models typically begins with a relatively narrow, deep channel. The channel subsequently widens, aggrades, and eventually stabilizes. Schumm et al. (1984) propose empirical relations between top width and drainage area (e.g. top width = 46.77 [drainage area]$^{0.39}$), derived from observations of stabilized channels, which can be used to identify whether incising channels are close to a stability threshold at which widening stops. A specific form of this relation can be useful within a limited geographic region, but the coefficient and exponent vary between regions.

Schumm et al. (1984) propose a five-stage model of channel evolution. This has been modified to six or more stages (Simon and Rinaldi 2013), which reflect shifts in the dominant adjustments and associated rates of sediment transport, bank stability, and cross-sectional geometry (Figure 6.5). The time necessary to develop each stage of evolution and the relative magnitudes of vertical and lateral adjustments vary widely between different streams and stream segments (Simon and Rinaldi 2006), and the

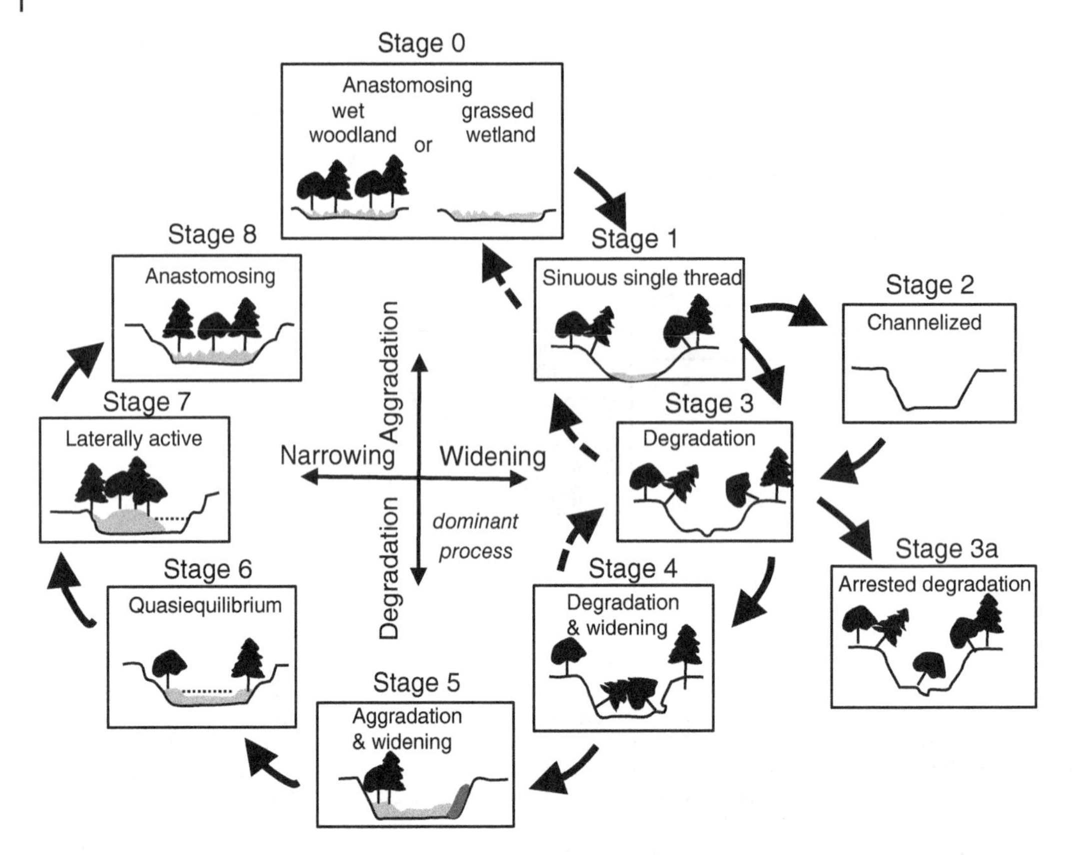

Figure 6.5 Illustration of a stream evolution model. Dashed arrows indicate alternatives to the normal progression (e.g. a Stage 1 stream can alter to a Stage 2 or Stage 0 configuration). Areas outside the circle represent construction and maintenance of stable end-points (Stage 2) or stabilization of incised channel banks by an erosionally resistant layer (Stage 3a). Source: After Cluer and Thorne (2014), Figure 4.

entire sequence can require 10^2–10^3 years (Simon et al. 2016) to complete. Cluer and Thorne (2014) conceptualize channel evolution as a cyclical, rather than a linear, process, and explicitly recognize different potential precursor (prior to disturbance) channel configurations, as well as different trajectories of channel adjustment. Like Lane's balance and complex response, channel evolution models are useful conceptualizations rather than quantitative predictions.

Incised channels of the type described in channel evolution models can form anywhere but are particularly common in arid and semiarid regions. Channels in these regions undergo repeated episodes of alternating incision and aggradation in response to internal thresholds, as described in Section 1.5.1 (Schumm and Hadley 1957), or to external changes in flood magnitude and frequency (Webb and Baker 1987; Hereford 2002), other aspects of climate (Leopold 1976), or land uses such as grazing and ground-water withdrawal (Cooke and Reeves 1976). Artificially channelized streams also typically go through the stages described in channel evolution models following completion of channelization.

Adjustments to cross-sectional geometry represent an intermediate level of channel response to changing water and sediment dynamics. Bed grain-size distribution and bedform configuration are more likely to change first as water and sediment supplies fluctuate, as described in Chapter 4. Adjustments to channel *w*/*d* ratio typically represent the next level, in response to larger-magnitude or more prolonged fluctuations in water and sediment. Changes in channel planform represent a third level of channel adjustment.

6.2 Channel Planform

Several classification schemes have been proposed to describe the wide variety of forms assumed by rivers when viewed on a two-dimensional planar surface such as a map (Buffington and Montgomery 2013). An obvious distinction is between rivers with single channels and those with multiple channels. Single channels are typically differentiated on the basis of *sinuosity*, the ratio defined by actual flow path downstream to straight-line distance between two points. Multiple channels are typically differentiated based on the lateral mobility of secondary channels.

Leopold and Wolman (1957) propose a tripartite classification of straight, meandering, and braided channels. Although this classification is still used, there are many channel planforms that do not fit well within these categories. Schumm (1985) proposes a broader range of 14 channel

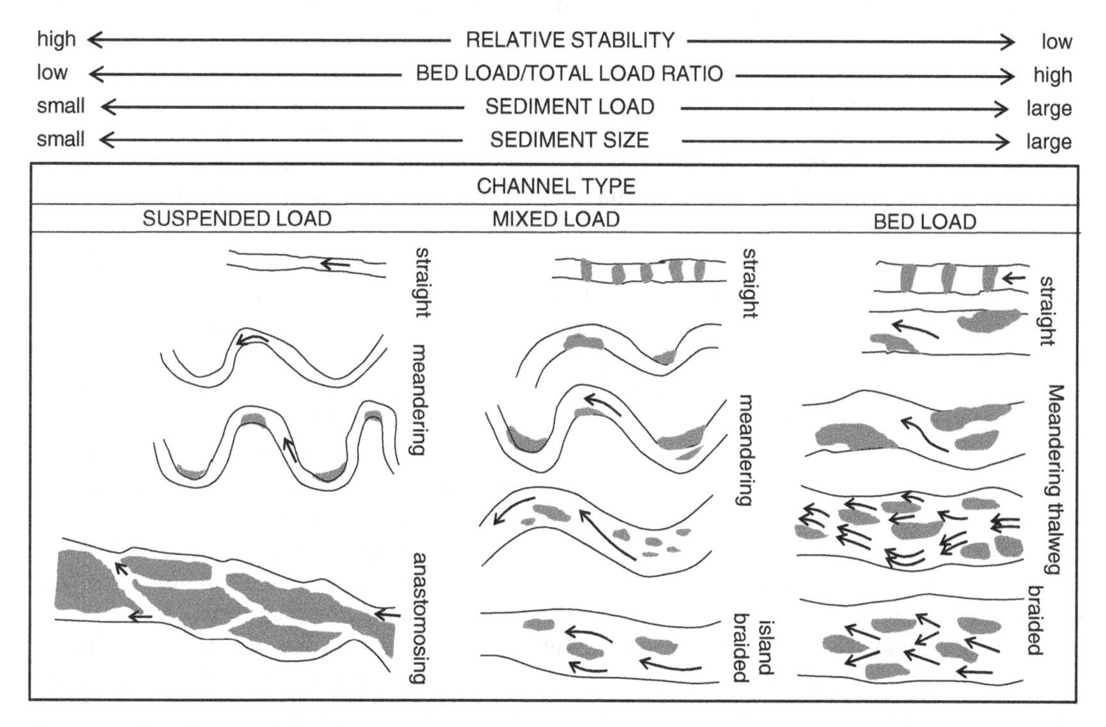

Figure 6.6 Classification of alluvial channel pattern (after Schumm 1981). Gray shading indicates depositional areas in the form of islands, bars, or riffles. Arrows indicate flow paths.

types (Figure 6.6), and other investigators have described additional categories such as wandering gravel-bed rivers (Carson 1984), anabranching rivers (Brice and Blodgett 1978), and compound rivers (Graf 1988) that regularly alternate between two or more planforms over time.

Any classification imposes more or less arbitrary boundaries on a continuum of channel form. Ideally, the boundaries reflect thresholds in the processes that create channel form. As with substrate, bedforms, and cross-sectional geometry, channel planform reflects an adjustment of the rate of energy expenditure in relation to water, sediment, and large wood supplied to the channel and the erosional resistance of the channel boundaries. Any channel planform that deviates from a single, straight channel – as many planforms do – represents an adjustment of resistance to flow. A channel with meanders, for example, effectively has a longer flow path for a given downstream length and gradient, and thus greater boundary resistance compared to an otherwise equivalent straight channel. The most commonly occurring planforms are described in more detail in the following sections.

6.2.1 Straight Channels

Straight channels have a single channel with sinuosity less than 1.5. They can be straight because they are closely confined by steep valley walls or other uplands such as terraces, or they can occur in erodible, alluvial boundaries. Even straight channels in erodible alluvial boundaries, however, remain straight because the banks have sufficient erosional resistance relative to available flow energy to limit bank erosion (Paola 2001). Straight channels can be further distinguished as those with or without mobile bedforms, those with alternating bars and a sinuous thalweg, and slightly sinuous channels with point bars (Schumm 1985).

Straight alluvial channels without mobile bedforms have mostly suspended sediment, a low gradient, and a deep, narrow cross-section. Such channels are rare (Schumm 1985). Straight alluvial channels with mobile bedforms are typically sand-bed channels in which the sequence of immobile bed, ripples, dunes, mobile plane bed, and antidunes described in Section 4.3.1 occurs as flow changes.

The majority of straight alluvial channels in mixed substrates or substrates coarser than sand have pool–riffle sequences of greater or lesser mobility, depending on the substrate and sediment supply. The downstream alternation between pools and riffles is associated with cross-sectional asymmetry and undulations in bed elevation, so that a regular channel planform does not equate to regularity in other dimensions or to uniformity of boundary roughness. Pools typically occur on alternate sides of the channel in a downstream direction, with intervening depositional features such as riffles and alternate bars (Figure 6.7). The *thalweg* – the line of deepest flow – is thus sinuous even though the channel boundaries are straight. The flow alternately converges passing through pools and diverges passing over depositional zones (e.g. MacWilliams et al. 2006; Sawyer et al. 2010).

The ubiquity of either a sinuous thalweg or a sinuous channel planform presumably reflects an inherent tendency in water flowing over a rough surface to develop a rotational component of flow as a result of greater hydraulic resistance along the channel boundaries than in the center of the channel. This rotational component, described as *helical flow*, creates downstream alternations in the location of greatest velocity that are expressed in differences in boundary erosion, sediment transport and deposition, and channel geometry. Sinuosity in flow and channel planform scales consistently across features as diverse as channels less than a meter wide on glacial ice and channels more than a kilometer wide in very large rivers (Leopold 1994).

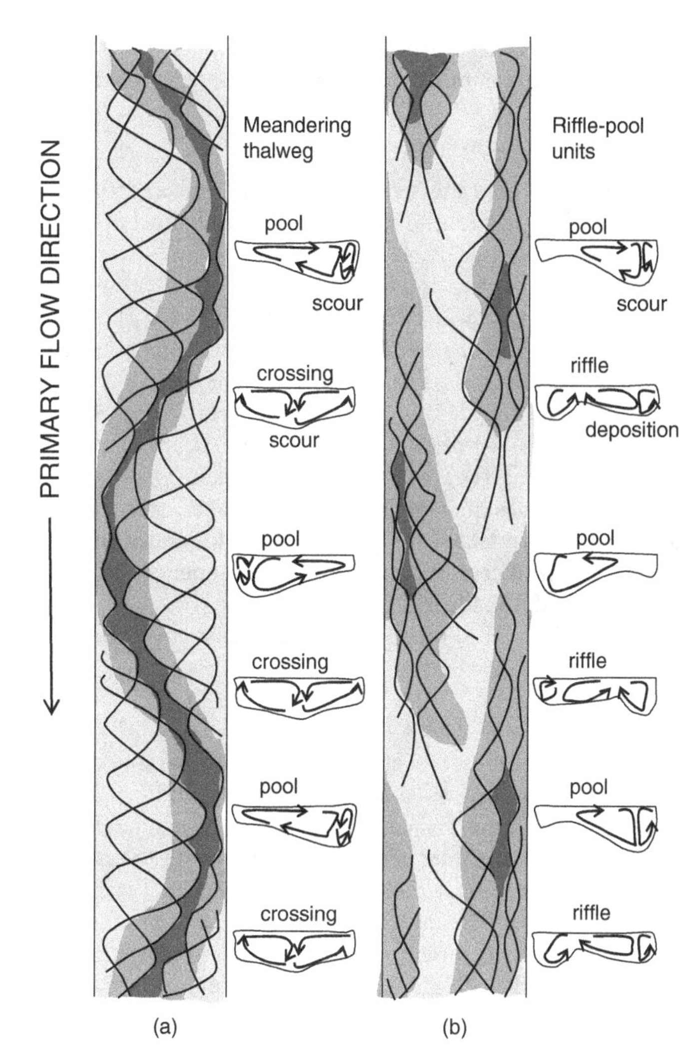

Figure 6.7 Plan view showing a sinuous thalweg and alternate bars within a straight channel, with associated helical flow and cross-sectional asymmetry shown by small channel cross-sectional views at right. (a) Model of twin, periodically reversing, surface-convergent helical cells based on work by Einstein and Shen (1964). (b) Model of surface-convergent flow produced by interactions between the flow and a mobile bed, creating pool–riffle units of alternate asymmetry, based on work by Thompson (1986). Black lines indicate currents, while progressively darker gray shading indicates progressively deeper portions of the channel. Source: After Knighton (1998), Figure 5.15, p. 199.

6.2.2 Meandering Channels

Single channels with a sinuosity greater than 1.5 appear to be the most widespread and common type of channel planform (Leopold 1994). Meandering channels can be differentiated based on the regularity of bends (Kellerhals et al. 1976) into:

- *irregular meanders* with a weakly repeated downstream pattern;
- *regular meanders* with a repeated pattern and a maximum deviation angle between the channel and down-valley axis <90°; and
- *tortuous meanders* with a less clearly repeated pattern and a maximum deviation angle >90°.

This differentiation reflects the fact that naturally formed meanders are seldom perfectly regular, but instead include randomness that reflects local controls on the erodibility of the channel boundaries.

In any given meander, the outer banks are commonly steep and eroding, and a pool is present at the bend apex. Cross-sectional bed topography slopes downward from the point bar on the opposite inner bank. Riffles are present in the inflection regions of the bend, in the straight limbs between bend apices where cross-sectional and bank geometry are most symmetrical (Hooke 2013).

Meander geometry is typically characterized using meander wavelength, λ, and radius of curvature, r_c, for individual bends. Path direction, θ, and change of direction, $\Delta\theta$, characterize meander geometry for multiple bends (Figure 6.8). Several parameters are necessary to adequately characterize meander form (Ferguson 1975). Standard approaches include: (i) assumption of a regular wave form and measurement of standard bend parameters such as wavelength, amplitude, sinuosity, and radius of curvature; and (ii) curve fitting using digitized center lines of meander sequences (Hooke 1977; Coulthard and Van De Wiel 2006; Güneralp and Rhoads 2009).

Langbein and Leopold (1966) introduce the idea of modeling meander geometry using the equation for a sine-generated curve

$$\theta = \omega \sin k x \tag{6.6}$$

where θ is channel direction expressed as a sinusoidal function of distance x, with parameters ω for the maximum angle between a channel segment and the mean down-valley axis and $k = 2\pi/\lambda$. A sine-generated curve represents a path in which the sum of squares of changes in channel direction per unit length is minimized, which effectively distributes stress uniformly along the curve. A sine-generated curve is a good approximation of the geometry of regular, symmetrical meanders (Leopold 1994). Many – perhaps the majority of – natural meanders, however, are not symmetrical (Carson and Lapointe 1983). Measurements of meander geometry across a range of environments suggest that λ is typically 10–14 times the channel width and r_c is typically 2–3 times the channel width (Knighton 1998).

Just as channel width is approximately proportional to $Q^{0.5}$, so meander wavelength typically varies as $\sim Q^{0.5}$, whether Q is mean annual flood, mean annual discharge, mean monthly maximum

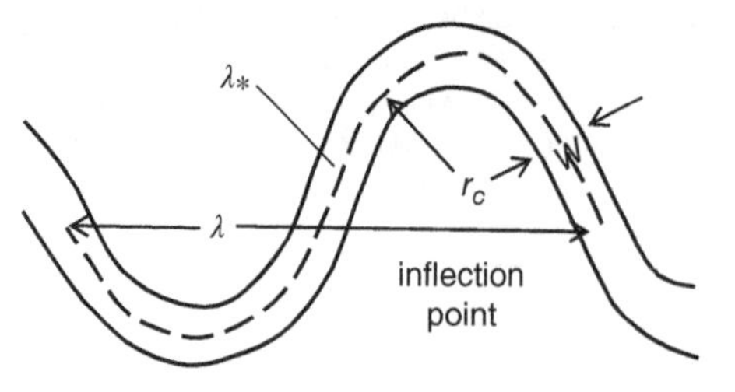

Figure 6.8 Components of meander geometry.

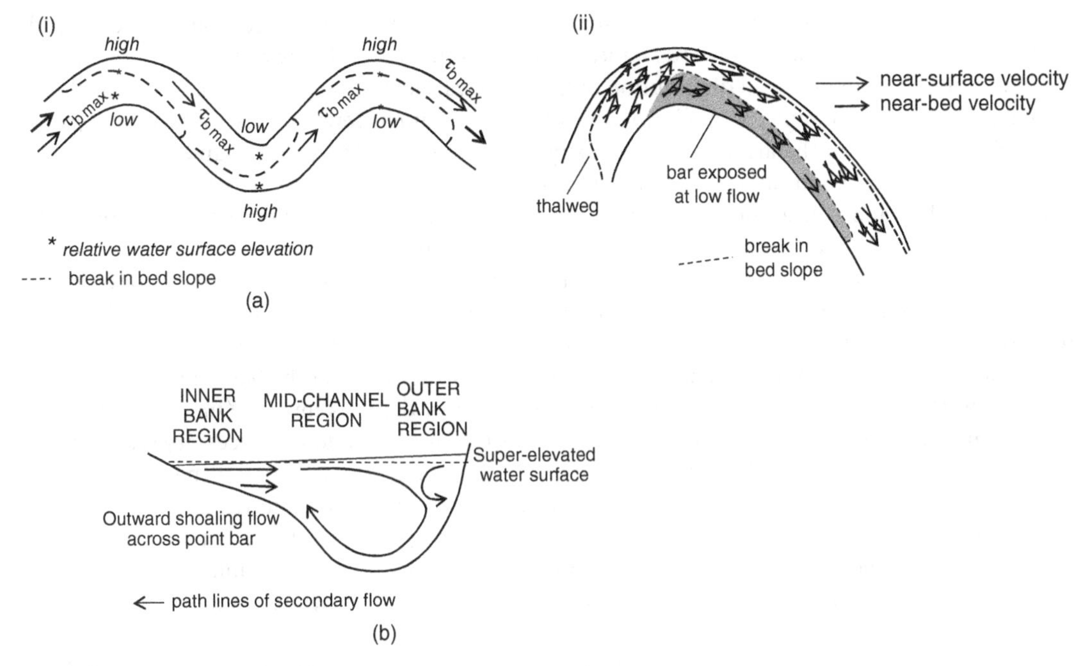

Figure 6.9 Sinuosity of the thalweg and associated velocity and shear stress in meandering channels. (a) (i) Location of maximum boundary shear stress, τ_b, and (ii) flow field in a bend with a well-developed bar (Source: After Dietrich 1987). (b) Secondary flow at a bend apex showing the outer bank cell and shoaling-induced outward flow over the point bar (Source: After Markham and Thorne 1992).

discharge, or some other measure of flow that is likely to transport the majority of sediment and thus strongly influence channel form parameters (Dury 1965; Knighton 1998). For a given discharge, meander wavelength also varies with boundary cohesion and gradient (Schumm 1963, 1967). The rates of vegetation establishment and growth relative to the rate of channel erosion further influence meander geometry and migration (Perucca et al. 2007). Wavelength decreases as boundary cohesion and gradient increase (e.g. Schwendel et al. 2015).

The thalweg of a meandering river does not maintain a central location along the channel in a downstream progression. Instead, the thalweg migrates across the channel through each bend from near the inner bank at the bend entrance to near the outer bank at the bend exit (Figure 6.9). The strongly helical flow in a meandering channel is expressed as superelevation of the water surface against the outer bank of each bend and a transverse current directed toward the outer bank at the surface and toward the inner bank at the bed. Helical flow facilitates preferential erosion of the outer bank and deposition on a point bar along the inner bank.

The details of the strength and location of the transverse, secondary currents reflect flow stage, meander geometry, and channel cross-sectional geometry. Secondary currents weaken with increasing stage and the flow follows a straighter path. Tighter bends with a lower ratio of radius of curvature to channel width (r_c/w) have stronger secondary circulation. A large width-to-depth ratio is associated with more extensive development of point bars. Point bars also reflect grain-size distribution (Hooke 2013).

Although models, in particular, typically assume relatively symmetrical and regular meander bend geometry, natural meanders can have substantial deviations in form and flow resistance, with associated deviations from an idealized distribution of hydraulic variables. The presence of instream wood along a meander, for example, strongly modifies the three-dimensional flow structure in a manner highly dependent on the arrangement, density, and mobility of the wood (Daniels and Rhoads 2004). In some bends, wood can create sufficient resistance to reduce flow velocity and constrain the high-velocity flow and helical motion to the channel center, whereas in other bends the wood can strongly deflect the flow away from the banks (Daniels and Rhoads 2004).

Hypotheses regarding the initiation of meandering tend to focus either on inherent properties of flow or on interactions between the flow and a mobile channel boundary. The argument for flow properties rests on the fact that water flowing down a rough, inclined surface – even a relatively smooth plate in a flume – in the absence of sediment develops sinuosity at the base of the flow where frictional resistance is greatest. Progressively more of the flow depth is involved in sinuous flow as discharge and velocity increase. The forces involved in this type of flow do not necessary scale well to natural channels (Hooke 2013), however, which typically have at least partly mobile beds.

Arguments for interactions between the flow and the boundary as the underlying cause of meandering emphasize deformation of the bed leading to the development of alternate bars that initiate meandering by deflecting flow toward the opposite banks (e.g. Seminara and Tubino 1989; van Dijk et al. 2012). Sinuous channels develop in materials without bars, however, such as those formed atop glacial ice or in bedrock.

Most likely, meanders are not the outcome of a single cause. Helicoidal, *curvature-driven secondary flow* is generated by superelevation of the water surface at the outside of a bend (Güneralp and Marston 2012). This creates substantial cross-stream variation in velocity, which redistributes downstream momentum, leading to a decrease in bed shear stress and deposition of a point bar along the inner bend, as well as a downstream increase in bed shear stress and erosion along the outer bend. The point bar deflects flow laterally toward the cutbank, creating *topographically driven secondary flow*. Seminara (2006), Güneralp and Marston (2012), and Hooke (2013) review theoretical and mechanical explanations for meandering in more detail.

Attention has also focused on why meanders form. Extremal hypotheses focus on the influence of meanders on energy expenditure. The reasoning is that meanders either (i) create the most uniform rate of energy expenditure along a channel by minimizing variance in hydraulic variables (Langbein and Leopold 1966) or (ii) establish a minimum channel slope for given input conditions (Chang 1988). Subsequent research, however, emphasizes the absence of equilibrium (Hooke 2013), focusing instead on continuous evolution and instability (e.g. Eaton et al. 2006). Discharge, sediment size and supply, bank resistance, and gradient are still acknowledged to influence meander migration and morphology, implying that any change in these variables results in a response in meander form and process (Hooke 2013).

Meanders can migrate very slowly, but even deeply incised bedrock channels that are sinuous show evidence of meander migration during incision (Harden 1990). Migration of individual meanders (Daniel 1971) can occur through:

- translation, which is downstream shifting of the bend without alteration in basic shape;
- extension, during which the bend increases its amplitude by migrating across the valley;
- rotation, in which the bend axis changes orientation; or
- lobing and compound growth, during which the bend become less regular and symmetrical.

Individual bends along a river can have different styles and rates of migration, and bends typically deform and become asymmetrical with migration. The fastest migration tends to occur when r_c/w is between 2 and 3 (Hickin and Nanson 1984). Ratios outside of this range lead to preferential migration in the upstream or downstream limb of the bend, with associated changes in curvature rather than rapid migration of the entire bend (Knighton 1998). Rates of meander migration can also reflect sediment supply: with other factors held constant, higher sediment loads correspond to faster annual migration rates and more frequent meander cutoff (Constantine et al. 2014). Parker et al. (2011) present a model of meander migration with separate relations for each bank. Migration of the eroding bank reflects the balance between flow forces, bank resistance, and armoring of slump blocks from past bank failure. Migration of the depositing bank depends on rates of sediment delivery and vegetation establishment.

Meander migration that increases the amplitude and tightness of bends can exceed a stability threshold and trigger a *chute cutoff* that creates a shorter channel across the inside of the point bar or a *neck cutoff* at the base of the bend (Constantine et al. 2010). The probability of chute initiation increases for bends that elongate more rapidly in a direction perpendicular to the valley axis trend (Grenfell et al. 2012). Because a cutoff increases channel gradient and local transport energy, the cutoff becomes the main channel and the longer flow path becomes a secondary or overflow channel that accumulates sediment during high flows.

Secondary channels can persist for decades to centuries, depending on the rates of sediment filling (Lewis and Lewin 1983; Grenfell et al. 2012). While present, secondary channels can form floodplain wetlands and increase habitat diversity for aquatic and riparian organisms. Secondary channels gradually fill with sediment settling from suspension during overbank flows. Because this sediment is commonly finer-grained than adjacent depositional areas of the floodplain, even secondary channels that completely filled with sediment decades or centuries earlier create persistent differences in ground water, soil moisture, and plant communities (e.g. Lewin and Ashworth 2014).

Field measurements conducted on meandering rivers typically focus on flow processes, bank erosion, and deposition and bar formation. Hooke (2013) notes that earlier measurements that included hydraulics, sediment dynamics, and channel changes (e.g. Jackson 1975; Bridge and Jarvis 1976; Dietrich et al. 1979; Thorne and Lewin 1979) remain standard datasets for whole meander bends, along with more recent work such as that by Frothingham and Rhoads (2003) and Parsapour-Moghaddam and Rennie (2018). The datasets developed during the 1970s come from sand- or gravel-bed rivers in the United States and United Kingdom. These data include some of the diversity present in alluvial rivers of the temperate zone, but likely do not adequately characterize sinuous rivers formed in bedrock or meandering channels in tropical or boreal environments.

Frothingham and Rhoads (2003) also describe a temperate zone river, but the methods they use to measure form and process along a single meander bend of the Embarras River in Illinois, USA illustrate more recent approaches to field measurements. These methods include an acoustic Doppler velocimeter mounted on a topset wading rod, with measurements examining the structure of time-averaged three-dimensional fluid motion and relating it to channel change via bank erosion. Parsapour-Moghaddam and Rennie (2018) use spatially dense acoustic Doppler current profiler data to calibrate a three-dimensional hydrodynamic river meander model for a small meandering channel in southeastern Canada.

Earlier hydraulic measurements with current meters and flow tracers have generally now been replaced by diverse acoustic Doppler instruments that facilitate very rapid measurement of flow direction and intensity (Dinehart and Burau 2005). Bank erosion is measured as reviewed in

Section 4.6. Sediment dynamics have been characterized using spatial and temporal variation in particle size and mapping of sedimentary structures in bars (e.g. Bennett et al. 1998). Bed topography is mapped using various surveying techniques, including bathymetry from boats (Hooke 2013), terrestrial scanning (Heritage and Hetherington 2007), differential GPS (Brasington et al. 2000), and, most recently, airborne LiDAR (Tonina et al. 2019).

Detailed process measurements typically cover a limited spatial area and a few years' duration (e.g. Hooke 1979), at most. Measurements of larger spatial areas and longer time spans rely on remote sensing data such as aerial photographs (Hickin and Nanson 1975; Gurnell 1997) and satellite imagery (Seker et al. 2005; Horton et al. 2017). Measurements of change over longer time intervals rely on historical records (Hooke 1977; Dort 2009) or inference of meander location and age from sedimentary and botanical evidence (Hickin and Nanson 1975; Malik 2006).

Initial attempts to replicate meanders in physical experiments had difficulty creating sufficient bank resistance to produce a single, meandering channel (e.g. Friedkin 1945). Consequently, early studies of hydraulics and bed topography relied on physical models with rigid walls. Mobile meanders have now been successfully produced in flumes by using vegetation such as alfalfa sprouts to enhance bank resistance (Tal and Paola 2007; Braudrick et al. 2009).

Advances in theoretical work and numerical simulations of meandering rivers during the 1980s are summarized in Ikeda and Parker (1989). Despite subsequent continuing work that ranges from one-dimensional approaches to three-dimensional modeling using computational fluid dynamics (Ruther and Olsen 2007), there is no standard method of predicting meander migration (Lagasse et al. 2004). Many theoretical and simulation models now exist, with diverse theoretical bases and assumptions, but most lack field validation or testing (Ferguson et al. 2003 provides an exception; Hooke 2013).

Most numerical models of meanders start with the theoretical development of flow patterns. Some include boundary interactions and the formation and movement of bars, others include curvature-induced patterns of flow, but most assume that bank erosion and bed movement reflect excess near-bank velocity (Ikeda et al. 1981; Hooke 2013). Some models simulate the formation and development of meanders, others simulate the detailed bed topography and hydraulics. The simplest typically focus on sequences of meander bends, whereas more computationally demanding models focus on single bends or very limited channel lengths. The latter approach can be used to model migration and chute cutoffs (Ruther and Olsen 2007), as well as meander migration via downstream translation, lateral extension, expansion, and downstream and upstream rotation (Chen and Duan 2006). Güneralp and Marston (2012) and Hooke (2013) provide more detail on measuring and modeling meandering rivers.

6.2.3 Wandering Channels

Wandering gravel-bed rivers have rapid bend migration, numerous bars, and frequent dissection of point bars. Wandering rivers are probably transitional between meandering and braided planforms (Carson 1984; Kidova et al. 2016). Wandering channels have also been subdivided into those with single channels, high channel migration rates, and frequent dissection of point bars, and those with multiple channels, a large supply of bed sediment, and low to moderate bank erodibility (Carson 1984). Wandering channels are most commonly described for mountainous regions with headwater glaciers or found downstream from large terraces, but this channel planform can occur in any environment (Church 1983; Burge 2005). Although the number of papers using the category of wandering channels is limited, the designation of this channel type reflects the fact that distinguishing meandering, braided, and anabranching channels is not always straightforward (Nanson and Knighton 1996).

6.2.4 Braided Channels

Braided rivers are multithread channels in which flow is separated by bars within a defined channel. Some of the bars can be submerged at high flows, but all are typically exposed at low flows. The degree to which bars are stabilized by vegetation varies, but a common distinction between multithread braided rivers and multithread anabranching rivers is that bars in braided rivers have less vegetation, are narrow relative to the width of the channel, and are relatively mobile (e.g. Carling et al. 2014). When bars and islands are distinguished, mid-channel bars are unvegetated and submerged at bankfull stage, whereas islands are vegetated and emergent (Brice 1964). Bars can be:

- longitudinal bars formed of crudely bedded gravel sheets;
- linguoid bars that are lobate in shape, made of sand or gravel deposited by downstream avalanche-face progradation; or
- point or side bars formed by coalescence of smaller bedforms such as dunes and linguoid bars at sites of lower energy (Miall 1977).

Alternatively, bars can be categorized as mid-channel, bank-attached, or compound (Wheaton et al. 2013).

Flume experiments with initially straight channels indicate that braiding can develop from the formation of single alternate bars, single mid-channel bars, or multiple mid-channel bars (Ashmore 2013). Such bars are low-amplitude bedforms occupying most of the channel width and connected to upstream scour pools. They deflect flow and start the development of channel sinuosity. Braiding develops either by cutoff of single bars at a critical bend amplitude or by bifurcation (splitting of flow) around mid-channel bars (Ashmore 2013). No single process leads to the division of flow and evolution of mid-channel bars. Instead, multiple depositional (e.g. bar building) and erosional (e.g. bar dissection) processes operate over time to create and maintain braiding (Wheaton et al. 2013).

Braiding is maintained by repetition of these initial bar-scale processes within the individual channels of the braided network. The temporal sequence starts with simple, well-defined pool–bar units from the pool head to the downstream bar margin in a single channel. With time, these are replaced by confluence–bar/bifurcation units, defined by the distance over which two channels join and then split downstream. These units form a basic morphological element of braided rivers (Figure 6.10) (Bridge 2003; Ashmore 2013).

Repeated division and joining of channels, and the associated divergence and convergence of flow, correspond to rapid shifts in channel position and the size and number of bars, particularly during floods. Braiding is produced by processes active at higher flows, rather than resulting solely from dissection of bars during low flows (Ashmore 2013). Flume experiments and numerical models indicate a strong, positive relationship between sediment supply and the frequency of channel avulsion (lateral movement and formation of a new channel), such that increased sediment supply and sediment heterogeneity cause greater channel mobility, bifurcation, and avulsion (Ashworth et al. 2007; Singh et al. 2017). Bars influence bifurcation processes more strongly as the bar height above the bed (bar amplitude) increases (Bertoldi et al. 2009). Adjustments to braided channel geometry alter flow resistance and the degree of braiding tends to increase with slope (Parker 1976) and stream power (Peirce et al. 2018).

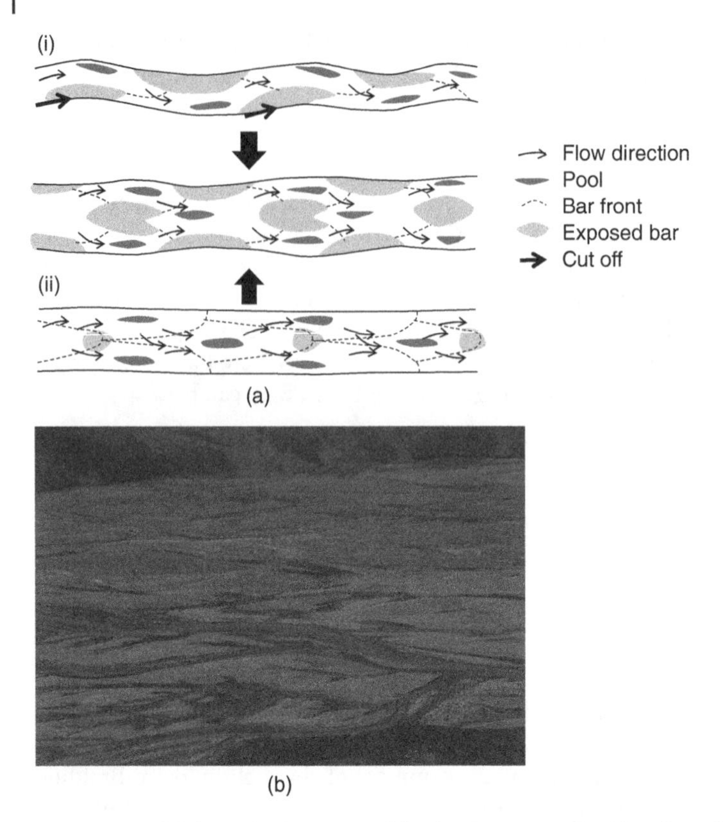

Figure 6.10 (a) Schematic illustration of the development of braiding from (i) initial alternate bars, through channel widening and chute cutoff, or (ii) central or higher-mode bars (Source: After Ashmore 2013, Figure 3). (b) Photograph of a wide (~1 km) braided river in northern Alaska, USA during low flow. Flow is from left to right. (*See color plate section for color representation of this figure*).

The degree of braiding has been quantified using various braiding indices. The most common approach is to count the mean number of active channels or braid bars per transect across the channel belt (Bridge 2003; Egozi and Ashmore 2008). Another measure is total sinuosity, which is the ratio of the total channel length of all active channels to the valley length (Hong and Davies 1979; Egozi and Ashmore 2008). Both of these measures are stage-dependent, in that higher stages can submerge bars and reduce the number of distinct subchannels. Redolfi et al. (2016) propose an index, α, derived from the frequency distribution of bed elevation. This index characterizes a statistical width–depth curve averaged longitudinally over multiple channel widths and defines a synthetic channel that includes information on river morphological complexity.

Braided channels are much less common than meandering channels. However, both the rock record prior to the evolution of land plants (Montgomery et al. 2003a; Davies and Gibling 2011) and flume experiments suggest that braiding is the default channel planform in rivers lacking sufficient riparian vegetation or cohesive bank sediments to substantially increase bank resistance to erosion (Paola 2001). Where woody riparian vegetation expands, commonly as a result of changes in flow regime, braided channels can metamorphose to a meandering or anabranching planform (Nadler and Schumm 1981; Piégay and Salvador 1997). Field studies indicate that the rate and location of growth of woody riparian vegetation strongly influence the formation and erosional resistance of islands (Gurnell and Petts 2006; Gurnell et al. 2019) and the stability of outer banks along a braided channel. Instream wood can also strongly influence the location and stability of bars, and the establishment of riparian forests on the bars, when sediment is deposited around a bar–apex logjam (Abbe and Montgomery 2003; Collins et al. 2012). The tendency toward braiding may be influenced by a river's ability to turn over its bed within the characteristic time required for riparian vegetation to establish and grow to a mature, scour-resistant state (Paola 2001). This dimensionless time-scale parameter can predict whether a channel will braid (Hicks et al. 2008).

Braided channels occur in diverse environments and across a broad range of scales. They are particularly common in arid and semiarid regions, downstream from glaciers, and in mountainous environments with abundant coarse sediment supply and limited riparian vegetation (Ashmore 2013). Proglacial braided channels are known by the Icelandic word *sandurs* at the point where the channel system expands freely, and as *valley sandurs* where development of the channel network is confined by valley walls (Krigstrom 1962).

Braiding tends to be associated with four conditions, although no single one is either sufficient or necessary to create a braided channel (Knighton 1998).

(1) Abundant bed load can cause braiding if the channel lacks capacity to transport the volume of sediment supplied or competence to move the size of sediment supplied (Griffiths 1979; Williams et al. 2015). Locally reduced transport capacity can facilitate sediment deposition sufficient to allow a bar to form and grow, deflecting the current toward the adjacent banks, creating local bank erosion, and introducing more sediment into the flow. This is the *central bar mechanism* that Leopold and Wolman (1957) invoke to explain the initiation of braiding. Related to this is the *transverse bar conversion mechanism* of initiation of braiding, in which flow convergence through a pool scours the bed and provides sufficient bedload for deposition downstream from the pool where the flow diverges, eventually causing the flow to be deflected around an elevated bar (Ashmore 1991b). Increased bedload supply causes aggradation and an increase in the degree of braiding, whereas decreased supply has the opposite effect (Germanoski and Schumm 1993; Thomas et al. 2007).

(2) Erodible banks facilitate continued channel widening and the development of multiple bars in wide, shallow flow with heterogeneous transport capacity. This corresponds to the argument presented above: single-thread or anabranching channels tend to form where cohesive banks result from silt and clay or from riparian vegetation (Eaton and Church 2004; Braudrick et al. 2009).

(3) Rapid fluctuations in discharge contribute to bank erosion and heterogeneous bedload movement. Large floods can initiate braiding, in part by removing stabilizing riparian vegetation and dramatically increasing bank erosion and channel width (Figure 6.11) (Burkham 1972; Friedman and Lee 2002; Jaquette et al. 2005). Some types of channels alternate repeatedly between a braided planform immediately after a large flood and a meandering planform that develops gradually during lower discharges over years to decades following a large flood.

(4) Steep valley gradients appear to promote braiding, although high stream power, QS, may be a better measure than simply gradient (Knighton 1998). Different threshold values have been proposed, but braided rivers tend to have higher values of stream power than meandering rivers of similar grain size (Ashmore 2013). The physical explanation for this observed correlation remains under debate.

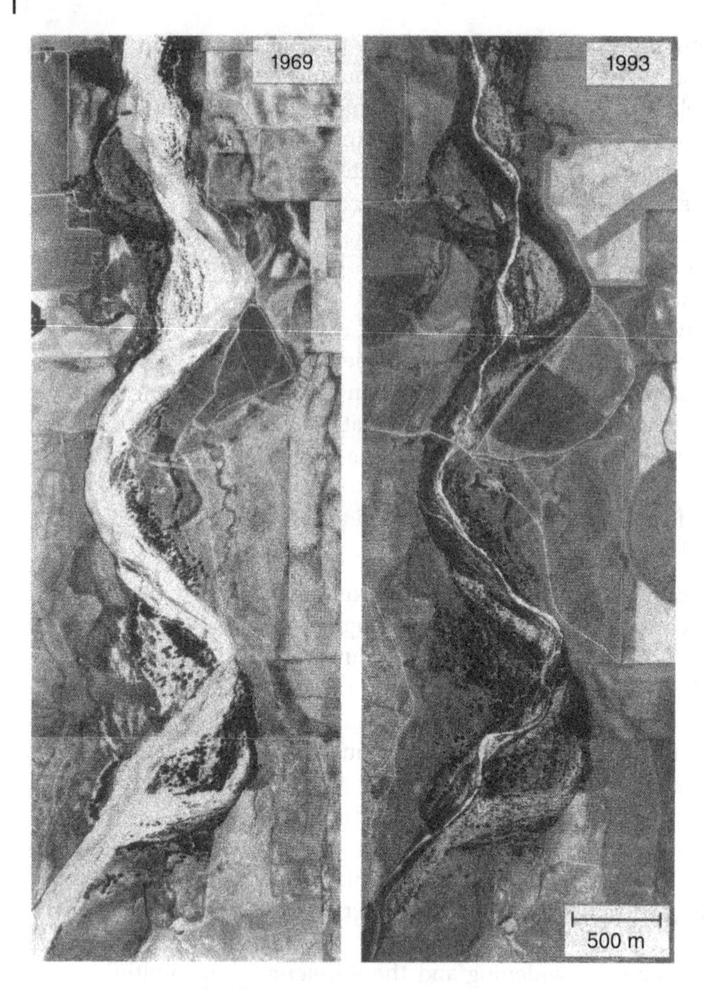

Figure 6.11 Example of a river (West Bijou Creek, Colorado, USA) that alternates between meandering and braided planform through time in relation to the magnitude of its floods. A large flood in 1965 removed much of the woody vegetation present along the channel, leading to a braided planform. The absence of large floods during succeeding decades allowed vegetation to regrow along the channel margins, and the active channel narrowed to a single-thread, slightly sinuous planform. Source: After Friedman and Lee (2002), Figure 2.

In addition to initiation through primarily depositional processes, as already described, braiding may also initiate in response to erosion of bars. Flume experiments indicate that dissection or the formation of cutoff channels across various types of bars occurs when flow follows a steeper route across the bar, causing headward incision, which increases the braiding index (Ashmore 1991b; Germanoski and Schumm 1993). The presence of bars is still basic to developing a braided channel, indicating that the fundamental underlying process is local deposition in response to loss of transport capacity.

Once braiding starts, all of the erosional and depositional processes just described may operate together to maintain braiding, as long as short-term transience in bed-load transport is maintained. Such transience is closely tied to longitudinally alternating convergent flow zones at confluences, in which sediment from upstream is transported and bed scour occurs, and to divergent flow zones at bifurcations, in which sediment is deposited (Ashmore 2013).

Confluence geometry reflects the relative discharge, junction angle, and orientation of the confluent channels. Maximum depth of scour increases as junction angle increases and as discharges in the confluent channels approach equal magnitude (Mosley 1976; Best and Rhoads 2008). *Avulsion frequency* is the frequency of formation of a new channel capable of carrying ≥50% of the discharge of the old channel. Avulsion frequency scales with the time necessary for bed sedimentation to produce a deposit equal to one channel depth, at which time the channel is likely to avulse. This is reflected in a dimensionless mobility number based on the relative rates of bank erosion and channel sedimentation (Jerolmack and Mohrig 2007).

As in the case of sinuosity in meandering channels, braiding represents a morphological adjustment of channel gradient to a stable configuration; in this case, by progressive channel subdivision, which effectively alters the channel gradient by altering the length of the flow path for a given vertical drop. Mid-channel bars exhibit similar width-to-length ratios over a wide range of spatial scales (Kelly 2006). This suggests that braiding is a self-organized, emergent property of the interaction between flow and a noncohesive sediment bed that must be constrained in some way if morphology other than braiding is to develop (Paola 2001; Ashmore 2013).

Development of new techniques for measuring flow and bed morphology has greatly improved the ability to quantify braided river process and form. One of the challenges to direct measurements in braided rivers has always been the spatial complexity of multiple channels that shift rapidly. Sambrook Smith et al. (2006) review techniques that use some type of energy signal to quickly collect large amounts of data. These techniques include flow measurement with acoustic Doppler current profiling (Bridge and Lunt 2006) and direct measurement of bed morphology with multibeam echo sounding (Best and Ashworth 1997). Remote sensing of braided river morphology using synoptic digital photogrammetric and airborne laser survey methods can provide event- to decadal-scale information on morphological changes (Lane 2006; Hicks et al. 2008), especially when imagery is analyzed using Structure from Motion software (e.g. Javernick et al. 2014). Characterization of deposits using ground-penetrating radar can provide information on thickness, lateral extent, and sedimentary structures in the subsurface (Bridge and Lunt 2006; Okazaki et al. 2015).

Physical experiments involving braided rivers now typically focus on specific processes or process–form interactions. Examples include particle travel distance (Kasprak et al. 2015), morphological active width in relation to hydrograph characteristics (Peirce et al. 2018), and storage and mobilization of large wood (Bertoldi et al. 2014).

Numerical modeling of braided rivers has also advanced significantly during recent decades (Lane 2006; Guin et al. 2010; Ramanathan et al. 2010; Williams et al. 2016). The first numerical representation of the spatially distributed, time-dependent evolution of river braiding produced what resembles a braided river without much of the hydraulic complexity considered necessary in other models (Murray and Paola 1994). Other studies that use two- and three-dimensional flow computational fluid dynamics modeling (Nicholas and Sambrook Smith 1999) and cellular automaton models (Coulthard et al. 2007; Van De Wiel et al. 2007), sometimes coupled with bedload transport models (McArdell and Faeh 2001), suggest that the presence or absence of significant local redeposition distinguishes braiding from other channel planforms (Murray and Paola 1997). For any type of numerical model, model output is increasingly compared to real river data (e.g. Javernick et al. 2016; Williams et al. 2016).

Figure 6.12 Anabranching planform along a portion of the Yukon River in central Alaska, USA. View is looking upstream. (*See color plate section for color representation of this figure*).

6.2.5 Anabranching Channels

Anabranching channels are multithread channels in which individual threads are separated by vegetated or otherwise stable bars and islands that are broad and long relative to the width of the channel and that divide flows at discharges up to bankfull (Figure 6.12) (Makaske 2001; Nanson 2013). The islands persist for decades to centuries and are similar in elevation to the floodplain (Knighton 1998). Individual anabranching channels can be straight, meandering, or braided but, unlike in distributary networks, the channels in an anabranchnig network eventually rejoin. Anabranching planforms can also include blind channels, however, which form by attachment of a lateral bar to a channel bank at the upstream end of an anabranch, creating a channel that is closed at the upstream end (Leli et al. 2018). A minimum width of floodplain relative to active channel width is necessary for anabranching channels to form (Morón et al. 2017). Anabranching channels have received less attention than meandering or braided channels, despite being found in very diverse environments.

Anabranching occurs in bedrock channels, and the characteristics of individual channels appear to be influenced by joint geometry in the bedrock. Anabranching reaches are particularly common immediately upstream of waterfalls (Kale et al. 1996; Tooth and McCarthy 2004), but can also occur in lower-gradient reaches without knickpoints (Heritage et al. 2001; Milan et al. 2018).

Anabranching is also particularly common in very large alluvial rivers (Jansen and Nanson 2004; Latrubesse 2008), although many of the historically anabranching segments of these rivers have been channelized to single-thread channels (Pišút 2002). Anabranching is the dominant channel pattern of the Amazon, Congo, Orinoco, Parana, and Brahmaputra, among other large rivers, and was historically present along rivers such as the Danube and Rhine (Latrubesse 2008, 2015).

Alluvial anabranches can form via three processes. In the first, anabranches develop as erosional channels scour into the floodplain during channel avulsion. Avulsion can also be triggered by sediment accumulation within a channel, particularly where an obstruction such as a channel-spanning logjam creates a backwater effect (Abbe and Montgomery 2003; O'Connor et al. 2003; Wohl 2011b). As a mid-channel bar develops, bank erosion is enhanced and multiple, subparallel channels may develop. Rapid aggradation can produce frequent avulsions and a network of channels in various stages of formation and abandonment, which can be wandering or anabranching (Nanson 2013). Finally, anabranches can develop from delta progradation and modification of the distributary network (Nanson 2013). In this situation, anabranching may reflect an efficient means of redistributing and storing excess sediment across a wide valley, because the lower w/d ratios within individual anabranches enhance bed shear stress and sediment transport (Huang and Nanson 2000). Conversely, anabranching increases boundary resistance and may effectively consume surplus energy along rivers with very low sediment loads (Nanson and Huang 2008). Along a high-energy gravel-bed river in northern England, for example, peak flows are distributed among the interconnected network of anabranches, dissipating flow energy, reducing shear stress, and aggrading, despite scour at channel bifurcation and confluence sites (Entwistle et al. 2018).

Definitions and characterizations of multichannel rivers, including anabranching, anastomosing, and braided types, remain unclear because of a lack of consensus among geomorphologists (Carling et al. 2014). Nanson and Knighton (1996) distinguish six types of anabranching rivers:

(1) *Anastomosing rivers*, which form in cohesive sediment and at low gradients, and have very low stream power. This type has been described in diverse environments, including the Rocky Mountains of Canada; arid central Australia, Africa, and Wyoming, USA; and tropical South America.
(2) Sand-dominated, island-form anabranching rivers, which rely on riparian vegetation to provide bank cohesion. This type has been described for Australia, and has relatively low values of stream power.
(3) Mixed-load, laterally active anabranching rivers, which have moderate values of stream power. This type ranges from small channels in Australia to very large rivers in South America and Asia.
(4) Sand-dominated, ridge-forming anabranching rivers, which have long, narrow, parallel ridges stabilized by riparian vegetation and straight anabranches with steep banks. This type also has moderate stream power and has mostly been described in Australia.
(5) Gravel-dominated, laterally active anabranching rivers, which are wandering channels with higher stream power. Although the basal coarse sediment is overlain by finer sediment on the islands, vegetation provides most of the bank stability. These channels, in particular, can be facilitated by obstructions such as logjams.
(6) Gravel-dominated, stable anabranching rivers, which exhibit higher stream power. Like type 5 channels, riparian vegetation is critical to enhancing bank stability in these channels, and obstructions appear to facilitate anabranching. The steeper, confined valleys in which type 6 channels occur limit lateral mobility.

6.2.6 Compound Channels

The term *compound channel* can be used to describe any channel that has distinct low- and high-flow portions, including those that simply overflow onto a floodplain (Graf 1988; Marston et al. 2005). Compound channel can also refer specifically to low- and high-flow portions with distinctly different planforms. Examples of the latter include proglacial streams (Fahnestock 1963) and tropical rivers (Gupta and Dutt 1989) that switch seasonally between meandering and braided as discharge fluctuates.

An intriguing example of a compound stream is Cooper Creek in central Australia. This creek has a clay-bed anastomosing planform at low flow, then dries completely, allowing the clay particles to aggregate into sand-sized pellets. At the start of the wet season, the pellets are transported as bedload in braided rivers, but they disaggregate as flow continues, transitioning to the anastomosing clay rivers as flows recede (Nanson et al. 1986).

6.2.7 Karst Channels

Channels developed in karst terrains form a unique subset of rivers in that they exhibit the properties of open-channel flow, but exist in subterranean environments. As noted earlier, karst processes and forms are associated with rocks that are readily soluble under surface or near-surface conditions, typically carbonates and evaporates. Rivers in karst terrains can flow underground for substantial distances and then abruptly resurface as a spring where the water table and the karst aquifer intersect the surface, or as a river in a pocket valley where an impermeable substrate forces flow to the surface (Lipar and Ferk 2015). Conversely, surface streams can abruptly disappear underground in a *blind valley* (White and White 2018). Or, a surface stream in a *dry valley* may contain flow only during large runoff inputs, despite existing in a wet climate, because only large runoff inputs effectively fill subsurface conduits and force flow to the surface (White and White 2018).

Subterranean rivers in cave systems can behave similarly to surface rivers in the sense of transporting sediment from clay to boulder-size clasts and developing channel geometry that reflects adjustments between available energy and substrate resistance (Springer et al. 2003). Karst rivers differ from most surface streams formed in bedrock in that a large portion of the sediment load can be carried in solution, and erosion of resistant channel boundaries can produce distinctive sculpted forms such as scallops and pockets via abrasion and solution (Springer and Wohl 2002). Along surface channels, a floodplain can attenuate the greater discharge and energy present during a flood. Subterranean karst rivers may respond to enhanced discharge with much greater flow depths and velocities if the cave passage occupied by the river channel is sufficiently large, and may also develop conditions of pipe flow (Springer 2004). (Pipe flow does not have the free surface found in open-channel flow.) Rates of incision in subterranean karst channels are likely to be strongly influenced by water-table dynamics (e.g. Springer et al. 2015). In many respects, subterranean karst channels are analogous to laterally confined bedrock channels at the surface, with downstream changes in substrate erodibility strongly influencing channel geometry.

6.2.8 Continuum Concept

As noted earlier, classifications impose boundaries on continuous variations in channel planform. Leopold and Wolman (1957) recognize this explicitly, arguing that channel pattern reflects interactions among continuous variables such as sediment grain size and volume, gradient, boundary erodibility, and flow energy. They propose that a continuum of channel patterns exists, with each pattern being defined by a combination of control variables. Subsequent research reiterates this point (e.g. Lewin and Ashworth 2014; Church and Ferguson 2015), such as work in Gaurav et al. (2017) demonstrating statistically homogenous distributions of channel width relative to discharge for individual meandering and braided subchannels.

Efforts to identify the most important control variables and consistent correlations with channel pattern include, but are not limited to:

- slope versus discharge (Figure 6.13) (Leopold and Wolman 1957; Beechie et al. 2006), under the assumption that discharge is an independent variable and the slope of alluvial channels reflects roughness, particle size, and drainage area;
- the percentage of silt and clay in the channel boundaries (Schumm 1963), because the presence of cohesive sediment strongly influences channel stability, shape, and sinuosity;
- the ratios of depth to width (d/w) and slope to Froude number (S/Fr) (Parker 1976), as a means of including channel form and flow energy parameters; and
- unit stream power versus median bed grain size (D_{50}) (Van den Berg 1995), as two boundary conditions that are nearly independent of channel pattern. Subsequent analyses, however, have rejected this basis for discriminating channel planform (Lewin and Brewer 2001).

Brotherton (1979) and subsequent authors (e.g. Church 2006) argue for the importance of the relative ease of eroding the channel banks versus transporting bank material. Where transport dominates, the channel remains straight, whereas erodible banks facilitate braiding, with meandering as an intermediate scenario (Knighton 1998).

In general, the progression from laterally stable (straight) channels through meandering to braided channels tends to correlate with increasing stream power, increasing w/d ratio (and hence increasing bank erodibility), and increasing volume and grain size of bed load (Ferguson 1987; Church 2006). Although plots such as Figure 6.13 oversimplify the complexity of channel planforms, they do provide useful insights regarding a given channel segment's proximity to a planform threshold and its likely response to relatively small changes in external variables (Schumm 1985; Knighton 1998).

Transitions between different planforms and the ability of channels to repeatedly cross thresholds as conditions of boundary resistance and flow energy change are particularly well represented using a cusp catastrophe model such as that in Figure 6.13c (Thom 1975; Francis et al. 2009). By representing channel boundary resistance on the x and z axes, and magnitude, frequency, and duration of flow energy on the y axis, such a model incorporates the balance between hydraulic driving forces and substrate resistance in a manner that facilitates understanding of the relative importance of individual variables influencing channel planform.

6.2.9 River Metamorphosis

The idea of continuity between individual channel patterns is also reflected in the recognition that pattern can change abruptly in space and time. *River metamorphosis* describes the "almost complete transformation of river morphology" in response to natural or human-induced changes (Schumm 1969). River metamorphosis typically refers to a rapid change in channel planform (Table 6.1; Figure 6.14).

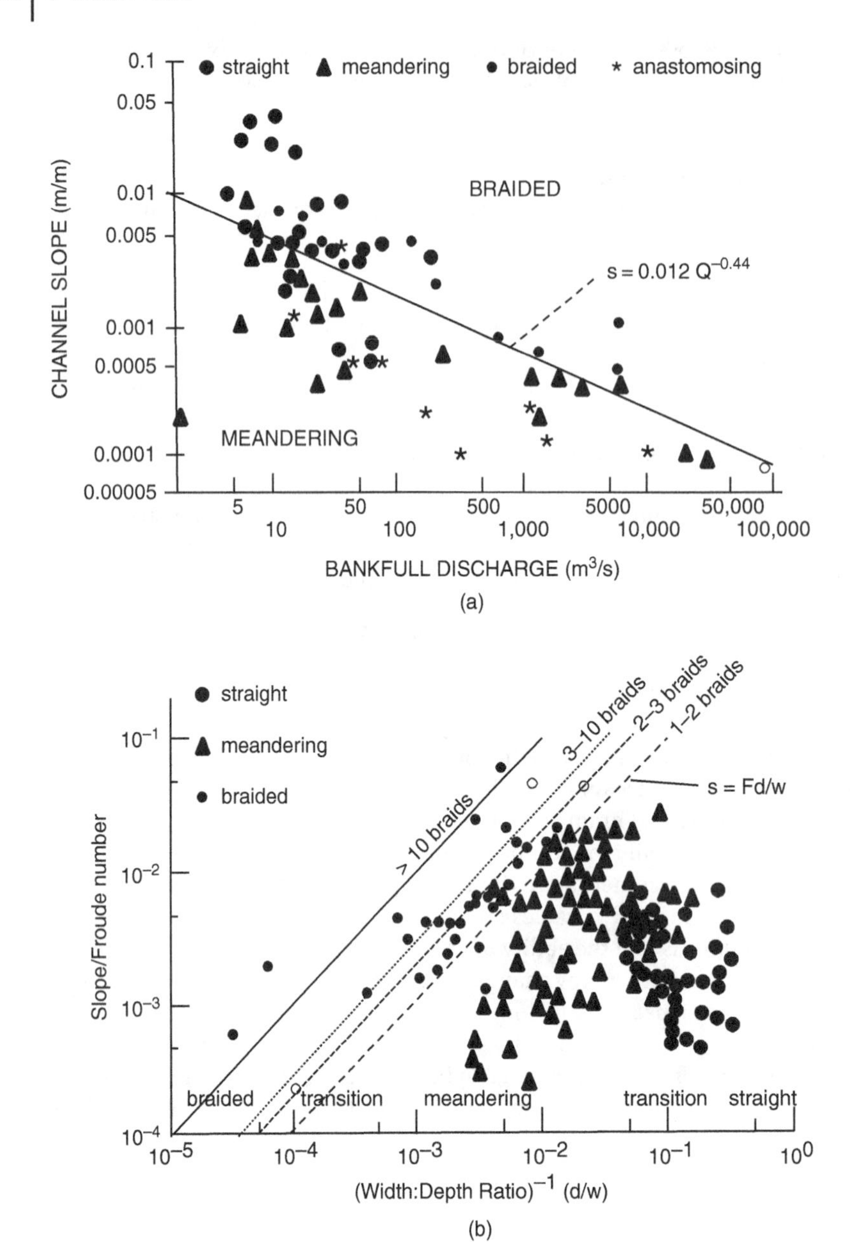

Figure 6.13 Different approaches to distinguishing channel planforms. (a) Braided and meandering with respect to slope versus discharge (Source: After Leopold and Wolman 1957). (b) Multiple channel types distinguished via slope/Froude number versus depth/width (Source: After Parker 1976). (c) Schematic representation of the continuity among different channel planforms, shown here in terms of a cusp catastrophe model (Thom 1975). This type of representation highlights how channels close to a threshold can have very different planforms (e.g. island braided versus meandering), with a small change in the controlling variables, such as hydrogeomorphic disturbance or vegetation dynamics, producing an abrupt change in channel planform (Source: After Francis et al. 2009, Figure 5).

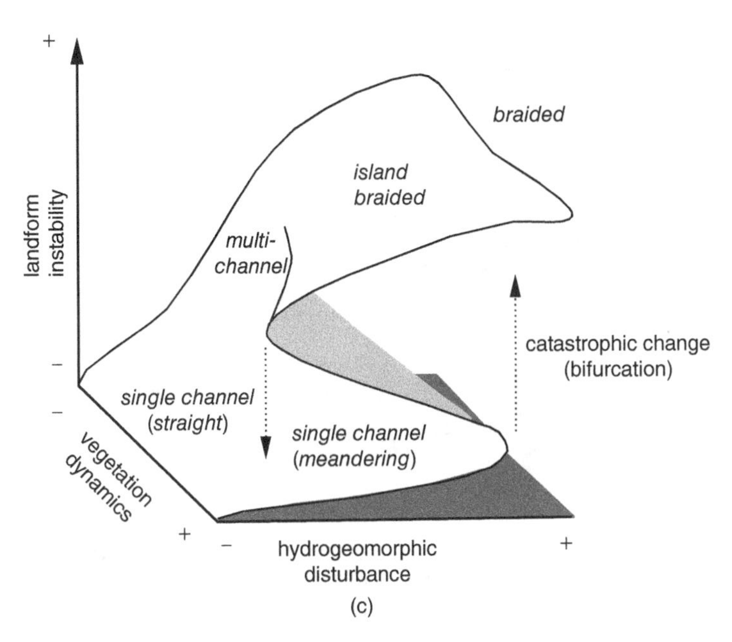

Figure 6.13 *(Continued)*

Table 6.1 Examples of river metamorphosis.

Description	References
Metamorphosis of a Polish river. Native vegetation was cleared across much of the drainage basin, and dramatic increases in sediment yield caused the meandering river to become braided.	Klimek (1987)
Metamorphosis of rivers in the US Southwest. Flow regulation, along with the introduction of exotic riparian vegetation that grows more densely along river banks, resulted in increased bank stability and sediment trapping. This took the form of narrowing, deepening, and the alteration of braided channels to meandering channels.	Birken and Cooper (2006); Reynolds et al. (2012)
Metamorphosis along rivers in the Great Plains, USA. Flow regulation caused a decrease in peak flows and an increase in base flows. Changes in flow regime allowed native riparian vegetation to grow more densely along the channel banks. The combined effects of loss of peak flows and increased bank resistance caused braided rivers to narrow to a meandering planform.	Williams (1978b); Nadler and Schumm (1981)
Metamorphosis of rivers due to the presence or absence of beaver dams. When present in a river, beaver build dams that promote channel–floodplain connectivity, overbank sedimentation and floodplain wetlands, and an anabranching planform. Beaver have been removed by trapping throughout much of their historic range. In parts of the western United States, they have also been outcompeted by large herbivores such as elk that have dramatically increased in population density following twentieth-century removal of their predators. Elk eat the same riparian woody species that support beaver. The combined loss of beaver dams and heavy elk grazing of riparian vegetation can result in metamorphosis of an anabranching channel to a single incised channel with unstable banks within the space of two to three decades. Where the effects of elk are somehow reduced (e.g. riparian grazing exclosures, reintroduction of predators) and beaver return to a site, the channel can again metamorphose to a multithread planform within a decade.	John and Klein (2004); Wolf et al. (2007); Beschta and Ripple (2012); Polvi and Wohl (2012)

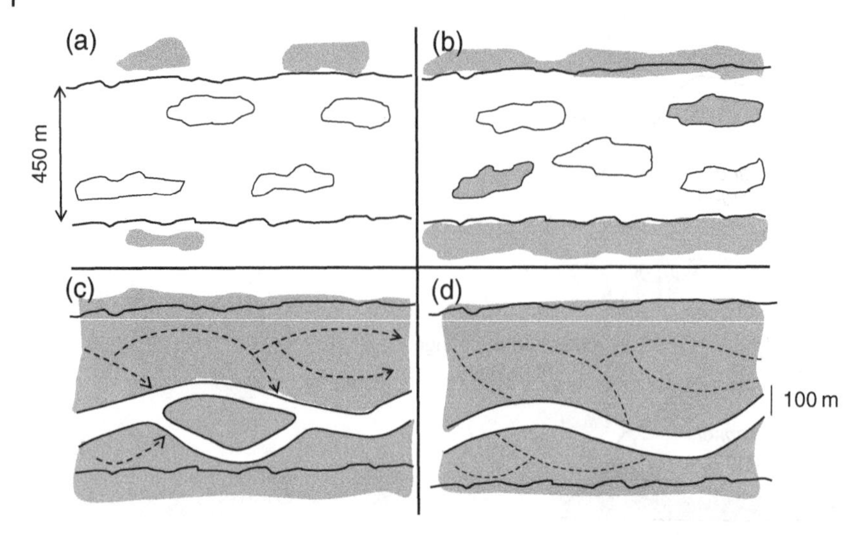

Figure 6.14 Schematic plan view illustration of river metamorphosis on the Great Plains of the United States. (a) Early 1800s: Highly seasonal discharge maintains a broad, shallow, braided channel with unvegetated bars. (b) Late 1800s: Flow regulation creates more consistent flows with higher base flow and lower flood peaks, allowing riparian vegetation (gray shading) to establish along the channel banks and on some of the braid bars. (c) Early 1900s: Droughts and flow regulation allow vegetation to establish below mean annual high water level. Bars become islands, and single thalweg is dominant. Dashed lines indicate vestiges of historic channels that remain on floodplain. (d) Contemporary channel: Bars and islands have become vegetated and attached to the floodplain, creating a wide, forested riparian corridor with a single narrow channel. Source: After Nadler and Schumm (1981), Figure 10.

River metamorphosis is not so much a model of how rivers adjust to changing external controls as the concept that pronounced channel change can occur very rapidly and in response to limited external change – such as loss of flood peaks or a change in biota – which then triggers numerous, nonlinear responses in river form.

6.3 Confluences

A specific aspect of changing channel planform is the presence, geometry, and erosional and depositional processes occurring where two segments of channelized flow meet at a confluence. Confluences are inherent in multithread planforms such as braided and anabranching rivers but can also be very important where tributary channels join single or multithread main channels or where secondary floodplain channels rejoin the main channel.

Confluences between tributaries and larger channels are ubiquitous in drainage networks, and confluences between subchannels are widespread in braided and anabranching channels. Confluences have received increasing attention in recent years because these highly turbulent locations strongly influence processes as diverse as contaminant dispersal and navigation (Gaudet and Roy 1995; Rice et al. 2008; Umar et al. 2018).

Average and maximum values of flow velocity typically increase at confluences because the cross-sectional area downstream is commonly smaller than the sum of the areas of the contributing channels. The presence of a flow separation zone downstream from the junction can further reduce effective cross-sectional area. The higher velocity can be associated with a scour zone in the bed. The size and location of the scour zone, along with hydraulic patterns, reflect the junction angle at the confluence and the ratio of discharge in the confluent channels (Lewis and Rhoads 2018): dimensionless scour depth increases as the junction angle increases and as the ratio of the minor tributary discharge to the mainstem discharge increases (Best 1988), as noted for braided channels. A shear layer develops at the margin of the flow-separation zone because of the low velocity in this zone relative to the main flow. A shear layer with significant vorticity also develops near the middle of the channel where tributary and mainstem flows merge (Figure 6.15). Turbulence associated with this shear layer may significantly influence the bed scour zone. Strong secondary circulation and helical flow also characterize confluences (Constantinescu et al. 2011). Velocity typically decreases beyond the confluence zone as flow merges into the single receiving channel (Best 1987).

Bars commonly form at confluences. The location and type of bar varies in relation to planform geometry. Symmetrical confluences with similar angles between each tributary and the receiving channel commonly have a mid-channel bar downstream from the confluence that reflects deposition of sediment eroded from the scour zone (Best 1986). Asymmetrical confluences typically have bars associated with the zone of flow separation, in which low velocity and recirculation promote sediment deposition (Figure 6.15). Among the most well-studied examples are the sand bars that form downstream from tributary junctions along the Colorado River in the Grand Canyon, USA (e.g. Wright and Kaplinski 2011; Mueller et al. 2018).

The Colorado in the Grand Canyon exemplifies a river within a laterally confined bedrock canyon. Smaller, steeper tributaries entering the river deposit bars and fans, commonly with episodic deposition during flash floods and debris flows on the tributaries (Webb et al. 1989). Coarse-grained tributary deposits create rapids and constrict the mainstem, leading to development of a shear layer, strong secondary circulation, and finer-grained separation and reattachment deposits (Figure 6.16). The separation and reattachment deposits are typically sand bars that form important riparian habitat and recreational sites for thousands of people who float down the Colorado through the Grand Canyon each year, and the eddy return-current channel creates a backwater that provides important habitat for endangered native fish (Schmidt and Graf 1990). Flume experiments and hydraulic modeling indicate that the length of the separation zone reflects the hydraulics and topography of the channel bed downstream from the confluence (Schmidt et al. 1993; Wiele et al. 1996). Aggradation within the separation zone effectively decreases the length of this zone (Schmidt et al. 1993). The great majority of marginal deposition in a channel configuration such as that in the Grand Canyon occurs in recirculation zones (Wiele et al. 1996).

Confluences between tributary and main channels in mountainous environments can also be strongly influenced by mass movements coming down the steeper tributaries. Low-order tributaries prone to debris flows can introduce abundant coarse sediment and wood into the main channel, resulting in more heterogeneous channel morphology and greater habitat diversity (Benda et al. 2003a). Tributary inputs can have minimal effects where the transport capacity of the mainstem is sufficient to rapidly redistribute tributary sediments (Rice 1998, 2017). The likelihood that tributary inputs will significantly influence channel morphology on the mainstem increases with the size of the tributary relative to the mainstem (Rice 1998, 1999; Benda et al. 2004a).

Confluences can be characterized as *concordant* when two channels of equal depth join, or more commonly as *discordant* when channels of different depths join (Robert 2014). Discordant confluences reflect differences in confluent channel dimensions as a result of differences in discharge and

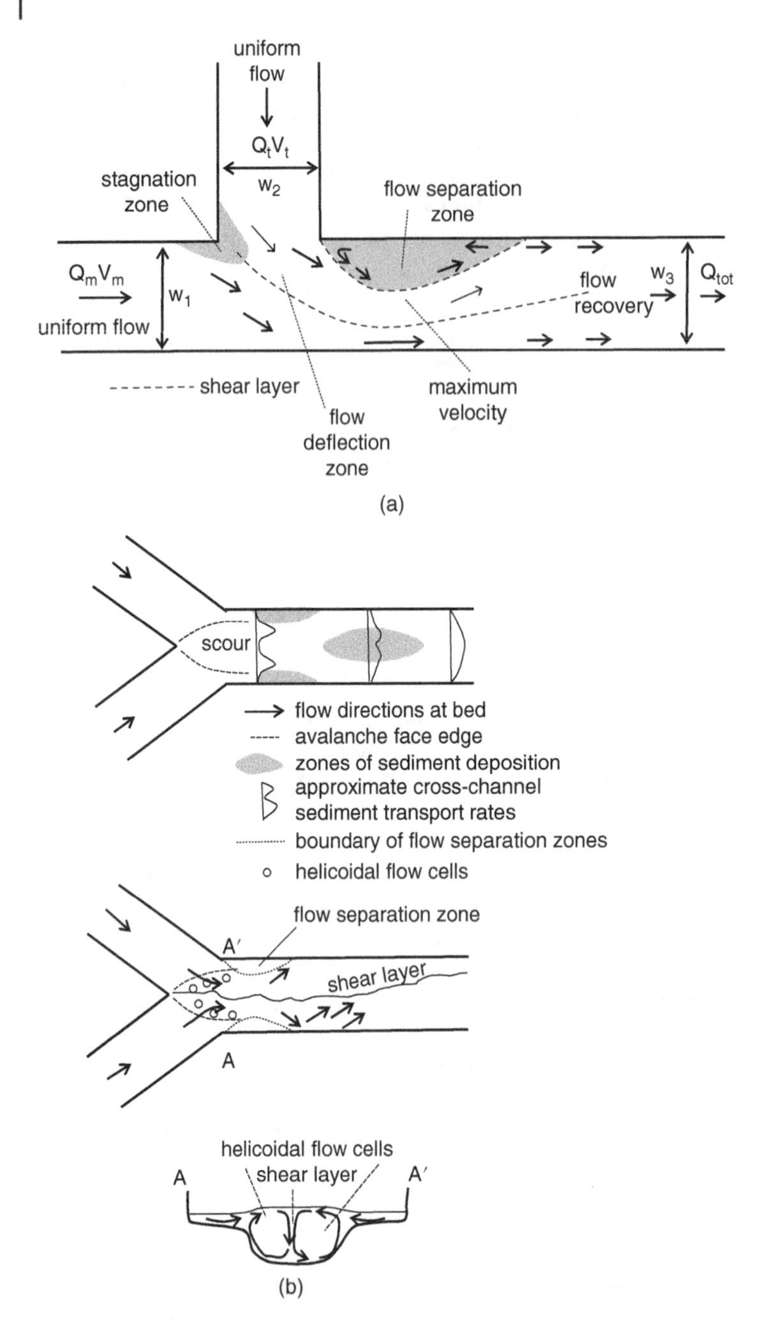

Figure 6.15 Hydraulics, cross-sectional geometry, and depositional features associated with channel confluences. (a) Conceptual model of flow dynamics at channel confluences, as derived from laboratory observations for an asymmetrical, 90° confluence with main channel and tributary of equal width and depth. *Q* is discharge, *V* is average velocity, and *w* is channel width; subscript *m* indicates the main channel and subscript *t* indicates the tributary. Source: After Best (1987), Figure 1. (b) Features of flow dynamics, bed morphology, and sediment transport at a symmetrical planform confluence. Source: After Mosley (1976), Figure 3 and Best (1987), Figure 5.

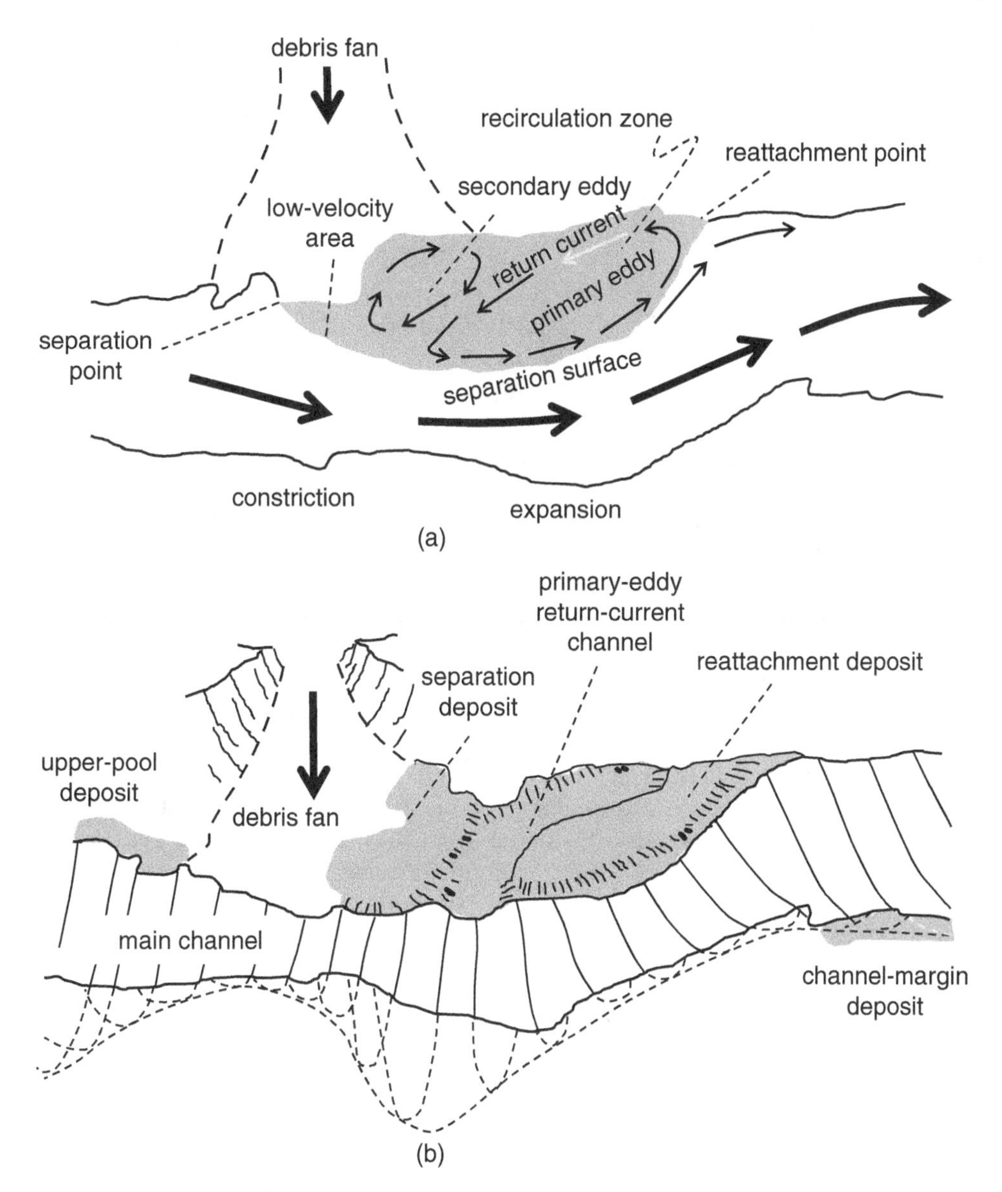

Figure 6.16 Separation and reattachment deposits in the Grand Canyon, USA. (a) Flow patterns. (b) Configuration of bed deposits. Gray shading indicates sand deposition. Source: After Schmidt and Graf (1990), Figure 3, p. 5.

boundary resistance; that is, differences in hydraulic geometry. The vertical mixing layer present in concordant confluences becomes more complicated at a discordant confluence, where the sudden drop in bed elevation from the tributary to the mainstem creates complex three-dimensional flow and vertical upwelling (Best and Roy 1991). The greater the bed discordance, the shorter the mixing length for the converging flows (Gaudet and Roy 1995).

In regions with active tectonic uplift or relative base-level fall, tributaries may not be able to incise as rapidly as the mainstem, leading to a pronounced drop at the tributary–mainstem confluence. In the most extreme case, this can result in a hanging valley (Section 6.5.3), with a substantial waterfall between the tributary and the mainstem (Wobus et al. 2006).

6.4 Bedrock Channels

Specific aspects of bedrock channels, including distinctive processes of erosion (Chapter 4), are treated in earlier chapters. Chapter 8 examines bedrock channels in the context of landscape evolution. The present discussion focuses on how channel form and processes of adjustment differ in bedrock channels from those in alluvial channels.

Bedrock channels have been variously defined. Bedrock channels have a substantial proportion (≥50%) of the channel boundary as exposed bedrock or covered by an alluvial veneer that is largely mobilized during high flows, so that underlying bedrock geometry strongly influences patterns of flow hydraulics and sediment movement (Tinkler and Wohl 1998). Bedrock channels lack a continuous alluvial sediment cover and reflect excess transport capacity over the long term (Whipple 2004). Bedrock channels are also defined as being unable to substantially widen, lower, or shift the bed without eroding bedrock (Turowski et al. 2008). The exposure of bedrock along a significant portion of the channel boundary indicates that transport capacity exceeds sediment supply, and this, along with the commonly substantial erosional resistance of the exposed bedrock, strongly influences channel form and channel changes through time. The spatial distribution of bedrock channels can reflect at least three controls. The first is tectonic uplift, which increases channel gradient and transport capacity beyond the available supply of sediment. The second is resistant lithology, which limits weathering and sediment supply to the river corridor (O'Connor et al. 2014) and results in deep, narrow cross-sectional geometry and an associated high capacity for sediment transport. Finally, bedrock channels in forested regions can reflect the absence of large wood, which when present can trap sufficient sediment to create forced alluvial reaches (Massong and Montgomery 2000).

There is no widely used classification of bedrock channel form analogous to the straight–meandering–braided–anabranching classification for alluvial channels. Bedrock channels can have straight, meandering, or anabranching planforms, but other forms of classification have been proposed based on cross-sectional and planform geometry (Wohl 1998; Ortega-Becerril et al. 2017b), the types of sculpted forms present (Richardson and Carling 2005), and morphology of waterfalls (Ortega et al. 2013).

The cross-sectional geometry and planform of bedrock channels reflect interactions among four factors. The first is hydraulic force (e.g. Carling et al. 2019). The second is sediment supply and attrition rate, as these influence the tendency of sediment to cover the bedrock bed or act as a tool of abrasion (Sklar and Dietrich 2004; Inoue et al. 2014; O'Connor et al. 2014). The third factor is living riparian vegetation and stationary large wood, which influence bedrock channel geometry to the extent that they trap and store otherwise mobile sediment and create alluvial patches within the bedrock channel (Massong and Montgomery 2000). The fourth factor is the characteristics of the bedrock, including joint geometry, rock tensile strength, and rates of chemical weathering, which also control bedrock geometry via their influence on the erosional resistance of bedrock exposed along the channel boundaries (Hancock et al. 2011; Ortega-Becerril et al. 2017a; Scott and Wohl 2019).

An actively incising bedrock channel typically has a relatively deep, narrow cross-section, which has been interpreted to maximize erosive force exerted on the bed (e.g. Baker 1988). Flume experiments using cohesive substrate illustrate how the form of an incising bedrock channel adjusts through time in response to fluctuations in sediment supply, discharge, bed gradient, and relative base level. As gradient increases, for example, channels become progressively narrower and deeper, and develop undulating wallforms (Wohl et al. 1999) and bedforms with longitudinally alternating patches of alluvial cover (Wohl and Ikeda 1997). Channel width increases as sediment supply increases (Finnegan et al. 2007; Johnson and Whipple 2010). Increasing hydraulic roughness associated with sculpted forms eventually leads to sediment deposition on the bed (Finnegan et al. 2007; Johnson and Whipple 2010). This concentrates sediment transport and bedrock abrasion along the margins of the alluvial deposit, facilitating channel widening. As bed roughness increases, rates of lateral abrasion in rough sections of channel can become up to five times higher than in smooth sections (Fuller 2014). Abrasion by suspended sediment can continue to widen the channel even if the bed is fully covered by sediment (Hartshorn et al. 2002; Scheingross et al. 2014).

A common difference between bedrock and alluvial channels is the rate of channel change, which is typically slower in bedrock channels because much of the change occurs during relatively high-magnitude, low-frequency floods (Baker 1987; Baynes et al. 2015). Long-term average rates of bedrock channel incision range from less than 1 cm per thousand years to 100 m per thousand years (Tinkler and Wohl 1998; Whipple et al. 2000b).

Like alluvial channels, bedrock channels exhibit statistically significant DHG relations, such that channel width increases with discharge, although comparison of relations within a region indicates that bedrock channels tend to be narrower and deeper (Montgomery and Gran 2001; Wohl and David 2008). As might be expected, bedrock channel width-to-depth ratio also varies locally with substrate resistance and sediment flux (e.g. Whitbread et al. 2015).

One of the challenges of working in bedrock channels is how to develop quantitative metrics of erosional resistance. Among those used to date is intact rock strength as measured using a Schmidt hammer (Aydin and Basu 2005). Originally developed to test concrete during the 1940s, the Schmidt hammer measures uniaxial compressive strength. Rock tensile strength has been measured using the Brazilian splitting tensile strength test (ISRM 1978).

6.5 River Gradient

The final level of adjustment in channel form occurs via changes in gradient. These can be changes in reach-scale gradient over lengths tens to hundreds of times the average channel width, or across all or much of the longitudinal profile of a river.

Like channel planform, gradient can be largely imposed on a river by external constraints such as changing base level and erosionally resistant substrate. Alternatively, gradient can be a response to existing supplies of water, sediment, and large wood. In the latter case, gradient represents another dimension of channel adjustment and exhibits changes in response to varying water, sediment, and wood supplies to the channel. The idea of gradient as a reflection of external inputs to the river is formally expressed in the concept of a graded stream.

The most widely cited definition of a *graded stream* was proposed by Mackin (1948, p. 64): "A graded stream is one in which, over a period of years, slope is delicately adjusted to provide, with available discharge and the prevailing channel characteristics, just the velocity required for transportation of all of the load supplied from above." Mackin (1948) traces this idea back to Gilbert (1877) and Davis (1902a), although Davis (1902a) views a graded stream as a condition developed only over very long periods of geologic time. Mackin (1948), following Gilbert's perception, identifies a graded stream as reflecting equilibrium, and applies this idea to the entire longitudinal profile of a river, as well as to more limited segments. This definition implies that stream geometry is relatively constant over the time period of interest, although fluctuations about a consistent mean can occur.

A graded stream can develop relatively quickly – in some cases, over a matter of hours to a few days – in channels with readily erodible substrate. A graded condition typically requires longer to develop over an entire longitudinal profile or in more erosionally resistant substrate. Developing a graded condition over the length of a river requires that all portions of the river adjust to base level and that bed sediment along the river is distributed in adjustment with discharge. Before these adjustments can be accomplished along substantial portions of a river, however, boundary conditions (base level, sediment and water yield, substrate resistance) are likely to change. Developing a graded condition in more erosionally resistant substrate requires a sufficient frequency and duration of flows of high magnitude that exceed the threshold of boundary erodibility, and such flows are likely to be of longer recurrence interval. Rivers in the northwest Indian Himalaya, for example, display increased concavity downstream from glacially modified reaches and require more than 500 000 years to recover a graded condition (Hobley et al. 2010).

Discontinuities in process and form along a channel can occur in association with the longitudinal transition from glacial to fluvial process domains (e.g. Lane et al. 2017) or at tributary junctions. These junctions can represent a substantial increase in discharge or an increase in the volume or average grain size of sediment (Section 4.5) supplied to the channel. The main channel gradient can decline or steepen in connection with the junction, depending on whether discharge or sediment supply increases more.

Discontinuities in channel process and form can also occur at transitions between bedrock, coarse-grained alluvial substrates, and fine-grained alluvial substrates (Howard 1980). The gradient of alluvial channels is commonly interpreted as reflecting hydraulic regime (Moshe et al. 2008). This condition is referred to as *transport-limited* (Howard 1994) because the transport of sediment is limited primarily by flow energy rather than sediment supply. In contrast, sediment transport in *supply-limited* channels is constrained primarily by the availability of sediment. Bed material grain size in a supply-limited channel adjusts by coarsening until the remaining exposed sediment is moved only in proportion to the supply of that grain size. If sufficient supply is not available, bedrock is exposed along the channel, creating a subset of supply-limited conditions that are referred to as *detachment-limited*. In detachment-limited channels, gradient may be less dependent on hydraulic regime because weathering must precede erosion (Howard 1994, 1998).

Alluvial channel gradient shows a weak inverse correlation with discharge. Similarly, the relationship between gradient and median bed material size is not simple, but can be significant when sites of similar drainage area or discharge are compared (Hack 1957). This indicates that gradient reflects both discharge and sediment supply (Knighton 1998). This is logical, given that bed material size and discharge influence particle mobility, channel-boundary roughness, rate of energy expenditure, and thus ability to adjust gradient. In channels with mixed grain sizes, questions arise as to which grain-size fraction exerts the strongest influence on the ability to adjust gradient. Most investigators use some grain-size fraction coarser than D_{50} as better representing the influence of grain size on gradient in gravel-bed streams.

A threshold grain size of 10 mm separates sand-bed and gravel-bed streams (Howard 1987). Gradient reflects the quantity and grain size of load below this threshold. On channels with bed material coarser than 10 mm, gradient reflects the threshold of motion of large grains, rather than the quantity of sediment. Sand-bed channels are referred to as *live-bed* or *regime channels* because sediment transport occurs at all but the lowest flows. Gravel-bed channels are known as *threshold* or *stable channels* because sediment moves only near bankfull discharge or during extreme flows (Howard 1980). These channels commonly have gradients near the threshold of motion for the coarse grain fraction (Howard et al. 1994).

Where large wood is present, it can act in a hybrid fashion, increasing hydraulic resistance during transport and when stable (analogous to sediment), and creating local base levels when stable and accumulated in sufficient quantity (analogous to resistant bedrock). In either scenario, the wood influences bed material size and reach-scale gradient and can thus alter correlations between gradient, discharge, and sediment size.

6.5.1 Longitudinal Profile

Longitudinal profile typically refers to gradient along the entire length of a river from the channel head to the base level at which a river enters a larger river or a body of standing water. The concept of *base level* is particularly important in understanding adjustment of river longitudinal profile. First articulated by Powell (1875, 1876), base level can be conceptualized as a lever arm that influences channel-bed elevation and gradient upstream. If base level increases in elevation, upstream gradient declines, transport energy declines, and sediment is typically deposited, causing the channel to aggrade to the new base level (Leopold and Bull 1979). If base level falls, gradient increases and the channel incises to the new base level starting at the point of base-level drop and propagating upstream (e.g. Dente et al. 2019).

Local base level can occur partway along a river, where the river enters a lake that has river outflow downstream. Local base level also refers to an erosionally resistant layer that limits upstream transmission of ultimate base-level fall or to the elevation at the mouth of a channel where it joins a larger river or a lake (e.g. Yanites et al. 2017). Sea level forms the ultimate base level for rivers.

Relative base-level change refers to a scenario in which the elevation difference between base level and a reference portion of a river changes. This can occur when tectonic uplift raises the drainage basin, for example, even if the ultimate base level of sea level remains constant during the period of uplift (e.g. Kemp et al. 2018). Sea level has, of course, fluctuated dramatically during the Quaternary in association with the advance and retreat of continental ice sheets, causing ultimate and relative base level to change repeatedly.

In the description of longitudinal profiles, as in many other areas of fluvial geomorphology, G.K. Gilbert (1877) led the way with his three laws of land sculpture, which included the *law of divides*: the gradient of a river steepens with proximity to the drainage divide. Although longitudinal profiles can be straight or convex (Figure 6.17), the overall gradient of medium- to large-sized rivers typically decreases downstream, creating a concave longitudinal profile that Hack (1957) describes as

$$S = k\,L^{n} \tag{6.7}$$

where S is gradient, k incorporates mean bed particle size, L is distance downstream from the drainage divide, and the exponent n is an index of profile concavity. S in this equation is a tangent to the curve that defines the relation between fall H and length L along a river (i.e. a longitudinal profile).

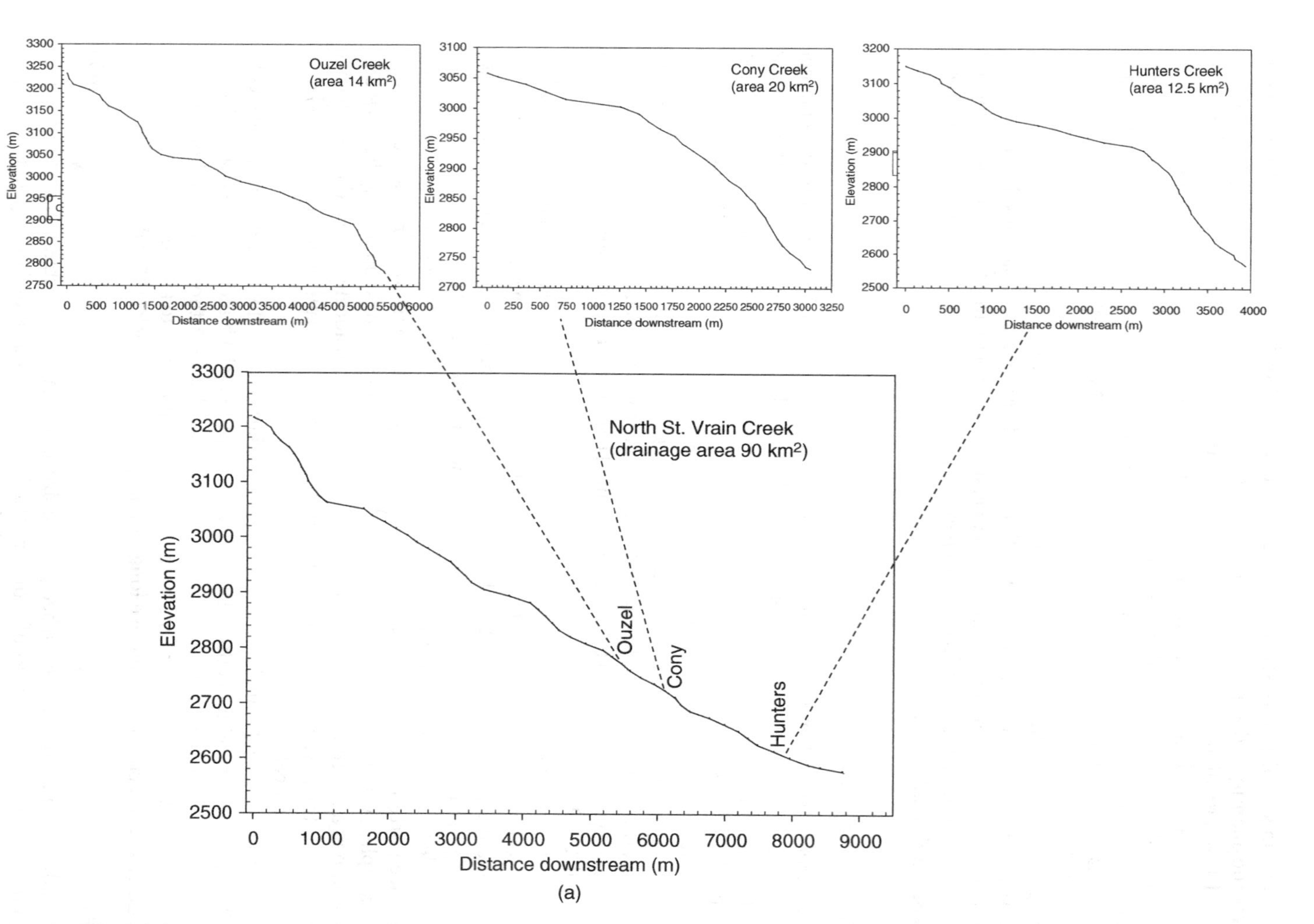

Figure 6.17 Sample longitudinal profiles for diverse rivers. (a) A mountainous river network in Colorado, USA. Ouzel, Cony, and Hunters creeks are each tributaries to North St. Vrain Creek at the points indicated on its longitudinal profile. Each is relatively straight or convex, and they steepen as they enter the main valley of North St. Vrain Creek, which was glaciated during the Pleistocene. (b) The Amazon and Rhine rivers, which display much more concave profiles. In each of these large rivers, the majority of elevation loss occurs in the mountainous headwaters.

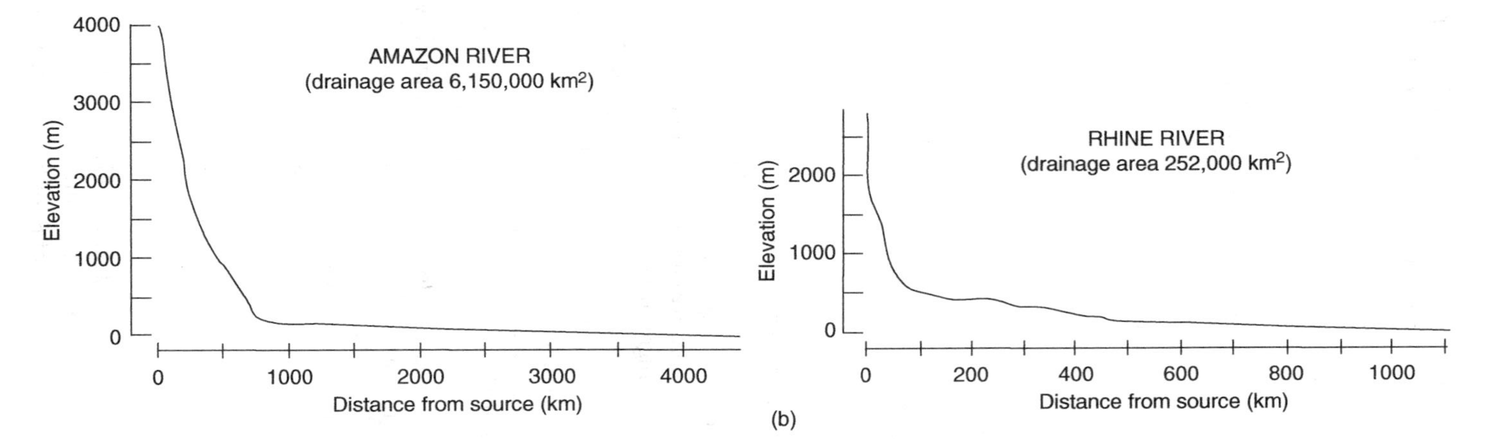

Figure 6.17 (*Continued*)

Based on field data from the eastern United States, Hack (1957) proposes an empirical form of Eq. (6.7) using mean bed particle size (D^{50}, in mm) and distance downstream from the drainage divide (L, in miles)

$$S = 25\left(\frac{D_{50}^{0.6}}{L}\right) \tag{6.8}$$

For channel segments with constant D_{50}, integration of Eq. (6.7) produces

$$H = k\, ln\, L + c \tag{6.9}$$

where c and k are empirically derived constants.

(The integration is

$$S = kL^n$$

$$H = kL^n dL$$

$$= klog_e L + c, \text{where } n \text{ equals} - 1)$$

In the much more common scenario in which particle size changes systematically downstream (i.e. n does not equal -1), the equation takes the form of

$$H = \left(\frac{k}{n+1}\right) L^{(n+1)} + c \tag{6.10}$$

Snow and Slingerland (1987) use numerical simulations to demonstrate that the specific form of mathematical function that best approximates a river's longitudinal profile depends on the relative rates of change in water and sediment discharge and grain size downstream. Exponential, logarithmic, and power functions all create smooth, concave-upward longitudinal profiles, but the profiles of real rivers typically contain irregularities (Knighton 1998).

The index of profile concavity (exponent n in Eq. (6.7)), as suggested by the correlations between reach-scale gradient and discharge and grain size, reflects rates of downstream change in discharge and substrate resistance (e.g. Roe et al. 2002; Gasparini et al. 2004), as well as ongoing response to relative base-level change (e.g. Martins et al. 2017). Profiles tend to be more concave where discharge increases rapidly downstream or grain size decreases rapidly. Increasing discharge translates to ability to transport the same bed-material load over progressively lower slopes. Similarly, decreasing grain size implies the ability to transport the available load over progressively lower slopes. Adjustments to increasing discharge or decreasing grain size can also occur in terms of bed roughness, cross-sectional geometry, and planform (Dust and Wohl 2012b).

Considerations of river longitudinal profile have thus far mostly ignored the potential influence of large wood. This remains an area that should be investigated, however – not least because recent work suggests that widespread wood removal triggered catchment-scale incision and formation of strath terraces (Collins et al. 2016; see Section 7.2).

The longitudinal profile of a river can also be characterized in terms of the concavity index, θ

$$S = k_s A^{-\theta} \tag{6.11}$$

where S is channel gradient, k_s is the steepness index, and A is drainage area (Flint 1974). This relation holds downstream of a critical drainage area. Critical drainage area varies, but is typically in the range of $0.2–0.9 \times 10^5$ m^2 (Whipple 2004). Values of θ for bedrock channel segments vary from 0.3

to 1.2 and can be negative over short reaches (Whipple 2004). Concavity values <0.4 are associated with short, steep drainages strongly influenced by debris flows, or with downstream increases in incision rate or rock strength that result in knickpoints. Moderate values of 0.4–0.7 correlate with active uplift and homogeneous substrate. High values of 0.7–1.0 correlate with downstream decreases in rock uplift rate or rock strength (Whipple 2004). The concavity index indicates the rate of decline in channel-bed gradient with increasing drainage area: the higher the index, the faster the rate of decline in channel-bed gradient.

As already noted, even predominantly concave longitudinal profiles tend to exhibit irregularities in the form of local steepening. These irregularities can result from more resistant substrate (e.g. Wohl and Ikeda 1998), large point inputs of sediment (e.g. Korup 2006), tectonic activity (e.g. Ambili and Narayana 2014), glacial history (Livers and Wohl 2015; Thayer et al. 2016), or continuing response to base-level fall (e.g. Zhang et al. 2017). Although irregularities generated by any of these mechanisms can persist in alluvial channels, they are more likely to persist in bedrock channels. Alluvial channels respond more quickly than bedrock channels to local perturbations, and alluvial channels are more likely to adjust channel parameters such as cross-sectional geometry and planform, as well as gradient (Schumm et al. 1987).

Irregularities in longitudinal profile have the potential to provide important insight into river history and adjustment. Consequently, they have been the subject of much attention, including how to quantify them using *SL* or *DS* indices.

6.5.2 Stream Gradient Index

A semi-log plot of the longitudinal profile of channel segments with a constant value of D_{50} should be a straight line, the slope of which (k) Hack (1957) refers to as the *stream gradient*, or *SL, index*. *SL* index can be calculated as the product of gradient (S) and total stream length (L) from the divide. This is the most common theoretical form of an equilibrium longitudinal profile against which actual profiles are assessed (Goldrick and Bishop 2007; Troiani et al. 2014).

Abrupt spatial changes in stream gradient, as indicated by discontinuities in the longitudinal profile of a river, can be an equilibrium response to substrate variations or a disequilibrium response caused by local deformation or upstream propagation of a knickpoint due to relative base-level fall (Hack 1973). An equilibrium response to spatial variations in substrate erodibility should persist because steeper channel segments form in more resistant substrate. A disequilibrium response should be transient because reach-scale gradient decreases after the knickpoint migrates upstream. There is no inherent mechanism, however, for differentiating equilibrium steepening associated with greater substrate resistance from disequilibrium steepening associated with relative base-level fall on an *SL* plot. Consequently, Goldrick and Bishop (2007) propose the *DS* approach, derived from the power relationship between discharge and downstream distance and the dependence of stream incision on stream power

$$H = H_o - k\left(\frac{L^{1-\lambda}}{1-\lambda}\right) \tag{6.12}$$

where H_0 is an estimate of the theoretical elevation of the drainage divide if hydraulic processes were active right to the drainage head, k reflects factors such as equilibrium incision rate and stream hydraulic geometry, as well as lithology, and λ is the exponent of the relationship between discharge and downstream distance. Eq. (6.12) is essentially a reparameterized version of Eq. (6.10), with $-k$

for k and $-\lambda$ for n. The *DS* form may be more appropriate for tectonically quiet areas than the *SL* form because the DS form differentiates equilibrium steepening, which appears as a parallel shift in a *DS* plot, from disequilibrium steepening, which appears as disordered outliers (Goldrick and Bishop 2007).

6.5.3 Knickpoints

As noted earlier, irregularities are more likely to persist in bedrock channels than in alluvial ones. Consequently, bedrock longitudinal profiles have been used as an index of rock uplift rate in *steady-state landscapes* that maintain statistically invariant topography and constant denudation rate (Whipple 2001), and in landscapes experiencing transient increases in erosion rates (e.g. Snyder et al. 2000; Whittaker et al. 2008; Roberts and White 2010; Vanacker et al. 2015). The assumption is that steeper portions of the profile reflect greater uplift. The ability of a river to maintain profile concavity in response to uplift, greater substrate erosional resistance, large sediment inputs, or base-level fall can be a function of discharge, with larger rivers capable of responding more quickly and maintaining smoother, more concave profiles (Merritts and Vincent 1989). Knickpoints, in particular, reflect an inability to maintain longitudinally continuous profile concavity.

A *knickpoint* is a step-like discontinuity in a river's longitudinal profile. A *knickzone* is a river segment steeper than upstream and downstream segments but which does not have a pronounced vertical discontinuity such as a waterfall. Knickpoints can occur in weakly consolidated alluvium, but they are best developed in cohesive alluvium or bedrock. They can be stepped, buttressed, or undercut, with headward erosion via parallel retreat or rotation of the face (Figure 6.18) (Holland and Pickup 1976; Gardner 1983). Flume experiments also suggest that knickpoints in homogeneous bedrock can propagate upstream as a result of vertical drilling of successive plunge pools (Scheingross and Lamb 2017; Scheingross et al. 2017). Knickpoints can migrate upstream from a site of base-level fall (Crosby and Whipple 2006) at a rate that reflects the rate of fall, rock resistance, and sediment supply (Anton et al. 2015; Faulkner et al. 2016; Grimaud et al. 2016; DiBiase et al. 2018). Knickpoints can also migrate upstream because of an increase in discharge relative to sediment supply, although increases in channel width at the knickpoint lip may dampen the effect of increased discharge on knickpoint retreat rate (Baynes et al. 2018). Knickpoints can form at a local base level created by more resistant substrate (Ortega et al. 2013), in which case they are likely to disappear once the river incises an inner channel through the resistant material (Wohl et al. 1994). Knickpoints can also form where a large point source of sediment such as a landslide overwhelms fluvial transport capacity (Korup et al. 2006).

Rates of knickpoint retreat can be two orders of magnitude greater than erosion rates elsewhere along the channel (Seidl et al. 1997). Knickpoints are the geomorphic hot spots of incision along a river's longitudinal profile. Rates of knickpoint retreat can correlate with drainage area, and thus follow the stream power law, in which incision is proportional to channel gradient and drainage area (Bishop et al. 2005) (Section 8.1.3). Rates of retreat also reflect spatial variations in rock erodibility (Harbor et al. 2005; Anton et al. 2015) and sediment flux passing over the knickpoint lip (Lamb et al. 2007). Cosmogenic ^{10}Be dating of strath terraces downstream from headward-retreating knickpoints in western Scotland, for example, indicates that knickpoint retreat rates have declined since the mid-Holocene. Jansen et al. (2011) attribute this to a depletion of paraglacial sediment supply and a deficiency of abrasive tools capable of eroding bedrock knickpoints.

The morphology and alluvial fill of the plunge pool below a knickpoint reflect interactions among discharge, flow velocity, waterfall drop height, and sediment supply. Flume experiments suggest that

(a)

(b)

Figure 6.18 Persistent and ephemeral knickpoints along channels. (a) Bedrock knickpoint where a tributary enters the Wulik River in Alaska, USA, representing the inability of the tributary to incise through the bedrock as rapidly as the mainstem river. (The Wulik River flows right to left in the foreground of this view.) (b) Knickpoint along an ephemeral channel tributary to the South Fork Poudre River in Colorado, USA, formed in response to enhanced water yield after a wildfire burned the catchment 4 months before the photograph was taken. (*See color plate section for color representation of this figure*).

Figure 6.19 Example of a tributary hanging valley in a glaciated river network in Yosemite National Park, California, USA. The mainstem river flows left to right in the foreground.

plunge pools initially erode vertically at high rates. Vertical erosion slows and ceases as pools deepen and sediment accumulates on the pool floor. Lateral erosion can continue after sediment accumulation, although rates decline as pools widen (Scheingross and Lamb 2017). As with other types of pools, increased sediment supply causes plunge pool aggradation (Scheingross and Lamb 2016).

At the network scale, knickpoints at tributary junctions can create *hanging valleys*, in which the tributary junction is perched well above the main valley bottom. Hanging tributary valleys are common in regions with alpine glaciation (e.g. Anderson et al. 2006c). Smaller tributary glaciers do not erode a valley as effectively as larger glaciers, leaving a tributary valley hanging when the glacial ice recedes (Figure 6.19). Hanging valleys can also form in tectonically active and incising landscapes with no history of glaciations. Under these conditions, rapid mainstem incision oversteepens tributary junctions beyond a threshold slope, or low tributary sediment flux during mainstem incision limits the tributary's ability to incise as rapidly as the mainstem (Crosby et al. 2007). The amount of oversteepening needed to form a hanging valley increases as tributary drainage area increases up to some maximum value, above which large tributaries can keep pace with base-level fall and mainstem incision.

Crosby et al. (2007) predict the maximum drainage area at which a tributary hanging valley, A_{temp}, can form

$$A_{temp} = \left(\frac{k_w k_q^b}{K_{GA}\beta} \frac{I_{\max}}{U_{initial}} \right)^{\frac{1}{1-bc}} \tag{6.13}$$

where k_w is a coefficient for the relation between channel width and discharge, k_q is a coefficient for the relation between discharge and drainage area, K_{GA} is a dimensional constant equal to (r/L_s), with r as the fraction of the volume detached off the bed with each collision and L_s as saltation hop length, β is the percentage of eroded material transported as bedload, I_{max} is maximum incision rate, and $U_{initial}$ is background rate of base-level fall.

Eq. (6.13) describes a transient instability in a river with a profile that reflects incision dependent on sediment flux. Crosby et al. (2007) assume that changes in sediment flux lag behind profile adjustment because the hillslope response that determines sediment flux depends on the transmission of base-level fall through the network. Temporary hanging valleys form after base-level fall if the mainstem transient incision rate exceeds the tributary maximum incision rate associated with the initial sediment flux. The terms in Eq. (6.13) describe how the oversteepened reach A_{temp} adjusts as the mainstem incision rate returns to just balancing the background rate of base-level fall ($U_{initial}$) while the tributary continues to incise at the rate I_{max}, which is dependent on abrasion reflected in the variables of K_{GA} and β.

6.6 Adjustment of Channel Form

Feedbacks between channel processes and various aspects of channel geometry are implicit in much of the material covered thus far in this chapter and in preceding chapters. The magnitude and duration of a flow hydrograph influence the energy available to perform work in the channel, and thus influence channel form, for example, but also respond to channel form, as the form influences the travel time of flood pulses. Hydraulic forces influence sediment mobility and bedforms, but also respond to changes in boundary roughness caused by mobile sediment and bedforms. Analogously, hydraulic forces influence the mobility of large wood, but also respond to changes in boundary roughness and configuration caused by mobile and stable large wood. Rate of erosion governs whether a longitudinal profile can maintain concavity during tectonic uplift, but profile concavity influences energy available for erosion.

Channel forms are typically described at the spatial scales of cross-sectional geometry, reach-scale planform and gradient, and basin-scale longitudinal profile. Adjustment of channel form can similarly occur at various spatial scales, including changes in w/d ratio or the grain size and quantity of bed material, changes in sinuosity or number of subparallel channels (if the channel is not laterally confined), and changes in the gradient and shape of the longitudinal profile.

Figure 6.20 illustrates the temporal and spatial scales over which various form components of rivers adjust to changing inputs and processes. This figure provides a very useful conceptual framework for thinking about feedbacks and adjustments in rivers, because the diagram illustrates that:

- smaller spatial-scale components of a river network (e.g. bed configuration) can change relatively quickly compared to larger-scale components such as planform;

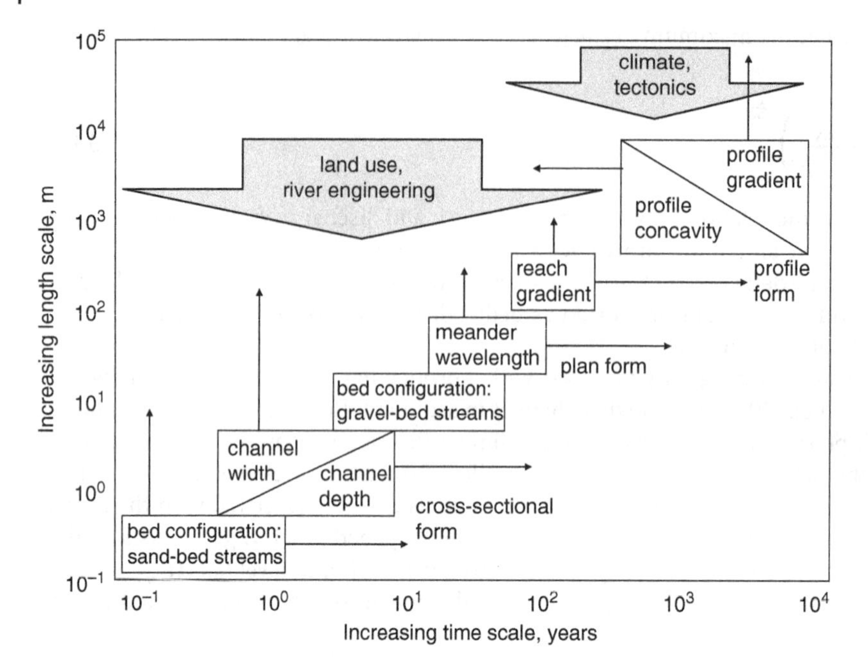

Figure 6.20 Temporal and spatial scales of river adjustment. Source: Modified from Knighton (1998), Figure 5.3, p 158.

- for a given spatial scale or type of component, such as bedforms, the rate of adjustment increases as the erosional resistance of the material increases – in other words, live-bed channels and sand bedforms adjust more quickly than threshold channels and boulder-bed step–pool sequences;
- rivers include multiple components that can adjust to changes in boundary conditions (water, sediment, and large wood inputs, substrate erodibility, base level), making it difficult to predict precisely the nature of river adjustment; and
- the space and time scales over which adjustments in diverse river features occur overlap, implying that multiple aspects of a river are likely to respond to a change in boundary conditions, as implied by the expanded Lane's balance.

Channel adjustment fundamentally represents some alteration of channel form in an attempt to reach equilibrium, as reflected in channel geometry and rate of flow energy expenditure, and as controlled by independent variables such as sediment, large wood, and water yield, substrate resistance, and base level. The final two sections of this chapter examine conceptual models developed to explain the physical processes that limit channel adjustment, and the high flows that are responsible for a substantial proportion of channel adjustment in many channels.

6.6.1 Extremal Hypotheses of Channel Adjustment

Channel adjustment is sometimes conceptualized using *extremal hypotheses*, which are models based on the assumption that equilibrium channel morphology corresponds to the morphology that maximizes or minimizes the value of a specific parameter (Darby and Van De Wiel 2003; Van De Wiel et al. 2016). Examples include minimization of rate of energy dissipation (Yang 1976; Tranmer et al. 2015) or stream power (Chang 1988), maximization of sediment transport rate (White et al. 1982) or friction factor (Davies and Sutherland 1983; Abrahams et al. 1995), and the maximum flow efficiency underlying DHG relations (Huang and Nanson 2000; Gleason 2015; Nanson and Huang 2018). This approach stems from Langbein and Leopold's (1964) conceptualization of river form as the most probable state between the opposing tendencies of minimum total rate of work and uniform distribution of energy expenditure.

Extremal hypotheses have been criticized as being teleological and lacking explanatory power (Ferguson 1986), but they do explain a wide range of observations (Darby and Van De Wiel 2003; Van De Wiel et al. 2016). The foundational assumption in extremal hypotheses is typically that a high rate of energy expenditure at a specified point in a channel will eventually result in boundary deformation, which will continue either until the rate of energy expenditure declines or the boundary is no longer adjustable by the energy available.

An example for steep, mobile-bed channels is the hypothesis that interactions between hydraulics and bed configuration prevent the Froude number from exceeding 1 for more than short distances or periods of time (Grant 1997). Observed cyclical patterns of creation and destruction of bedforms (particularly antidunes) effectively maintain critical flow in these channels. Extremal hypotheses have also been used to explain the development of bank roughness in extremely deep, narrow bedrock slot canyons (Wohl et al. 1999), cross-sectional geometry (Eaton et al. 2004), and anabranching channel planform (Huang and Nanson 2007), among other characteristics. In the Grand Canyon, USA, tributary inputs create lateral constrictions of the Colorado River that correspond to supercritical flow during extreme floods. These inputs progressively increase constriction of the main channel through time, until a large flood occurs that widens the constrictions to the point that flow remains subcritical (Kieffer 1989). Episodic channel adjustment to maintain subcritical flow during large floods is another example of an extremal tendency.

6.6.2 Nonlinear Behavior and Alternative States

As described in Chapter 1, nonlinear behavior describes a situation in which output is not directly proportional to input because of interactions among variables (Phillips 2003). The existence of hysteresis in sediment and large wood transport, complex responses of channels to base-level change, thresholds, self-organization, and alternative states in river corridors all provide evidence that rivers are most appropriately characterized as exhibiting nonlinear behavior.

Ecologists describe a system with alternative states as being present when self-reinforcing feedbacks are capable of creating two or more stable, persistent configurations under a given set of environmental conditions (Holling 1973; May 1977). These systems can change between alternative states in response to changing conditions or severe perturbation (Scheffer et al. 2001). This conceptual model is closely related to the idea of river metamorphosis, described in Section 6.2.9. The beaver meadow–elk grassland metamorphosis described in that section (Table 6.1) is also an example of alternative states.

Alternative states with respect to large wood were described in Section 5.6. Another example of alternative states comes from warm, dryland rivers of the southwestern United States that alternate between shallow channels with extensive riverine wetlands and incised, gravel-bed channels in response to the magnitude of large floods. Floods above a threshold magnitude can remove riparian vegetation and initiate channel incision, which subsequent smaller floods can maintain. In the absence of extremely large floods, riparian vegetation stabilizes channel and floodplain surfaces, traps and stores fine sediment, and promotes the development of shallow swamps (Heffernan 2008). Alternative states driven by changes in the magnitude and frequency of flooding have also been described for semiarid grasslands (e.g. Friedman and Lee 2002).

6.6.3 Geomorphic Effects of Floods

The geomorphic importance of floods of varying magnitude differs widely among rivers. Along rivers with relatively low hydrologic variability through time, the largest floods are unlikely to have persistent effects on channel and valley morphology because subsequent smaller flows quickly modify the erosional and depositional features created during large floods. Large floods are more likely to create persistent effects as erosional resistance of channel boundaries increases and as hydrologic variability increases (Kochel 1988) because subsequent smaller flows are less able to modify the geomorphic effects of large floods.

Within a region, large floods are typically more important in highland rivers than in lowland rivers (Froehlich and Starkel 1987; Patton 1988; Eaton et al. 2003). Highland rivers have steep gradients, narrow valley bottoms, coarse bedload, relatively flashy hydrographs, and relatively unerodible channel boundaries, all of which magnify the geomorphic effects of floods (Kochel 1988). Lowland rivers have larger buffering capacity associated with well-developed floodplains and greater drainage area, as well as finer bedload and channel boundaries that can be mobilized by smaller discharges. Floods can thus be of varying *geomorphic effectiveness* – defined as the ability of a flow to modify channel morphology (Wolman and Gerson 1978) – across different segments of a river network (e.g. Miller 1990; Fryirs et al. 2015).

The geomorphic effectiveness of a particular flood typically varies spatially along a river and between neighboring rivers in response to spatial variations in flood hydraulics, sediment supply, and erodibility of the channel boundaries (Miller 1995; Cenderelli and Wohl 2003; Procter et al. 2010). Antecedent conditions and the magnitude of a flood relative to earlier floods also influence geomorphic effectiveness (Eaton and Lapointe 2001; Cenderelli and Wohl 2003; Fryirs et al. 2015). The greater the difference in magnitude between a particular flood and previous flows, the more likely that particular flood is to create substantial channel change.

Erosional features created by large floods over cohesive substrates (Baker 1988) include sculpted features, inner channels, and knickpoints. Flood erosional features in unconsolidated materials (Figure 6.21) (Miller and Parkinson 1993) include:

- longitudinal grooves – elongate linear grooves parallel or subparallel to the local direction of flood flow, tens to hundreds of meters long, form in groups, with individual grooves spaced 0.5–3 m apart; width and depth values range from centimeters to greater than a meter;
- channel widening and incision (e.g. Wicherski et al. 2017);
- stripped floodplains, which occur where general scouring that is not restricted to a well-defined scour mark or channel removes vegetation and fine-grained alluvium to depths of up to 1.5 m (e.g. Dean and Schmidt 2013);
- anabranching erosion channels, which reflect incomplete channel widening that creates remnant islands in expanded channels;
- cutoff chutes in the form of well-defined channels that are typically several hundred meters long; and
- erosion of tributary fans impinging on the floodplain and main channel (e.g. Kieffer 1989; Stokes and Mather 2015).

Figure 6.21 Examples of flood erosional features along the Big Thompson River in Colorado, USA during an extreme flood in 1976. (a) Channel widening along the upper, alluvial reaches of the river. (b) Channel widening in a bedrock-constrained canyon, which has resulted in the destruction of a bridge. Source: Both photographs courtesy of Stanley A. Schumm.

Depositional features created by large floods include: gravel bars within and along the margins of the channel; wake deposits in the lee of a large obstacle to flow; slackwater deposits of sediment deposited from suspension in areas of flow separation; terrace-like boulder berms; levees and jams of large wood; and aggradation within channels. Depositional features in overbank areas include gravel splays and gravel and sand sheets (Miller and Parkinson 1993; Lucía et al. 2018). Gravel splays are lobate features in planform, with a convex profile. They are associated with severe channel or floodplain erosion and are deposited where confined flow becomes unconfined, such as main-channel flow breaching a levee and entering the floodplain. Gravel and sand sheets are typically broader and thinner than splays. Floods can also drive substantial changes in channel planform, typically from single-channel to braided (Scott and Gravlee 1968; Cenderelli and Cluer 1998; Friedman and Lee 2002).

The magnitude of channel change during a flood can be difficult to predict, but the longitudinal distribution of predominantly erosional and depositional flood features correlates strongly with pre-flood valley and channel geometry (as these influence the distribution of hydraulic forces during a flood) and with sediment supply (e.g. Dean and Schmidt 2013). The longitudinal distribution of flood-induced changes is thus predictable (e.g. Surian et al. 2016; Righini et al. 2017). Steep, narrow

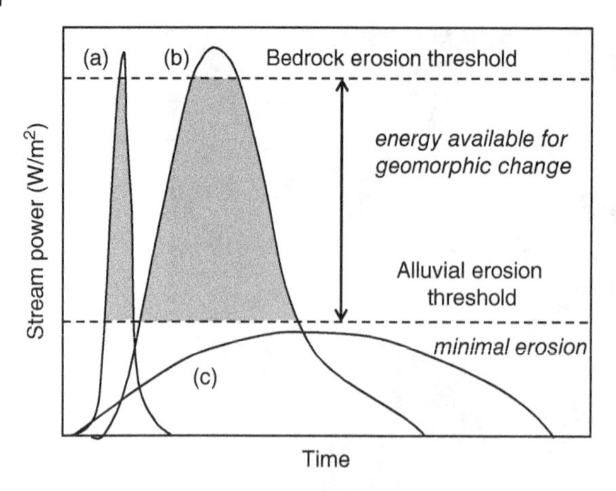

Figure 6.22 Schematic illustration of how the energy available for geomorphic change during a flood varies in relation to flood magnitude and duration. Hydrograph (a) represents a peaked, short-duration flood of the type that results from a convective storm or damburst. Hydrograph (b) represents a longer-duration, high-magnitude flood, such as might result from a cyclone. Hydrograph (c) represents a sustained, low-magnitude flood of the type that might occur during snowmelt. Source: After Costa and O'Connor (1995), Figure 11, p. 54.

reaches are likely to experience predominantly erosion during floods, whereas relatively low-gradient, wide reaches become sites of deposition (e.g. Shroba et al. 1979; Wohl 1992).

Despite the site-specific nature of erosion and deposition, various thresholds have been proposed for significant channel modification during floods, including a minimum stream power per unit area of 300 W/m^2 for low-gradient alluvial channels (Magilligan 1992) and an empirical power relation between stream power per unit area and drainage area ($\omega = 21A^{0.36}$) (Wohl et al. 2001). Costa and O'Connor (1995) propose distinct thresholds for different types of substrates and channel modification during floods (Figure 6.22). An example of a threshold in mixed bedrock–coarse alluvial rivers is floods large enough to mobilize boulders that otherwise prevent erosion of the bedrock channel boundaries (e.g. Cook et al. 2018). If such thresholds can be quantified for a channel reach, and a measure of the flow energy available to perform geomorphic work (e.g. excess shear stress or stream power) can be quantified over some time interval based on flood magnitude, duration, and frequency, then flood effectiveness can be quantitatively estimated.

6.7 Human Influences on Channel Form

In addition to regulating flow, humans have been directly altering channel form for centuries. The intent behind these alterations is as diverse as the alterations themselves (Table 6.2). Early, well documented examples of human influences on channel form include Li Ping's extensive system of irrigation canals and flood-control structures built more than 2100 years ago in the Szechuan region of China (Yao 1943). Levees were constructed along the Yodo River in Japan during the fourth century AD. Check dams were built along mountain rivers in Japan at least as early as 806 AD (Japanese Ministry of Construction 1993). Wood has been cleared from channels since the twelfth century in France (Piégay and Gurnell 1997). Flood embankments were built along Italy's Po River during the fourteenth century (Braga and Gervasoni 1989). In North and South America, Australia, New Zealand, and other areas colonized by Europeans, alterations to channel form typically start soon after European settlement (Brierley et al. 2005).

As with flow regulation, an extensive literature documents numerous case studies of altered channel form and associated changes in longitudinal, lateral, and vertical connectivity. This section briefly reviews some of the most widespread types of direct channel alteration.

Table 6.2 Examples of human influences on channel form.

Description of human modification of channel	Example references
Increasing channel conveyance by dredging the channel or building levees to reduce overbank flooding	Pinter (2005)
Straightening sinuous channels and stabilizing banks to reduce channel mobility and bank erosion	Brookes (1985)
Removing instream wood or beaver dams to increase conveyance and downstream water supplies, to reduce overbank flooding, or to enhance fish passage	Gurnell et al. (2002); Wohl (2014)
Building check dams and weirs to limit bed incision or downstream movement of coarse bed load	Garcia and Lenzi (2010)
Extending channel networks via canals that connect the headwaters of two networks that share a common drainage divide, or that shift water across tens to hundreds of kilometers to a different river network	Yevjevich (2001)
Burying or laterally shifting channels perceived as inconveniently placed with respect to urban areas, transportation corridors, or other land uses	Elmore and Kaushal (2008)
Removing sediment from channels to mine placer metals or construction aggregate	Kondolf (1997)
Reconfiguring a channel to a form considered more esthetically pleasing	Le Lay et al. (2013a)
Directly (grazing, deforestation) or indirectly (invasive exotic species) altering riparian vegetation	Trimble and Mendel (1995); González et al. (2015)

Levees are as ubiquitous along lowland (as opposed to mountain) rivers as are dams along all rivers. And, like dams, levees have a very long history. Linear mounds constructed along rivers to limit overbank flooding are known as levees, dikes or dykes, embankments, and floodbanks, among other things (Petroski 2006). Levees can be permanent or temporary, but they have been constructed at least since 2600 BC, when they were built in the Indus River valley (Clark 1982). At present, particularly extensive permanent levees line the Danube, Po, Rhine, Meuse, and Rhone Rivers of Europe, the Mississippi and Sacramento river systems in the United States, and most of the large rivers of China (Petts 1984, 1989). The Mississippi levee system is one of the world's largest. Begun by French settlers in Louisiana during the eighteenth century, it now includes more than 5600 km of levees along the middle and lower portions of the river, and most major tributaries such as the Illinois, Ohio, and Missouri Rivers also have extensive levees (Winkley 1994).

Levees reduce or eliminate channel–floodplain exchanges of water, solutes, sediment, large wood, and organisms. Levees facilitate higher-magnitude, shorter-duration floods and exacerbate flooding in downstream areas without levees (Tobin 1995; Pinter 2005). As levees cause flood peak discharge and shear stress to increase, the river bed between the levees coarsens and may incise (Frings et al. 2009). Levees severely reduce lateral connectivity between the channel and floodplain, leading to loss of habitat, animal abundance, and biodiversity in channel and floodplain environments (Hohensinner et al. 2004). The floodplain can be a major sediment source as a result of bank erosion along sinuous rivers (e.g. Dunne et al. 1998), but levees and associated bank protection truncate this sediment exchange, commonly leading to channel incision (Kesel 2003).

Along rivers with extremely high suspended sediment loads, including the Huang He in China, levees limit lateral channel movement and facilitate sediment deposition within the channel. With time, such channels become elevated above the surrounding floodplain – a situation that the Chinese describe as "hanging rivers" (Xu 2004). Levees have been built along the lower Huang He since 475 BC, and the bed of the river is now mostly 3–5 m above the floodplain beyond the levees. In some places, the river bed is 10 m above its surroundings. Water seeps into the ground from this elevated river, reducing discharge by an estimated 83 million m^3 each year (Xu 2004). Along with other changes in the drainage and consumptive water use, this seepage loss caused the lower Huang He to become ephemeral starting in 1972, a condition for which there is no earlier evidence in the historical record. With time, periods of no flow in this large river have become more frequent and the point of no-flow has moved farther upstream (Xu 2004).

Dredging, channelization (sometimes known as canalization in Europe), and bank stabilization commonly occur together and in association with levees, the intent being to make the channel more physically uniform in dimensions, including flow depth, and to limit overbank flooding, bank erosion, and lateral channel movements. *Dredging* involves physically removing sediment from the channel in order to increase cross-sectional area and flow conveyance, usually for purposes of navigation or flood control. *Channelization* involves making the channel straighter and larger in cross-sectional area, typically via dredging and other forms of sediment removal, such as digging out the streambanks with heavy machinery (e.g. Lennox and Rasmussen 2016). *Bank stabilization* (also known as *bank hardening*) involves increasing the erosional resistance of the streambanks using methods that range from planting riparian vegetation to covering the banks in large boulders (riprap) or concrete (e.g. Allen and Leech 1997; Biedenharn et al. 1997; Florsheim et al. 2008).

Collectively, dredging, channelization, and bank stabilization are so ubiquitous and of such antiquity in many regions that most people have no idea what altered rivers looked like historically. Channels at opposite ends of the drainage basin in the Danube River of Europe provide an example.

Alpine headwater tributaries of the Danube have been extensively "trained" since the sixteenth century. *Training* refers to engineering channel form and mobility, and includes bank stabilization, check dams, and confinement of braided channels into a single straightened and stabilized channel (e.g. Skublics et al. 2016) (Figure 6.23). Larger, mid-basin tributaries have been similarly engineered since the nineteenth century to alter braided channels to single channels (Habersack and Piégay 2008; Muhar et al. 2008; Hohensinner et al. 2018). Sediment introduced to channels from hillslopes and formerly transported downstream has been stored in check dams and sediment detention basins in the headwaters. Sediment temporarily stored in floodplains and alluvial fans along mid-network valley bottoms has been stabilized to limit mobilization during floods. The net effect of these alterations has been to substantially reduce sediment supply to downstream river segments, exacerbating incision and bank erosion (Habersack and Piégay 2008).

Floodplain habitat has also been reduced. The Upper Danube drainage, for example, has lost 95% of historically present floodplain habitat (Bloesch 2003). Lateral connectivity between the main channel and secondary channels has declined, along with associated river complexity, aquatic habitat and productivity (Hohensinner et al. 2004), and the abundance and diversity of fish (Aarts et al. 2004).

Downstream from the Alps, the Danube alternately flows through narrow canyons and broad alluvial basins that now contain cities such as Vienna, Austria and Bratislava, Slovakia. Historically, the Danube had a braided or anastomosing planform in these basins, with multiple channels separated by side arms, backwaters, and forested floodplain. River training began along these segments of the river in the eighteenth century and accelerated during the nineteenth, resulting in a single, straight,

(a)

(b)

Figure 6.23 Examples of trained alpine rivers. (a) In this view upstream in the town of St. Jakob, Austria, the river is channelized, with stabilized banks and regularly spaced concrete steps. (b) This river in the Italian Dolomites flows between stabilized banks and over pronounced artificial steps (closed check dams).

highly stabilized channel relative to historical, multithread channel planforms (Pišút 2002). As in the headwater tributaries, these alterations have resulted in diverse problems, from bed and bank erosion that undermines engineering infrastructure along the stabilized channel to dramatic losses in the abundance and diversity of aquatic and riparian species (Bloesch 2003). Surface connectivity between the main channel, secondary channels, and floodplain has been reduced by flow regulation, and habitat and biodiversity within the alluvial reaches have declined (e.g. Tockner et al. 1998; Schiemer et al. 1999). Problems associated with flow regulation and river engineering in the Danube drainage basin have propagated downstream to the delta, where ecosystems have declined as a result of excess nutrients, pesticide contamination, altered flow regime, and loss of floodplains and delta plains (Pringle et al. 1995), and to the Black Sea, where pollution, eutrophication (Humborg et al. 1997), and overfishing caused the collapse of commercial fisheries until nutrient fluxes were reduced (Kideys 2002).

Channelization of lowland alluvial channels, typically undertaken to increase conveyance and reduce overbank flooding, was particularly widespread in portions of the United States until the 1970s (Gillette 1972; Schoof 1980; Wohl 2004a) and continues elsewhere in the world (e.g. Dutta et al. 2018). By making channels straighter, steeper, and less complex, channelization triggers responses that vary from incision and widening in the channelized reaches and upstream segments of the drainage to aggradation and exacerbated overbank flooding in downstream reaches (Schoof 1980; Simon 1994; Wyżga 2001). Channel evolution models (Section 6.1.6) were developed to describe the successive responses through time of channelized streams.

Check dams are a type of channel alteration common in steep, mountain channels (Lenzi 2002; Itoh et al. 2013), and are primarily designed to retain sediment. Check dams share some similarities with weirs. *Weirs* are designed to provide grade control, or local base level, in lowland alluvial channels that are actively incising, sometimes because they have been artificially straightened or channelized. Check dams are also used to create a local base level that limits upstream migration of headcuts in steep channels. By storing sediment, check dams are designed to limit downstream deposition, overbank flooding, and channel avulsion, and to enhance profile irregularity and dissipation of flow energy. Check dams are particularly abundant in steep channels of Europe and Asia (Castillo et al. 2007; Shieh et al. 2007; Garcia and Lenzi 2010; Fortugno et al. 2017).

Closed check dams built of concrete or rock are designed to trap all sediment until the structure fills, at which point the sediment can be removed or another structure can be built. *Open check dams* built of some type of very coarse mesh can pass finer sediment downstream. Check dams are typically built in series, rather than singly, along a channel.

Although check dams typically do not strongly affect flow regime, they greatly disrupt downstream fluxes of sediment and large wood. Disruption of sediment fluxes commonly causes exacerbated channel erosion downstream (Wyżga 1991), as well as altering channel geometry, substrate characteristics, and aquatic and riparian communities (Nakamura et al. 2000; Bombino et al. 2009; Fortugno et al. 2017). New approaches to check dams include designing structures that mimic naturally occurring step–pool bedforms in both appearance and function (Lenzi 2002; Comiti et al. 2009b).

Although people have built hundreds of thousands of dams and check dams along rivers throughout the world, they have also actively removed naturally occurring channel-spanning obstructions, including large wood and beaver dams. Beavers, once extremely abundant and widespread throughout Europe and North America, are now highly restricted in distribution and abundance, and are effectively absent from much of their former range (Naiman et al. 1988; Pollock et al. 2015). Loss of beaver populations results in loss of beaver dams. This causes increases in conveyance, flow velocity, sediment erosion, and transport, as well as decreases in overbank flooding, riparian water tables, nutrient storage, and aquatic and riparian habitat diversity and abundance (Naiman et al. 1994; Westbrook et al. 2006; Burchsted et al. 2010; Pollock et al. 2014; Wegener et al. 2017). (Conversely, beavers introduced to Chile in 1946 have locally eliminated riparian forests and greatly expanded meadow environments; Anderson et al. 2006b; Tadich et al. 2018).

As discussed in Chapter 5, wood has been removed from channels for centuries to reduce flooding, to enhance river navigation and fish passage, and because it is considered esthetically unattractive. Active removal of instream wood, and passive loss of such wood because of reduced recruitment from riparian zones and hillslopes, has caused reduced stability and complexity along a wide variety of channels. Attempts to actively reintroduce instream wood in the form of engineered logjams (Southerland and Reckendorf 2010; Abbe and Brooks 2011; Roni et al. 2015) or individual pieces remain limited by the lack of a detailed, quantitative understanding of how wood characteristics (abundance, size, spatial distribution along a channel or river network) relate to channel form and process.

Another form of channel alteration involves actively or passively altering riparian vegetation (González et al. 2015). Active alteration includes removal and replanting (e.g. Mackay et al. 2016). Passive alteration includes changes associated with riparian grazing or the spread of invasive, exotic species (e.g. Hultine and Bush 2011). Removing or replanting riparian vegetation alters bank erodibility, as well as near- and overbank flow resistance and sediment deposition (Section 4.6).

Riparian grazing affects the density, type, and spatial extent of riparian vegetation, exposes the bank substrate, and directly erodes the banks via trampling (Cooke and Reeves 1976; Trimble and Mendel 1995). Enhanced bank and overbank erosion can result in deposition of fine sediments on the bed, reducing pool volume, spawning and macroinvertebrate habitat, and hyporheic exchange (Myers and Swanson 1996). Several studies indicate that these effects can be rapidly reduced or eliminated if short-duration grazing or grazing exclosures replace continuous grazing of the riparian corridor (Magilligan and McDowell 1997; Magner et al. 2008).

Invasive, exotic riparian species with different characteristics than native species can alter the density of riparian plants and thus influence streambank resistance to erosion and near- and overbank sedimentation (Graf 1978; Allred and Schmidt 1999). Exotic plants can also change patterns of water uptake and transpiration, thereby altering streamflow and riparian water tables, as well as nutrient cycling and riparian habitat for other species (Hultine and Bush 2011). Invasive, exotic riparian species are particularly widespread and well documented in southeastern Australia (e.g. McInerney et al. 2016) and the western United States (e.g. Nagler et al. 2011).

Although *instream mining* for sand and gravel used in construction and for placer deposits of precious metals has occurred for millennia in some regions, systematic studies of the physical and ecological effects of these activities began only during the twentieth century (Gilbert 1917). By altering substrate grain-size distribution and abundance, all forms of instream mining disrupt sediment dynamics and cause a variety of channel adjustments (e.g. Béjar et al. 2018). Large, localized excavations can initiate a knickpoint (Wishart et al. 2008). Bed and bank erosion downstream from mining can be exacerbated by reduced sediment supply (Lagasse et al. 1980). Sediment deficit at the mining site can result in bed coarsening and loss of aquatic habitat (Kondolf 1997). Disruption of a coarse surface layer can increase sediment mobility and downstream turbidity, aggradation, overbank deposition, and lateral channel mobility (James 1997; Parker et al. 1997). These changes can be so severe that they cause a meandering channel to become braided (Knighton 1989; Hilmes and Wohl 1995). Mining-related changes in channel form and process can also significantly disrupt aquatic communities, from primary production by algae to survival of fish (Van Nieuwenhuyse and LaPerriere 1986).

Diversion of flow to process placer deposits further disrupts river form and process, and toxic material such as mercury associated with placer mining can contaminate aquatic and riparian systems for decades to centuries after mining ceases (Marcus et al. 2001; Taylor and Kesterton 2002; Macklin et al. 2006). Mining and associated river contamination in parts of Europe, in particular, began centuries ago (Macklin et al. 2006; Cortizas et al. 2016).

In summary, it is important to once more emphasize that the cumulative, combined effects of direct and indirect human alteration of rivers are ubiquitous (Gregory 2006; James and Lecce 2013). River networks in even seemingly remote areas of polar regions, the tropics, vast inland deserts, and sparsely settled mountains have been directly and indirectly altered by a variety of human activities (Wohl 2006, 2011a; Comiti 2012). In many cases, these activities and the resulting changes in river systems occurred so long ago that there is little or no historical record, let alone collective awareness, of them.

Increasing scientific attention to prehistoric or historical activities that altered rivers has led to the use of the phrases *legacy effects* and *legacy sediments* (Walter and Merritts 2008; James 2013; Wohl 2015a). A legacy can be defined as "something received from a predecessor or from the past." In the context of rivers, legacy refers to sediments or channel form resulting from historical land uses. An example comes from a study of streams in the mid-Atlantic Piedmont of the eastern United States, where streams that historically had small, anabranching channels with extensive vegetated wetlands and low sediment accumulation were transformed by the construction of tens of thousands of milldams during the seventeenth to nineteenth-centuries (Walter and Merritts 2008). After dam construction, 1–5 m of fine, slackwater sediment buried the original channel network and wetlands upstream from each dam. As the dams were abandoned, they fell into disrepair and were breached, causing channel incision down to Pleistocene basal gravels. The resulting incised, meandering gravel-bed streams were assumed to represent a channel form with little human influence until Walter and Merritts (2008) undertook detailed historical and stratigraphic reconstructions of the region and revealed the legacy of historical alteration. These insights have proved controversial because of differing views of how to protect and restore river corridors in the region. At present there is a dichotomy between those who seek to protect single-thread channels with riparian forests and those who seek to restore multithread channels with floodplain wetlands (Merritts et al. 2012; Forshay and Mayer 2012). Similarly dense networks of mill dams were once present throughout Britain and much of northern Europe (e.g. Kaiser et al. 2018).

6.8 Summary

Water, sediment, and large wood are the drivers of channel form. We study hydraulics and sediment and wood dynamics to understand channel form – to predict the channel form likely to result from specific engineering manipulations such as altered flow regime downstream from a dam or stabilization of banks through an urban area and to interpret the environment that has resulted in an existing natural channel form. The fundamental questions in understanding channel form through time and space include:

- What magnitude and frequency of flow are most important in creating and maintaining channel form?
- How do sediment and large wood fluxes and storage interact with flow and configuration of the river corridor to determine stability of channel form?
- What is the relative importance of characteristics such as valley geometry; substrate erosional resistance as a function of grain-size distribution, living riparian vegetation, and stationary large wood; and channel configuration in governing the magnitude, rate, frequency, and duration of channel change?
- How can thresholds of channel change be most effectively quantified and predicted?
- Through what processes, and how frequently, do changes in channel form occur? The relations between water, sediment, large wood, and channel form are not a simple, one-way process.

Channel form responds to discharges of water, sediment, and large wood, but form also influences the distribution of hydraulic variables and the dynamics of sediment and wood in the river corridor.

Channel forms – from cross-sectional geometry to longitudinal profiles – represent the intersection of physics and history. The physical balance between hydraulic driving forces and substrate erodibility governs the creation and maintenance of river form. But the history of past channel adjustments and of larger factors as diverse as tectonics, climate, and land use, which influence driving forces and resistance, constrain the manner in which the balance is expressed. A river is a physical system with a history.

7

Extra-Channel Environments

Extra-channel environments refer to fluvial erosional and depositional processes and the resulting fluvial landforms created outside of active channels. These processes and landforms are contrasted with those within the boundaries of the active channel (Chapters 3–6). This is an arbitrary distinction because of the close coupling between active channels and marginal areas of fluvial erosion and deposition, but the distinction serves to emphasize some of the unique aspects of floodplains, terraces, alluvial fans, deltas, and estuaries. Each of these features can be laterally and longitudinally extensive and persistent landforms. As terrestrial–aquatic interfaces, they can also be chemically complex and biologically diverse and productive environments that are strongly influenced by biotic activities. Globally, these riverine environments are also heavily altered by human activities, with consequences for hazards and for ecological sustainability.

7.1 Floodplains

Floodplains are low-relief sedimentary surfaces adjacent to the active channel that are constructed by fluvial processes and frequently inundated (Nanson and Croke 1992). Floodplain relief is typically ~0.1 to 0.5 times the bankfull depth of the river (Dunne and Aalto 2013). Engineers designate floodplains based on average recurrence interval of flooding, as in the 10-year floodplain or 100-year floodplain (Graf 1988). Nanson and Croke (1992) refer to such surfaces as the *hydraulic floodplain*, because the designation does not imply anything about geomorphic history or fluvial influence. Geomorphologists traditionally refer to floodplains as those surfaces that are flooded at least once every 2 years, with the assumption that such surfaces are composed largely of fluvial sediments deposited under the current flow regime, rather than relict sediments deposited under very different conditions. Nanson and Croke (1992) describe this as a *genetic floodplain*. Even a genetic floodplain can be flooded less frequently than approximately once every 2 years, however, depending on the hydroclimatic regime. Data from 28 gaged sites on rivers in the western United States indicate that the most frequent recurrence interval for floodplain inundation is about 1.5 years, varying from 1 to 32 years (Williams 1978a). Floodplains can form along river courses in valley bottoms, on alluvial fans, and on deltas (Bridge 2003).

Floodplains along large, lowland rivers such as the Amazon or the Yukon can be extremely extensive (Figure 7.1). Floodplains extend for approximately 4000 km along the Amazon and the Nile, and can extend up to 100 km across. Dunne and Aalto (2013) distinguish the modern, most recently active *channel belt landforms* on a floodplain from the wider floodplain, which can be of Holocene or even

Rivers in the Landscape, Second Edition. Ellen Wohl.

Figure 7.1 The floodplain of the Yukon River in central Alaska, USA. In this view taken from a small airplane, the extent of the floodplain and the complexity of soil moisture, plant communities, small-scale topography, and depositional history are apparent. (*See color plate section for color representation of this figure*).

Pleistocene age. The channel belt along a meandering or braided river can be up to several times the width of the channel bends, and such bends can up to 100 times the bankfull channel width. Channel belts on multithread large rivers can be even wider (Dunne and Aalto 2013).

7.1.1 Floodplain Functions

Floodplains can be examined in many contexts. They exert an important influence on lateral and longitudinal connectivity within a drainage network, buffering hillslope inputs and commonly slowing the downstream movement of water, sediment, large wood, nutrients, contaminants, and organisms. Floodplains thus serve as reservoirs, in the sense of at least temporarily storing diverse materials. Storage time is likely to range from hours to weeks for flood waters and up to thousands of years for sediment (Wohl 2015b). On the 16 km^2 floodplain of the Middle Fork Flathead River in Montana, USA, for example, mean hydrologic residence time for water in the channel is 1–1.2 days (decreasing with river discharge), whereas the residence time of water on the floodplain surface is 1.3–1.9 days (Helton et al. 2014). Floodplain aquifer residence time is 194–238 days and, unlike with surface waters, does not decrease as discharge increases (Helton et al. 2014).

Floodplains act as a safety valve during large floods. As flood water spills over the channel banks and across the floodplain, the water typically encounters greater flow resistance and moves more slowly, facilitating sediment deposition. Turbulence and flow separation between the overbank and channelized flows also dissipate energy and reduce channel bed and bank erosion (e.g. Shiono et al. 1999). Slower overbank flow attenuates the flood, creating a peak discharge of lower magnitude and longer duration (Lininger and Latrubesse 2016). This effect can be substantial, with up to 30% reduction in peak discharge (Lininger and Latrubesse 2016). Greater floodplain length, width, and hydraulic roughness, and the presence of detention areas such as lakes and wetlands, all increase attenuation (O'Sullivan et al. 2012). Valley-bottom morphology, floodplain size, valley-bottom downstream gradient, and hydraulic roughness can also strongly influence peak discharge, especially for moderate-magnitude (5- to 50-year return interval) floods (Woltemade and Potter 1994). The potential for peak flow attenuation relates to both the flood volume and the ratio of peak discharge to total flow volume. Floods of moderate magnitude with relatively high peak-to-volume ratios are likely to be attenuated more than either smaller or larger floods (Woltemade and Potter 1994).

Storage of surface water is not simply a function of floodplain surface configuration and hydraulic roughness. Subsurface water can well up to the surface and surface water can infiltrate and follow subsurface flow paths of varying depth and complexity (e.g. Helton et al. 2014). Consequently, although many hydrologic models of flood inundation treat a floodplain as though it has an impermeable surface, this simplifying assumption can lead to errors in estimating flood peak attenuation.

Storage of surface and subsurface water results in time lags of water release from the floodplain that can be very difficult to predict. As flood waters recede, overbank flows can be trapped behind natural levees and remain on the floodplain until the water infiltrates or evaporates. Dry, wooded floodplains can attenuate flood peaks more than wet, grassy floodplains because of differences in soil moisture and hydraulic roughness caused by vegetation (Acreman and Holden 2013). Rates of change in floodplain storage can be lowest at sites farthest from a channel, indicating greater storage times at these sites (Alsdorf et al. 2005). Not all floodplain sediment functions similarly. Permeable sediment favors subsurface storage, whereas impermeable sediment deflects influent groundwater through aquifers below the alluvium or across the floodplain surface, greatly restricting the buffering capacity of the floodplain (Burt 1996). Burt et al. (2002) describe a case study in which a reverse groundwater ridge develops in the floodplain subsurface during overbank flows, resulting in strong groundwater flux velocities directed toward the base of the hillslopes adjoining the floodplain. The impact of the groundwater ridge is to switch off hillslope inputs to the floodplain. The reverse groundwater ridging process also occurs during in-channel flows when the flood stage is high. Scenarios of this type indicate the complexity of floodplain surface–subsurface flow interactions through time and across space. Subsurface water can return to the channel via flow paths that range from diffuse, matrix flow that takes weeks to months, to concentrated flow in soil pipes or macropores, which may occur within minutes to hours.

Floodplains typically enhance the storage of sediment, organic matter, and solutes over diverse time spans that depend on factors such as the frequency of overbank flow, the width of the floodplain relative to the rate of lateral channel migration, and mass balance between bank erosion, sediment transport, and in-channel and overbank deposition (e.g. Lauer and Parker 2008a,b). The mean turnover time of individual floodplains varies widely, with published rates ranging from 20 years to more than 7000 years (Wittmann and von Blanckenburg 2009; Wohl 2015b). Turnover time is probably best described as a population of numbers rather than a single number, because both field studies and numerical models indicate that channels preferentially reoccupy recently abandoned locations (Konrad 2012), which equates to a decreasing probability of reoccupation with time since abandonment.

Sediment and solutes entering the floodplain can come from upland environments, upstream or tributary portions of the river network, or, particularly in the case of solutes, ground-water sources (e.g. Krause et al. 2007). The spatial heterogeneity of these materials in the floodplain reflects the history of floodplain erosion and deposition, and, in the case of nutrients and organic matter, the history of biological uptake and autochthonous production on the floodplain (e.g. organic matter from plant litterfall). The resulting mosaic of vegetation, soil texture, and soil moisture creates spatial heterogeneity in carbon and nitrogen cycling (e.g. Appling et al. 2014), surface–subsurface water fluxes (Mouw et al. 2009), and habitat. Floodplains and riparian zones are widely recognized to be efficient sites of nitrate uptake (e.g. Wollheim et al. 2014) and organic carbon storage (e.g. Hoffmann et al. 2009; Wohl et al. 2012c; Hanberry et al. 2015; Sutfin et al. 2016), and the spatial and temporal characteristics of these floodplain functions largely reflect the history of floodplain development as expressed in the three-dimensional spatial mosaic of floodplain sediments. Analogously, floodplain spatial heterogeneity both reflects erosional and depositional history and influences continuing erosion and deposition by creating patches of more resistant substrate that influence channel migration (e.g. Schwendel et al. 2015).

Between inputs and outputs, particulate organic matter can be stored on or in a floodplain for tens of thousands of years. Buried logs within floodplain sediments have been dated to 3300 years of age in Poland (Kukulak et al. 2002), 13 500 years in Missouri, USA (Guyette et al. 2008), and 17 000 years in Tasmania, Australia (Nanson et al. 1995). Typical mean ages of soil carbon in the humic substances component of soil exceed 1000 years (MacCarthy et al. 1990). Although large wood can form a substantial reservoir of particulate organic matter and organic carbon where a dry climate limits rates of decay (Wohl et al. 2012c; Lininger et al. 2017), the largest organic matter and organic carbon stocks are typically found in floodplain soils (Sutfin et al. 2016). Soil moisture exerts a primary influence on the stability of soil organic matter and carbon. Cold temperatures, saturated soils, and a reducing environment limit decomposition and biological uptake, and the highest values of soil organic matter and carbon typically occur in wetlands (Eswaran et al. 1993; Sanders et al. 2017) and permafrost (Tarnocai et al. 2009).

One of the implications of floodplain turnover is the potential for the storage and subsequent release of contaminants. Several studies document preferential floodplain storage of mining-related contaminants (e.g. Lecce and Pavlowsky 2014). Floodplain storage results from dispersal of tailings during large floods caused by tailings-dam failures (Marcus et al. 2001). Floodplain storage also reflects the tendency of mining-related contaminants such as mercury to travel adsorbed to the fine sediment that accumulates in floodplains via vertical accretion (Miller et al. 1999). Once deposited on the floodplain, the contaminants can slowly leak into the river via bank erosion over a period of centuries (Macklin et al. 1997).

Excess nutrients can also form a contaminant with delayed release from a floodplain. Wang et al. (2013) describe the "nitrate time bomb" created by the fact that nitrate can be retained for decades in the unsaturated and saturated zones of a floodplain before being leached into surface and ground water. Application of numerical models of these storage and release processes to watersheds in the United Kingdom indicates that peak nitrate loading from existing and historical inputs will not occur for another 30 years (Wang et al. 2013).

Floodplains tend to be biologically rich habitats. Many aquatic organisms use the floodplain during periods of inundation as breeding, nursery, and feeding areas (Junk et al. 1989; Hurd et al. 2016; McInerney et al. 2017). The flush of nutrients that return to the channel as flood waters recede helps to support aquatic communities in the channel (Kreiling et al. 2013), although the original source of this floodwater (e.g. overbank flow versus groundwater) influences nutrient content and productivity (Keizer et al. 2014). The diversity of floodplain habitat and moisture levels commonly supports greater biodiversity than is found in the adjacent uplands (Ward et al. 1999), and this diversity depends on continued interactions with the channel and underlying hyporheic zone (e.g. Hughes 1997; Stella et al. 2011). Large wood can exert a particularly important influence on floodplain process and form in forested regions, including helping to create and maintain floodplain spatial heterogeneity (e.g. Collins et al. 2012) and providing habitat and resources for floodplain biota (e.g. Pettit and Naiman 2006).

7.1.2 Floodplain Hydrology

The water inundating a floodplain does not necessarily come exclusively from overbank flow. Flooding can occur before natural levees on the main channel overtop because flood waters can enter a floodplain from rising ground water, overland flow from adjacent slopes, secondary channels across floodplains, and tributaries, as well as overbank flow from the main channel (e.g. Rudorff et al. 2014a). Mertes (2000) distinguishes inundation patterns on dry floodplains, where most water overflows channel banks, from those on saturated floodplains, where rising ground water can contribute significantly. Flooding associated with rising ground water is evidenced by increasing depth of standing water before the flood wave arrives and by clear water that appears distinctly different from the turbid flood waters of the main channel (Mertes 1997). The *perirheic zone* refers to the zone of mixing surrounding the flowing river water (Mertes 2000). Mixing between flooding river water and locally derived water is governed by characteristics such as the spatial density of floodplain channels: high density limits mixing because the floodplain channels contain flood waters leaving the main channel.

The Amazon River in some respects exemplifies floodplain dynamics. As the largest river in the world, the Amazon has enormous floodplains. A largely unregulated river without human-built levees, the Amazon still experiences a natural flooding regime and high channel–floodplain connectivity. Limited mixing occurs in the upper reaches of the river, which have a high density of floodplain channels, whereas the middle reaches have enhanced mixing from overbank flows across levees along the main channel. The lateral extent of the incursion of river water into the floodplain varies from about 5 km in upstream reaches of the Amazon to 5–15 km in middle reaches. Mixing is also limited in the lower reaches of the Amazon, where large floodplain lakes have sufficient hydraulic head to prevent river water from entering large portions of the floodplain (Mertes 2000). The dominant sources of inflow to the floodplain also vary seasonally and between years. On average, river-to-floodplain discharge represents less than 1% of the Amazon River discharge at Óbidos in the lower Amazon, but 82% of annual water fluxes to the floodplain (Rudorff et al. 2014b). Consequently, relatively small changes in mainstem peak discharge cause disproportionately large changes in water flux to the floodplain (Rudorff et al. 2014b).

7.1.3 Depositional Processes and Floodplain Stratigraphy

Floodplain form reflects a history of erosion and deposition, but the majority of floodplains are predominantly depositional environments that store large volumes of sediment for varying lengths of time. Floodplain deposits can be categorized as *vertical accretion* deposits, which form when sediment settles from suspension in the lower-energy environment of the floodplain, or as *lateral accretion* deposits. Vertical accretion deposits include levee and backswamp or flood basin sediments, whereas lateral accretion deposits include islands, bars, and channel lag sediments originally deposited in the active channel and then accreted to the floodplain as the channel migrates laterally.

Sedimentation rates from vertical accretion typically decrease across the floodplain with distance from the channel (e.g. Swanson et al. 2008), and active sedimentation extends no farther than a few kilometers from the channel even on very large floodplains (Dunne and Aalto 2013). Rates of vertical accretion reported for diverse rivers range from 0 to nearly 6000 mm per year (Knighton 1998) but are typically at most a few centimeters per year when averaged (e.g. Moody and Troutman 2000). Deposition can be episodic, however, and dominated by large floods. Floods occurring on average every 8 years in association with La Niña climatic episodes, for example, deposit 20–80 cm-thick layers of sediment on the floodplains of the Beni and Mamore rivers in the Amazon basin (Aalto et al. 2003). Alternatively, deposition can be dominated by annual floods, such as the snowmelt peak, on rivers with less hydrological variability (Moody 2019). Sedimentation is commonly extremely slow at distances of more than a kilometer from the main channel, and these lower-elevation portions of the floodplain can be covered by lakes or swamps (Dunne and Aalto 2013). In regions wet enough to support at least partial vegetation cover, floodplain deposits commonly include organic matter transported by flood waters and varying in size from fine particles (~0.5–1 mm in diameter) to large tree trunks. Vegetation increases organic matter in floodplain sediments both by trapping fluvially transported material and by dropping plant litter in the form of leaves and branches (Gonzalez et al. 2014). Floodplain sediments are also typically bioturbated by plant roots and animals (Bridge 2003).

Natural *levees*, as distinct from levees constructed by humans, are discontinuous, linear features immediately adjacent and parallel to the channel banks that form where the coarsest suspended sediment is deposited as velocity drops in the mixing zone between channelized and overbank flows (Figure 7.2). Sediment accumulates faster along the levee than it accumulates as flow spreads farther from the channel. This results in a continuous berm that gradually declines in elevation with distance from the channel (Dunne and Aalto 2013) and can be up to four channel widths across (Bridge 2003). Levee height commonly decreases downstream as the grain size of sediments in suspension

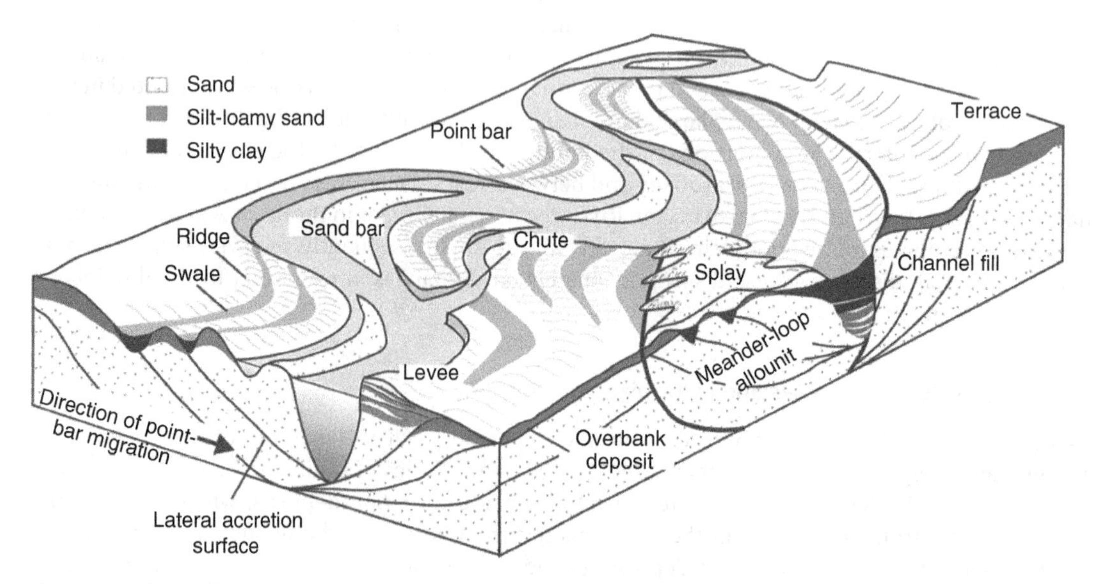

Figure 7.2 Idealized diagram of a large, sinuous river floodplain, showing the types of lateral and vertical accretion deposits discussed in the text. Source: After Holbrook et al. (2006), Figure 2 and Dunne and Aalto (2013), Figure 21.

decreases (Dunne and Aalto 2013). Levee height and deposition on the levee during a flood can be longitudinally discontinuous, with greater height and deposition immediately downstream of zones of channel enlargement during the flood (Smith and Perez-Arlucea 2008). Levee morphology and grain-size distribution can also change with age: as continued deposition builds up a levee, only the finer component of suspended load may be transported onto or across the levee (Cazanacli and Smith 1998).

Backswamp or *flood basin* deposits are typically finer-grained sediment – wash load – settling from suspension across the portions of the floodplain between the levees and the valley walls or the outer limit of flooding (e.g. Kesel et al. 1974). Flood basins are typically much longer than wide, and segmented by alluvial ridges and crevasse splays (Bridge 2003). Sediment deposited in this environment can fill abandoned or secondary channels (known by different names in different parts of the world, e.g. slough, billabong), or can be deposited more uniformly across the floodplain flats. Very large depressions on the flood basin can contain lacustrine (lake) deposits and deltas. Ephemeral lakes in arid regions can contain evaporite minerals. Backswamp deposits are typically rich in organic matter (e.g. Aslan and Autin 1998), which can come from terrestrial or aquatic sources. Flood basins form the lower elevation areas of floodplains.

Alluvial ridges are slightly higher areas formed of active and abandoned channels and bars. The ridges include accretionary features, levees, and crevasse channels and splays (Bridge 2003).

Splay or *crevasse splay deposits* occur when a levee is breached by piping or overtopping. The breaching process creates a narrow cut through the levee that contains higher-velocity flow capable of transporting coarser suspended sediment into the backswamp area. Sand splays formed in this manner can be quite extensive. The 1993 flood on the Mississippi River created sand-splay deposits over 60 cm thick that extended hundreds of meters in length and breadth (Jacobson and Oberg 1997).

The other major category of floodplain deposits is associated with lateral accretion. *Lateral accretion* deposits result from sediment deposition along the margins of the active channel, which can become part of the floodplain as the channel migrates laterally. Lateral accretion deposits include islands and bars, as well as lag deposits of the channel-bed sediment. The classic example is point-bar sediment incorporated into the floodplain as a meander bend migrates toward the outer bank (the cutbank) of the bend. Lateral accretion deposits can form the bulk of floodplain sediment in gravel-bed rivers that are laterally mobile (e.g. Page et al. 2003).

Well-developed bars can concentrate flow along the outer margins of bends and accelerate bank erosion and bend growth (Legleiter et al. 2011). Bend migration results in lower shear stress and sediment transport across the bar, which facilitates growth of vegetation and accumulation of finer sediment from suspension, raising the bar surface and blocking minor channels that might have developed across the bar (Braudrick et al. 2009). As bars are incorporated into the floodplain, sequences of arcuate bars with intervening, poorly drained depressions create *scroll-bar topography*, which can persist for centuries as differences in soil grain size and moisture support different plant communities (Figure 7.3) (e.g. Rodnight et al. 2005).

The lateral accretion of bars in a braided river creates a *braid plain* of channel-margin sediments. This plain, which is analogous to the channel belt of a sinuous river, has an irregular surface and thin, discontinuous, vertically accreted sedimentary covers in depressions and former channels (Dunne and Aalto 2013).

The juxtaposition of different types of deposits creates vertical relief on floodplains. A floodplain created primarily by lateral accretion is sometimes known as a *flat floodplain* (Butzer 1976) and consists mainly of bed-material deposits – either bedload in gravel-bed rivers or suspendible sandy or

(a)

(b)

Figure 7.3 Examples of scroll bars on the floodplains of sinuous rivers in Alaska, USA. (a) Scroll bars highlighted by differences in vegetation along a river on the coastal plain near Kotzebue in western Alaska. (b) Sandy scroll bars on the inside of a meander bend along a river in central Alaska. As in (a), differences in vegetation highlight the locations of the bars. (*See color plate section for color representation of this figure*).

silty bed material in large, lowland rivers (Dunne and Aalto 2013). This type of floodplain contrasts with a *convex floodplain*, which is found along rivers dominated by vertical accretion (Butzer 1976). Here, the thickest and coarsest deposits close to the channel create a slightly higher floodplain elevation adjacent to the channel than at the margins of the floodplain.

Floodplains can also include nonfluvial sediments intermixed with the fluvial material. Eolian sediments can be interbedded with fluvial floodplain deposits in arid regions as a result of eolian input from adjacent regions or reworking of unvegetated floodplain sediments (e.g. Spooner et al. 2001). Colluvial sediments can be deposited on the outer margins of a floodplain with steep valley walls or in the floodplain of very small channels (e.g. Knox 2006).

Floodplains thus form from a combination of within-channel and overbank deposits. The relative importance of these two basic depositional environments varies in relation to factors such as the rate of lateral channel movement and the frequency, extent, duration, and sediment concentration of overbank flows. In an influential paper on floodplain deposition, Wolman and Leopold (1957) propose that lateral accretion deposits dominate most floodplains, accounting for up to 90% of the total sediment deposited. Subsequent research, however, suggests that the situation is more complicated and varies greatly among different rivers.

The time scales of vertical and lateral floodplain accretion can differ, and rapid lateral channel migration can remove gradually accumulated vertical accretion deposits (e.g. Nanson 1986). The relative importance of lateral and vertical accretion can alter through time in response to:

- changes in the volume and grain size of sediment available during floods – finer-grained sediment carried in suspension is more likely to be associated with vertical accretion, and coarser material moving in contact with the bed is more likely to be associated with lateral accretion;
- the transport energy of the flood waters – higher transport energy can result in overbank velocities too high to permit settling from suspension and vertical accretion; and
- the geometry of depositional sites – floodplains with numerous depressions that enhance locally low velocity and settling from suspension can promote vertical accretion.

Differences in the relative importance of lateral and vertical accretion through time and space underlie the Nanson and Croke (1992) floodplain classification (Section 7.1.6). Although this classification has since been partly expanded (e.g. Thayer and Ashmore 2016 for partly alluvial channels), the original form continues to provide an effective mechanism for distinguishing the relative importance of later and vertical accretion at the reach scale in diverse river settings.

The relative importance of vertical and lateral accretion can also vary downstream. Along the Brazilian portion of the Amazon River, floodplain deposition in the upstream reaches results from lateral accretion in floodplain channels. Overbank deposition dominates downstream portions of the floodplain (Mertes et al. 1996).

As a relatively well-studied and still-natural river, the Amazon provides insight into some of the aspects of floodplain sedimentation characteristic of very large rivers. The huge drainage area and long translation times for passage of a flood wave along large rivers create gradually rising and falling hydrographs. Overbank flow can last from many weeks to as long as 6 months (Richey et al. 1989). The combination of long-duration overbank flows and abundant fine-grained particles with low settling velocity can result in large transfers of sediment into the floodplain (Dunne and Aalto 2013).

Recent work emphasizes biotic influences on floodplain form and process. Riparian vegetation growing on bars can help to stabilize and enlarge the bars, facilitating lateral accretion to the floodplain (Gurnell and Petts 2006; Zen et al. 2017). Formation of channel-spanning obstructions such as

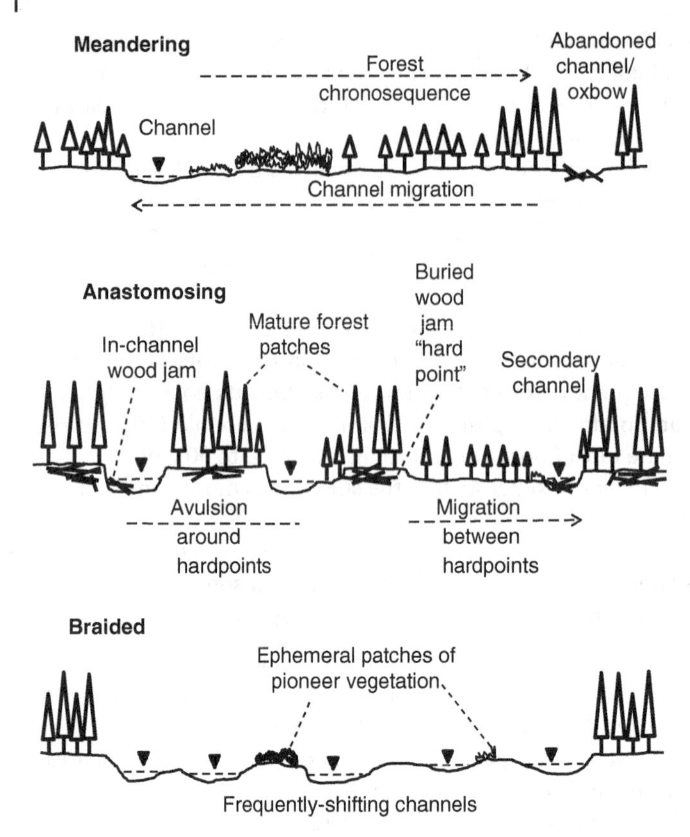

Figure 7.4 Cross-sectional views of three idealized models of floodplain landforms and forest development for river valleys of moderate gradient in the US Pacific Northwest. The meandering river model is dominated by lateral migration of meanders and associated meander cutoffs. These processes create scroll-bar topography, oxbow ponds, and sloughs. Plant germination on newly accreted sediments, and continued lateral migration and forest succession, create chronosequences of surface and forest ages. In the anastomosing river model, stable alluvial deposits associated with woodjams can resist lateral channel erosion for hundreds of years, providing sites for tree germination. Channels avulse around stable patches and migrate as in the meandering river model, creating a mosaic of multiple channels and floodplain elevations, and greater diversity in forest patch age. Braided rivers are dominated by multiple, frequently shifting channels. Mature forest vegetation is limited to the channel margins, although ephemeral patches of pioneer vegetation can grow on braid bars within the channel network. Source: After Collins et al. (2012), Figure 3.

beaver dams (Westbrook et al. 2006; Polvi and Wohl 2012, 2013) and logjams (Brummer et al. 2006; Oswald and Wohl 2008; Lombardo 2017) can lead to local aggradation and loss of conveyance that enhances overbank flow. This can in turn lead to the formation of multithread channels, as well as enhancing floodplain vertical accretion (Collins and Montgomery 2002; Sear et al. 2010; Westbrook et al. 2011; Wohl 2011b; Polvi and Wohl 2012). Collins et al. (2012) conceptualize these influences in terms of the *floodplain large wood cycle* described earlier (Figure 7.4). Historical examples such as the Graft Raft on the Red River in Louisiana, USA (Triska 1984) reflect the influence of wood rafts on channel–floodplain connectivity in large rivers (Wohl 2013b, 2014).

A sedimentary facies suite attributed to low-energy, organic-rich rivers with multiple anabranching or anastomosing channels and stable alluvial islands first appeared during the Carboniferous, when tree-like plants could create channel blockage via logjams (Davies and Gibling 2011). In other words, large wood has been influencing flow resistance, channel conveyance, and overbank flows for a very long time, across all the forested regions of the world. Wohl (2013b) proposes that widespread removal of instream wood during the past two centuries resulted in a fundamental change in floodplain dynamics because of the associated decrease in overbank flows and sedimentation, as well as in avulsion and formation of multithread planform.

The balance through time among vertical accretion, lateral accretion, and lateral channel migration is reflected in floodplain stratigraphy. Sinuous channels typically create a characteristic *fining-upward sequence* (e.g. Pettijohn 1957). As a meander bend migrates laterally, a geographic location that starts as a deeper portion of the channel near the outside of the bend transitions to a point-bar environment and eventually to a floodplain. The stratigraphic sequence at such a location thus fines upward from relatively coarse channel-lag deposits to slightly finer-grained point-bar deposits, which are in turn capped by even finer overbank deposits. Floodplains created by sinuous channels also commonly have predictable lateral stratigraphy associated with the migration of numerous meander bends.

Braided channels and sinuous channels subject to frequent, large avulsions are likely to have much less predictable stratigraphy (e.g. Miall 1977). Floodplains along these rivers can be described as a three-dimensional mosaic in which different grain sizes representing diverse depositional environments can be abruptly juxtaposed laterally and vertically. Arid-region braided channels, for example, repeatedly avulse across a broad floodplain. Historical information about channel location – including floodplain stratigraphy – can be used to develop a locational probability map that provides a statistical view of a river's most likely location at any point in time (Graf 2001).

Lateral accretion of bars dominates the floodplain stratigraphy of braided rivers, creating an irregular surface capped by a thin, discontinuous cover of finer, vertical accretion sediment accumulated in former channels (Dunne and Aalto 2013). Levees along braided rivers tend to be less tall as a result of lateral channel movement. These levees are also easily eroded, so that sand splays are an important mechanism of floodplain sedimentation. Relatively steep gradients and sparse vegetation facilitate chute cutoffs and avulsions that can be rapidly filled if suspended sediment concentrations are high (Dunne and Aalto 2013).

Although avulsion is particularly characteristic of braided rivers, this process can occur in any type of river. *Avulsion* refers to the shifting of a channel or channel belt across a floodplain (Allen 1965). Avulsion typically occurs during floods, although downstream blockage of a channel by wood or ice, for example, can create sufficiently high water levels to promote avulsion in the absence of flooding (Jones and Schumm 1999). The new channel belt follows the zone of maximum floodplain slope while migrating to the lowest part of the floodplain. This process can occur over a period of years to centuries (Bridge 2003), or it can happen during a single, very large flood.

The frequency of avulsion increases with increasing deposition rate (e.g. Bryant et al. 1995). Deposition enhances topographic relief and cross-valley slope relative to down-valley slope of the channel belt. Frequency of avulsion also increases with base-level rise, which causes aggradation and growth of alluvial ridges (Bridge 2003).

7.1.4 Erosional Processes and Floodplain Turnover Times

Floodplain sediment can be removed by localized erosion during extreme floods. Examples of localized erosion include longitudinal grooves, scour marks, stripped floodplains, chutes, and anastomosing erosion channels (Miller and Parkinson 1993). Longitudinal grooves are elongate linear grooves parallel or subparallel to the local direction of flood flow that extend down the valley floor tens to hundreds of meters, with depths and widths from a few centimeters to over a meter. Miller and Parkinson (1993) interpret these features as the early stages of floodplain erosion, because the rest of the valley bottom is largely unaltered. Scour marks vary in shape from small circular or elliptical pits to elongate parabolic or spindle-shaped and irregular marks that appear to be associated with flow separation and formation of vortices. Individual marks can be less than 0.3 m deep and 1 m in diameter, but the largest reach tens to hundreds of meters long and tens of meters wide. Stripped floodplains occur where flood waters remove a veneer of finer overbank deposit to reveal underlying cobble and boulder sediments across most or all of a floodplain. Concentrated flow on the floodplain produces a well-defined chute channel comparable in dimensions to the pre-flood river channel, rather than stripping the entire floodplain surface. Anastomosing erosion channels across a floodplain are associated with incomplete channel widening that leaves remnant islands in the expanded channel (Miller and Parkinson 1993). Removal of floodplain vegetation by a flood strongly influences the ability of flood waters to create such localized erosional features (Constantine et al. 2010).

Gradual change in floodplain morphology can facilitate abrupt, localized erosion during floods, a phenomenon Nanson (1986) describes as catastrophic stripping. Along high-energy coastal Australian rivers, gradual overbank deposition over periods of 10^2–10^3 years builds floodplains of fine-textured alluvium with large levees. Growth of the levees and adjacent floodplain surfaces progressively restricts peak flows to the main channel and floodplain back-channels, concentrating erosional energy until widespread scour of the channel boundaries and floodplain occurs. This catastrophic stripping is discontinuous along a channel (Nanson 1986; Fryirs et al. 2009).

Longitudinal variation in valley geometry and channel gradient can also result in localized floodplain erosion at the reach scale. Patton (1988) found that Widespread flooding in the northeastern United States indicated that sedimentation during large floods is important in constructing floodplains within highland drainages (Patton, 1988). In these drainages, reaches of lateral confinement and steep gradient experienced erosion during flooding. Less confined, lower-gradient reaches within the highland drainages experienced primarily deposition. Lowland drainages, in contrast, were not as strongly influenced because the magnitudes of flood erosion and deposition did not exceed those during a succession of more moderate flows.

More continuous floodplain erosion occurs as a result of channel lateral migration and bank erosion. Along the Brazilian portion of the Amazon River, sediment exchanges between the channel and floodplain in each direction exceed the annual flux of sediment out of the river in the lower portion of the drainage, although there is a net accumulation of sediment on the floodplain (Dunne et al. 1998). Quantitative studies of floodplain dynamics along the Strickland River of Papua New Guinea similarly indicate that ~50% of the total sediment load is recycled between the channel and floodplain via cutbank erosion and point-bar deposition (Aalto et al. 2008). In gravel-bed channels such as the Fraser River of Canada, sediment inputs from bank erosion can exceed net downstream sediment flux by an order of magnitude, partly because bank sediment is largely transferred to adjacent bars, creating short sediment step lengths (Ham 2005).

The continual exchange between sediment within the channel and that stored in the floodplain results in *floodplain turnover*: erosion of existing floodplain sediment and deposition of new sediment. Turnover times typically increase downstream, for two reasons. First, increasing floodplain width downstream means that longer time spans are needed for lateral channel migration to completely cross the floodplain. Second, even large floods are less likely to completely erode the floodplain than in smaller, laterally confined portions in upstream reaches (Patton 1988). Estimated turnover times are ~1000 years in upstream portions of the Brazilian Amazon and >4000 years in the downstream portions of the river near Óbidos (Mertes et al. 1996). Based on cosmogenic ^{26}Al

and ^{10}Be, some of the sediment in transport within the Amazon channel network has been stored in distal, deeply buried portions of floodplains for as long as 5 million years (Wittmann et al. 2011).

Because of the slow turnover times of sediment in many large floodplains, floodplain stratigraphy can provide a variety of paleoenvironmental information. The dimensions and stratigraphy of paleochannels can be used to infer past flow regimes (e.g. Sousa de Morais et al. 2016) (Section 3.2.2). Pollen and fossils such as gastropods in the sediments filling abandoned floodplain lakes and channels can be used to infer past climates (e.g. Fayó et al. 2018). Archeological sites are also common in floodplains and can provide useful information on floodplain chronology and local plants and animals (e.g. Schott 2017).

7.1.5 Downstream Trends in Floodplain Form and Process

The general pattern of floodplain form is a progressive downstream increase in longitudinal continuity and lateral extent. Exceptions can occur in low-relief terrains where even headwater channels have little lateral confinement, allowing wider, longitudinally continuous floodplains to develop. Exceptions also occur in rivers that cross mountain ranges. The Danube River of Europe, for example, heads in mountainous regions, but the middle and lower sections of its course alternate between wide alluvial basins with multithread channels and extensive floodplains, and narrowly confined bedrock gorges with little floodplain development (Wohl 2011a).

Valley and floodplain width within regions of high relief can also exhibit substantial longitudinal variation as a result of local changes in bedrock erodibility, glacial history, sediment supply, network structure, and biota (Wohl 2010b). Changes in lithology and jointing can influence bedrock erodibility, so that a river alternates between wider, lower-gradient segments where bedrock is more readily eroded and a floodplain develops, and steep, narrow gorges largely lacking floodplains where bedrock is more resistant (Ehlen and Wohl 2002; Schanz and Montgomery 2016). Portions of a valley immediately upstream from a glacial moraine can be wider and of lower gradient, allowing more extensive floodplain development (e.g. Montgomery 2002; Livers and Wohl 2015). Local point sources of coarse sediment, such as debris flows and landslides, can laterally constrict floodplains, as well as creating at least a temporary local base level that alters channel form (Korup 2013). Network structure, as expressed in the arrangement and relative size of tributaries entering a main channel, influences fluxes of water, sediment, and large wood into the mainstem and affects the form and processes of the mainstem floodplain (Benda et al. 2004a; Czuba and Foufoula-Georgiou 2014; Gran and Czuba 2017). Biota, including beavers that build dams and forests that allow sufficient wood recruitment to support channel-spanning logjams, can create channel segments of lower gradient and greater overbank flow (Westbrook et al. 2011; Polvi and Wohl 2012).

Longitudinal variations in lateral valley confinement and floodplain width correspond to spatial variations in floodplain functions. Valley confinement and elevation explain nearly 90% of the variability in the field-delineated width of the riparian zone, as defined based on plant species present, in headwater rivers of the Colorado Rocky Mountains, USA (Polvi et al. 2011). In mountain streams of the western United States, network-scale storage of organic carbon in downed, dead wood and floodplain soils is dominated by relatively short segments of the river corridor with less lateral confinement (Wohl et al. 2012c; Sutfin et al. 2016; Scott and Wohl 2018a,b).

Exceptions to the general downstream trend of increasing floodplain width can also result from structural controls or changes in substrate resistance. The Amazon River is entrenched where it crosses each of four structural arches, resulting in restricted channel movement and narrower floodplains through these segments (Mertes et al. 1996). Similar effects have been described along the

Mississippi (Schumm et al. 1994) and other large alluvial rivers (Bridge 2003). The very low gradients of large alluvial rivers allow these rivers to be modified by even slow, small tectonic movements (Dunne and Aalto 2013), resulting in downstream changes in gradient, sediment transport, channel planform, and floodplain width and processes.

Regional- to continental-scale tectonic setting can also influence floodplain extent on very large rivers. The largest rivers with long, low-gradient reaches that facilitate extensive sedimentation and floodplain development drain the tectonically passive margins of continents (Potter 1978). Rivers that head in orogens (belts of deformed rocks, typically associated with higher topographic relief) have greater sediment supplies than those that drain cratons (tectonically stable continental interiors, typically with lower topographic relief) (Milliman and Meade 1983). If a river flows away from the orogen, like the Amazon, it can create extensive floodplains, whereas rivers that flow parallel to the orogen and receive sediment from tributaries along much of their length, like the Orinoco, tend to have narrower and asymmetrical floodplains where tributary fans of coarse sediment shift the main river away from the orogen (Dunne and Aalto 2013).

7.1.6 Classification of Floodplains

Nanson and Croke (1992) suggests a three-part classification for floodplains, based on specific stream power and sediment texture as an influence on substrate resistance. The intent of this classification is to distinguish floodplains with respect to processes of origin and resulting morphology.

(1) High-energy, noncohesive floodplains with specific stream power >300 W/m^2 at bankfull are disequilibrium landforms that partly or completely erode during infrequent extreme floods. These floodplains typify steep upland portions of a drainage, in which bedrock or very coarse alluvium limits lateral channel migration. Relatively coarse vertical accretion deposits dominate these floodplains.

(2) Medium-energy noncohesive floodplains with bankfull specific stream power 10–300 W/m^2 are in dynamic equilibrium with annual to decadal flow regime. Geomorphic change is limited during extreme floods because energy is dissipated by overbank flow. Lateral point-bar accretion and braid-channel accretion tend to dominate deposition on these floodplains.

(3) Low-energy cohesive floodplains with bankfull specific stream power < 10 W/m^2 form along laterally stable channels with low gradients. Energy is dissipated by overbank flow during extreme floods and cohesive bank sediments limit lateral channel migration, so vertical accretion and infrequent channel avulsion dominate floodplain deposition.

Because floodplains do not necessarily progressively increase in width downstream, and because channel and floodplain characteristics are so closely related, some river classifications emphasize the degree of valley confinement and floodplain development. The river styles framework in Brierley and Fryirs (2005), for example, begins by differentiating valley setting as being confined, partly confined, or unconfined. Similarly, designations of river geomorphic spatial differentiation in mountainous settings emphasize the degree of valley confinement.

7.1.7 Human Influences on Floodplains

Floodplains physically and chemically connected to the active channel and performing the functions described in Section 7.1.1 were historically much more extensive. Humans have systematically reduced the extent and duration of overbank flow by regulating the flow and removing flood peaks as much as possible, constructing levees or transportation corridors (Blanton and Marcus 2009) that physically block access to the floodplain, and enlarging channels so that higher flows remain within the channel.

Floodplains have been particularly targeted by these activities because extensive and prolonged overbank flooding is regarded as a nuisance or a hazard that limits human access to the flooded areas or travel across these areas and because floodplains have long been regarded as highly desirable locations for human communities because of their fertile soils and easy access to a water supply and water-based transport. Many of the world's great cities are largely or wholly located within floodplains, including London (capital of the United Kingdom), Seoul (capital of South Korea), Paris (capital of France), Tokyo (capital of Japan), Dhaka (capital of Bangladesh), Caracas (capital of Venezuela), and Kinshasa (capital of the Democratic Republic of Congo). Obviously, governments on every continent think they can control flooding sufficiently to continue to base themselves within a floodplain. The location of major cities within floodplains can result in tremendous damage and loss of life when flow regulation, levees, or channelization fail for some reason. In addition to the obvious hazard of flooding, lateral channel migration and avulsion can create substantial change, causing hazards for people and structures.

Relatively few large rivers with largely unaltered flood regimes and floodplains remain. Among them are the Amazon and the Congo, the two largest rivers in the world, and the great northern rivers the Lena and the Mackenzie, which are rated as moderately affected by dams, and the Yukon, which is minimally affected (Nilsson et al. 2005). The commonality among these rivers is that they have low average population density within their drainage basin. In contrast, large rivers such as the Mississippi, Danube, Nile, Murray–Darling, and Yenisey, which historically had large floods and extensive floodplains, have lost flood peaks due to flow regulation and lost floodplains due to disconnection from the river and to human settlement and land use within the floodplain.

Loss of floodplain–channel connectivity and floodplain function has resulted in decreases in storage of fine sediment (e.g. Kroes et al. 2015) and organic carbon (Hanberry et al. 2015), retention and biological uptake of excess nutrients (Kroes et al. 2015), and flood peak attenuation (e.g. Jacobson et al. 2015). Alterations of floodplains have also caused loss of habitat abundance and diversity and associated biomass and biodiversity (e.g. Moyle and Mount 2007). Globally, these declines in floodplain function likely contribute significantly to substantial increases in nutrient fluxes to the oceans and disproportionately high rates of extinction for riparian and freshwater organisms (e.g. Ricciardi and Rasmussen 1999; Rahel 2007).

7.2 Terraces

A floodplain surface can remain active for thousands of years along a river with stable base level and relatively consistent water and sediment inputs. Along many rivers, however, the longevity of a particular floodplain surface is limited because incision or aggradation reduces the lateral connectivity between the channel and the floodplain surface. When this occurs, a terrace can form. *Terraces* are relict channel–floodplain features that have been isolated from contemporary river processes (Davis 1902b, 1909). Each terrace consists of a *tread*, the flat portion that represents the former floodplain surface, and a *riser*, the steep portion that separates the terrace from adjacent surfaces.

Terraces can be created by aggradation of the floodplain to a level no longer accessed by relatively frequent flows or by aggradation followed by incision. Terraces can also be created by incision of the channel. These various combinations can result from a change in base level or altered water, sediment, or large wood supplies resulting from changes in climate, land cover, or tectonics (Leopold and Bull 1979; Collins et al. 2016). Terraces can also result from intrinsic processes associated with sediment transport that is discontinuous in time and space (e.g. Schick 1974; Womack and Schumm 1977) (Section 1.5).

Terraces are not necessarily present along a river, but where they do exist, they can be used to infer past longitudinal profiles, as well as paleoenvironmental conditions (Wolman and Leopold 1957; Ryder and Church 1986; Kolb et al. 2017). Where terraces are present in a drainage basin, they typically do not extend fully into the headwaters because of lack of formation or subsequent erosion.

Terraces have been widely used by humans because they provide flat, in many cases stable surfaces, which are close to rivers and suitable for agriculture and cities. Terraces containing coarse-grained channel lag deposits are also mined for placer metals and construction aggregate (e.g. James 1989).

7.2.1 Terrace Classifications

Numerous classifications are used in connection with terraces. Some are simple and descriptive. *Paired terraces* have equivalent surfaces on both sides of a river valley and indicate that lateral erosion prevailed over vertical incision for a period of time. *Unpaired terraces* do not match across the valley and occur in such situations as continued incision while a sinuous river migrates laterally across the valley (Davis 1902b).

Other terrace classifications focus on the dominant process creating the terrace. The names of *erosional terraces* and *depositional terraces*, for example, are self-explanatory, although in practice it can be difficult to assign a primary cause of formation because both erosion and deposition are involved in the formation of many terraces. Similarly, *tectonic terraces* and *climatic terraces* (Bull 1990) can be difficult to distinguish in a region with active tectonic uplift or relative base-level fall that has also undergone climatically driven changes in water and sediment yield over the period of terrace formation (e.g. Woolderink et al. 2018).

One classification differentiates *event terraces* associated with a single extreme perturbation such as a very large flood or debris flow and *sustained terraces*, which represent a persistent change in water and sediment yield. Bull (1990) also distinguishes major terraces associated with tectonic or climatic changes and sustained conditions from minor terraces associated with a complex response, internal thresholds, and single extreme perturbations.

One of the more straightforward classifications focuses on terrace composition and differentiates strath and fill terraces (Figure 7.5) (Leopold and Miller 1954; Howard 1959). *Strath terraces* have low-relief terrace treads formed in bedrock or other cohesive materials such as glacial till,

Figure 7.5 Examples of river terraces. (a) Two small fill terraces (top outer edge of each indicated by a dashed white line) are visible in this view of a small creek in northern California, USA. Successive waves of aggradation and incision along the creek created the higher fill terrace just visible at the base of the forest, then the younger fill terrace about 1 m in elevation below it, and finally lowered and widened the channel, killing the riparian trees and leaving the stumps visible in the photograph. Flow in the channel is from left to right. (b) Aerial view of numerous fill terraces along a river in Nepal. (c) A strath terrace along the Mattole River in California, USA. The white material exposed in the lower portion of the cutbank is bedrock. An upward-fining alluvial sequence of cobbles to loamy soil overlies the bedrock strath. Flow is from the foreground toward the rear in this view. The cutbank is approximately 6 m tall. (d) Distant view of a very tall strath terrace along the Duke River in northwestern Canada. The strath surface is approximately 100 m above the active channel. The lower view shows a close-up of the bedrock–fill contact, which is indicated by a dashed line in both photographs. Arrow indicates flow direction on the main channel. *(See color plate section for color representation of this figure)*.

(a)

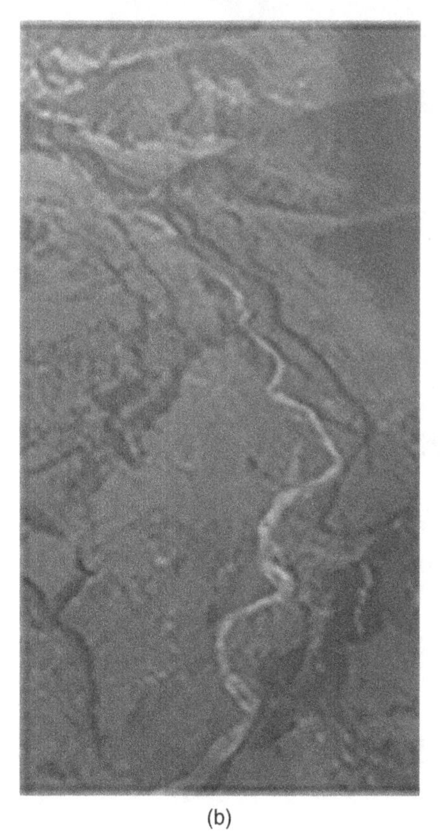

(b)

(c)

(d)

Figure 7.5 (*Continued*)

overlain by a veneer of alluvium thin enough to be mobilized throughout its entire depth by the river (Bull 1991). Existence of a strath terrace implies a period of vertical stability during which the river formed a relatively planar bedrock valley bottom via lateral erosion, followed by a period of vertical incision as transport capacity increased beyond sediment supply (Merritts et al. 1994). Strath terraces are less likely to form as uplift and rate of incision increase, because the river cuts downward too fast to form a strath tread (Merritts et al. 1994). Strath terraces are also less likely to form where resistant bedrock lithology limits valley-bottom width (Schanz and Montgomery 2016). Strath terraces tend to be more extensive where rivers flow over bedrock that is less resistant to weathering and erosion as a result of lithology or jointing, and more poorly developed over more resistant rock (Montgomery 2004; Wohl 2008).

Fill terraces are alluvial sequences that are too thick to be mobilized throughout their depth by the river. Because fill terraces can form more rapidly than straths, they do not necessarily imply a period of vertical stability. They do imply a period of aggradation, followed by incision. Alluvium in fill terraces is commonly topped by 0.1–1 m of fine overbank sediments and, in drier regions, eolian deposits (Pazzaglia 2013).

A relatively common scenario among fill terraces associated with high-latitude climatic fluctuations during the Quaternary is illustrated by the Maas River of northern Europe. Terrace alluvium aggraded along the Maas during periods of cold climate and glacial advance, whereas the river incised during periods of warm climate and glacial retreat (van den Berg and van Hoof 2001). Alternating glacial and interglacial periods during the Quaternary produced 21 paired and unpaired fill terraces and ~100 m of incision along the Maas (van den Berg and van Hoof 2001).

7.2.2 Mechanisms of Terrace Formation and Preservation

Terraces can be interpreted as deviations from a graded condition. A graded river can aggrade in response to base-level rise or incise in response to base-level fall, as long as the channel remains fixed to the changing base level such that the river maintains a steady-state profile with uniform valley and channel geometry and constant concavity and steepness (Gilbert 1877; Mackin 1937; Pazzaglia 2013). Creation and preservation of a terrace reflects unsteadiness in the profile because the channel aggrades or incises and geometry, concavity, and steepness can be altered. This unsteadiness is a transient response to changes in water and sediment input or base level.

Strath terraces have been attributed to a wide variety of changes in river dynamics (Table 7.1) (Wohl 2010b; Pazzaglia 2013). In the Pacific Northwest region of North America, a relatively recent (c. 1890 AD) episode of strath terrace formation in weak bedrock may reflect widespread removal of large wood and associated reductions in hydraulic roughness and sediment retention in channels (Collins et al. 2016).

Numerical modeling suggests that formation of strath terraces depends on input variability that creates a changing ratio of vertical to lateral erosion rates (Hancock and Anderson 2002). A wide variety of external and internal factors can create input variability, which explains how strath terraces can result from different causes in diverse river catchments or within a catchment through time. However, other numerical models suggest that the intrinsic unsteadiness of lateral migration in rivers may generate terraces under constant rates of incision without external forcing (Limaye and Lamb 2016). Again, these results support the diverse field-based interpretations for causes of terrace formation.

Similarly, fill terraces have been attributed to diverse changes (Table 7.2) (Wohl 2010b; Pazzaglia 2013). Fill terraces differ from strath terraces in that fill terraces can be event-based features resulting

Table 7.1 Causes of strath terrace formation.

Inferred cause	Example reference
Periods of balanced sediment supply	-
Altered sediment supply	Fuller et al. (2009)
Glacial–interglacial transitions and associated changes in water and sediment supply	Pan et al. (2003)
Tectonically induced changes in rock uplift	Lavé and Avouac (2001)
Falling local base level or eustatic base-level fall (global sea-level change)	Pazzaglia and Gardner (1993), Finnegan and Balco (2013)
Autocyclic oscillations in erosion rate in laterally migrating channels	Finnegan and Dietrich (2011)

Table 7.2 Causes of fill terrace formation.

Inferred cause	Example reference
Fluctuating water and sediment discharge during glacial cycles, volcanic eruptions, climatic change, and changing land use	Brocard et al. (2003)
Repetitive lateral shifting and stillstands during continuous downcutting	Mizutani (1998)
Fluctuating base level	Faulkner et al. (2016)
Complex response to base-level change	Schumm (1973)
Sediment waves migrating down catchments over periods of tens to hundreds of years in response to hillslope mass movements or land use	Oldknow and Hooke (2017)

from a single flood, debris flow, or landslide that overwhelms sediment transport capacity and causes floodplain aggradation before river flow incises the deposit (Miller and Benda 2000; Strecker et al. 2003). Fill terraces are also more likely to be reduced in planform size during lateral erosion after the terrace forms (Moody and Meade 2008).

The mechanisms by which terraces form, along with terrace geochronology, have been the subject of much research, because a terrace can record some change in external control variables, such as water and sediment yield or base level, and thus provide paleoenvironmental information. Terrace studies began in the nineteenth century with the conceptual model that river terraces reflect an externally imposed change in river dynamics associated with glaciation, uplift, or changing base level (e.g. Davis 1902b). Terraces continued to be studied under this assumption until Schumm's work on semiarid alluvial channels (Schumm and Hadley 1957) and physical experiments (Schumm 1973) demonstrated that terraces could result from the crossing of internal thresholds while external inputs to the river network remained unchanging. Schumm and Hadley (1957) described a cycle of erosion for ephemeral rivers in semiarid regions that involves longitudinally discontinuous incision. Infiltration into the streambed during brief periods of flow causes downstream decreases in discharge, local aggradation, steepened channel gradients, and incision, all without any change in climate or land cover that alters water and sediment inputs to the river network. This represented a major shift in thinking about how some types of terrace can form and led to the idea that event-based terraces might result from temporary fluctuations in water and sediment supply without major shifts in climate, tectonics, or land use. The formation of terraces as a result of the crossing of internal thresholds, rather than external forcing, has subsequently been documented from other field sites (e.g. Schick 1974; Malatesta et al. 2017). Diverse flume experiments have produced strath and fill terraces during periods of constant discharge and base level (Mizutani 1998; Hasbargen and Paola 2000).

Physical experiments and numerical modeling have been used to understand the mechanisms of terrace formation and preservation. Examples include physical experiments on knickpoint formation and migration, and the formation of gorges and terraces downstream from the knickpoint in both alluvial and cohesive (simulated bedrock) substrates (e.g. Schumm et al. 1987; Lamb et al. 2015; Grimaud et al. 2016; Baynes et al. 2018). These experiments indicate that knickpoints can retreat headward through a combination of parallel retreat and vertical channel incision, resulting in upstream-dipping strath terraces analogous to those found along the Susquehanna River in Pennsylvania, USA (Frankel et al. 2007).

Early numerical simulations of terrace formation, such as that by Boll et al. (1988), parameterized discharge, sediment transport, and erosion within a cross-section. Each parameter was varied through time to mimic the effects of changing water and sediment supply and base level. Subsequent simulations of terraces have included adaptations of models such as CAESAR, a cellular automaton model of landscape evolution (van de Wiel et al. 2007), and the centerline evolution model of Howard and Knutson (1984), as adapted by Limaye and Lamb (2016). One-dimensional models greatly simplify river complexity, but can nonetheless produce terraces (Veldkamp and Tebbens 2001; Hancock and Anderson 2002).

7.2.3 Terraces as Paleoprofiles and Paleoenvironmental Indicators

Terrace grain size and stratigraphy can be used to infer sediment supply, flow conditions, and climate (e.g. Baker 1974; Bridgland 2000). The longitudinal profile of a terrace can approximate the paleoprofile of the river (e.g. Pastor et al. 2015). The age of the terrace can be used to constrain the timing of the events that resulted in terrace formation (e.g. Delmas et al. 2015). However, inferring paleoenvironmental conditions from terrace characteristics is complicated by several factors.

First, a given change in climate that alters water and sediment yield to a river network can create very different responses within that network, depending on the climatic conditions present prior to the change. This is illustrated by the Charwell River basin of central New Zealand (Bull 1991), in which the lower portion of the drainage, from the basin mouth to the contemporary tree line, changed from periglacial conditions to greater vegetation density and soil thickness. This reduced sediment yield, increased peak discharge, and resulted in stream incision. The upper portion of the drainage above timberline experienced an increase in periglacial processes and increased sediment yield, so that channels in this zone did not incise (Bull 1991).

Second, a change in base level, such as that caused by tectonic uplift, can result in different responses across a river network. For example, larger rivers in the vicinity of a triple junction in northern California, USA can maintain uniform incision into bedrock during relatively rapid uplift, creating sequences of strath terraces along the rivers (Merritts and Vincent 1989; Merritts et al. 1994). Smaller headwater rivers cannot incise rapidly enough to keep pace with uplift and instead become steeper with time rather than creating strath terraces. Studies elsewhere indicate that, even on larger rivers, terrace formation can be longitudinally discontinuous if the river lacks the power to create strath surfaces along segments of more resistant bedrock (Wohl 2008).

Third, event-based terraces that result from local disturbances such as landslides create fill terraces of limited longitudinal extent (Montgomery and Abbe 2006; Wohl et al. 2009). Even more sustained changes such as base-level rise or fall may only produce terraces in the lower portion of a river network (Merritts et al. 1994). Strath terraces, in particular, may require some minimum drainage area to form (Garcia 2006), because of the stream energy required to create a strath surface in bedrock. The formation and preservation of straths can also vary with differences in bedrock erosional resistance (Montgomery 2004; Wohl 2008). Terraces may be thickest, widest, and best preserved near tributary junctions, where tributary sediment widens the mainstem and helps to build a terrace sufficiently large to be preserved (Pazzaglia 2013).

Fourth, a single perturbation such as base-level fall can result in multiple terraces if complex response occurs (Schumm 1973). Each terrace along a river therefore cannot necessarily be correlated with an external change.

Fifth, the duration of time represented in terrace alluvium can vary widely. The alluvium capping a strath terrace is synchronous with formation of a strath, for example, whereas the alluvium of a fill terrace is younger than the underlying surface, and can accumulate over long periods of time (Pazzaglia 2013).

Finally, the timing of terrace formation can substantially lag the timing of any external perturbation in water and sediment yield or base level (Pazzaglia 2013). Base level on the Drac River in the French Alps dropped 800 m following removal of an ice-dam during glacial retreat. As a knickpoint moved upstream through the basin in response to base-level lowering, terrace formation lagged behind base-level drop by 2000–5000 years, depending on location within the drainage (Brocard et al. 2003).

These complicating factors do not by any means render terraces useless as paleoenvironmental indicators. The presence of spatial and temporal differences and lag times in terrace formation within a river network only means that terraces must be interpreted carefully and in a broader context than a single terrace exposure (e.g. Merritts et al. 1994), as with any other river feature.

Strath and fill terraces can be used to reconstruct river longitudinal profile at the time of strath or floodplain formation (Paola and Mohrig 1996), although creation of straths, in particular, can lag by several thousand years the input changes that cause terrace formation (Merritts and Vincent 1989). Also, both types of terraces can subsequently be tectonically deformed (Finnegan 2013). The terrace surface is the easiest proxy to use for paleoprofile. This surface can be deformed by weathering and inputs of colluvium from adjacent hillslopes, however, so the more difficult-to-discern underlying bedrock strath provides a better indicator of original longitudinal profile (Pazzaglia 2013).

Any paleoenvironmental inferences drawn from terraces rely on establishing a chronology of terrace formation (Table 7.3) (Wohl 2010b; Pazzaglia 2013). Terrace chronology can be based on absolute methods that provide a numerically precise age, although the accuracy of this age is not necessarily greater than that achieved using relative methods. Relative methods infer differences in age between terraces by characterizing parameters that change through time, although the rate of change in a parameter is not necessarily linear. Many studies utilize multiple techniques (e.g. Guralnik et al. 2010; Sancho et al. 2016). The actual event or period of time dated can be either the phase of tread construction (erosion of a strath or deposition of a fill) or the incision that produces

Table 7.3 Geochronological methods applied to terraces.

Method	Sample references
Absolute	
^{14}C dating of organic matter in terrace sediments	Merritts et al. (1994)
Dendrochronology of terrace vegetation	Pierson (2007)
Luminescence techniques – luminescence measures energy stored in sediment as a result of natural, background radioactive decay; the energy stored is a function of the time over which the sediment has been buried, as well as background decay rate	Pederson et al. (2006), Lyons et al. (2015)
U-series dating of pedogenic carbonate in terrace sediments	Candy et al. (2004)
Cosmogenic isotope dating of terrace sediments or bedrock in strath terraces	Riihimaki et al. (2006), Delmas et al. (2015)
Paleomagnetic dating of terrace sediments	Pan et al. (2003)
Tephrochronology of volcanic ashes incorporated in terrace sediments	Dethier (2001)
Radiometric ages of bounding lithologic units such as basalt flows	Maddy et al. (2005)
Relative	
Development of weathering rinds on coarse clasts in terrace sediments	Pazzaglia and Brandon (2001)
Soil characteristics – thickness, accumulation of translocated clays or soluble salts, color, and iron-oxide speciation	Tsai et al. (2007)
Amino acid racemization of organic compounds such as shells	Penkman et al. (2007)
Lichenometry – the diameter of lichen colonies, which increases with time	Baumgart-Kotarba et al. (2003)

a terrace scarp (Ritter et al. 2011). The phase of tread construction is time-transgressive and of relatively long duration, whereas incision can be widespread and rapid (Pazzaglia 2013).

As numerous detailed case studies accumulate in the scientific literature on fluvial geomorphology, diverse groups have begun to establish digital catalogs and databases. The Fluvial Archives Group (FLAG) is actively documenting the global record of terrace stratigraphy (Vandenberghe and Maddy 2000; Cordier et al. 2017) and creating databases for Pleistocene and Holocene sedimentary archives.

As with other types of fluvial deposits, LiDAR and DEMs facilitate the designation and mapping of terraces (Jones et al. 2007; Johnson et al. 2019). Shallow geophysical techniques such as ground-penetrating radar and electrical resistivity (Froese et al. 2005; Bábek et al. 2018) are used to image terrace subsurface geometry. These images can substantially enhance the information available from limited surface exposures.

7.3 Alluvial Fans

Alluvial fans are primarily depositional environments that are shaped by the combined effects of erosion and deposition, similar to the other features addressed in this chapter. Channels exiting high-gradient, laterally confined canyons in regions with high relief commonly create a fan-shaped deposit known generically as an *alluvial fan*, although colluvial processes can contribute to sediment accumulation. Water flow, hyperconcentrated flow, debris flow, rockfall, landslide, and snow avalanches can all contribute sediment to alluvial fans. Fans can be differentiated based on primary depositional process and morphology, resulting in debris cones dominated by rockfalls, debris fans dominated by debris flow, and so forth. Fans classified in this way can be differentiated as those dominated by *inertial transport* (debris flows, rockfalls) versus those dominated by *traction transport* (hyperconcentrated and fluvial flows) (Stock 2013).

Where parallel channels drain from a mountain range into a basin, adjacent alluvial fans can coalesce to form an *alluvial apron* or *bajada* (e.g. Owen et al. 2006). *Fluvial megafans* are unusually large alluvial fans that mostly occur between 15° and 35° north and south of the equator where rivers that undergo moderate to extreme discharge fluctuations enter actively aggrading basins (Leier et al. 2005; Latrubesse 2015). Australians use the term *floodout* to describe sites where a marked reduction in channel capacity creates increased overbank flows and associated deposition (Tooth 1999). Floodouts bear some similarities to fans, but are not necessarily associated with a change in lateral channel confinement. *Paraglacial fans* are alluvial fans composed mainly of reworked glacial deposits (Ryder 1971; Ballantyne 2002). *Fan-deltas* form at the margins of the ocean, where rivers flow from a mountainous region across a narrow coastal zone (Nemec and Steel 1988). Fan-deltas can grade into deltas that form in freshwater or marine subaqueous environments. Small alluvial fans form at crevasse splays below levees on floodplains.

Alluvial fans can form in any climatic region. Fans in arid and semiarid environments are by far the most well-studied, however, not least because the lack of continuous vegetation cover results in good exposure of fan features (Figure 7.6).

Lower transport capacity compared to upstream channel reaches promotes aggradation, overbank flows, and frequent avulsion on alluvial fans, which creates numerous hazards for structures built on these surfaces (e.g. Cavalli and Marchi 2008; Croissant et al. 2017). Fans are also heterogeneous with respect to process, form, grain size and stratigraphy, and soils and vegetation (e.g. Weissmann and Fogg 1999; Wang et al. 2017). Fan sediments tend to be relatively porous and permeable, which facilitates infiltration and subsurface flow, allowing fans to hydrologically buffer a catchment (Herron and Wilson 2001; Vivoni et al. 2006). Wetlands can occur at the downstream ends of fans in humid climates, as lower gradients promote deposition of finer, less permeable sediments and subsurface flow rises back toward the surface (Woods et al. 2006).

Fans can also buffer sediment delivery by at least temporarily storing sediment. Fans thus lower the connectivity of sediment movement and the coupling between hillslopes and channels or between tributaries and larger rivers (Fryirs et al. 2007a). The volume of sediment stored can fluctuate substantially through time, as processes of fan erosion – incision of the main or distributary channels or erosion of the toe of a tributary fan by a larger river – alternate with processes of deposition.

Figure 7.6 Examples of alluvial fans. (a) A fan created at the junction of a tributary with the Colorado River in the Grand Canyon, USA. The older fan deposits are outlined with a dashed white line, the most recent deposits with a dotted white line at far right center. Vegetated deposits in the foreground are the floodplain. Mainstem flow is left to right. (b) An event-based alluvial fan created by a damburst flood down the Roaring River in Colorado, USA. The Roaring River enters the Fall River valley; the Fall River flows from lower left to upper right in this view. The outburst flood occurred in 1982, and the resulting alluvial fan appears lighter in color (outlined in dashed white line) in this photograph, taken nearly 30 years later, because of slow growth of vegetation on the fan surface. (*See color plate section for color representation of this figure*).

7.3.1 Erosional and Depositional Processes

Alluvial fans are geomorphically complex and dynamic environments. Channelized water flows, hyperconcentrated flows, and debris flows alternate spatially and temporally with unconfined sheet flow or with rapid infiltration that results in sieve deposits and subsurface flow (Griffiths et al. 2006). Deposits on many fans are coarse-grained and poorly sorted as a result of relatively short transport distances, inputs by debris flows and flash floods, and rapid loss of flow capacity as a result of infiltration or overtopping of shallow channels (Blair and McPherson 2009).

Episodic aggradation at the fanhead causes oversteepening that initiates incision. Consequently, an incised channel, or *fanhead trench*, is present near the apex of many alluvial fans, with active deposition concentrated farther down the fan (Blair and McPherson 2009). Channels also frequently avulse or are abandoned through stream capture (e.g. de Haas et al. 2018).

Downstream from the zone of most active incision, channels are commonly braided and can decrease in size downstream as a result of discharge infiltration, so that channelized flow can give way to sheetflow (Bridge 2003). Although deposition can produce relatively smooth surfaces, secondary channel networks can dissect these surfaces once deposition shifts elsewhere on the fan. Airborne laser swath mapping, LiDAR, and other remote, high-resolution imaging can be used to map fan surface roughness (e.g. Frankel and Dolan 2007; Cavalli and Marchi 2008).

Blair and McPherson (2009) distinguish primary and secondary depositional processes on alluvial fans. Primary processes transport sediment from the catchment onto an alluvial fan and cause fan construction. Secondary processes modify previously deposited sediment via overland flow, wind erosion or deposition, bioturbation, soil development, weathering, and toe erosion. Secondary processes dominate fan surfaces away from the usually limited areas of recent deposition. Blair and McPherson (2009) also distinguish three classes and 13 different types of alluvial fans (Table 7.4)

Table 7.4 Types of alluvial fans.

Type BR alluvial fans (sediment sourced from bedrock)
BR-1: fans dominated by rockfalls
BR-2: fans dominated by rock slides
BR-3: fans dominated by rock avalanches
BR-4: fans dominated by earth flows
Type CC alluvial fans (sediment sourced from cohesive colluviums)
CC-1: fans dominated by debris-flow levees
CC-2: fans dominated proximally by debris-flow levees and distally by lobes
CC-3: fans dominated by clast-rich debris-flow lobes
CC-4: fans dominated by clast-poor debris-flow lobes
CC-5: fans dominated by colluvial slips or slides
Type NC alluvial fans (sediment sourced from noncohesive colluviums)
NC-1: fans dominated by sandy, clast-poor, sheetflood deposits
NC-2: fans dominated by sand-poor and clast-poor sheetflood deposits
NC-3: fans dominated by gravel-poor and sandy sheetflood deposits
NC-4: fans dominated by noncohesive, sediment-gravity flows

Source: After Blair and McPherson (2009), Table 14.1.

based on dominant primary processes and textures. The classes (bedrock, cohesive colluvium, non-cohesive colluvium) reflect the catchment slope material from which primary processes are triggered.

Very few direct observations of transport and depositional processes on fluvially dominated fans have been published (Stock 2013). These observations consistently indicate fluctuations in water and sediment supply during a single storm, as well as shallow, hydraulically rough flow and unstable flows with large concentrations of suspended and bed load. Other consistent patterns include the formation of coarse bars by particle accretion and subsequent diversion of flow and sediment around these bedforms on gravel-bed surfaces. These processes of channel bifurcation and avulsion are analogous to those in braided rivers. Processes of sediment accumulation seem to be consistently associated with bifurcation, avulsion, and surges of water and sediment on alluvial fans.

Alluvial fans are typically built over time by continuing deposition, but individual fans can result from a single extreme event, such as massive erosion associated with a damburst flood (Figure 7.6b) (Blair 2001). Fans can also be dominated by a specific and sometimes repetitive type of disturbance, such as tributary glaciation, volcanic eruption, wildfire-induced mass movements (e.g. Meyer and Pierce 2003; Pierce and Meyer 2008), or landslides (e.g. Korup et al. 2004; Torres et al. 2004). An example comes from the Southern Alps of New Zealand, where multiple case studies document a series of alluvial fans dominated by sediment inputs from large ($>10^6$ m^3) landslides (Korup et al. 2004).

7.3.2 Fan Geometry and Stratigraphy

An alluvial fan approximates a segment of a cone radiating downslope from a point where a channel flows out from a high-relief catchment (Blair and McPherson 2009). In plan view, a fan is arcuate, creating a 180° semicircle unless the toe of the fan is truncated by erosional processes or the sides of the fan are restricted by adjacent fans. Longitudinal fan profiles within and away from the channel are typically segmented, with the steepest slopes along the flanks (Hooke and Rohrer 1979). Segmentation can reflect intermittent uplift, channel incision, and episodic deposition of various types (Bull 1964; Arboleya et al. 2008). Channel gradient typically decreases downstream from 0.10–0.04 to 0.01 m/m on fluvially dominated fans (Stock et al. 2007). This decrease in gradient reflects progressive deposition and declining sediment loads and grain size down-fan. Radial fan profiles tend to be concave or planar, and cross-fan profiles are convex (Bull 1977; Blair and McPherson 2009). Fans typically extend 0.5–10 km from a mountain front, but large fans can extend 20 km (Blair and McPherson 2009).

The volume of sediment available for deposition on a fan and the processes of deposition exert first-order controls on fan geometry. Climate, lithology, tectonics, and drainage area all influence sediment delivery to a fan. Climate influences weathering and the caliber of sediment supplied to the fan, as well as the flow available for sediment transport. Oguchi and Ohmori (1994) compare alluvial fans formed in the wet climate of Japan to those formed in the arid US Southwest, across a range of lithologies. Although fan area in each region is proportional to the catchment sediment output per unit time, sediment yield per unit area is higher in the Japanese basins for the same basin slope.

Lithologies that weather to very coarse sediment deposited by mass movements typically create the steepest fans. Lithologies that weather rapidly and release very fine sediments result in moderately steep fans of larger volumes, whereas lithologies that weather to sand-sized sediment create low-gradient fans of smaller volume per unit drainage area (Bull 1964; Blair 1999).

Tectonics influence rates of sediment production and delivery to rivers in upstream catchments and the accommodation space in depositional sites. A basin that is subsiding relative to the source area can develop a thicker fan of more limited spatial extent, whereas a tectonically stable basin can develop a thinner but more areally extensive fan (Whipple and Trayler 1996).

Tectonics can dominate fan formation in regions with active extension (divergence of plates), compression (convergence of plates), or transtension (lateral plate movement) in which confined steep valleys are juxtaposed with unconfined depositional regions such as piedmonts (Stock 2013). These areas favor large fans because of the lack of confinement on depositional space. Extensional regions such as the US Basin and Range, the Tibetan Plateau, and the Andean back-arc are characterized by mountain ranges and intervening subsiding basins along the margins of which alluvial fans form. These regions tend to be arid because of rain-shadow effects from bounding mountain ranges and current global circulation patterns, which facilitates identification and study of their alluvial fans. Some of the largest fans, such as the Kosi fan, form in convergent tectonic settings such as along the Himalayan or Taiwan thrust fronts (Stock 2013). In this setting, active uplift provides abundant sources of sediment, but the depositional area can be less constrained than in an extensional setting with numerous mountain ranges. Other examples of formerly tectonically active settings conductive to fan formation are retreating escarpments such as South Africa's Drakensberg or the southeastern Australia passive margin, which juxtapose sediment sources in higher-relief areas with relatively flat piedmonts.

Some of the earliest quantitative work on alluvial fans demonstrated that drainage area corresponds to fan size, A_f, and fan slope, S_f, as

$$A_f = aA_d{}^n \tag{7.1}$$

$$S_f = cA_d{}^d \tag{7.2}$$

with a varying slope of 0.7–1.1 for the regression line, n, which averages 0.9 for fans in the United States (Bull 1962). Coefficient a varies by more than an order of magnitude as a result of lithology, climate, and the space available for deposition on the fan (Viseras et al. 2003). Although these equations continue to be widely applied, limitations exist (Blair and McPherson 2009). These include the comparison of plan-view areas for three-dimensional features (Eq. 7.1) that are not of constant thickness and for which thickness is commonly unknown, as well as the lack of clear guidelines for defining area, which may not be straightforward (Stock 2013).

Various morphometric indices, including relief and ruggedness, as well as particle size and roundness and the fabric of deposits, can be used to distinguish spatial differences in depositional processes on fans (de Scally and Owens 2004; Stock 2013) and dominant depositional processes on individual fans (Al-Farraj and Harvey 2005). Debris flows with high sediment concentration can create steep and rugged topography on the upper fan, for example, whereas low sediment concentration can create smoother topography on the lower fan by filling channels and other depressions (Whipple and Dunne 1992).

Spatial and temporal heterogeneity in depositional processes on alluvial fans make the resulting stratigraphic records challenging to interpret. Nonetheless, an extensive literature (Wohl 2010b; Stock 2013) documents how changes in the volume and style of fan deposition through time have been used to infer environmental changes. These include climatically driven changes in sediment supply to alluvial fans (Oguchi and Oguchi 2004; Waters et al. 2010) and changes in sediment supply associated with alpine glacier retreat (Hornung et al. 2010). Fan deposition can be used to infer the magnitude and frequency of event-based sedimentation associated with storms or wildfires (Pierce and Meyer 2008), human-induced and tectonically driven changes in sediment supply (Chiverrell et al. 2007) or deposition (Spaliviero 2003), and short- and long-term sediment budgets for the upstream catchment (McEwen et al. 2002).

7.3.3 Mapping, Studying, and Living on Fans

As with other fluvial features, the ability to differentiate geomorphic units and map surficial features on alluvial fans is rapidly improving with the increasing availability of high-resolution LiDAR (Jones et al. 2007; Regmi et al. 2014) and airborne laser swath mapping (Staley and Wasklewicz 2006) data. The increasing availability of frequently repeated satellite imagery also facilitates change detection on fans (Torres et al. 2004), as well as the calibration of parameters for numerical simulations of flow processes on fans (Catani et al. 2003). Fan stratigraphy can be accurately assessed using shallow geophysical techniques, including ground-penetrating radar (Ekes and Friele 2003; Gu et al. 2018) and seismic refraction (Schrott et al. 2003; Maraio et al. 2018).

Surficial maps of alluvial fans are especially useful when combined with absolute or relative age dating. Mapped features and chronologies of change through time can be used to understand how climate and tectonics interact to shape fans and how soil-forming processes influence hydrologic processes and biological communities on fans, as well as to map hazards across fan surfaces (Stock 2013). A wide variety of geochronologic methods have been applied to alluvial fans, mostly with the intent of estimating how long a portion of a fan has been stable (Table 7.5).

Partly because of the episodic and unpredictable timing of processes on alluvial fans, investigators have relied heavily on physical experiments and numerical simulations to examine fan processes. Physical experiments have released water into a sloping inlet box filled with sandy sediment (e.g. Hooke 1967, 1968; Hooke and Rohrer 1979) in order to examine how changing water and sediment discharge and down-fan gradient influence process and form. Some experiments have used overhead sprinklers to simulate the rainfall driving basin erosion and fan deposition across a relatively broad and sloping surface (e.g. Schumm et al. 1987). Other experiments have used experimental basins that are broader than they are long to examine how deposition rate influences avulsion frequency (Bryant et al. 1995) and how fluctuating base level influences fan process and form (Whipple et al. 1998). The eXperimental EarthScape basin at the National Center for Earth-Surface Dynamics in the United States (Paola et al. 2001) has been used to investigate the creation and preservation of different types of deposits on fans (Sheets et al. 2002) and the rate at which flow occupies an experimental fan surface (Cazanacli et al. 2002). Stock (2013) provides a thorough review of physical experiments involving alluvial fans.

Numerical simulations of fan processes and morphology have grown progressively more complex since the Price (1974) random-walk model of fan deposition (Stock 2013). For example, Salcher et al. (2010) uses a numerical model to investigate climate-induced aggradation and degradation cycles on a fan, as well as the influences of tectonic subsidence and erosion by an axial river over 25 000 years. Studies since Price's have used sediment transport models with thresholds to examine longitudinal profiles on fans (Parker et al. 1998a,b; Whipple et al.1998; Stock et al. 2007) and a diffusivity model to estimate surface profiles (De Chant et al. 1999). Coulthard et al. (2002) and Nicholas and Quine (2007) use a cellular automaton model to couple upland river basin and fan evolution. Clevis et al. (2003) use a forward numerical model to simulate the effects of pulsed tectonic activity on fan dynamics. Two- and three-dimensional numerical models are used to simulate sediment dynamics and resulting morphology on floodplains, deltas, and fans (Karssenberg and Bridge 2008; Croissant et al. 2017).

Complex and intermittent erosion and deposition on alluvial fans also create special challenges for hazard zoning and mitigation. Fans are commonly the only gently sloping surfaces available in

Table 7.5 Geochronological methods applied to alluvial fans.

Method	Sample references
Historical records – these include, for example, floods recorded since the fifteenth century on alluvial fans in northern Italy and debris flows on alluvial fans in Slovakia	Crosta and Frattini (2004); Kotarba (2005)
Dendrochronology – the ages of trees growing on forested fans can be used to provide a minimum time over which the surface has been stable or to determine the timing of stem burial during aggradation or root exposure during erosion	May and Gresswell (2004); Wilford et al. (2005); Turk et al. (2008); Schürch et al. (2016)
Luminescence techniques – luminescence can reveal the time elapsed since finer fan sediments were last transported and exposed to solar radiation	Robinson et al. (2005); Yeats and Thakur (2008); Singh et al. (2016)
Clast weathering – the degree to which cobble-size and larger clasts on a fan surface are weathered can indicate the time elapsed since the clasts were last transported (and presumably abraded to expose fresh rock)	Kotarba (2005)
Rock varnish – the degree of development can be used to estimate the duration of surface exposure of larger clasts	Hooke and Dorn (1992)
Cosmogenic isotopes – accumulation can be used to estimate the duration of surface exposure of larger clasts	Barnard et al. (2004, 2006); Frankel et al. (2007)
Radiocarbon dating – dating organic matter incorporated in fan deposits can provide a maximum estimate of time of deposition	Lewis and Birnie (2001); Sanborn et al. (2006)
Bomb isotopes – accumulation of isotopes such as ^{7}Be, ^{137}Cs, and ^{210}Pb that were disseminated throughout the global atmosphere with the start of nuclear bomb testing during the 1950s can be used to determine the age of sediments deposited during the 1950s or later	Wallbrink et al. (2005)
Soil properties – relative dating of alluvial fans relies on the degree to which soil has developed or pedogenic carbon has accumulated in arid-region fan soils	Machete (1985); McFadden et al. (1989); Kochel et al. (1997); McDonald et al. (2003); Rockwell et al. (2019)

high-relief environments, and hazards associated with erosion and deposition are intermittent in time. Consequently, urbanization concentrates on fans (Zorn et al. 2006). Approaches used to mitigate hazards (Kellerhals and Church 1990; Pelletier et al. 2005; Wohl 2010b; Stock 2013) include channelization and bank stabilization to limit overbank flows, sheetwash, and channel avulsion. Debris interception barriers and detention basins are used to contain sediment entering the fan, rather than allow the sediment to disperse and bury infrastructure or fill active channels. Hazard mapping is used to guide the development of infrastructure, and warning devices are used to facilitate rapid evacuation in case of an erosional or depositional event.

7.4 Deltas

A delta is analogous to an alluvial fan in that it is a primarily depositional feature that results from a loss of transport capacity – in the case of a delta, where a river enters a body of standing water. Just as an alluvial fan is roughly triangular in shape, the word "delta" was first applied to a fluvial deposit 2500 years ago by the historian Herodotus because of the Nile delta's resemblance to the Greek letter delta, Δ (Ritter et al. 2011).

A delta can protrude well beyond the adjacent coastline when a river carries large volumes of sediment and currents in the receiving body have limited ability to rework the sediment. Examples of such deltas include the Mississippi River delta in the Gulf of Mexico, the Danube in the Black Sea, the Nile in the Mediterranean Sea, Italy's Po-Adige delta in the Adriatic Sea, Russia's Volga River delta in the Caspian Sea, the Ganges-Brahmaputra and Irrawaddy in the Bay of Bengal, and Africa's Niger in the Gulf of Guinea.

Alternatively, delta deposits can be primarily upstream from the adjacent coastline, so that the delta does not protrude into the ocean as a result of limited particulate sediment transport from the river basin or thorough reworking and transport of fluvial sediment by currents in the receiving water body. Examples include the Amazon and the Rio de la Plata in South America, Australia's Murray–Darling, the Congo in Africa, and the Columbia in North America.

Regardless of the shape, deltas are biologically diverse and productive environments that typically support commercial fisheries (e.g. Bănăduc et al. 2016). Deltas also host vegetative communities that can help to reduce the velocity and associated damage of incoming oceanic storm surges (Kirwan et al. 2016). Also, as with inland river channels, vegetation on and around deltas – including salt marshes, seagrass beds, and mangrove forests – can affect hydrodynamics, sediment transport, and delta morphology (e.g. Fagherazzi et al. 2017). Delta features such as the area of islands and bars and type and the extent of vegetation influence the *tidal prism* – the volume of water exchanged in a tidal cycle – as well as values of bed shear stress and thus sediment dynamics (Canestrelli et al. 2010). Also, as with inland river channels, the details of delta form and process influence biogeochemical processing such as nutrient concentrations (e.g. Henry and Twilley 2014) and organic carbon stock (e.g. Siewert et al. 2016).

As river flow entering the body of standing water expands and decelerates, sediment is deposited. Various types of currents in the receiving water body rework the river sediment, sometimes with the result that no delta forms. The morphology and stratigraphy of a delta reflect the balance between river inputs and the reworking of sediment by the receiving water body. In general, the coarsest sediment is deposited closest to the river mouth and finer sediment in suspension is carried farther into the receiving body. Large databases containing data from numerous deltas indicate that the area of a delta is best predicted from average discharge, total sediment load entering, and offshore accommodation space (Syvitski and Saito 2007).

Many of the world's great cities are located on deltas along marine shorelines. This is typically a mixed blessing (e.g. for Shanghai, New Orleans, Calcutta, Rotterdam, Marseille, and Lisbon). On the plus side, deltas provide ease of transport and navigation of goods from the terrestrial interior and across the ocean, fertile soils and rich fisheries, and reserves of oil and natural gas. These benefits are offset by some of the same hazards common on alluvial fans, such as overbank flooding and channel avulsion. Deltas also include hazards associated with subsidence of delta sediments (Wang et al. 2012) and flooding of low-lying portions during marine storm surges (e.g. Neumann et al. 2015). Hazards on deltas can result from marine erosion if river sediment supply decreases or sea level rises (Poulos et al. 2009). Urban areas risk loss of water supply, transport, or waste disposal if river flow shifts location across the delta (e.g. Rudra 2014). Agricultural regions and urban areas dependent on ground water can be particularly affected by salinization of delta soils and ground water if river flow decreases, sea level rises, or ground water is pumped at a rate faster than natural recharge (Rodríguez-Rodríguez et al. 2011). Like alluvial fans, deltas exemplify *terra* ***non****-firma*.

7.4.1 Processes of Erosion and Deposition

Deposition on deltas, as on alluvial fans, constantly shifts location across or down the delta in response to changes in river water and sediment discharge, as well as changes in delta morphology resulting from deposition, subsidence, and reworking by waves and tides. The classic model of delta deposition involves deposition of a longitudinal bar where river transport capacity declines because of mixing between river water and the water of the receiving basin (e.g. Bierman and Montgomery 2014). The initial bar forces the river to bifurcate and the process continues, developing a *distributary channel* network. Distributary channels formed at least in part by avulsion are lined with natural levees. With continued deposition, the distributary network progrades into the receiving body.

Slingerland and Smith (1998) developed a model predicting avulsion stability based on the ratio of the water surface slopes of the two branches of a bifurcation. Ratios <1 (where the secondary channel slope is the numerator and the main channel slope is the denominator) result in a failed avulsion in which the incipient avulsion fills with sediment. Ratios >5 create avulsions in which the river abandons its original course in favor of the steeper channel.

Avulsion frequency scales with the time required for sedimentation on the streambed to create a deposit equal to one channel depth. Jerolmack and Mohrig (2007) use relative rates of bank erosion and channel sedimentation to derive a dimensionless mobility number to predict the conditions under which distributary channels form.

An active delta includes *subdeltas* created when sediment diverted through breached levees accumulates as crevasse splays (e.g. Coleman et al. 1998). Channels on the subdelta also bifurcate and prograde until they develop gradients similar to those of the main channel and deposition shifts to a new subdelta. On the Mississippi River, the process of forming and abandoning subdeltas occurs over a few decades (Morgan 1970). Physical experiments suggest that the time scale for these processes on river-dominated deltas reflects the time over which a river mouth bar that is prograding basinward reaches a critical size and stops prograding (Edmonds et al. 2009). This triggers a wave of bed aggradation that moves upstream, increasing overbank flows and shear stresses exerted on the levees, and triggering avulsion and the growth of subdeltas.

In high-latitude rivers, ice and ice-jam floods can exert an important control on processes of delta erosion and deposition. Canada's 12 000 km^2 Mackenzie River delta on the shore of the Arctic Ocean provides a well-studied example (Goulding et al. 2009b). The northerly flow direction of the Mackenzie results in the spring snowmelt flow peak to progressing downstream with the seasonal advance of warm weather. The flood wave commonly encounters an intact and resistant ice cover downstream, causing the formation of ice-jams and enhanced overbank flooding (Prowse and Beltaos 2002). The spatial extent and duration of overbank flooding influence erosion and deposition; habitat abundance, diversity, and connectivity; biogeochemical processes; lake flooding; and even surface albedo as floodwaters hasten the melting of snow and ice (Goulding et al. 2009a). Ice transported by the floodwaters into overbank areas enhances localized erosion. Decreases in the severity of river-ice break-up during recent decades have lessened the flooding of some delta lakes, which may lead to loss of these water bodies, as well as changes in the biogeochemical interactions between river water and the floodplain ecosystem (Prowse et al. 2011).

As a delta front progrades farther into the receiving basin, shorter routes for water flow into the basin become available along the sides of the delta. These shorter routes can also access topographically lower portions of the coastline than the delta, which has been progressively built up as well as out. River flows commonly access such shorter routes via crevasses developed in levees well upstream from the delta, causing *delta switching* (also known as channel switching), a relatively abrupt shift in river course and delta location (e.g. Roberts 1998; Zheng et al. 2018). Over time, delta switching results in distinct depositional lobes within the general depositional complex (Aslan and Autin 1999). Delta switching occurs approximately every 1000–2000 years, for example, along the Mississippi River (Aslan and Autin 1999) and Spain's Ebro River (Somoza et al. 1998).

Interactions between inputs to a delta, processes of erosion and deposition, and delta form can be studied within the context of connectivity (e.g. Sendrowski and Passalacqua 2017). Passalacqua (2017) proposes the framework of a Delta Connectome, in which a delta is represented as a network composed of nodes (e.g. bifurcations and junctions) and links (e.g. channels) – a leaky network that continuously exchanges fluxes of matter and energy with its surroundings and evolves over time. Focusing on these exchanges can inform a basic understanding of deltas, as well as their management and restoration (Passalacqua 2017).

7.4.2 Delta Morphology and Stratigraphy

Deltas reflect the changing balance through time between at least four factors (Bridge 2003): the volume and grain-size distribution of river sediments; the relative density of river water and the receiving water body; the currents of the receiving water body; and subsidence and water-level change in the receiving body, which influence relative change in base level.

(1) *River sediments.* The characteristics of the river sediments supplied to the delta integrate everything in the river catchment that influences sediment supply – lithology, relief, tectonics, climate, land cover, and land use – as well as river discharge and transport capacity. The greater the sediment load coming down the river, the more likely the river is to build a delta. Conversely, the stronger the currents in the receiving water body, the more likely they are to rework the river sediments and dominate the size and form of the delta or to completely remove the river sediments.

(2) *Relative water density.* The relative densities of river and receiving waters reflect salinity, temperature, and concentration of suspended sediment. When the densities are similar (*homopycnal flow*), the river and basin water mix in three dimensions close to the shoreline, and much of the sediment is deposited close to the river mouth (Bridge 2003). This is most likely to occur where a river enters a freshwater lake. River water denser than basin water (*hyperpycnal*) – such as very cold or turbid water entering a freshwater lake – will flow as a bottom current, mixing only along the lateral margins and carrying sediment farther into the basin. Hyperpycnal flows can also occur where a river enters an ocean, as on China's Huanghe (Yellow River) (Yang et al. 2018). River water of lower density than basin water (*hypopycnal*) – such as freshwater entering saline water – can flow over the basin water as a buoyant plume, mixing along the bottom and sides, and carrying sediment farther into the basin than does homopycnal flow. This type of flow is common on the delta of the Mississippi River (e.g. Keller et al. 2017).

(3) *Currents.* Lakes and enclosed seas typically experience lower energy waves and tidal currents, although storm-generated winds can temporarily enhance wave energy. Deltas in these

lower-energy environments are likely to reflect predominantly river processes. River mouths entering long, narrow embayments such as the Bay of Fundy on Canada's southeastern coastline, or semi-enclosed seas such as the North Sea in the Atlantic or the Sea of Cortez off northwestern Mexico, can have very strong tidal currents that cause tidal bores to travel up the bay and into the entering river, strongly influencing delta shape and size (Bonneton et al. 2015).

(4) *Relative change in base level.* Changes in relative base level influence river gradient and transport capacity, the location of the shoreline, and the energy of currents in the receiving basin. Either rise or fall of relative base level can shift the locations of active erosion and deposition across a delta. An example comes from a Pleistocene-age lake delta in northwestern Germany, in which periods of lake-level rise correspond to vertically stacked delta systems as depositional centers shifted upslope, and periods of fall correspond to the development of a single incised valley with coarse-grained delta lobes in front of the valley (Winsemann et al. 2011).

One of the more commonly used classifications of deltas emphasizes the relative importance of river processes or reworking by the receiving body (Figure 7.7) (Ritter et al. 2011). *High-constructive deltas* reflect predominantly river deposition; these are sometimes known as *river-dominated deltas*

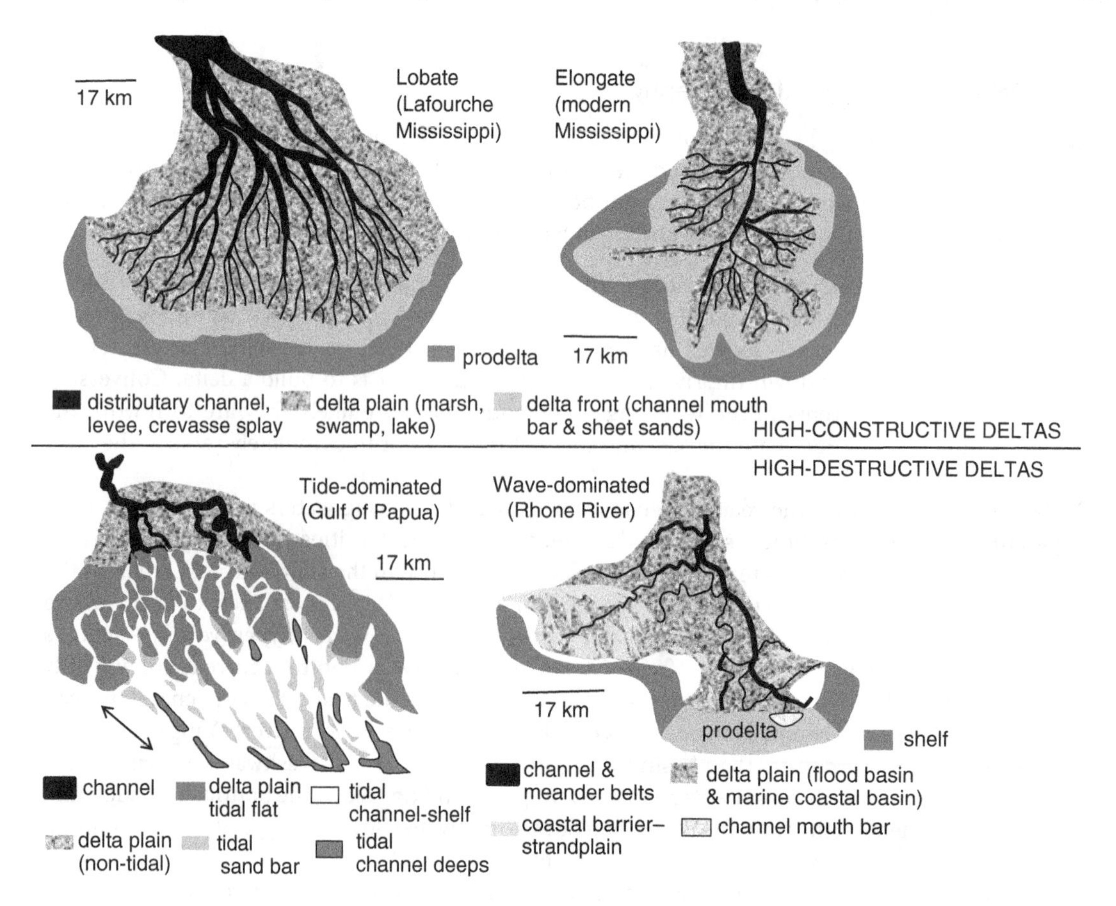

Figure 7.7 Basic delta types. Source: After Ritter et al. (2011), Figure 7.35, p. 299.

(Fagherazzi 2008). River-dominated deltas can be *elongate*, with greater length (in the downstream direction) than width, like the contemporary delta of the Mississippi River. Elongate deltas have more silt and clay, and subside rapidly once deposition ceases or shifts elsewhere on the delta. River-dominated deltas can also be *lobate*, with greater width than length and slightly coarser sediment that subsides more slowly upon abandonment, allowing time for waves and tides to rework the sandy sediment. The now-abandoned Holocene deltas of the Mississippi River are lobate.

High-destructive deltas are shaped predominantly by currents in the receiving water body. Sediment is moved alongshore to form beach ridges and arcuate sand barriers in *wave-dominated deltas* such as those of the Rhone and Nile Rivers. *Tide-dominated deltas* have strongly developed tidal channels, bars, and tidal flats, with sand deposits that radiate linearly from the river mouth, such as those in the Gulf of Papua (Ritter et al. 2011).

Any type of delta can be subdivided into delta plains and a delta front. For most deltas, the *delta plain* will be much larger. This is an extensive area of low slope crossed by distributary channels. Delta plains include subaerial portions of floodplains, levees, and crevasse splays, and subaqueous portions of marshes, lakes, interdistributary bays, tidal drainage channels, and tidal flats (Bridge 2003). The gradient of the delta plain correlates with the ratio of sediment supply to sediment retention on the delta, sediment concentration (a proxy for delta plain sedimentation), and mean water discharge (Syvitski and Saito 2007). The delta plain gradient increases as sediment accumulates on the plain, but decreases as river discharge increases. Delta plains are more affected by river currents than by wave and tidal currents, and thus resemble inland floodplains (Bridge 2003). Deposition along the lower portions of tributary channels can be enhanced during periods of high tides or storm surges, leading to increased frequency of avulsion (Bridge 2003).

Delta plains can have extensive vegetation cover that varies with climatic setting, rate of deposition, and – down the delta – salinity gradient. Some of the world's great deltas are covered in dense subtropical forest and mangrove swamps (e.g. the Sundarbans at the mouth of the Ganges in India) (Rahman et al. 2015). Other deltas have primarily sedges, rushes, and marsh plants (e.g. the Mississippi River delta in the United States, the Pearl River delta in China), rather than woody vegetation (e.g. Baustian et al. 2017). Regardless of the specific plant communities, vegetation cover strongly increases the resistance of delta sediments to erosion (e.g. Hood 2010). Removal of vegetation thus typically results in much faster and more severe delta erosion, as well as less efficient trapping and storage of incoming sediment.

The *delta front* is the steeper, basinward portion of a delta, where sediment-carrying river water enters open water. High rates of deposition and limited reworking by currents in the receiving basin can lead to steepening of the delta front beyond the angle of repose of the sediment, causing a sediment gravity flow that can develop into a turbidity current. A *turbidity current* is analogous to a subaqueous debris flow, in that it is a relatively dense, flowing slurry of sediment that follows bottom topography and can travel farther into the receiving basin than suspended sediment (e.g. Heezen et al. 1954; Symons et al. 2017).

Deposition on a delta front is episodic, as is deposition on the basin floor beyond the delta. Stratigraphy in a prograding delta that is progressively building into the receiving basin is characterized by *bottomsets* – fine-grained, low-inclination deposits of the basin floor; *foresets* – deposits of the delta front at the angle of repose; and *topsets* – relatively flat-lying and coarse deposits of the delta plain (Figure 7.8) (Bridge 2003). Gilbert (1885) introduced these terms, so deltas with this type of stratigraphy are known as *Gilbert-type deltas*. Edmonds et al. (2011) refer to Gilbert-type deltas as

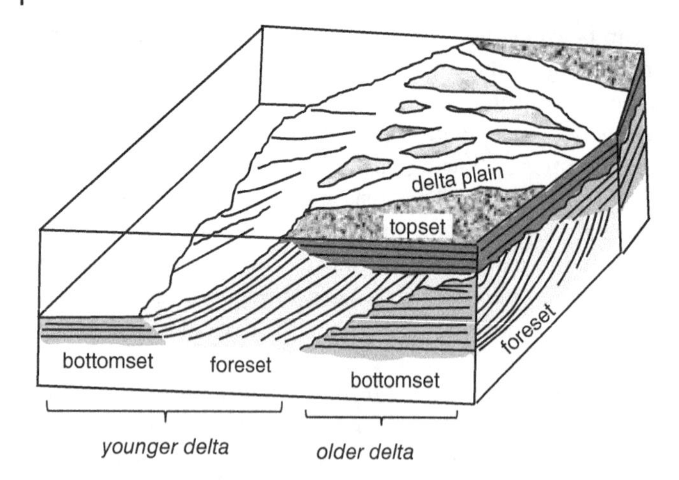

Figure 7.8 Schematic block diagram of a Gilbert-type or foreset-dominated delta, showing the three primary types of beds, the delta plain, bars, and distributary channels. River flow enters a body of standing water from the right.

foreset-dominated deltas, and distinguish them from *topset-dominated deltas*, in which the distributary channels incise into pre-delta sediment.

As with other fluvial processes, delta dynamics can be studied using physical experiments (e.g. Edmonds et al. 2009; Muto et al. 2012) and numerical simulations (e.g. Canestrelli et al. 2010; Geleynse et al. 2011; Martyr-Koller et al. 2017). Fagherazzi and Overeem (2007) provide a thorough review of numerical models of delta dynamics. Numerical models vary in computational complexity from reduced-complexity models that use a weighted random walk method to route water and sediment flux (Liang et al. 2015), through two-dimensional cross-shelf models that focus on the evolution of the continental shelf profile and ignore along-shelf variations in morphology, and two-dimensional along-shelf models that focus on the evolution of the along-shelf morphology and consider either a constant cross-shelf profile or integrate across the shelf. Three-dimensional models include pseudo-three-dimensional shelf-area models based on depth-averaged formations of hydrodynamics and sediment transport and fully three-dimensional models that reproduce the full structure of the flow field coupled with sediment transport.

7.4.3 Paleoenvironmental Records

The depositional environment that is a delta integrates upstream changes in yields of water, sediment, and other materials. Consequently, delta morphology and stratigraphy are commonly examined as a source of information on catchment paleoenvironments. The length of paleoenvironmental record that can be accessed in this manner depends partly on the age and erosional history of the delta (in the sense of erosion removing portions of the depositional record) and partly on the techniques used to examine deltas. Changes in sedimentary facies, mineralogy, and isotopic ratios, as well as in the location of delta sublobes, are used to infer changes in depositional processes driven by either upstream changes in water and sediment yield or changes in coastal wave and tide energy (Krom et al. 2002; Hori et al. 2004; Hori and Saito 2007; Liu et al. 2010). Chronologies of delta deposition can be constrained using luminescence and radiocarbon techniques (Hori and Saito 2007; Tamura et al. 2012; Chamberlain et al. 2017).

Over shorter time periods, changes in delta morphology and process can be studied using aerial photographs, satellite images, and repeat ground-based surveys (e.g. Tamura et al. 2010; Anthony et al. 2015). Distinct spatial patterns within a delta, reflected in statistical differences in parameters such as island size, shape factor, and aspect ratio, suggest that planform information extracted from satellite imagery carries the signature of processes responsible for vegetation and delta formation and change (Passalacqua et al. 2013).

7.4.4 Deltas in the Anthropocene

As with other aspects of river networks, many deltas have recently undergone major changes as a result of direct and indirect human alterations of incoming water and sediment yield and of the delta distributary network (e.g. Syvitski et al. 2009). Many deltas are densely populated and heavily farmed – close to half a billion people live on or near deltas, and an estimated 25% of the world's population lives on deltaic lowlands (Tamura et al. 2012). These people are increasingly vulnerable to changes in delta morphology and dynamics.

One of the primary changes is sediment compaction and delta subsidence, so that deltas are sinking more rapidly than sea level is rising (Syvitski et al. 2009; Wang et al. 2012; Anthony et al. 2015). Delta subsidence results from several factors. Sediment consolidates with time as pore water is expelled. Trapping of sediments and river flow in reservoirs upstream from a delta reduces ongoing inputs that would otherwise offset progressive compaction and subsidence (e.g. English et al. 1997; Jun et al. 2010; Nageswara Rao et al. 2010; Yang et al. 2011; Stanley and Clemente 2017). Removal of oil, gas, and water from delta sediments reduces pore pressure and facilitates compaction. Floodplain engineering on deltas, including channelization, limits overbank deposition and conveys sediment more efficiently beyond the delta and into the ocean (Giosan et al. 2013). Diversion of water and sediment flows either away from a delta or to a different portion of the coastline exacerbates delta shifting, causing the abandoned delta to compact, subside, and erode (Jabaloy-Sanchez et al. 2010; Restrepo and Kettner 2012). And, rising sea level subjects deltas to more intense erosion by wave and tidal currents, as well as storm surges. On many deltas, multiple human activities affect morphodynamics (e.g. Canuel et al. 2009; Syvitski et al. 2013; Tessler et al. 2016).

Data from 33 deltas around the world indicate that 85% experienced severe flooding during the first decade of the twenty-first century, causing the temporary submergence of 260 000 km^2 (Syvitski et al. 2009). Conservative estimates indicate that the amount of delta surface area vulnerable to flooding could increase by 50% under current projections of sea-level rise during the twenty-first century. Coastal wetlands associated with deltas are also being lost to submergence or erosion. A 2005 survey of 42 deltas across the globe indicated the loss of nearly 16 000 km^2 of wetlands during the preceding 20 years, with an average rate of loss of 95 km^2 per year (Coleman et al. 2005). Diverse assessments of coastal wetland loss estimate that between 46 and 63% has been lost globally, predominantly during the past century (Spencer et al. 2016).

Mining of construction aggregate, land reclamation, and changes in delta morphology can also influence the pathways of water and sediment across a delta, as well as delta submergence and the ability of waves and tides to rework delta sediments. Large-scale sand excavation from China's Pearl River delta since the mid-1980s, for example, has resulted in increased water depth and lowered streambed elevation (Zhang et al. 2010). This has led to an increased tidal prism and upstream movement of the tidal limit, which have facilitated increased saltwater intrusion into the estuary.

Some deltas changed little during the twentieth century, and their aggradation rate remains in balance with, or exceeds, subsidence or relative sea-level rise. Examples include deltas on the Amazon (Brazil), Congo (Democratic Republic of Congo), Orinoco (Venezuela), Fly (Papua New Guinea), and Mahaka (Borneo) rivers (Syvitski et al. 2009). These are rivers with a lower density of human settlement in the upstream drainage basin and minimal flow and sediment regulation. Deltas with minimal human manipulation provide a control against which to evaluate the effects of direct human alteration of deltas within the context of changing climate and sea level.

Human-induced changes are exacerbated where very little or no river flow now reaches a delta as a result of consumptive water uses upstream. Deltas subject to such a severe change in flow regime include those on the Indus River in southern Asia (McManus 2002) and the Colorado River in the United States and Mexico (Shafroth et al. 2017). Where mitigation is being undertaken, experimental high-flow releases (Shafroth et al. 2017) and controlled diversions of water and sediment (Nittrouer et al. 2012) can be used to replenish water and sediment supplies to deltas.

In addition to changes in morphology caused by human alterations, delta sediments can be locally contaminated by toxic synthetic chemicals such as pesticides (Feo et al. 2010) and petroleum byproducts (Wang et al. 2011). Any portion of a river network can be affected by such contaminants, but deltas are particularly vulnerable because they integrate diverse upstream sources of contamination. Contaminants can affect the delta environment and, when remobilized by wave or tidal erosion, can be spread into the nearshore environment. Naturally occurring contaminants, such as arsenic associated with highly reducing deltaic sediments in the Bengal and Mekong deltas in southern and southeastern Asia, can also strongly influence delta environments (Berg et al. 2007).

7.5 Estuaries

An *estuary* represents an incised river valley along the coastline of an ocean. The river valley incised during a period of lower sea level that caused base-level lowering of the river, typically during a period of continental ice-sheet formation. Subsequent sea-level rise during ice-sheet melting and retreat backflooded the estuary. The typical scenario is that the estuary grows progressively shallower and less subject to marine influence as terrestrial and marine sediments accumulate. An estuary that is a sediment sink is also referred to as a *microtidal estuary* (Cooper 2001). Usually, estuarine filling occurs during or shortly after backflooding on sediment-rich rivers. Estuary filling can require time periods longer than the Holocene on rivers with very low sediment loads, where river floods periodically scour the estuary, or where sediment is primarily transported through the estuary to the marine environment (Cooper 2001; Cooper et al. 2012).

Numerous factors influence the form and processes of estuaries (Dalrymple and Choi 2007). Among the most fundamental is the bathymetry, which ranges from a relatively shallow-water, channelized environment landward of the coast to deeper, unconfined settings on the shelf. Another influence comes from the source of the energy that moves sediment, which can range from predominantly river currents to tidal, wave, and oceanic shelf currents. The frequency, rate, and direction of sediment movement influence estuarine form and process. Sediment movements vary from unidirectional and continuous to seasonal or episodic when driven by rivers. Sediment movements can repeatedly reverse in tidal settings, and can be episodic and coast-parallel when driven by waves. Sediment movements also alternate between onshore and offshore in tide-dominated shelf environments. Water salinity, which ranges from fresh, through brackish, to fully marine, also influences estuarine process and form.

Given this breadth of river and coastal influences, estuaries exhibit significant diversity. Most estuaries, however, possess a three-part structure composed of (i) an outer, marine-dominated portion where the net bedload transport is headward, (ii) a relatively low-energy central zone with net bedload convergence, and (iii) an inner, river-dominated (but still marine-influenced) portion where net sediment transport is seaward (Dalrymple et al. 1992). Each zone is developed to differing degrees among diverse estuaries as a reflection of site-specific sediment availability, gradient of the coastal zone, and the stage of estuary development.

Various classifications exist for estuaries. Examining estuaries in South Africa, for example, Cooper (2001) distinguishes five types based on contemporary morphodynamics. *Normally open estuaries* maintain a semipermanent connection with the open sea. The three primary categories of *open estuaries* are barrier-inlet systems maintained by river discharge (river-dominated estuaries) and tidal discharge (tide-dominated estuaries), and open estuaries that lack a supratidal barrier because of inadequate availability of marine sediment. *Closed estuaries* are separated from the sea for long periods by a continuous supratidal barrier. Perched closed estuaries develop behind high berms and maintain a water level above high-tide level, whereas nonperched closed estuaries develop behind barriers of lower elevation with wide beach profiles.

Dalrymple et al. (1992) distinguish simply wave- and tide-dominated estuaries. *Wave-dominated estuaries* typically have three well-defined portions: a marine sand body composed of barrier, washover, tidal inlet, and tidal delta deposits; a fine-grained central basin; and a bay-head delta that experiences tidal or salt-water influences (Figure 7.9). Examples of wave-dominated estuaries include the Hawkesbury Estuary and Lake Macquarie in Australia, eastern shore estuaries in Novia

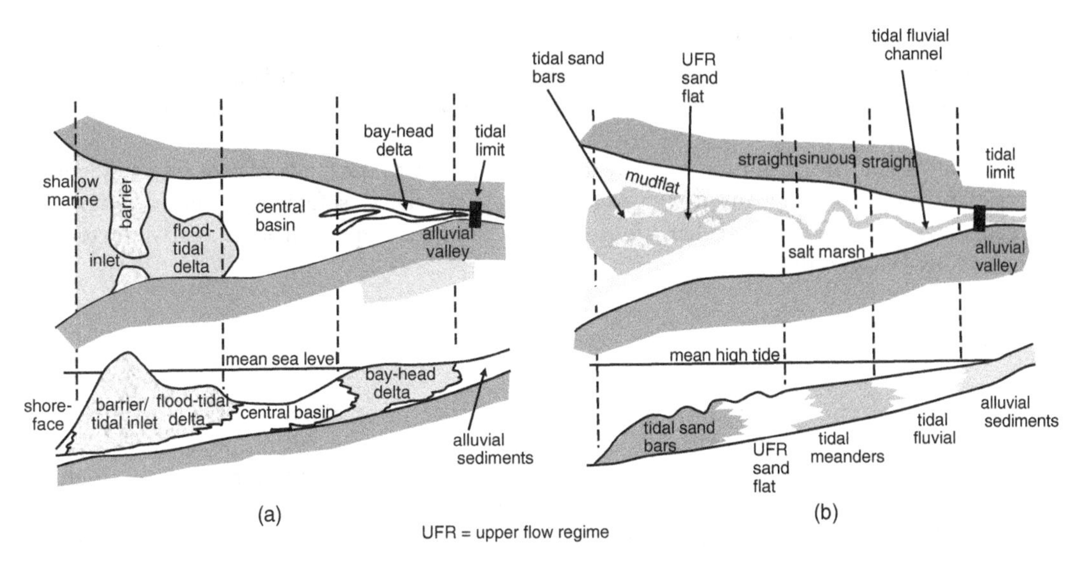

Figure 7.9 Distribution of morphological elements in plan view and sedimentary facies in longitudinal section within an idealized (a) wave-dominated and (b) tide-dominated estuary. For (a), the barrier/sand plug is shown as attached to the headland, but it may not be connected on low-gradient coasts, where it can be separated from the mainland by a lagoon. The longitudinal section represents the onset of estuary filling after a period of sea-level rise. For (b), the longitudinal section is along the axis of the channel and does not show all sedimentary facies present. Source: After Dalrymple et al. (1992), Figures 4 and 7.

Scotia and the Miramichi River in Canada, and San Antonio Bay and Lavaca Bay in the United States (Dalrymple et al. 1992). *Tide-dominated estuaries* have a marine sand body composed of elongate sand bars and broad sand flats; a central zone of tight meanders where bedload is transported by flood-tidal and river currents; and an inner, river-dominated zone with a single, low-sinuosity channel. Examples of tide-dominated estuaries include the Adelaide and Ord Rivers in Australia, the Cumberland Basin and Avon River in Canada, and Alaska's Cook Inlet in the United States (Dalrymple et al. 1992).

Estuaries share some characteristics with deltas in that they include depositional surfaces on which tidally influenced channels form. Numerous studies demonstrate that these tidal channels are funnel- or trumpet-shaped, such that channel width tapers upstream in an approximately exponential fashion (Chappell and Woodroffe 1994; Fagherazzi and Furbish 2001; Savenije 2005). The shape of the exponential width profile relates to the mouth width and river output. Channels and estuaries with larger mouths have a more pronounced funnel shape than do systems with narrower mouths, and higher river water and sediment discharge create a longer funnel (Davies and Woodroffe 2010).

As in the case of deltas, human coastal settlement and resource use concentrate around estuaries. Estuaries play an important role in the cycling of carbon and other nutrients through the exchange and modification of terrestrial organic matter transported by rivers to the ocean (Canuel et al. 2012; Alongi 2014). Because estuaries typically retain organic matter, these environments have high rates of primary production, and support abundant and diverse wetlands and fisheries (Boerema and Meire 2017). Dramatically increased fluxes of nitrogen to estuaries as a result of livestock wastes, industrial-scale application of agricultural fertilizers, and other human activities have caused many estuaries to experience eutrophication (Green et al. 2004; Paerl et al. 2014). *Eutrophication* is an ecosystem response to increased nutrients, typically involveing a substantial increase in phytoplankton and associated decrease in dissolved oxygen, which together cause *hypoxia*, or so-called "dead zones." At least 375 hypoxic coastal zones have been identified around the world, primarily around western Europe, the eastern and southern coasts of the United States, and Asia (CENR 2000).

Many of the same techniques used to study deltas are applied to the study of process and form in estuaries. Bathymetric mapping of submerged features is particularly useful in studying estuaries (e.g. Ganju et al. 2017), and navigational charts for individual estuaries can date from the early nineteenth century (van der Wal and Pye 2003).

7.6 Summary

The fluvial landforms discussed in this chapter – floodplains, terraces, alluvial fans, deltas, and estuaries – are inherently dynamic and rapidly changing environments. Ever-shifting balances among different processes of deposition and erosion can reflect changes external to the landform, such as variation in climate or base level. Shifting erosion and deposition can also reflect internal adjustments within the environment of the landform even as the external setting remains relatively stable, as when continued development of levee crevasses results in delta shifting. This continual variability can make fluvial landforms difficult to interpret, but floodplains, terraces, fans, and deltas are nonetheless valuable repositories of information on river dynamics over time spans of tens to hundreds of thousands of years.

Human activities increasingly alter process and form in these extra-channel environments. Indirect effects associated with climate change and rising sea level combine with direct effects including flow regulation, levee construction, and channelization to change fluxes of water, sediment, solutes, and large wood between channels and extra-channel environments. Altered fluxes – altered connectivity – have in many cases led to reduced water quality, lower biological productivity and ecological sustainability, and increased hazards to human populations and infrastructure as a result of eroding landforms. The great challenge of the future is to understand how past human actions have changed river networks and how undesirable changes can be mitigated or reversed as we go forward.

8

Rivers in the Landscape

This final chapter looks at interactions between rivers and landscapes across varying time and space scales, starting with geological time scales of 10^3–10^7 years and entire continents. The second section explores distinctive climatic signatures of river process and form associated with high latitudes, low latitudes, and warm drylands. The third section examines the significance of spatial context and the potential for relatively abrupt spatial transitions in process and form within a drainage basin. The fourth section returns to the idea of connectivity, which was introduced in Chapter 1, and explores the implications of connectivity for the diverse river processes and forms discussed throughout the book. The fifth section examines contemporary river management in the context of the diverse topics introduced in earlier chapters.

Preceding chapters have introduced the basic processes of water, solute, sediment, and large wood movement into and through channel networks and the river forms and adjustments that result through time. This final chapter returns to the idea that river process and form reflect not only physics, but also distinctive processes and forms associated with a specific geographic location and its history.

8.1 Rivers and Topography

Having examined the details of how water, sediment, and large wood move down hillslopes and into channels, and then through a river network, it is useful to step back and consider the larger-scale distribution of rivers across continents. Important insights into interactions between rivers and topography can be gained by examining river configuration.

The manner in which rivers both respond to and shape surrounding topography has been investigated systematically for more than a century. Early work explored why some rivers cut through mountain ranges rather than simply flowing downward from high points in the landscape. Significant questions at the time of this late-nineteenth-century research included, What is the role of rivers (as opposed to glaciers or other processes) in cutting deep river canyons?, and How do large-scale structures (mountains, deep canyons) relate to movements of Earth's crust? Subsequent research has emphasized (i) how redistribution of mass at the surface by river erosion can influence redistribution of subsurface mass via movements of molten material in the crust and tectonic movements, and (ii) how river gradient and channel width can be used as indicators of spatial variations in rock uplift.

The most obvious topographic influences on river process and form occur in mountainous environments where rivers cut deep, narrow canyons as they flow down toward adjacent lowlands. For

Rivers in the Landscape, Second Edition. Ellen Wohl.

more than a century, however, investigators have recognized that rivers do not always follow topography. Among rivers that cut across mountain ranges in the western United States, Powell (1875, 1876) distinguishes *antecedent* drainage networks in which pre-existing channels maintained their spatial arrangement while the underlying landmass was deformed and uplifted, and *superimposed* channels, which incised downward to a buried structure. In both cases, the river flows through or across the mountain range, rather than being a consequence of the topography.

Today, these two types of drainages are commonly referred to as *transverse drainages*, which cut across bedrock topographic highs such as anticlines and upwarps (Douglass and Schmeeckle 2007). Transverse drainage subsumes antecedent and superimposed drainages, overflow, and drainage piracy. In *drainage piracy*, a drainage network on one side of a bedrock high erodes headward sufficiently to lower or breach the drainage divide and divert flow from a network on the other side. An ongoing example is the Casiquiare Canal, which is not a canal but rather a natural channel, a tributary to the Rio Negro in the northern extent of the Amazon River drainage in South America. The canal connects the Rio Negro and the Amazon with the adjoining Rio Orinoco drainage as a result of drainage piracy, with the canal capturing a portion of the Orinoco's headwaters (Eden 1971).

Research since the nineteenth century has developed tools that illuminate the influences of tectonics on the spatial arrangement of rivers and the geometry of individual river channels, as well as the influences of river incision on tectonics and topography. The manner in which tectonic forcing creates topographic patterns and spatial variability of relief in river networks gives rise to the *inverse problem*, in which topographic features are used to infer tectonic uplift rates. This approach to understanding uplift is challenging because river response time to tectonic perturbations governs which tectonic events are preserved in topography (Goren et al. 2014). In addition, rates of river incision into bedrock exist in non-steady-state even over measurement intervals of 10^4–10^7 years as a result of episodic interruptions in river incision triggered by alluvial deposition (Finnegan et al. 2014).

The gradient and width of rivers incised into bedrock are the geometric parameters most commonly used to infer the spatial distribution and relative magnitude of tectonic forces. A river being incised into bedrock, rather than alluvium, implies that the channel's capacity to transport sediment exceeds the sediment supply (Howard 1980). Regardless of where they occur along a river's course, bedrock river segments typically have smaller width-to-depth ratios and steeper gradients than alluvial segments (Montgomery and Gran 2001; Wohl and David 2008). These differences in geometry effectively enhance the flow's limited ability to incise the channel bed and enlarge the channel cross-section relative to alluvial segments.

Bedrock river segments can be interpreted as geologically transient features that are not confined to headwaters. The middle and lower Danube River of Europe alternately flows across large alluvial basins and cuts through mountain ranges. The lower Mississippi River in the United States, commonly thought of as a fully alluvial river, is better described as a mixed bedrock–alluvial channel because of the presence of a cohesive, Pleistocene-age clay unit that influences river process and form in a manner analogous to bedrock (Schumm et al. 1994; Nittrouer et al. 2011).

The relative lack of erosive ability that produces a bedrock river segment can reflect greater erosional resistance where the river crosses a different lithology (Wohl 2000b) or changes in relative base level associated with base-level fall or with uplift of the drainage (Howard 1980; Seidl et al. 1994). Where the presence of bedrock river segments reflects enhanced incision in response to relative base-level fall, river incision is the primary nonglacial mechanism of transmitting base-level changes across the landscape (Hancock et al. 1998). Bedrock river incision steepens adjacent

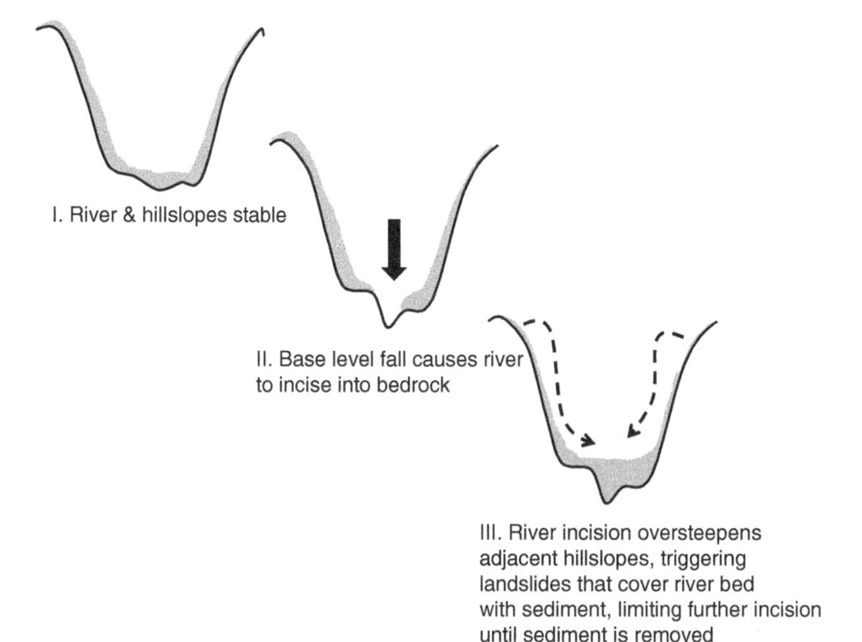

Figure 8.1 Illustration of the effects of bedrock-channel incision on landscapes. In these schematic views looking upstream within a river valley, base-level change triggers river incision, which affects the stability of adjacent hillslopes and tributaries. Gray shading represents hillslope regolith and valley alluvium.

hillslopes, increases topographic relief between summits and valley bottoms, removes mass from the landscape, and ultimately sets the rate at which the entire landscape evolves (Figure 8.1) (Howard 1994; Burbank et al. 1996; Hancock et al. 1998; Whipple et al. 2013).

The idea that hillslopes and rivers mutually adjust was first expressed by Gilbert (1877) and subsequently formally stated in the context of *dynamic equilibrium* as a condition in which "every slope and every channel in an erosional system is adjusted to every other. When the topography is in equilibrium and erosional energy remains the same all elements of the topography are downwasting at the same rate" (Hack 1960, p. 80).

8.1.1 Tectonics, Topography, and Large Rivers

Although much of the research summarized in the preceding section focused on rivers in mountainous terrain, investigators have also examined the spatial arrangement of very large rivers in the context of tectonic history and topography (Figure 8.2). The location and configuration of most of Earth's largest river basins reflect the tectonic assembly and deformation of continental land masses (Potter 1978). The basic configuration of some major rivers has been extremely persistent. The Mississippi drainage, for example, has existed about 250 million years, or about 1/16 of Earth's history (Potter 1978). This persistence partly reflects the control of deeper crustal structures.

The 28 largest rivers discharge across trailing-edge coasts without compressional deformation, which reflects the continental asymmetry of many watersheds (Inman and Nordstrom 1971). The mouths of many large rivers occupy grabens or crustal downwarps. The alignment of several major

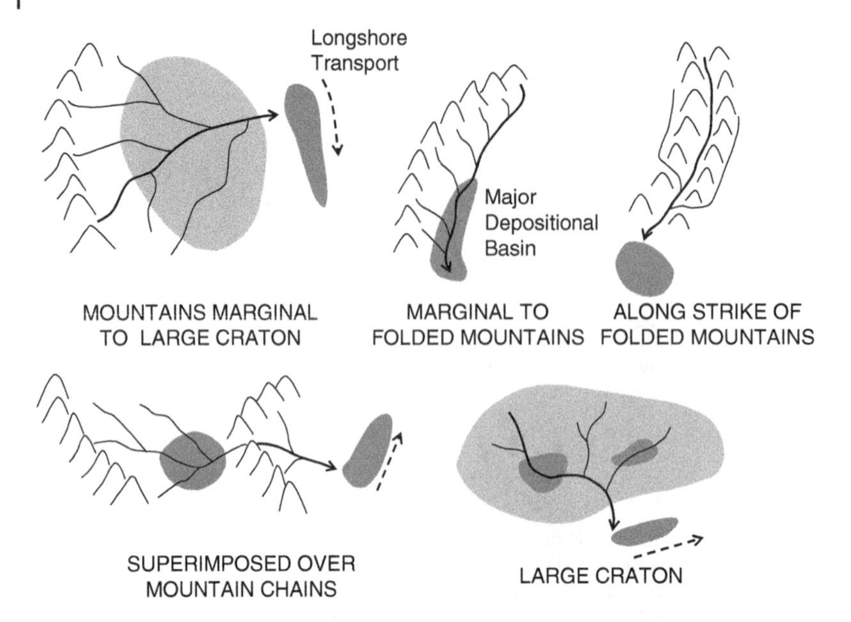

Figure 8.2 Types of big river settings. Lighter gray shading indicates basement bedrock; darker gray shading indicates major depositional areas. Source: After Potter (1978), Figure 9.

rivers, including the Amazon in South America (Potter 1978), the Nile in Africa (Schumm and Galay 1994), and the Rio Grande in North America, correlate with large-scale crustal fracture patterns.

The Amazon River provides an example of the three scales at which tectonics can influence large-scale features of a drainage network (Mertes and Dunne 2007; Dunne and Aalto 2013). At the continental-scale (5×10^3 km), the assembly of orogen (mountains), foreland basin, cratons, and grabens influences production of runoff, sediment supply, and accommodation space. Specifically, the Amazon heads on the Andean arc at the leading edge of the South American Plate. The Andes provide the majority of the sediment that the Amazon carries across a very broad lowland, before entering a graben that localizes the mouth of the river on the trailing edge of the continent.

At an intermediate scale (10^2–10^3 km), the spacing of crustal warping transverse to the river course influences gradient, valley width, channel sinuosity, accommodation space, and sediment distribution across the floodplain. Four major structural arches lie transverse to the main course of the Amazon between the Peru–Brazil border and the Atlantic Ocean (Mertes and Dunne 2007). As the Amazon crosses each of these arches, channel gradient increases, the valley grows narrower, and the channel becomes less sinuous.

At the local scale (10^1–10^2 km), brittle crustal fracturing influences channel orientation and gradient. Channel alignment follows fractures because these create localized zones of more readily eroded bedrock (Latrubesse and Franzinelli 2002; Roy et al. 2016).

These examples illustrate that, when examining river process and form, it is important to remember that deeper crustal structures and geologic processes occurring over millions of years can strongly influence river characteristics. These influences may be most readily detected in mountainous portions of a river network, but can also influence the world's largest lowland rivers, such as the Amazon and the Mississippi.

8.1.2 Indicators of Relations Between Rivers and Landscape Evolution

W.M. Davis (1899) first attempted to relate river geometry to landscape evolution in a conceptual model known as the *cycle of erosion*. This model posited high topographic relief and rivers with steep gradients in geologically young landscapes. As erosion gradually transferred mass to lower elevations, topographic relief and river gradients progressively decreased from mature to old landscapes.

The cycle of erosion, despite the name, assumed a highly linear landscape evolution with time and implied that an observer could readily interpret the relative geologic age of distinct landscapes based on their topography. This conceptualization is typically contrasted with the ideas of Davis' contemporary, G.K. Gilbert. Gilbert (1877) emphasized nonlinear change with periods of little net change, or equilibrium, as a result of feedbacks such as those subsequently recognized as tectonic aneurysms or isostatic rebound. *Isostatic rebound* is delayed upward flexure of Earth's crust in response to removal of mass such as a continental ice sheet that had previously depressed the crust. The implication of tectonic aneurysms or isostatic rebound is that elevation or relief may change relatively little over time spans of 10^3–10^4 years, despite continued erosion and transfer of mass to lower elevations.

Many subsequent investigators have demonstrated that rates of landscape change fluctuate substantially through time and space (e.g. Bierman and Nichols 2004; Hahm et al. 2014). Conceptual models now tend to emphasize that most landscape change occurs during relatively short periods of time and is concentrated in relatively small portions of a drainage basin, although net change in elevation or relief may be minor.

Topographic metrics are still used to infer relative rates or stages of landscape evolution. Among these metrics are *hypsometric curves*, which illustrate the distribution of mass within a basin by plotting proportion of total basin height against proportion of total basin area. Strahler (1952) proposes that these curves could be used to distinguish relative basin age as a function of decreasing hypsometric integral – the area under the hypsometric curve – with increasing age. Subsequent research suggests that hypsometric curves can be used to infer the history and processes of basin development. The distribution of mass within a basin reflects uplift rates and variations in erodibility of different lithologic units (Walcott and Summerfield 2008; Pérez-Peña et al. 2009), as well as differences in diffusive (hillslope) versus fluvial sediment transport (Willgoose and Hancock 1998) and glacial versus fluvial erosion (Sternai et al. 2011). Hypsometric curves are likely to be concave-down everywhere, for example, within landscapes dominated by diffusive transport (Willgoose and Hancock 1998). Glacial valleys are more likely to have concave-up curves than are fluvial valleys (Sternai et al. 2011).

8.1.3 Tectonic Influences on River Geometry

Increased availability of topographic data in the form of electronic digital elevation models (DEMs) greatly enhanced the ability to detect irregularities in river longitudinal profile starting in the 1990s. Profile irregularities can reflect downstream variations in lithology and erodibility (Valla et al. 2010), glacial history (Hobley et al. 2010; Gran et al. 2013), sediment inputs (Cowie et al. 2008), and rock uplift (Snyder et al. 2000; Li et al. 2019), so interpreting the significance of irregularities typically requires knowledge of other characteristics of the river environment (e.g. Marrucci et al. 2018). Where investigators have independent evidence of uplift rate, as in the central Apennines of Italy (Whittaker et al. 2008) or the Santa Ynez Mountains of California, USA (Duvall et al. 2004), steeper gradients strongly correlate with greater rates of rock uplift (Whipple et al. 2013).

Spatial variations in channel width-to-depth ratio are not as readily detected using remote information as are changes in river gradient, but variations in the width of bedrock channels can also reflect differential uplift (Whittaker et al. 2007a,b; Attal et al. 2008; Yanites et al. 2010), as well as changes in rock erodibility (Wohl and Merritt 2001; Spotila et al. 2015). Typically, segments of higher uplift or more resistant rock have deeper, narrow cross-sectional geometry. Modeling suggests that incorporating channel-width adjustment or sediment-transport dynamics decreases the sensitivity of a river profile to rate of rock uplift (Yanites 2018), indicating the interconnectedness of bedrock-channel process and form as they influence response to uplift.

Adjustments of gradient and width in response to increasing substrate resistance or uplift are typically tightly coupled (Whipple 2004; Stark 2006). The most commonly used approach is to interpret downstream variations in scaling relations among channel width w, drainage area A, gradient S, and discharge Q – in other words, downstream hydraulic geometry – as reflecting changes in rock erodibility or uplift rate (Duvall et al. 2004; Cowie et al. 2006; Jansen 2006; Whitbread et al. 2015). Scaling laws change along bedrock channels crossing an active fault in the central Italian Apennines, for example (Whittaker et al. 2007a), and along rivers crossing growing folds in New Zealand (Amos and Burbank 2007).

8.1.4 Effects of River Incision on Tectonics

Decades of research indicate that rivers can convey tremendous volumes of sediment from high-relief landscapes. Rivers remove up to five times more sediment per unit area from mountainous basins than from lowland basins (Corbel 1959; Willenbring et al. 2013). An estimated 96% of the approximately 7819 million tons of sediment delivered to the oceans by rivers each year originates in mountainous settings (Milliman and Syvitski 1992).

By the 1990s, investigators realized that one implication of this ability to remove mass from mountainous regions is that river incision can affect crustal structure in mountain belts by changing the distribution of stress in the crust (Molnar and England 1990; Hoffman and Grotzinger 1993; Beaumont and Quinlan 1994; Small and Anderson 1998). Local rheological variations arise in a deforming orogen as a result of deep and rapid incision by glaciers or rivers (Zeitler et al. 2001). The crust weakens as the strong upper crust is locally stripped from above by erosion. This causes the local geotherm (or rate of change in temperature with depth below the surface) to steepen from below as a focused, rapid uplift of hot rock occurs. In other words, incision by glaciers or rivers removes enough mass that molten material rises preferentially from Earth's interior into the eroding area. If efficient erosion continues, material continues to flow into the weakened zone, maintaining local elevation and relief (Koons et al. 2002; Booth et al. 2009a,b). This conceptualization of the interactions between river erosion, uplift, and topography is known as the *tectonic aneurysm model* (Zeitler et al. 2001).

A river's ability to incise depends partly on discharge and the climate that supplies runoff. Contemporary research emphasizes strong coupling among climate, erosion, and tectonics. Gradients in climate (Bookhagen and Burbank 2010) and tectonic forcing influence erosional intensity, which governs the development of topography, which in turn influences climate and tectonics (Roe et al. 2002). This is demonstrated in the Himalaya in southern Asia (Montgomery and Stolar 2006), where erosion along major rivers causes focused rock uplift. The uplift creates anticlines, and the anticlines correlate with local rainfall maxima because monsoon precipitation is advected up the river valleys. The greater rainfall in turn increases the erosive capability of the rivers (Figure 8.3). In some regions, however, climate can be decoupled from topography, so that topography reflects predominantly tectonic forces (e.g. Forte et al. 2016).

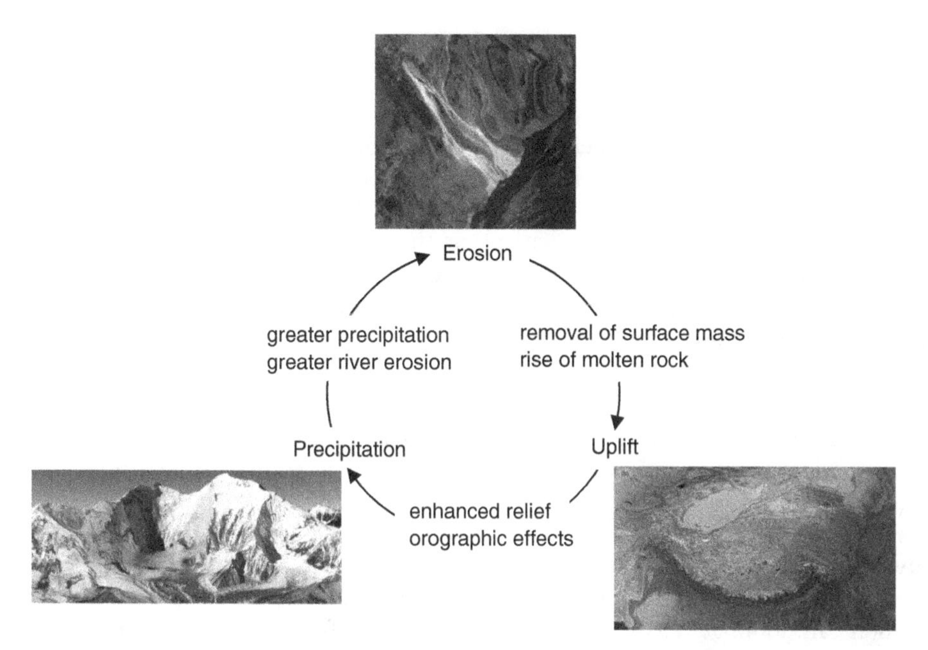

Figure 8.3 Schematic illustration of the interactions among tectonics, topography and climate, as illustrated by research in the Himalaya in southern Asia. The inset photos illustrate (clockwise from top) a landslide along a valley wall in the Nepalese Himalaya, the Tibetan Plateau as seen from space, and snow, ice, and glaciers around Mount Everest. Source: Inset photographs courtesy of Google Earth. (*See color plate section for color representation of this figure*).

8.1.5 Bedrock-Channel Incision and Landscape Evolution

The processes and rates of bedrock-channel incision are integral to landscape evolution. As noted previously, upstream transmission of relative base-level fall can be limited by the rate at which a river incises into bedrock, and the rate of channel incision at least partly governs the stability of adjacent hillslopes. Consequently, the rate and spatial distribution of bedrock-channel incision in response to relative base-level lowering limit the rate of adjustment for the entire drainage basin (e.g. DiBiase et al. 2018).

Most investigations of bedrock-channel erosion infer rates of downcutting based on terrace ages or cosmogenic nuclide dating of bedrock surfaces. Such approaches integrate all erosive processes over time spans of hundreds of thousands to millions of years. Limited studies have directly measured bedrock erosion using high-precision repeat surveys (Hartshorn et al. 2002; Beer et al. 2015) and erosion pins (Stock et al. 2005; Turowski and Cook 2017). Directly measured short-term rates range from 4 to 400 mm/y (Tinkler and Wohl 1998) and greatly exceed long-term estimates of erosion rates, which range from 0.005 to 10 mm/y (Tinkler and Wohl 1998). This discrepancy suggests that bedrock-channel erosion is episodic over longer time periods.

Numerous equations have been proposed to model reach- to basin-scale bedrock erosion. Tests of these equations come from field and flume data. The simplest and earliest formulation, which remains widely used, is known as the *stream-power incision law* (Howard 1980) because it equates average erosion rate, E, with total stream power, substituting the more readily measured drainage area, A, for discharge

$$E = k\,A^m\,S^n \tag{8.1}$$

where S is gradient and k is a constant that includes the inherent bed erodibility and magnitude and frequency characteristics of the flow. If erosion rates were directly proportional to bed shear stress, m would equal 0.38 and n would equal 0.81 (Howard 1980). Channels described by Eq. (8.1) are known as *detachment-limited* because erosion depends on the erodibility of the bedrock. Under *transport-limited* conditions, volumetric transport capacity, Q_{eq}, is a function of stream power, sediment flux is equal to transport capacity, and erosion or deposition rate equals the downstream divergence of sediment flux (Willgoose et al. 1991)

$$Q_{eq} = K_t\,A^m\,S^n \tag{8.2}$$

where K_t is a sediment transport coefficient and m and n are area and slope exponents as in Eq. (8.1). The distinction between detachment- and transport-limited conditions is incorporated in some numerical models (e.g. Davy and Lague 2009).

Numerous field studies now suggest that a single rate law cannot approximate incision across an entire channel network or among diverse networks because of differences in the dominant mechanism of erosion (Seidl and Dietrich 1992). Debris-flow abrasion can dominate steep tributaries, knickpoint propagation can dominate channels with rapid base-level fall, and fluvial abrasion or macroabrasion described by stream-power incision laws can dominate channels with relatively stable base level (Stock and Montgomery 1999; van der Beek and Bishop 2003; Cook et al. 2009). Ongoing work also calls into question some of the assumptions in the stream-power law (e.g. Lague 2014). A stochastic-threshold stream-power model, for example, in which a stochastic distribution of discharges is used to define a discharge threshold for channel erosion, better explains observed erosion rates in the Himalaya than does a simple stream-power model using drainage area or mean annual runoff (Scherler et al. 2017). A global compilation of more than 1400 basin-averaged erosion rates using ^{10}Be indicates that the slope exponent in the stream-power law is generally >1, indicating that the relationship between erosion rate and channel gradient is nonlinear and supporting the hypothesis that bedrock-channel incision is a threshold-controlled process (Harel et al. 2016).

Equations for bedrock-channel erosion are commonly incorporated as one component of landscape-evolution models, which also include equations for processes such as bedrock weathering and diffusive sediment transport on hillslopes (Figure 8.4). Tucker and Hancock (2010) review the evolution of landscape-evolution models from late-nineteenth-century qualitative conceptual models such as W.M. Davis' cycle of erosion through contemporary quantitative numerical simulations, emphasizing the rapid advances in numerical simulation since circa the 1980s. "Model" is now used to refer to both the underlying theory and the computer programs that calculate approximate solutions to the governing equations of landscape evolution.

A landscape-evolution model includes representations of:

- continuity of mass;
- the production and transport of dissolved and particulate sediment on hillslopes;
- runoff generation and routing of water across the landscape;
- erosion and transport by water and water–sediment mixtures; and
- numerical methods used to discretize space and iterate forward in time to solve the governing equations (Tucker and Hancock 2010).

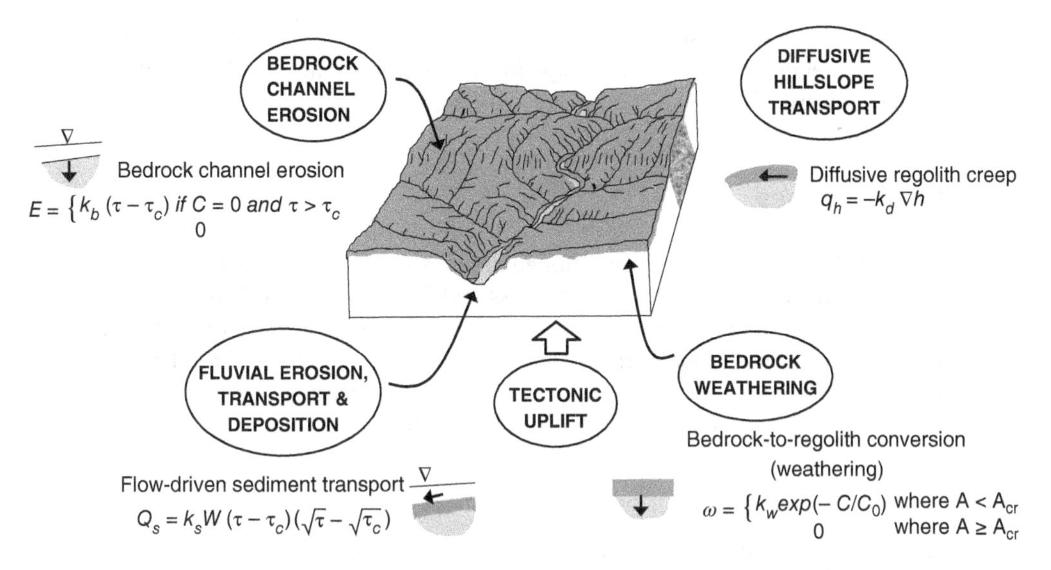

Figure 8.4 Schematic illustration of the components of a landscape-evolution model, including primary processes as described by equations in the model. In the equations, h is elevation, C is regolith thickness, ω is the lowering rate of the bedrock surface due to weathering, q_h is hillslope sediment transport rate per unit slope width, Q_s is volumetric overland or channelized sediment transport rate, τ is shear stress, τ_c is critical shear stress for erosion by flow, E is bedrock erosion rate, and C_0, k_b, k_d, k_w, and k_s are constants. Source: After Tucker and Slingerland (1997), Figures 1 and 2.

Initial and boundary conditions on the system, such as climate forcing, base-level control, and substrate characteristics, must also be specified.

The development of landscape-evolution models has required the development of governing equations, sometimes known as *geomorphic transport laws*, for weathering of bedrock, hillslope mass movements, hillslope sediment diffusion, overland and channelized flow and associated erosion, sediment transport, and channel geometry. These equations are solved using finite-difference (e.g. DELIM; Howard 1994), finite-volume (Tucker et al. 2001; Campforts et al. 2017), finite-element (Simpson and Schlunegger 2003), or cellular automaton (CA) methods (e.g. CAESAR; Coulthard et al. 1997; Hancock et al. 2015).

A finite difference is a mathematical expression of the difference between two points that is used to numerically solve differential equations describing boundary value problems, such as fluxes between the points. A finite-volume method is based on computing fluxes in and out along the boundaries of a finite volume of space. A finite-element method is based on computing fluxes in and out along the boundaries of a small subdomain or finite element, which can be two-dimensional rather than a volume. CA models are based on a discrete universe made up of cells. Each cell has an internal state consisting of a finite number of information bits. The system of cells evolves in discrete time steps as automata that follow rules to compute their new internal state. The rule governing evolution of the system is the same for all cells and is a function of the states of the neighbor cells.

Regardless of the specific method, the basic idea underlying a landscape-evolution model is that the time rate of change of height in a spatial unit of the landscape reflects the difference between sediment inflows and outflows along the unit's boundaries. The unit is defined by discretizing the land surface into a regular or irregular grid (Tucker and Hancock 2010). Different types of mathematical expressions are then used to describe fluxes between grid elements. Models can be tested by comparing model output against landscape form and process measured over diverse time and space scales (e.g. Howard 1994).

8.2 Climatic Signatures

Many of the preceding chapters have briefly mentioned differences in river process and form in relation to climate. Climatically induced diversity in process and form has been relatively neglected in scientific river studies, however, because the great majority of studies have been conducted in the temperate latitudes where many scientists live. This imbalance and relative neglect of high latitudes, low latitudes, and drylands has been changing in recent decades, and this section reviews the implications of climatic differences for interactions between process and form.

8.2.1 High Latitudes

The salient feature of high latitudes with respect to interactions between river process and form is the occurrence of very cold temperatures that maintain seasonal ice cover on rivers and permanently frozen ground. As discussed in Section 3.2.8, the formation, presence, and break-up of ice create distinct hydrologic and hydraulic effects on stage and discharge. The presence of ice can also strongly influence channel and floodplain erosion, deposition, and connectivity (Figure 8.5).

Ice cover and ice-jams that form during break-up create backwater effects that enhance overbank flooding. When ice-jams break, the resulting surge can enhance overbank flooding downstream. Changing hydraulic forces during ice-jams and break-up, along with the mechanical effects of large chunks of moving ice, can enhance bank erosion and the overbank flooding and scouring that structure riparian vegetation (Beltaos 2002; Shen 2016). Avulsion induced by ice-jams has been described for both meandering and braided channels. Channel ice can facilitate bank erosion by gouging the banks, increasing the bank loading, and reducing vegetation growth along the banks (Ettema and Kempema 2012). Ice cover can alter lateral variations in flow depth and boundary shear stress within a channel (Ettema and Kempema 2012). Ice-backwater effects can alter the direction of flow within a channel and the connectivity between the main channel, secondary channels, and the floodplain (Prowse and Beltaos 2002). Ice that retards flow can decrease bed-material transport, but ice can also increase sediment transport by directly moving the bed sediment via ice rafting (transport of sediment by floating ice), ice gouging, or ice push (Ettema and Kempema 2012). Warming climate is causing a decrease in duration of ice cover and changes in ice thickness in at least some high-latitude rivers (Shiklomanov and Lammers 2014).

Ice-jams and the surges that result from their release can cause habitat degradation or loss, species stress or mortality, and deposition of fines and deterioration of spawning grounds. The highest suspended sediment concentrations occur during freeze-over and break-up, however, and ice-jams and surges also replenish adjacent floodplains with sediment and nutrients (Beltaos 2002).

Permanently frozen ground, known as permafrost, is sediment or bedrock that has a temperature at or below 0 °C for at least two consecutive years (Wright et al. 2009). Permafrost underlies an estimated 24% of land in the northern hemisphere (Zhang et al. 2003) and strongly influences rainfall–runoff relations, erosional resistance of channel banks and floodplains (Figure 8.6), and thus river process and form. Permafrost is overlain by an active layer that thaws during the warm season and can vary

(a)

(b)

Figure 8.5 Aufeis along the Kongakut River, which flows north to the Arctic Ocean in northern Alaska, USA. Aufeis is shelf ice that forms along river margins as ground water continues to flow from adjacent uplands into the river corridor during autumn after air temperatures have dropped below freezing. (a) View of aufeis remaining along the braided channel in late June. Flow is toward the rear in this view, and the valley bottom is approximately 700 m wide. (b) Closer view of the river-side edge of the aufeis. Person in inflatable raft at right center for scale. (*See color plate section for color representation of this figure*).

Figure 8.6 Permafrost exposed in a cutbank along the Yukon River in the interior of Alaska, USA. White stripes of ice alternate with frozen sediment in the cutbank exposure. Cutbank is approximately 3 m tall. (*See color plate section for color representation of this figure*).

from a few centimeters to greater than a meter in thickness. Warming air temperatures are causing permafrost boundaries to recede poleward, as well as leading to thinning of permafrost layers and increasing lateral disconnectivity in frozen ground. Permafrost degradation in turn results in hillslope erosion and development of thermokarst, which increases sediment yield to rivers (Gooseff et al. 2009; Lamoureux and Lafrenière 2009). Because the thickness of the active layer governs the depth to which plant roots can penetrate and thus what tree species can survive in permafrost terrain, permafrost degradation will change the species composition and distribution of vegetation within high-latitude watersheds, with associated changes in runoff, hydraulic roughness, and channel banks in high-latitude rivers (Osterkamp et al. 2009). Changes in permafrost thickness and distribution will also affect runoff, sediment, and organic matter delivered to rivers (Walvoord and Kurylyk 2016), river hydrology (Overeem and Syvitski 2010), and sediment transport (Kokelj et al. 2013). The magnitude and rate of change currently occurring in hydrologic and geomorphic processes in the high latitudes as a result of warming climate (Rowland et al. 2010) will significantly alter high-latitude river process and form in ways that investigators are still struggling to understand.

Although there is no unique type of cold-region river geometry, *sandur* (a valley segment undergoing rapid aggradation, with a downstream decrease in particle size), braided channels, meandering channels, and anastomosing channels in wetland environments are particularly common in cold regions (Vandenberghe and Woo 2002). For all of these channel types, the periods of dynamic change in seasonal ice cover – freeze-over and break-up – tend to be the periods of greatest channel and floodplain geomorphic change in cold-region rivers (Prowse and Beltaos 2002), even though these periods do not coincide with the greatest seasonal discharge. The enhanced erosion associated with ice also increases the frequency of changes in channel cross-sectional geometry, planform, and channel–floodplain connectivity (Ettema and Kempema 2012).

8.2.2 Low Latitudes

The salient features of low latitudes are the magnitude and speed with which various fluxes occur (Wohl et al. 2012a). Low latitudes here are synonymous with the tropics, the area of surplus radiative energy bounded by anticyclonic circulations near 30° N and S (Scatena and Gupta 2013). This region includes the humid tropics, where average annual rainfall is greater than potential evapotranspiration and precipitation is sufficient to support evergreen or semideciduous forests, and the dry tropics. The dry tropics are sufficiently similar to temperate dry regions to be treated in the next subsection.

The humid tropics can be further distinguished as seasonal and aseasonal (Gupta 1995). The aseasonal humid tropics typically have mean annual rainfall between 2000 and 4000 mm/y, whereas mean annual rainfall in the seasonal humid tropics varies between 1000 and 6000 mm/y and interannual variability in runoff can be large (Scatena and Gupta 2013). The seasonal humid tropics have a marked seasonal concentration of rainfall and runoff, typically reflecting the Intertropical Convergence Zone (ITCZ) or monsoonal circulation patterns, and 80% of the annual runoff can occur in 4 or 5 months of the year (Scatena and Gupta 2013). This can result in substantial variations in river process and form between wet and dry seasons, including distinctly different wet- and dry-season channel geometry (Wohl 1992; Gupta 1995).

Extremely intense rainfall and preferential shallow flow paths such as macropores (Section 2.2.1) result in large hydrologic inputs to tropical channels, which tend to have a very flashy flow regime in smaller drainages (Wohl et al. 2012a) and a prolonged peak flood (exceeding 3 months) in very large basins such as the Amazon and Congo (e.g. Rudorff et al. 2014b). Smaller catchments are more likely to have basin-wide intense storms than are equivalently sized catchments at higher latitudes. Monsoons, hurricanes, and ITCZ-related storms tend to cover sufficiently large areas that even larger catchments receive geomorphically significant rainfalls simultaneously across their area (Scatena and Gupta 2013). Widespread intense rainfall results in high percentages of contributing area and channel-modifying discharges that occur simultaneously throughout the catchment. These characteristics contrast with the more spatially restricted precipitation inputs and channel modifications of equivalently sized catchments at higher latitudes.

Continual high air temperatures and abundant vegetation combine with high values of precipitation to create rapid weathering of rock and soil and of organic inputs such as wood. Large inputs of material to channels occur in high-relief tropical environments when cyclones or hurricanes trigger widespread landslides that strip hillslopes of weathered rock and vegetation (Figure 8.7a) (Scatena and Lugo 1995; Goldsmith et al. 2008; Hilton et al. 2011b; Wohl et al. 2012b). Instream wood does not persist in low-latitude rivers. Although individual pieces of wood and large jams can create important geomorphic and ecological effects (Figure 8.7b) (Wohl et al. 2009, 2012b; Martín-Vide et al. 2014), the transience of wood as a result of combined rapid decay and high transport rates is particularly noticeable (Spencer et al. 1990; Soldner et al. 2004; Cadol and Wohl 2010; Wohl and Ogden 2013).

Although the hydrology of low-latitude rivers has distinctive characteristics, and process–form interactions occur more rapidly and frequently than in higher latitudes, Scatena and Gupta (2013) conclude that tropical rivers do not have diagnostic landforms that can be solely attributed to their low-latitude location. A key distinction relative to rivers in higher latitudes is the high frequency of geomorphically significant flows and channel changes.

Figure 8.7 The Upper Rio Chagres in Panama. (a) Low-level aerial view of landslides that occurred during widespread intense rainfall shortly before the photo was taken. The landslides introduced substantial sediment and large wood into the channel. White arrow indicates flow direction. Active channel is approximately 35 m wide. (b) View up the mouth of a tributary channel. A logjam approximately 7 m tall formed at the mouth of the tributary and created a thick wedge of sediment and large wood (not visible here) on its upstream side, which extended more than 100 m up the tributary. Within 2 years, this jam was breached and much of the sediment wedge had been eroded. Person at upper left within white oval for scale. White arrow indicates flow direction on main channel. (*See color plate section for color representation of this figure*).

8.2.3 Warm Drylands

The salient feature of interactions between process and form within rivers in warm dryland regions is spatial and temporal discontinuities. As with tropical rivers, Tooth (2013) concludes that there are no features unique to warm dryland rivers, although certain characteristics are more common in drylands than elsewhere. Warm dryland here includes hyperarid, arid, semiarid, and dry subhumid environments, but excludes cold, high-latitude, and high-altitude regions. Warm dryland regions are particularly prevalent within the subtropical, high-pressure belts of the northern and southern hemispheres, and much of the scientific research on rivers in warm drylands has been conducted in the western United States, Australia, the Middle East, and southern Africa.

Warm drylands share high, but variable, degrees of aridity that reflect low precipitation and high potential evapotranspiration. Long periods with little or no precipitation and stream flow are interrupted by intense rainfall and runoff, which can generate short-duration flash floods (Figure 8.8). Irregular precipitation and low water tables keep many dryland channels ephemeral or intermittent, although rivers that originate in wetter mountainous highlands before flowing into dry regions can be perennial (Tooth 2013). Ephemeral and intermittent rivers are now increasingly referred to as temporary rivers (e.g. Datry et al. 2011, 2014).

Short periods of high water and sediment connectivity between uplands and channels in warm drylands can result from overland flow and limited upland vegetation (Bracken and Croke 2007). Limited riparian vegetation and sediment cohesion can create unstable banks that promote a braided

Figure 8.8 Downstream view along Bumblebee Creek in central Arizona, USA. Although the wetted channel is only about 1.5 m wide under the base flow conditions shown here, the white oval highlights organic material deposited high along the channel margins by a flash flood. (*See color plate section for color representation of this figure*).

planform (Merritt and Wohl 2003), and floodplains (defined as frequently inundated) are limited in extent (Reid et al. 1998).

Flash floods within channels or across piedmonts are particularly characteristic of catchments less than 100 km^2 in size (Tooth 2013). Flash floods have steep rising and recessional limbs and may last only minutes to hours, but can generate very high values of discharge per unit drainage area and correspondingly substantial erosion and deposition (e.g. Reid et al. 1998; Segura-Beltran et al. 2016). Longitudinal declines in flow resulting from infiltration and evaporation can correspond to a lack of longitudinal continuity in a distinct channel form, as when channels end in floodouts (e.g. Sutfin et al. 2014; Larkin et al. 2017). Downstream transmission losses are a particularly important mechanism for recharging alluvial and regional aquifers in drylands (e.g. Morin et al. 2009; Mvandaba et al. 2018). Downstream transmission losses in dryland channels, along with limited contributing area where small convective storms affect only a limited portion of a large drainage basin, create large spatial variability in discharge and temporal variability between storms (Tooth 2013).

Larger river networks are likely to have more sustained flows, especially if they head in wetter uplands and then flow down into dry lowlands, as in much of the interior western United States. Because these perennial rivers provide vital water supplies for agriculture and urbanization in the lowlands, the natural flow regime is especially likely to have been altered by flow regulation, resulting in significant changes in river process, form, and connectivity (Williams 1978b; Nadler and Schumm 1981; Chin et al. 2017). Large dryland rivers are not typically considered to have substantial channel and floodplain large wood loads, for example, but perennial rivers in deserts historically had extensive woody riparian vegetation. Riparian deforestation and flow regulation in these rivers have significantly reduced historical wood loads (Minckley and Rinne 1985; Stout et al. 2018), contributing to loss of river ecosystem function, including habitat for endangered native fish species (e.g. Crook and Robertson 1999).

Although ephemeral dryland channels can have less dense riparian vegetation than perennial rivers, dryland vegetation can be concentrated in and along channels and can generate substantial flow resistance. Resistance can increase with stage as woody vegetation becomes immersed, particularly along the tops of banks and bars (Knighton and Nanson 2002; Griffin et al. 2005). Resistance associated with riparian vegetation can lead to a scenario of streambed scour at lower stages during a flood, when vegetation is not inundated and thus contributes little resistance, and to one of fill during the higher stages when bank flow resistance is strongly affected by vegetation (Merritt and Wohl 2003).

Dryland rivers typically transport large quantities of suspended and bed load and display strong hysteresis of bed scour during the rising limb and fill during the falling limb (e.g. Alexandrov et al. 2003). Scour is facilitated by the absence or poor development of coarse surface layers. Poorly developed coarse surface layers may reflect abundant upland sediment supplies, enhanced sediment mixing during scour and fill, high rates of bedload transport, and short-duration flows that minimize winnowing of fine particles from the bed (Tooth 2013). Lack of coarse surface layers also facilitates high rates of bed-load transport that increase more consistently with increasing flow because particles across a large size range are available for entrainment at the start of flow and the bed becomes highly mobile at even modest flows (Tooth 2013).

Much of the research on small- to medium-size dryland channels emphasizes abrupt transitions between incising and aggrading conditions across a channel network and through time (Schumm and Hadley 1957; Prosser and Slade 1994; Tucker et al. 2006; Nichols et al. 2016). This emphasis is exemplified by the extensive literature on arroyos in the US Southwest (Graf 1983, 1988; Harvey 2008: Miller 2017).

Repeated large floods appear to dominate process and form in many dryland channels, as evidenced by numerous studies of flood-related channel changes and by recovery times of decades or longer (Tooth 2013). Dryland channels may be particularly susceptible to change during floods because of limited bank cohesion in the absence of dense riparian vegetation and abundant silt and clay, and because of the lack of intervening smaller flows that could modify flood-generated erosional and depositional features (Tooth 2013).

In contrast to the emphasis on equilibrium conditions along perennial rivers, the ability of rare floods to cause substantial change along dryland rivers has led some investigators to describe dryland channels as non- or disequilibrium systems (Graf 1988; Tooth and Nanson 2000). Tooth (2013), however, emphasizes the global diversity of dryland river process and form as a result of varying degrees of aridity, tectonic activity, and structural and lithological settings (Tooth 2000; Nanson et al. 2002). In addition, the identification of equilibrium or disequilibrium is highly dependent on temporal and spatial frames of reference, so that individual dryland rivers can exhibit both conditions (Tooth and Nanson 2000).

Dryland river networks are especially vulnerable to consumptive uses of surface and ground water. The aquatic and riparian plants and animals living in dryland river networks have evolved distinctive adaptations to limited water supplies and to high temporal variability in water availability (Dodds et al. 2004). These organisms, however, cannot survive beyond some limit of hydrological alteration (Fausch and Bestgen 1997) and mechanical (e.g. grade control or irrigation intake structures) disconnectivity (Ficke and Myrick 2011). In the semiarid Great Plains of the western United States, for example, small-bodied native fish rely on brief periods of longitudinal flow connectivity and on the persistence of refuge pools that retain water year-round (Falke et al. 2011). As rates of ground-water withdrawal have exceeded recharge in the region, the duration and extent of longitudinal hydrological dysconnectivity have increased and the number and size of refuge pools have declined, severely stressing or eliminating populations of native fish (e.g. Falke et al. 2010).

The hydrologic regime of larger rivers in the western United States that head in mountains and are perennial because of snowmelt has also been altered by flow regulation, which tends to reduce the temporal variability of the hydrograph by storing the snowmelt peak flow and releasing it more gradually throughout the year. This has resulted in extensive narrowing of historically braided channels, which have become anastomosing or meandering within wooded floodplains (Williams 1978b; Nadler and Schumm 1981). Analogous examples of substantial hydrologic alterations causing changes in channel connectivity and channel and floodplain morphology come from dryland regions in Australia (Page et al. 2005; Crook et al. 2015), the southwestern United States (Figure 8.9) (Jaeger and Olden 2012; Goodrich et al. 2018), and the Mediterranean portion of Europe (Skoulikidis et al. 2017).

An extreme illustration of hydrological alteration in a dryland river is provided by large rivers that no longer flow to the ocean because of consumptive water use and flow regulation, such as the Colorado River of the western United States, the Indus River of Pakistan, Australia's Murray River, and China's Huanghe (Yellow River) (Postel 1999; Famiglietti 2014).

Water quality in dryland rivers is also especially vulnerable to human activities because even rivers that drain relatively large regions can have such low flow that their ability to dilute contaminants is limited. Water quality in Australia's Murray–Darling River, for example, is degraded by salts and excess nutrients (e.g. Holland et al. 2015).

Temporary rivers may also be especially vulnerable to urban encroachment because, if a channel is incised and remains dry much of the time, there is a tendency to assume that infrequent, relatively

Figure 8.9 A portion of the Rio Grande in Big Bend National Park, Texas, USA. Here, the active channel has narrowed substantially as a result of upstream flow regulation and dense growth of invasive exotic plants (*Tamarix* spp., *Arundo donax*). Wetted channel is approximately 35 m wide. (*See color plate section for color representation of this figure*).

short-duration flows will be contained within the channel, rather than flooding overbank areas. This assumption can lead to enormous damage when incised channels widen or avulse during large floods within urban areas (e.g. Kresan 1988).

Finally, dryland rivers can provide unique challenges to management designed to sustain river integrity because river form and process are so changeable through time and space and because systematic records of water and sediment discharge are less likely to exist for these river networks. Water-quality standards developed for perennial rivers may be applicable only under certain circumstances or not at all. The Mediterranean Intermittent River Management (MIRAGE) project addresses many of these difficulties and has developed a toolbox designed for sequential use in characterizing the hydrological, ecological, and chemical status of temporary rivers (Prat et al. 2014). Analogously, specification of environmental flow regimes to sustain native biota in dryland rivers can be hampered by lack of systematic flow records. Under these circumstances, the community composition of macroinvertebrates can be used to differentiate rivers characterized in a natural state by longitudinal connectivity versus disconnected pools (Cid et al. 2016).

8.3 Spatial Differentiation Along a River

Turning to smaller spatial scales such as the subwatershed or reach level, distinct suites of geologic and climatic processes can create spatial differentiation within river networks. Schumm (1977) conceptualizes drainage basins as consisting of an upstream zone of production from which water and sediment are derived, a central zone of transfer in which inputs can equal outputs in a stable river, and a downstream zone of deposition. Although acknowledging that production, transfer, and deposition occur continuously throughout a drainage basin, this organization recognizes the existence of spatial zonation in dominant processes within a catchment.

Subsequent conceptual frameworks have also emphasized spatial zonation. Montgomery and Buffington (1997), for example, distinguish source, transport, and response segments in reach-scale classification of mountain channel morphology (Figure 4.15). Sklar and Dietrich (1998) hypothesize consistent changes in dominant incision mechanism (debris flow, fluvial) and substrate type (coarse-bed alluvial, fine-bed alluvial) at threshold slopes, regardless of drainage area (Figure 2.11).

Montgomery (1999) builds on this work in describing *process domains*, defined as spatially identifiable areas of a landscape or drainage basin characterized by distinct suites of geomorphic processes (Figure 2.8). The existence of process domains implies that river networks can be divided into discrete regions in which ecological community structure and dynamics respond to distinctly different physical disturbance regimes (Montgomery 1999). The delineation of process domains has subsequently proven useful in understanding spatial patterns of riparian vegetation (Polvi et al. 2011), sediment dynamics (Wohl 2010a), organic carbon stock in river corridors (Wohl et al. 2012c; Sutfin and Wohl 2017), aquatic ecosystem dynamics and biodiversity (Bellmore and Baxter 2014), channel geometry (Livers and Wohl 2015), and connectivity (Wohl et al. 2019a) within mountainous river networks.

As noted earlier, some river geomorphic parameters exhibit progressive downstream trends whereas others exhibit so much local variation that any systematic longitudinal trends which might be present are obscured (Wohl 2010b). Local variation that overwhelms progressive trends is particularly characteristic of mountainous terrain, where spatially abrupt longitudinal zonation in substrate resistance, gradient, valley geometry, and sediment sources can create substantial variability in river process and form. Under these conditions, characterizing river dynamics at reach scales can be more accurate than assuming that parameters will change progressively downstream. Examples of geomorphic parameters for which spatial variation is better explained by process domain classifications than by drainage area or discharge in mountainous drainage basins include riparian zone width (Polvi et al. 2011), floodplain volume and carbon storage (Wohl et al. 2012c), connectivity (Wohl et al. 2019a), instream wood load (Wohl and Cadol 2011), and biomass and biodiversity (Bellmore and Baxter 2014; Herdrich et al. 2018; Venarsky et al. 2018).

Process domains can also apply to very large drainage basins that have distinct spatial differences associated with topography or climate. The wet, high-relief, sediment-producing Ethiopian Highlands portion of the Blue Nile, for example, is distinctly different than the dry, low-relief mainstem Nile lower in the drainage (Wohl 2011a). Similarly, the steep, narrowly confined segments of the Danube that cut through mountainous terrain in Europe are distinctly different than the intervening anastomosing or braided segments in broad alluvial basins (Wohl 2011a).

Process domains can provide a useful organizational framework for delineating the spatial distribution and relative abundance of different valley and channel types (Wohl et al. 2007), and this can facilitate identification of sensitive or rare areas and formulation of different management strategies for distinct physical settings (Buffington and Tonina 2009b). The concept of process domains can be readily applied at the reach scale at which most river management occurs. With even minimal field calibration, process domains can also provide a framework for remotely predicting at least relative variations in numerous valley-bottom characteristics.

Other conceptual frameworks that emphasize spatial zonation include the *River Styles framework* (Brierley and Fryirs 2005), structured around the five spatial scales of:

- catchment;
- landscape units of relatively homogeneous topography within the catchment;
- reaches with consistent channel planform, assemblage of channel and floodplain geomorphic units, and bed-material texture;
- channel and floodplain geomorphic units such as pools; and
- hydraulic units of homogeneous flow and substrate characteristics.

River style is classified at the reach scale using a procedural tree that starts with lateral valley confinement and continues with characteristics such as river planform, geomorphic units, and bed-material texture, through to a descriptive classification of categories such as "bedrock-confined discontinuous floodplain" and "multichannel sand bed" (Brierley and Fryirs 2005). The catchment-wide distribution of river styles can be used to understand spatial distribution of controls on river process and form, relative abundance of different river styles, and potential sensitivity and resilience of different portions of the river network, in a way analogous to the application of process domains.

Stream biologists also emphasize zonation at differing spatial scales. This is exemplified by the widely used hierarchy of stream system (10^3 m length), stream segment (10^2 m), reach (10^1 m), pool/riffle (10^0 m), and microhabitat (10^{-1} m) of Frissell et al. (1986).

Aquatic and riparian ecologists have also examined the relative importance of progressive trends versus local controls. One of the primary conceptual models of aquatic ecology, the *river continuum concept*, emphasizes continuous longitudinal gradients in the structure and function of ecological communities along a river network (Vannote et al. 1980). In contrast, the *serial discontinuity concept* focuses on how dams disrupt longitudinal gradients along river courses (Ward and Stanford 1983, 1995). *Hierarchical patch dynamics* (Pringle et al. 1988; Poole 2002) emphasizes the existence of relatively homogeneous units from the scale of microhabitat up to channel reaches, with distinct changes in process and form between patches. Reach-scale patches, like process domains or river styles, can result from changes in physical processes such as glaciation or differential rock erodibility (Ehlen and Wohl 2002) and from biotic influences such as beaver dams (Burchsted et al. 2010).

The appropriateness of conceptual models that emphasize progressive downstream changes versus patchiness varies depending on the river characteristic being described and the specific drainage basin. Recognition that a variety of river processes and forms can exhibit abrupt spatial transitions, however, illustrates the importance of considering landscape context when examining river process and form (Wohl 2018a). Although much of the work emphasizing patchiness and relatively abrupt changes in process domains and river styles comes from relatively high-relief environments in which rock structure, tectonics, or glacial history can create strong longitudinal changes in valley geometry, the presence of spatial differentiation can also be useful in explaining process and form in low-relief river networks (e.g. Jones et al. 2008). The presence and characteristics of connectivity exert an important influence on both spatially continuous and spatially discontinuous processes and forms.

8.4 Connectivity

Although the first chapter introduced the idea of diverse forms of connectivity, it is worth returning to this concept and exploring the implications of connectivity for several of the processes and forms

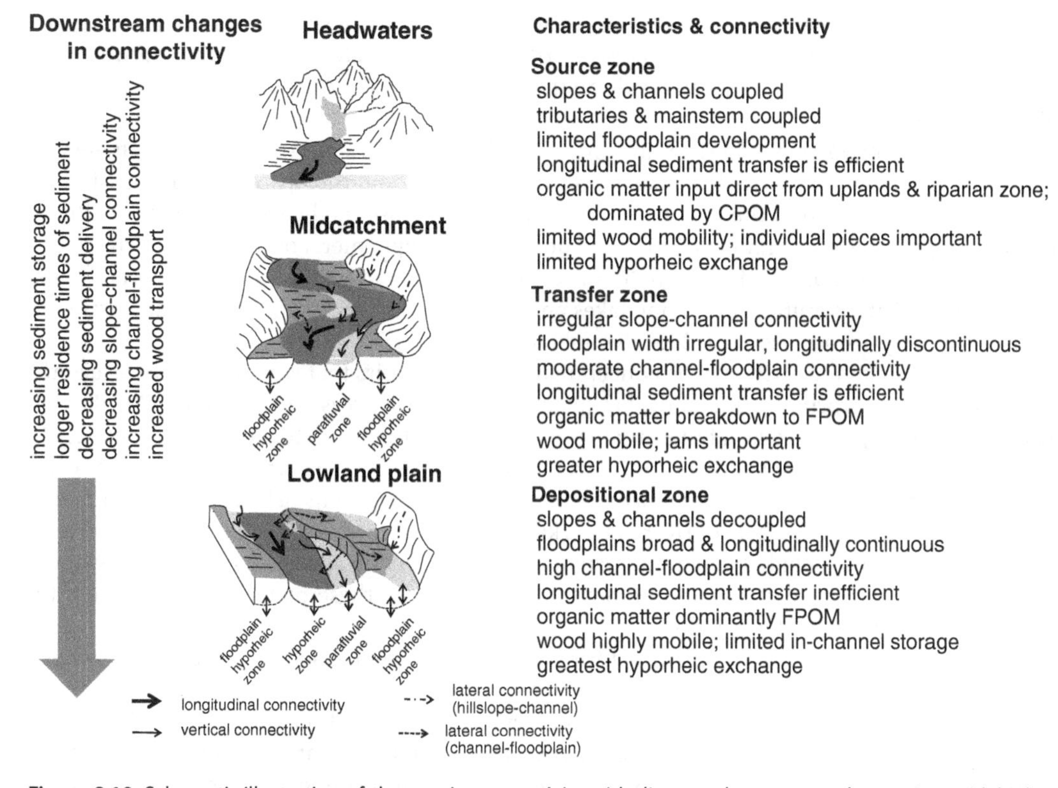

Figure 8.10 Schematic illustration of changes in connectivity with distance downstream along a river with high-relief headwaters. Moving downstream, the river flows through headwater valleys with relatively thin, narrow alluvial veneers over bedrock and then through progressively wider and deeper alluvial valleys with greater floodplain development and hyporheic exchange. The presence of a floodplain buffers the mainstem river from hillslope and tributary inputs by creating depositional zones along its length, and progressively more extensive floodplains typically equate to greater average residence time of sediment, surface flow during overbank floods, and subsurface flow. CPOM is coarse particulate organic matter (>1 mm in diameter), FPOM is fine particulate organic matter (0.45 μm–1 mm). Source: After Brierley and Fryirs (2005), Figure 2.10, p. 44. (*See color plate section for color representation of this figure*).

discussed to this point, including: interactions between hillslope, floodplain, channel, and hyporheic environments; sources, transport, and residence times of water, solutes, sediment, and large wood; and spatial zonation within a drainage basin. Figure 8.10 illustrates how various forms of connectivity change throughout a basin.

High connectivity implies that matter and organisms move rapidly and easily within a river network. Landscapes typically include some characteristics that create at least temporary storage and limit connectivity. Subsurface units of low permeability can limit the downslope transmission of water from hillslopes to channels, or limit hyporheic and ground-water exchanges along channels (e.g. Gooseff et al. 2017). Lakes, broad floodplains with extensive wetlands, and numerous channel-spanning obstructions such as beaver dams and logjams can substantially decrease the rate at which floods move through a river network (e.g. Lininger and Latrubesse 2016; Wegener et al. 2017). Analogously, depositional features such as alluvial fans can limit the rate at which sediment is introduced from hillslopes to channels (Fryirs et al. 2007a). Extensive floodplains can increase the time necessary for sediment entering a river network to move completely through that network. As described in Section 7.1.4, sediment can reside on the floodplains of the Amazon River for thousands to millions of years.

Erosionally resistant portions of a river network can influence landscape connectivity. Segments of bedrock channel commonly act as local base levels, for example, and the upstream transmission of base-level change is limited to the rate at which the river can incise the bedrock segment (DiBiase et al. 2018). Large waterfalls (May et al. 2017) and portions of an ephemeral or intermittent river network (Jaeger and Olden 2012; Cid et al. 2016) that are dry can limit migration of organisms and thus biological connectivity. Naturally occurring lakes and artificial reservoirs can limit the downstream transmission of large wood (Seo et al. 2008; Kramer and Wohl 2015; Senter et al. 2017).

Features that limit connectivity can be conceptualized as reservoirs that store materials, as exemplified by alluvial fans storing sediment or floodplains storing peak flows during a flood. An alluvial fan can also be conceptualized as a buffer that restricts sediment delivery to a channel (Fryirs et al. 2007a). Features that limit connectivity can be conceptualized as barriers, as in the case of a local base level that limits profile adjustment or a dry stream segment that limits fish dispersal. Whether a reservoir or a barrier, these aspects of river networks exert critical controls on fluxes of material and organisms, and must be included when understanding or quantifying all aspects of river networks, from production of water, solutes, and sediment, to movement of these materials downslope into channels and through channel networks.

Connectivity is not *a priori* good or bad in terms of river ecosystem functionality. Some river networks naturally have high levels of connectivity, whereas others include many features that limit connectivity (e.g. Burchsted et al. 2010; Mould and Fryirs 2017). The three dimensions of connectivity commonly have different relations to reach-scale characteristics: channel obstructions such as logjams and beaver dams, for example, promote lateral and vertical connectivity for water, solutes, and particulate organic matter, but limit longitudinal connectivity for these materials. High sediment inputs that promote channel avulsion and high rates of lateral migration may increase lateral connectivity for water, solutes, sediment, and large wood, but restrict longitudinal connectivity for these materials.

There is no single method that adequately measures diverse forms of connectivity (Wohl et al. 2019a). Methods used to date include those that (i) quantify fluxes and use the measurements to infer the degree of connectivity in the transport system at a single point in space (e.g. Jaeger and Olden 2012) or numerically model fluxes at multiple points simultaneously (e.g. Coulthard and Van De Wiel 2017); (ii) infer fluxes of materials by measuring the key drivers that govern connectivity (e.g. inferred sediment connectivity based on drainage area, mean slope, and travel distance; e.g. Cavalli et al. 2013); (iii) infer fluxes of materials by measuring sediment storage and assessing topographic parameters that influence sediment storage and flux (e.g. Nicoll and Brierley 2017); and (iv) represent the geomorphic system under consideration as a network composed of source or storage elements (nodes) connected by pathways of potential transport (links) and then use analytical techniques such as network or graph theory to quantitatively estimate connectivity (e.g. Tejedor et al. 2015).

The issue of scale is critical when quantifying connectivity, because the degree of connectivity varies with the temporal and spatial scales under consideration. Most transport processes are intermittent, for example, so connectivity should increase when measured over longer characteristic time scales but decrease as spatial scale increases, because transport occurs over finite distances during any transport event (Bracken et al. 2013, 2015). This suggests that measurements should be made at a sufficiently large multiple of the fundamental temporal and spatial scales of the phenomenon of interest in order to include a representative sample of transport events (Wohl et al. 2019a). Because the available tools and methods used to collect data constrain the scales at which connectivity can be analyzed, technological advances such as terrestrial LiDAR and structure-from-motion photogrammetry (e.g. Smith and Vericat 2015) are enhancing the ability to quantify connectivity.

Among the challenges in managing rivers are those of quantifying connectivity and understanding how human activities have increased or decreased connectivity within a landscape (Kondolf et al. 2006). As discussed in Chapter 1, most human activities decrease hydrological, sediment, biological, and landscape connectivity within a river network, although a few alterations such as flow diversions and removal of naturally occurring obstructions such as beaver dams can increase connectivity. In contrast, many human activities (e.g. tile drains, impervious surfaces) increase connectivity between hillslopes and river corridors (Covino 2017).

Unanticipated side effects of altered connectivity can require many decades to become apparent. Before the 1970 completion of the Aswan High Dam on the Nile in Egypt, the river annually carried more than a hundred million tons of silt to the Nile Delta. The dam now traps much of this sediment, causing subsidence and erosion in the delta: former delta villages are now 2 km out to sea. Plankton formerly nourished by nutrients in the river flow have dramatically decreased in abundance, contributing to a collapse of the sardine populations that fed on them. Reduction of sediment connectivity along this major river has thus altered physical, chemical, and biological characteristics of the lower river and the nearshore zone (Wohl 2011a).

In Siberia, the Novosibirsk Dam reduces seasonal peak flows along more than a thousand kilometers of the Ob River. The Ob historically provided an important commercial fishery, but many species of fish require access to the floodplains for spawning and nursery habitat during the spring peak flows. Fish in the Ob need at least 20 days of flooding in order to spawn, hatch, and grow. The Novosibirsk Dam has reduced floodplain habitat by half during years of average flow, and eliminated this habitat during dry years. This dam, along with several others in the Ob catchment, also limits longitudinal movements by fish, creating genetic isolation and constraining the ability of each fish population to find appropriate habitat during dry periods. The commercial fishery along the Ob has largely collapsed since construction of the dams (Wohl 2011a).

The effects of altered connectivity along the Nile and the Ob may appear obvious in retrospect, but the challenge of anticipating not only the type, but also the magnitude of altered connectivity in connection with river engineering remains substantial. Numerical simulation can be particularly useful in this context by facilitating the ability to evaluate alternative scenarios, but any model must be accurately parameterized and underpinned by a solid understanding of the sources and characteristics of connectivity within a drainage basin. Again, knowledge of landscape context is critical to effective understanding and management of river process and form (Wohl et al. 2019a).

Connectivity receives so much attention in river science at present because it ultimately reflects geomorphic context and governs the extent to which a river network or a reach of a river is integrated into the greater landscape. Geomorphic context includes spatial dimensions of river corridor geometry, location within a drainage basin, and location within a global context. Context also includes temporal dimensions of the frequency and duration of specific processes influencing the river corridor and the historical sequence of natural and human-induced processes that continue to influence the river corridor (Wohl 2018a).

8.5 River Management in an Environmental Context

This portion of the chapter focuses on river management undertaken specifically to restore rivers in an environmental context. People have been managing – or at least attempting to manage – rivers for millennia. Past management actions were typically undertaken with the intent of making rivers or associated resources more conveniently accessible to humans: damming rivers to ensure water supply, for example, or building levees and channelizing rivers to enhance agricultural use or settlement on floodplains. Although river restoration and rehabilitation are sometimes viewed as being fundamentally different than past river management, they can also be viewed as the latest iteration of the attempt to reconfigure rivers to conform to human expectations; in the case of restoration, expectations of more natural or ecologically functional rivers.

Restoration activities such as the planting of riparian vegetation to stabilize streambanks date to the seventeenth century in Europe (Evette et al. 2009). Projects designed to restore rivers have increased dramatically in number and scope since the 1990s (Bernhardt et al. 2005), particularly in the United States, western Europe (e.g. Muhar et al. 2016), and Australia (Brooks and Lake 2007). As with any form of river management, some projects have largely achieved their original purpose whereas others have been thorough failures. The rationales that underpin river restoration and the factors that result in success or failure are worth examining because the ability to restore rivers provides an effective measure of our understanding of river process and form.

8.5.1 Reference Conditions

Any river restoration is undertaken to achieve some desired end result of river process and form. Restoration is undertaken for a variety of reasons, including those related to recreation, water quality, esthetics, protection of aquatic and riparian species, bank stabilization, fish passage, flow modification and dam removal, and the creation of a more natural environment (Table 8.1) (Bernhardt et al. 2005). The latter point is perhaps the most difficult, because achieving a more natural environment entails addressing at least one fundamental question: What is natural? (Graf 1996; Wohl 2011c).

"Natural" is commonly assumed to imply minimal human alteration, although natural is increasingly being defined in terms of *physical integrity* (Graf 2001) or the ability of a river to adjust to existing conditions (Fryirs and Brierley 2009). Humans have been manipulating natural landscapes for many thousands of years, by using fire to alter land cover, selectively hunting some animals to extinction, domesticating plants and animals and then altering ecosystems to favor domesticated species, and cutting trees for fuel and building materials (Wohl et al. 2017c). So, at what point in history do we consider a given ecosystem to have last been natural: prior to agriculture, prior to the Industrial Revolution, or prior to some arbitrary human population density?

Whatever (pre) historical period is chosen, the characteristics of rivers during that period are typically known as *reference conditions*, which can also be defined as the best available conditions that could be expected at a site (Norris and Thoms 1999). The latter definition can be highly problematic, however, because great disagreement or uncertainty can arise as to what constitutes "best available." Reference conditions encompass all of the aspects of river process and form discussed to this point, including: flow, sediment, and wood regimes; water chemistry; substrate; bedforms; channel morphology, planform and gradient; and longitudinal, lateral, and vertical connectivity.

In a region where some river basins have undergone minimal human alteration, contemporary rivers can provide reference conditions for altered river basins (e.g. Larned et al. 2008). This approach

Table 8.1 National River Restoration Science Synthesis (NRRSS) working group list of goal categories and operational definitions for river restoration projects.

Category	Description
Esthetics/recreation/education	Activities that increase community value: use, appearance, access, safety, and knowledge
Bank stabilization	Practices designed to reduce or eliminate erosion of banks
Channel reconfiguration	Alteration of channel geometry: includes restoration of meanders and in-channel structures that alter river thalweg
Dam removal/retrofit	Removal of dams and weirs or modifications to existing dams to reduce negative ecological impacts (excludes dam modifications solely intended for the improvement of fish passage)
Fish passage	Removal of barriers to longitudinal migration of fishes: includes physical removal of barriers, construction of alternative pathways, and construction of barriers to prevent access by undesirable species
Floodplain reconnection	Practices that increase overbank flows and the flux of organisms and materials between channel and floodplain
Flow modification	Practices that alter the timing and delivery of water quantity (does not include stormwater management)
Instream habitat improvement	Alteration of structural complexity (bedforms, cross-sectional geometry, substrate, hydraulics) to increase habitat availability and diversity for target organisms and provide breeding habitat and refugia from disturbance and predation
Instream species management	Practices that directly alter aquatic native species distribution and abundance through the addition (stocking) or translocation of plant and animal species or the removal of exotic species
Land acquisition	Practices that obtain lease, title, or easements for streamside land for the explicit purpose of preservation or removal of impacting agents or the facilitation of future restoration projects
Riparian management	Revegetation of riparian zone or removal of exotic species of plants and animals
Stormwater management	Special case of flow modification that includes the construction and management of structures (ponds, wetlands, flow regulators) in urban areas to modify the release of storm runoff
Water-quality management	Practices that protect existing water quality or change the chemical composition or suspended load, including remediation of acid mine drainage

Source: After Bernhardt et al. (2007), Table 1.

must be used with caution, however, because contemporary conditions constitute a snapshot in time that reflects only a single state or a limited portion of the fluctuations that naturally occur in rivers (SER 2002).

In many regions of the world, there are no relatively unaltered rivers, so reference conditions must be inferred from historical, botanical, and geologic records (Morgan et al. 1994; Nonaka and Spies 2005; Stoddard et al. 2006; Wohl 2011c). Lack of information on reference conditions, as well as continuing change in catchment parameters, can limit the usefulness of reference conditions (Hughes et al. 2005; Newson and Large 2006; Dufour and Piégay 2009). Consequently, reference conditions may be most appropriate as an ideal rather than as a goal for restoration (Osborne et al. 1993). This thinking underlies balanced sediment regimes (Wohl et al. 2015b) and target wood regimes (Wohl et al. 2019), both of which describe regimes that are not natural but that can create and sustain desired conditions within the river corridor.

A tremendous amount of effort may be necessary to characterize reference conditions, not least because river process and form can vary substantially across even a relatively small drainage basin and because rivers are never static in time. Even in the absence of human manipulation, rivers undergo fluctuations in process and form associated with natural events such as floods or droughts, landslides, wildfires, tectonic uplift or subsidence, and continuing adjustment to long-term changes in climate or tectonics. Consequently, a key component of understanding reference conditions is being able to quantify the *natural or historical range of variability* (NRV) for a given parameter or set of parameters (Morgan et al. 1994; Nonaka and Spies 2005; Wohl 2011c; Rubin et al. 2012; Brierley and Fryirs 2016; Grimsley et al. 2016).

Ongoing climate change, abundant and widespread invasive species, and human population growth and resource use cause some scientists and managers to question the relevance of NRV (Safford et al. 2008; Dufour and Piégay 2009). If the world already looks fundamentally different than it did prior to human manipulation, and if it will grow increasingly different in the future, what do past river process and form matter? Other scientists and managers contend that, even if a river cannot be restored to NRV, detailed, quantitative understanding of prior and existing river characteristics can inform management by constraining the range of potential river process and form and providing insight into the conditions necessary to sustain native species (e.g. Koel and Sparks 2002).

Knowledge of NRV provides insight into the conditions that native riverine species or communities might require for survival, as well as the thresholds or minimum values of process or form that must be maintained in order to sustain biological communities (e.g. flood threshold for overbank flooding that provides fish access to the floodplain for spawning and nursery habitat). Knowledge of NRV also facilitates delineation of the spatial distribution of different suites of geomorphic processes, such as portions of a mountainous headwater catchment dominated by debris flows versus portions dominated by fluvial processes. This facilitates evaluation of the location, relative rarity, and connectivity of sensitive stream segments that are likely to respond to alterations or that contain biologically unique communities (McDonald et al. 2004; Brierley and Fryirs 2005; Wohl et al. 2007). Insight into NRV provides information on the relative magnitude of variation in specific river attributes among process domains (Wohl 2011c). Knowledge of NRV thus underpins our understanding of process and form in any river network (Wohl 2018a).

8.5.2 Restoration

Restoration is undertaken for many reasons and at many scales, from a single river segment only a few hundred meters in length to entire large drainage basins. The National River Restoration Science Synthesis database includes more than 37 000 restoration projects in the United States. Most of the projects in the database were implemented on river segments less than 1 km in length (Bernhardt et al. 2005), but the most high-profile are those involving much larger segments, such as the Grand Canyon of the Colorado River in Arizona (Melis 2011; Melis et al. 2015) or the Kissimmee River in Florida (Toth et al. 1993; Warne et al. 2000; Wohl et al. 2008; Koebel and Bousquin 2014), both in the United States, and the Danube River in western Europe (Tockner et al. 1998, 1999; Bloesch and Sieber 2003; Hein et al. 2016). These basin-scale restoration projects overseen by governmental agencies are more likely to include independent oversight and monitoring and to have integrated restoration plans that span diverse spatial scales and river processes than are projects that are smaller in spatial extent.

Restoration in the context of this chapter is used to include both *restoration* – a return to a close approximation of the river condition prior to disturbance (however disturbance may be defined) – and *rehabilitation* – improvements of a visual nature, sometimes described as "putting the channel back into good condition" (however good condition may be defined) (National Academy 1992).

River restoration can be described as following one of three approaches (Palmer et al. 1997; McDonald et al. 2004). The *field of dreams approach* uses traditional engineering techniques to modify river form to a desired condition, with the expectation that this will create the processes necessary to maintain that form (e.g. Lepori et al. 2005). This is the most widely used approach to restoration on relatively short river segments. This approach is named in reference to the movie *Field of Dreams*, which became famous for the phrase, "If you build it, he will come." In the case of river restoration, the implication is that restoring river form will also restore function, so that organisms such as fish will return and thrive. This approach to restoration can also be characterized as involving active intervention within the river corridor, in contrast to passive restoration that focuses on processes outside of the corridor, such as upland reforestation to reduce sediment yields to the river.

The *system function approach* to river restoration identifies and alters the initial conditions required to achieve restoration goals. This could involve modifying fine sediment inputs to a targeted segment of river by establishing riparian buffer strips, for example, or adding large wood to increase flow resistance and sediment retention. The underlying rationale is that modifying initial conditions such as water or sediment inputs will cause river process and form to adjust in a desired manner.

The *keystone method* identifies and incorporates crucial components of process and form and recognizes uncertainty in the resulting river responses. Riffle–pool sequences might be the keystones of river form and process in a project designed to restore fish habitat, for example, so that restoration would focus on the parameters necessary to create and maintain riffles and pools.

River restoration can also be distinguished as focusing on reconfiguration or reconnection (Bernhardt and Palmer 2011; Wohl et al. 2015c). As the name implies, reconfiguration focuses on changing channel form and is analogous to the field of dreams approach. Reconnection emphasizes connectivity, as in removal of artificial bank protection and grade controls and widening of the portion of the floodplain actively connected to the channel in order to restore braided planforms on European rivers (Tockner et al. 1998, 1999; Habersack and Piégay 2008; Muhar et al. 2008; Theule et al. 2015).

River restoration as currently practiced is commonly not scientific because hypotheses regarding river response to a given restoration action are not posed and tested. A large-scale survey of river restoration in the United States found that fewer than 10% of projects included any form of monitoring or assessment, although projects with higher costs were more likely to be monitored (Bernhardt et al. 2005). Less than half of all projects set measurable objectives, but nearly two-thirds of project managers felt that their projects were "completely successful" (Bernhardt et al. 2007). This reflects the fact that perceived ecological degradation typically motivated the projects. Post-project appearance and positive public opinion were the most commonly used metrics of success (Bernhardt et al. 2007), however, rather than more objective metrics or metrics grounded in scientific understanding of river process and form. At present, monitoring is still not routinely included in small-scale restoration projects, but is at least becoming more common than during the first decade of the twenty-first century.

The lack of monitoring for restoration effectiveness is highlighted by a consideration of the history of river restoration. The design of instream structures such as rock and log dams and deflectors used for habitat improvement goes back to at least the 1880s in the United States, and even earlier in Europe (Thompson and Stull 2002). Many of these structures are still used today with very little modification of initial designs, although systematic examination indicates that such structures do not necessarily guarantee demonstrable benefits for fish communities (Thompson 2006), and may in fact decrease habitat abundance and diversity over a period of many decades (Thompson 2002) (Figure 8.11). In other words, because we typically do not objectively and systematically evaluate the

Figure 8.11 Instream structures used for restoration. (a) Vortex weir along a river in Charlotte, North Carolina, USA. A vortex weir is u- or v-shaped, with the apex pointing upstream. The structure is designed to deflect flow toward the channel center and promote bed scour, which forms a pool. (b) Collapsed lunker in the Catskills region of New York State, USA. Lunkers are designed to stabilized stream banks and promote edge cover for fish. Source: Both photographs courtesy of Douglas M. Thompson.

success of restoration projects over periods of several years following project completion, we are not learning from our mistakes.

In this context, the relatively new practice of stream mitigation banking has the potential to facilitate continued degradation of rivers. Stream mitigation banking, as practiced in the United States, gives developers the option to offset construction impacts to streams by purchasing credits. The credits are generated by for-profit companies that restore streams on a speculative basis and are approved by federal regulatory agencies (Lave et al. 2008). Critical issues such as the location and proper amount of compensation, as well as how stream credits should be measured and certified, remain unresolved (Lave et al. 2008). This strongly suggests that at least some stream mitigation banking practices will provide a cover for stream degradation and lead to net loss of river health (Lave 2018).

Small- to medium-scale river restoration has become an industry dominated by consulting firms with a background in civil engineering rather than in river science, with designs developed and implemented by those with relatively little knowledge of river process and form. The scientific community has become increasingly vocal in criticizing restoration practices (e.g. Pasternack 2013; Palmer et al. 2014; Wohl et al. 2015c). Numerous papers emphasize that river restoration must be based on or include five factors (Kondolf and Larson 1995; Hughes et al. 2001; Kondolf et al. 2001; Ward et al. 2001; Hilderbrand et al. 2005; Wohl et al. 2005; Kondolf et al. 2006; Sear et al. 2008; Brierley and Fryirs 2009; Hester and Gooseff 2010).

First, restoration should be designed with explicit recognition of complexity and uncertainty regarding river process and form, including the historical context of variations in process and form through time. A well-documented example of failed restoration imposed a stabilized single-thread channel on a river segment that had repeatedly alternated between multi- and single-thread planforms during previous decades in response to fluctuations in flood magnitude and frequency (Kondolf et al. 2001). The restored channel was completely altered by a flood within 3 months of project completion.

Second, restoration should emphasize processes that create and sustain river form, rather than imposition of rigid forms that are unlikely to be sustainable under existing water and sediment regimes. Perhaps the most egregious and common example is braided river segments that are restored to sinuous, single-thread rivers without addressing the water and sediment yields that produced a braided planform, and without any consideration of meander dynamics (the constructed bends are commonly stabilized to prevent migration) (e.g. Kondolf et al. 2001). This practice likely reflects a cultural preference for single-thread, meandering channels in North America and Europe (Kondolf 2006; Le Lay et al. 2013a). Sometimes, the re-meandered rivers were historically straightened, but in other cases the rivers have a braided planform reflects historical conditions of flow and sediment regime. Cuneo and Uvas Creeks are two well-publicized examples in California, USA in which restoration that imposed a meandering planform failed spectacularly within a few years as moderately sized floods converted the restored river reaches back to a braided form (Kondolf et al. 2001; Kondolf 2006). In each case, practitioners ignored historical evidence of large sediment fluxes and braided planform, and imposed a single-thread, sinuous channel with stabilized banks.

Examples of restoration strategies designed to initiate processes are forms of stage 0 restoration (see Section 5.7) and beaver dam analogs. Stage 0 restoration refers to the reestablishment of a multichannel planform. This approach can involve introducing substantial quantities of large wood and allowing flows to redistribute the wood, which is likely to result in logjams and formation of secondary channels. Stage 0 restoration can also involve excavation of legacy sediments and the introduction of widely spaced, temporary obstacles (e.g. large wood pieces, hay bales) that deflect flow in a manner likely to facilitate formation of swale-shaped channels and a multichannel planform (e.g. Booth and Loheide 2012). Stage 0 restoration can also employ beavers, with beaver dam analogs built in locations in which real beavers can enhance and maintain the dams (Pollock et al. 2014, 2015).

Third, projects should be monitored after completion, using the set of variables most effective for evaluating achievement of objectives, and at the correct scale of measurement (Comiti et al. 2009b provides an example of effective monitoring). If the primary objective of restoration is to enhance biodiversity, for example, then monitoring habitat heterogeneity under the assumption that habitat heterogeneity always correlates with biodiversity is not as appropriate as directly monitoring metrics of biodiversity. A synthesis of restoration project designed to restore habitat and biodiversity found that most projects did restore heterogeneity, but almost none resulted in greater macroinvertebrate biodiversity (Palmer et al. 2010b). Comiti et al. (2009b) provide an example of the monitoring of changes in channel process along mountain streams in Europe, where organic matter retention and benthic macroinvertebrate biodiversity were higher in river segments with morphologically based artificial steps constructed of logs and boulders than in river segments with traditional concrete check dams.

Fourth, consideration of the watershed context, rather than an isolated segment of river, is crucial because of the influences of physical, chemical, and biological connectivity on alterations undertaken for river restoration. The Carmel River in California, USA provides an example where restoration using native riparian vegetation was not initially successful because decades of ground-water withdrawal had lowered the water table below a depth that could be accessed by the vegetation (Kondolf and Curry 1986). Consequently, the native plants had to be artificially irrigated to ensure their survival.

Fifth, accomodation of the heterogeneity and spatial and temporal variations inherent in rivers is necessary for successful restoration (Brierley and Fryirs 2009). Rivers continually adjust parameters such as bedform configuration, bed grain-size distribution, and channel width-to-depth ratio in response to fluctuations in water, sediment, and wood yields to the channel. These adjustments are commonly not synchronous or of exactly the same magnitude between distinct reaches of the river. Allowing the channel some freedom to adjust to changes imposed during restoration, as well as changes that will inevitably occur afterward, increases the likelihood that the objectives of restoration will continue to be met over a period of many years.

Numerous papers examine how numerical simulations can be used to predict restoration outcomes prior to project implementation (e.g. Brooks and Brierley 2004; Singer and Dunne 2006; Dixon et al. 2016). However, as Bernhardt et al. (2007) emphasize in a survey of river practitioners, publishing more scientific studies of river restoration will not by itself change the existing situation. River restoration can only improve through direct, collaborative involvement among scientists, managers, and practitioners.

Such collaborations appear to be an obvious next step, but can be very difficult to achieve. Interdisciplinary scientific teams face significant challenges because of differences in terminology, conceptual models, qualitative versus quantitative knowledge, and temporal and spatial scales of interest (Benda et al. 2002). These challenges multiply when the pool of participants is broadened beyond the scientific community. There is no question, however, that river restoration requires an interdisciplinary approach. As reviewed by Pasternack (2013), wetland rehabilitation provides a model. Wetland rehabilitation is facilitated by a certification program hosted by the Society of Wetland Scientists, which includes research scientists, governmental regulators, and practitioners. There is no equivalent for river restoration, partly because the river science community is diverse and oriented toward specific academic disciplines, as well as being strongly divided between research scientists and practitioners and poorly organized (Castro 2008; Pasternack 2013). Although regulators and practitioners would like to establish a universal restoration approach that would standardize methods, research scientists remain highly skeptical that such a "cookbook" technique can be effective. At present, there is no scientific consensus about the scientific foundations for restoration, what the practice should entail, or who should be allowed to undertake restoration (Darby and Sear 2008; Pasternack 2013; Wohl et al. 2015c).

An important consideration in river restoration is that it is not an all-or-nothing process. A river does not have to be – and typically cannot be – restored to some completely natural condition that existed prior to intensive resource use. Partial restoration within the constraints existing in a watershed can nonetheless restore a great deal of physical and ecological form and function. Small watersheds in the central United States that are effectively completely devoted to agriculture, for example, have highly channelized streams. These streams cannot be restored in the traditional sense. River engineering is ubiquitous, and land cover has been altered throughout the watershed, so that water and sediment yields are completely altered (Rhoads et al. 1999). The physical and ecological function of these streams can be improved with *naturalization*, however, which defines a viable management goal for watersheds within landscapes intensively modified by humans. A naturalized channel has sustainable hydraulic and morphologic diversity that supports greater biodiversity than in an unrestored channel, even though the naturalized channel may still be completely altered relative to its natural state. A naturalized channel might be deepened but not straightened, for example, allowing the development of limited sinuosity and associated physical diversity in hydraulics, substrate, and channel geometry (Rhoads et al. 1999), or limited secondary channels might be constructed or allowed to form to create a multichannel planform (e.g. Škarpich et al. 2019).

Primary considerations in restoring selected reaches of a river or river network are, how many and where? Bernhardt and Palmer (2011) discuss the "fuzzy logic" of attempting to reverse catchment-scale degradation by restoring small segments of an entire river network. In scenarios where catchment-scale restoration is limited, however, identifying portions of the river corridor that can be disproportionately influential in retaining excess nutrients, for example, or increasing habitat abundance and diversity, can enhance the effects of restoration. The concept of river beads becomes important in this context.

Because river restoration is commonly spatially constrained by existing land ownership and use, scientists undertaking basin-scale restoration in the Missouri–Mississippi River drainage of the United States proposed a *string of beads approach*. In this approach, land acquisition and restoration activities focus on key floodplain habitats such as floodprone areas near tributary confluences or remnant backwaters that form beads along the string of the otherwise-altered river corridor (Galat et al. 1998; Lemke et al. 2017). This spatially discontinuous approach to river restoration yields demonstrated improvement in water quality, flood hazard mitigation, and biodiversity. Questions remain, however, regarding the number and location of beads that will maximize desired river responses to restoration (Wohl et al. 2018b).

The European Water Framework Directive provides an example of a governmental approach that may help to organize efforts toward common objectives and enhance river restoration at a national or transnational scale (European Commission 2000). The directive was designed primarily to improve and protect water quality, with a set deadline for achieving "good status" for all surface waters within member nations by 2015. This status was to be achieved by meeting requirements for ecological protection and minimum water quality standards analogous to those enforced by the US Environmental Protection Agency. These goals, although laudable, were difficult to implement: 95% of England's rivers, for example, were at risk of failing legislated environmental objectives prior to 2015 (Green and Fernández-Bilbao 2006; Pasternack 2013). Now that the first 6-year planning cycle has finished (as of 2015), river basin management plans are required to establish measurement programs for bodies of water falling below the benchmark in order to improve their status and become compliant over subsequent planning cycles (Hughes et al. 2016).

One relatively recent and high-profile approach to restoration via reconnection comes from dam removal and experimental high flow releases (Sections 3.2.9 and 4.9). High flow releases can be used to restore river process and, indirectly, form downstream from a dam. Dam removal effectively restores all forms of longitudinal connectivity (i.e. water, solutes, sediment, large wood, and organisms) and, indirectly, lateral and vertical connectivity within the river corridor. The restoration of natural sediment and wood regimes, however, can require years to decades as the downstream portion of the river adjusts to the substantial pulse of sediment that typically enters following dam removal (e.g. East et al. 2015). Dam removal and experimental high flow releases grew from a progressive recognition of the importance of environmental flows.

8.5.3 Instream, Channel Maintenance, and Environmental Flows

A vital aspect of river restoration at many sites is preserving or restoring a natural, as opposed to regulated, flow regime. As discussed earlier, the natural flow regime (Poff et al. 1997) outlines the importance of magnitude, frequency, duration, timing, and rate of change of flow in river networks for physical process and form and for ecological communities. When flow regime is altered directly by flow regulation or indirectly by changes in land cover that alter water yield to a river, channel characteristics and riverine biota are affected. Subsequent papers have documented how flow regulation tends to homogenize river flow regimes, with the consequence that riverine physical characteristics and biotic communities also become more homogeneous with time (Moyle and Mount 2007; Poff et al. 2007; Peipoch et al. 2015). Growing awareness of the importance of all aspects of a river's flow regime led to the current emphasis on environmental flows.

Initial efforts to protect river flow focused on minimum flows. In arid and semiarid regions such as the western United States, dams and diversions designed to manipulate water for consumptive uses including agricultural irrigation can result in river segments that are completely dewatered for some or all of the year. As two fish biologists wrote in one of their papers about rivers in such regions, "it is obvious that without water, there can be no fish" (Fausch and Bestgen 1997).

The concept *of instream flows* developed as a means to preserve some minimum flow level within the channel. An early version of this was based on the Instream Flow Incremental Methodology (IFIM) (Bovee and Milhous 1978). IFIM uses a biological model that describes the habitat preference of individual fish species in terms of depth, velocity, and substrate, and a hydraulic model that estimates how habitat availability varies with discharge. The intent behind this method is to be able to specify the minimum flows below which individual species likely cannot be sustained in a river segment, as well as the inferred gains in habitat and potentially in fish biomass as flow increases. Although the method has been criticized for being overly simplistic – the model does not account for biological interactions such as competition or predation, for example – IFIM remains widely used in evaluating alternative water management options (Stalnaker et al. 1995; Macura et al. 2017).

Minimum flows can allow river organisms to survive for a period of time, but a river that in essence has only continual base flows will eventually lose its capacity to support a diverse aquatic community. Periodic high flows are indirectly necessary to biotic communities because higher flows maintain habitat by performing functions such as scouring pools, winnowing fine sediments from the bed, and limiting channel narrowing through encroachment of riparian vegetation. High flows are also directly necessary because they provide a window of opportunity during which organisms can disperse longitudinally and laterally, accessing new habitat for breeding and feeding, as well as maintaining lateral connectivity between the channel and floodplain.

Recognition of the importance of higher flows first led to the concept of *channel maintenance flows*, typically defined as the components of a river's flow regime necessary to maintain specific physical channel characteristics, such as sediment transport or flood conveyance (e.g. Andrews and Nankervis 1995). Channel maintenance flows are an applied equivalent of the concepts of bankfull, effective, or dominant discharge. In this applied context, channel maintenance flows can specify a particular magnitude and frequency of flow to achieve a limited objective, such as pool scour, or they can incorporate a broader range of flow magnitudes designed to maintain a physically diverse channel (Andrews and Nankervis 1995). Analogous to instream flows, the intent behind channel maintenance flows is to quantify and then legally establish or protect the magnitude and frequency of flow necessary to maintain specific components of river process and form.

The latest iteration in this progressively expanding view of the components of the flow regime necessary to preserve physical and ecological integrity in a river is *environmental flows*. The concept of environmental flows grew out of experimental flow releases from dams, such as those on the Colorado River through the Grand Canyon, USA in 1996, 2004, 2008, 2010, and continuing at present (Melis 2011; Melis et al. 2015; Mueller et al. 2018). An experimental flood is tied to a qualitative or quantitative model of a river ecosystem that predicts some beneficial effect from the flood. In the case of the Colorado River, the floods are designed primarily to deposit finer sediment (silt and sand) along the channel margins in order to restore riparian and backwater habitat that has been lost through progressive erosion of sand bars and backwater habitats since construction of Glen Canyon Dam in 1963. Ideally, the experimental flow release is conducted in an *adaptive management* context in which the results of the flood are systematically assessed and the underlying model of the river ecosystem is modified as needed (Melis 2011). Experimental releases from dams are now documented for several rivers in diverse settings (e.g. Mürle et al. 2003; Konrad et al. 2011; Mueller et al. 2018). For example, experimental flow releases designed to restore channel–floodplain connectivity have been conducted along multiple rivers in southeastern Australia, although these have had limited success in reversing environmental degradation because the flows are relatively small and of short duration (e.g. Lind et al. 2007; Rolls et al. 2013).

Environmental flows now refer both to such experimental releases, which are typically limited in duration, and to an annual hydrograph that specifies magnitude, frequency, timing, duration, and rate of change in flow, commonly for diverse conditions such as wet years and dry years (Figure 8.12) (e.g. Rathburn et al. 2009). As with other forms of river restoration, developing the guidelines for environmental flows is time-consuming and challenging because this process forces geomorphologists and riverine ecologists to specify flow thresholds related to targets such as winnowing fine sediment, mobilizing the entire bed, creating overbank flows for channel–floodplain connectivity, and maintaining diversity of species and individual ages within a riparian forest. The complexity and uncertainty associated with river process and form, as discussed at length in Chapters 3, 4, and 5, mean that this task is sometimes straightforward, but more often includes a great deal of uncertainty. Uncertainty that can be acceptable in scientific research becomes more challenging in a management context in which every cubic meter of water released from a hydroelectric dam during an experimental flood, for example, equates to a loss of revenue for the entity operating the dam.

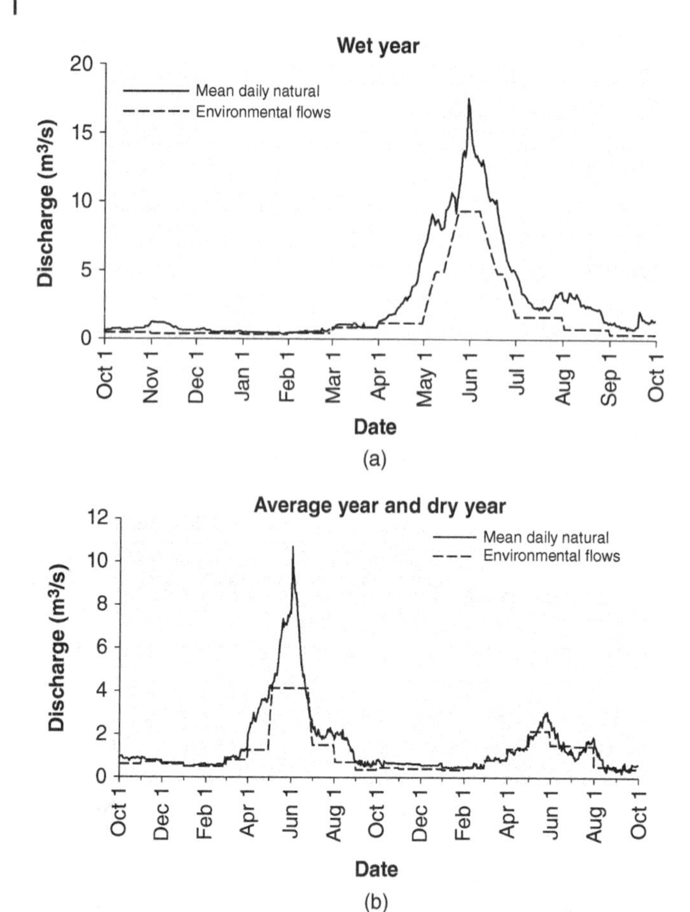

Figure 8.12 Suggested annual hydrographs for (a) wet and (b) dry or average years in the North Fork Poudre River drainage basin, Colorado, USA. Each hydrograph shows the average flows that would occur under natural (unregulated) conditions and the minimum environmental flows recommended to protect desired ecosystem attributes including winnowing fine sediment from spawning gravels, maintaining pool volume, and creating overbank inundation for riparian vegetation. Source: After Rathburn et al. (2009), Figures 13 and 14.

The procedure of assessing environmental flow requirements has gradually assumed characteristic steps of (i) quantifying natural and altered stream flows and the changes in the flow regime in terms of relevant hydrologic metrics (Richter et al. 1996; Gao et al. 2009; Mackay et al. 2014) and (ii) quantifying relationships between hydrologic metrics and physical and biological river attributes, which essentially involves coupling physical and biological models (Sanderson et al. 2011; Webb et al. 2015). A large-scale example comes from the Okavango River system in southern Africa, where multiple response curves for individual metrics of river function and biota were combined using Decision Support Software to model potential river ecosystem response to different scenarios of flow regulation (King et al. 2014).

Once recommendations are developed for environmental flows, the process moves into the policy arena, in which a community much broader than scientists typically weighs in on water availability and use. As Arthington and Pusey (2003, p. 377) describe the process in an Australian context, the two vital questions are: "How much water does a river need? and How can this water be clawed back from other users?" In southern Africa, predictions of the tradeoffs between water development and river ecosystem response are being used in a social context to define *development space* – the acceptable balance between flow regulation and loss of river health (King and Brown 2010). An extensive literature has come into being during the past decade that describes environmental flow assessments and recommendations, as well as a variety of case studies (e.g. Tharme 2003; Arthington et al. 2006; Shafroth et al. 2010; Foster et al. 2018). However, Poff (2018) emphasizes the need to move beyond static, regime-based flow metrics to dynamic, time-varying flow characteristics, as well as the importance of expanding the ecological metrics used to design and assess environmental flows. Poff (2018) also highlights the importance of non-flow parameters such as water temperature and sediment dynamics.

Environmental flows are in many cases driven by the need to preserve endangered species, and ecologists tend to focus on flow regime. Geomorphologists increasingly emphasize the equal importance of sediment dynamics in maintaining channel complexity, habitat heterogeneity, and nutrient cycling (Pitlick and Wilcock 2001; Wohl et al. 2015b). Altered flow regimes are commonly accompanied by altered sediment dynamics as a result of sediment trapping behind dams or changed ability of flows to entrain and transport sediment present along the river corridor. A body of scientific literature on sediment dynamics in the context of environmental flows is just beginning to appear (Rubin et al. 1998; Wiele et al. 2007; Wohl et al. 2015b). The conceptualization of critical inputs and fluxes of materials in river corridors has recently expanded to include large wood (Wohl et al. 2019).

Environmental flows are also closely connected to ecohydrology and ecohydraulics. *Ecohydrology* is the investigation of integrated biological and hydrologic processes, at the scale of a river network or river reach, and the resulting changes to hydrologic, physical, chemical, and ecological attributes of river corridors (Zalewski 2000). Ecohydrology can be used to guide river restoration by quantifying the relationships between hydrologic drivers, other environmental stressors (e.g. connectivity, water quality), and stream organisms (e.g. DuBowy 2013). *Ecohydraulics* focuses on how the magnitude and spatial distribution of hydraulic variables such as flow depth and velocity vary in relation to discharge and channel morphology, under the assumption that hydraulic characteristics provide a useful means of predicting, evaluating, and designing habitat conditions favorable to targeted species (Leclerc et al. 1996; Clifford et al. 2006). Ecohydrology and ecohydraulics are increasingly used to develop specific prescriptions for river restoration on a wide range of rivers (e.g. Mallen-Cooper and Zampatti 2018; Entwistle et al. 2019).

8.5.4 River Health

Global and regional syntheses indicate that aquatic and riparian species are becoming increasingly homogeneous – a few hardy generalist species tend to dominate many communities – because of flow regulation (Dynesius and Nilsson 1994; Moyle and Mount 2007; Poff et al. 2007; Braatne et al. 2008). A common theme among diverse case studies of regulated rivers is the loss of physical and ecological complexity (Surian 1999; Graf et al. 2002; Peipoch et al. 2015). This can be conceptualized with changes in hydrology, sediment supply, and large wood dynamics as first-order effects, changes in hydraulics, substrate mobility, and channel form as second-order effects, and changes in biota as third-order effects (Burke et al. 2009).

River restoration in any form is driven by the perception that a river is to some extent unhealthy and can be improved. The concept of river health is intuitively appealing to many people and easy to communicate at a general level to nonscientists (Karr 1999). Scientists debate whether such a conceptualization is useful or appropriate, however, as well as how to quantify river health (Boulton 1999; Fairweather 1999; Harris and Silveira 1999; Gippel et al. 2017). Much of this debate occurs in the biological literature, partly because river health is an example of ecosystem health (Norris and Thoms 1999).

River health is related to *ecological integrity*, which is the ability of an ecosystem to support and maintain a community of organisms with species composition, diversity, and functional organization similar to those within natural habitats in the same region (Parrish et al. 2003). This definition of ecological integrity emphasizes biota, but implicitly includes physical and chemical processes that sustain the biota.

Biologists typically explicitly include physical and chemical aspects of rivers in the consideration of river health, as exemplified by defining *river health* as the degree to which a river's energy source, water quality, and flow regime, plus its biota and their habitats, match the natural condition at all scales (Karr 1991; Harris and Silveira 1999). Numerous qualitative and quantitative metrics of river health have been developed. These are typically focused on biological metrics (Harris and Silveira 1999; Karr 1999; Everall et al. 2017), although cumulative metrics sometimes include measures of water quality (Bunn et al. 1999), habitat (Maddock 1999; Norris and Thoms 1999; Kim and An 2015), or flow regime (Richter et al. 1996; Poff et al. 2010; Mackay et al. 2014).

Many of the biological metrics used to characterize river health focus on some aspect of biodiversity. *Biodiversity* is typically defined in terms of number of species within a given ecosystem, but can be quantified in a wide variety of ways, each of which provides specific information about the ecosystem under consideration. Biodiversity reflects biological influences such as competition and predation, as well as physical influences such as the diversity, abundance, and stability of habitat, and the connectivity of habitat (Gaston and Spicer 2004).

Geomorphologists have been slower to develop metrics of physical river condition to facilitate quantification of difference between contemporary and reference conditions for a river, as well as evaluation of river health. Geomorphic conceptualizations of rivers emphasize diversity of form and process through space and through time (McDonald et al. 2004; Brierley and Fryirs 2005; Pasternack 2013). Diversity of form and process reflects the hydrology, sediment supply, large wood, hydraulics, substrate, geomorphic history, and biota at the site. A desirable level of physical diversity can constitute *physical integrity*, which Graf (2001) defines as a set of active fluvial processes and landforms such that the river maintains dynamic equilibrium, with adjustments not exceeding limits of change defined by societal values. In other words, a river has physical integrity when river process and form are actively connected under the current hydrologic and sediment regime.

A geomorphic perspective on river health would characterize a healthy river as having two basic characteristics. First, a healthy river has the ability to adjust form and process in response to changes in water, sediment, and wood inputs, whether these changes occur over many decades to centuries (e.g. climate variability) or over relatively short time periods (e.g. a large flood or landslide). Second, a healthy river has spatial and temporal ranges of water, sediment, and large wood inputs and river geometry similar to those present under natural conditions. Discussions of how to evaluate river health have taken on increased importance as broad governmental regulations such as the European Union's Water Framework and Habitats Directive have mandated delineation of diverse aspects of river health (Newson and Large 2006; Wernersson et al. 2015).

One danger involved in this type of national or transnational evaluation is that of oversimplifying what constitutes a natural or healthy river. Some rivers are naturally depauperate in species, for example, because of harsh physical or chemical conditions or a history of geographic isolation (e.g. Tolkinnen et al. 2015). Some rivers receive large sediment inputs and exhibit substantial channel instability because of natural factors such as semiarid climate or erodible lithology in the watershed (e.g. Constantine et al. 2014; Rickenmann et al. 2016). Some rivers have low levels of longitudinal connectivity as a result of natural flow intermittency or large numbers of in-channel obstructions such as logjams and beaver dams (e.g. Mould and Fryirs 2017). Recognition of complexity and diversity as inherent properties within a river network and between river networks remains crucial for research and management of rivers. This takes us back to viewing rivers in the context of the greater landscape.

8.6 Rivers with a History

As explained in Chapter 6, a river is a physical system with a history. The influence of previous climatic and tectonic regimes on river form and process can extend back in time beyond the Quaternary because of the slow response of some aspects of river networks to change. Among these aspects are:

- topography;
- the spatial arrangement of river channels within a network;
- relief ratio;
- drainage density for river segments larger than first- and second-order channels;
- river longitudinal profiles; and
- valley geometry.

Consequently, inherited characteristics of these features can continue to influence contemporary process and form, as illustrated by continental-scale controls on large rivers (Figure 8.2) (Potter 1978).

At smaller spatial scales, a single river channel flowing across different lithologies, structural features, or tectonic zones can exhibit striking differences in river and valley geometry and rate of incision in response to geological controls that formed millions to hundreds of millions of years ago. Examples include the lower Mississippi River in the eastern United States. This sinuous river can be subdivided into reaches that differ in gradient and sinuosity where the river crosses more erosionally resistant Tertiary-age sediments and fault zones, even though many of these faults show relatively little recent activity (Schumm et al. 2000). Rivers with drainage areas larger than 10 km^2 in the Central Apennines of Italy have long-profile convexities where they cross faults that have undergone an increase in displacement during the past million years, whereas those crossing faults with constant displacement rates lack such convexities (Whittaker et al. 2008).

Landscape configurations or persistent erosional and depositional features relict from Quaternary glaciation provide another example of how past events continue to influence river form and process. Glaciated mountains can have distinctly different valley geometry above and below the elevation limits of Pleistocene valley glaciers, with greater cross-sectional area and steeper valley walls in glacial valleys relative to fluvial valleys (Montgomery 2002; Amerson et al. 2008; Livers and Wohl 2015). Glaciated and fluvial portions of a mountain range can also be eroding at different rates in response to the effects of differing sediment supply and base-level controls (Anderson et al. 2006c). Tributary glacial valleys that eroded to a base level defined by the upper level of the main valley glacier can persist as hanging valleys with large vertical drops between the tributary valley mouth and the main valley floor for thousands of years after glacial ice retreats.

Recessional moraines are persistent depositional features perpendicular to valley orientation that create local base levels and valley segments with lower river gradient and finer substrate than segments immediately up- and downstream, even long after the moraine is incised by a river. Recessional moraines that fill with meltwater as valley glaciers retreat can fail catastrophically. Although individual moraines fail only once, failure of successive moraines along a valley can create numerous outburst floods along a river network over periods of decades to centuries as moraines successively up-valley are abruptly drained. The discharge and stream power of these *outburst floods* typically greatly exceed the discharge and stream power generated during annual floods induced by rainfall or snowmelt (Cenderelli 2000; O'Connor et al. 2013), and annual floods may be largely incapable of modifying the outburst-flood erosional and depositional features (Cenderelli and Wohl 2001). Portions of a valley subject to repeated outburst floods can also become less responsive in that earlier floods have already modified valley morphology to convey exceptionally large discharges (Cenderelli and Wohl 2003).

The continental-scale ice sheets that covered portions of North America, northern Europe, and northern Asia also left enduring signatures on river networks, diverting existing channels, altering water and sediment supply to channels beyond the ice margins, and changing local river gradients via isostatic flexure of the crust. One of the most spectacular categories of ice-sheet effects on river networks is the occurrence of megafloods during periods of glacial retreat. *Megafloods* are relatively short-duration flows that constitute the largest known freshwater floods, with discharges that generally exceed 1 million m^3/s (Baker 2013). Although other mechanisms, such as failure of rock dams or caldera lake impoundments, can generate megafloods, most were associated with ice-marginal lakes or water released from within the ice sheet. Among the megafloods documented thus far (Baker 2013) are those of:

- the Channeled Scabland in Washington, USA;
- the Laurentide Ice Sheet ice-marginal lakes in north-central North America, which flowed down the Mississippi, St. Lawrence, Mackenzie, and Hudson rivers;
- the Patagonian Ice Sheet of southern Argentina and Chile;
- Icelandic jökulhlaups;
- the Fennoscandian Ice Sheet, which drained southward and influenced the English Channel and the North Sea; and
- the northern mountain areas of central Asia, including Kirgizstan, Mongolia, and Siberia.

These exceptionally large floods created erosional and depositional features of such large magnitude and extent that subsequent geomorphic processes during the Holocene have only partially – or in some cases, little – modified megaflood terrains.

More recent Holocene history can also influence river process and form. River response to disturbance partly depends on the time elapsed since the last disturbance of a similar magnitude. Along mountainous headwater catchments in which wildfires and subsequent rainfall induce debris flows, such flows strongly influence channel morphology only if a minimum period has passed since the last debris flow (Wohl and Pearthree 1991). This minimum period is necessary for sufficient sediment to accumulate in the channel to be eroded by the next debris flow. Another example comes from the Drôme River in France, where a flood in 1978 caused avulsion, channel straightening, and incision (Toone et al. 2014); 53 years had passed since a flood of equivalent or larger magnitude. In contrast, a hundred-year flood in 1994 caused relatively minor channel changes, partly in response to inherent channel structural features imposed by the 1978 flood.

Human use of resources can also continue to influence river process and form long after the relevant human activity has ceased. Earlier chapters provided numerous examples of such influences, including mill dams along rivers in the eastern United States (Walter and Merritts 2008). A 1998 study in the southern Appalachian Mountains found that whole-watershed land use in the 1950s was the best predictor of contemporary invertebrate and fish diversity because of persistent effects on aquatic habitat (Harding et al. 1998). A century after cut logs for railroad ties were floated down streams in the Medicine Bow National Forest of Wyoming, USA, the streams used for log floating had less instream wood, lower densities of large riparian trees, lower channel complexity, a greater proportion of riffles, and fewer pools than did otherwise analogous streams that were not used for log floating (Young et al. 1994; Ruffing et al. 2015). As noted earlier, 200 years may be required before instream wood volumes completely recover following timber harvest (Bragg et al. 2000; Stout et al. 2018).

The cumulative effects of historical processes occurring over different temporal and spatial scales continue to influence contemporary river process and form (Wohl 2018a). Failure to recognize this influence can limit understanding of rivers and lead to erroneous predictions of river response to ongoing climate change and management actions.

8.7 The Greater Context

Isaac Asimov once wrote – perhaps in a moment of frustration – that the only constant is change. This phrase can be adapted to rivers in at least two contexts. First, natural rivers are continually adjusting process and form – spatial distribution of hydraulic forces; water chemistry; sediment entrainment, transport, and deposition; instream and floodplain wood; bed configuration; channel cross-sectional geometry; channel planform; reach gradient and longitudinal profile – in response to changing inputs of water, sediment, and wood, or to continuing development of the river. Indeed, one definition of a natural – as opposed to a completely engineered – river is that a natural river possesses physical integrity because it is able to adjust to changing inputs. River process and form reflect some balance between continually changing external inputs of water, solutes, sediment, and large wood, and ongoing adjustments within the river. Under these circumstances, considerable insight can be gained by asking why, under variable external forcing, rivers do not change even more.

The second context for adapting Asimov's phrase to rivers is that, although any river or river segment follows the basic laws of physics and chemistry, characterizing feedbacks between channel process and form using a numerical equation or qualitative conceptual model that applies to all rivers is limited by the place-specific effects of lithology, tectonics, weathering regime, landscape history, river history, seasonal presence of ice cover or cyclones, and so forth. The only constant within a river network is changes through time and space. The only constant among river networks is river-specific changes in the interactions between process and form. Recognition of these characteristics further emphasizes the importance of understanding landscape context for any particular river network or river segment.

The nineteenth-century Scottish-American conservationist John Muir famously wrote in 1911, "When we try to pick out anything by itself, we find it hitched to everything else in the Universe." Many other thinkers have stated the same concept in different words, recognizing the interconnectedness of natural systems and people. Rivers are no exception to this rule, as I have emphasized from the opening pages of this book. As the intensity and extent of human alteration of river form and process have accelerated globally since the mid-twentieth century, abundant evidence has appeared that reflects riverine influences on the entire critical zone. Changes in coastal environments provide a vivid example.

Human activities now dominate nitrogen budgets in regions such as Asia, Europe, and North America (Boyer et al. 2006). Substantial increases in nitrogen yields to rivers result from industrial-scale agriculture, feedlots, and septic systems. Riverine corridors have been simplified via channelization, removal of riparian vegetation, levees, and flow regulation, all of which reduce channel–floodplain connectivity. This simplification reduces nitrogen retention and processing by rivers. Greater inputs and less storage create a "one-two punch," resulting in substantially increased nitrogen fluxes down rivers to coastal areas. This has created eutrophication of estuaries and other nearshore environments. Consequently, fluxes of nitrogen now exceed planetary boundaries for sustainability and resilience (Steffen et al. 2015).

For the most part, we cannot yet predict in detail how diverse changes in rivers resulting directly and indirectly from human activities will affect the greater landscape. Synthesizing studies in the northern high latitudes, for example, Woo (2010) and Carmack et al. (2016) explain how freshwater discharge during spring snowmelt influences numerous and diverse processes in the nearshore zone and greater Arctic Ocean (Figure 8.13). Freshwater discharge influences the dynamics of coastal sea ice, terrestrial sediment and organic matter plumes into the ocean, and thermohaline circulation in polar seas. The complex effects of global warming on these interactions remain largely unknown. Less sea ice may modify existing energy and moisture fluxes, for example, and thus alter coastal storm patterns and inland water balances (Woo 2010; Carmack et al. 2016).

Unfortunately, we too commonly recognize the importance of landscape context and connectivity within and between river networks once our activities have altered connectivity and caused unintended negative consequences. Examples span spatial scales from headwaters to the world's largest rivers, climatic zones from hot and cold deserts to rainforests, and tectonically active to passive terrains with varying lithology and structure. The point is not to induce despair as we contemplate past failures to account for the importance of connectivity, but rather to highlight the importance of being fully cognizant of various forms of connectivity as we move forward with river management. Rivers

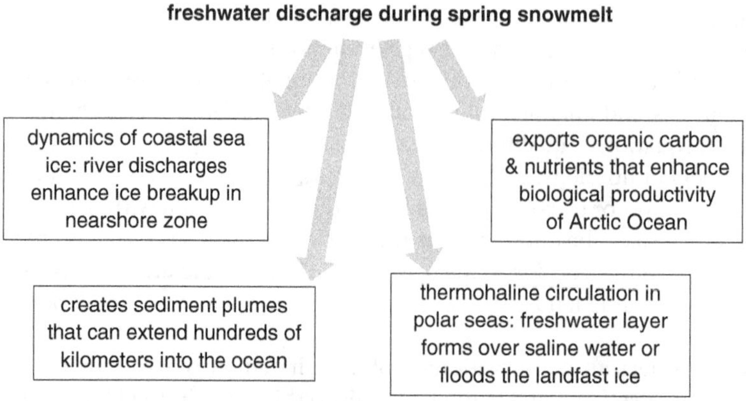

Figure 8.13 Schematic diagram of the effects of freshwater discharge during spring snowmelt from rivers draining to the Arctic Ocean.

are physical systems with a history and rivers exist within a global landscape that includes the atmosphere, land masses, oceans, and ground water, as well as all the wondrously diverse organisms that live on our planet. Rivers are at the heart of nearly every landscape on Earth, and river form and process are more integral to human communities than is any other single landscape component. The challenge of developing ways to coexist with healthy, functional rivers is integral to both a scientific understanding of rivers and the application of that understanding through river management. This challenge can only be met by treating rivers as part of the greater landscape. If we do not view rivers in this integrative, holistic context, we make the same mistakes over and over, risking nothing less than the survival of our own societies.

References

Aalto, R., Maurice-Bourgoin, L., Dunne, T. et al. (2003). Episodic sediment accumulation on Amazonian floodplains influenced by El Niño/Southern Oscillation. *Nature* 425: 493–497.

Aalto, R., Lauer, J.W., and Dietrich, W.E. (2008). Spatial and temporal dynamics of sediment accumulation and exchange along Strickland River floodplains (Papua New Guinea) over decadal-to-centennial timescales. *Journal of Geophysical Research* 113, F01S04.

Aarts, B.G.W., van den Brink, F.W.B., and Nienhuis, P.H. (2004). Habitat loss as the main cause of the slow recovery of fish faunas of regulated large rivers in Europe: the transversal floodplain gradient. *River Research and Applications* 20: 3–23.

Abbe, T. and Brooks, A. (2011). Geomorphic, engineering, and ecological considerations when using wood in river restoration. In: *Stream Restoration in Dynamic Fluvial Systems: Scientific Approaches, Analyses, and Tools* (eds. A. Simon, S.J. Bennett and J.M. Castro), 419–451. Washington, DC: American Geophysical Union Press.

Abbe, T.B. and Montgomery, D.R. (2003). Patterns and processes of wood debris accumulation in the Queets River basin, Washington. *Geomorphology* 51: 81–107.

Aberle, J. and Smart, G.M. (2003). The influence of roughness structure on flow resistance on steep slopes. *Journal of Hydraulic Research* 41: 259–269.

Aberle, J., Nikora, V., Henning, M. et al. (2010). Statistical characterization of bed roughness due to bed forms: a field study in the Elbe River at Aken, Germany. *Water Resources Research* 46, W03521.

Abrahams, A.D., Li, G., and Atkinson, J.F. (1995). Step-pool streams: adjustment to maximum flow resistance. *Water Resources Research* 31: 2593–2602.

Ackers, P. and White, W.R. (1973). Sediment transport: new approach and analysis. *Journal of the Hydraulics Division ASCE* 99: 204–254.

Acreman, M. and Holden, J. (2013). How wetlands affect floods. *Wetlands* 33: 773–786.

Acuña, V., Datry, T., Marshall, J. et al. (2014). Why should we care about temporary waterways? *Science* 343: 1080–1081.

Adams, R.K. and Spotila, J.A. (2005). The form and function of headwater streams based on field and modeling investigations in the southern Appalachian Mountains. *Earth Surface Processes and Landforms* 30: 1521–1546.

Aizen, V., Aizen, E., and Melack, J. (1995). Characteristics of runoff formation at the Kirgizskiy Alatoo, Tien Shan. In: *Biogeochemistry of Seasonally Snow-Covered Catchments* (eds. K.A. Tonnessen, M.W. Williams and M. Tranter), 413–430. Wallingford: IAHS Publications no. 228.

Albertson, L.K. and Daniels, M.D. (2018). Crayfish ecosystem engineering effects on riverbed disturbance and topography are mediated by size and behavior. *Freshwater Science* 37: 836–844.

Rivers in the Landscape, Second Edition. Ellen Wohl.

Alexander, J. and Cooker, M.J. (2016). Moving boulders in flash floods and estimating flow conditions using boulders in ancient deposits. *Sedimentology* 63: 1582–1595.

Alexandrov, Y., Laronne, J.B., and Reid, I. (2003). Suspended sediment concentration and its variation with water discharge in a dryland ephemeral channel, northern Negev, Israel. *Journal of Arid Environments* 53: 73–84.

Al-Farraj, A. and Harvey, A.M. (2005). Morphometry and depositional style of late Pleistocene alluvial fans; Wadi Al-Bih, northern UAE and Oman. In: *Alluvial Fans: Geomorphology, Sedimentology, Dynamics*, vol. 251 (eds. A.M. Harvey, A.E. Mather and M. Stokes), 85–94. Geological Society Special Publications.

Alho, P., Kukko, A., Hyyppä, H. et al. (2009). Application of boat-based laser scanning for river survey. *Earth Surface Processes and Landforms* 34: 1831–1838.

Alin, S.R., Aalto, R., Goni, M. et al. (2008). Biogeochemical characterization of carbon sources in the Strickland and Fly Rivers, Papua New Guinea. *Journal of Geophysical Research: Earth Surface* 113, F01S05.

Allan, J.D. (1995). *Stream Ecology: Structure and Function of Running Waters*. London: Chapman and Hall.

Allan, A.F. and Frostick, L. (1999). Framework dilation, winnowing, and matrix particle size: the behavior of some sand-gravel mixtures in a laboratory flume. *Journal of Sedimentary Research* 69: 21–26.

Allen, J.R.L. (1965). A review of the origin and characteristics of recent alluvial sediments. *Sedimentology* 5: 89–191.

Allen, J.R.L. (1982). *Sedimentary Structures: Their Character and Physical Basis*, Developments in Sedimentology, vol. 1, 30. Amsterdam: Elsevier.

Allen, J.R.L. (1985). *Principles of Physical Sedimentology*. London: Allen and Unwin.

Allen, H.H. and Leech, J.R. (1997). *Bioengineering for Streambank Erosion Control. Report 1 - Guidelines*. Vicksburg, MS: US Army Corps of Engineers http://www.dtic.mil/docs/citations/ADA326294.

Alley, W.M., Healy, R.W., LaBaugh, J.W., and Reilly, T.E. (2002). Flow and storage in groundwater systems. *Science* 296: 1985–1990.

Allmendinger, N.E., Pizzuto, J.E., Potter, N. et al. (2005). The influence of riparian vegetation on stream width, eastern Pennsylvania, USA. *Geological Society of America Bulletin* 117: 229–243.

Allred, T.M. and Schmidt, J.C. (1999). Channel narrowing by vertical accretion along the Green River near Green River, Utah. *Geological Society of America Bulletin* 111: 1757–1772.

Alongi, D.M. (2014). Carbon cycling and storage in mangrove forests. *Annual Review of Marine Science* 6: 195–219.

Alonso, C.V. (2004). Transport mechanics of stream-borne logs. In: *Riparian Vegetation and Fluvial Geomorphology* (eds. S.J. Bennett and A. Simon), 59–69. Washington, DC: American Geophysical Union Press.

Alsdorf, D., Dunne, T., Melack, J. et al. (2005). Diffusion modeling of recessional flow on central Amazonian floodplains. *Geophysical Research Letters* 32, L21405.

Ambili, V. and Narayana, A.C. (2014). Tectonic effects on the longitudinal profiles of the Chaliyar River and its tributaries, Southwest India. *Geomorphology* 217: 37–47.

Amerson, B.E., Montgomery, D.R., and Meyer, G. (2008). Relative size of fluvial and glaciated valleys in Central Idaho. *Geomorphology* 93: 537–547.

Amos, C.B. and Burbank, D.W. (2007). Channel width response to differential uplift. *Journal of Geophysical Research* 112, F02010.

Anderson, R.S. and Anderson, S.P. (2010). *Geomorphology: The Mechanics and Chemistry of Landscapes*. Cambridge: Cambridge University Press.

Anderson, S.P. and Dietrich, W.E. (2001). Chemical weathering and runoff chemistry in a steep headwater catchment. *Hydrological Processes* 15: 1791–1815.

Anderson, S.P., Longacre, S.A., and Kraal, E.R. (2003). Patterns of water chemistry and discharge in the glacier-fed Kennicott River, Alaska: evidence for subglacial water storage cycles. *Chemical Geology* 202: 297–312.

Anderson, J.K., Wondzell, S.M., Gooseff, M.N., and Haggerty, R. (2005). Patterns in stream longitudinal profiles and implications for hyporheic exchange flow at the H.J. Andrews Experimental Forest, Oregon, USA. *Hydrological Processes* 19: 2931–2949.

Anderson, B.G., Rutherfurd, I.D., and Western, A.W. (2006a). An analysis of the influence of riparian vegetation on the propagation of flood waves. *Environmental Modelling and Software* 21: 1290–1296.

Anderson, C.B., Griffith, C.R., Rosemond, A.D. et al. (2006b). The effects of invasive North American beavers on riparian plant communities in Cape Horn, Chile: do exotic beavers engineer differently in sub-Antarctic ecosystems? *Biological Conservation* 128: 467–474.

Anderson, R.S., Riihimaki, C.A., Safran, E.B., and MacGregor, K.R. (2006c). Facing reality: late Cenozoic evolution of smooth peaks, glacially ornamented valleys, and deep river gorges of Colorado's Front Range. In: *Tectonics, Climate, and Landscape Evolution* (eds. S.D. Willett, N. Hovius, M.T. Brandon and D.M. Fisher), 397–418. Boulder, CO: Geological Society of America Special Paper 398.

Anderson, S.W., Anderson, S.P., and Anderson, R.S. (2015). Exhumation by debris flows in the 2013 Colorado Front Range storm. *Geology* 41: 391–394.

Andreasson, J., Bergstrom, S., Carlsson, B., and Graham, L.P. (2003). The effect of downscaling techniques on assessing water resources impacts from climate change scenarios. In: *Water Resources Systems – Water Availability and Global Change* (eds. S. Franks, G. Bloschl, M. Kumagai, et al.), 160–164. Wallingford: IAHS Publications no. 280.

Andreoli, A., Comiti, F., and Lenzi, M.A. (2007). Characteristics, distribution and geomorphic role of large woody debris in a mountain stream of the Chilean Andes. *Earth Surface Processes and Landforms* 32: 1675–1692.

Andrews, E.D. (1980). Entrainment of gravel from naturally sorted riverbed material. *Geological Society of America Bulletin* 94: 1225–1231.

Andrews, E.D. (1984). Bed-material entrainment and hydraulic geometry of gravel-bed rivers in Colorado. *Geological Society of America Bulletin* 95: 371–378.

Andrews, E.D. and Erman, D.C. (1986). Persistence in the size distribution of surficial bed material during an extreme snowmelt flood. *Water Resources Research* 22: 191–197.

Andrews, E.D. and Nankervis, J.M. (1995). Effective discharge and the design of channel maintenance flows for gravel-bed rivers. In: *Natural and Anthropogenic Influences in Fluvial Geomorphology* (eds. J.E. Costa, A.J. Miller, K.W. Potter and P.R. Wilcock), 151–164. Washington, DC: American Geophysical Union.

Andrews, E.D. and Smith, J.D. (1992). A theoretical model for calculating marginal bed load transport rates of gravel. In: *Dynamics of Gravel-Bed Rivers* (eds. P. Billi, R.D. Hey, C.R. Thorne and P. Tacconi), 41–52. Chichester: Wiley.

Anthony, D.J. and Harvey, M.D. (1991). Stage-dependent cross-section adjustments in a meandering reach of Fall River, Colorado. *Geomorphology* 4: 187–203.

Anthony, E.J., Brunier, G., Besset, M. et al. (2015). Linking rapid erosion of the Mekong River delta to human activities. *Scientific Reports* 5: 14745.

Anton, L., Mather, A.E., Stokes, M. et al. (2015). Exceptional river gorge formation from unexceptional floods. *Nature Communications* 6, 7963.

Appling, A.P., Bernhardt, E.S., and Stanford, J.A. (2014). Floodplain biogeochemical mosaics: a multidimensional view of alluvial soils. *Journal of Geophysical Research: Biogeosciences* 119: 1538–1553.

Arboleya, M.-L., Babault, J., Owen, L.A. et al. (2008). Timing and nature of quaternary fluvial incision in the Ouarzazate foreland basin, Morocco. *Journal of the Geological Society of London* 165: 1059–1073.

Arrigoni, A.S., Poole, G.C., Mertes, L.A.K. et al. (2008). Buffered, lagged, or cooled? Disentangling hyporheic influences on temperature cycles in stream channels. *Water Resources Research* 44, W09418.

Arthington, A.H. and Pusey, B.J. (2003). Flow restoration and protection in Australian rivers. *River Research and Applications* 19: 377–395.

Arthington, A.H., Bunn, S.E., Poff, N.L., and Naiman, R.J. (2006). The challenge of providing environmental flow rules to sustain river ecosystems. *Ecological Applications* 16: 1311–1318.

Ashley, G.M. (1990). Classification of large-scale subaqeous bedforms: a new look at an old problem. *Journal of Sedimentary Petrology* 60: 160–172.

Ashmore, P. (1991a). Channel morphology and bed load pulses in braided, gravel-bed streams. *Geografiska Annaler, Series A* 73: 37–52.

Ashmore, P.E. (1991b). How do gravel-bed rivers braid? *Canadian Journal of Earth Sciences* 28: 326–341.

Ashmore, P. (2013). Morphology and dynamics of braided rivers. In: *Treatise on Fluvial Geomorphology* (ed. E. Wohl), 290–312. Amsterdam: Elsevier.

Ashmore, P.E. and Day, T.J. (1988). Effective discharge for suspended sediment transport in streams of the Saskatchewan River basin. *Water Resources Research* 24: 864–870.

Ashworth, P.J. and Ferguson, R.I. (1989). Size-selective entrainment of bed load in gravel bed streams. *Water Resources Research* 25: 627–634.

Ashworth, P.J., Best, J.L., and Jones, M.A. (2007). The relationship between channel avulsion, flow occupancy and aggradation in braided rivers: insights from an experimental model. *Sedimentology* 54: 497–513.

Aslan, A. and Autin, W.J. (1998). Holocene flood-plain soil formation in the southern lower Mississippi Valley: implications for interpreting alluvial paleosols. *Geological Society of America Bulletin* 110: 433–449.

Aslan, A. and Autin, W.J. (1999). Evolution of the Holocene Mississippi River floodplain, Ferriday, Louisiana: insights on the origin of fine-grained floodplains. *Journal of Sedimentary Research* 69: 800–815.

Attal, M. and Lavé, J. (2006). Changes of bedload characteristics along the Marsyandi River (central Nepal): implications for understanding hillslope sediment supply, sediment load evolution along fluvial networks, and denudation in active orogenic belts. In: *Tectonics, Climate, and Landscape Evolution* (eds. S.D. Willett, N. Hovius, M.T. Brandon and D.M. Fisher), 143–171. Geological Society of America Special Paper 398.

Attal, M., Tucker, G.E., Whittaker, A.C. et al. (2008). Modeling fluvial incision and transient landscape evolution: influence of dynamic channel adjustment. *Journal of Geophysical Research* 113, F03013.

Aufdenkampe, A.K., Mayorga, E., Raymond, P.A. et al. (2011). Riverine coupling of biogeochemical cycles between land, oceans, and atmosphere. *Frontiers in Ecology and the Environment* 9: 53–60.

Aydin, A. and Basu, A. (2005). The Schmidt hammer in rock material characterization. *Engineering Geology* 81: 1–14.

Azari, M., Moradi, H.R., Saghafian, B., and Faramarzi, M. (2016). Climate change impacts on streamflow and sediment yield in the north of Iran. *Hydrological Sciences Journal* 61: 123–133.

Baartman, J.E.M., Masselink, R., Keesstra, S.D., and Temme, A.J.A.M. (2013). Linking landscape morphological complexity and sediment connectivity. *Earth Surface Processes and Landforms* 38: 1457–1471.

Bábek, O., Sedlacek, J., Novak, A., and Letal, A. (2018). Electrical resistivity imaging of anastomosing river subsurface stratigraphy and possible controls of fluvial style change in a graben-like basin, Czech Republic. *Geomorphology* 317: 139–156.

Bagnold, R.A. (1977). Bedload transport by natural waters. *Water Resources Research* 13: 303–312.

Baker, V.R. (1974). Paleohydraulic interpretation of quaternary alluvium near Golden, Colorado. *Quaternary Research* 4: 94–112.

Baker, V.R. (1978). Large-scale erosional and depositional features of the Channeled Scabland. In: *The Channeled Scabland* (eds. V.R. Baker and D. Nummedal), 81–115. Washington, DC: National Aeronautics and Space Administration.

Baker, V.R. (1987). Paleoflood hydrology and extraordinary flood events. *Journal of Hydrology* 96: 79–99.

Baker, V.R. (1988). Flood erosion. In: *Flood Geomorphology* (eds. V.R. Baker, R.C. Kochel and P.C. Patton), 81–95. New York: Wiley.

Baker, V.R. (2013). Global late Quaternary fluvial paleohydrology: with special emphasis on paleofloods and megafloods. In: *Treatise on Fluvial Geomorphology* (ed. E. Wohl), 512–527. Amsterdam: Elsevier.

Baker, V.R. and Kochel, R.C. (1988). Flood sedimentation in bedrock fluvial systems. In: *Flood Geomorphology* (eds. V.R. Baker, R.C. Kochel and P.C. Patton), 123–137. New York: Wiley.

Baker, V.R. and Pickup, G. (1987). Flood geomorphology of the Katherine Gorge, Northern Territory, Australia. *Geological Society of America Bulletin* 98: 635–646.

Baker, D.B., Richards, P.R., Loftus, T.T., and Kramer, J.W. (2004). A new flashiness index: characteristics and applications to Midwestern rivers and streams. *Journal of the American Water Resources Association* 40: 503–522.

Baker, D.W., Bledsoe, B.P., Albano, C.M., and Poff, N.L. (2011). Downstream effects of diversion dams on sediment and hydraulic conditions of Rocky Mountain streams. *River Research and Applications* 27: 388–401.

Ballantyne, C.K. (2002). Paraglacial geomorphology. *Quaternary Science Reviews* 21: 1935–2017.

Ballesteros-Cánova, J.A., Stoffel, M., George, S.S., and Hirschboeck, K. (2015a). A review of flood records from tree rings. *Progress in Physical Geography* 39: 794–816.

Ballesteros-Cánova, J.A., Marquez-Penaranda, J.F., Sanchez-Silva, M. et al. (2015b). Can tree tilting be used for paleoflood discharge estimations? *Journal of Hydrology* 529: 480–489.

Bănăduc, D., Rey, S., Trichkova, T. et al. (2016). The lower Danube River-Danube Delta-North West Black Sea: a pivotal area of major interest for the past, present and future of its fish fauna – a short review. *Science of the Total Environment* 545–546: 137–151.

Barnard, P.L., Owen, L.A., Sharma, M.C., and Finkel, R.C. (2004). Late Quaternary (Holocene) landscape evolution of a monsoon-influenced high Himalayan valley, Gori Ganga, Nanda Devi, NE Garhwal. *Geomorphology* 61: 91–110.

Barnard, P.L., Owen, L.A., and Finkel, R.C. (2006). Quaternary fans and terraces in the Kumbu Himal south of Mount Everest; their characteristics, age and formation. *Journal of the Geological Society of London* 163: 383–399.

Barnes, H.L. (1956). Cavitation as a geological agent. *American Journal of Science* 254: 493–505.

Barnes, H.H. 1967. Roughness characteristics of natural channels. US Geological Survey Water-Supply Paper 1849.

Barnes, C. and Bonell, M. (2005). How to choose an appropriate catchment model. In: *Forests, Water and People in the Humid Tropics* (eds. M. Bonnell and L.A. Bruijnzeel), 717–741. Cambridge: Cambridge University Press.

Barnett, T.P., Adam, J.C., and Lettenmaier, D.P. (2005). Potential impacts of a warming climate on water availability in snow-dominated regions. *Nature* 438: 303–309.

Barry, R.G. (2008). *Mountain Weather and Climate*, 3e. Cambridge: Cambridge University Press.

Barry, R.G. and Chorley, R.J. (1987). *Atmosphere, Weather and Climate*, 5e. Methuen, London.

Bar-Yam, B. (1997). *Dynamics of Complex Systems*. Reading, MA: Addison Wesley.

Bathurst, J.C. (1985). Flow resistance estimation in mountain rivers. *Journal of Hydraulic Engineering* 111: 625–641.

Bathurst, J.C. (1990). Tests of three discharge gauging techniques in mountain rivers. In: *Hydrology of Mountainous Areas* (ed. L. Molnar), 93–100. Wallingford: IAHS Publications no. 190.

Bathurst, J.C. (2002). At-a-site variation and minimum flow resistance for mountain rivers. *Journal of Hydrology* 269: 11–26.

Battin, T.J., Kaplan, L.A., Findlay, S. et al. (2008). Biophysical controls on organic carbon fluxes in fluvial networks. *Nature Geoscience* 1: 95–100.

Battin, T.J., Luyssaert, S., Kaplan, L.A. et al. (2009). The boundless carbon cycle. *Nature Geoscience* 2: 598–600.

Baumgart-Kotarba, M., Bravard, J.-P., Chardon, M. et al. (2003). High-mountain valley floors evolution during recession of alpine glaciers in the Massif des Ecrins, France. *Geographia Polonica* 76: 65–87.

Baustian, M.M., Stagg, C.L., Perry, C.L. et al. (2017). Relationships between salinity and short-term soil carbon accumulation rates from marsh types across a landscape in the Mississippi River Delta. *Wetlands* 37: 313–324.

Bayazit, M. (2015). Nonstationarity of hydrological records and recent trends in trend analysis: a state-of-the-art review. *Environmental Processes* 2: 527–542.

Bayley, P.B. (1991). The flood-pulse advantage and the restoration of river-floodplain systems. *Regulated Rivers: Research & Management* 6: 75–86.

Baynes, E.R.C., Attal, M., Niedermann, S. et al. (2015). Erosion during extreme flood events dominates Holocene canyon evolution in Northeast Iceland. *Proceedings of the National Academy of Sciences of the United States of America* https://doi.org/10.1073/pnas.1415443112.

Baynes, E.R.C., Lague, D., and Kermarrec, J.J. (2018). Supercritical river terraces generated by hydraulic and geomorphic interactions. *Geology* 46: 499–502.

Beaumont, C. and Quinlan, G. (1994). A geodynamic framework for interpreting crustal-scale seismic reflectivity patterns in compressional orogens. *International Journal of Geophysics* 116: 754–783.

Beauvais, A.A. and Montgomery, D.R. (1997). Are channel networks statistically self-similar? *Geology* 25: 1063–1066.

Becker, B. and Kromer, B. (1993). The continental tree-ring record – absolute chronology, ^{14}C calibration, and climatic change at 11 ka. *Palaeogeography, Palaeoclimatology, Palaeocology* 103: 67–71.

Beckman, N.D. and Wohl, E. (2014a). Carbon storage in mountainous headwater streams: the role of old-growth forest and logjams. *Water Resources Research* 50: 2376–2393.

Beckman, N.D. and Wohl, E. (2014b). Effects of forest stand age on the characteristics of logjams in mountainous forest streams. *Earth Surface Processes and Landforms* 39: 1421–1431.

Bednarek, A.T. and Hart, D.D. (2005). Modifying dam operations to restore rivers: ecological responses to Tennessee River dam mitigation. *Ecological Applications* 15: 997–1008.

Beechie, T.J., Liermann, M., Pollock, M.M. et al. (2006). Channel pattern and river-floodplain dynamics in forested mountain river systems. *Geomorphology* 78: 124–141.

Beer, A.R., Turowski, J.M., Fritschi, B., and Rieke-Zapp, D.H. (2015). Field instrumentation for high-resolution parallel monitoring of bedrock erosion and bedload transport. *Earth Surface Processes and Landforms* 40: 530–541.

Beer, A.R., Turowski, J.M., and Kirchner, J.W. (2017). Spatial patterns of erosion in a bedrock gorge. *Journal of Geophysical Research: Earth Surface* 122: 191–214.

Béjar, M., Vericat, D., Batalla, R.J., and Gibbins, C.N. (2018). Variation in flow and suspended sediment transport in a montane river affected by hydropeaking and instream mining. *Geomorphology* 310: 69–83.

Bellmore, J.R. and Baxter, C.V. (2014). Effects of geomorphic process domains on river ecosystems: a comparison of floodplain and confined valley segments. *River Research and Applications* 30: 617–630.

Belmont, P., Gran, K.B., Schottler, S.P. et al. (2011). Large shift in source of fine sediment in the upper Mississippi River. *Environmental Science and Technology* 45: 8804–8810.

Beltaos, S. (2002). Effects of climate on mid-winter ice jams. *Hydrological Processes* 16: 789–804.

Beltaos, S., Carter, T., Rowsell, R., and DePalma, S.G.S. (2018). Erosion potential of dynamic ice breakup in lower Athabasca River. Part I: field measurements and initial quantification. *Cold Regions Science and Technology* 149: 16–28.

Bencala, K.E. and Walters, R.A. (1983). Simulations of solute transport in a mountain pool-and-riffle stream: a transient storage model. *Water Resources Research* 19: 718–724.

Benda, L.E. (1990). The influence of debris flows on channels and valley floors in the Oregon Coast Range, USA. *Earth Surface Processes and Landforms* 15: 457–466.

Benda, L.E. and Dunne, T. (1987). Sediment routing by debris flow. In: *Erosion and Sedimentation in the Pacific Rim* (eds. W.G. Wells II, P.M. Wohlgemuth and A.G. Campbell), 213–223. Wallingford: IAHS Publications no. 165.

Benda, L.E. and Sias, J.C. (2003). A quantitative framework for evaluating the mass balance of in-stream organic debris. *Forest Ecology and Management* 172: 1–16.

Benda, L.E., Poff, N.L., Tague, C. et al. (2002). How to avoid train wrecks when using science in environmental problem solving. *BioScience* 52: 1127–1136.

Benda, L., Veldhuisen, C., and Black, J. (2003a). Debris flows as agents of morphological heterogeneity at low-order confluences, Olympic Mountains, Washington. *Geological Society of America Bulletin* 115: 1110–1121.

Benda, L.E., Miller, D., Sais, J. et al. (2003b). Wood recruitment processes and wood budgeting. In: *The Ecology and Management of Wood in World Rivers* (eds. S.V. Gregory, K.L. Boyer and A.M. Gurnell), 49–73. Bethesda, MD: American Fisheries Society.

Benda, L., Andras, K., Miller, D., and Bigelow, P. (2004a). Confluence effects in rivers: interactions of basin scale, network geometry, and disturbance regimes. *Water Resources Research* 40, W05402.

Benda, L., Poff, N.L., Miller, D. et al. (2004b). The Network Dynamics Hypothesis: how channel networks structure riverine habitats. *BioScience* 54: 413–427.

Benda, L.E., Hassan, M.A., Church, M., and May, C.L. (2005). Geomorphology of steepland headwaters: the transition from hillslopes to channels. *Journal of the American Water Resources Association* 41: 835–851.

Benito, G. and O'Connor, J. (2013). Quantitative paleoflood hydrology. In: *Treatise on Fluvial Geomorphology* (ed. E. Wohl), 459–474. Amsterdam: Elsevier.

Benito, G., Machado, M.J., and Perez-Gonzalez, A. (1996). Climate change and flood sensitivity in Spain. In: *Global Continental Changes: The Context of Palaeohydrology*, vol. 115 (eds. J. Branson, A.G. Brown and K.J. Gregory), 85–98. Geological Society Special Publication.

Benke, A.C. and Wallace, J.B. (2003). Influence of wood on invertebrate communities in streams and rivers. In: *The Ecology and Management of Wood in World Rivers*, vol. 37 (eds. S.V. Gregory, K.L. Boyer and A.M. Gurnell), 149–177. Bethesda, MD: American Fisheries Society Symposium.

Bennett, S.J., Simon, A., and Kuhnle, R. (1998). Temporal variations in point bar morphology within two incised river meanders, Goodwin Creek, Mississippi. *Water Resources Engineering* 98: 1422–1427.

Berg, M., Stengel, C., Trang, P.T.K. et al. (2007). Magnitude of arsenic pollution in the Mekong and Red River deltas – Cambodia and Vietnam. *Science of the Total Environment* 372: 413–425.

Berghuijs, W.R., Woods, R.A., and Hrachowitz, M. (2014). A precipitation shift from snow towards rain leads to a decrease in streamflow. *Nature Climate Change* 4: 583–586.

Bergkamp, G. (1998). A hierarchical view of the interactions of runoff and infiltration with vegetation and microtopography in semiarid shrublands. *Catena* 33: 201–220.

Berner, E.K. and Berner, R.A. (1987). *The Global Water Cycle: Geochemistry and Environment*. Englewood Cliffs, NJ: Prentice Hall.

Bernhardt, E.S. and Palmer, M.A. (2011). River restoration: the fuzzy logic of repairing reaches to reverse catchment scale degradation. *Ecological Applications* 21: 1926–1931.

Bernhardt, E.S., Palmer, M.A., Allan, J.D. et al. (2005). Synthesizing US river restoration efforts. *Science* 308: 636–637.

Bernhardt, E.S., Sudduth, E.B., Palmer, M.A. et al. (2007). Restoring rivers one reach at a time: results from a survey of US river restoration practitioners. *Restoration Ecology* 15: 482–493.

Bertoldi, W., Zanoni, L., Miori, S. et al. (2009). Interaction between migrating bars and bifurcations in gravel bed rivers. *Water Resources Research* 45, W06418.

Bertoldi, W., Zanoni, L., and Tubino, M. (2010). Assessment of morphological changes induced by flow and flood pulses in a gravel bed braided river: the Tagliamento River (Italy). *Geomorphology* 114: 348–360.

Bertoldi, W., Welber, M., Mao, L. et al. (2014). A flume experiment on wood storage and remobilization in braided river systems. *Earth Surface Processes and Landforms* 39: 804–813.

Bertoldi, W., Welber, M., Gurnell, A. et al. (2015). Physical modelling of the combined effect of vegetation and wood on river morphology. *Geomorphology* 246: 178–187.

Beschta, R.L. and Ripple, W.J. (2012). The role of large predators in maintaining riparian plant communities and river morphology. *Geomorphology* 157–158: 88–98.

Best, J.L. (1986). The morphology of river channel confluences. *Progress in Physical Geography* 10: 157–174.

Best, J.L. (1987). Flow dynamics at river channel confluences: implications for sediment transport and bed morphology. In: *Recent Developments in Fluvial Sedimentology* (eds. F.G. Ethridge, R.M. Flores and M.D. Harvey), 27–35. SEPM Special Publication 39.

Best, J.L. (1988). Sediment transport and bed morphology at river channel confluences. *Sedimentology* 35: 481–498.

Best, J.L. (1992). On the entrainment of sediment and initiation of bed defects: insights from recent developments within turbulent boundary layer research. *Sedimentology* 39: 797–811.

Best, J.L. (1993). On the interactions between turbulent flow structure, sediment transport and bedform development. In: *Turbulence: Perspectives on Flow and Sediment Transport* (eds. N.J. Clifford, J.R. French and J. Hardistry), 61–92. Chichester: Wiley.

Best, J.L. (2005). The fluid dynamics of river dunes: a review and some future research directions. *Journal of Geophysical Research* 110, F04S02.

Best, J.L. and Ashworth, P.J. (1997). Scour in large braided rivers and the recognition of sequence stratigraphic boundaries. *Nature* 387: 275–277.

Best, J.L. and Rhoads, B.L. (2008). Sediment transport, bed morphology and the sedimentology of river channel confluences. In: *River Confluences, Tributaries and the Fluvial Network* (eds. S. Rice, A. Roy and B.L. Rhoads), 45–72. Chichester: Wiley.

Best, J.L. and Roy, A.G. (1991). Mixing-layer distortion at the confluence of channels of different depth. *Nature* 350: 411–413.

Beven, K. (2012). *Rainfall-Runoff Modelling: The Primer*, 2e. Chichester: Wiley-Blackwell.

Beven, K.J. and Binley, A. (1992). The future of distributed models: model calibration and uncertainty prediction. *Hydrological Processes* 6: 279–298.

Beven, K. and Germann, P. (1982). Macropores and water flow in soils. *Water Resources Research* 18 (5): 1311–1325.

Beven, K.J. and Kirkby, M.J. (1979). A physically-based variable contributing area model of basin hydrology. *Hydrology Science Bulletin* 24: 43–69.

Biedenharn, D.S. and Thorne, C.R. (1994). Magnitude-frequency analysis of sediment transport in the lower Mississippi River. *Regulated Rivers: Research and Management* 9: 237–251.

Biedenharn, D.S., Elliott, C.M., and Watson, C.C. (1997). *The WES Stream Investigation and Streambank Stabilization Handbook*. Vicksburg, MS: US Army Engineers Waterways Experiment Station (WES).

Bieger, K., Rathjens, H., Allen, P.M., and Arnold, J.G. (2015). Development and evaluation of bankfull hydraulic geometry relationships for the physiographic regions of the United States. *Journal of the American Water Resources Association* 51: 842–858.

Bierman, P.R. and Montgomery, D.R. (2014). *Key Concepts in Geomorphology*. New York: W.H. Freeman.

Bierman, P.R. and Nichols, K.K. (2004). Rock to sediment – slope to sea with ^{10}Be – rates of landscape change. *Annual Review of Earth and Planetary Sciences* 32: 215–255.

Bierman, P.R., Reuter, J.M., Pavich, M. et al. (2005). Using cosmogenic nuclides to contrast rates of erosion and sediment yield in a semi-arid, arroyo-dominated landscape, Rio Puerco Basin, New Mexico. *Earth Surface Processes and Landforms* 30: 935–953.

Bilby, R.E. (1981). Role of organic debris dams in regulating the export of dissolved and particulate matter from a forested watershed. *Ecology* 62: 1234–1243.

Bilby, R.E. (2003). Decomposition and nutrient dynamics of wood in streams and rivers. In: *The Ecology and Management of Wood in World Rivers*, vol. 37 (eds. S.V. Gregory, K.L. Boyer and A.M. Gurnell), 135–147. Bethesda, MD: American Fisheries Society Symposium.

Bilby, R.E. and Ward, J.W. (1989). Changes in characteristics and function of woody debris with increasing size of streams in western Washington. *Transactions of the American Fisheries Society* 118: 368–378.

Bintanja, R. and Selten, F.M. (2014). Future increases in Arctic precipitation linked to local evaporation and sea-ice retreat. *Nature* 509: 479–482.

Birken, A.S. and Cooper, D.J. (2006). Processes of *Tamarix* invasion and floodplain development along the lower Green River, Utah. *Ecological Applications* 16: 1103–1120.

Bishop, P., Hoey, T.B., Jansen, J.D., and Artza, I.L. (2005). Knickpoint recession rate and catchment area: the case of uplifted rivers in eastern Scotland. *Earth Surface Processes and Landforms* 30: 767–778.

Bisson, P.A., Bilby, R.E., Bryant, M.D. et al. (1987). Large woody debris in forested streams in the Pacific Northwest: past, present, and future. In: *Streamside Management: Forestry and Fishery Implications* (eds. E.O. Salo and T.W. Cundy), 143–190. Seattle, WA: University of Washington, Institute of Forest Resources, Contribution No. 57.

Bisson, P.A., Wondzell, S.M., Reeves, G.H., and Gregory, S.V. (2003). Trends in using wood to restore aquatic habitats and fish communities in western North American rivers. *American Fisheries Society Symposium* 37: 391–406.

Blair, T.C. (1999). Alluvial fan and catchment initiation by rock avalanching, Owens Valley, California. *Geomorphology* 28: 201–221.

Blair, T.C. (2001). Outburst flood sedimentation on the proglacial Tuttle Canyon alluvial fan, Owens Valley, California, USA. *Journal of Sedimentary Research* 71: 657–679.

Blair, T.C. and McPherson, J.G. (2009). Processes and forms of alluvial fans. In: *Geomorphology of Desert Environments*, 2e (eds. A.J. Parsons and A.D. Abrahams), 413–467. Berlin: Springer.

Blann, K.L., Anderson, J.L., Sands, G.R., and Vondracek, B. (2009). Effects of agricultural drainage on aquatic ecosystems: a review. *Critical Reviews in Environmental Science and Technology* 39: 909–1001.

Blanton, P. and Marcus, W.A. (2009). Railroads, roads and lateral disconnection in the river landscapes of the continental United States. *Geomorphology* 112: 212–227.

Bledsoe, B.P. and Watson, C.C. (2001). Effects of urbanization on channel instability. *Journal of the American Water Resources Association* 37: 255–270.

Bloesch, J. (2003). Flood plain conservation in the Danube River basin, the link between hydrology and limnology. *Archiv für Hydrobiologie Supplement: Large Rivers* 147: 347–362.

Bloesch, J. and Sieber, U. (2003). The morphological destruction and subsequent restoration programmes of large rivers in Europe. *Large Rivers* 14: 363–385.

Blois, G., Best, J.L., Sambrook Smith, G.H., and Hardy, R.J. (2014). Effect of bed permeability and hyporheic flow on turbulent flow over bed forms. *Geophysical Research Letters* 41: 6435–6442.

Bocchiola, D., Rulli, M., and Rosso, R. (2006). Transport of large woody debris in the presence of obstacles. *Geomorphology* 76: 166–178.

Boerema, A. and Meire, P. (2017). Management for estuarine ecosystem services: a review. *Ecological Engineering* 98: 172–182.

Boivin, M., Buffin-Belanger, T., and Piegay, H. (2017). Interannual kinetics (2010–2013) of large wood in a river corridor exposed to a 50-year flood event and fluvial ice dynamics. *Geomorphology* 279: 59–73.

Boll, J., Thewessen, T.J.M., Meijer, E.L., and Kroonenberg, S.B. (1988). A simulation of the development of river terraces. *Zeitschrift für Geomorphologie* 32: 31–45.

Bollati, M., Pellegrini, L., Rinaldi, M. et al. (2014). Reach-scale morphological adjustments and stages of channel evolution: the case of the Trebbia River (northern Italy). *Geomorphology* 221: 176–186.

Bombino, G., Gurnell, A.M., Tamburino, V. et al. (2009). Adjustments in channel form, sediment caliber and vegetation around check-dams in the headwater reaches of mountain torrents, Calabria, Italy. *Earth Surface Processes and Landforms* 34: 1011–1021.

Bonnell, M. and Gilmour, D.A. (1978). The development of overland flow in a tropical rain-forest catchment. *Journal of Hydrology* 39: 365–382.

Bonneton, P., Bonneton, N., Parisot, J.P., and Castelle, B. (2015). Tidal bore dynamics in funnel-shaped estuaries. *Journal of Geophysical Research Oceans* 120: 923–941.

Bookhagen, B. and Burbank, D.W. (2010). Toward a complete Himalayan hydrological budget: spatiotemporal distribution of snowmelt and rainfall and their impact on river discharge. *Journal of Geophysical Research* 115, F03019.

Booth, E.G. and Loheide, S.P. (2012). Hydroecological model predictions indicate wetter and more diverse soil water regimes and vegetation types following floodplain restoration. *Journal of Geophysical Research: Biogeosciences* 117, G02011.

Booth, A.L., Chamberlain, C.P., Kidd, W.S.F., and Zeitler, P.K. (2009a). Constraints on the metamorphic evolution of the eastern Himalayan syntaxis from geochronologic and petrologic studies of Namche Barwa. *Geological Society of America Bulletin* 121: 385–407.

Booth, A.M., Roering, J.J., and Perron, J.T. (2009b). Automated landslide mapping using spectral analysis and high-resolution topographic data: Puget Sound lowlands, Washington, and Portland Hills, Oregon. *Geomorphology* 109: 132–147.

Booth, D.B., Roy, A.H., Smith, B., and Capps, K.A. (2016). Global perspectives on the urban stream syndrome. *Freshwater Science* 35: 412–420.

Botter, G., Basso, S., Porporato, A. et al. (2010). Natural streamflow regime alterations: damming of the Piave River basin (Italy). *Water Resources Research* 46, W06522.

Boucher, E., Bégin, Y., Arsenault, D., and Ouarda, T.B.M.J. (2012). Long-term and large-scale river-ice processes in cold-region watersheds. In: *Gravel-Bed Rivers: Processes, Tools, Environments* (eds. M. Church, P.M. Biron and A.G. Roy), 546–554. Hoboken, NJ: Wiley.

Boulton, A.J. (1999). An overview of river health assessment: philosophies, practice, problems and prognosis. *Freshwater Biology* 41: 469–479.

Bovee, K.D., R. Milhous. 1978. Hydraulic simulation in instream flow studies: theory and techniques. Instream Flow Information Paper nr 5, FWS/OBS-78/33, US Fish and Wildlife Service, Fort Collins, Colorado.

Bowman, D., Svoray, T., Devora, S. et al. (2010). Extreme rates of channel incision and shape evolution in response to a continuous, rapid base-level fall, the Dead Sea, Israel. *Geomorphology* 114: 227–237.

Boyer, E.W., Hornberger, G.M., Bencala, K.E., and McKnight, D.M. (2000). Response characteristics of DOC flushing in an alpine catchment. *Hydrological Processes* 11: 1635–1647.

Boyer, E.W., Howarth, R.W., Galloway, J.N. et al. (2006). Riverine nitrogen export from the continents to the coasts. *Global Biogeochemical Cycles* 20, GB1S91.

BR and ERDC (Bureau of Reclamation and US Army Engineer Research and Development Center) (2016). *National Large Wood Manual: Assessment, Planning, Design, and Maintenance of Large Wood in Fluvial Ecosystems: Restoring Process, Function, and Structure*. Boise, ID.

Braatne, J.H., Rood, S.B., Goater, L.A., and Blair, C.L. (2008). Analyzing the impacts of dams on riparian ecosystems: a review of research strategies and their relevance to the Snake River through Hells Canyon. *Environmental Management* 41: 267–281.

Bracken, L.J. and Croke, J. (2007). The concept of hydrological connectivity and its contribution to understanding runoff-dominated geomorphic systems. *Hydrological Processes* 21: 1749–1763.

Bracken, L.J., Wainwright, J., Ali, G.A. et al. (2013). Concepts of hydrological connectivity: research approaches, pathways and future agendas. *Earth-Science Reviews* 119: 17–34.

Bracken, L.J., Turnbull, L., Wainwright, J., and Bogaart, P. (2015). Sediment connectivity: a framework for understanding sediment transfer at multiple scales. *Earth Surface Processes and Landforms* 40: 177–188.

Bradley, R.S. (1985). *Quaternary Paleoclimatology: Methods of Paleoclimatic Reconstruction*. London: Chapman and Hall.

Bradley, D.N. and Tucker, G.E. (2012). Measuring gravel transport and dispersion in a mountain river using passive radio tracers. *Earth Surface Processes and Landforms* 37: 1034–1045.

Bradshaw, P. (1985). *An Introduction to Turbulence and Its Measurement*. Oxford: Pergamon Press.

Braga, G. and Gervasoni, S. (1989). Evolution of the Po River: an example of the application of historic maps. In: *Historical Change of Large Alluvial Rivers: Western Europe* (eds. G.E. Petts, H. Möller and A.L. Roux), 113–126. Chichester: Wiley.

Bragg, D.C. (2000). Simulating catastrophic and individualistic large woody debris recruitment for a small riparian stream. *Ecology* 81: 1383–1394.

Bragg, D.C., J.L. Kershner, D.W. Roberts. 2000. Modeling large woody debris recruitment for small streams of the central Rocky Mountains. USDA Forest Service General Technical Report RMRS-GTR-55, Fort Collins, CO.

Brandt, S.A. (2000). Classification of geomorphological effects downstream of dams. *Catena* 40: 375–401.

Brantley, S.L., Eissenstat, D.M., Marshall, J.A. et al. (2017). Reviews and syntheses: on the roles trees play in building and plumbing the critical zone. *Biogeosciences* 14: 5115–5142.

Brasington, J., Rumsby, B.T., and McVey, R.A. (2000). Monitoring and modeling morphological change in a braided gravel-bed river using high resolution GPS-based survey. *Earth Surface Processes and Landforms* 25: 973–990.

Braudrick, C.A. and Grant, G.E. (2000). When do logs move in rivers? *Water Resources Research* 36: 571–583.

Braudrick, C.A. and Grant, G.E. (2001). Transport and deposition of large woody debris in streams: a flume experiment. *Earth Surface Processes and Landforms* 22: 669–683.

Braudrick, C.A., Grant, G.E., Ishikawa, Y., and Ikeda, H. (1997). Dynamics of wood transport in streams: a flume experiment. *Earth Surface Processes and Landforms* 22: 669–683.

Braudrick, C.A., Dietrich, W.E., Leverich, G.T., and Sklar, L.S. (2009). Experimental evidence for the conditions necessary to sustain meandering in coarse-bedded rivers. *Proceedings of the National Academy of Sciences of the United States of America* 106: 16936–16941.

Bray, D.I. (1979). Estimating average velocity in gravel-bed rivers. *Journal of Hydraulics Division, ASCE* 105: 1103–1122.

Brayshaw, A.C. (1984). Characteristics and origin of cluster bedforms in coarse-grained alluvial channels. In: *Sedimentology of Gravels and Conglomerates* (eds. E.H. Koster and R.J. Steel), 77–85. Canadian Society of Petroleum Geologists, Memoir 10.

Brice, J.C. 1964. Channel patterns and terraces of the Loup Rivers in Nebraska. US Geological Survey Professional Paper 422D.

Brice, J.C. and Blodgett, J.C. 1978. Countermeasures for hydraulic problems at bridges. Federal Highway Administration Report FHWA-RD-78-162, vols. 1, 2, Washington, DC.

Bridge, J.S. (2003). *Rivers and Floodplains: Forms, Processes, and Sedimentary Record*. Malden, MA: Blackwell Publishing.

Bridge, J.S. and Jarvis, J. (1976). Flow and sedimentary processes in meandering river South Esk, Glen-Cova, Scotland. *Earth Surface Processes and Landforms* 1: 303–336.

Bridge, J.S. and Lunt, I.A. (2006). Depositional models of braided rivers. In: *Braided Rivers: Process, Deposits, Ecology, and Management* (eds. G.H. Sambrook Smith, J.L. Best, C.S. Bristow and G.E. Petts) Special Publication of the International Association of Sedimentologistis 36, 11–50.

Bridgland, D.R. (2000). River terrace systems in north-west Europe: an archive of environmental change, uplift and early human occupation. *Quaternary Science Reviews* 19: 1293–1303.

Brierley, G.J. and Fryirs, K.A. (2005). *Geomorphology and River Management: Applications of the River Styles Framework*. Oxford: Blackwell.

Brierley, G.J. and Fryirs, K.A. (2009). Don't fight the site: three geomorphic considerations in catchment-scale river rehabilitation planning. *Environmental Management* 43: 1201–1218.

Brierley, G.J. and Fryirs, K.A. (2016). The use of evolutionary trajectories to guide "moving targets" in the management of river futures. *River Research and Applications* 32: 823–835.

Brierley, G.J., Brooks, A.P., and Fryirs, K.A. (2005). Did humid-temperate rivers in the old and new worlds respond differently to clearance of riparian vegetation and removal of woody debris? *Progress in Physical Geography* 29: 27–49.

Brierley, G.J., Fryirs, K.A., and Jain, V. (2006). Landscape connectivity: the geographic basis of geomorphic applications. *Area* 38: 165–174.

Briggs, M.A., Gooseff, M.N., Arp, C.D., and Baker, M.A. (2009). A method for estimating surface transient storage parameters for streams with concurrent hyporheic exchange. *Water Resources Research* 45, W00D27.

Briggs, M.A., Lautz, L.K., Hare, D.K., and Gonzalez-Pinzon, R. (2013). Relating hyporheic fluxes, residence times, and redox-sensitive biogeochemical processes upstream of beaver dams. *Freshwater Science* 32: 622–641.

Brocard, G.Y., van der Beek, P.A., Bourles, D.L. et al. (2003). Long-term fluvial incision rates and postglacial river relaxation time in the French western Alps from ^{10}Be dating of alluvial terraces with assessment of inheritance, soil development and wind ablation effects. *Earth and Planetary Science Letters* 209: 197–214.

Brooker, M.P. (1981). The impact of impoundments on the downstream fisheries and general ecology of rivers. In: *Advances in Applied Biology*, vol. 6 (ed. T.H. Coaker), 91–152. London: Academic Press.

Brookes, A. (1985). River channelization: traditional engineering methods, physical consequences and alternative practices. *Progress in Physical Geography* 9: 44–73.

Brooks, S.M. (2003). Slopes and slope processes: research over the past decade. *Progress in Physical Geography* 27: 130–141.

Brooks, A.P. and Brierley, G.J. (2004). Framing realistic river rehabilitation targets in light of altered sediment supply and transport relationships: lessons from East Gippsland, Australia. *Geomorphology* 58: 107–123.

Brooks, S.S. and Lake, P.S. (2007). River restoration in Victoria, Australia: change is in the wind, and none too soon. *Restoration Ecology* 15: 584–591.

Brooks, A.P., Brierley, G.J., and Millar, R.G. (2003). The long-term control of vegetation and woody debris on channel and flood-plain evolution: insights from a paired catchment study in southeastern Australia. *Geomorphology* 51: 7–29.

Brooks, R., Barnard, R., Coulombe, R., and McDonnell, J.J. (2010). Ecohydrologic separation of water between trees and streams in a Mediterranean climate. *Nature Geoscience* 3: 100–104.

Brotherton, D.I. (1979). On the origin and characteristics of river channel patterns. *Journal of Hydrology* 44: 211–230.

Brown, C.B. (1950). Sediment transportation. In: *Engineering Hydraulics* (ed. H. Rouse), 769–857. New York: Wiley.

Brown, L., Thorne, R., and Woo, M.-K. (2008). Using satellite imagery to validate snow distribution simulated by a hydrological model in large northern basins. *Hydrological Processes* 22: 2777–2787.

Bruijnzeel, L.A. (2005). Tropical montane cloud forest: a unique hydrological case. In: *Forests, Water and People in the Humid Tropics* (eds. M. Bonell and L.A. Bruijnzeel), 462–483. Cambridge: Cambridge University Press.

Brummer, C.J. and Montgomery, D.R. (2003). Downstream coarsening in headwater channels. *Water Resources Research* 39 https://doi.org/10.1029/2003WR001981.

Brummer, C.J., Abbe, T.B., Sampson, J.R., and Montgomery, D.R. (2006). Influence of vertical channel change associated with wood accumulations on delineating channel migration zones, Washington, USA. *Geomorphology* 80: 295–309.

Brunsden, D. and Thornes, J.B. (1979). Landscape sensitivity and change. *Transactions of the Institute of British Geographers* 4 (4): 463–484.

Bryan, R.B. and Jones, J.A.A. (1997). The significance of soil piping processes: inventory and prospect. *Geomorphology* 20: 209–218.

Bryant, M., Falk, P., and Paola, C. (1995). Experimental study of avulsion frequency and rate of deposition. *Geology* 23: 365–368.

Buffin-Bélanger, T., Roy, A.G., and Kirkbride, A.D. (2000). On large-scale flow structures in a gravel-bed river. *Geomorphology* 32: 417–435.

Buffin-Bélanger, T., Roy, A.G., and Demers, S. (2013). Turbulence in river flows. In: *Treatise on Fluvial Geomorphology* (ed. E. Wohl), 70–86. Amsterdam: Elsevier.

Buffington, J.M. and Montgomery, D.R. (1997). A systematic analysis of eight decades of incipient motion studies, with special reference to gravel-bedded rivers. *Water Resources Research* 33: 1993–2029.

Buffington, J.M. and Montgomery, D.R. (1999a). A procedure for classifying textural facies in gravel-bed rivers. *Water Resources Research* 35: 1903–1914.

Buffington, J.M. and Montgomery, D.R. (1999b). Effects of hydraulic roughness on surface textures of gravel-bed rivers. *Water Resources Research* 35: 2507–3521.

Buffington, J.M. and Montgomery, D.R. (2013). Geomorphic classification of rivers. In: *Treatise on Fluvial Geomorphology* (ed. E. Wohl), 730–766. Amsterdam: Elsevier.

Buffington, J.M. and Tonina, D. (2009a). Hyporheic exchange in mountain rivers I: mechanics and environmental effects. *Geographical Compass* 3: 1063–1086.

Buffington, J.M. and Tonina, D. (2009b). Hyporheic exchange in mountain rivers II: effects of channel morphology on mechanics, scales, and rates of exchange. *Geographical Compass* 3: 1038–1062.

Buffington, J.M., Lisle, T.E., Woodsmith, R.D., and Hilton, S. (2002). Controls on the size and occurrence of pools in coarse-grained forest rivers. *River Research and Applications* 18: 507–531.

Bull, W.B. 1962. Relations of alluvial-fan size and slope to drainage-basin and lithology in western Fresno County, California. US Geological Survey Professional Paper 450-B.

Bull, W.B. 1964. Geomorphology of segmented alluvial fans in western Fresno County, California. US Geological Survey Professional Paper 352-E.

Bull, W.B. (1977). The alluvial fan environment. *Progress in Physical Geography* 1: 222–270.

Bull, W.B. (1990). Stream-terrace genesis: implications for soil development. *Geomorphology* 3: 351–367.

Bull, W.B. (1991). *Geomorphic Responses to Climatic Change*. New York: Oxford University Press.

Bunn, S.E. and Arthington, A.H. (2002). Basic principles and ecological conseq uences of altered flow regimes for aquatic biodiversity. *Environmental Management* 30: 492–507.

Bunn, S.E., Davies, P.M., and Mosisch, T.D. (1999). Ecosystem measures of river health and their response to riparian and catchment degradation. *Freshwater Biology* 41: 333–345.

Bunte, K. 2004. Gravel mitigation and augmentation below hydroelectric dams: a geomorphological perspective. Stream Systems Technology Center, Rocky Mountain Research Station, Forest Service, US Department of Agriculture, Fort Collins, CO.

Bunte, K., S.R. Abt. 2001. Sampling surface and subsurface particle-size distributions in wadable gravel- and cobble-bed streams for analyses in sediment transport, hydraulics, and streambed monitoring. USDA Forest Service General Technical Report RMRS-GTR-74, Fort Collins, CO.

Bunte, K., Abt, S.R., Potyondy, J.P., and Ryan, S.E. (2004). Measurement of coarse gravel and cobble transport using portable bedload traps. *Journal of Hydraulic Engineering* 130: 879–893.

Bunte, K., K.W. Swingle, S.R. Abt. 2007. Guidelines for using bedload traps in coarse-bedded mountain streams: construction, installation, operation, and sample processing. USDA Forest Service General Technical Report RMRS-GTR-191, Rocky Mountain Research Station, Fort Collins, CO.

Burbank, D.W., Leland, J., Fielding, E. et al. (1996). Bedrock incision, rock uplift and threshold hillslopes in the northwestern Himalayas. *Nature* 379: 505–510.

Burchsted, D., Daniels, M., Thorson, R., and Vokoun, J. (2010). The river discontinuum: applying beaver modifications to baseline conditions for restoration of forested headwaters. *BioScience* 60: 908–922.

Burge, L.M. (2005). Wandering Miramichi rivers, New Brunswick, Canada. *Geomorphology* 69: 253–274.

Burke, B.C., Heimsath, A.M., and White, A.F. (2007). Coupling chemical weathering with soil production across soil-mantled landscapes. *Earth Surface Processes and Landforms* 32: 853–873.

Burke, M., Jorde, K., and Buffington, J.M. (2009). Application of a hierarchical framework for assessing environmental impacts of dam operation: changes in streamflow, bed mobility and recruitment of riparian trees in a western North American river. *Journal of Environmental Management* 90 (Suppl. 3): S224–S236.

Burkham, D.E. 1972. Channel changes of the Gila River in Safford Valley, Arizona 1846–1970. US Geological Survey Professional Paper 655-G.

Burns, D.A. and McDonnell, J.J. (1998). Effects of a beaver pond on runoff processes: comparison of two headwater catchments. *Journal of Hydrology* 205: 248–264.

Burns, D.A., McDonnell, J.J., Hooper, R.P. et al. (2001). Quantifying contributions to storm runoff through end-member mixing analysis and hydrologic measurements at the Panola Mountain research watershed (Georgia, USA). *Hydrological Processes* 15: 1903–1924.

Burt, T.P. (1996). The hydrological role of floodplains within the drainage basin system. In: *Buffer Zones: Their Processes and Potential in Water Protection* (eds. N. Haycock, T. Burt, K. Goulding and G. Pinay), 21–32. St. Albans: Haycock Associated Ltd.

Burt, T.P., Bates, P.D., Stewart, M.D. et al. (2002). Water table fluctuations within the floodplain of the River Severn, England. *Journal of Hydrology* 262: 1–20.

Burton, A. and Bathurst, J.C. (1998). Physically based modeling of shallow landslide sediment yield at a catchment scale. *Environmental Geology* 35: 89–99.

Burton, A., Arkell, T.J., and Bathurst, J.C. (1998). Field variability of landslide model parameters. *Environmental Geology* 35: 100–114.

Busato, L., Boaga, J., Perri, M.T. et al. (2019). Hydrogeophysical characterization and monitoring of the hyporheic and riparian zones: the Vermigliana Creek case study. *Science of the Total Environment* 648: 1105–1120.

Buscombe, D., Rubin, D.M., and Warrick, J.A. (2010). A universal approximation of grain size from images of noncohesive sediment. *Journal of Geophysical Research Earth Surface* 115, F02015.

Butler, D.R. (2006). Human-induced changes in animal populations and distributions, and the subsequent effects on fluvial systems. *Geomorphology* 79: 448–459.

Butler, D.R. and Malanson, G.P. (2005). The geomorphic influences of beaver dams and failures of beaver dams. *Geomorphology* 71: 48–60.

Butzer, K.W. (1976). *Geomorphology from the Earth.* New York: Harper and Row.

Buxton, T.H. (2010). Modeling entrainment of waterlogged large wood in stream channels. *Water Resources Research* 46, W10537.

Buxton, T.H., Buffington, J.M., Yager, E.M. et al. (2015). The relative stability of salmon redds and unspawned streambeds. *Water Resources Research* 51: 6074–6092.

Byrne, J.M., Berg, A., and Townshend, I. (1999). Linking observed and general circulation model upper air circulation patterns to current and future snow runoff for the Rocky Mountains. *Water Resources Research* 35: 3793–3802.

Caamaño, D., Goodwin, P., Buffington, J.M. et al. (2009). Unifying criterion for the velocity reversal hypothesis in gravel-bed rivers. *Journal of Hydraulic Engineering* 135: 66–70.

Cadol, D. and Wohl, E. (2010). Wood retention and transport in tropical, headwater streams, La Selva Biological Station, Costa Rica. *Geomorphology* 123: 61–73.

Caine, N. (1989). Hydrograph separation in a small alpine basin based on inorganic solute concentrations. *Journal of Hydrology* 112: 89–101.

Campbell, E.P. (2005). Physical-statistical models for predictions in ungauged basins. In: *Predictions in Ungauged Basins: International Perspectives on the State of the Art and Pathways Forward* (eds. S. Franks, M. Sivipalan, K. Takeuchi and Y. Tachikawa), 292–298. Wallingford: IAHS Publications no. 301.

Campforts, B., Schwanghart, W., and Govers, G. (2017). Accurate simulation of transient landscape evolution by eliminating numerical diffusion: the TTLEM 1.0 model. *Earth Surface Dynamics* 5: 47–66.

Candy, I., Black, S., and Sellwood, B.W. (2004). Interpreting the response of a dryland river system to late quaternary climate change. *Quaternary Science Reviews* 23: 2513–2523.

Canestrelli, A., Fagherazzi, S., Defina, A., and Lanzoni, S. (2010). Tidal hydrodynamics and erosional power in the Fly River delta, Papua New Guinea. *Journal of Geophysical Research* 115 https://doi.org/10.1029/2009JF001355.

Cannon, S.H., Kirkham, R.M., and Parise, M. (2001). Wildfire-related debris-flow initiation processes, Storm King Mountain, Colorado. *Geomorphology* 39: 171–188.

Canuel, E.A., Lerberg, E.J., Dickhut, R.M. et al. (2009). Changes in sediment and organic carbon accumulation in a highly-disturbed ecosystem: the Sacramento-San Joaquin River Delta (California, USA). *Marine Pollution Bulletin* 59: 154–163.

Canuel, E.A., Cammer, S.S., McIntosh, H.A., and Pondell, C.R. (2012). Climate change impacts on the organic carbon cycle at the land-ocean interface. *Annual Reviews Earth Planetary Science* 40: 685–711.

Carbonneau, P.E., Bergeron, N., and Lane, S.N. (2005). Automated grain size measurements from airborne remote sensing for long profile measurements of fluvial grain sizes. *Water Resources Research* 41, W11426.

Cardenas, M.S. and Gooseff, M.N. (2008). Comparison of hyporheic exchange under covered and uncovered channels based on linked surface and groundwater flow simulations. *Water Resources Research* 44, W03418.

Cardenas, M.B., Wilson, J.L., and Zlotnik, V.A. (2004). Impact of heterogeneity, bed forms, and stream curvature on subchannel hyporheic exchange. *Water Resources Research* 40, W08307.

Carey, R.O. and Migliaccio, K.W. (2009). Contribution of wastewater treatment plant effluents to nutrient dynamics in aquatic systems: a review. *Environmental Management* 44: 205–217.

Carey, M., Molden, O.C., Rasmussen, B. et al. (2017). Impacts of glacier recession and declining meltwater on mountain societies. *Annals of the American Association of Geographers* 107: 350–359.

Carling, P.A. (1988). Channel change and sediment transport in regulated UK rivers. *Regulated Rivers* 2: 369–388.

Carling, P.A. (1995). Flow-separation berms downstream of a hydraulic jump in a bedrock channel. *Geomorphology* 11: 245–253.

Carling, P.A. and Grodek, T. (1994). Indirect estimation of ungauged peak discharges in a bedrock channel with reference to design discharge selection. *Hydrological Processes* 8: 497–511.

Carling, P.A. and Reader, N.A. (1982). Structure, composition and bulk properties of upland stream gravels. *Earth Surface Processes and Landforms* 7: 349–365.

Carling, P.A. and Wood, N. (1994). Simulation of flow over pool-riffle topography: a consideration of the velocity-reversal hypothesis. *Earth Surface Processes and Landforms* 19: 319–332.

Carling, P.A., Orr, H., and Kelsey, A. (2006). The dispersion of magnetite bedload tracer across a gravel point bar and the development of heavy mineral placers. *Ore Geology Reviews* 28: 402–416.

Carling, P., Jansen, J., and Meshkova, L. (2014). Multichannel rivers: their definition and classification. *Earth Surface Processes and Landforms* 39: 26–37.

Carling, P.A., Huang, H.Q., Su, T., and Hornby, D. (2019). Flow structure in large bedrock channels: the example of macroturbulent rapids, lower Mekong River, Southeast Asia. *Earth Surface Processes and Landforms* https://doi.org/10.1002/esp.4537.

Carmack, E.C., Yamamoto-Kawai, M., Haine, T.W.N. et al. (2016). Freshwater and its role in the Arctic marine system: sources, disposition, storage, export, and physical and biogeochemical consequences in the Arctic and global oceans. *Journal of Geophysical Research Biogeosciences* 121: 675–717.

Carr, M.L. and Vuyovich, C.M. (2014). Investigating the effects of long-term hydro-climatic trends on Midwest ice jam events. *Cold Regions Science and Technology* 106–107: 66–81.

Carrara, P.E. and Carroll, T.R. (1979). The determination of erosion rates from exposed tree roots in the Piceance Basin, Colorado. *Earth Surface Processes* 4: 307–317.

Carson, M.A. (1984). Observations on the meandering-braided river transition, the Canterbury plains, New Zealand: part two. *New Zealand Geographer* 40: 89–99.

Carson, M.A. and Kirkby, M.J. (1972). *Hillslope Form and Process*. London: Cambridge University Press.

Carson, M.A. and Lapointe, M.F. (1983). The inherent asymmetry of river meander planform. *Journal of Geology* 91: 41–55.

Carson, E.C., Knox, J.C., and Mickelson, D.M. (2007). Response of bankfull flood magnitudes to Holocene climate change, Uinta Mountains, northeastern Utah. *Geological Society of America Bulletin* 119: 1066–1078.

Caruso, A., Ridolfi, L., and Boano, F. (2016). Impact of watershed topography on hyporheic exchange. *Advances in Water Resources* 94: 400–411.

Castillo, V.M., Mosch, W.M., Garcia, C.C. et al. (2007). Effectiveness and geomorphological impacts of check dams for soil erosion control in a semiarid Mediterranean catchment: El Carcavo (Murcia, Spain). *Catena* 70: 416–427.

Castro, J. (2008). Certificate in river restoration: one approach to training and professional development. *The Stream Restoration Networker* 2 (1): 2–4.

Castro, J.M. and Jackson, P.L. (2001). Bankfull discharge recurrence intervals and regional hydraulic geometry relationships: patterns in the Pacific northwest, USA. *Journal of the American Water Resources Association* 37: 1249–1262.

Catani, F., Farina, P., Moretti, S., and Nico, G. (2003). Spaceborne radar interferometry; a promising tool for hydrological analysis in mountain alluvial fan environments. In: *Erosion Prediction in Ungauged Basins: Integrating Methods and Techniques* (ed. D.H. de Boer), 241–248. Wallingford: IAHS Publications no. 279.

Cavalli, M. and Marchi, L. (2008). Characterisation of the surface morphology of an alpine alluvial fan using airborne LIDAR. *Natural Hazards and Earth System Sciences* 8: 323–333.

Cavalli, M., Trevisani, S., Comiti, F., and Marchi, L. (2013). Geomorphometric assessment of spatial sediment connectivity in small alpine catchments. *Geomorphology* 188: 31–41.

Cavalli, M., Goldin, B., Comiti, F. et al. (2017). Assessment of erosion and deposition in steep mountain basins by differencing sequential digital terrain models. *Geomorphology* 291: 4–16.

Cazanacli, D. and Smith, N.D. (1998). A study of morphology and texture of natural levees – Cumberland marshes, Saskatchewan, Canada. *Geomorphology* 25: 43–55.

Cazanacli, D., Paola, C., and Parker, G. (2002). Experimental steep, braided flow: application to flooding risk on fans. *Journal of Hydraulic Engineering* 128: 322–330.

Cenderelli, D.A. (2000). Floods from natural and artificial dam failures. In: *Inland Flood Hazards: Human, Riparian, and Aquatic Communities* (ed. E. Wohl), 73–103. Cambridge: Cambridge University Press.

Cenderelli, D.A. and Cluer, B.L. (1998). Depositional processes and sediment supply in resistant-boundary channels: examples from two case studies. In: *Rivers Over Rock: Fluvial Processes in Bedrock Channels* (eds. K.J. Tinkler and E.E. Wohl), 105–131. Washington, DC: American Geophysical Union Press.

Cenderelli, D.A. and Wohl, E.E. (2001). Peak discharge estimates of glacier-lake outburst floods and "normal" climatic floods in the Mount Everest region, Nepal. *Geomorphology* 40: 57–90.

Cenderelli, D.A. and Wohl, E.E. (2003). Flow hydraulics and geomorphic effects of glacial-lake outburst floods in the Mount Everest region, Nepal. *Earth Surface Processes and Landforms* 28: 385–407.

CENR (2000). *Integrated Assessment of Hypoxia in the Northern Gulf of Mexico*. Washington, DC: National Science and Technology Council Committee on Environment and Natural Resources.

Chamberlain, E.L., Walling, J., Reimann, T. et al. (2017). Luminescence dating of delta sediments: novel approaches explored for the Ganges-Brahmaputra-Meghna Delta. *Quaternary Geochronology* 41: 97–111.

Chang, H.H. (1978). *Fluvial Processes in River Engineering*. New York: Wiley.

Chang, H.H. (1988). On the cause of river meandering. In: *International Conference on River Regime* (ed. W.R. White), 83–93. Wallingford: Hydraulics Research.

Changnon, S.A. (2002). Frequency of heavy rainstorms on areas from 10 to 100 km^2, defined using dense rain gauge networks. *Journal of Hydrometeorology* 3: 220–223.

Chanson, H. (1996). Comment on "step-pool streams: adjustment to maximum flow resistance" by A.D. Abrahams, G. Li, and J.F. Atkinson. *Water Resources Research* 32: 3401–3402.

Chanson, H. (1999). *The Hydraulics of Open Channel Flow: An Introduction*. London: Arnold.

Chaponniere, A., Boulet, G., Chehbouni, A., and Aresmouk, M. (2008). Understanding hydrological processes with scarce data in a mountain environment. *Hydrological Processes* 22: 1908–1921.

Chappell, J. and Woodroffe, C. (1994). Macrotidal estuaries. In: *Coastal Evolution: Late Quaternary Shoreline Morphodynamics* (eds. C. Woodroffe and R. Carter), 187–218. Cambridge: Cambridge University Press.

Chappell, N.A., Bidin, K., Sherlock, M.D., and Lancaster, J.W. (2005). Parsimonious spatial representation of tropical soils within dynamic rainfall-runoff models. In: *Forests, Water and People in the Humid Tropics* (eds. M. Bonell and L.A. Bruijnzeel), 756–769. Cambridge: Cambridge University Press.

Chatanantavet, P. and Parker, G. (2009). Physically based modeling of bedrock incision by abrasion, plucking, and macroabrasion. *Journal of Geophysical Research* 114, F04018.

Chatanantavet, P., Lajeunesse, E., Parker, G. et al. (2010). Physically based model of downstream fining in bedrock streams with lateral input. *Water Resources Research* 46, W02581.

Chen, D. and Duan, J.D. (2006). Simulating sine-generated meandering channel evolution with an analytical model. *Journal of Hydraulic Research* 44: 363–373.

Chen, J. and Ohmura, A. (1990). On the influence of alpine glaciers on runoff. In: *Hydrology in Mountainous Regions I. Hydrological Measurements, the Water Cycle* (eds. H. Lang and A. Musy), 117–125. Wallingford: IAHS Publications no. 193.

Chen, X., Wei, X., Scherer, R., and Hogan, D. (2008). Effects of large woody debris on surface structure and aquatic habitat in forested streams, southern interior British Columbia, Canada. *River Research and Applications* 24: 862–875.

Cheng, L., AghaKouchak, A., Gilleland, E., and Katz, R.W. (2014). Non-stationary extreme value analysis in a changing climate. *Climatic Change* 127: 353–369.

Chiew, F.H. and McMahon, T.A. (2002). Modelling the impacts of climate change on Australian streamflow. *Hydrological Processes* 16: 1235–1245.

Chin, A. (2006). Urban transformation of river landscapes in a global context. *Geomorphology* 79: 460–487.

Chin, A. and Wohl, E. (2005). Toward a theory for step pools in stream channels. *Progress in Physical Geography* 29: 275–296.

Chin, A., Daniels, M.D., Urban, M.A. et al. (2008). Perceptions of wood in rivers and challenges for stream restoration in the United States. *Environmental Management* 41: 893–903.

Chin, A., O'Dowd, A.P., and Gregory, K.J. (2013). Urbanization and river channels. In: *Treatise on Fluvial Geomorphology* (ed. E. Wohl), 809–826. Amsterdam: Elsevier.

Chin, A., Laurencio, L.R., Daniels, M.D. et al. (2014). The significance of perceptions and feedbacks for effectively managing wood in rivers. *River Research and Applications* 30: 98–111.

Chin, A., Gridley, R., Tyner, L., and Gregory, K. (2017). Adjustment of dryland stream channels over four decades of urbanization. *Anthropocene* 20: 24–36.

Chiverrell, R.C., Harvey, A.M., and Foster, G.C. (2007). Hill slope gullying in the Solway Firth, Morecambe Bay region, Great Britain: responses to human impact and/or climatic deterioration? *Geomorphology* 84: 317–343.

Chorley, R.J. 1962. Geomorphology and general systems theory. US Geological Survey Professional Paper 500-B.

Chow, V.T. (1959). *Open-Channel Hydraulics*. New York: McGraw-Hill.

Church, M. (1983). Pattern of instability in a wandering gravel bed channel. In: *Modern and Ancient Fluvial Systems* (eds. J.D. Collinson and J. Lewin), 169–180. Oxford: International Association of Sedimentologists, Blackwell Scientific.

Church, M. (2006). Bed material transport and the morphology of alluvial river channels. *Annual Reviews of Earth and Planetary Science* 34: 325–354.

Church, M. and Ferguson, R.I. (2015). Morphodynamics: rivers beyond steady state. *Water Resources Research* 51: 1883–1897.

Church, M. and Hassan, M.A. (1992). Size and distance of travel of unconstrained clasts on a streambed. *Water Resources Research* 28: 299–303.

Church, M.A., McLean, D.G., and Wolcott, J.F. (1987). River bed gravels: sampling and analysis. In: *Sediment Transport in Gravel-Bed Rivers* (eds. C.R. Thorne, J.C. Bathurst and R.D. Hey), 43–88. Chichester: Wiley.

Cid, N., Verkaik, I., Garcia-Roger, E.M. et al. (2016). A biological tool to assess flow connectivity in reference temporary streams from the Mediterranean Basin. *Science of the Total Environment* 540: 178–190.

Clapp, E.M., Bierman, P.R., Schick, A.P. et al. (2000). Sediment yield exceeds sediment production in arid region drainage basins. *Geology* 28: 995–998.

Clark, C. (1982). *Flood*. Alexandria, VA: Time-Life Books.

Clark, D.B., Clark, D.A., Brown, S. et al. (2002). Stocks and flows of coarse woody debris across a tropical rain forest nutrient and topography gradient. *Forest Ecology and Management* 164: 237–248.

Clark, A.H., Shattuck, M.D., Ouellette, N.T., and O'Hern, C.S. (2017). Role of grain dynamics in determining the onset of sediment transport. *Physical Review Fluids* 2, 034305.

Clayton, J.A. and Pitlick, J. (2007). Spatial and temporal variations in bed load transport intensity in a gravel bed river bend. *Water Resources Research* 43, W02426.

Clayton, J.A. and Pitlick, J. (2008). Persistence of the surface texture of a gravel-bed river during a large flood. *Earth Surface Processes and Landforms* 33: 661–673.

Clevis, Q., de Boer, P., and Wachter, M. (2003). Numerical modelling of drainage basin evolution and three-dimensional alluvial fan stratigraphy. *Sedimentary Geology* 163: 85–110.

Clifford, N.J. (1993). Differential bed sedimentology and the maintenance of riffle-pool sequences. *Catena* 20: 447–468.

Clifford, N.J. and French, J.R. (1993a). Monitoring and modeling turbulent flows: historical and contemporary perspectives. In: *Turbulence: Perspectives on Flow and Sediment Transport* (eds. N.J. Clifford, J.R. French and J. Hardistry), 1–34. Chichester: Wiley.

Clifford, N.J. and French, J.R. (1993b). Monitoring and analysis of turbulence in geophysical boundaries: some analytical and conceptual issues. In: *Turbulence: Perspectives on Flow and Sediment Transport* (eds. N.J. Clifford, J.R. French and J. Hardistry), 93–120. Chichester: Wiley.

Clifford, N.J., Richards, K.S., and Robert, A. (1992). Estimation of flow resistance in gravel-bedded rivers: a physical explanation of the multiplier of roughness length. *Earth Surface Processes and Landforms* 17: 111–126.

Clifford, N.J., Harmar, O.P., Harvey, G., and Petts, G.E. (2006). Physical habitat, eco-hydraulics and river design: a review and re-evaluation of some popular concepts and methods. *Aquatic Conservation: Marine and Freshwater Ecosystems* 16: 389–408.

Clow, D.W., Ingersoll, G.P., Mast, M.A. et al. (2002). Comparison of snowpack and winter wet-deposition chemistry in the Rocky Mountains, USA: implications for winter dry deposition. *Atmospheric Environment* 36: 2337–2348.

Clow, D.W., Schrott, L., Webb, R. et al. (2003). Ground water occurrence and contributions to streamflow in an alpine catchment, Colorado front range. *Ground Water* 41: 937–950.

Cluer, B. and Thorne, C. (2014). A stream evolution model integrating habitat and ecosystem benefits. *River Research and Applications* 30: 135–154.

Cohn, T.A., Lane, W.L., and Stedinger, J.R. (2001). Confidence intervals for Expected Moments Algorithm flood quantile estimates. *Water Resources Research* 37: 1695–1706.

Colby, B.R. 1964. Discharge of sands and mean-velocity relationships in sand-bed streams. US Geological Survey Professional Paper 462-A.

Coleman, J.M., Roberts, H.H., and Stone, G.W. (1998). Mississippi River Delta: an overview. *Journal of Coastal Research* 14: 698–716.

Coleman, S.E., Melville, B.W., and Gore, L. (2003). Fluvial entrainment of protruding fractured rock. *Journal of Hydraulic Engineering ASCE* 129: 872–884.

Coleman, J., Huh, O., Braud, D., and Roberts, H.H. (2005). Major world delta variability and wetland loss. *Transactions of the Gulf Coast Association of Geological Societies* 55: 102–131.

Collier, M.P., Webb, R.H., and Andrews, E.D. (1997). Experimental flooding in Grand Canyon. *Scientific American* 276: 66–73.

Collins, D.N. (1998). Rainfall-induced high-magnitude events in highly glacierized Alpine basins. In: *Hydrology, Water Resources, and Ecology in Headwaters* (eds. K. Kovar, U. Tappeiner, N.E. Peters and R.G. Craig), 365–372. Wallingford: IAHS Publications no. 193.

Collins, D.N. (2006). Climatic variation and runoff in mountain basins with differing proportions of glacier cover. *Hydrology Research* 37: 315–326.

Collins, D.B.G. and Bras, R.L. (2010). Climatic and ecological controls of equilibrium drainage density, relief and channel concavity in dry lands. *Water Resources Research* 46, W04508.

Collins, B.D. and Montgomery, D.R. (2002). Forest development, wood jams and restoration of floodplain rivers in the Puget Lowland, Washington. *Restoration Ecology* 10: 237–247.

Collins, D.N. and Taylor, D.P. (1990). Variability of runoff from partially-glacierised Alpine basins. In: *Hydrology in Mountainous Regions I. Hydrological Measurements, the Water Cycle* (eds. H. Lang and A. Musy), 365–372. Wallingford: IAHS Publications no. 193.

Collins, M.E., Doolittle, J.A., and Rourke, R.V. (1989). Mapping depth to bedrock on a glaciated landscape with ground-penetrating radar. *Soil Science Society of America Journal* 53: 1806–1812.

Collins, B.D., Montgomery, D.R., Fetherston, K.L., and Abbe, T.B. (2012). The floodplain large-wood cycle hypothesis: a mechanism for the physical and biotic structuring of temperate forested alluvial valleys in the North Pacific coastal ecoregion. *Geomorphology* 139–140: 460–470.

Collins, B.D., Montgomery, D.R., Schanz, S.A., and Larsen, I.J. (2016). Rates and mechanisms of bedrock incision and strath terrace formation in a forested catchment, Cascade Range, Washington. *Geological Society of America Bulletin* 128: 926–943.

Comiti, F. (2012). How natural are Alpine mountain rivers? Evidence from the Italian Alps. *Earth Surface Processes and Landforms* 37: 693–707.

Comiti, F., Mao, L., Preciso, E. et al. (2008). Large wood and flash floods: evidence from the 2007 event in the Davča basin (Slovenia). In: *Monitoring, Simulation, Prevention and Retention of Dense and Debris Flow II*, vol. 60 (eds. D. de Wrachien, C.A. Brebbia and M.A. Lenzi), 173–182. WIT Transactions of Engineering Sciences.

Comiti, F., Cadol, D., and Wohl, E. (2009a). Flow regimes, bed morphology, and flow resistance in self-formed step-pool channels. *Water Resources Research* 45, W04424.

Comiti, F., Mao, L., Lenzi, M.A., and Siligardi, M. (2009b). Artificial steps to stabilize mountain rivers: a post-project ecological assessment. *River Research and Applications* 25: 639–659.

Comiti, F., Lucía, A., and Rickenmann, D. (2016). Large wood recruitment and transport during large floods: a review. *Geomorphology* 269: 23–39.

Constantine, C.R., Mount, J.F., and Florsheim, J.L. (2003). The effects of longitudinal differences in gravel mobility on the downstream fining pattern in the Cosumnes River, California. *Journal of Geology* 111: 233–241.

Constantine, J.A., McLean, S.R., and Dunne, T. (2010). A mechanism of chute cutoff along large meandering rivers with uniform floodplain topography. *Geological Society of America Bulletin* 122: 855–869.

Constantine, J.A., Dunne, T., Ahmed, J. et al. (2014). Sediment supply as a driver of river meandering and floodplain evolution in the Amazon Basin. *Nature Geoscience* 7: 899–903.

Constantinescu, G., Miyawaki, S., Rhoads, B. et al. (2011). Structure of turbulent flow at a river confluence with momentum and velocity ratios close to 1: insight provided by an eddy-resolving numerical simulation. *Water Resources Research* 47, W05507.

Constantz, J. (2008). Heat as a tracer to determine streambed water exchanges. *Water Resources Research* 44, W00D10.

Constantz, J. and Stonestrom, D.A. (2003). Heat as a tracer of water movement near streams. In: *Heat as a Tool for Studying the Movement of Ground Water Near Streams* (eds. D.A. Stonestrom and J. Constantz) US Geological Survey Circular 1260, 1–6.

Cook, K.L., Whipple, K.X., Heimsath, A.M., and Hanks, T.C. (2009). Rapid incision of the Colorado River in Glen Canyon – insights from channel profiles, local incision rates, and modeling of lithologic controls. *Earth Surface Processes and Landforms* 34: 994–1010.

Cook, K.L., Andermann, C., Gimbert, F. et al. (2018). Glacial lake outburst floods as drivers of fluvial erosion in the Himalaya. *Science* 362: 53–57.

Cooke, R.U. and Reeves, R.W. (1976). *Arroyos and Environmental Change in the American South-West.* Oxford: Clarendon Press.

Cooley, S.W. and Pavelsky, T.M. (2016). Spatial and temporal patterns in Arctic river ice breakup revealed by automated ice detection from MODIS imagery. *Remote Sensing of Environment* 175: 310–322.

Cooper, J.A.G. (2001). Geomorphological variability among micro-tidal estuaries from the wave-dominated South African coast. *Geomorphology* 40: 99–122.

Cooper, J.A.G., Green, A.N., and Wright, C.I. (2012). Evolution of an incised valley coastal plain estuary under low sediment supply: a "give-up" estuary. *Sedimentology* 59: 899–916.

Coopersmith, E., Yaeger, M.A., Ye, S. et al. (2012). Exploring the physical controls of regional patterns of flow duration curves – part 3: a catchment classification system based on regime curve indicators. *Hydrology and Earth System Sciences* 16: 4467–4482.

Corbel, J. (1959). Vitesse de l erosion. *Zeitschrift Geomorphologie* 3: 1–28.

Cordier, S., Briant, B., Bridgland, D. et al. (2017). The Fluvial Archives Group: 20 years of research connecting fuvial geomorphology and palaeoenvironments. *Quaternary Science Reviews* 166: 1–9.

Cortizas, A.M., Lopez-Merino, L., Bindler, R. et al. (2016). Early atmospheric metal pollution provides evidence for Chalcolithiic/Bronze Age mining and metallurgy in southwestern Europe. *Science of the Total Environment* 545–546: 398–406.

Costa, J.E. (1984). Physical geomorphology of debris flows. In: *Developments and Applications of Geomorphology* (eds. J.E. Costa and P.J. Fleisher), 268–317. Berlin: Springer.

Costa, M.H. (2005). Large-scale hydrological impacts of tropical forest conversion. In: *Forests, Water and People in the Humid Tropics* (eds. M. Bonell and L.A. Bruijnzeel), 590–597. Cambridge: Cambridge University Press.

Costa, J.E. and O'Connor, J.E. (1995). Geomorphically effective floods. In: *Natural and Anthropogenic Influences in Fluvial Geomorphology* (eds. J.E. Costa, A.J. Miller, K.W. Potter and P.R. Wilcock), 45–56. Washington, DC: American Geophysical Union Press.

Costa, J.E. and Schuster, R.L. (1988). The formation and failure of natural dams. *Geological Society of America Bulletin* 100: 1054–1068.

Cote, D., Kehler, D.J., Bourne, C., and Wiersma, Y.F. (2009). A new measure of longitudinal connectivity for stream networks. *Landscape Ecology* 24: 101–113.

Coulthard, T.J. and Van De Wiel, M.J. (2006). A cellular model of river meandering. *Earth Surface Processes and Landforms* 31: 123–132.

Coulthard, T.J. and Van de Wiel, M.J. (2013). Numerical modeling in fluvial geomorphology. In: *Treatise on Fluvial Geomorphology* (ed. E. Wohl), 694–710. Amsterdam: Elsevier.

Coulthard, T.J. and Van De Wiel, M.J. (2017). Modelling long term basin scale sediment connectivity, driven by spatial land use changes. *Geomorphology* 277: 265–281.

Coulthard, T.J., Kirkby, M.J., and Macklin, M.G. (1997). Modelling hydraulic, sediment transport and slope processes, at a catchment scale, using a cellular automaton approach. In: *Proceedings of GeoComputation '97 and SIRC '97*, 309–318.

Coulthard, T.J., Macklin, M.G., and Kirkby, M.J. (2002). A cellular model of Holocene upland river basin and alluvial fan evolution. *Earth Surface Processes and Landforms* 27: 269–288.

Coulthard, T.J., Lewin, J., and Macklin, M.G. (2005). Modelling differential catchment response to environmental change. *Geomorphology* 69: 222–241.

Coulthard, T.J., Hicks, D.M., and Van De Wiel, M.J. (2007). Cellular modeling of river catchments and reaches: advantages, limitations and prospects. *Geomorphology* 90: 192–207.

Covino, T. (2017). Hydrologic connectivity as a framework for understanding biogeochemical flux through watersheds and along fluvial networks. *Geomorphology* 277: 133–144.

Cowie, P.A., Attal, M., Tucker, G.E. et al. (2006). Investigating the surface process response to fault interaction and linkage using a numerical modelling approach. *Basin Research* 18: 23–266.

Cowie, P.A., Whittaker, A.C., Attal, M. et al. (2008). New constraints on sediment-flux-dependent river incision: implications for extracting tectonic signals from river profiles. *Geology* 36: 535–538.

Cramer, M., ed. 2012. Stream habitat restoration guidelines. Co-published by the Washington Departments of Fish and Wildlife, Natural Resources, Transportation and Ecology, Washington State Recreation and Conservation Office, Puget Sound Partnership, and the US Fish and Wildlife Service, Olympia, WA.

Cristan, R., Aust, W.M., Bolding, M.C. et al. (2016). Effectiveness of forestry best management practices in the United States: literature review. *Forest Ecology and Management* 360: 133–151.

Croissant, T., Lague, D., Davy, P. et al. (2017). A precipiton-based approach to model hydro-sedimentary hazards induced by large sediment supplies in alluvial fans. *Earth Surface Processes and Landforms* 42: 2054–2067.

Croke, J. and Mockler, S. (2001). Gully initiation and road-to-stream linkage in a forested catchment, southeastern Australia. *Earth Surface Processes and Landforms* 26: 205–217.

Crook, D.A. and Robertson, A.I. (1999). Relationships between riverine fish and woody debris: implications for lowland rivers. *Marine and Freshwater Research* 50: 941–953.

Crook, D.A., Lowe, W.H., Allendorf, F.W. et al. (2015). Human effects on ecological connectivity in aquatic ecosystems: integrating scientific approaches to support management and mitigation. *Science of the Total Environment* 534: 52–64.

Crosby, B.T. and Whipple, K.X. (2006). Knickpoint initiation and distribution within fluvial networks: 236 waterfalls in the Waipaoa River, North Island, New Zealand. *Geomorphology* 82: 16–38.

Crosby, B.T., Whipple, K.X., Gasparini, N.M., and Wobus, C.W. (2007). Formation of fluvial hanging valleys: theory and simulation. *Journal of Geophysical Research* 112, F03S10.

Crossland, C.J., Kremer, H.H., Lindeboom, H.J. et al. (eds.) (2005). *Coastal Fluxes in the Anthropocene*. Berlin: Springer.

Crosta, G.B. and Frattini, P. (2004). Controls on modern alluvial fan processes in the Central Alps, northern Italy. *Earth Surface Processes and Landforms* 29: 267–293.

Crowder, D.W. and Knapp, H.V. (2005). Effective discharge recurrence intervals of Illinois streams. *Geomorphology* 64: 167–184.

Cui, Y. and Parker, G. (2005). Numerical model of sediment pulses and supply disturbances in mountain rivers. *Journal of Hydraulic Engineering* 131: 646–656.

Culp, J.M., Scrimgeour, G.J., and Townsend, G.D. (1996). Simulated fine woody debris accumulations in a stream increase rainbow trout fry abundance. *Transactions of the American Fisheries Society* 125: 472–479.

Curran, J.C. and Hession, W.C. (2013). Vegetative impacts on hydraulics and sediment processes across the fluvial system. *Journal of Hydrology* 505: 364–376.

Curran, J.C. and Tan, L. (2014). The effect of cluster morphology on the turbulent flows over an armored gravel bed surface. *Journal of Hydro-Environment Research* 8: 129–142.

Curran, J.H. and Wohl, E.E. (2003). Large woody debris and flow resistance in step-pool channels, Cascade Range, Washington. *Geomorphology* 51: 141–147.

Curtis, K.E., Renshaw, C.E., Magilligan, F.J., and Dade, W.B. (2010). Temporal and spatial scales of geomorphic adjustments to reduced competency following flow regulation in bedload-dominated systems. *Geomorphology* 118: 105–117.

Curtis, P.G., Slay, C.M., Harris, N.L. et al. (2018). Classifying drivers of global forest loss. *Science* 361: 1108–1111.

Czuba, J.A. and Foufoula-Georgiou, E. (2014). A network-based framework for identifying potential synchronizations and amplifications of sediment delivery in river basins. *Water Resources Research* 50: 3826–3851.

Czuba, J.A. and Foufoula-Georgiou, E. (2015). Dynamic connectivity in a fluvial network for identifying hotspots of geomorphic change. *Water Resources Research* 51: 1401–1421.

Czuba, J.A. and Foufoula-Georgiou, E. (2017). Dynamic connectivity in a fluvial network for identifying hot spots of geomorphic change. *Water Resources Research* 51: 1401–1421.

Dadson, S.J., Hovius, N., Chen, H. et al. (2004). Earthquake-triggered increase in sediment delivery from an active mountain belt. *Geology* 32: 733–736.

D'Agostino, V. and Michelini, T. (2015). On kinematics and flow velocity prediction in step-pool channels. *Water Resources Research* 51: 4650–4667.

Dalrymple, T. 1960. Flood-frequency analyses. US Geological Survey Water-Supply Paper 1543-A.

Dalrymple, T., M.A. Benson. 1967. Measurement of peak discharge by the slope-area method. Techniques of Water-Resources Investigations of the US Geological Survey, Book 3, Chapter A1, 1–12.

Dalrymple, R.W. and Choi, K. (2007). Morphologic and facies trends through the fluvial-marine transition in tide-dominated depositional systems: a schematic framework for environmental and sequence stratigraphic interpretation. *Earth-Science Reviews* 81: 135–174.

Dalrymple, R.W., Zaitlin, B.A., and Boyd, R. (1992). Estuarine facies models: conceptual basis and stratigraphic implications. *Journal of Sedimentary Petrology* 62: 1130–1146.

Daniel, J.F. 1971. Channel movement of meandering Indiana streams. US Geological Survey Professional Paper 732-A.

Daniels, M.D. and Rhoads, B.L. (2004). Effect of large woody debris configuration on three-dimensional flow structure in two low-energy meander bends at varying stages. *Water Resources Research* 40, W11302.

D'Aoust, S.G. and Millar, R.G. (2000). Stability of ballasted woody debris habitat structures. *Journal of Hydraulic Engineering* 126: 810–817.

Darby, S. and Sear, D.A. (2008). *River Restoration: Managing the Uncertainty in Restoring Physical Habitat*. Chichester: Wiley.

Darby, S.E. and Van De Wiel, M.J. (2003). Models in fluvial geomorphology. In: *Tools in Fluvial Geomorphology* (eds. G.M. Kondolf and H. Piégay), 503–537. Chichester: Wiley.

Darcy, H. (1856). *Les fontaines publiques de la ville de Dijon*. Paris: Victor Dalmont.

Das, N.N. and Mohanty, B.P. (2006). Root zone soil moisture assessment using remote sensing and vadose zone modeling. *Vadose Zone Journal* 5: 296–307.

Datry, T., Arscott, D.B., and Sabater, S. (2011). Recent perspectives on temporary river ecology. *Aquatic Sciences* 73: 453–457.

Datry, T., Larned, S.T., and Tockner, K. (2014). Intermittent rivers: a challenge for freshwater ecology. *BioScience* 64: 229–235.

David, G.C.L., Bledsoe, B.P., Merritt, D.M., and Wohl, E. (2009). The impacts of ski slope development on stream channel morphology in the White River National Forest, Colorado, USA. *Geomorphology* 103: 375–388.

David, G.C.L., Wohl, E., Yochum, S.E., and Bledsoe, B.P. (2010). Controls on at-a-station hydraulic geometry in steep headwater streams, Colorado, USA. *Earth Surface Processes and Landforms* 35: 1820–1837.

David, G.C.G., Legleiter, C.J., Wohl, E., and Yochum, S.E. (2013). Characterizing spatial variability in velocity and turbulence intensity using 3-D acoustic Doppler velocimeter data in a plane-bed reach of East St. Louis Creek, Colorado, USA. *Geomorphology* 183: 28–44.

Davidson, S.L. and Eaton, B.C. (2013). Modeling channel morphodynamic response to variations in large wood: implications for stream rehabilitation in degraded watersheds. *Geomorphology* 202: 59–73.

Davidson, S., MacKenzie, L., and Eaton, B. (2015). Large wood transport and jam formation in a series of flume experiments. *Water Resources Research* 51: 10065–10077.

Davies, N.S. and Gibling, M.R. (2011). Evolution of fixed-channel alluvial plains in response to Carboniferous vegetation. *Nature Geoscience* 4: 629–633.

Davies, T.R.H. and Sutherland, A.J. (1983). Extremal hypotheses for river behavior. *Water Resources Research* 19: 141–148.

Davies, G. and Woodroffe, D.C. (2010). Tidal estuary width convergence: theory and form in north Australian estuaries. *Earth Surface Processes and Landforms* 35: 737–749.

Davis, W.M. (1899). The geographical cycle. *Geographical Journal* 14: 481–504.

Davis, W.M. (1902a). Base level, grade, and peneplain. *Journal of Geology* 10: 77–111.

Davis, W.M. (1902b). River terraces in New England. *Bulletin of the Museum of Comparative Zoology at Harvard College* 38: 281–346.

Davis, W.M. (1909). *Geographical Essays*. Dover Publications.

Davis, J.A. (2006). *Naples and Napoleon: Southern Italy and the European Revolutions 1780–1860*. New York: Oxford University Press.

Davy, P. and Lague, D. (2009). Fluvial erosion/transport equation of landscape evolution models revisited. *Journal of Geophysical Research Earth Surface* 114, F03007.

De Chant, L.J., Pease, P.P., and Tchakerian, V.P. (1999). Modelling alluvial fan morphology. *Earth Surface Processes and Landforms* 24: 641–652.

De Haas, T., Densmore, A.L., Stoffel, M. et al. (2018). Avulsions and the spatio-temporal evolution of debris-flow fans. *Earth-Science Reviews* 177: 53–75.

De Jong, C. (2015). Challenges for mountain hydrology in the third millennium. *Frontiers in Environmental Science* 3 https://doi.org/10.3389/fenvs.2015.00038.

De Scally, F.A. and Owens, I.F. (2004). Morphometric controls and geomorphic responses on fans in the Southern Alps, New Zealand. *Earth Surface Processes and Landforms* 29: 311–322.

De Scally, F.E.S., Slaymaker, O., and Owens, I. (2001). Morphometric controls and basin response in the Cascade Mountains. *Geografiska Annaler* 83A: 117–130.

De Vente, J., Poesen, J., Arabkhedri, M., and Verstraeten, G. (2007). The sediment delivery problem revisited. *Progress in Physical Geography* 31: 155–178.

Dean, D.J. and Schmidt, J.C. (2011). The role of feedback mechanisms in historic channel changes of the lower Rio Grande in the Big Bend region. *Geomorphology* 126: 333–349.

Dean, D.J. and Schmidt, J.C. (2013). The geomorphic effectiveness of a large flood on the Rio Grande in the Big Bend region: insights on the geomorphic controls and post-flood geomorphic response. *Geomorphology* 201: 183–198.

Delcaillau, B., Amrhar, M., Namous, M. et al. (2011). Transpressional tectonics in the Marrakech High Atlans: insight by the geomorphic evolution of drainage basins. *Geomorphology* 134: 344–362.

Dell'Agnese, A., Brardinoni, F., Toro, M. et al. (2015). Bedload transport in a formerly glaciated mountain catchment constrained by particle tracking. *Earth Surface Dynamics* 3: 527–542.

Delmas, M., Braucher, R., Gunnell, Y. et al. (2015). Constraints on Pleistocene glaciofluvial terrace age and related soil chronosequences features from vertical ^{10}Be profiles in the Ariége River catchment (Pyrenees, France). *Global and Planetary Change* 132: 39–53.

Dente, E., Lensky, N.G., Morin, E. et al. (2019). Sinuosity evolution along an incising channel: new insights from the Jordan River response to the Dead Sea level fall. *Earth Surface Processes and Landforms* https://doi.org/10.1002/esp.4530.

Derksen, C. and Brown, R. (2012). Spring snow cover extent reductions in the 2008-2012 period exceeding climate model projections. *Geophysical Research Letters* 39, L19504.

Deroanne, C. and Petit, F. (1999). Longitudinal evaluation of the bed load size and of its mobilization in a gravel bed river. In: *Floods and Landslides* (eds. R. Casale and C. Margottini), 335–342. Berlin: Springer.

Detert, M., Klar, M., Wenka, T., and Jirka, G.H. (2008). Pressure- and velocity-measurements above and within a porous gravel bed at the threshold of stability. In: *Gravel-Bed Rivers VI: From Process Understanding to River Restoration* (eds. H. Habersack, H. Piégay and M. Rinaldi), 85–107. Amsterdam: Elsevier.

Dethier, D.P. (2001). Pleistocene incision rates in the western United States calibrated using Lava Creek B tephra. *Geology* 29: 783–786.

Dezileau, L., Terrier, B., Berger, J.F. et al. (2014). A multidating approach applied to historical slackwater flood deposits of the Gardon River, SE France. *Geomorphology* 214: 56–68.

DiBiase, R.A., Denn, A.R., Bierman, P.R. et al. (2018). Stratigraphic control of landscape response to base-level fall, Young Womans Creek, Pennsylvania, USA. *Earth and Planetary Science Letters* 504: 163–173.

Dietrich, W.E. (1987). Mechanics of flow and sediment transport in river bends. In: *River Channels: Environment and Process* (ed. K.S. Richards), 179–227. Oxford: Blackwell.

Dietrich, W.E. and Dunne, T. (1978). Sediment budget for a small catchment in mountainous terrain. *Zeitschrift Geomorphologie* 29: 191–206.

Dietrich, W.E. and Dunne, T. (1993). The channel head. In: *Channel Network Hydrology* (eds. K. Beven and M.J. Kirkby), 175–219. Hoboken, NJ: Wiley.

Dietrich, W.E., Smith, J.D., and Dunne, T. (1979). Flow and sediment transport in a sand bedded meander. *Journal of Geology* 87: 305–315.

Dietrich, W.E., Kirchner, J.W., Ikeda, H., and Iseya, F. (1989). Sediment supply and the development of the coarse surface layer in gravel-bedded rivers. *Nature* 340: 215–217.

Dietrich, W.E., Wilson, C.J., Montgomery, D.R. et al. (1992). Erosion thresholds and land surface morphology. *Geology* 20: 675–679.

Dietrich, W.E., Wilson, C.J., Montgomery, D.R., and McKean, J. (1993). Analysis of erosion thresholds, channel networks, and landscape morphology using a digital terrain model. *Journal of Geology* 101: 259–278.

Dietrich, W.E., Bellugi, D.G., Sklar, L.S. et al. (2003). Geomorphic transport laws for predicting landscape form and dynamics. In: *Prediction in Geomorphology* (eds. P.R. Wilcock and R.M. Iverson), 103–132. Washington, DC: American Geophysical Union Press.

Dinehart, R.L. and Burau, J.R. (2005). Averaged indicators of secondary flow in repeated acoustic Doppler current profiler crossings of bends. *Water Resources Research* 41, W09405.

Dingwall, P.R. (1972). Erosion by overland flow on an alpine debris slope. In: *Mountain Geomorphology: Geomorphological Processes in the Canadian Cordillera* (eds. H.O. Slaymaker and H.J. McPherson), 113–120. Vancouver: Tantalus Research.

Diplas, P. and Sutherland, A.J. (1988). Sampling techniques for gravel sized sediments. *Journal of Hydraulic Engineering* 114: 484–501.

Diplas, P., Dancey, C.L., Celik, A.O. et al. (2008). The role of impulse on the initiation of particle movement under turbulent flow conditions. *Science* 322: 717–720.

Dixon, S.J. and Sear, D.A. (2014). The influence of geomorphology on large wood dynamics in a low gradient headwater stream. *Water Resources Research* 50: 9194–9210.

Dixon, S.J., Sear, D.A., Odoni, N.A. et al. (2016). The effects of river restoration on catchment scale flood risk and flood hydrology. *Earth Surface Processes and Landforms* 41: 997–1008.

Do Carmo, J.S.A. (2007). The environmental impact and risks associated with changes in fluvial morphodynamic processes. In: *Water in Celtic Countries: Quantity, Quality and Climate Variability* (eds. J.P. Lobo Ferreira and J.M.P. Viera), 307–319. Wallingford: IAHS Publications no. 310.

Dodds, W.K., Gido, K., Whiles, M.R. et al. (2004). Life on the edge: the ecology of Great Plains prairie streams. *BioScience* 54: 205–216.

Dolan, R., Howard, A., and Trimble, D. (1978). Structural control of the rapids and pools of the Colorado River in the grand canyon. *Science* 202: 629–631.

Domenico, P.A. and Schwartz, F.W. (1998). *Physical and Chemical Hydrogeology*. New York: Wiley.

Doody, T. and Benyon, R. (2011). Quantifying water savings from willow removal in Australian streams. *Journal of Environmental Management* 92: 926–935.

Dort, W. (2009). *Historical Channel Changes of the Kansas River and its Major Tributaries*. New York: American Geographical Society.

Dosseto, A. and Schaller, M. (2016). The erosion response to Quaternary climate change quantified using uranium isotopes and in situ-produced cosmogenic nuclides. *Earth-Science Reviews* 155: 60–81.

Douglas, I. (1967). Man, vegetation and the sediment yields of rivers. *Nature* 215: 925–928.

Douglas, I. (2009). Hydrological investigations of forest disturbance and land cover impacts in South-East Asia: a review. *Philosophical Transactions of the Royal Society B* 354: 1725–1738.

Douglass, J. and Schmeeckle, M. (2007). Analogue modeling of transverse drainage mechanisms. *Geomorphology* 84: 22–43.

Downs, P.W. and de Asua, R.R. (2016). Modelling catchment processes. In: *Tools in Fluvial Geomorphology* (eds. G.M. Kondolf and H. Piegay), 159–179. Chichester: Wiley-Blackwell.
Downs, P.W. and Priestnall, G. (2003). Modelling catchment processes. In: *Tools in Fluvial Geomorphology* (eds. G.M. Kondold and H.M. Piégay), 205–230. Chichester: Wiley.
Doyle, M.W. and Ensign, S.H. (2009). Alternative reference frames in river system science. *BioScience* 59: 499–510.
Drake, T.W., Raymond, P.A., and Spencer, R.G.M. (2018). Terrestrial carbon inputs to inland waters: a current synthesis of estimates and uncertainty. *Limnology and Oceanography* 3: 132–142.
Draut, A.E., Logan, J.B., and Mastin, M.C. (2011). Channel evolution on the dammed Elwha River, Washington, USA. *Geomorphology* 127: 71–87.
Drever, J.I. (1988). *The Geochemistry of Natural Waters*, 2e. Englewood Cliffs, NJ: Prentice-Hall.
Drigo, R. (2005). Trends and patterns of tropical land use change. In: *Forests, Water and People in the Humid Tropics* (eds. M. Bonell and L.A. Bruijnzeel), 9–39. Cambridge: Cambridge University Press.
Dubinski, I.M. and Wohl, E. (2013). Relationships between block quarrying, bed shear stress, and stream power: a physical model of block quarrying of a jointed bedrock channel. *Geomorphology* 180–181: 66–81.
DuBowy, P.J. (2013). Mississippi River ecohydrology: past, present and future. *Ecohydrology and Hydrobiology* 13: 73–83.
Dufour, S. and Piégay, H. (2009). From the myth of a lost paradise to targeted river restoration: forget natural references and focus on human benefits. *River Research and Applications* 25: 568–581.
Dugdale, S.J., Carbonneau, P.E., and Campbell, D. (2010). Aerial photosieving of exposed gravel bars for the rapid calibration of airborne grain size maps. *Earth Surface Processes and Landforms* 35: 627–639.
Dunne, T. (1978). Field studies of hillslope flow processes. In: *Hillslope Hydrology* (ed. M.J. Kirkby), 227–293. Toronto: Wiley.
Dunne, T. (1980). Formation and controls of channel networks. *Progress in Physical Geography* 4: 211–239.
Dunne, T. (1990). Hydrology, mechanics, and geomorphic implications of erosion by subsurface flow. In: *Groundwater Geomorphology; The Role of Subsurface Water in Earth-surface Processes and Landforms* (eds. C.G. Higgins and D.R. Coates), 1–28. Boulder, CO, Special Paper 252,: Geological Society of America.
Dunne, T. and Aalto, R.E. (2013). Large river floodplains. In: *Treatise on Fluvial Geomorphology* (ed. E. Wohl). Amsterdam: Elsevier.
Dunne, R. and Black, R.D. (1970a). An experimental investigation of runoff production in permeable soils. *Water Resources Research* 6 (2): 478–490.
Dunne, R. and Black, R.D. (1970b). Partial area contributions to storm runoff in a small New England watershed. *Water Resources Research* 6 (5): 1296–1311.
Dunne, T. and Leopold, L.B. (1978). *Water in Environmental Planning*. New York: W.H. Freeman.
Dunne, T., Zhang, W., and Aubry, B.F. (1991). Effects of rainfall, vegetation, and microtopography on infiltration and runoff. *Water Resources Research* 27 (9): 2271–2285.
Dunne, T., Whipple, K.X., and Aubry, B.F. (1995). Microtopography of hillslopes and initiation of channels by Horton overland flow. In: *Natural and Anthropogenic Influences in Fluvial Geomorphology* (eds. J.E. Costa, A.J. Miller, K.W. Potter and P.R. Wilcock), 27–44. Washington, DC: American Geophysical Union Press.
Dunne, T., Mertes, L.A.K., Meade, R.H. et al. (1998). Exchanges of sediment between the flood plain and channel of the Amazon River in Brazil. *Geological Society of America Bulletin* 110: 450–467.

Dunne, T., Malmon, D.V., and Mudd, S.M. (2010). A rain splash transport equation assimilating field and laboratory measurements. *Journal of Geophysical Research* 115, F01001.

Dury, G.H. 1965. Theoretical implications of underfit streams. US Geological Survey Professional Paper 452-C.

Dust, D. and Wohl, E. (2012a). Characterization of the hydraulics at natural step crests via weir flow concepts. *Water Resources Research* 48, W09542.

Dust, D. and Wohl, E. (2012b). Conceptual model for complex river responses using an expanded Lane's relation. *Geomorphology* 139-140: 109–121.

Dutta, V., Sharma, U., Iqbal, K. et al. (2018). Impact of river channelization and riverfront development on fluvial habitat: evidence from Gomti River, a tributary of Ganges, India. *Environmental Sustainability* 1: 167–184.

Duvall, A., Kirby, E., and Burbank, D. (2004). Tectonic and lithologic controls on bedrock channel profiles and processes in coastal California. *Journal of Geophysical Research* 109, F03002.

Dykes, A.P. and Thornes, J.B. (2000). Hillslope hydrology in tropical rainforest steeplands in Brunei. *Hydrological Processes* 14: 215–235.

Dynesius, M. and Nilsson, C. (1994). Fragmentation and flow regulation of river systems in the northern third of the world. *Science* 266: 753–762.

East, A.E., Pess, G.R., Bounty, J.A. et al. (2015). Large-scale dam removal on the Elwha River, Washington, USA: river channel and floodplain geomorphic change. *Geomorphology* 228: 765–786.

Eaton, B.C. (2006). Bank stability analysis for regime models of vegetated gravel bed rivers. *Earth Surface Processes and Landforms* 31: 1438–1444.

Eaton, B.C. (2013). Hydraulic geometry: empirical investigations and theoretical approaches. In: *Treatise on Fluvial Geomorphology* (ed. E. Wohl), 313–329. Amsterdam: Elsevier.

Eaton, B.C. and Church, M. (2004). A graded stream response relation for bed load-dominated streams. *Journal of Geophysical Research Earth Surface* 109, F03011.

Eaton, B.C. and Lapointe, M.F. (2001). Effects of large floods on sediment transport and reach morphology in the cobble-bed Sainte Marguerite River. *Geomorphology* 40: 291–309.

Eaton, B.C. and Millar, R.G. (2004). Optimal alluvial channel width under a bank stability constraint. *Geomorphology* 62: 35–45.

Eaton, L.S., Morgan, B.A., Kochel, R.C., and Howard, A.D. (2003). Role of debris flows in long-term landscape denudation in the central Appalachians of Virginia. *Geology* 31: 339–342.

Eaton, B.C., Church, M., and Millar, R.G. (2004). Rational regime model of alluvial channel morphology and response. *Earth Surface Processes and Landforms* 29: 511–529.

Eaton, B.C., Church, M., and Davies, T.R.H. (2006). A conceptual model for meander initiation in bedload-dominated streams. *Earth Surface Processes and Landforms* 31: 875–891.

Eckhardt, K. and Ulbrich, U. (2003). Potential impacts of climate change on groundwater recharge and streamflow in a central European low mountain range. *Journal of Hydrology* 284: 244–252.

Eckley, M.S. and Hinchliff, D.L. (1986). Glen Canyon Dam's quick fix. *Civil Engineering/ASCE* 56: 46–48.

Eden, M.J. (1971). Scientific exploration in the Venezuelan Amazonas. *The Geographical Journal* 137: 149–156.

Edmonds, D.A., Hoyal, D.C.J.D., Sheets, B.A., and Slingerland, R.L. (2009). Predicting delta avulsions: implications for coastal wetland restoration. *Geology* 37: 759–762.

Edmonds, D.A., Shaw, J.B., and Mohrig, D. (2011). Topset-dominated deltas: a new model for river delta stratigraphy. *Geology* 39: 1175–1178.

Edwards, T.K. and Glysson, G.D. (1999). Field methods for measurement of fluvial sediment. In: *US Geological Survey Techniques of Water Resources Investigations*, Book 3, Chapter C2,. Denver, CO.

Egli, M., Brandová, D., Böhlert, R. et al. (2010). ^{10}Be inventories in Alpine soils and their potential for dating land surfaces. *Geomorphology* 119: 62–73.

Egozi, R. and Ashmore, P. (2008). Defining and measuring braiding intensity. *Earth Surface Processes and Landforms* 33: 2121–2138.

Ehlen, J. and Wohl, E. (2002). Joints and landform evolution in bedrock canyons. *Transactions, Japanese Geomorphological Union* 23: 237–255.

Einstein, H.A. (1950). The bed-load function for sediment transportation in open channel flows. US Department of Agriculture, Technical Bulletin No. 1026, Washington, DC.

Einstein, H.A. and Banks, R.B. (1950). Fluid resistance of composite roughness. *Transactions of the American Geophysical Union* 31: 603–610.

Einstein, H.A. and Shen, H.W. (1964). A study of meandering in straight alluvial channels. *Journal of Geophysical Research* 69: 5239–5247.

Ekes, C. and Friele, P. (2003). Sedimentary architecture and post-glacial evolution of Cheekye fan, southwestern British Columbia, Canada. In: *Ground Penetrating Radar in Sediments* (eds. C.S. Bristow and H.M. Jol) Geological Society Special Publication 211, 87–98.

Elder, K., Kattelmann, R., and Ferguson, R. (1990). Refinements in diluting gauging for mountain streams. In: *Hydrology in Mountainous Regions. I. Hydrological Measurements: The Water Cycle* (eds. H. Lang and A. Musy), 247–254. Wallingford: IAHS Publications no. 193.

Elfström, A. (1987). Large boulder deposits and catastrophic floods. A case study of the Båldakatj area, Swedish Lapland. *Geografiska Annaler* 69A: 101–121.

Ellis, L.M., Molles, M.C., and Crawford, C.S. (1999). Influence of experimental flooding on litter dynamics in a Rio Grande riparian forest, New Mexico. *Restoration Ecology* 7: 193–204.

Elmore, A.J. and Kaushal, S.S. (2008). Disappearing headwaters: patterns of stream burial due to urbanization. *Frontiers in Ecology and the Environment* 6: 308–312.

Elsenbeer, H. (2001). Hydrological flowpaths in tropical rainforest soilscapes – a review. *Hydrological Processes* 15: 1751–1759.

Elsenbeer, H. and Vertessy, R.A. (2000). Stormflow generation and flowpath characteristics in an Amazonian rainforest catchment. *Hydrological Processes* 14: 2367–2381.

Elshorbagy, A., Wagener, T., Razavi, S., and Sauchyn, D. (2016). Correlation and causation in tree-ring-based reconstruction of paleohydrology in cold semiarid regions. *Water Resources Research* 52: 7053–7069.

Ely, L.L., Webb, R.H., and Enzel, Y. (1992). Accuracy of post-bomb ^{137}Cs and ^{14}C in dating fluvial deposits. *Quaternary Research* 38: 196–204.

Ely, L.L., Enzel, Y., Baker, V.R., and Cayan, D.R. (1993). A 5000-year record of extreme floods and climate change in the southwestern United States. *Science* 262: 410–412.

Emmett, W.W. 1980. A field calibration of the sediment-trapping characteristics of the Helley-Smith bedload sampler. US Geological Survey Professional Paper 1139.

Engelund, F. and Hansen, E. (1967). *A Monograph on Sediment Transport in Alluvial Streams.* Copenhagen: TEKNISKFORLAG.

England, J.F., Jarrett, R.D., and Salas, J.D. (2003). Data-based comparisons of moments estimators using historical and paleoflood data. *Journal of Hydrology* 278: 172–196.

England, J.F., Godaire, J.E., Klinger, R.E. et al. (2010). Paleohydrologic bounds and extreme flood frequency of the upper Arkansas River, Colorado, USA. *Geomorphology* 124: 1–16.

England, J.F., Julien, P.Y., and Velleux, M.L. (2014). Physically-based extreme flood frequency with stochastic storm transposition and paleoflood data on large watersheds. *Journal of Hydrology* 510: 228–245.

English, M.C., Hill, R.B., Stone, M.A., and Ormson, R. (1997). Geomorphological and botanical change on the outer Slave River delta, NWT, before and after impoundment of the Peace River. *Hydrological Processes* 11: 1707–1724.

Engstrom, D.R., Almendinger, J.E., and Wolin, J.A. (2009). Historical changes in sediment and phosphorus loading to the upper Mississippi River: mass-balance reconstructions from the sediments of Lake Pepin. *Journal of Paleolimnology* 41: 563–588.

Ensign, S.H. and Doyle, M.W. (2005). In-channel transient storage and associated nutrient retention: evidence from experimental manipulations. *Limnology and Oceanography* 50: 1740–1751.

Entwistle, N., Heritage, G., and Milan, D. (2018). Flood energy dissipation in anabranching channels. *River Research and Applications* 34: 709–720.

Entwistle, N., Heritage, G.L., and Milan, D. (2019). Ecohydraulic modelling of anabranching rivers. *River Research and Applications* 35: 353–364.

Ergenzinger, P. and de Jong, C. (2002). Perspectives on bed load measurement. In: *Erosion and Sediment Transport Measurement in Rivers: Technological and Methodological Advances* (ed. J. Bogen), 113–125. Wallingford: IAHS Publcations no. 44.

Erwin, S.O., Schmidt, J.C., and Nelson, N.C. (2011). Downstream effects of impounding a natural lake; the Snake River downstream from Jackson Lake Dam, Wyoming, USA. *Earth Surface Processes and Landforms* 36: 1421–1434.

Eswaran, H., Van Den Berg, E., and Reich, P. (1993). Organic carbon in soils of the world. *Soil Science Society of America Journal* 57: 192–194.

Ettema, R., S.F. Daly. 2004. Sediment transport under ice. US Army Corps of Engineers, Cold Regions Research and Engineering Laboratory, Hanover, New Hampshire, ERDC/CRREL TR-04-20.

Ettema, R. and Kempema, E.W. (2012). River-ice effects on gravel-bed channels. In: *Gravel-bed Rivers: Processes, Tools, Environments* (eds. M. Church, P.M. Biron and A.G. Roy), 525–540. Hoboken, NJ: Wiley.

European Commission. 2000. Directive 2000/60/EC of the European Parliament and of the Council of 23 October 2000 establishing a framework for community action in the field of water policy. Official Journal of the European Communities, L 327.

Everall, N.C., Johnson, M.F., Wood, P. et al. (2017). Comparability of macroinvertebrate biomonitoring indices of river health derived from semi-quantitative and quantitative methodologies. *Ecological Indicators* 78: 437–448.

Everitt, B.L. (1968). Use of the cottonwood in an investigation of the recent history of a flood plain. *American Journal of Science* 266: 417–439.

Evette, A., Labonne, S., Rey, F. et al. (2009). History of bioengineering techniques for erosion control in rivers in western Europe. *Environmental Management* 43: 972–984.

Ewen, J., Parkin, G., and O'Connell, P.E. (2000). SHETRAN: a coupled surface/subsurface modelling system for 3D water flow and sediment and solute transport in river basins. *ASCE Journal of Hydraulic Engineering* 5: 250–258.

Fagherazzi, S. (2008). Self-organization of tidal deltas. *Proceedings of the National Academy of Sciences* 105: 18692–18695.

Fagherazzi, S. and Furbish, D.J. (2001). On the shape and widening of salt marsh creeks. *Journal of Geophysical Research* 106: 991–1003.

Fagherazzi, S. and Overeem, I. (2007). Models of deltaic and inner continental shelf landform evolution. *Annual Review of Earth and Planetary Sciences* 35: 685–715.

Fagherazzi, S., Bryan, K.R., and Nardin, W. (2017). Buried alive or washed away: the challenging life of mangroves in the Mekong Delta. *Oceanography* 30: 48–59.

Fahnestock, R.K. 1963. Morphology and hydrology of a glacial stream – White River, Mount Rainier, Washington. US Geological Survey Professional Paper 422A.

Fairweather, P.G. (1999). State of environment indicators of "river health": exploring the metaphor. *Freshwater Biology* 41: 211–220.

Falke, J.A., Bestgen, K.R., and Fausch, K.D. (2010). Streamflow reductions and habitat drying affect growth, survival, and recruitment of brassy minnow across a Great Plains riverscape. *Transactions of the American Fisheries Society* 139: 1566–1583.

Falke, J.A., Fausch, K.D., Magelky, R. et al. (2011). The role of groundwater pumping and drought in shaping ecological futures for stream fishes in a dryland river basin of the western Great Plains, USA. *Ecohydrology* 4: 682–697.

Famiglietti, J.S. (2014). The global groundwater crisis. *Nature Climate Change* 4: 945–948.

Fanelli, R.M. and Lautz, L.K. (2008). Patterns of water, heat, and solute flux through streambeds around small dams. *Ground Water* 46: 671–687.

Farnsworth, K.L. and Milliman, J.D. (2003). Effects of climatic and anthropogenic change on small mountainous rivers; the Salinas River example. *Global and Planetary Change* 39: 53–64.

Faulkner, D.J., Larson, P.H., Jol, H.M. et al. (2016). Autogenic incision and terrace formation resulting from abrupt late-glacial base-level fall, lower Chippewa River, Wisconsin, USA. *Geomorphology* 266: 75–95.

Fausch, K.D. and Bestgen, K.R. (1997). Ecology of fishes indigenous to the central and southwestern Great Plains. In: *Ecology and Conservation of Great Plains Vertebrates* (eds. F.L. Knopf and F.B. Samson), 131–166. New York: Springer-Verlag.

Fayó, R., Espinosa, M.A., Velez-Agudelo, C.A. et al. (2018). Diatom-based reconstruction of Holocene hydrological changes along the Colorado River floodplain (northern Patagonia, Argentina). *Journal of Paleolimnology* 60: 427–443.

Fearnside, P.M. (2015). Emissions from tropical hydropower and the IPCC. *Environmental Science and Policy* 50: 225–239.

Felder, G., Zischg, A., and Weingartner, R. (2017). The effect of coupling hydrologic and hydrodynamic models on probable maximum flood estimation. *Journal of Hydrology* 550: 157–165.

Fenn, C.R. (1987). Sediment transfer processes in alpine glacier basins. In: *Glacio-Fluvial Sediment Transfer* (eds. A.M. Gurnell and M.J. Clark), 59–85. Chichester: Wiley.

Feo, M.L., Ginebreda, A., Eljarrat, E., and Barcelo, D. (2010). Presence of pyrethroid pesticides in water and sediments of Ebro River delta. *Journal of Hydrology* 393: 156–162.

Ferguson, R.I. (1975). Meander irregularity and wavelength estimation. *Journal of Hydrology* 26: 315–333.

Ferguson, R.I. (1986). Hydraulics and hydraulic geometry. *Progress in Physical Geography* 10: 1–31.

Ferguson, R.I. (1987). Hydraulic and sedimentary controls of channel pattern. In: *River Channels: Environment and Process* (ed. K.S. Richards), 129–158. Oxford: Blackwell.

Ferguson, R.I. (2003). Emergence of abrupt gravel to sand transitions along rivers through sorting processes. *Geology* 31: 159–162.

Ferguson, R.I. (2007). Flow resistance equations for gravel- and boulder-bed streams. *Water Resources Research* 43, W05427.

Ferguson, R.I. (2012). River channel slope, flow resistance, and gravel entrainment thresholds. *Water Resources Research* 48, W05517.

Ferguson, R. (2013). Reach-scale flow resistance. In: *Treatise on Fluvial Geomorphology* (ed. E. Wohl), 50–68. Amsterdam: Elsevier.

Ferguson, R.I., Parsons, D.R., Lane, S.N., and Hardy, R.J. (2003). Flow in meander bends with recirculation at the inner bank. *Water Resources Research* 39 https://doi.org/10.1029/2003WR001965.

Ferguson, R.I., Sharma, B.P., Hodge, R.A. et al. (2017a). Bed load tracer mobility in a mixed bedrock/alluvial channel. *Journal of Geophysical Research Earth Surface* 122: 807–822.

Ferguson, R.I., Sharma, B.P., Hardy, R.J. et al. (2017b). Flow resistance and hydraulic geometry in contrasting reaches of a bedrock channel. *Water Resources Research* 53: 2278–2293.

Ferrier, K.L., Kirchner, J.W., and Finkel, R.C. (2005). Erosion rates over millennial and decadal timescales at Caspar Creek and Redwood Creek, northern California Coast Ranges. *Earth Surface Processes and Landforms* 30: 1025–1038.

Ferro, V. and Porto, P. (2012). Identifying a dominant discharge for natural rivers in southern Italy. *Geomorphology* 139–140: 313–321.

Ficke, A.D. and Myrick, C. (2011). The swimming and jumping ability of three small Great Plains fishes: implications for fishway design. *Transactions of the American Fisheries Society* 140: 521–531.

Figueroa, A.M. and Knott, J.R. (2010). Tectonic geomorphology of the southern Sierra Nevada Mountains (California): evidence for uplift and basin formation. *Geomorphology* 123: 34–45.

Finley, J.B., Drever, J.I., and Turk, J.T. (1995). Sulfur isotope dynamics in a high-elevation catchment, West Glacier Lake, Wyoming. *Water, Air, & Soil Pollution* 79: 227–241.

Finnegan, N.J. (2013). Interpretation and downstream correlation of bedrock river terrace treads created from propagating knickpoints. *Journal of Geophysical Research: Earth Surface* 118 https://doi.org/10.1029/2012JF002534.

Finnegan, N.J. and Balco, G. (2013). Sediment supply, base level, braiding, and bedrock river terrace formation: Arroyo Seco, California, USA. *Geological Society of America Bulletin* 125: 1114–1124.

Finnegan, N.J. and Dietrich, W.E. (2011). Episodic bedrock strath terrace formation due to meander migration and cutoff. *Geology* 39: 143–146.

Finnegan, N.J., Roe, G., Montgomery, D.R., and Hallet, B. (2005). Controls on the channel width of rivers: implications for modeling fluvial incision of bedrock. *Geological Society of America Bulletin* 33: 229–232.

Finnegan, N.J., Sklar, L.S., and Fuller, T.K. (2007). Interplay of sediment supply, river incision, and channel morphology revealed by the transient evolution of an experimental bedrock channel. *Journal of Geophysical Research* 112, F03S11.

Finnegan, N.J., Schumer, R., and Finnegan, S. (2014). A signature of transience in bedrock river incision rates over timescales of 10^4–10^7 years. *Nature* 505: 391–394.

Flemming, B.W. (1988). Zur klassifikation subaquatischer, stromungstrans versaler transportkorper. *Bochumer Geologische und Geotechnische Arbeit* 29: 44–47.

Flener, C., Lotsari, E., Alho, P., and Käyhkö, J. (2012). Comparison of empirical and theoretical remote sensing based bathymetry models in river environments. *River Research and Applications* 28: 118–133.

Flint, J. (1974). Stream gradient as a function of order, magnitude, and discharge. *Water Resources Research* 10: 969–973.

Flint, A.L., Flint, L.E., and Dettinger, M.D. (2008). Modeling soil moisture processes and recharge under a melting snowpack. *Vadose Zone Journal* 7: 350–257.

Florsheim, J.L. and Dettinger, M.D. (2015). Promoting atmospheric-river and snowmelt-fueled biogeomorphic processes by restoring river-floodplain connectivity in California's Central Valley. In: *Geomorphic Approaches to Integrated Floodplain Management of Lowland Fluvial Systems in North America and Europe* (eds. P.F. Hudson and H. Middelkoop), 119–141. New York: Springer.

Florsheim, J.L., Keller, E.A., and Best, D.W. (1991). Fluvial sediment transport in response to moderate storm flows following chaparral wildfire, Ventura County, southern California. *Geological Society of America Bulletin* 103: 504–511.

Florsheim, J.L., Mount, J.F., and Chin, A. (2008). Bank erosion as a desirable attribute of rivers. *BioScience* 58: 519–529.

Foley, J.A., DeFries, R., Asner, G.P. et al. (2005). Global consequences of land use. *Science* 309: 570–574.

Folk, R.L. (1980). *Petrology of Sedimentary Rocks*. Austin, TX: Hemphill.

Fonstad, M.A. (2003). Spatial variation in the power of mountain streams in the Sangre de Cristo Mountains, New Mexico. *Geomorphology* 55: 75–96.

Fonstad, M.A. and Marcus, W.A. (2005). Remote sensing of stream depths with hydraulically assisted bathymetry (HAB) models. *Geomorphology* 72: 320–339.

Ford, D. and Williams, P. (2007). *Karst Hydrogeology and Geomorphology*. Chichester: Wiley.

Forshay, K., P. Mayer. 2012. Effects of watershed restoration on nitrogen in a stream impacted by legacy sediments: big spring run stream restoration project as a case study. National Risk Management Research Laboratory, US Environmental Protection Agency, Fact Sheet.

Forte, A.M., Whipple, K.X., Bookhagen, B., and Rossi, M.W. (2016). Decoupling of modern shortening rates, climate, and topography in the Caucasus. *Earth and Planetary Science Letters* 449: 282–294.

Fortugno, D., Boix-Fayos, C., Bombino, G. et al. (2017). Adjustments in channel morphology due to land-use changes and check dam installation in mountain torrents of Calabria (southern Italy). *Earth Surface Processes and Landforms* 42: 2469–2483.

Foster, S.G., Mahoney, J.M., and Rood, S.B. (2018). Functional flows: an environmental flow regime benefits riparian cottonwoods along the Waterton River, Alberta. *Restoration Ecology* 26: 921–932.

Fox, M. and Bolton, S. (2007). A regional and geomorphic reference for quantities and volumes of instream wood in unmanaged forested basins of Washington State. *North American Journal of Fisheries Management* 27: 342–359.

Fox, R.W. and McDonald, A.T. (1978). *Introduction to Fluid Mechanics*, 2e. New York: Wiley.

Francis, R.A., Corenblit, D., and Edwards, P.J. (2009). Perspectives on biogeomorphology, ecosystem engineering and self-organisation in island-braided fluvial ecosystems. *Aquatic Sciences* 71: 290–304.

Frankel, K.L. and Dolan, J.F. (2007). Characterizing arid region alluvial fan surface roughness with airborne laser swath mapping digital topographic data. *Journal of Geophysical Research Earth Surface* 112 https://doi.org/10.1029/2006JF000644.

Frankel, K., Brantley, K.S., Dolan, J.F. et al. (2007). Comsogenic ^{10}Be and ^{36}Cl geochronology of offset alluvial fans along the northern Death Valley fault zone: implications for transient strain in the eastern California shear zone. *Journal of Geophysical Research* 122 https://doi.org/10.1029/2006JB004350.

Fredlund, D.G., Morgenstern, N.R., and Widger, R.A. (1978). The shear strength of unsaturated soils. *Canadian Geotechnical Journal* 15: 313–321.

Freeman, M.C., Pringle, C.M., and Jackson, C.R. (2007). Hydrologic connectivity and the contribution of stream headwaters to ecological integrity at regional scales. *Journal of the North American Water Resources Association* 43: 5–14.

Frey, P. and Church, M. (2012). Gravel transport in granular perspective. In: *Gravel-Bed Rivers: Processes, Tools, Environments* (eds. M. Church, P.M. Biron and A.G. Roy), 39–55. Chichester: Wiley.

Friedkin, J.F. (1945). *A Laboratory Study of the Meandering of Alluvial Rivers*. Vicksburg, MI: US Waterways Experiment Station.

Friedman, J.M. and Lee, V.J. (2002). Extreme floods, channel change, and riparian forests along ephemeral streams. *Ecological Monographs* 72: 409–425.

Friedman, J.M., Vincent, K.R., and Shafroth, P.B. (2005). Dating floodplain sediment using tree-ring response to burial. *Earth Surface Processes and Landforms* 30: 1077–1091.

Frings, R.M., Berbee, B.M., Erkens, G. et al. (2009). Human-induced changes in bed shear stress and bed grain size in the River Waal (The Netherlands) during the past 900 years. *Earth Surface Processes and Landforms* 34: 503–514.

Fripp, J.B. and Diplas, P. (1993). Surface sampling in gravel streams. *Journal of Hydraulic Engineering* 119: 473–490.

Frissell, C.A., Liss, W.J., Warren, C.E., and Hurley, M.D. (1986). A hierarchical framework for stream habitat classification: viewing streams in a watershed context. *Environmental Management* 10: 199–214.

Froehlich, W. and Starkel, L. (1987). Normal and extreme monsoon rains – their role in the shaping of the Darjeeling Himalaya. *Studia Geomorphologica Carpatho-Balcanica* 21: 129–158.

Froese, D.G., Smith, D.G., and Clement, D.T. (2005). Characterizing large river history with shallow geophysics; middle Yukon River, Yukon Territory and Alaska. *Geomorphology* 67: 391–406.

Frothingham, K.M. and Rhoads, B.L. (2003). Three-dimensional flow structure and channel change in an asymmetrical compound meander loop, Embarras River, Illinois. *Earth Surface Processes and Landforms* 28: 625–644.

Fryirs, K.A. (2013). (Dis)connectivity in catchment sediment cascades: a fresh look at the sediment delivery problem. *Earth Surface Processes and Landforms* 38: 30–46.

Fryirs, K.A. (2017). River sensitivity: a lost foundation concept in fluvial geomorphology. *Earth Surface Processes and Landforms* 42: 55–70.

Fryirs, K. and Brierley, G.J. (2001). Variability in sediment delivery and storage along river courses in Bega catchment, NSW, Australia: implications for geomorphic river recovery. *Geomorphology* 38: 237–265.

Fryirs, K. and Brierley, G.J. (2009). Naturalness and place in river rehabilitation. *Ecology and Society* 14 http://www.ecologyandsociety.org/vol14/iss1/art20.

Fryirs, K.A., Brierley, G.J., Preston, N.J., and Kasai, M. (2007a). Buffers, barriers and blankets: the (dis)connectivity of catchment-scale sediment cascades. *Catena* 70: 49–67.

Fryirs, K.A., Brierley, G.J., Preston, N.J., and Spencer, J. (2007b). Catchment scale (dis)connectivity in sediment flux in the upper Hunter catchment, New South Wales, Australia. *Geomorphology* 84: 297–316.

Fryirs, K., Spink, A., and Brierley, G. (2009). Post-European settlement response gradients of river sensitivity and recovery across the upper Hunter catchment, Australia. *Earth Surface Processes and Landforms* 34: 897–918.

Fryirs, K., Brierley, G.J., and Erskine, W.D. (2012). Use of ergodic reasoning to reconstruct the historical range of variability and evolutionary trajectory of rivers. *Earth Surface Processes and Landforms* 37: 763–773.

Fryirs, K., Lisenby, P., and Croke, J. (2015). Morphological and historical resilience to catastrophic flooding: the case of Lockyer Creek, SE Queensland, Australia. *Geomorphology* 241: 55–71.

Fryirs, K.A., Brierley, G.J., Hancock, F. et al. (2018). Tracking geomorphic recovery in process-based river management. *Land Degradation and Development* 29: 3221–3244.

Fuchs, M. and Lang, A. (2009). Luminescence dating of hillslope deposits – a review. *Geomorphology* 109: 17–26.

Fuller, T.K. 2014. Field, experimental and numerical investigations into the mechanisms and drivers of lateral erosion in bedrock channels. Unpublished PhD dissertation, University of Minnesota, Minneapolis, MN.

Fuller, T.K., Perg, L.A., Willenbring, J.K., and Lepper, K. (2009). Field evidence for climate-driven changes in sediment supply leading to strath terrace formation. *Geology* 37: 467–470.

Furbish, D.J., Childs, E.M., Haff, P.K., and Schmeeckle, M.W. (2009a). Rain splash of soil grains as a stochastic advection-dispersion process, with implications for desert plant-soil interactions and land-surface evolution. *Journal of Geophysical Research* 114, F00A03.

Furbish, D.J., Haff, P.K., Dietrich, W.E., and Heimsath, A.M. (2009b). Statistical description of slope-dependent soil transport and the diffusion-like coefficient. *Journal of Geophysical Research* 114, F00A05.

Furbish, D.J., Schmeeckle, M.W., Schumer, R., and Fathel, S.L. (2016). Probability distributions of bed load particle velocities, accelerations, hop distances, and travel times informed by Jaynes's principle of maximum entropy. *Journal of Geophysical Research: Earth Surface* 121: 1373–1390.

Gabet, E.J. and Mudd, S.M. (2010). Bedrock erosion by root fracture and tree throw: a coupled biogeomorphic model to explore the humped soil production function and the persistence of hillslope soils. *Journal of Geophysical Research: Earth Surface* 115, F04005.

Galat, D.L., Fredrickson, L.H., Humburg, D.D. et al. (1998). Flooding to restore connectivity of regulated, large-river wetlands. *BioScience* 48: 721–733.

Galay, V.J. (1983). Causes of river bed degradation. *Water Resources Research* 19: 1057–1090.

Gallaway, J.M., Martin, Y.E., and Johnson, E.A. (2009). Sediment transport due to tree root throw: integrating tree population dynamics, wildfire, and geomorphic response. *Earth Surface Processes and Landforms* 34: 1255–1269.

Gallisdorfer, M.S., Bennett, S.J., Atkinson, J.F. et al. (2014). Physical-scale model designs for engineered log jams in rivers. *Journal of Hydro-Environment Research* 8: 115–128.

Galloway, J.N., Dentener, F.J., Capone, D.G. et al. (2004). Nitrogen cycles: past, present, and future. *Biogeochemistry* 70: 153–226.

Galy, A., France-Lanord, C., and Lartiges, B. (2008). Loading and fate of particulate organic carbon from the Himalay to the Ganges-Brahmaputra delta. *Geochimica et Cosmochimica Acta* 72: 1767–1787.

Galy, V., Peucker-Ehrenbrink, B., and Eglinton, T. (2015). Global carbon export from the terrestrial biosphere controlled by erosion. *Nature* 521: 204–207.

Galy-Lacaux, C., Delmas, R., Kouadio, G. et al. (1999). Long-term greenhouse gas emissions from hydroelectric reservoirs in tropical forest regions. *Global Biogeochemical Cycles* 13: 503–517.

Ganju, N.K., Suttles, S.E., Beudin, A. et al. (2017). Quantification of storm-induced bathymetric change in a back-barrier estuary. *Estuaries and Coasts* 40: 22–36.

Gannett, M.W., Manga, M., and Lite, K.E. (2003). Groundwater hydrology of the Upper Deschutes basin and its influence on streamflow. In: *A Peculiar River: Geology, Geomorphology, and Hydrology of the Deschutes River, Oregon* (eds. J.E. O'Connor and G.E. Grant), 31–49. Washington, DC: American Geophysical Union Press.

Gao, Y., Vogel, R.M., Kroll, C.N. et al. (2009). Development of representative indicators of hydrologic alteration. *Journal of Hydrology* 374: 136–147.

Garcia, A.F. (2006). Thresholds of strath genesis deduced from landscape response to stream piracy by Pancho Rico Creek in the Coast Ranges of central California. *American Journal of Science* 306: 655–681.

Garcia, C.C. and Lenzi, M.A. (eds.) (2010). *Check Dams, Morphological Adjustments and Erosion Control in Torrential Streams*. New York: Nova Science.

Garcia Parra, B., Pena Rojas, L.E., Barrios, M., and Munera Estrada, J.C. (2016). Uncertainty of discharge estimation in high-grade Andean streams. *Flow Measurement and Instrumentation* 48: 42–50.

Garcin, Y., Schildgen, T.F., Acosta, V.T. et al. (2017). Short-lived increase in erosion during the African Humid Period: evidence from the northern Kenya Rift. *Earth and Planetary Science Letters* 459: 58–69.

Gardner, T.W. (1983). Experimental study of knickpoint and longitudinal profile evolution in cohesive, homogeneous material. *Geological Society of America Bulletin* 94: 664–672.

Gasparini, N.M., Tucker, G.E., and Bras, R.L. (2004). Network-scale dynamics of grain-size sorting: implications for downstream fining, stream-profile concavity, and drainage basin morphology. *Earth Surface Processes and Landforms* 29: 401–421.

Gaston, K.J. and Spicer, J.I. (2004). *Biodiversity: An Introduction*, 2e. New York: Blackwell.

Gaudet, J.M. and Roy, A.G. (1995). Effects of bed morphology on flow mixing length at river confluences. *Nature* 373: 138–139.

Gaurav, K., Tandon, S.K., Devauchelle, O. et al. (2017). A single width-discharge regime relationship for individual threads of braided and meandering rivers from the Himalayan Foreland. *Geomorphology* 295: 126–133.

Geleynse, N., Storms, J.E.A., Walstra, D.-J.R. et al. (2011). Controls on river delta formation; insights from numerical modeling. *Earth and Planetary Science Letters* 302: 217–226.

Gellis, A.C. and Walling, D.E. (2011). Sediment source fingerprinting (tracing) and sediment budgets as tools in targeting river and watershed restoration programs. In: *Stream Restoration in Dynamic Fluvial Systems: Scientific Approaches, Analyses, and Tools* (eds. A. Simon, S.J. Bennett and J.M. Castro), 263–291. Washington, DC: American Geophysical Union Press.

Gellis, A.C., Pavich, M.J., Bierman, P.R. et al. (2004). Modern sediment yield compared to geologic rates of sediment production in a semi-arid basin, New Mexico: assessing the human impact. *Earth Surface Processes and Landforms* 29: 1359–1372.

Gellis, A.C., Webb, R.M.T., McIntyre, S.C., and Wolfe, W.J. (2006). Land-use effects on erosion, sediment yields, and reservoir sedimentation: a case study in the Lago Loíza basin, Puerto Rico. *Physical Geography* 27: 39–69.

Genereux, D.P. and Jordan, M. (2006). Interbasin groundwater flow and groundwater interaction with surface water in a lowland rainforest, Costa Rica: a review. *Journal of Hydrology* 320: 385–399.

Gerbersdorf, S.U. and Wieprecht, S. (2015). Biostabilization of cohesive sediments: revisiting the role of abiotic conditions, physiology and diversity of microbes, polymeric secretion, and biofilm architecture. *Geobiology* 13: 68–97.

Germann, P.F. (1990). Preferential flow and the generation of runoff 1. Boundary layer flow theory. *Water Resources Research* 26: 3055–3063.

Germanoski, G. and Schumm, S.A. (1993). Changes in braided river morphology resulting from aggradation and degradation. *Journal of Geology* 101: 451–466.

Gerrard, J. (1990). *Mountainous Environments: An Examination of the Physical Geography of Mountains*. Cambridge, MA: The MIT Press.

Geza, M., Poeter, E.P., and McCray, J.E. (2009). Quantifying predictive uncertainty for a mountain-watershed model. *Journal of Hydrology* 376: 170–181.

Ghasemizade, M. and Schirmer, M. (2013). Subsurface flow contribution in the hydrological cycle: lessons learned and challenges ahead: a review. *Environmental Earth Sciences* 69: 707–718.

Ghinassi, M., Moody, J., and Martin, D. (2018). Influence of extreme and annual floods on point-bar sedimentation: inferences from Powder River, Montana, USA. *Geological Society of America Bulletin* https://doi.org/10.1130/B31990.1.

Gilbert, G.K. (1877). *Report on the Geology of the Henry Mountains*. Washington, DC: US Geological Survey.

Gilbert, G.K. 1885. The topographic features of lake shores. US Geological Survey Annual Report 5.

Gilbert, G.K. 1914. Transportation of debris by running water. US Geological Survey Professional Paper 86.

Gilbert, G.K. 1917. Hydraulic-mining debris in the Sierra Nevada. US Geological Survey Professional Paper 105.

Gillan, B.J., Harper, J.T., and Moore, J.N. (2010). Timing of present and future snowmelt from high elevations in Northwest Montana. *Water Resources Research* 46, W01507.

Gillette, R. (1972). Stream channelization: conflict between ditchers, conservationists. *Science* 176: 890–894.

Gilman, K. and Newson, M.D. (1980). *Soils Pipes and Pipeflow – A Hydrological Study in Upland Wales*. Norwich: Geobooks.

Gimbert, F., Fuller, B.M., Lamb, M.P. et al. (2019). Particle transport mechanics and induced seismic noise in steep flume experiments with accelerometer-embedded tracers. *Earth Surface Processes and Landforms* https://doi.org/10.1002/esp.4495.

Giorgi, F., Shields Brodeur, C., and Bates, G.T. (1994). Regional climate change scenarios over the United States produced with a nested regional climate model. *Journal of Climate* 7: 375–399.

Giosan, L., Constantinescu, S., Filip, F., and Deng, B. (2013). Maintenance of large deltas through channelization: nature vs. humans in the Danube delta. *Anthropocene* 1: 35–45.

Gippel, C.J. (1995). Environmental hydraulics of large woody debris in streams and rivers. *Journal of Environmental Engineering* 121: 388–394.

Gippel, C.J., O'Neill, I.C., and Finlayson, B.L. (1992). *The Hydraulic Basis of Snag Management*. New South Wales: Land and Water Resources Development Corporation, Department of Water Resources.

Gippel, C., Zhang, Y., Qu, X.D. et al. (2017). Design of a national river health assessment program for China. In: *Decision Making in Water Resources Policy and Management: An Australian Perspective* (eds. B.T. Hart and J. Doolan), 321–339. Amsterdam: Elsevier.

Gleason, C.J. (2015). Hydraulic geometry of natural rivers: a review and future directions. *Progress in Physical Geography* 39: 337–360.

Gleason, C.J. and Smith, L.C. (2014). Toward global mapping of river discharge using satellite images and at-many-stations hydraulic geometry. *Proceedings of the National Academy of Sciences of the United States of America* 111: 4788–4791.

Gleason, C.J. and Wang, J. (2015). Theoretical basis for at-many-stations hydraulic geometry. *Geophysical Research Letters* 42: 7107–7114.

Glock, W.S. (1931). The development of drainage systems: a synoptic view. *Geographical Review* 21: 475–482.

Goldrick, G. and Bishop, P. (2007). Regional analysis of bedrock stream long profiles: evaluation of Hack's SL form, and formulation and assessment of an alternative (the DS form). *Earth Surface Processes and Landforms* 32: 649–671.

Goldsmith, E. and Hildyard, N. (1984). *The Social and Environmental Effects of Large Dams*. San Francisco, CA: Sierra Club.

Goldsmith, S.T., Carey, A.E., Lyons, W.B. et al. (2008). Extreme storm events, landscape denudation, and carbon sequestration: Typhoon Mindulle, Choshui River, Taiwan. *Geology* 36: 483–486.

Goldstein, J. (1999). Emergence as a construct: history and issues. *Emergence* 1: 49–72.

Gomez, B. and Troutman, B.M. (1997). Evaluation of process errors in bed load sampling using a dune model. *Water Resources Research* 33: 2387–2398.

Gomez, B., Naff, R.L., and Hubbell, D.W. (1989). Temporal variations in bedload transport rates associated with the migration of bedforms. *Earth Surface Processes and Landforms* 14: 135–156.

Gomez-Velez, J.D. and Harvey, J.W. (2014). A hydrogeomorphic river network model predicts where and why hyporheic exchange is important in large basins. *Geophysical Research Letters* 41: 6403–6412.

Gomez-Velez, J.D., Harvey, J.W., Cardenas, M.B., and Kie, B. (2015). Denitrification in the Mississippi River network controlled by flow through river bedforms. *Nature Geoscience* 8: 941–945.

Gonzalez, R.L. and Pasternack, G.B. (2015). Reenvisioning cross-sectional at-a-station hydraulic geometry as spatially explicit hydraulic topography. *Geomorphology* 246: 394–406.

González, E., Cabezas, A., Corenblit, D., and Steiger, J. (2014). Autochthonous versus allochthonous organic matter in recent soil C accumulation along a floodplain biogeomorphic gradient: an exploratory study. *Journal of Environmental Geography* 7: 29–38.

González, E., Sher, A.A., Tabacchi, E. et al. (2015). Restoration of riparian vegetation: a global review of implementation and evaluation approaches in the international, peer-reviewed literature. *Journal of Environmental Management* 158: 85–94.

Gonzalez-Pinzon, R., Ward, A.S., Hatch, C.E. et al. (2015). A field comparison of multiple techniques to quantify groundwater-surface-water interactions. *Freshwater Science* 34: 139–160.

Goode, J.R. and Wohl, E. (2010a). Coarse sediment transport in a bedrock channel with complex bed topography. *Water Resources Research* 46, W11524.

Goode, J.R. and Wohl, E. (2010b). Substrate controls on the longitudinal profile of bedrock channels: implications for reach-scale roughness. *Journal of Geophysical Research: Earth Surfaces* 115, F03018.

Goode, J.R., Luce, C.H., and Buffington, J.M. (2012). Enhanced sediment delivery in a changing climate in semi-arid mountain basins: implications for water resource management and aquatic habitat in the northern Rocky Mountains. *Geomorphology* 139–140: 1–15.

Goodrich, D.C., Kepner, W.G., Levick, L.R., and Wigington, P.J. (2018). Southwestern intermittent and ephemeral stream connectivity. *Journal of the American Water Resources Association* 54: 400–422.

Goolsby, D.A., W.A. Battaglia, G.B. Lawrence, R.S. Artz, B.T. Aulenbach, R.P. Hooper, et al. 1999. Flux and sources of nutrients in the Mississippi-Atchafalaya river basin. Topic 3 report for the integrated assessment on hypoxia in the Gulf of Mexico. NOAA Coastal Ocean Program Decision Analysis Series No. 17. NOAA Coastal Ocean Program, Silver Spring, MD.

Gooseff, M.N. (2010). Defining hyporheic zones – advancing our conceptual and operational definitions of where stream water and groundwater meet. *Geographical Compass* 4: 945–955.

Gooseff, M.N., Wondzell, S.M., Haggerty, R., and Anderson, J. (2003). Comparing transient storage modeling and residence time distribution (RTD) analysis in geomorphically varied reaches in the Lookout Creek basin, Oregon, USA. *Advances in Water Resources* 26: 925–937.

Gooseff, M.N., LaNier, J., Haggerty, R., and Kokkeler, K. (2005). Determing in-channel (dead zone) transient storage by comparing solute transport in a bedrock channel-alluvial channel sequence, Oregon. *Water Resources Research* 41, W06014.

Gooseff, M.N., Anderson, J.K., Wondzell, S.M. et al. (2006). A modeling study of hyporheic exchange pattern and sequence, size, and spacing of stream bedforms in mountain stream networks, Oregon, USA. *Hydrological Processes* 20: 2443–2457.

Gooseff, M.N., Hall, R.O., and Tank, J.L. (2007). Relating transient storage to channel complexity in streams of varying land use in Jackson Hole, Wyoming. *Water Resources Research* 43 W01417.

Gooseff, M.N., Balser, A., Bowden, W.B., and Jones, J.B. (2009). Effects of hillslope thermokarst in northern Alaska. *Eos, Transactions of the American Geophysical Union* 90: 29–30.

Gooseff, M.N., Wlostowski, A., McKnight, D.M., and Jaros, C. (2017). Hydrologic connectivity and implications for ecosystem processes – lessons from naked watersheds. *Geomorphology* 277: 63–71.

Goren, L., Fox, M., and Willett, S.D. (2014). Tectonics from fluvial topography using formal linear inversion: theory and applications to the Inyo Mountains, California. *Journal of Geophysical Research: Earth Surface* 119: 1651–1681.

Gorrick, S. and Rodríguez, J.F. (2012). Sediment dynamics in a sand bed stream with riparian vegetation. *Water Resources Research* 48, W02505.

Götzinger, J.R., Barthel, J., and Jagelke, A.B. (2008). The role of groundwater recharge and baseflow in integrated models. In: *Groundwater–Surface Water Interaction: Process Understanding, Conceptualization and Modeling* (eds. C. Abesser, T. Wagener and G. Nuetzmann), 103–109. Wallingford: IAHS Publications no. 321.

Gough, L.P. 1993. Understanding our fragile environment: lessons from geochemical studies. US Geological Survey Circular 1005.

Goulding, H.L., Prowse, T.D., and Bonsal, B. (2009a). Hydroclimatic controls on the occurrence of break-up and ice-jam flooding in the Mackenzie Delta, NWT, Canada. *Journal of Hydrology* 379: 251–267.

Goulding, H.L., Prowse, T.D., and Beltaos, S. (2009b). Spatial and temporal patterns of break-up and ice-jam flooding in the Mackenzie Delta, NWT. *Hydrological Processes* 23: 2654–2670.

Graf, W.L. (1978). Fluvial adjustments to the spread of tamarisk in the Colorado Plateau region. *Geological Society of America Bulletin* 89: 1491–1501.

Graf, W.L. (1983). Flood-related channel change in an arid region river. *Earth Surface Processes and Landforms* 8: 125–139.

Graf, W.L. (1988). *Fluvial Processes in Dryland Rivers.* Berlin: Springer-Verlag.

Graf, W.L. (1996). Geomorphology and policy for restoration of impounded American rivers: what is "natural?". In: *The Scientific Nature of Geomorphology* (eds. B.L. Rhoads and C.E. Thorn), 443–473. New York: Wiley.

Graf, W.L. (2001). Damage control: restoring the physical integrity of America's rivers. *Annals of the Association of American Geographers* 91: 1–27.

Graf, W.L. (2006). Downstream hydrologic and geomorphic effects of large dams on American rivers. *Geomorphology* 79: 336–360.

Graf, W.L., Stromberg, J., and Valentine, B. (2002). Rivers, dams, and willow flycatchers: a summary of their science and policy connections. *Geomorphology* 47: 169–188.

Grahame, T.J. and Schlesinger, R.B. (2007). Health effects of airborne particulate matter: do we know enough to consider regulating specific particle types or sources? *Inhalation Toxicology* 19: 457–481.

Gramlich, A., Stoll, S., Stamm, C. et al. (2018). Effects of artificial land drainage on hydrology, nutrient and pesticide fluxes from agricultural fields – a review. *Agriculture, Ecosystems and Environment* 266: 84–99.

Gran, K.B. (2012). Strong seasonality in sand loading and resulting feedbacks on sediment transport, bed texture, and channel planform at Mount Pinatubo, Philippines. *Earth Surface Processes and Landforms* 37: 1012–1022.
Gran, K.B. and Czuba, J.A. (2017). Sediment pulse evolution and the role of network structure. *Geomorphology* 277: 17–30.
Gran, K.B. and Montgomery, D.R. (2005). Spatial and temporal patterns in fluvial recovery following volcanic eruptions: channel response to basin-wide sediment loading at Mount Pinatubo, Philippines. *Geological Society of America Bulletin* 117: 195–211.
Gran, K.B., Montgomery, D.R., and Sutherland, D.G. (2006). Channel bed evolution and sediment transport under declining sand inputs. *Water Resources Research* 42 W10407.
Gran, K.B., Montgomery, D.R., and Halbur, J.C. (2011). Long-term elevated post-eruption sedimentation at Mount Pinatubo, Philippines. *Geology* 39: 367–370.
Gran, K.B., Finnegan, N., Johnson, A.L. et al. (2013). Landscape evolution, valley excavation, and terrace development following abrupt postglacial base-level fall. *Geological Society of America Bulletin* 125: 1851–1864.
Gran, K.B., Tal, M., and Wartman, E.D. (2015). Co-evolution of riparian vegetation and channel dynamics in an aggrading braided river system, Mount Pinatubo, Philippines. *Earth Surface Processes and Landforms* 40: 1101–1115.
Grant, G.E. (1997). Critical flow constrains flow hydraulics in mobile-bed streams: a new hypothesis. *Water Resources Research* 33: 349–358.
Grant, G.E., Schmidt, J.C., and Lewis, S.L. (2003). A geological framework for interpreting downstream effects of dams on rivers. In: *A Peculiar River: Geology, Geomorphology, and Hydrology of the Deschutes River, Oregon* (eds. J.E. O'Connor and G.E. Grant), 203–219. Washington, DC: American Geophysical Union Press.
Grant, G.E., O'Connor, J.E., and Wolman, M.G. (2013). A river runs through it: conceptual models in fluvial geomorphology. In: *Treatise on Fluvial Geomorphology* (ed. E. Wohl), 6–20. Amsterdam: Elsevier.
Green, C. and Fernández-Bilbao, A. (2006). Implementing the water framework directive: how to define a "competent authority." *Journal of Contemporary Water Research and Education* 135: 65–73.
Green, K.C. and Westbrook, C.J. (2009). Changes in riparian area structure, channel hydraulics, and sediment yield following loss of beaver dams. *BC Journal of Ecosystems and Management* 10: 68–79.
Green, P.A., Vörösmarty, C.J., Meybeck, M. et al. (2004). Pre-industrial and contemporary fluxes of nitrogen through rivers: a global assessment based on typology. *Biogeochemistry* 68: 71–105.
Gregory, K.J. (1976). Lichens and the determination of river channel capacity. *Earth Surface Processes* 1: 273–285.
Gregory, K.J. (2006). The human role in changing river channels. *Geomorphology* 79: 172–191.
Gregory, K.J. and Park, C.C. (1974). Adjustment of river channel capacity downstream from a reservoir. *Water Resources Research* 10: 870–873.
Gregory, S.V., Swanson, F.J., McKee, W.A., and Cummins, K.W. (1991). An ecosystem perspective of riparian zones. *BioScience* 41: 540–551.
Gregory, S.V., Meleason, M.A., and Sobota, D.J. (2003). Modeling the dynamics of wood in streams and rivers. In: *The Ecology and Management of Wood in World Rivers* (eds. S.V. Gregory, K.L. Boyer and A.M. Gurnell) American Fisheries Society Symposium 37, 315–335. Bethesda, MD.
Grenfell, M., Aalto, R., and Nicholas, A. (2012). Chute channel dynamics in large, sand-bed meandering rivers. *Earth Surface Processes and Landforms* 37: 315–331.

Griffin, E.R., Kean, J.W., Vincent, K.R. et al. (2005). Modeling effects of bank friction and woody bank vegetation on channel flow and boundary shear stress in the Rio Puerco, New Mexico. *Journal of Geophysical Research* 110 F04023.

Griffiths, G.A. (1979). Recent sedimentation history of the Waimakariri River, New Zealand. *Journal of Hydrology New Zealand* 18: 6–28.

Griffiths, G.A. (1981). Flow resistance in coarse gravel bed rivers. *Journal of Hydraulics Division, ASCE* 107: 899–918.

Griffiths, G.A. (1987). Form resistance in gravel channels with mobile beds. *Journal of Hydraulic Engineering* 115: 340–355.

Griffiths, G.A. and Carson, M.A. (2000). Channel width for maximum bedload transport capacity in gravel-bed rivers, South Island, New Zealand. *Journal of Hydrology (New Zealand)* 39: 107–126.

Griffiths, P.G., Hereford, R., and Webb, R.H. (2006). Sediment yield and runoff frequency of small drainage basins in the Mojave Desert, USA. *Geomorphology* 74: 232–244.

Grill, G., Dallaire, C.O., Chouinard, E.F. et al. (2014). Development of new indicators to evaluate river fragmentation and flow regulation at large scales: a case study for the Mekong River basin. *Ecological Indicators* 45: 148–159.

Grill, G., Lehner, B., Lumsden, A.E. et al. (2015). An index-based framework for assessing patterns and trends in river fragmentation and flow regulation by global dams at multiple scales. *Environmental Research Letters* 10 https://doi.org/10.1088/1748-9326/10/1/015001.

Grimaud, J.L., Paola, C., and Voller, V. (2016). Experimental migration of knickpoints: influence of style of base-level fall and bed lithology. *Earth Surface Dynamics* 4: 11–123.

Grimm, N.B. and Fisher, S.G. (1992). Responses of arid-land streams to changing climate. In: *Global Climate Change and Freshwater Ecosystems* (eds. P. Firth and S.G. Fisher), 211–233. New York: Springer.

Grimsley, K.J., Rathburn, S.L., Friedman, J.M., and Mangano, J.F. (2016). Debris flow occurrence and sediment persistence, Upper Colorado River Valley, Colorado. *Environmental Management* 58: 76–92.

Gu, Z., Shi, C., Yang, H., and Yao, H. (2018). Analysis of dynamic sedimentary environments in alluvial fans of some tributaries of the upper Yellow River of China based on ground penetrating radar (GPR) and sediment cores. *Quaternary International*.

Guin, A., Ramanathan, R., Ritzi, R.W. et al. (2010). Simulating the heterogeneity in braided channel belt deposits: 2. Examples of results and comparison to natural deposits. *Water Resources Research* 46 W04516.

Güneralp, İ. and Marston, R. (2012). Process-form linkages in meander morphodynamics: bridging theoretical modeling and real world complexity. *Progress in Physical Geography* 35: 718–746.

Güneralp, İ. and Rhoads, B.L. (2009). Empirical analysis of the planform curvature-migration relation of meandering rivers. *Water Resources Research* 45 W09424.

Gupta, A. (1995). Magnitude, frequency, and special factors affecting channel form and processes in the seasonal tropics. In: *Natural and Anthropogenic Influences in Fluvial Geomorphology* (eds. J.E. Costa, A.J. Miller, K.W. Potter and P.R. Wilcock), 125–136. Washington, DC: American Geophysical Union Press.

Gupta, A. and Dutt, A. (1989). The Auranga: description of a tropical river. *Zeitschrift für Geomorphologie* 33: 73–92.

Guralnik, B., Matmon, A., Avni, Y. et al. (2010). Constraining the evolution of river terraces with integrated OSL and cosmogenic nuclide data. *Quaternary Geochronology* https://doi.org/10.1016/j.quageo.2010.06.002.

Gurnell, A.M. (1995). Sediment yield from alpine glacier basins. In: *Sediment and Water Quality in River Catchments* (eds. I. Foster, A. Gurnell and B. Webb), 407–435. Chichester: Wiley.

Gurnell, A.M. (1997). Channel change on the River Dee meanders, 1946–1992, from the analysis of air photographs. *Regulated Rivers: Research & Management* 13: 13–26.

Gurnell, A.M. (2003). Wood storage and mobility. In: *The Ecology and Management of Wood in World Rivers* (eds. S.V. Gregory, K.L. Boyer and A.M. Gurnell) American Fisheries Society Symposium 37, 75–91. Bethesda, MD.

Gurnell, A.M. (2007). Analogies between mineral sediment and vegetative particle dynamics in fluvial systems. *Geomorphology* 89: 9–22.

Gurnell, A.M. (2013). Wood in fluvial systems. In: *Treatise on Fluvial Geomorphology* (ed. E. Wohl), 163–188. Amsterdam: Elsevier.

Gurnell, A.M. (2014). Plants as river system engineers. *Earth Surface Processes and Landforms* 39: 4–25.

Gurnell, A.M. and Grabowski, R.C. (2016). Vegetation-hydrogeomorphology interactions in a low-energy, human-impacted river. *River Research and Applications* 32: 202–215.

Gurnell, A.M. and Petts, G. (2006). Trees as riparian engineers: the Tagliamento River, Italy. *Earth Surface Processes and Landforms* 31: 1558–1574.

Gurnell, A.M., Petts, G.E., Hannah, D.M. et al. (2001). Riparian vegetation and island formation along the gravel-bed Fiume Tagliamento, Italy. *Earth Surface Processes and Landforms* 26: 31–62.

Gurnell, A.M., Piegay, H., Swanson, F.J., and Gregory, S.V. (2002). Large wood and fluvial processes. *Freshwater Biology* 47: 601–619.

Gurnell, A.M., Peiry, J.-L., and Petts, G.E. (2003). Using historical data in fluvial geomorphology. In: *Tools in Geomorphology* (eds. G.M. Kondolf and H. Piégay), 77–101. Chichester: Wiley.

Gurnell, A., Tockner, K., Edwards, P., and Petts, G. (2005). Effects of deposited wood on biocomplexity of river corridors. *Frontiers in Ecology and the Environment* 3: 377–382.

Gurnell, A., Lee, M., and Souch, C. (2007). Urban rivers: hydrology, geomorphology, ecology and opportunities for change. *Geography Compass* 1: 1118–1137.

Gurnell, A.M., Bertoldi, W., and Corenblit, D. (2012). Changing river channels: the roles of hydrological processes, plants and pioneer fluvial landforms in humid temperate, mixed load, gravel bed rivers. *Earth-Science Reviews* 111: 129–141.

Gurnell, A.M., Corenblit, D., Gracia de Jalon, D. et al. (2016a). A conceptual model of vegetation-hydrogeomorphology interactions within river corridors. *River Research and Applications* 32: 142–163.

Gurnell, A.M., Bertoldi, W., Tockner, K. et al. (2016b). How large is a river? Conceptualizing river landscape signatures and envelopes in four dimensions. *WIREs Water* 3: 313–325.

Gurnell, A.M., Bertoldi, W., Francis, R.A. et al. (2019). Understanding processes of island development on an island braided river over timescales from days to decades. *Earth Surface Processes and Landforms* 44: 624–640.

Guthrie, R.H. and Evans, S.G. (2007). Work, persistence, and formative events; the geomorphic impact of landslides. *Geomorphology* 88: 266–275.

Guyette, R.P., Dey, D.C., and Stambaugh, M.C. (2008). The temporal distribution and carbon storage of large oak wood in streams and floodplain deposits. *Ecosystems* 11: 643–653.

Gyalistras, D., Schar, C., Davies, H.C., and Wanner, H. (1998). Future Alpine climate. In: *Views from the Alps: Regional Perspectives on Climate Change* (eds. P. Cebon, U. Dahinden, H.C. Davies, et al.), 171–224. Cambridge, MA: The MIT Press.

Habersack, H.M. and Piégay, H. (2008). River restoration in the Alps and their surroundings: past experience and future challenges. In: *Gravel-Bed Rivers VI: From Process Understanding to River Restoration* (eds. H. Habersack, H. Piégay and M. Rinaldi), 703–737. Amsterdam: Elsevier.

Habersack, H., Jäger, E., and Hauer, C. (2013). The status of the Danube River sediment regime and morphology as a basis for future basin management. *International Journal of River Basin Management* 11: 153–166.

Hack, J.T. 1957. Studies of longitudinal profiles in Virginia and Maryland. US Geological Survey Professional Paper 294-B.

Hack, J.T. (1960). Interpretation of erosional topography in humid temperate regions. *American Journal of Science* 258-A: 80–97.

Hack, J.T. (1973). Stream-profile analysis and stream-gradient index. *Journal of Research of the US Geological Survey* 1: 421–429.

Hagg, W. and Braun, L. (2005). The influence of glacier retreat on water yield from high mountain areas: comparison of Alps and Central Asia. In: *Climate and Hydrology in Mountain Areas* (eds. C. de Jong, D. Collins and R. Ranzi), 263–275. Chichester: Wiley.

Haggerty, R. and Reeves, R. (2002). *STAMMT-L Version 1.0 User's Mannual*. Albuquerque, NM: Sandia National Laboratory.

Hagstrom, C.A., Leckie, D.A., and Smith, M.G. (2018). Point bar sedimentation and erosion produced by an extreme flood in a sand and gravel-bed meandering river. *Sedimentary Geology* 377: 1–16.

Hahm, W.J., Riebe, C.S., Lukens, C.E., and Araki, S. (2014). Bedrock composition regulates mountain ecosystems and landscape evolution. *Proceedings of the National Academy of Science* 111: 3338–3343.

Hales, R.C. and Roering, J.J. (2007). Climatic controls on frost cracking and implications for the evolution of bedrock landscapes. *Journal of Geophysical Research* 112 F02033.

Hales, T.C. and Roering, J.J. (2009). A frost "buzzsaw" mechanism for erosion of the eastern Southern Alps, New Zealand. *Geomorphology* 107: 241–253.

Hall, R.O. and Tank, J.L. (2003). Ecosystem metabolism controls nitrogen uptake in streams in Grand Teton National Park, Wyoming. *Limnology and Oceanography* 48: 1120–1128.

Ham, D.G. 2005. Morphodynamics and sediment transport in a wandering gravel-bed channel: Fraser River, British Columbia. PhD dissertation, University of British Columbia, Vancouver.

Hamilton, L.S. and Bruijnzeel, L.A. (1997). Mountain watersheds – integrating water, soils, gravity, vegetation, and people. In: *Mountains of the World: A Global Priority* (eds. B. Messerli and J.D. Ives), 337–370. London: The Parthenon.

Hanberry, B.B., Kabrick, J.M., and He, H.S. (2015). Potential tree and soil carbon storage in a major historical floodplain forest with disrupted ecological function. *Perspectives in Plant Ecology, Evolution and Systematics* 17: 17–23.

Hancock, G.S. and Anderson, R.S. (2002). Numerical modeling of fluvial strath-terrace formation in response to oscillating climate. *Geological Society of America Bulletin* 114: 1131–1142.

Hancock, G.S., Anderson, R.S., and Whipple, K.X. (1998). Beyond power: bedrock river incision process and form. In: *Rivers Over Rock: Fluvial Processes in Bedrock Channels* (eds. K.J. Tinkler and E.E. Wohl), 35–60. Washington, DC: American Geophysical Union Press.

Hancock, P.J., Boulton, A.J., and Humphreys, W.F. (2005). Aquifers and hyporheic zones: towards an ecological understanding of groundwater. *Hydrogeology Journal* 13: 98–111.

Hancock, G.R., Lowry, J.B.C., Coulthard, T.J. et al. (2010). A catchment scale evaluation of the SIBERIA and CAESAR landscape evolution models. *Earth Surface Processes and Landforms* 35: 863–875.

Hancock, G.S., Small, E.E., and Wobus, C. (2011). Modelling the effects of weathering on bedrock-floored channel geometry. *Journal of Geophysical Research* 116 F03018.
Hancock, G.R., Lowry, J.B.C., and Coulthard, T.J. (2015). Catchment reconstruction – erosional stability at millennial time scales using landscape evolution models. *Geomorphology* 231: 15–27.
Harbor, D.J., Bacastow, A., Heath, A., and Rogers, J. (2005). Capturing variable knickpoint retreat in the central Appalachians, USA. *Geografia Fisica e Dinamica Quaternaria* 28: 23–36.
Harden, D.R. (1990). Controlling factors in the distribution and development of incised meanders in the central Colorado Plateau. *Geological Society of America Bulletin* 102: 233–242.
Harden, C.P. (2006). Human impacts on headwater fluvial systems in the northern and central Andes. *Geomorphology* 79: 249–263.
Harden, C.P. and Scruggs, P.D. (2003). Infiltration on mountain slopes: a comparison of three environments. *Geomorphology* 55: 5–24.
Harding, J.S., Benfield, E.F., Bolstad, P.V. et al. (1998). Stream biodiversity: the ghost of land use past. *Proceedings of the National Academy of Sciences of the United States of America* 95: 14843–14847.
Harel, M.A., Mudd, S.M., and Attal, M. (2016). Global analysis of the stream power law parameters based on worldwide ^{10}Be denudation rates. *Geomorphology* 268: 184–196.
Harmon, M.E. (1982). Decomposition of standing dead trees in the Southern Appalachian Mountains. *Oecologia* 52: 214–215.
Harmon, M.E., Franklin, J.F., Swanson, F.J. et al. (1986). Ecology of coarse woody debris in temperate ecosystems. *Advances in Ecological Resources* 15: 133–302.
Harris, J.H. and Silveira, R. (1999). Large-scale assessments of river health using an Index of Biotic Integrity with low-diversity fish communities. *Freshwater Biology* 41: 235–252.
Hart, D.R. (1995). Parameter estimation and stochastic interpretation of the transient storage model for solute transport in streams. *Water Resources Research* 31: 323–328.
Hart, D.D., Johnson, T.E., Bushaw-Newton, K.L. et al. (2002). Dam removal: challenges and opportunities for ecological research and river restoration. *BioScience* 52: 669–681.
Hartman, G. (1996). Habitat selection by European beaver (*Castor fiber*) colonizing a boreal landscape. *Journal of Zoology (London)* 240: 317–325.
Hartshorn, K., Hovius, N., Dade, W.B., and Slingerland, R.L. (2002). Climate-driven bedrock incision in an active mountain belt. *Science* 297: 2036–2038.
Harvey, A.M. (1997). Coupling between hillslope gully systems and stream channels in the Howgill Fells, northwest England: temporal implications. *Geomorphologie: Relief, Processus, Environnement* 3: 3–19.
Harvey, A.M. (2002). Effective timescales of coupling within fluvial systems. *Geomorphology* 44: 175–201.
Harvey, A.M. (2007). Differential recovery from the effects of a 100 year storm: significance of long term hillslope channel coupling: Howgill Fells, northwest England. *Geomorphology* 84: 192–208.
Harvey, A.M. (2008). Developments in dryland geomorphology. In: *The History of the Study of Landforms or the Development of Geomorphology*, Quaternary and Recent Processes and Forms (1890–1965) and The Mid-Century Revolutions, vol. 4 (eds. T.P. Burt, R.J. Chorley, D. Brunsden, et al.), 729–765. London: The Geological Society.
Harvey, J.W. and Bencala, K.E. (1993). The effect of streambed topography on surface-subsurface water exchange in mountain catchments. *Water Resources Research* 29: 89–98.
Harvey, J.W. and Fuller, C.C. (1998). Effect of enhanced manganese oxidation in the hyporheic zone on basin-scale geochemical mass balance. *Water Resources Research* 34: 623–636.
Harvey, J. and Gooseff, M. (2015). River corridor science: hydrologic exchange and ecological consequences from bedforms to basins. *Water Resources Research* 51: 6893–6922.

Harvey, M.D., Mussetter, R.A., and Wick, E.J. (1993). A physical process-biological response model for spawning habitat formation for the endangered Colorado squawfish. *Rivers* 4: 114–131.

Hasbargen, L.E. and Paola, C. (2000). Landscape instability in an experimental drainage basin. *Geology* 28: 1067–1070.

Haschenburger, J.K. (2006). Observations of event-based streambed deformation in a gravel bed channel. *Water Resources Research* 42 W11412.

Haschenburger, J.K. (2013). Bedload kinematics and fluxes. In: *Treatise on Fluvial Geomorphology* (ed. E. Wohl), 104–123. Amsterdam: Elsevier.

Hasnain, S.I. (2002). Himalayan glaciers meltdown: impact on South Asian rivers. In: *FRIEND 2002: Regional Hydrology; Bridging the Gap Between Research and Practice* (eds. H.A.J. van Lanen and S. Demuth), 417–423. Wallingford: IAHS Publications no. 274.

Hassan, M.A. and Bradley, D.N. (2017). Geomorphic controls on tracer particle dispersion in gravel-bed rivers. In: *Gravel-Bed Rivers: Processes and Disasters* (eds. D. Tsutsumi and J.B. Laronne), 159–184. Chichester: Wiley.

Hassan, M.A. and Ergenzinger, P. (2003). Use of tracers in fluvial geomorphology. In: *Tools in Fluvial Geomorphology* (eds. G.M. Kondolf and H. Piégay), 397–423. Chichester: Wiley.

Hassan, M.A. and Reid, I. (1990). The influence of microform bed roughness elements on flow and sediment transport in gravel bed rivers. *Earth Surface Processes and Landforms* 15: 739–750.

Hassan, M.A. and Roy, A.G. (2016). Coarse particle tracing in fluvial geomorphology. In: *Tools in Fluvial Geomorphology* (eds. G.M. Kondolf and H. Piegay), 306–323. Chichester: Wiley-Blackwell.

Hassan, M.A. and Woodsmith, R.D. (2004). Bed load transport in an obstruction-formed pool in a forest, gravelbed stream. *Geomorphology* 58: 203–221.

Hassan, M.A., Church, M., and Ashworth, P.J. (1992). Virtual rate and mean distance of travel of individual clasts in gravel-bed channels. *Earth Surface Processes and Landforms* 17: 617–627.

Hassan, M.A., Hogan, D.L., Bird, S.A. et al. (2005). Spatial and temporal dynamics of wood in headwater streams of the Pacific Northwest. *Journal of the American Water Resources Association* 41: 899–919.

Hassan, M.A., Egozi, R., and Parker, G. (2006). Experiments on the effect of hydrograph characteristics on vertical grain sorting in gravel bed rivers. *Water Resources Research* 42 W09408.

Hassan, M.A., Gottesfeld, A.S., Montgomery, D.R. et al. (2008). Salmon-driven bed load transport and bed morphology in mountain streams. *Geophysical Research Letters* 35 L04405.

Hassan, M.A., Church, M., Rempel, J., and Enkin, R.J. (2009). Promise, performance and current limitations of a magnetic Bedload Movement Detector. *Earth Surface Processes and Landforms* 34: 1022–1032.

Hassan, M.A., Petticrew, E.L., Montgomery, D.R. et al. (2011). Salmon as biogeomorphic agents in gravel bed rivers: the effect of fish on sediment mobility and spawning habitat. In: *Stream Restoration in Dynamic Fluvial Systems: Scientific Approaches, Analyses, and Tools* (eds. A. Simon, S.J. Bennett and J.M. Castro), 337–352. Washington, DC: American Geophysical Union Press.

Hassan, M.A., Tonina, D., Beckie, R.D., and Kinnear, M. (2015). The effects of discharge and slope on hyporheic flow in step-pool morphologies. *Hydrological Processes* 29: 419–433.

Hatfield, R.G. and Maher, B.A. (2009). Fingerprinting upland sediment sources: particle size-specific magnetic linkages between soils, lake sediments and suspended sediments. *Earth Surface Processes and Landforms* 34: 1359–1373.

Hauer, F.R., Baron, J.S., Campbell, D.H. et al. (1997). Assessment of climate change and freshwater ecosystems of the Rocky Mountains, USA and Canada. *Hydrological Processes* 11: 903–924.

He, S., Hipel, K.W., and Kilgour, D.M. (2014). Water diversion conflicts in China: a hierarchical perspective. *Water Resources Management* 28: 1823–1837.

Heckmann, T. and Vericat, D. (2018). Computing spatially distributed sediment delivery ratios: inferring functional sediment connectivity from repeat high-resolution digital elevation models. *Earth Surface Processes and Landforms* 43: 1547–1554.

Heckmann, T., Schwanghart, W., and Phillips, J.D. (2015). Graph theory – recent developments of its application in geomorphology. *Geomorphology* 243: 130–146.

Hedman, C.W., Van Lear, D.H., and Swank, W.T. (1996). In-stream large woody debris loading and riparian forest seral stage associations in the southern Appalachian Mountains. *Canadian Journal of Forest Research* 26: 1218–1227.

Heezen, B.C., Ericson, D.B., and Ewing, M. (1954). Further evidence for a turbidity current following the 1929 Grand Banks earthquake. *Deep Sea Research* 1: 193–202.

Heffernan, J.B. (2008). Wetlands as an alternative stable state in desert streams. *Ecology* 89: 1261–1271.

Heimsath, A.M., Dietrich, W.E., Nishiizumi, K., and Finkel, R.C. (1997). The soil production function and landscape equilibrium. *Nature* 388: 358–361.

Heimsath, A.M., Dietrich, W.E., Nishiizumi, K., and Finkel, R.C. (1999). Cosmogenic nuclides, topography, and the spatial variation of soil depth. *Geomorphology* 27: 151–172.

Hein, T., Schwarz, U., Habersack, H. et al. (2016). Current status and restoration options for floodplains along the Danube River. *Science of the Total Environment* 543: 778–790.

Heine, R.A. and Pinter, N. (2012). Levee effects upon flood levels: an empirical assessment. *Hydrological Processes* 26: 3225–3240.

Helton, A.M., Poole, G.C., Payn, R.A. et al. (2014). Relative influences of the river channel, floodplain surface, and alluvial aquifer on simulated hydrologic residence time in a montane river floodplain. *Geomorphology* 205: 17–26.

Helton, A.M., Hall, R.O., and Bertuzzo, E. (2018). How network structure can affect nitrogen removal by streams. *Freshwater Biology* 63: 128–140.

Hendrick, R.R., Ely, L.L., and Papanicolaou, A.N. (2010). The role of hydrologic processes and geomorphology on the morphology and evolution of sediment clusters in gravel-bed rivers. *Geomorphology* 114: 483–496.

Henkle, J.E., Wohl, E., and Beckman, N. (2011). Locations of channel heads in the semiarid Colorado Front Range, USA. *Geomorphology* 129: 309–319.

Henry, K.M. and Twilley, R.R. (2014). Nutrient biogeochemistry during the early stages of delta development in the Mississippi River deltaic plain. *Ecosystems* 17: 327–343.

Herdrich, A.T., Winkelman, D.L., Venarsky, M.P. et al. (2018). The loss of large wood affects Rocky Mountain trout populations. *Ecology of Freshwater Fish* 27: 1023–1036.

Hereford, R. (2002). Valley-fill alluviation during the Little Ice Age (ca. AD 1400–1880), Paria River Basin and southern Colorado Plateau, United States. *Geological Society of America Bulletin* 114: 1550–1563.

Heritage, G.L. and Hetherington, D. (2005). The use of high-resolution field laser scanning for mapping surface topography in fluvial systems. In: *Sediment Budgets I* (eds. D.E. Walling and A.J. Horowitz), 269–277. Wallingford: IAHS Publications no. 291.

Heritage, G. and Hetherington, D. (2007). Towards a protocol for laser scanning in fluvial geomorphology. *Earth Surface Processes and Landforms* 32: 66–74.

Heritage, G.L., Charlton, M.E., and O'Regan, S. (2001). Morphological classification of fluvial environments: an investigation of the continuum of channel types. *Journal of Geology* 109: 21–33.

Herron, N. and Wilson, C. (2001). A water balance approach to assessing the hydrologic buffering potential of an alluvial fan. *Water Resources Research* 37: 341–351.

Hester, E.T. and Doyle, M.W. (2008). In-stream geomorphic structures as drivers of hyporheic exchange. *Water Resources Research* 44 W03417.

Hester, E.T. and Gooseff, M.N. (2010). Moving beyond the banks: hyporheic restoration is fundamental to restoring ecological services and functions of streams. *Environmental Science and Technology* 44: 1521–1525.

Heuer, K., Tonnessen, K.A., and Ingersoll, G.P. (2000). Comparison of precipitation chemistry in the central Rocky Mountains, Colorado, USA. *Atmospheric Environment* 34: 1713–1722.

Hewitt, K. (1998). Catastrophic landslides and their effects on the Upper Indus streams, Karakoram Himalaya, northern Pakistan. *Geomorphology* 26: 47–80.

Hewlett, J.D. and Hibbert, A.R. (1967). Factors affecting the response of small watersheds to precipitation in humid areas. In: *Forest Hydrology* (eds. W.E. Sopper and H.W. Lull), 275–290. New York: Pergamon Press.

Hey, R.D. (1979). Flow resistance in gravel-bed rivers. *Journal of Hydraulics Division, ASCE* 105: 365–379.

Hey, R.D. (1988). Bar form resistance in gravel-bed rivers. *Journal of Hydraulic Engineering* 114: 1498–1508.

Hey, R.D. and Thorne, C.R. (1986). Stable channels with mobile gravel beds. *Journal of Hydraulic Engineering* 112: 671–689.

Hickin, E.J. (1984). Vegetation and river channel dynamics. *Canadian Geographer* 28: 111–126.

Hickin, E.J. and Nanson, G.C. (1975). The character of channel migration on the Beaton River, northeast British Columbia, Canada. *Geological Society of America Bulletin* 86: 487–494.

Hickin, E.J. and Nanson, G.C. (1984). Lateral migration rates of river bends. *Journal of Hydraulic Engineering* 110: 1557–1567.

Hicks, F. (2009). An overview of river ice problems: CRIPE07 guest editorial. *Cold Regions Science and Technology* 55: 175–185.

Hicks, D.M. and Gomez, B. (2016). Sediment transport. In: *Tools in Fluvial Geomorphology* (eds. G.M. Kondolf and H. Piegay), 324–356. Chichester: Wiley-Blackwell.

Hicks, D.M., Duncan, M.J., Lane, S.N. et al. (2008). Contemporary morphological change in braided gravel-bed rivers: new developments from field and laboratory studies, with particular reference to the influence of riparian vegetation. In: *Gravel-bed Rivers VI: From Process Understanding to River Restoration* (eds. H. Habersack, H. Piégay and M. Rinaldi), 557–586. Amsterdam: Elsevier.

Hildebrandt, A., Al Aufi, M.A., Amerjeed, M. et al. (2007). Ecohydrology of a seasonal cloud forest in Dhofar: 1. Field experiment. *Water Resources Research* 43 W10411.

Hilderbrand, R.H., Watts, A.C., and Randle, A.M. (2005). The myths of restoration ecology. *Ecology and Society* 10 (1): 19.

Hillman, E.J., Bigelow, S.G., Samuelson, G.M. et al. (2016). Increasing river flow expands riparian habitat: influences of flow augmentation on channel form, riparian vegetation and birds along the little Bow River, Alberta. *River Research and Applications* 32: 1687–1697.

Hilmes, M.M. and Wohl, E.E. (1995). Changes in channel morphology associated with placer mining. *Physical Geography* 16: 223–242.

Hilton, R.G., Galy, A., and Hovius, N. (2008a). Riverine particulate organic carbon from an active mountain belt: importance of landslides. *Global Biogeochemical Cycles* 22, GB1017.

Hilton, R.G., Galy, A., Hovius, N. et al. (2008b). Tropical-cyclone-driven erosion of the terrestrial biosphere from mountains. *Nature Geoscience* 1: 759–762.

Hilton, R.G., Galy, A., Hovius, N. et al. (2011a). Efficient transport of fossil organic carbon to the ocean by steep mountain rivers: an orogenic carbon sequestration mechanism. *Geology* 39: 71–74.
Hilton, R.G., Meunier, P., Hovius, N. et al. (2011b). Landslide impact on organic carbon cycling in a temperate montane forest. *Earth Surface Processes and Landforms* 36: 1670–1679.
Hirsch, R.M., Walker, J.F., Day, J.C., and Kallio, R. (1990). The influence of man on hydrologic systems. In: *Surface Water Hydrology* (eds. M.G. Wolman and H.C. Riggs), 329–359. Boulder, CO: Geological Society of America.
Hirschboeck, K.K. (1987). Hydroclimatically-defined mixed distributions in partial duration flood series. In: *Hydrologic Frequency Modeling* (ed. V.P. Singh), 199–212. Dordrecht: D. Reidel.
Hirschboeck, K.K. (1988). Flood hydroclimatology. In: *Flood Geomorphology* (eds. V.R. Baker, R.C. Kochel and P.C. Patton), 27–49. New York: Wiley.
Hjulström, F. (1935). Studies of the morphological activity of rivers as illustrated by the River Fyris. *Bulletin Mineral. Geologic Institute, University of Uppsala* 25: 221–258.
Hobley, D.E.J., Sinclair, H.D., and Cowie, P.A. (2010). Processes, rates, and time scales of fluvial response in an ancient postglacial landscape of the northwest Indian Himalaya. *Geological Society of America Bulletin* 122: 1569–1584.
Hodge, R.A. and Hoey, T.B. (2016). A Froude-scaled model of a bedrock-alluvial channel reach: 2. Sediment cover. *Journal of Geophysical Research: Earth Surface* 121: 1597–1618.
Hodge, R., Brasington, J., and Richards, K. (2009). In situ characterization of grain-scale fluvial morphology using Terrestrial Laser Scanning. *Earth Surface Processes and Landforms* 34: 954–968.
Hodge, R.A., Hoey, T.B., and Sklar, L.S. (2011). Bed load transport in bedrock rivers: the role of sediment cover in grain entrainment, translation, and deposition. *Journal of Geophysical Research* 116 F04028.
Hodge, R., Hoey, T., Maniatis, G., and Lepretre, E. (2016). Formation and erosion of sediment cover in an experimental bedrock-alluvial channel. *Earth Surface Processes and Landforms* 41: 1409–1420.
Hodgkins, R., Cooper, R., Wadham, J., and Tranter, M. (2009). The hydrology of the proglacial zone of a high-Arctic glacier (Finsterwalderbreen, Svalbard): atmospheric and surface water fluxes. *Journal of Hydrology* 378: 150–160.
Hoey, T. (1992). Temporal variations in bed load transport rates and sediment storage in gravel-bed rivers. *Progress in Physical Geography* 16: 319–338.
Hoffman, P.F. and Grotzinger, J.P. (1993). Orographic precipitation, erosional unloading, and tectonic style. *Geology* 21: 195–198.
Hoffmann, T., Glatzel, S., and Dikau, R. (2009). A carbon storage perspective on alluvial sediment storage in the Rhine catchment. *Geomorphology* 108: 127–137.
Hohensinner, S., Habersack, H., Jungwirth, M., and Zauner, G. (2004). Reconstruction of the characteristics of a natural alluvial river-floodplain system and hydromorphological changes following human modifications: the Danube River (1812–1991). *River Research and Applications* 20: 25–41.
Hohensinner, S., Hauer, C., and Muhar, S. (2018). River morphology, channelization, and habitat restoration. In: *Riverine Ecosystem Management* (eds. S. Schmutz and J. Sendzimir), 41–65. Cham: Springer.
Hohermuth, B. and Weitbrecht, V. (2018). Influence of bed-load transport on flow resistance of step-pool channels. *Water Resources Research* 54: 5567–5583.
Holbrook, J., Kliem, G., Nzewunwah, C. et al. (2006). Surficial alluvium and topography of the Overton Bottoms North Unit, Big Muddy National Fish and Wildlife Refuge in the Missouri River Valley and its potential influence on Environmental Management. In: *Science to Support Adaptive*

Management – Overton Bottoms North Unit, Big Muddy National Fish and Wildlife Refuge (ed. R.B. Jacobson) US Geological Survey Scientific Investigations Report 2006-5086, Missouri, 17–31.

Holland, W.N. and Pickup, G. (1976). Flume study of knickpoint development in stratified sediment. *Geological Society of America Bulletin* 87: 76–82.

Holland, J.E., Luck, G.W., and Finlayson, C.M. (2015). Threats to food production and water quality in the Murray-Darling Basin of Australia. *Ecosystem Services* 12: 55–70.

Holling, C.S. (1973). Resilience and stability of ecological systems. *Annual Review of Ecology and Systematics* 4: 1–23.

Holmes, R.M., McClelland, J.W., Peterson, B.J. et al. (2012). Seasonal and annual fluxes of nutrients and organic matter from large rivers to the Arctic Ocean and surrounding seas. *Estuaries and Coasts* 35: 369–382.

Hong, L.B. and Davies, T.R.H. (1979). A study of stream braiding. *Geological Society of America Bulletin* 90: 1839–1859.

Hood, W.G. (2010). Tidal channel meander formation by depositional rather than erosional processes: examples from the prograding Skagit River delta (Washington, USA). *Earth Surface Processes and Landforms* 35: 319–330.

Hood, G.A. and Bayley, S.E. (2008). Beaver (*Castor canadensis*) mitigate the effects of climate on the area of open water in boreal wetlands in western Canada. *Biological Conservation* 141: 556–567.

Hooke, R.L.B. (1967). Processes on arid-region alluvial fans. *Journal of Geology* 75: 438–460.

Hooke, R.L.B. (1968). Steady-state relationships on arid-region alluvial fans in closed basins. *American Journal of Science* 266: 609–629.

Hooke, J.M. (1977). The distribution and nature of changes in river channel pattern. In: *River Channel Changes* (ed. K.J. Gregory), 265–280. Chichester: Wiley.

Hooke, J.M. (1979). Analysis of the processes of river bank erosion. *Journal of Hydrology* 42: 39–62.

Hooke, R.L.B. (2000). On the history of humans as geomorphic agents. *Geology* 28: 843–846.

Hooke, J. (2003). Coarse sediment connectivity in river channel systems: a conceptual framework and methodology. *Geomorphology* 56: 79–94.

Hooke, J.M. (2013). Meandering rivers. In: *Treatise on Fluvial Geomorphology* (ed. E. Wohl), 260–287. Amsterdam: Elsevier.

Hooke, R.L.B. and Dorn, R.I. (1992). Segmentation of alluvial fans in Death Valley, California: new insights from surface exposure dating and laboratory modeling. *Earth Surface Processes and Landforms* 17: 557–574.

Hooke, R.L.B. and Rohrer, W.L. (1979). Geometry of alluvial fans: effect of discharge and sediment size. *Earth Surface Processes* 4: 147–166.

Hooshyar, M., Wang, D., Kim, S. et al. (2016). Valley and channel networks extraction based on local topographic curvature and *k*-means clustering of contours. *Water Resources Research* 52: 8081–8102.

Hopp, L. and McDonnell, J.J. (2009). Connectivity at the hillslope scale: identifying interactions between storm size, bedrock permeability, slope angle and soil depth. *Journal of Hydrology* 376: 378–391.

Hori, K. and Saito, Y. (2007). An early Holocene sea-level jump and delta initiation. *Geophysical Research Letters* 34 https://doi.org/10.1029/2007GL031029.

Hori, K., Tanabe, S., Saito, Y. et al. (2004). Delta initiation and Holocene sea-level change; example from the Song Hong (Red River) delta, Vietnam. *Sedimentary Geology* 164: 237–249.

Hori, M., Sugiura, K., Kobayashi, K. et al. (2017). A 38-year (1978–2015) Northern Hemisphere daily snow cover extent product derived using consistent objective criteria from satellite-borne optical sensors. *Remote Sensing of Environment* 191: 402–418.

Hornung, J., Pflanz, D., Hechler, A. et al. (2010). 3-D architecture, depositional patterns and climate triggered sediment fluxes of an alpine alluvial fan (Samedan, Switzerland). *Geomorphology* 114: 202–214.

Horton, R.E. (1932). Drainage-basin characteristics. *Transactions of the American Geophysical Union* 13: 350–361.

Horton, R.E. (1945). Erosional development of streams and their drainage basins: hydrophysical approach to quantitative morphology. *Geological Society of America Bulletin* 56: 275–370.

Horton, A.J., Constantine, J.A., Hales, T.C. et al. (2017). Modification of river meandering by tropical deforestation. *Geology* 45: 511–514.

Hotchkiss, E.R., Hall, R.O., Sponseller, R.A. et al. (2015). Sources of and processes controlling CO_2 emissions change with the size of streams and rivers. *Nature Geoscience* 8: 696–699.

Houjou, K., Shimizu, Y., and Ishii, C. (1990). Calculation of boundary shear stress in open channel flow. *Journal of Hydroscience and Hydraulic Engineering* 8: 21–37.

Houssais, M. and Jerolmack, D.J. (2017). Toward a unifying constitutive relation for sediment transport across environments. *Geomorphology* 277: 251–264.

Houssais, M., Ortiz, C.P., Durian, D.J., and Jerolmack, D.J. (2015). Onset of sediment transport is a continuous transition driven by fluid shear and granular creep. *Nature Communications* 6 6527.

Hovius, N., Stark, C.P., Tutton, M.A., and Abbott, L.D. (1998). Landslide-driven drainage network evolution in a pre-steady-state mountain belt: Finisterre Mountains, Papua New Guinea. *Geology* 26: 1071–1074.

Howard, A.D. (1959). Numerical systems of terrace nomenclature: a critique. *Journal of Geology* 67: 239–243.

Howard, A.D. (1980). Thresholds in river regimes. In: *Thresholds in Geomorphology* (eds. D.R. Coates and J.D. Vitek), 227–258. London: George Allen and Unwin.

Howard, A.D. (1987). Modelling fluvial systems: rock-, gravel- and sand-bed channels. In: *River Channels: Environment and Process* (ed. K. Richards), 69–94. New York: Blackwell.

Howard, A.D. (1994). A detachment-limited model of drainage basin evolution. *Water Resources Research* 30: 2261–2285.

Howard, A.D. (1998). Long profile development of bedrock channels: interaction of weathering, mass wasting, bed erosion, and sediment transport. In: *Rivers Over Rock: Fluvial Processes in Bedrock Channels* (eds. K.J. Tinkler and E.E. Wohl), 297–319. Washington, DC: American Geophysical Union Press.

Howard, A.D. and Knutson, T.R. (1984). Sufficient conditions for river meandering – a simulation approach. *Water Resources Research* 20: 1659–1667.

Howard, A.D., Dietrich, W.E., and Seidl, M.A. (1994). Modelling fluvial erosion on regional to continental scales. *Journal of Geophysical Research* 99 (B7): 13971–13986.

Hsu, L., Finnegan, N.J., and Brodsky, E.E. (2011). A seismic signature of river bedload transport during storm events. *Geophysical Research Letters* 38 L13407.

Huang, H.Q. and Nanson, G.C. (2000). Hydraulic geometry and maximum flow efficiency as products of the principle of least action. *Earth Surface Processes and Landforms* 25: 1–16.

Huang, H.Q. and Nanson, G.C. (2007). Why some alluvial rivers develop an anabranching pattern. *Water Resources Research* 43 W07441.

Hughes, F.M.R. (1997). Floodplain biogeomorphology. *Progress in Physical Geography* 21: 501–529.

Hughes, F.M.R., Adams, W.M., Muller, E. et al. (2001). The importance of different scale processes for the restoration of floodplain woodlands. *Regulated Rivers: Research and Management* 17: 325–345.

Hughes, F.M.R., Colston, A., and Mountford, J.O. (2005). Restoring riparian ecosystems: the challenge of accommodating variability and designing restoration trajectories. *Ecology and Society* 10 https://doi .org/10.5751/ES-01292-100112.

Hughes, S.J., Cabral, J.A., Bastos, R. et al. (2016). A stochastic dynamic model to assess land use change scenarios on the ecological status of fluvial water bodies under the Water Framework Directive. *Science of the Total Environment* 565: 427–439.

Hultine, K.R. and Bush, S.E. (2011). Ecohydrological consequences of non-native riparian vegetation in the southwestern United States: a review from an ecophysiological perspective. *Water Resources Research* 47 W07542.

Humborg, C., Ittekkot, V., Cociasu, A., and Bodungen, B.V. (1997). Effect of Danube River dam on Black Sea biogeochemistry and ecosystem structure. *Nature* 386: 385–388.

Humphries, M.S., Kindness, A., Ellery, W.N., and Hughes, J.C. (2011). Water chemistry and effect of evapotranspiration on chemical sedimentation on the Mkuze River floodplain, South Africa. *Journal of Arid Environments* 75: 555–565.

Hupp, C.R. (1988). Plant ecological aspects of flood geomorphology and paleoflood history. In: *Flood Geomorphology* (eds. V.R. Baker, R.C. Kochel and P.C. Patton), 335–356. New York: Wiley.

Hupp, C.R. and Bornette, G. (2003). Vegetation as a tool in the interpretation of fluvial geomorphic processes and landforms in humid temperate areas. In: *Tools in Geomorphology* (eds. G.M. Kondolf and H. Piégay), 269–288. Chichester: Wiley.

Hupp, C.R. and Simon, A. (1991). Bank accretion and the development of vegetated depositional surfaces along modified alluvial channels. *Geomorphology* 4: 111–124.

Hurd, L.E., Sousa, R.G.C., Siqueira-Souza, F.K. et al. (2016). Amazon floodplain fish communities: habitat connectivity and conservation in a rapidly deteriorating environment. *Biological Conservation* 195: 118–127.

Hyatt, T.L. and Naiman, R.J. (2001). The residence time of large woody debris in the Queets River, Washington, USA. *Ecological Applications* 11: 191–202.

Ijjász-Vásquez, E.I. and Bras, R.L. (1995). Scaling regimes of local slope versus contributing area in digital elevation models. *Geomorphology* 12: 299–311.

Ikeda, S. and Parker, G. (eds.) (1989). *River Meandering*. Washington, DC: American Geophysical Union Press.

Ikeda, S., Parker, G., and Sawai, K. (1981). Bend theory of river meanders. 1. Linear development. *Journal of Fluid Mechanics* 112: 363–377.

Inman, D.L. and Nordstrom, C.E. (1971). On the tectonic and morphologic classification of coasts. *Journal of Geology* 79: 1–21.

Inoue, T., Izumi, N., Shimizu, Y., and Parker, G. (2014). Interaction among alluvial cover, bed roughness, and incision rate in purely bedrock and alluvial-bedrock channel. *Journal of Geophysical Research – Earth Surface* 119: 2123–2146.

Inoue, T., Iwasaki, T., Parker, G., and Shimizu, Y. (2016). Numerical simulation of effects of sediment supply on bedrock channel mophorlogy. *Journal of Hydraulic Engineering* 142 04016014.

IPCC (Intergovernmental Panel on Climate Change) (2008). *Climate Change 2007*. IPCC Fourth Assessment Report. Cambridge: Cambridge University Press.

ISRM (International Society for Rock Mechanics) (1978). Suggested methods for determining tensile strength of rock materials. *International Journal Mechanics Minerals Science Geomechanics Abstracts* 21: 145–153.

Istanbulluoglu, E., Tarboton, D.G., Pack, R.T., and Luce, C. (2002). A probabilistic approach for channel initiation. *Water Resources Research* 38 https://doi.org/10.1029/2001WR000782.

Itoh, T., Horiuchi, S., Mizuyama, T., and Kaitsuka, K. (2013). Hydraulic model tests for evaluating sediment control function with a grid-type Sabo dam in mountainous torrents. *International Journal of Sediment Research* 28: 511–522.

Iverson, R.J. (2005). Debris-flow mechanics. In: *Debris-Flow Hazards and Related Phenomena* (eds. M. Jakob and O. Hungr), 105–134. Berlin: Springer.

Iverson, R.M., Logan, M., LaHusen, R.G., and Berti, M. (2010). The perfect debris flow? Aggregated results from 28 large-scale experiments. *Journal of Geophysical Research – Earth Surface* 115 F03005.

Ives, R.L. (1942). The beaver-meadow complex. *Journal of Geomorphology* 5: 191–203.

Jabaloy-Sanchez, A., Lobo, F.J., Azor, A. et al. (2010). Human-driven coastline changes in the Adra River deltaic system, Southeast Spain. *Geomorphology* 119: 9–22.

Jackson, R.G. (1975). Velocity-bed-form-texture patterns of meander bends in lower Wabash River of Illinois and Indiana. *Geological Society of America Bulletin* 86: 1511–1522.

Jackson, C.R. and Sturm, C.A. (2002). Woody debris and channel morphology in first- and second-order forested channels in the Washington Coast Ranges. *Water Resources Research* 38 https://doi.org/10.1029/2001WR001138.

Jackson, J.R., Pasternack, G.B., and Wheaton, J.M. (2015). Virtual manipulation of topography to test potential pool-riffle maintenance mechanisms. *Geomorphology* 228: 617–627.

Jacobson, R.B. and Gran, K.B. (1999). Gravel sediment routing from widespread, low-intensity landscape disturbance, Current River basin, Missouri. *Earth Surface Processes and Landforms* 24: 897–917.

Jacobson, R.B., K.A. Oberg. 1997. Geomorphic changes on the Mississippi River flood plain at Miller City, Illinois, as a result of the flood of 1993. US Geological Survey Circular 1120-J.

Jacobson, R.B., McGeehin, J.P., Cron, E.D. et al. (1993). Landslides triggered by the storm of November 3–5, 1985, Wills Mountain anticline, West Virginia and Virginia. *US Geological Survey Bulletin* C: C1–C33.

Jacobson, P.J., Jacobson, K.M., Angermeier, P.L., and Cherry, D.S. (1999). Transport, retention, and ecological significance of woody debris within a large ephemeral river. *Journal of the North American Benthological Society* 18: 429–444.

Jacobson, R.B., Lindner, G., and Bitner, C. (2015). The role of floodplain restoration in mitigating flood risk, Lower Missouri River, USA. In: *Geomorphic Approaches to Integrated Floodplain Management of Lowland Fluvial Systems in North America and Europe* (eds. P.F. Hudson and H. Middelkoop), 203–243. New York: Springer.

Jacobson, R.B., O'Connor, J.E., and Oguchi, T. (2016). Surficial geologic tools in fluvial geomorphology. In: *Tools in Geomorphology* (eds. G.M. Kondolf and H. Piégay), 15–39. Chichester: Wiley.

Jaeger, K.L. and Olden, J.D. (2012). Electrical resistance sensor arrays as a means to quantify longitudinal connectivity of rivers. *River Research and Applications* 28: 1843–1852.

Jaeger, K.L., Montgomery, D.R., and Bolton, S.M. (2007). Channel and perennial flow initiation in headwater streams: management implications of variability in source-area size. *Environmental Management* 40: 775–786.

James, L.A. (1989). Sustained storage and transport of hydraulic gold mining sediment in the Bear River, California. *Annals of the Association of American Geographers* 79: 570–592.

James, L.A. (1997). Channel incision on the lower American River, California, from streamflow gage records. *Water Resources Research* 33: 485–490.

James, L.A. (2013). Legacy sediment: definitions and processes of episodically produced anthropogenic sediment. *Anthropocene* 2: 16–26.

James, L.A. and Lecce, S.A. (2013). Impacts of land-use and land-cover change on river systems. In: *Treatise on Fluvial Geomorphology* (ed. E. Wohl), 768–793. Amsterdam: Elsevier.

Jansen, J.D. (2006). Flood magnitude-frequency and lithologic control on bedrock river incision in post-orogenic terrain. *Geomorphology* 82: 39–57.

Jansen, J.D. and Nanson, G.C. (2004). Anabranching and maximum flow efficiency in Magela Creek, northern Australia. *Water Resources Research* 40 W04503.

Jansen, J.D., Fabel, D., Bishop, P. et al. (2011). Does decreasing paraglacial sediment supply slow knickpoint retreat? *Geology* 39: 543–546.

Jansson, R., Nilsson, C., Dynesius, M., and Andersson, E. (2000). Effects of river regulation on river-margin vegetation: a comparison of eight boreal rivers. *Ecological Applications* 10: 203–224.

Japanese Ministry of Construction (1993). *Sabo*. Kobe: Ministry of Construction.

Jaquette, C., Wohl, E., and Cooper, D. (2005). Establishing a context for river rehabilitation, North Fork Gunnison River, Colorado. *Environmental Management* 35: 593–606.

Jarrett, R.D. (1987). Errors in slope–area computations of peak discharges in mountain streams. *Journal of Hydrology* 96: 53–67.

Jarrett, R.D. (1992). Hydraulics of mountain rivers. In: *Channel Flow Resistance: Centennial of Manning's Formula* (ed. B.C. Yen), 287–298. Littleton, CO: Water Resources Publications.

Jarrett, R.D. and England, J.F. (2002). Reliability of paleostage indicators for paleoflood studies. In: *Ancient Floods, Modern Hazards: Principles and Applications of Paleoflood Hydrology* (eds. P.K. House, R.H. Webb, V.R. Baker and D.R. Levish), 91–109. Washington, DC: American Geophysical Union Press.

Jasechko, S., Sharp, Z.D., Gibson, J.J. et al. (2013). Terrestrial water fluxes dominated by transpiration. *Nature* 496: 347–351.

Javernick, L., Brasington, J., and Caruso, B. (2014). Modelling the topography of shallow braided rivers using structure-from-motion photogrammetry. *Geomorphology* 213: 166–182.

Javernick, L., Hicks, D.M., Measures, R. et al. (2016). Numerical modelling of braided rivers with structure-from-motion-derived terrain models. *River Research and Applications* 32: 1071–1081.

Jefferson, A., Grant, G.E., Lewis, S.L., and Lancaster, S.T. (2010). Coevolution of hydrology and topography on a basalt landscape in the Oregon Cascade Range, USA. *Earth Surface Processes and Landforms* 35: 803–816.

Jeffries, R., Darby, S.E., and Sear, D.A. (2003). The influence of vegetation and organic debris on flood-plain sediment dynamics: case study of a low-order stream in the new Forest, England. *Geomorphology* 51: 61–80.

Jerde, C.L., Chadderton, W.L., Mahon, A.R. et al. (2013). Detection of Asian carp DNA as part of a Great Lakes basin-wide surveillance program. *Canadian Journal of Fisheries and Aquatic Sciences* 70: 522–526.

Jerolmack, D.J. and Mohrig, D. (2007). Conditions for branching in depositional rivers. *Geology* 35: 463–466.

Jimenez Sanchez, M. (2002). Slope deposits in the upper Nalon River basin (NW Spain): an approach to a quantitative comparison. *Geomorphology* 43: 165–178.

Jochner, M., Turowski, J.M., Badoux, A. et al. (2015). The role of log jams and exceptional flood events in mobilizing coarse particulate organic matter in a steep headwater stream. *Earth Surface Dynamics* 3: 311–320.

John, S. and Klein, A. (2004). Hydrogeomorphic effects of beaver dams on floodplain morphology: avulsion processes and sediment fluxes in upland valley floors (Spessart, Germany). *Quaternaire* 15: 219–231.

Johnson, J.P.L. (2014). A surface roughness model for predicting alluvial cover and bed load transport rate in bedrock channels. *Journal of Geophysical Research: Earth Surface* 119: 2147–2173.

Johnson, J.P.L. (2017). Clustering statistics, roughness feedbacks, and randomness in experimental step-pool morphodynamics. *Geophysical Research Letters* 44: 3653–3662.

Johnson, A.M. and Rodine, J.R. (1984). Debris flows. In: *Slope Instability* (eds. D. Brunsden and D.B. Prior), 257–361. New York: Wiley.

Johnson, J.P.L. and Whipple, K.X. (2007). Feedbacks between erosion and sediment transport in experimental bedrock channels. *Earth Surface Processes and Landforms* 32: 1048–1062.

Johnson, J.P.L. and Whipple, K.X. (2010). Evaluating the controls of shear stress, sediment supply, alluvial cover, and channel morphology on experimental bedrock incision rate. *Journal of Geophysical Research: Earth Surface* 115, F02018.

Johnson, S.L., Swanson, F.J., Grant, G.E., and Wondzell, S.M. (2000). Riparian forest disturbances by a mountain flood – the influence of floated wood. *Hydrological Processes* 14: 3031–3050.

Johnson, J.P.L., Whipple, K.X., and Sklar, L.S. (2010). Contrasting bedrock incision rates from snowmelt and flash floods in the Henry Mountains, Utah. *Geological Society of America Bulletin* 122: 1600–1615.

Johnson, Z.C., Warwick, J.L., and Schumer, R. (2014). Factors affecting hyporheic and surface transient storage in a western US river. *Journal of Hydrology* 510: 325–339.

Johnson, K.M., Snyder, N.P., Castle, S. et al. (2019). Legacy sediment storage in New England river valleys: anthropogenic processes in a postglacial landscape. *Geomorphology* 327: 417–437.

Jones, J.A.A. (1981). *The Nature of Soil Piping: A Review of Research.* Norwich: Geobooks.

Jones, J.A.A. (2010). Soil piping and catchment response. *Hydrological Processes* 24: 1548–1566.

Jones, L.S. and Schumm, S.A. (1999). Causes of avulsion: an overview. In: *Fluvial Sedimentology VI*, vol. 28 (eds. N.D. Smith and J. Rogers) Special Publications of the International Association of Sedimentologists, 171–178.

Jones, J.A., Swanson, F.J., Wemple, B.C., and Snyder, K.U. (2000). Effects of roads on hydrology, geomorphology, and disturbance patches in stream networks. *Conservation Biology* 14: 76–86.

Jones, W.K., Culver, D.C., and Herman, J.S. (2003). *Epikarst.* Proceedings of the 2003 Epikarst Symposium. Special Publication 9,. Charles Town, WV: Karst Waters Institute.

Jones, A.F., Brewer, P.A., Johnstone, E., and Macklin, M.G. (2007). High-resolution interpretative geomorphological mapping of river valley environments using airborne LiDAR data. *Earth Surface Processes and Landforms* 32: 1574–1592.

Jones, K.L., Poole, G.C., O'Daniel, S.J. et al. (2008). Surface hydrology of low-relief landscapes: assessing surface water flow impedance using LIDAR-derived digital elevation models. *Remote Sensing of Environment* 112: 4148–4158.

Jorde, K., Burke, M., Scheidt, N. et al. (2008). Reservoir operations, physical processes, and ecosystem losses. In: *Gravel-Bed Rivers VI: From Process Understanding to River Restoration* (eds. H. Habersack, H. Piégay and M. Rinaldi), 607–636. Amsterdam: Elsevier.

Julien, P.Y. (1998). *Erosion and Sedimentation.* Cambridge: Cambridge University Press.

Jun, P., Chen, S., and Dong, P. (2010). Temporal variation of sediment load in the Yellow River basin, China, and its impacts on the lower reaches and the river delta. *Catena* 83: 136–147.

Jungers, M.C., Bierman, P.R., Matmon, A. et al. (2009). Tracing hillslope sediment production and transport with in situ and meteoric ^{10}Be. *Journal of Geophysical Research* 114, F04020.

Junk, W.J., Bayley, P.B., and Sparks, R.E. (1989). The flood pulse concept in river-floodplain systems. *Canadian Special Publication of Fisheries and Aquatic Sciences* 106: 110–127.

Kaiser, K., Guggenberger, G., and Haumeier, L. (2004). Changes in dissolved lignin-derived phenols, netural sugars, uronic acids, and amino sugars with depth in forested Haplic Arenosols and Rendzic Leptosols. *Biogeochemistry* 70: 135–151.

Kaiser, K., Keller, N., Brande, A. et al. (2018). A large-scale medieval dam-lake cascade in central Europe: water level dynamics of the Havel River, Berlin-Brandenburg region, Germany. *Geoarchaeology* 33: 237–259.

Kale, V.S. and Hire, P.S. (2007). Temporal variations in the specific stream power and total energy expenditure of a monsoonal river: the Tapi River, India. *Geomorphology* 92: 134–146.

Kale, V.S., Baker, V.R., and Mishra, S. (1996). Multi-channel patterns of bedrock rivers: an example from the central Narmada basin, India. *Catena* 26: 85–98.

Kampf, S.K. and Mirus, B.B. (2013). Subsurface and surface flow leading to channel initiation. In: *Treatise on Fluvial Geomorphology* (ed. E. Wohl), 23–42. Amsterdam: Elsevier.

Kampf, S.K., Brogan, D.J., Schmeer, S. et al. (2016). How do geomorphic effects of rainfall vary with storm type and spatial scale in a post-fire landscape? *Geomorphology* 273: 39–51.

Kang, K. and Merwade, V. (2011). Development and application of a storage-release based distributed hydrologic model using GIS. *Journal of Hydrology* 403: 1–13.

Kao, S.J. and Milliman, J.D. (2008). Water and sediment discharge from small mountainous rivers, Taiwan: the roles of lithology, episodic events, and human activities. *Journal of Geology* 116: 431–448.

Kao, S.J., Hilton, R.G., Selvaraj, K. et al. (2014). Preservation of terrestrial organic carbon in marine sediments offshore Taiwan: mountain building and atmospheric carbon dioxide sequestration. *Earth Surface Dynamics* 2: 127–139.

Karr, J.R. (1991). Biological integrity: a long-neglected aspect of water resource management. *Ecological Applications* 1: 66–84.

Karr, J.R. (1999). Defining and measuring river health. *Freshwater Biology* 41: 221–234.

Karssenberg, D. and Bridge, J.S. (2008). A three-dimensional numerical model of sediment transport, erosion and deposition within a network of channel belts, floodplain and hill slope; extrinsic and intrinsic controls on floodplain dynamics and alluvial architecture. *Sedimentology* 55: 1717–1745.

Kasprak, A., Wheaton, J.M., Ashmore, P.E. et al. (2015). The relationship between particle travel distance and channel morphology: results from physical models of braided rivers. *Journal of Geophysical Research Earth Surface* 120: 55–74.

Kasprak, A., Hough-Snee, N., Beechie, T. et al. (2016). The blurred line between form and process: a comparison of stream channel classification frameworks. *PLoS One* 11, e0150293.

Kattenberg, A., Giorgi, F., Grassl, H., and Meehl, G.A. (1996). Climate change 1995: the science of climate change. In: *Climate Change 1995. Intergovernmental Panel on Climate Change* (eds. R.T. Watson, M.C. Zinyowera and R.H. Moss). Cambridge: Cambridge University Press.

Katz, G.L., Friedman, J.M., and Beatty, S.W. (2005). Delayed effects of flood control on a flood-dependent riparian forest. *Ecological Applications* 15: 1019–1035.

Kaushal, S.S., Likens, G.E., Pace, M.L. et al. (2018). Freshwater salinization syndrome on a continental scale. *Proceedings of the National Academy of Sciences of the United States of America* https://doi.org/10.1073/pnas.1711234115.

Kean, J.W. and Smith, J.D. (2006a). Form drag in rivers due to small-scale natural topographic features: 1. Regular sequences. *Water Resources Research* 111, F04009.

Kean, J.W. and Smith, J.D. (2006b). Form drag in rivers due to small-scale natural topographic features: 2. Irregular sequences. *Water Resources Research* 111, F04010.

Keizer, F.M., Schot, P.P., Okruszko, T. et al. (2014). A new look at the flood pulse concept: the (ir)relevance of the moving littoral in temperate zone rivers. *Ecological Engineering* 64: 85–99.

Keller, E.A. (1971). Areal sorting of bedload material, the hypothesis of velocity reversal. *Geological Society of America Bulletin* 82: 279–280.

Keller, E.A. and Melhorn, W.N. (1978). Rhythmic spacing and origin of pools and riffles. *Geological Society of America Bulletin* 89: 723–730.

Keller, E.A. and Swanson, F.J. (1979). Effects of large organic material on channel form and fluvial processes. *Earth Surface Processes* 4: 361–380.

Keller, E.A., A. MacDonald, T. Tally, N.J. Merrit. 1995. Effects of large organic debris on channel morphology and sediment storage in selected tributaries of Redwood Creek, northwestern California. US Geological Survey Professional Paper 1454.

Keller, T., Pielmeier, C., Rixen, C. et al. (2004). Impact of artificial snow and ski-slope grooming on snowpack properties and soil thermal regime in a sub-alpine ski area. *Annals of Glaciology* 38: 314–318.

Keller, G., Bentley, S.J., Georgiou, I.Y. et al. (2017). River-plume sedimentation and $^{210}Pb/^{7}Be$ seabed delivery on the Mississippi River delta front. *Geo-Marine Letters* 37: 259–272.

Kellerhals, R. and Church, M. (1990). Hazard management on fans, with examples from British Columbia. In: *Alluvial Fans: A Field Approach* (eds. A.H. Rachocki and M. Church), 335–354. Chichester: Wiley.

Kellerhals, R., Church, M., and Bray, D.I. (1976). Classification of river processes. *Journal of the Hydraulics Division American Society of Civil Engineers* 102: 813–829.

Kellerhals, R., Church, M., and Davies, L.B. (1979). Morphological effects of interbasin river diversions. *Canadian Journal of Civil Engineering* 6: 18–31.

Kelly, S. (2006). Scaling and hierarchy in braided rivers and their deposits: examples and implications for reservoir modeling. In: *Braided Rivers: Process, Deposits, Ecology and Management* (eds. G.H. Sambrook Smith, J.L. Best, C.S. Bristow and G.E. Petts), 75–106. Oxford: Blackwell.

Kemp, P., Sear, D., Collins, A. et al. (2011). The impacts of fine sediment on riverine fish. *Hydrological Processes* 25: 1800–1821.

Kemp, A.C., Wright, A.J., Edwards, R.J. et al. (2018). Relative sea-level change in Newfoundland, Canada during the past ~3000 years. *Quaternary Science Reviews* 201: 89–110.

Kendall, C., McDonnell, J.J., and Gu, W. (2001). A look inside "black box" hydrograph separation models: a study at the Hydrohill catchment. *Hydrological Processes* 15: 1877–1902.

Kesel, R.H. (2003). Human modifications to the sediment regime of the Lower Mississippi River flood plain. *Geomorphology* 56: 325–334.

Kesel, R.H., Dunne, K.C., McDonald, R.C. et al. (1974). Lateral erosion and overbank deposition on the Mississippi River in Louisiana caused by 1973 flooding. *Geology* 2: 461–464.

Kideys, A.E. (2002). Fall and rise of the Black Sea ecosystem. *Science* 297: 1482–1484.

Kidova, A., Lehotsky, M., and Rusnak, M. (2016). Geomorphic diversity in the braided-wandering Bela River, Slovak Carpathians, as a response to flood variability and environmental changes. *Geomorphology* 272: 137–149.

Kieffer, S.W. (1989). Geologic nozzles. *Reviews of Geophysics* 27: 3–38.

Kim, J.Y. and An, K.G. (2015). Integrated ecological river health assessments, based on water chemistry, physical habitat quality and biological integrity. *Water* 7: 6378–6403.

Kim, H.J., Sidle, R.C., Moore, R.D., and Hudson, R. (2004). Throughflow variability during snowmelt in a forested mountain catchment, coastal British Columbia, Canada. *Hydrological Processes* 18: 1219–1236.

King, J. and Brown, C. (2010). Integrated basin flow assessments: concepts and method development in Africa and South-East Asia. *Freshwater Biology* 55: 127–146.

King, J., Beuster, H., Brown, C., and Joubert, A. (2014). Pro-active management: the role of environmental flows in transboundary cooperative planning for the Okavango River system. *Hydrological Sciences Journal* 59: 786–800.

Kirchner, J.W. (1993). Statistical inevitability of Horton's laws and the apparent randomness of stream channel networks. *Geology* 21: 591–594.

Kirchner, J. (2003). A double paradox in catchment hydrology and geochemistry. *Hydrological Processes* 17: 871–874.

Kirchner, J.W. (2006). Getting the right answers for the right reasons: linking measurements, analysis, and models to advance the science of hydrology. *Water Resources Research* 42, W03S04.

Kirkby, M.J. (1967). Measurement and theory of soil creep. *Journal of Geology* 75: 359–378.

Kirkby, M. (1988). Hillslope runoff processes and models. *Journal of Hydrology* 100: 315–339.

Kirwan, M.L., Temmerman, S., Skeehan, E.E. et al. (2016). Overestimation of marsh vulnerability to sea level rise. *Nature Climate Change* 6: 253–260.

Kjeldsen, T.R., Macdonald, N., Lang, M. et al. (2014). Documentary evidence of past floods in Europe and their utility in flood frequency estimation. *Journal of Hydrology* 517: 963–973.

Klavon, K., Fox, G., Guertault, L. et al. (2017). Evaluating a process-based model for use in streambank stabilization: insights on the Bank Stability and Toe Erosion Model (BSTEM). *Earth Surface Processes and Landforms* 42: 191–213.

Kleinhans, M.G., Wilbers, A.W.E., De Swaaf, A., and Van Den Berg, J.H. (2002). Sediment supply-limited bedforms in sand-gravel bed rivers. *Journal of Sedimentary Research* 72: 629–640.

Klimek, K. (1987). Man's impact on fluvial processes in the Polish western Carpathians. *Geografiska Annaler* 69A (1): 221–226.

Klingeman, P.C., R.T. Milhous. 1970. Oak Creek vortex bedload sampler. American Geophysical Union, 17th Annual Pacific Northwest Regional Meeting, Tacoma, WA.

Knighton, A.D. (1974). Variation in width-discharge relation and some implications for hydraulic geometry. *Geological Society of America Bulletin* 85: 1069–1076.

Knighton, D. (1984). *Fluvial Forms and Processes*. London: Edward Arnold.

Knighton, D. (1989). River adjustment to changes in sediment load: the effects of tin mining on the Ringarooma River, Tasmania, 1875–1984. *Earth Surface Processes and Landforms* 14: 333–359.

Knighton, D. (1998). *Fluvial Forms and Processes: A New Perspective*. London: Arnold.

Knighton, A.D. (1999). Downstream variation in stream power. *Geomorphology* 29: 293–306.

Knighton, A.D. and Nanson, G.C. (2002). Inbank and overbank velocity conditions in an arid zone anastomosing river. *Hydrological Processes* 16: 1771–1791.

Knox, J.C. (2006). Floodplain sedimentation in the Upper Mississippi Valley: natural versus human accelerated. *Geomorphology* 79: 286–310.

Knutson, T.R., McBride, J.I., Chan, J. et al. (2010). Tropical cyclones and climate change. *Nature Geoscience* 3: 157–163.

Koboltschnig, G.R., Schoener, W., Zappa, M., and Holmann, H. (2007). Contribution of glacier melt to stream runoff; if the climatically extreme summer of 2003 had happened in 1979. *Annals of Glaciology* 46: 303–308.

Kochel, R.C. (1988). Geomorphic impact of large floods: review and new perspectives on magnitude and frequency. In: *Flood Geomorphology* (eds. V.R. Baker, R.C. Kochel and P.C. Patton), 169–187. New York: Wiley.

Kochel, R.C., Miller, J.R., and Ritter, D.F. (1997). Geomorphic response to minor cyclic climate changes, San Diego County, California. *Geomorphology* 19: 277–302.

Koebel, J.W. and Bousquin, S.G. (2014). The Kissimmee River restoration project and evaluation program, Florida, USA. *Restoration Ecology* 22: 345–352.

Koel, T.M. and Sparks, R.E. (2002). Historical patterns of river stage and fish communities as criteria for operations of dams on the Illinois River. *River Research and Applications* 18: 3–19.

Kokelj, S.V., Lacelle, D., Lantz, T.C. et al. (2013). Thawing of massive ground ice in mega slumps drives increases in stream sediment and solute flux across a range of watershed scales. *Journal of Geophysical Research Earth Surface* 118: 681–692.

Kokelj, S.V., Tunnicliffe, J., Lacelle, D. et al. (2015). Increased precipitation drives mega slump development and destabilization of ice-rich permafrost terrain, northwestern Canada. *Global and Planetary Change* 129: 56–68.

Kolb, T., Fuchs, M., Moine, O., and Zoller, L. (2017). Quaternary river terraces and hillslope sediments as archives for paleoenvironmental reconstruction: new insights from the headwaters of the Main River, Germany. *Zeitschrift für Geomorphologie* 61: 53–76.

Komar, P.D. and Carling, P.A. (1991). Grain sorting in gravel-bed streams and the choice of particle sizes for flow-competence evaluations. *Sedimentology* 38: 489–502.

Kondolf, G.M. (1997). Hungry water: effects of dams and gravel mining on river channels. *Environmental Management* 21: 533–551.

Kondolf, G.M. (2001). Planning approaches to mitigate adverse human impacts on land/water ecotones. *International Journal of Ecohydrology and Hydrobiology* 1: 111–116.

Kondolf, G.M. (2006). River restoration and meanders. *Ecology and Society* 11 (2): 42.

Kondolf, G.M. and Curry, R.R. (1986). Channel erosion along the Carmel River, Monterey County, California. *Earth Surface Processes and Landforms* 11: 307–319.

Kondolf, G.M. and Larson, M. (1995). Historical channel analysis and its application to riparian and aquatic habitat restoration. *Aquatic Conservation: Marine and Freshwater Ecosystems* 5: 109–126.

Kondolf, G.M., Smeltzer, M.W., and Railsback, S.F. (2001). Design and performance of a channel reconstruction project in a coastal California gravel-bed stream. *Environmental Management* 28: 761–776.

Kondolf, G.M., Piegay, H., and Landon, N. (2002). Channel response to increased and decreased bedload supply from land use change: contrasts between two catchments. *Geomorphology* 45: 35–51.

Kondolf, G.M., Boulton, A.J., O'Daniel, S. et al. (2006). Process-based ecological river restoration: visualizing three-dimensional connectivity and dynamic vectors to recover lost linkages. *Ecology and Society* 11 (2): 5.

Kondolf, G.M., Gao, Y., Annandale, G.W. et al. (2014). Sustainable sediment management in reservoirs and regulated rivers: experiences from five continents. *Earth's Future* 2: 256–280.

Konrad, C.P. (2012). Reoccupation of floodplains by rivers and its relation to age structure of floodplain vegetation. *Journal of Geophysical Research* 117, F00N13.

Konrad, C.P., Olden, J.D., Lytle, D.A. et al. (2011). Large-scale flow experiments for managing river systems. *BioScience* 61: 948–959.

Koons, P.O., Zeitler, P.K., Chamberlain, C.P. et al. (2002). Mechanical links between erosion and metamorphism in Nanga Parbat, Pakistan Himalaya. *American Journal of Science* 302: 749–773.

Korup, O. (2006). Rock-slope failure and the river long profile. *Geology* 34: 45–48.
Korup, O. (2013). Landslides in the fluvial system. In: *Treatise on Fluvial Geomorphology* (ed. E. Wohl), 244–259. Amsterdam: Elsevier.
Korup, O., McSaveney, M.J., and Davies, T.R.H. (2004). Sediment generation and delivery from large historic landslides in the Southern Alps, New Zealand. *Geomorphology* 61: 189–207.
Korup, O., Strom, A.L., and Weidinger, J.T. (2006). Fluvial response to large rock-slope failures; examples from the Himalayas, the Tien Shan, and the Southern Alps in New Zealand. *Geomorphology* 78: 3–21.
Korup, O., Densmore, A.L., and Schlunegger, F. (2010). The role of landslides in mountain range evolution. *Geomorphology* 120: 77–90.
Koster, E.H. (1978). Transverse ribs: their characteristics, origin and paleohydraulic significance. In: *Fluvial Sedimentology* (ed. A.D. Miall), 161–186. Canadian Society of Petroleum Geologists Memoir 5.
Kotarba, A. (2005). Geomorphic processes and vegetation pattern changes case study in the Zelene Pleso Valley, High Tatra, Slovakia. *Studia Geomorphologica Carpatho-Balcanica* 39: 39–47.
Kovacs, A. and Parker, G. (1994). A new vectorial bedload formulation and its application to the time evolution of straight river channels. *Journal of Fluid Mechanics* 267: 153–183.
Kraft, C.E. and Warren, D.R. (2003). Development of spatial pattern in large woody debris and debris dams in streams. *Geomorphology* 51: 127–139.
Kramer, P.J. and Boyer, J.S. (1995). *Water Relations of Plants and Soils*. San Diego, CA: Academic Press.
Kramer, N. and Wohl, E. (2014). Estimating fluvial wood discharge using time-lapse photography with varying sampling intervals. *Earth Surface Processes and Landforms* 39: 844–852.
Kramer, N. and Wohl, E. (2015). Driftcretions: the legacy impacts of driftwood on shoreline morphology. *Geophysical Research Letters* 4: 5855–5864.
Kramer, N. and Wohl, E. (2017). Rules of the road: a qualitative and quantitative synthesis of large wood transport through drainage networks. *Geomorphology* 276: 74–97.
Kramer, N., Wohl, E.E., and Harry, D.L. (2012). Using ground penetrating radar to "unearth" buried beaver dams. *Geology* 40: 43–46.
Kramer, N., Wohl, E., Hess-Homeier, B., and Leisz, S. (2017). The pulse of driftwood export from a very large forested river basin over multiple time scales, Slave River, Canada. *Water Resources Research* 53: 1928–1947.
Kratzer, C.R., R.N. Biagtan. 1997. Determination of travel times in the lower San Joaquin River basin, California, from dye-tracer studies during 1994–1995. US Geological Survey Water-Resources Investigations Report 97-4018.
Krause, S., Bronstert, A., and Zehe, E. (2007). Groundwater-surface water interactions in a north German lowland floodplain – implications for the river discharge dynamics and riparian water balance. *Journal of Hydrology* 347: 404–417.
Kreiling, R.M., Schubauer-Berigan, J.P., Richardson, W.B. et al. (2013). Wetland management reduces sediment and nutrient loading to the Upper Mississippi River. *Biogeochemistry* 42: 573–583.
Kresan, P.L. (1988). The Tucson, Arizona, flood of October 1983: implications for land management along alluvial river channels. In: *Flood Geomorphology* (eds. V.R. Baker, R.C. Kochel and P.C. Patton), 465–489. New York: Wiley.
Krigstrom, A. (1962). Geomorphological studies of sandur plains and their braided rivers in Iceland. *Geografiska Annaler* 44: 328–346.
Kroes, D.E., Schenk, E.R., Noe, G.B., and Benthem, A.J. (2015). Sediment and nutrient trapping as a result of a temporary Mississippi River floodplain restoration: the Morganza Spillway during the 2011 Mississippi River flood. *Ecological Engineering* 82: 91–102.

Krom, M.D., Stanley, J.D., Cliff, R.A., and Woodward, J.C. (2002). Nile River sediment fluctuations over the past 7000 yr and their key role in sapropel development. *Geology* 30: 71–74.

Kueppers, L.M., Southon, J., Baer, P., and Harte, J. (2004). Dead wood biomass and turnover time, measured by radiocarbon, along a subalpine elevation gradient. *Oecologia* 141: 641–651.

Kuhnle, R.A. (2013). Suspended load. In: *Treatise on Fluvial Geomorphology* (ed. E. Wohl), 124–136. Amsterdam: Elsevier.

Kuhnle, R.A. and Southard, J.B. (1988). Bed load transport fluctuations in a gravel bed laboratory channel. *Water Resources Research* 24: 247–260.

Kuhnle, R.A., Bingner, R.L., Foster, G.R., and Grissinger, E.H. (1996). Effect of land use changes on sediment transport in Goodwin Creek. *Water Resources Research* 32: 3189–3196.

Kukulak, J., Pazdur, A., and Kuc, T. (2002). Radiocarbon dated wood debris in floodplain deposits of the San River in the Bieszczady Mountains. *Geochronometria* 21: 129–136.

Kundzewicz, Z.W. (2011). Nonstationarity in water resources – Central European perspective. *Journal of the American Water Resources Association* 47: 550–562.

Kundzewicz, Z.W., Mata, L.J., Arnell, N.W. et al. (2007). Freshwater resources and their management. Climate change 2007: impacts, adaptation, and vulnerability. In: *Contribution of Working Group II to the Fourth Assessment Report of the Intergovernmental Panel on Climate Change* (eds. M.L. Parry, O.F. Canziani, J.P. Palutikof, et al.), 173–210. Cambridge: Cambridge University Press.

Kuo, C.-W. and Brierley, G.J. (2013). The influence of landscape configuration upon patterns of sediment storage in a highly connected river system. *Geomorphology* 180–181: 255–266.

Kupfer, J.A., Meitzen, K.M., and Gao, P. (2014). Flooding and surface connectivity of Taxodium-Nyssa stands in a southern floodplain forest ecosystem. *River Research and Applications* 31: 1299–1310.

Kurashige, Y. (1999). Monitoring of thickness of river-bed sediment in the Pankenai River, Hokkaido, Japan. *Transactions, Japanese Geomorphological Union* 20: 21–33.

Kurashige, Y., Kibayashi, H., and Nakajima, G. (2003). Chronology of alluvial sediment using the date of production of buried refuse: a case study in an ungauged river in central Japan. In: *Erosion Prediction in Ungauged Basins: Integrating Methods and Techniques* (ed. D. de Boer), 43–50. Wallingford: IAHS Publications no. 279.

Lacey, R.W.J., Legendre, P., and Roy, A.G. (2007). Spatial-scale partitioning of in situ turbulent flow data over a pebble cluster in a gravel-bed river. *Water Resources Research* 43, W03416.

Laenen, A., J.C. Risley. 1997. Precipitation-runoff and streamflow-routing models for the Willamette River Basin, Oregon. US Geological Survey Water Resources Investigations Report 95-4284.

Lagasse, P.F., Winkley, B.R., and Simons, D.B. (1980). Impact of gravel mining on river system stability. *Journal of Waterway Port Coastal Ocean Division American Society of Civil Engineering* 106: 389–404.

Lagasse, P.F., L.W. Zevenbergen, W.J. Spitz, C.R. Thorne. 2004. Methodology for predicting channel migration. NCHRP Web-Only Document 67 (Project 24-16). Report prepared for Transportation Research Board of the US National Academies.

Lague, D. (2014). The stream power river incision model: evidence, theory and beyond. *Earth Surface Processes and Landforms* 39: 38–61.

Lai, C. and Katul, G. (2000). The dynamic role of root-water uptake in coupling potential to actual transpiration. *Advances in Water Resources* 23: 427–439.

Lamarre, H. and Roy, A.G. (2008). The role of morphology on the displacement of particles in a step-pool river system. *Geomorphology* 99: 270–279.

Lamb, M.P. and Fonstad, M.A. (2010). Rapid formation of a modern bedrock canyon by a single flood event. *Nature Geoscience* 3: 477–481.

Lamb, M.P., Howard, A.D., Johnson, J. et al. (2006). Can springs cut canyons into rock? *Journal of Geophysical Research* 111, E07002.

Lamb, M.P., Howard, A.D., Dietrich, W.E., and Perron, J.T. (2007). Formation of ampitheater-headed valleys by waterfall erosion after large-scale slumping on Hawai'i. *Geological Society of America Bulletin* 119: 805–822.

Lamb, M.P., Dietrich, W.E., and Venditti, J.G. (2008). Is the critical Shields stress for incipient sediment motion dependent on channel-bed slope? *Journal of Geophysical Research* 113, F02008.

Lamb, M.P., Finnegan, N.J., Scheingross, J.S., and Sklar, L.S. (2015). New insights into the mechanics of fluvial bedrock erosion through flume experiments and theory. *Geomorphology* 244: 33–55.

Lamb, M.P., Brun, F., and Fuller, B.M. (2017). Hydrodynamics of steep streams with planar coarse-grained beds: turbulence, flow resistance, and implications for sediment transport. *Water Resources Research* 53: 2240–2263.

Lamoureux, S.F. and Lafrenière, M.J. (2009). Fluvial impacts of extensive active layer detachments, Cape Bounty, Melville Island, Canada. *Arctic, Antarctarctic, and Alpine Research* 41: 59–68.

Lana-Renault, N., Regüés, D., Latron, J. et al. (2006). A volumetric approach to estimated bed load transport in a mountain stream (Central Spanish Pyrenees). In: *Sediment Dynamics and the Hydromorphology of Fluvial Systems* (eds. J.S. Rowan, R.W. Duck and A. Werritty), 89–95. Wallingford: IAHS Publications no. 306.

Lancaster, S.T. and Casebeer, N.E. (2007). Sediment storage and evacuation in headwater valleys at the transition between debris-flow and fluvial processes. *Geology* 35: 1027–1030.

Lancaster, S.T., Hayes, S.K., and Grant, G.E. (2003). Effects of wood on debris flow runout in small mountain watersheds. *Water Resources Research* 39: 1168.

Lancaster, S.T., Underwood, E.F., and Frueh, W.T. (2010). Sediment reservoirs at mountain stream confluences: dynamics and effects of tributaries dominated by debris-flow and fluvial processes. *Geological Society of America Bulletin* 122: 1775–1786.

Lane, E.W. (1955). The importance of fluvial morphology in hydraulic engineering. *American Society of Civil Engineers Proceedings Separate* 81: 1–17.

Lane, S.N. (2006). Approaching the system-scale understanding of braided river behavior. In: *Braided Rivers: Process, Deposits, Ecology and Management*, vol. 36 (eds. G.H. Sambrook Smith, J.L. Best, C.S. Bristow and G.E. Petts) Special Publications., 107–135. International Association of Sedimentology.

Lane, S.N. (2017). Natural flood management. *WIREs Water* 4, e1211.

Lane, S.N. and Richards, K.S. (1998). High resolution, two-dimensional spatial modelling of flow processes in a multi-thread channel. *Hydrological Processes* 12: 1279–1298.

Lane, S.N., Chandler, J.H., and Richards, K.S. (1994). Development in monitoring and modelling small scale river bed topography. *Earth Surface Processes and Landforms* 19: 349–368.

Lane, S.N., Hardy, R.J., Ferguson, R.I., and Parsons, D.R. (2005). A framework for model verification and validation of CFD schemes in natural open channel flows. In: *Computational Fluid Dynamics: Applications in Environmental Hydraulics* (eds. P.D. Bates, S.L. Lane and R.I. Ferguson), 169–192. Chichester: Wiley.

Lane, S.N., Bakker, M., Gabbud, C. et al. (2017). Sediment export, transient landscape response and catchment-scale connectivity following rapid climate warming and Alpine glacier recession. *Geomorphology* 277: 210–227.

Langbein, W.B. (1964). Geometry of river channels. *Journal of the Hydraulics Division* 90: 301–312.

Langbein, W.B. and Leopold, L.B. (1964). Quasi-equilibrium states in channel morphology. *American Journal of Science* 262: 782–794.

Langbein, W.B., L.B. Leopold. 1966. River meanders – theory of minimum variance. US Geological Survey Professional Paper 422H.

Langbein, W.B., L.B. Leopold. 1968. River channel bars and dunes – theory of kinematic waves. US Geological Survey Professional Paper 422L.

Langbein, W.B. and Schumm, S.A. (1958). Yield of sediment in relation to mean annual precipitation. *Transactions, American Geophysical Union* 39: 1076–1084.

Larkin, Z., Ralph, T.J., Tooth, S., and McCarthy, T.S. (2017). The interplay between extrinsic and intrinsic controls in determining floodplain wetland characteristics in the South African drylands. *Earth Surface Processes and Landforms* 42: 1092–1109.

Larned, S.T., Hicks, D.M., Schmidt, J. et al. (2008). The Selwyn River of New Zealand: a benchmark system for alluvial plain rivers. *River Research and Applications* 24: 1–21.

Laronne, J.B., Alexandrov, Y., Bergman, N. et al. (2003). The continuous monitoring of bed load flux in various fluvial environments. In: *Erosion and Sediment Transport Measurement in Rivers: Technological and Methodological Advances* (eds. J. Bogen, T. Fergus and D.E. Walling), 134–145. Wallingford: IAHS Publications no. 283.

Larsen, I.J. and MacDonald, L.H. (2007). Predicting postfire sediment yields at the hillslope scale: testing RUSLE and Disturbed WEPP. *Water Resources Research* 43, W11412.

Larsen, M.C. and Román, A.S. (2001). Mass wasting and sediment storage in a small montane watershed: an extreme case of anthropogenic disturbance in the humid tropics. In: *Geomorphic Processes and Riverine Habitat* (eds. J.M. Dorava, D.R. Montgomery, B.B. Palcsak and F.A. Fitzpatrick), 119–138. Washington, DC: American Geophysical Union Press.

Larsen, L.G., Choi, J., Nungesser, M.K., and Harvey, J.W. (2012). Directional connectivity in hydrology and ecology. *Ecological Applications* 22: 2204–2220.

Larsen, I.J., Montgomery, D.R., and Greenberg, H.M. (2014). The contribution of mountains to global denudation. *Geology* 42: 527–530.

Larsen, A., May, J.H., Moss, P., and Hacker, J. (2016). Could alluvial knickpoint retreat rather than fire drive the loss of alluvial wet monsoon forest, tropical northern Australia? *Earth Surface Processes and Landforms* 41: 1583–1594.

Latocha, A. and Migoń, P. (2006). Geomorphology of medium-high mountains under changing human impact, from managed slopes to nature restoration: a study from the Sudetes, SW Poland. *Earth Surface Processes and Landforms* 31: 1657–1673.

Latrubesse, E.M. (2008). Patterns of anabranching channels: the ultimate end-member adjustment of mega rivers. *Geomorphology* 10: 130–145.

Latrubesse, E.M. (2015). Large rivers, megafans and other Quaternary avulsive fluvial systems: a potential "who's who" in the geologic record. *Earth-Science Reviews* 146: 1–30.

Latrubesse, E.M. and Franzinelli, E. (2002). The Holocene alluvial plain of the middle Amazon River. *Geomorphology* 44: 241–257.

Latrubesse, E.M., Arima, E.Y., Dunne, T. et al. (2017). Damming the rivers of the Amazon basin. *Nature* 546: 363–369.

Latterell, J.J. and Naiman, R.J. (2007). Sources and dynamics of large logs in a temperate floodplain river. *Ecological Applications* 17: 1127–1141.

Lauer, J.W. and Parker, G. (2008a). Modeling framework for sediment deposition, storage, and evacuation in the floodplain of a meandering river: theory. *Water Resources Research* 44, W04425.

Lauer, J.W. and Parker, G. (2008b). Net local removal of floodplain sediment by river meander migration. *Geomorphology* 96: 123–149.

Laurel, D. and Wohl, E. (2018). The persistence of beaver-induced geomorphic heterogeneity and organic carbon stock in river corridors. *Earth Surface Processes and Landforms*.

Lautz, L.K. (2010). Impacts of nonideal field conditions on vertical water velocity estimates from streambed temperature time series. *Water Resources Research* 46, W01509.

Lautz, L.K. and Fanelli, R.M. (2008). Seasonal biogeochemical hotspots in the streambed around restoration structures. *Biogeochemistry* 91: 85–104.

Lautz, L.K. and Siegel, D.I. (2007). The effect of transient storage on nitrate uptake lengths in streams: an inter-site comparison. *Hydrological Processes* 21: 3533–3548.

Lave, R. (2018). Stream mitigation banking. *WIREs Water* 5, e1279.

Lavé, J. and Avouac, J.P. (2001). Fluvial incision and tectonic uplift across the Himalayas of central Nepal. *Journal of Geophysical Research: Solid Earth* 106: 26561–26591.

Lave, R., Robertson, M.M., and Doyle, M.W. (2008). Why you should pay attention to stream mitigation banking. *Ecological Restoration* 26: 2878–2289.

Lawler, D.M. (1991). A new technique for the automatic monitoring of erosion and deposition rates. *Water Resources Research* 27: 2125–2128.

Lawler, D.M. (1992). Process dominance in bank erosion systems. In: *Lowland Floodplain Rivers: Geomorphological Perspectives* (eds. P.A. Carling and G.E. Petts), 117–143. Chichester: Wiley.

Lawler, D.M. (1993). The measurement of river bank erosion and lateral channel change: a review. *Earth Surface Processes and Landforms* 18: 777–821.

Lawler, D.M. (2005). The importance of high-resolution monitoring in erosion and deposition dynamics studies: examples from estuarine and fluvial systems. *Geomorphology* 64: 1–23.

Lawler, D.M. (2008). Advances in the continuous monitoring of erosion and deposition dynamics: develpments and applications of the new PEEP-3T system. *Geomorphology* 93: 17–39.

Le Bouteiller, C. and Venditti, J.G. (2015). Sediment transport and shear stress partitioning in a vegetated flow. *Water Resources Research* 51: 2901–2922.

Le Lay, Y.F., Piegay, H., and Riviere-Honegger, A. (2013a). Perception of braided river landscapes: implications for public participation and sustainable management. *Journal of Environmental Management* 119: 1–12.

Le Lay, Y.F., Piegay, H., and Moulin, B. (2013b). Wood entrance, deposition, transfer, and effects on fluvial forms and processes: problem statements and challenging issues. In: *Treatise on Geomorphology, v. 12, Ecogeomorphology* (eds. J. Shroder, D.R. Butler and C.R. Hupp), 20–36. San Diego, CA: Academic Press.

Lebedeva, M.I., Fletcher, R.C., and Brantley, S.L. (2010). A mathematical model for steady-state regolith production at constant erosion rate. *Earth Surface Processes and Landforms* 35: 508–524.

Lecce, S.A. and Pavlowsky, R.T. (2014). Floodplain storage of sediment contaminated by mercury and copper from historic gold mining at Gold Hill, North Carolina, USA. *Geomorphology* 206: 122–132.

Leclerc, M., Capra, H., Valentin, S. et al. (1996). *Ecohydraulics 2000: Proceedings of the Second IAHR International Conference on Habitat Hydraulics*. Quebec: International Association for Hydro-Environment Engineering and Research.

Leeder, M.R. (1983). On the interactions between turbulent flow, sediment transport and bedform mechanics in channelized flows. In: *Modern and Ancient Fluvial Systems*, vol. 6 (eds. J.D. Collinson and D. Lewin), 5–18. International Association of Sedimentologists Special Publication.

Legleiter, C.J., Phelps, T.L., and Wohl, E.E. (2007). Geostatistical analysis of the effects of stage and roughness on reach-scale spatial patterns of velocity and turbulence intensity. *Geomorphology* 83: 322–345.

Legleiter, C.J., Roberts, D.A., and Lawrence, R.L. (2009). Spectrally based remote sensing of river bathymetry. *Earth Surface Processes and Landforms* 34: 1039–1059.

Legleiter, C.J., Harrison, L.R., and Dunne, T. (2011). Effect of point bar development on the local force balance governing flow in a simple, meandering gravel bed river. *Journal of Geophysical Research: Earth Surface* 116, F01005.

Legleiter, C.J., Mobley, C.D., and Overstreet, B.T. (2017a). A framework for modeling connections between hydraulics, water surface roughness, and surface reflectance in open channel flows. *Journal of Geophysical Research: Earth Surface* 122: 1715–1741.

Legleiter, C.J., Kinzel, P.J., and Nelson, J.M. (2017b). Remote measurement of river discharge using thermal particle image velocimetry (PIV) and various sources of bathymetric information. *Journal of Hydrology* 554: 490–506.

Lehner, B., Reidy Liermann, C., Revenga, C. et al. (2011). High-resolution mapping of the world's reservoirs and dams for sustainable river-flow management. *Frontiers in Ecology and the Environment* 9: 494–502.

Leibowitz, S.G., Wigington, P.J., Rains, M.C., and Downing, D.M. (2008). Non-navigable streams and adjacent wetlands: addressing science needs following the Supreme Court's *Rapanos* decision. *Frontiers in Ecology and the Environment* 6: 364–371.

Leier, A.L., DeCelles, P.G., and Pelletier, J.D. (2005). Mountains, monsoons, and megafans. *Geology* 33: 289–292.

Leithold, E.L., Blair, N.E., and Wegmann, K.W. (2016). Source-to-sink sedimentary systems and global carbon burial: a river runs through it. *Earth-Science Reviews* 153: 30–42.

Leli, I.T., Stevaux, J.C., and Assine, M.L. (2018). Genesis and sedimentary record of blind channel and islands of the anabranching river: an evolution model. *Geomorphology* 302: 35–45.

Lemke, A.M., Herkert, J.R., Walk, J.W., and Blodgett, K.D. (2017). Application of key ecological attributes to assess early restoration of river floodplain habitats: a case study. *Hydrobiologia* 804: 19–33.

Lennox, P.A. and Rasmussen, J.B. (2016). Long-term effects of channelization on a cold-water stream community. *Canadian Journal of Fisheries and Aquatic Sciences* 73: 1530–1537.

Lenzi, M.A. (2001). Step-pool evolution in the Rio Cordon, northeastern Italy. *Earth Surface Processes and Landforms* 26: 991–1008.

Lenzi, M.A. (2002). Stream bed stabilization using boulder check dams that mimic step-pool morphology features in northern Italy. *Geomorphology* 45: 243–260.

Lenzi, M.A. (2004). Displacement and transport of marked pebbles, cobbles and boulders during floods in a steep mountain stream. *Hydrological Processes* 18: 1899–1914.

Lenzi, M.A., Marchi, L., and Scussel, G.R. (1990). Measurement of coarse sediment transport in a small Alpine stream. In: *Hydrology in Mountainous Regions I. Hydrological Measurements: The Water Cycle* (eds. H. Lang and A. Musy), 283–290. Wallingford: IAHS Publications no. 193.

Lenzi, M.A., Mao, L., and Comiti, F. (2004). Magnitude-frequency analysis of bed load data in an alpine boulder bed stream. *Water Resources Research* 40, W07201.

Lenzi, M.A., Mao, L., and Comiti, F. (2006). Effective discharge for sediment transport in a mountain river: computational approaches and geomorphic effectiveness. *Journal of Hydrology* 326: 257–276.

Leopold, L.B. (1976). Reversal of erosion cycle and climatic change. *Quaternary Research* 6: 557–562.

Leopold, L.B. (1994). *A View of the River*. Cambridge, MA: Harvard University Press.

Leopold, L.B. and Bull, W.B. (1979). Base level, aggradation, and grade. *Proceedings of the American Philosophical Society* 123: 168–202.

Leopold, L.B. and Emmett, W.E. (1976). Bedload measurements, East Fork River, Wyoming. *Proceedings of the National Academy of Sciences of the United States of America* 73: 1000–1004.

Leopold, L.B., W.B. Langbein. 1962. The concept of entropy in landscape evolution. US Geological Survey Professional Paper 500-A.

Leopold, L.B., T. Maddock. 1953. The hydraulic geometry of stream channels and some physiographic implications. US Geological Survey Professional Paper 252.

Leopold, L.B., J.P. Miller. 1954. A post-glacial chronology for some alluvial valleys in Wyoming. US Geological Survey Water-Supply Paper 1261.

Leopold, L.B., M.G. Wolman. 1957. River channel patterns – braided, meandering and straight. US Geological Survey Professional Paper 282B.

Leopold, L.B., R.A. Bagnold, M.G. Wolman, L.M. Brush. 1960. Flow resistance in sinuous or irregular channels. US Geological Survey Professional Paper 282-D.

Leopold, L.B., Wolman, M.G., and Miller, J.P. (1964). *Fluvial Processes in Geomorphology*. San Francisco: W.H. Freeman.

Lepori, F., Palm, D., Brännäs, E., and Malmqvist, B. (2005). Does restoration of structural heterogeneity in streams enhance fish and macroinvertebrate diversity? *Ecological Applications* 15 (6): 2060–2071.

Levia, D.F. and Germer, S. (2015). A review of stemflow generation dynamics and stemflow-environment interactions in forests and shrublands. *Reviews of Geophysics* 53: 673–714.

Lewin, J. and Ashworth, P.J. (2014). Defining large river channel patterns: alluvial exchange and plurality. *Geomorphology* 215: 83–98.

Lewin, J. and Ashworth, P.J. (2014). The negative relief of large river floodplains. *Earth-Science Reviews* 129: 1–23.

Lewin, J. and Brewer, P.A. (2001). Predicting channel patterns. *Geomorphology* 40: 329–339.

Lewin, J. and Hughes, D. (1976). Assessing channel change on Welsh rivers. *Cambria* 3: 1–10.

Lewis, S.G. and Birnie, F. (2001). Little Ice Age alluvial fan development in Langedalen, western Norway. *Geografiska Annaler* 83A: 179–190.

Lewis, G.W. and Lewin, J. (1983). Alluvial cutoffs in Wales and the Borderlands. In: *Modern and Ancient Fluvial Systems*, vol. 6 (eds. J.D. Collinson and J. Lewin), 145–154. Special Publications International Association of Sedimentologists.

Lewis, Q.W. and Rhoads, B.L. (2018). LSPIV measurements of two-dimensional flow structure in streams using small unmanned aerial systems: 2. Hydrodynamic mapping at river confluences. *Water Resources Research* 54: 7981–7999.

Leyland, J., Hackney, C.R., Darby, S.E. et al. (2017). Extreme flood-driven fluvial bank erosion and sediment loads: direct process measurements using integrated Mobile Laser Scanning (MLS) and hydro-acoustic techniques. *Earth Surface Processes and Landforms* 42: 334–346.

Li, Q., Pan, B., Gao, H. et al. (2019). Differential rock uplift along the northeastern margin of the Tibetan Plateau inferred from bedrock channel longitudinal profiles. *Journal of Asian Earth Sciences* 169: 182–198.

Lian, O.B. (2007). Optically-stimulated luminescence. In: *Encyclopedia of Quaternary Science* (ed. S.A. Elias), 1491–1505. Amsterdam: Elsevier.

Liang, M., Geleynse, N., Edmonds, D.A., and Passalacqua, P. (2015). A reduced-complexity model for river delta formation – Part 2: assessment of the flow routing scheme. *Earth Surface Dynamics* 3: 87–104.

Liébault, F., Clément, P., Piégay, H. et al. (2002). Contemporary channel changes in the Eygues basin, southern French Prealps: the relationship of subbasin variability to watershed characteristics. *Geomorphology* 45: 53–66.

Liebe, J.R., van de Giesen, N., Andreini, M. et al. (2009). Determining watershed response in data poor environments with remotely sensed small reservoirs as runoff gauges. *Water Resources Research* 45, W07410.

Lienkaemper, G.W. and Swanson, F.J. (1987). Dynamics of large woody debris in streams in old-growth Douglas-fir forests. *Canadian Journal of Forest Research* 17: 150–156.

Ligon, F.K., Dietrich, W.E., and Trush, W.J. (1995). Downstream ecological effects of dams. *BioScience* 45: 183–192.

Limaye, A.B.S. and Lamb, M.P. (2016). Numerical model predictions of autogenic fluvial terraces and comparison to climate change expectations. *Journal of Geophysical Research Earth Surface* 121: 512–544.

Limerinos, J.T. 1970. Determination of the Manning coefficient from measured bed roughness in natural channels. US Geological Survey Water Supply Paper 1898-B.

Lind, P.R., Robson, B.J., and Mitchell, B.D. (2007). Multiple lines of evidence for the beneficial effects of environmental flows in two lowland rivers in Victoria, Australia. *River Research and Applications* 23: 933–946.

Lininger, K.B. and Latrubesse, E.M. (2016). Flooding hydrology and peak discharge attenuation along the middle Araguaia River in central Brazil. *Catena* 143: 90–101.

Lininger, K.B., Wohl, E., Sutfin, N.A., and Rose, J.R. (2017). Floodplain downed wood volumes: a comparison across three biomes. *Earth Surface Processes and Landforms* 42: 1248–1261.

Link, T.E., Flerchinger, G.N., Unsworth, M., and Marks, D. (2005). Water relations in an old-growth Douglas fir stand. In: *Climate and Hydrology in Mountainous Areas* (eds. C. de Jong, D. Collins and R. Ranzi), 147–159. Chichester: Wiley.

Lipar, M. and Ferk, M. (2015). Karst pocket valleys and their implications on Pliocene–Quaternary hydrology and climate: examples from the Nullarbor Plain, southern Australia. *Earth-Science Reviews* 150: 1–13.

Lisle, T.E. (1982). Effects of aggradation and degradation on riffle-pool morphology in natural gravel channels, northwestern California. *Water Resources Research* 18: 1643–1651.

Lisle, T.E. (1986). Stabilization of a gravel channel by large streamside obstructions and bedrock bends, Jacoby Creek, northwestern California. *Geological Society of America Bulletin* 97: 999–1011.

Lisle, T.E. (2008). The evolution of sediment waves influenced by varying transport capacity in heterogeneous rivers. In: *Gravel-bed Rivers VI: From Process Understanding to River Restoration* (eds. H. Habersack, H. Piégay and M. Rinaldi), 443–472. Amsterdam: Elsevier.

Lisle, T.E., R.E. Eads. 1991. Methods to measure sedimentation of spawning gravels. USDA Forest Service Research Note PSW-411, Berkeley, CA.

Lisle, T.E. and Hilton, S. (1992). The volume of fine sediment in pools: an index of sediment supply in gravel-bed streams. *Water Resources Bulletin* 28: 371–383.

Lisle, T.E., Ikeda, H., and Iseya, F. (1991). Formation of stationary alternate bars in a steep channel with mixed-size sediment: a flume experiment. *Earth Surface Processes and Landforms* 16: 463–469.

Lisle, T.E., Pizzuto, J.E., Ikeda, H. et al. (1997). Evolution of a sediment wave in an experimental channel. *Water Resources Research* 33: 1971–1981.

Lisle, T.E., Cui, Y., Parker, G. et al. (2001). The dominance of dispersion in the evolution of bed material waves in gravelbed rivers. *Earth Surface Processes and Landforms* 26: 1409–1420.

Little, P.J., Richardson, J.S., and Alila, Y. (2013). Channel and landscape dynamics in the alluvial forest mosaic of the Carmanah River valley, British Columbia, Canada. *Geomorphology* 202: 86–100.

Liu, J., Saito, Y., Kong, X. et al. (2010). Sedimentary record of environmental evolution off the Yangtze River estuary, East China Sea, during the last ~13 000 years, with special reference to the influence of the Yellow River on the Yangtze River delta during the last 600 years. *Quaternary Science Reviews* 29: 2424–2438.

Livers, B. and Wohl, E. (2015). An evaluation of stream characteristics in glacial versus fluvial process domains in the Colorado Front Range. *Geomorphology* 231: 72–82.

Livers, B. and Wohl, E. (2016). Sources and interpretation of channel complexity in forested subalpine streams of the Southern Rocky Mountains. *Water Resources Research* 52: 3910–3929.

Livers, B., Wohl, E., Jackson, K.J., and Sutfin, N.A. (2018). Historical land use as a driver of alternative states for stream form and function in forested mountain watersheds of the Southern Rocky Mountains. *Earth Surface Processes and Landforms* 43: 669–684.

Lombardo, U. (2017). River logjams cause frequent large-scale forest die-off events in southwestern Amazonia. *Earth Surface Dynamics* 8: 565–575.

Longfield, S.A., Faulkner, D., Kjeldsen, T.R. et al. (2018). Incorporating sedimentological data in UK flood frequency estimation. *Journal of Flood Risk Management* 12, e12449.

López-Moreno, J.I., Goyette, S., and Beniston, M. (2009). Impact of climate change on snowpack in the Pyrenees: horizontal spatial variability and vertical gradients. *Journal of Hydrology* 374: 384–396.

López-Vicente, M., Nadal-Romero, E., and Cammeraat, E.L.H. (2017). Hydrological connectivity does change over 70 years of abandonment and afforestation in the Spanish Pyrenees. *Land Degradation and Development* 28: 1298–1310.

Lorang, M.S. and Aggett, G. (2005). Potential sedimentation impacts related to dam removal: Icicle Creek, Washington, USA. *Geomorphology* 71: 182–201.

Lowrance, R., Todd, R., Fail, J. et al. (1984). Riparian forests as nutrient filters in agricultural watersheds. *BioScience* 34: 374–377.

Loye, A., Pedrazzini, A., Theule, J.I. et al. (2012). Influence of bedrock structures on the spatial pattern of erosional landforms in small alpine catchments. *Earth Surface Processes and Landforms* 37: 1407–1423.

Lu, H., Moran, C.J., and Sivapalan, M. (2005). A theoretical exploration of catchment-scale sediment delivery. *Water Resources Research* 41, W09415.

Lu, X.X., Zhang, S., and Xu, J. (2010). Climate change and sediment flux from the Roof of the World. *Earth Surface Processes and Landforms* 35: 732–735.

Lucía, A., Comiti, F., Borga, M. et al. (2015). Dynamics of large wood during a flash flood in two mountain catchments. *Natural Hazards and Earth System Sciences* 15: 1741–1755.

Lucía, A., Schwientek, M., Eberle, J., and Zarfl, C. (2018). Planform changes and large wood dynamics in two torrents during a severe flash flood in Braunsbach, Germany 2016. *Science of the Total Environment* 640–641: 315–326.

Lugt, H.J. (1983). *Vortex Flow in Nature and Technology*. New York: Wiley.

Luhar, M. and Nepf, H.M. (2013). From the blade scale to the reach scale: a characterization of aquatic vegetative drag. *Advances in Water Resources* 51: 305–316.

Luke, A., Vrugt, J.A., AghaKouchak, A. et al. (2017). Predicting nonstationary flood frequencies: evidence supports an updated stationarity thesis in the United States. *Water Resources Research* 53: 5469–5494.

Lundquist, J.D., Dettinger, M.D., and Cayan, D.R. (2005). Snow-fed streamflow timing at different basin scales: case study of the Tuolumne River above Hetch Hetchy, Yosemite, California. *Water Resources Research* 41, W07005.

Lyons, W.B., Nezat, C.A., Carey, A.E., and Hicks, D.M. (2002). Organic carbon fluxes to the ocean from high-standing islands. *Geology* 30: 443–446.

Lyons, R., Tooth, S., and Duller, G.A.T. (2015). Late Quaternary climatic changes revealed by luminescence dating, mineral magnetism and diffuse reflectance spectroscopy of river terrace palaeosols: a new form of geoproxy data for the southern African interior. *Quaternary Science Reviews* 95: 43–59.

MacCarthy, P., Bloom, P.R., Clapp, C.E., and Malcom, R.L. (eds.) (1990). *Humic Substances in Soil and Crop Sciences: Selected Readings*. Madison, WI: Soil Science Society of America.

MacFarlane, W.A. and Wohl, E. (2003). Influence of step composition on step geometry and flow resistance in step-pool streams of the Washington Cascades. *Water Resources Research* 39: 1037.

Machete, M.N. (1985). Calcic soils of the southwestern United States. In: *Soils and Quaternary Geology of the Southwestern United States* (ed. D.L. Weide), 1–21. Boulder, CO: Geological Society of America Special Paper 203.

Mackay, S.J., Arthington, A.H., and James, C.S. (2014). Classificaiton and comparison of natural and altered flow regimes to support an Australian trial of the Ecological Limits of Hydrologic Alteration framework. *Ecohydrology* 7: 1485–1507.

Mackay, J.E., Cunningham, S.C., and Cavagnaro, T.R. (2016). Riparian reforestation: are there changes in soil carbon and soil microbial communities? *Science of the Total Environment* 566–567: 960–967.

Mackin, J.H. (1937). Erosional history of the Big Horn Basin, Wyoming. *Geological Society of America Bulletin* 48: 813–893.

Mackin, J.H. (1948). Concept of the graded river. *Geological Society of America Bulletin* 59: 463–512.

Macklin, M.G. and Lewin, J. (2008). Alluvial responses to the changing Earth system. *Earth Surface Processes and Landforms* 33: 1374–1395.

Macklin, M.G., Hudson-Edwards, K.A., and Dawson, E.J. (1997). The significance of pollution from historic metal mining in the Pennine orefields on river sediment contaminant fluxes to the North Sea. *The Science of the Total Environment* 194/195: 391–397.

Macklin, M.G., Brewer, P.A., Hudson-Edwards, K.A. et al. (2006). A geomorphological approach to the management of rivers contaminated by metal mining. *Geomorphology* 79: 423–447.

Macura, V., Stefunkova, Z.S., Majorosova, M. et al. (2017). Influence of discharge on fish habitat suitability curves in mountain watercourses in IFIM methodology. *Journal of Hydrology and Hydromechanics* 66: 12–22.

MacVicar, B. and Piégay, H. (2012). Implementation and validation of video monitoring for wood budgeting in a wandering piedmont river, the Ain River (France). *Earth Surface Processes and Landforms* 37: 1272–1289.

MacVicar, B.J., Rennie, C.D., and Roy, A.G. (2010). Discussion of "Unifying criterion for the velocity reversal hypothesis in gravel-bed rivers" by D. Caamaño, P. Goodwin, J.M. Buffington, J.C.P. Liou, and S. Daley-Laursen. *Journal of Hydraulic Engineering* 136: 550–552.

MacWilliams, M.L., Wheaton, J.M., Pasternack, G.B. et al. (2006). Flow convergence routing hypothesis for pool-riffle maintenance in alluvial rivers. *Water Resources Research* 42, W10427.

Maddock, I. (1999). The importance of physical habitat assessment for evaluating river health. *Freshwater Biology* 41: 373–391.

Maddy, D., Demir, T., Bridgland, D.R. et al. (2005). An obliquity-controlled early Pleistocene river terrace record from western Turkey? *Quaternary Research* 63: 339–346.

Madej, M.A. (2001). Development of channel organization and roughness following sediment pulses in single-thread, gravel bed rivers. *Water Resources Research* 37: 2259–2272.

Madej, M.A. and Ozaki, V. (2009). Channel response to sediment wave propagation and movement, Redwood Creek, California, USA. *Earth Surface Processes and Landforms* 21: 911–927.

Magilligan, F.J. (1992). Thresholds and the spatial variability of flood power during extreme floods. *Geomorphology* 5: 373–390.

Magilligan, F.J. and McDowell, P.F. (1997). Stream channel adjustments following elimination of cattle grazing. *Journal of the American Water Resources Association* 33: 867–878.

Magilligan, F.J. and Nislow, K.H. (2001). Long-term changes in regional hydrologic regime following impoundment in a humid-climate watershed. *Journal of the American Water Resources Association* 37: 1551–1569.

Magilligan, F.J., Nislow, K.H., and Renshaw, C.E. (2013). Flow regulation by dams. In: *Treatise on Fluvial Geomorphology* (ed. E. Wohl), 794–807. Amsterdam: Elsevier.

Magilligan, F.J., Graber, B.E., Nislow, K.H. et al. (2016). River restoration by dam removal: enhancing connectivity at watershed scales. *Elementa: Science of the Anthropocene* 4 https://doi.org/10.12952/journal.elementa.000108.

Magirl, C.S., Hilldale, R.C., Curran, C.A. et al. (2015). Large-scale dam removal on the Elwha River, Washington, USA: fluvial sediment load. *Geomorphology* 246: 669–686.

Magner, J.A., Vondracek, B., and Brooks, K.N. (2008). Grazed riparian management and stream channel response in southeastern Minnesota (USA) streams. *Environmental Management* 42: 377–390.

Maitre, V., Cosandey, A.-C., Desagher, E., and Parriaux, A. (2003). Effectiveness of groundwater nitrate removal in a river riparian area; the importance of hydrogeological conditions. *Journal of Hydrology* 278: 76–93.

Major, J.J., O'Connor, J.E., Grant, G.E. et al. (2008). Initial fluvial response to the removal of Oregon's Marmot Dam. *Eos, Transactions of the American Geophysical Union* 89: 241–242.

Makaske, B. (2001). Anastomosing rivers: a review of their classification, origin and sedimentary products. *Earth-Science Reviews* 53: 149–196.

Malatesta, L.C., Prancevic, J.P., and Avouac, J.P. (2017). Autogenic entrenchment patterns and terraces due to coupling with lateral erosion in incising alluvial channels. *Journal of Geophysical Research: Earth Surface* 122: 335–355.

Malik, I. (2006). Contribution to understanding the historical evolution of meandering rivers using dendrochronological methods: example of the Mala Panew River in southern Poland. *Earth Surface Processes and Landforms* 31: 1227–1245.

Malik, I. and Matyja, M. (2008). Bank erosion history of a mountain stream determined by means of anatomical changes in exposed tree roots over the last 100 years (Bilá Opava River – Czech Republic). *Geomorphology* 98: 126–142.

Mallen-Cooper, M. and Zampatti, B.P. (2018). History, hydrology and hydraulics: rethinking the ecological management of large rivers. *Ecohydrology* 11, e1965.

Manners, R.B., Doyle, M.W., and Small, M.J. (2007). Structure and hydraulics of natural woody debris jams. *Water Resources Research* 43, W06432.

Mao, L. and Lenzi, M.A. (2007). Sediment mobility and bedload transport conditions in an alpine stream. *Hydrological Processes* 21: 1882–1891.

Mao, L., Andreoli, A., Comiti, F., and Lenzi, M.A. (2008). Geomorphic effects of large wood jams on a sub-Antarctic mountain stream. *River Research and Applications* 24: 249–266.

Mao, L., Dell'Agnese, A., Huincache, C. et al. (2014). Bedload hysteresis in a glacier-fed mountain river. *Earth Surface Processes and Landforms* 39: 964–976.

Maraio, S., Bruno, P.P.G., Picotti, V. et al. (2018). High-resolution seismic imaging of debris-flow fans, alluvial valley fills and hosting bedrock geometry in Vinschgau/Val Venosta, eastern Italian Alps. *Journal of Applied Geophysics* 157: 61–72.

Marburg, A.E., Turner, M.G., and Kratz, T.K. (2006). Natural and anthropogenic variation in coarse wood among and within lakes. *Journal of Ecology* 94: 558–568.

Marchenko, S.S., Gorbunov, A.P., and Romanovsky, V.E. (2007). Permafrost warming in the Tien Shan Mountains, central Asia. *Global and Planetary Change* 56: 311–327.

Marcus, W.A., Ladd, S.C., Stoughton, J.A., and Stock, J.W. (1995). Pebble counts and the role of user-dependent bias in documenting sediment size distributions. *Water Resources Research* 31: 2625–2631.

Marcus, W.A., Meyer, G.A., and Nimmo, D.R. (2001). Geomorphic control of persistent mine impacts in a Yellowstone Park stream and implications for the recovery of fluvial systems. *Geology* 29: 355–358.

Marcus, W.A., Marston, R.A., Colvard, C.R., and Gray, R.D. (2002). Mapping the spatial and temporal distributions of woody debris in streams of the Greater Yellowstone ecosystem, USA. *Geomorphology* 44: 323–335.

Marinucci, M.R., Giorgi, F., Beniston, M. et al. (1995). High resolution simulations of January and July climate over the western Alpine region with a nested regional modeling system. *Theoretical and Applied Climatology* 51: 119–138.

Marion, A., Bellinello, M., Guymer, I., and Packman, A. (2002). Effect of bed form geometry on the penetration of nonreactive solutes into a streambed. *Water Resources Research* 38: 1209.

Marion, A., Packman, A.I., Zaramella, M., and Bottacin-Busolin, A. (2008a). Hyporheic flows in stratified beds. *Water Resources Research* 44: W09433.

Marion, A., Zaramella, M., and Bottacin-Busolin, A. (2008b). Solute transport in rivers with multiple storage zones: the STIR model. *Water Resources Research* 44: W10406.

Markham, A.J. and Thorne, C.R. (1992). Geomorphology of gravel-bed river bends. In: *Dynamics of Gravel-Bed Rivers* (eds. P. Billi, R.D. Hey, C.R. Thorne and P. Tacconi), 433–450. Chichester: Wiley.

Marks, K. and Bates, P. (2000). Integration of high-resolution topographic data with floodplain flow models. *Hydrological Processes* 14: 2109–2122.

Marks, J.C., Parnell, R., Carter, C. et al. (2006). Interactions between geomorphology and ecosystem processes in travertine streams: implications for decommissioning a dam on Fossil Creek, Arizona. *Geomorphology* 77: 299–307.

Marquis, G.A. and Roy, A.G. (2012). Using multiple bed load measurements: toward the identification of bed dilation and contraction in gravel-bed rivers. *Journal of Geophysical Research* 117, F01014.

Marrucci, M., Zeilinger, G., Ribolini, A., and Schwanghart, W. (2018). Origin of knickpoints in an alpine context subject to different perturbing factors, Stura Valley, Maritime Alps (north-western Italy). *Geosciences* 8: 443.

Marshall, J.A. and Roering, J.J. (2014). Diagenetic variation in the Oregon Coast Range: implications for rock strength, soil production, hillslope form, and landscape evolution. *Journal of Geophysical Research Earth Surface* 119: 1395–1417.

Marston, R.A., Mills, J.D., Wrazien, D.R. et al. (2005). Effects of Jackson Lake Dam on the Snake River and its floodplain, Grand Teton National Park, Wyoming, USA. *Geomorphology* 71: 79–98.

Martin, R.L. and Jerolmack, D.J. (2013). Origin of hysteresis in bed form response to unsteady flows. *Water Resources Research* 49: 1314–1333.

Martin, D.A. and Moody, J.A. (2001). Comparison of soil infiltration rates in burned and unburned mountainous watersheds. *Hydrological Processes* 15: 2893–2903.

Martin, R.L., Jerolmack, D.J., and Schumer, R. (2012). The physical basis for anomalous diffusion in bed load transport. *Journal of Geophysical Research: Earth Surface* 117, F01018.

Martinez-Carreras, N., Soler, M., Hernandez, E., and Gallart, F. (2007). Simulating badland erosion with KINEROS2 in a small Mediterranean mountain basin (Vallcebre, eastern Pyrenees). *Catena* 71: 145–154.

Martins, A.A., Cabral, J., Cunha, P.P. et al. (2017). Tectonic and lithological controls on fluvial landscape development in central-eastern Portugal: insights from long profile tributary stream analyses. *Geomorphology* 276: 144–163.

Martín-Vide, J.P., Amarilla, M., and Zarate, F.J. (2014). Collapse of the Pilcomayo River. *Geomorphology* 205: 155–163.

Martyr-Koller, C., Kemkamp, H.W.J., van Dam, A. et al. (2017). Application of an unstructured 3D finite volume numerical model to flows and salinity dynamics in the San Francisco Bay-Delta. *Estuarine, Coastal and Shelf Science* 192: 86–107.

Maser, C. and Sedell, J.R. (1994). *From the Forest to the Sea: The Ecology of Wood in Streams, Rivers, Estuaries and Oceans*. Delray Beach, FL: St. Lucie Press.

Massong, T.M. and Montgomery, D.R. (2000). Influence of sediment supply, lithology, and wood debris on the distribution of bedrock and alluvial channels. *Geological Society of America Bulletin* 112: 591–599.

Mast, M.A., Kendall, C., Campbell, D.H. et al. (1995). Determination of hydrologic pathways in an alpine-subalpine basin using isotopic and chemical tracers, Loch Vale Watershed, Colorado, USA. In: *Biogeochemistry of Seasonally Snow-Covered Catchments* (eds. K.A. Tonnessen, M.W. Williams and M. Tranter), 263–270. Wallingford: IAHS Publications no. 228.

Matheson, A., Thoms, M., and Reid, M. (2017). Does reintroducing large wood influence the hydraulic landscape of a lowland river system? *Geomorphology* 292: 128–141.

Mattikalli, N.M. and Engman, E.T. (1997). Microwave remote sensing and GIS for monitoring surface soil moisture and estimation of soil properties. In: *Remote Sensing and Geographic Information Systems for Design and Operation of Water Resource Systems* (eds. M.F. Baumgartner, G.A. Schultz and A.I. Johnson), 229–236. Wallingford: IAHS Publications no. 242.

May, R.M. (1977). Thresholds and breakpoints in ecosystems with a multiplicity of stable states. *Nature* 269: 471–477.

May, C.L. and Gresswell, R.E. (2003). Large wood recruitment and redistribution in headwater streams in the southern Oregon Coast Range, USA. *Canadian Journal of Forest Research* 33: 1352–1362.

May, C.L. and Gresswell, R.E. (2004). Spatial and temporal patterns of debris-flow deposition in the Oregon Coast Range, USA. *Geomorphology* 57: 135–149.

May, C., Roering, J., Snow, K. et al. (2017). The waterfall paradox: how knickpoints disconnect hillslope and channel processes, isolating salmonid populations in ideal habitats. *Geomorphology* 277: 228–236.

Mazzorana, B., Hubl, J., Zischg, A., and Largiader, A. (2011). Modelling woody material transport and deposition in alpine rivers. *Natural Hazards* 56: 425–449.

McArdell, B.W. and Faeh, R. (2001). A computational investigation of river braiding. In: *Gravel-Bed Rivers V* (ed. M.P. Mosley), 73–86. Wellington: New Zealand Hydrological Society.

McCabe, G.J., Clark, M.P., and Hay, L.E. (2007). Rain-on-snow events in the western United States. *Bulletin of the American Meteorological Society* 88: 319–328.

McCarthy, J.M. 2008. Factors influencing ecological recovery downstream of diversion dams in southern Rocky Mountain streams. Unpublished MS thesis, Colorado State University, Ft. Collins, CO.

McClain, M.E., G. Pinay, R.M. Holmes. 1999. Contrasting biogeochemical cycles of riparian forests in temperate, wet tropical, and arid regions. 1999 Annual Meeting Abstracts, Ecological Society of America.

McClain, M.E., Boyer, E.W., Dent, C.L. et al. (2003). Biogeochemical hot spots and hot moments at the interface of terrestrial and aquatic ecosystems. *Ecosystems* 6: 301–312.

McCluskey, A.H. and Grant, S.B. (2016). Flipping the thin film model: mass transfer by hyporheic exchange in gaining and losing streams. *Water Resources Research* 52: 7806–7818.

McColl, K.A., Alemohammad, S.H., Akbar, R. et al. (2017). The global distribution and dynamics of surface soil moisture. *Nature Geoscience* 10: 100–104.

McCord, V.A. (1996). Fluvial process dendrogeomorphology: reconstructions of flood events from the southwestern United States using flood-scarred trees. In: *Tree Rings, Environment, and Humanity* (eds. J.S. Dean, D.M. Meko and T.W. Swetnam), 689–699. Tucson, AZ: University of Arizona Press.

McCormick, B.C., Eshleman, K.N., Griffith, J.L., and Townsend, P.A. (2009). Detection of flooding responses at the river basin scale enhanced by land use change. *Water Resources Research* 45, W08401.

McDonald, E.V., McFadden, L.D., and Wells, S.G. (2003). Regional response of alluvial fans to the Pleistocene–Holocene climatic transition, Mojave Desert, California. In: *Paleoenvironments and Paleohydrology of the Mojave and Southern Great Basin Deserts* (eds. Y. Enzel, S.G. Wells and N. Lancaster), 189–205. Boulder, CO: Geological Society of America Special Paper 368.

McDonald, A., Lane, S.N., Haycock, N.E., and Chalk, E.A. (2004). Rivers of dreams: on the gulf between theoretical and practical aspects of an upland river restoration. *Transactions. Institute of British Geographers* 29: 257–281.

McDonnell, J.J. (2003). Where does water go when it rains? Moving beyond the variable source area concept of rainfall-runoff response. *Hydrological Processes* 17: 1869–1875.

McDonnell, J.J. (2014). The two water worlds hypothesis: ecological separation of water between streams and trees? *WIREs Water* 1: 323–329.

McDonnell, J.J., Stewart, M.K., and Owens, I.F. (1991). Effect of catchment-scale subsurface mixing on stream isotopic response. *Water Resources Research* 27: 3065–3073.

McDonnell, J.J., McGlynn, B., Kendall, K. et al. (1998). The role of near-stream riparian zones in the hydrology of steep upland catchments. In: *Hydrology, Water Resources and Ecology in Headwaters* (eds. K. Kovar, U. Tappeiner, N.E. Peters and R.G. Craig), 173–180. Wallingford: IAHS Publications no. 248.

McDonnell, J.J., McGlynn, B., Vache, K., and Tromp-Van Meerveld, I. (2005). A perspective on hillslope hydrology in the context of PUB. In: *Predictions in Ungauged Basins: International Perspectives on the State of the Art and Pathways Forward* (eds. S. Franks, M. Sivapalan, K. Takeuchi and Y. Tachikawa), 204–212. Wallingford: IAHS Publications no. 301.

McDonnell, J.J., Sivapalan, M., Vaché, K. et al. (2007). Moving beyond heterogeneity and process complexity: a new vision for watershed hydrology. *Water Resources Research* 43, W07031.

McElroy, B., Willenbring, J., and Mohrig, D. (2017). Addressing time-scale-dependent erosion rates from measurement methods with censorship. *Geological Society of America Bulletin* 130: 381–395.

McEwan, I., Sørensen, M., Heald, J. et al. (2004). Probabilistic modeling of bed-load composition. *Journal of Hydraulic Engineering* 130 (2): 129–139.

McEwen, L.J., Matthews, J.A., Shakesby, R.A., and Berrisford, M.S. (2002). Holocene gorge excavation linked to boulder fan formation and frost weathering in a Norwegian alpine periglaciofluvial system. *Arctic, Antarctic, and Alpine Research* 34: 345–357.

McFadden, L.D., Ritter, J.B., and Wells, S.G. (1989). Use of multiparameter relative-age methods for age estimations and correlation of alluvial fan surfaces on a desert piedmont, eastern Mojave Desert, California. *Quaternary Research* 32: 276–290.

McGinness, H.M., Thoms, M.C., and Southwell, M.R. (2002). Connectivity and fragmentation of flood plain-river exchanges in a semiarid, anabranching river system. In: *The Structure, Function, and Management Implications of Fluvial Sedimentary Systems* (eds. F.J. Dyer, M.C. Thoms and J.M. Olley), 19–26. Wallingford: IAHS Publications no. 276.

McGlynn, B. and McDonnell, J.J. (2003). The role of discrete landscape units in controlling catchment dissolved organic carbon dynamics. *Water Resources Research* 39: 3-1–3-18.

McGlynn, B.L., McDonnell, J.J., Seibert, J., and Kendall, C. (2004). Scale effects on headwater catchment runoff timing, flow sources, and groundwater-streamflow relations. *Water Resources Research* 40, W07504.

McInerney, P.J., Rees, G.N., Gawne, B. et al. (2016). Invasive willows drive instream community structure. *Freshwater Biology* 61: 1379–1391.

McInerney, P.J., Stoffels, R.J., Shackleton, M.E., and Davey, C.D. (2017). Flooding drives a macroinvertebrate biomass boom in ephemeral floodplain wetlands. *Freshwater Science* 36: 726–738.

McKean, J.A., Dietrich, W.E., Finkel, R.C. et al. (1993). Quantification of soil production and downslope creep rates from cosmogenic ^{10}Be accumulations on a hillslope profile. *Geology* 21: 343–346.

McLean, S.R., Nelson, J.M., and Wolfe, S.R. (1994). Turbulence structure over two-dimensional bed forms: implications for sediment transport. *Journal of Geophysical Research* 99: 12729–12747.

McManus, J. (2002). Deltaic responses to changes in river regimes. *Marine Chemistry* 79: 155–170.

McMillan, S.K., Tuttle, A.K., Jennings, G.D., and Gardner, A. (2014). Influence of restoration age and riparian vegetation on reach-scale nutrient retention in restored urban streams. *Journal of the American Water Resources Association* 50: 626–638.

Meade, R.H., ed. 1996. Contaminants in the Mississippi River, 1987–92. US Geological Survey Circular 1133.

Meade, R.H. (2007). Transcontinental moving and storage: the Orinoco and Amazon Rivers transfer the Andes to the Atlantic. In: *Large Rivers: Geomorphology and Management* (ed. A. Gupta), 45–63. Chichester: Wiley.

Meade, R.H., Dunne, T., Richey, J.E. et al. (1985). Storage and remobilization of suspended sediment in the lower Amazon River of Brazil. *Science* 228: 488–490.

Meade, R.H., Yuzyk, T.R., and Day, T.J. (1990). Movement and storage of sediment in rivers of the United States and Canada. In: *Surface Water Hydrology* (eds. M.G. Wolman and H.C. Riggs), 255–280. Boulder, CO: Geological Society of America.

Mei-e, R. and Xianmo, Z. (1994). Anthropogenic influences on changes in the sediment load of the Yellow River, China, during the Holocene. *The Holocene* 4: 314–320.

Meixner, T., Bales, R.C., Williams, M.W. et al. (2000). Stream chemistry modeling of two watersheds in the Front Range, Colorado. *Water Resources Research* 36: 77–87.

Meko, D.M., Friedman, J.M., Touchan, R. et al. (2015). Alternative standardization approaches to improving streamflow reconstructions with ring-width indices of riparian trees. *The Holocene* 25: 1093–1101.

Melcher, N.B., Costa, J.E., Haeni, F.P. et al. (2002). River discharge measurements by using helicopter-mounted radar. *Geophysical Research Letters* 29: 2084.

Melis, T.S., ed. 2011. Effects of three high-flow experiments on the Colorado River ecosystem downstream from Glen Canyon Dam, Arizona. US Geological Survey Circular 1366.

Melis, T.S., Korman, J., and Kennedy, T.A. (2012). Abiotic and biotic responses of the Colorado River to controlled floods at Glen Canyon Dam, Arizona, USA. *River Research and Applications* 28: 764–776.

Melis, T.S., Walters, C.J., and Korman, J. (2015). Surprise and opportunity for learning in Grand Canyon: the Glen Canyon Dam adaptive management program. *Ecology and Society* 20: 22.

Melis, T.S., Pine, W.E., Korman, J. et al. (2016). Using large-scale flow experiments to rehabilitate Colorado River ecosystem function in Grand Canyon: basis for an adaptive climate-resilient strategy. In: *Water Policy and Planning in a Variable and Changing Climate* (eds. K.A. Miller, A.F. Hamlet, D.S. Kenney and K.T. Redmond), 315–345. Boca Raton, FL: CRC Press.

Merrill, L. and Tonjes, D.J. (2014). A review of the hyporheic zone, stream restoration, and means to enhance denitrification. *Critical Reviews in Environmental Science and Technology* 44: 2337–2379.

Merritt, D.M. (2013). Reciprocal relations between riparian vegetation, fluvial landforms, and channel processes. In: *Treatise on Fluvial Geomorphology* (ed. E. Wohl), 220–243. Amsterdam: Elsevier.

Merritt, D.M. and Wohl, E.E. (2003). Downstream hydraulic geometry and channel adjustment during a flood along an ephemeral, arid-region drainage. *Geomorphology* 52: 165–180.

Merritt, D.M. and Wohl, E.E. (2006). Plant dispersal along rivers fragmented by dams. *River Research and Applications* 22: 1–26.

Merritts, D.J. and Vincent, K.R. (1989). Geomorphic response of coastal streams to low, intermediate, and high rates of uplift, Mendocino triple junction region, northern California. *Geological Society of America Bulletin* 100: 1373–1388.

Merritts, D.J., Vincent, K.R., and Wohl, E.E. (1994). Long river profiles, tectonism, and eustasy: a guide to interpreting fluvial terraces. *Journal of Geophysical Research* 99 (B7): 14031–14050.

Merritts, D., Walter, R., Rahnis, M. et al. (2012). Anthropocene streams and base-level controls from historic dams in the unglaciated mid-Atlantic region, USA. *Philosophical Transactions of the Royal Society A* 369: 976–1009.

Merten, E., Finlay, J., Johnson, L. et al. (2010). Factors influencing wood mobilization in streams. *Water Resources Research* 46, W10514.

Merten, E.C., Vaz, P.G., Decker-Fritz, J.A. et al. (2013). Relative importance of breakage and decay as processes depleting large wood from streams. *Geomorphology* 190: 40–47.

Mertes, L.A.K. (1997). Documentation and significance of the perirheic zone on inundated floodplains. *Water Resources Research* 33: 1749–1762.

Mertes, L.A.K. (2000). Inundation hydrology. In: *Inland Flood Hazards: Human, Riparian, and Aquatic Communities* (ed. E. Wohl), 145–166. Cambridge: Cambridge University Press.

Mertes, L.A.K. and Dunne, T. (2007). The effects of tectonics, climatic history, and sea-level history on the form and behavior of the modern Amazon River. In: *Large Rivers* (ed. A. Gupta), 115–144. Chichester: Wiley.

Mertes, L.A.K., Dunne, T., and Martinelli, L.A. (1996). Channel-floodplain geomorphology along the Solimões-Amazon River, Brazil. *Geological Society of America Bulletin* 108: 1089–1107.

Merz, J.E., Pasternack, G.B., and Wheaton, J.M. (2006). Sediment budget for a salmonid spawning habitat rehabilitation in a regulated river. *Geomorphology* 76: 207–228.

Meyer, G.A. and Pierce, J.L. (2003). Climatic controls on fire-induced sediment pulses in Yellowstone National Park and central Idaho: a long-term perspective. *Forest Ecology and Management* 178: 89–104.

Meyer, G.A. and Wells, S.G. (1997). Fire-related sedimentation events on alluvial fans, Yellowstone National Park, USA. *Journal of Sedimentary Research* 67: 776–791.

Meyer-Peter, E. and Mueller, R. (1948). Formulas for bedload transport. In: *International Association for Hydraulic Research Proceedings*, 2nd Congress, 39–65. Stockholm.

Miall, A.D. (1977). A review of the braided-river depositional environment. *Earth-Science Reviews* 13: 1–62.

Milan, D.J. (2012). Geomorphic impact and system recovery following an extreme flood in an upland stream: Thinhope Burn, northern England, UK. *Geomorphology* 138: 319–328.

Milan, D.J. and Large, A.R.G. (2014). Magnetic tracing of fine-sediment over pool-riffle morphology. *Catena* 115: 134–149.

Milan, D.J., Heritage, G.L., and Large, A.R.G. (2002). Tracer particle entrainment and deposition loci; influence of flow character and implications for riffle-pool maintenance. In: *Sediment Flux to Basins; Causes, Controls and Consequences*, vol. 191 (eds. S.J. Jones and L.E. Frostick), 133–148. London: Geological Society Special Publications vol.

Milan, D., Heritage, G., Entwistle, N., and Tooth, S. (2018). Morphodynamic simulation of sediment deposition patterns on a recently stripped bedrock anastomosed channel. *Proceedings, International Association of Hydrological Sciences* 377: 51–56.

Milhous, R.T. 1973. Sediment transport in a gravel-bottom stream. PhD dissertation, Oregon State University, Corvallis, OR.

Millar, R.G. and Quick, M.C. (1994). Flow resistance of high-gradient gravel channels. In: *Proceedings of the ASCE Hydraulic Engineering '94 Conference, August 1–5, Buffalo, NY* (eds. G.V. Cotroneo and R.R. Rumer), 717–721. New York: American Society of Civil Engineers.

Miller, A.J. (1990). Flood hydrology and geomorphic effectiveness in the central Appalachians. *Earth Surface Processes and Landforms* 15: 119–134.

Miller, A.J. (1995). Valley morphology and boundary conditions influencing spatial variations of flood flow. In: *Natural and Anthropogenic Influences in Fluvial Geomorphology* (eds. J.E. Costa, A.J. Miller, K.W. Potter and P.R. Wilcock), 57–81. Washington, DC: American Geophysical Union Press.

Miller, J.R. (2017). Causality of historic arroyo incision in the southwestern United States. *Anthropocene* 18: 69–75.

Miller, D.J. and Benda, L. (2000). Effects of punctuated sediment supply on valley-floor landforms and sediment transport. *Geological Society of America Bulletin* 112: 1814–1824.

Miller, J.R. and Friedman, J.M. (2009). Influence of flow variability on floodplain formation and destruction, Little Missouri River, North Dakota. *Geological Society of America Bulletin* 121: 752–759.

Miller, A.J. and Parkinson, D.J. (1993). Flood hydrology and geomorphic effects on river channels and floodplains: the flood of November 4–5, 1985, in the South Potomac River basin of West Virginia. In: *Geomorphic Studies of the Storm and Flood of November 3–4, 1985, in the Upper Potomac and Cheat River basins in West Virginia and Virginia* (ed. R.B. Jacobson) US Geological Survey Bulletin 1981, E1–E96.

Miller, A.J. and Zegre, N.P. (2014). Mountaintop removal mining and catchment hydrology. *Water* 6: 472–499.

Miller, E.K., Carson, C.D., Friedland, A.J., and Blum, J.D. (1995). Chemical and isotopic tracers of snowmelt flow paths in a subalpine watershed. In: *Biogeochemistry of Seasonally Snow-Covered Catchments* (eds. K.A. Tonnessen, M.W. Williams and M. Tranter), 349–353. Wallingford: IAHS Publications no. 228.

Miller, J., Barr, R., Grow, D. et al. (1999). Effects of the 1997 flood on the transport and storage of sediment and mercury within the Carson River valley, west-central Nevada. *Journal of Geology* 107: 313–327.

Miller, K.L., Szabo, T., Jerolmack, D.J., and Domokos, G. (2014). Quantifying the significance of abrasion and selective transport for downstream fluvial grain size evolution. *Journal of Geophysical Research: Earth Surface* 119: 2412–2429.

Milliman, J.D. and Meade, R.H. (1983). World-wide delivery of river sediment to oceans. *Journal of Geology* 91: 1–21.

Milliman, J.D. and Syvitski, J.P.M. (1992). Geomorphic/tectonic control of sediment discharge to the ocean: the importance of small mountainous rivers. *Journal of Geology* 100: 525–544.

Milly, P.C.D., Betancourt, J., Falkenmark, M. et al. (2008). Stationarity is dead: whither water management? *Science* 319: 573–574.

Milton, L.E. (1966). The geomorphic irrelevance of some drainage net laws. *Australian Geographical Studies* 4: 89–95.

Minckley, W.L. and Rinne, J.N. (1985). Large woody debris in hot-desert streams: an historical review. *Desert Plants* 7: 142–153.

Minshall, G.W. (1984). Aquatic insect-substratum relationships. In: *The Ecology of Aquatic Insects* (eds. V.H. Resh and D.M. Rosenberg), 358–400. New York: Praeger Publishers.

Mirus, B.B., Ebel, B.A., Loague, K., and Wemple, B.C. (2007). Simulated effect of a forest road on near-surface hydrologic response: redux. *Earth Surface Processes and Landforms* 32: 126–142.

Mitsch, W.J., Day, J.W., Gilliam, J.W. et al. (2001). Reducing nitrogen loading to the Gulf of Mexico from the Mississippi River basin: strategies to counter a persistent ecological problem. *BioScience* 51: 373–388.

Mizutani, T. (1998). Laboratory experiment and digital simulation of multiple fill-cut terrace formation. *Geomorphology* 24: 353–361.

Moir, H.C., Gibbins, C.N., Soulsby, C., and Webb, J. (2004). Linking channel geomorphic characteristics to spatial patterns of spawning activity and discharge use by Atlantic salmon (*Salmo salar* L.). *Geomorphology* 60: 21–35.

Molnár, P. (2013). Network-scale energy distribution. In: *Treatise on Fluvial Geomorphology* (ed. E. Wohl), 43–49. Amsterdam: Elsevier.

Molnar, P. and England, P. (1990). Late Cenozoic uplift of mountain ranges and global climate change: chicken or egg? *Nature* 346: 29–34.

Molnár, P. and Ramírez, J.A. (1998). An analysis of energy expenditure in Goodwin Creek. *Water Resources Research* 34: 1819–1829.

Molod, A., Takacs, L., Suarez, M., and Bacmeister, J. (2015). Development of the GEOS-5 atmospheric general circulation model: evolution from MERRA to MERRA2. *Geoscientific Model Development* 8: 1339–1356.

Monsalve, A., Yager, E.M., Turowski, J.M., and Rickenmann, D. (2016). A probabilistic formulation of bed load transport to include spatial variability of flow and surface grain size distributions. *Water Resources Research* 52: 3579–3598.

Montgomery, D.R. (1994). Road surface drainage, channel initiation, and slope instability. *Water Resources Research* 30: 1925–1932.

Montgomery, D.R. (1999). Process domains and the river continuum. *Journal of the American Water Resources Association* 35: 397–410.

Montgomery, D.R. (2001). Slope distributions, threshold hillslopes, and steady-state topography. *American Journal of Science* 301: 432–454.

Montgomery, D.R. (2002). Valley formation by fluvial and glacial erosion. *Geology* 30: 1047–1050.

Montgomery, D.R. (2004). Observations on the role of lithology in strath terrace formation and bedrock channel width. *American Journal of Science* 304: 454–476.

Montgomery, D.R. (2007). Is agriculture eroding civilization's foundation? *GSA Today* 17 (10): 4–9.

Montgomery, D.R. and Abbe, T.B. (2006). Influence of logjam-formed hard points on the formation of valley-bottom landforms in an old-growth forest valley, Queets River, Washington, USA. *Quaternary Research* 65: 147–155.

Montgomery, D.R. and Buffington, J.M. (1997). Channel-reach morphology in mountain drainage basins. *Geological Society of America Bulletin* 109: 596–611.

Montgomery, D.R. and Dietrich, W.E. (1988). Where do channels begin? *Nature* 336: 232–234.

Montgomery, D.R. and Dietrich, W.E. (1989). Source areas, drainage density, and channel initiation. *Water Resources Research* 25: 1907–1918.

Montgomery, D.R. and Dietrich, W.E. (1992). Channel initiation and the problem of landscape scale. *Science* 255: 826–830.

Montgomery, D.R. and Dietrich, W.E. (1994). A physically based model for the topographic control on shallow landsliding. *Water Resources Research* 30: 1153–1171.

Montgomery, D.R. and Dietrich, W.E. (2002). Runoff generation in a steep, soil-mantled landscape. *Water Resources Research* 38 (9): 1168.

Montgomery, D.R. and Foufoula-Georgiou, E. (1993). Channel network source representation using digital elevation models. *Water Resources Research* 29: 3925–3934.

Montgomery, D.R. and Gran, K.B. (2001). Downstream variations in the width of bedrock channels. *Water Resources Research* 37: 1841–1846.

Montgomery, D.R. and Stolar, D.B. (2006). Reconsidering Himalayan river anticlines. *Geomorphology* 82: 4–15.

Montgomery, D.R., Buffington, J.M., Smith, R.D. et al. (1995). Pool spacing in forest channels. *Water Resources Research* 31: 1097–1105.

Montgomery, D.R., Abbe, T.B., Buffington, J.M. et al. (1996). Distribution of bedrock and alluvial channels in forested mountain drainage basins. *Nature* 381: 587–589.

Montgomery, D.R., Dietrich, W.E., and Sullivan, K. (1998). The role of GIS in watershed analysis. In: *Landform Monitoring, Modeling and Analysis* (eds. S.N. Lane, K.S. Richards and J.H. Chandler), 241–261. Chichester: Wiley.

Montgomery, D.R., Dietrich, W.E., and Heffner, J.T. (2002). Piezometric response in shallow bedrock at CB1: implications for runoff generation and landsliding. *Water Resources Research* 38: 1274.

Montgomery, D.R., Collins, B.D., Buffington, J.M., and Abbe, T.B. (2003a). Geomorphic effects of wood in rivers. In: *The Ecology and Management of Wood in World Rivers* (eds. S.V. Gregory, K.L. Boyer and A.M. Gurnell), 21–47. Bethesda, MD: American Fisheries Society.

Montgomery, D.R., Massong, T.M., and Hawley, S.C.S. (2003b). Influence of debris flows and log jams on the location of pools and alluvial channel reaches, Oregon Coast Range. *Geological Society of America Bulletin* 115: 78–88.

Montgomery, D.R., Schmidt, K.M., Dietrich, W.E., and McKean, J. (2009). Instrumental record of debris flow initiation during natural rainfall: implications for modeling slope stability. *Journal of Geophysical Research* 114, F01031.

Moody, J.A. (2019). Dynamic relations for the deposition of sediment on floodplains and point bars of a freely-meandering river. *Geomorphology* 327: 585–597.

Moody, J.A. and Martin, R.G. (2015). Measurements of the initiation of post-wildfire runoff during rainstorms using in situ overland flow detectors. *Earth Surface Processes and Landforms* 40: 1043–1056.
Moody, J.A. and Meade, R.H. (2008). Terrace aggradation during the 1978 flood on Powder River, Montana, USA. *Geomorphology* 99: 387–403.
Moody, J.A. and Troutman, B.M. (2000). Quantitative model of the growth of floodplains by vertical accretion. *Earth Surface Processes and Landforms* 25: 115–133.
Morgan, J. (1970). Deltas – a résumé. *Journal of Geological Education* 18: 107–117.
Morgan, P., Aplet, G.H., Haufler, J.B. et al. (1994). Historical range of variability: a useful tool for evaluating ecosystem change. *Journal of Sustainable Forestry* 2: 87–111.
Morin, E., Grodek, T., Dahan, O. et al. (2009). Flood routing and alluvial aquifer recharge along the ephemeral arid Kuiseb River, Namibia. *Journal of Hydrology* 368: 262–275.
Morón, S., Edmonds, D.A., and Amos, K. (2017). The role of floodplain width and alluvial bar growth as a precursor for the formation of anabranching rivers. *Geomorphology* 278: 78–90.
Morse, P.D. and Wolfe, S.A. (2015). Geological and meteorological controls on icing (aufeis) dynamics (1985 to 2014) in subarctic Canada. *Journal of Geophysical Research* 120: 1670–1686.
Moshe, L.B., Haviv, I., Enzel, Y. et al. (2008). Incision of alluvial channels in response to continuous base level fall: field characterization, modeling, and validation along the Dead Sea. *Geomorphology* 93: 524–536.
Mosley, M.P. (1976). An experimental study of channel confluences. *Journal of Geology* 84: 535–562.
Mosley, M.P. and Tinsdale, D.S. (1985). Sediment variability and bed material sampling in gravel-bed rivers. *Earth Surface Processes and Landforms* 10: 465–482.
Mote, P.W., Hamlet, A.F., Clark, M.P., and Lettenmaier, D.P. (2005). Declining mountain snowpack in western North America. *Bulletin of the American Meteorological Society* 86: 39–49.
Mould, S. and Fryirs, K. (2017). The Holocene evolution and geomorphology of a chain of ponds, southeast Ausralia: establishing a physical template for river management. *Catena* 149: 349–362.
Moulin, B. and Piégay, H. (2004). Characteristics and temporal variability of large woody debris trapped trapped in a reservoir on the River Rhone (France): implications for river basin management. *River Research and Applications* 20: 79–97.
Mouw, J.E.B., Stanford, J.A., and Alaback, P.B. (2009). Influences of flooding and hyporheic exchange on floodplain plant richness and productivity. *River Research and Applications* 25: 929–945.
Moyle, P.B. and Mount, J.F. (2007). Homogenous rivers, homogenous faunas. *Proceedings of the National Academy of Sciences of the United States of America* 104: 5711–5712.
Muehlbauer, J.D., Collins, S.F., Doyle, M.W., and Tockner, K. (2014). How wide is a stream? Spatial extent of the potential "stream signature" in terrestrial food webs using meta-analysis. *Ecology* 95: 44–55.
Mueller, E.R., Grams, P.E., Hazel, J.E., and Schmidt, J.C. (2018). Variability in eddy sand bar dynamics during two decades of controlled flooding of the Colorado River in the Grand Canyon. *Sedimentary Geology* 363: 181–199.
Muhar, S., Jungwirth, M., Unfer, G. et al. (2008). Restoring riverine landscapes at the Drau River: successes and deficits in the context of ecological integrity. In: *Gravel-bed Rivers VI: From Process Understanding to River Restoration* (eds. H. Habersack, H. Piégay and M. Rinaldi), 779–807. Amsterdam: Elsevier.
Muhar, S., Januschke, K., Kail, J. et al. (2016). Evaluating good-practice cases for river restoration across Europe: context, methodological framework, selected results and recommendations. *Hydrobiologia* 769: 3–19.

Mul, M.L., Mutiibwa, R.K., Foppen, J.W.A. et al. (2007). Identification of groundwater flow systems using geological mapping and chemical spring analysis in south Pare Mountains, Tanzania. *Physics and Chemistry of the Earth* 32: 1015–1022.

Mürle, U., Ortlepp, J., and Zahner, M. (2003). Effects of experimental flooding on riverine morphology, structure and riparian vegetation: The River Spöl, Swiss National Park. *Aquatic Sciences* 65: 191–198.

Murphy, B.P., Johnson, J.P.L., Gasparini, N.M., and Skar, L.S. (2016). Chemical weathering as a mechanism for the climatic control of bedrock river incision. *Nature* 532: 223–227.

Murray, A.B. and Paola, C. (1994). A cellular model of braided rivers. *Nature* 371: 54–57.

Murray, A.B. and Paola, C. (1997). Properties of a cellular braided-stream model. *Earth Surface Processes and Landforms* 22: 1001–1025.

Muto, T., Yamagishi, C., Sekiguchi, T. et al. (2012). The hydraulic autogenesis of distinct cyclicity in delta foreset bedding: flume experiments. *Journal of Sedimentary Research* 82: 545–558.

Mvandaba, V., Hughes, D., Kapangaziwiri, E. et al. (2018). The delineation of alluvial aquifers towards a better understanding of channel transmission losses in the Limpopo River basin. *Physics and Chemistry of the Earth* 108: 60–73.

Myers, T.J. and Swanson, S. (1996). Long-term aquatic habitat restoration: Mahogany Creek, Nevada, as a case study. *Water Resources Bulletin* 32: 241–252.

Nadler, C.T. and Schumm, S.A. (1981). Metamorphosis of South Platte and Arkansas Rivers, eastern Colorado. *Physical Geography* 2: 95–115.

Nageswara Rao, K., Subraelu, P., Naga Kumar, K.C.V. et al. (2010). Impacts of sediment retention by dams on delta shoreline recession: evidences from the Krishna and Godavari deltas, India. *Earth Surface Processes and Landforms* 35: 817–827.

Nagler, P.L., Scott, R.L., Westenburg, C. et al. (2005). Evapotranspiration on western US rivers estimated using the Enhanced Vegetation Index from MODIS and data from eddy covariance and Bowen ratio flux towers. *Remote Sensing of Environment* 97: 337–351.

Nagler, P.L., Glenn, E.P., Jarnevich, C.S., and Shafroth, P.B. (2011). Distribution and abundance of Saltcedar and Russian olive in the western United States. *Critical Reviews in Plant Sciences* 30: 508–523.

Naiman, R.J., Melillo, J.M., and Hobbie, J.E. (1986). Ecosystem alteration of boreal forest streams by beaver (*Castor canadensis*). *Ecology* 67: 1254–1269.

Naiman, R.J., Johnston, C.A., and Kelley, J.C. (1988). Alteration of North American streams by beaver. *BioScience* 38: 753–762.

Naiman, R.J., Pinay, G., Johnston, C.A., and Pastor, J. (1994). Beaver influences on the long-term biogeochemical characteristics of boreal forest drainage networks. *Ecology* 75: 905–921.

Naiman, R.J., Décamps, H., and McClain, M.E. (2005). *Riparia: Ecology, Conservation, and Management of Streamside Communities*. Amsterdam: Elsevier.

Nakamura, F. and Swanson, F.J. (1993). Effects of coarse woody debris on morphology and sediment storage of a mountain stream system in western Oregon. *Earth Surface Processes and Landforms* 18: 43–61.

Nakamura, F., Swanson, F.J., and Wondzell, S.M. (2000). Disturbance regimes of stream and riparian systems – a disturbance-cascade perspective. *Hydrological Processes* 14: 2849–2860.

Namour, P., Schmitt, L., Eschbach, D. et al. (2015). Stream pollution concentration in riffle geomorphic units (Yzeron basin, France). *Science of the Total Environment* 532: 80–90.

Nanson, G.C. (1986). Episodes of vertical accretion and catastrophic stripping: a model of disequilibrium flood-plain development. *Geological Society of America Bulletin* 97: 1467–1475.

Nanson, G.C. (2013). Anabranching and anastomosing rivers. In: *Treatise on Fluvial Geomorphology* (ed. E. Wohl), 330–344. Amsterdam: Elsevier.
Nanson, G.C. and Croke, J.C. (1992). A genetic classification of floodplains. *Geomorphology* 4: 459–486.
Nanson, G.C. and Huang, H.Q. (2008). Least action principle, equilibrium states, iterative adjustment and the stability of alluvial channels. *Earth Surface Processes and Landforms* 33: 923–942.
Nanson, G.C. and Huang, H.Q. (2018). A philosophy of rivers: equilibrium states, channel evolution, teleomatic change and least action principle. *Geomorphology* 302: 3–19.
Nanson, G.C. and Knighton, A.D. (1996). Anabranching rivers: their cause, character and classification. *Earth Surface Processes and Landforms* 21: 217–239.
Nanson, G.C., Rust, B.R., and Taylor, G. (1986). Coexistent mud braids and anastomosing channels in an arid-zone river: Cooper Creek, central Australia. *Geology* 14: 175–178.
Nanson, G.C., Barbetti, M., and Taylor, G. (1995). River stabilisation due to changing climate and vegetation during the late Quaternary in western Tasmania, Australia. *Geomorphology* 13: 145–158.
Nanson, G.C., Tooth, S., and Knighton, A.D. (2002). A global perspective on dryland rivers: perceptions, misconceptions and distinctions. In: *Dryland Rivers: Hydrology and Geomorphology of Semi-Arid Channels* (eds. L.J. Bull and M.J. Kirkby), 17–54. Chichester: Wiley.
National Academy (1992). *Restoration of Aquatic Ecosystems: Science, Technology, and Public Policy.* Washington, DC: National Academy Press.
Navratil, O., Albert, M.-B., Hérouin, E., and Gresillon, J.-M. (2006). Determination of bankfull discharge magnitude and frequency: comparison of methods on 16 gravel-bed river reaches. *Earth Surface Processes and Landforms* 31: 1345–1363.
Nayak, A., Marks, D., Chandler, D.G., and Seyfried, M. (2010). Long-term snow, climate, and streamflow trends at the Reynolds Creek Experimental Watershed, Owyhee Mountains, Idaho, United States. *Water Resources Research* 46, W06519.
Neal, A. (2004). Ground-penetrating radar and its use in sedimentology: principles, problems, and progress. *Earth-Science Reviews* 66: 261–330.
Neal, E.G., Waler, M.T., and Coffeen, C. (2002). Linking the Pacific Decadal Oscillation to seasonal stream discharge patterns in southeast Alaska. *Journal of Hydrology* 263: 188–197.
Nelson, A. and Dubé, K. (2016). Channel response to an extreme flood and sediment pulse in a mixed bedrock and gravel-bed river. *Earth Surface Processes and Landforms* 41: 178–195.
Nelson, J.M., Shreve, R.L., McLean, S.R., and Drake, T.G. (1995). Role of near-bed turbulence structure in bed-load transport and bed form mechanics. *Water Resources Research* 31 (8): 2071–2086.
Nelson, P.A., Brew, A.K., and Morgan, J.A. (2015). Morphodynamic response of a variable-width channel to changes in sediment supply. *Water Resources Research* 51: 5717–5734.
Nemec, W. and Steel, R.J. (1988). What is a fan delta and how do we recognize it? In: *Fan Deltas: Sedimentology and Tectonic Settings* (eds. W. Nemec and R.J. Steel), 3–13. Glasgow: Blackie and Son.
Neumann, J.E., Emanuel, K.A., Ravela, S. et al. (2015). Risks of coastal storm surge and the effect of sea level rise in the red River Delta, Vietnam. *Sustainability* 7: 6553–6572.
Newman, M.E.J., Barabasi, A.L., and Watts, D.J. (2006). *The Structure and Dynamics of Networks.* Princeton, NJ: Princeton University Press.
Newson, M.D. and Large, A.R.G. (2006). "Natural" rivers, "hydromorphological quality" and river restoration: a challenging new agenda for applied fluvial geomorphology. *Earth Surface Processes and Landforms* 31: 1606–1624.
Nicholas, A.P. and Quine, T.A. (2007). Crossing the divide: representation of channels and processes in reduced-complexity river models at reach and landscape scales. *Geomorphology* 90: 318–339.

Nicholas, A.P. and Sambrook Smith, G.H. (1999). Numerical simulation of three-dimensional flow hydraulics in a braided channel. *Hydrological Processes* 13: 913–929.

Nichols, K., Bierman, P., Finkel, R., and Larsen, J. (2005). Sediment generation rates for the Upper Rio Chagres basin: evidence from cosmogenic ^{10}Be. In: *The Rio Chagres, Panama: A Multidisciplinary Profile of a Tropical Watershed* (ed. R.S. Harmon), 297–313. Dordrecht: Springer.

Nichols, M.H., Nearing, M., Hernandez, M., and Polyakov, V.O. (2016). Monitoring channel head erosion processes in response to an artificially induced abrupt base level change using time-lapse photography. *Geomorphology* 265: 107–116.

Nicoll, T. and Brierley, G. (2017). Within-catchment variability in landscape connectivity measures in the Garang catchment, upper Yellow River. *Geomorphology* 277: 197–209.

Niedzialek, J.M. and Ogden, F.L. (2005). Runoff production in the upper Rio Chagres watershed, Panama. In: *The Upper Rio Chagres, Panama: A Multidisciplinary Profile of a Tropical Watershed* (ed. R.S. Harmon), 149–168. Dordrecht: Springer.

Niemitz, J., Haynes, C., and Lasher, G. (2013). Legacy sediments and historic land use: Chemostratigraphic evidence for excess nutrient and heavy metal sources and remobilization. *Geology* 41: 47–50.

Nienow, P., Sharp, M., and Willis, I. (1998). Seasonal changes in the morphology of the subglacial drainage system, Haut Glacier d'Arolla, Switzerland. *Earth Surface Processes and Landforms* 23: 825–843.

Nihlgard, B.J., Swank, W.T., and Mitchell, M.J. (1994). Biological processes and catchment studies. In: *Biogeochemistry of Small Catchments: A Tool for Environmental Research* (eds. B. Moldan and J. Cerny), 133–161. Chichester: Wiley.

Nijssen, B., O'Donnell, G.M., Hamlet, A.F., and Lettenmaier, D.P. (2001). Hydrologic sensitivity of global rivers to climate change. *Climatic Change* 50: 143–175.

Nikora, V.I., Goring, D.G., and Biggs, B.J.F. (1998). On gravel-bed roughness characterization. *Water Resources Research* 34: 517–527.

Nilsson, C. and Berggren, K. (2000). Alterations of riparian ecosystems caused by river regulation. *BioScience* 50: 783–792.

Nilsson, C., Jansson, R., and Zinko, U. (1997). Long-term responses of river-margin vegetation to water-level regulation. *Science* 276: 798–800.

Nilsson, C., Reidy, C.A., Dynesius, M., and Revenga, C. (2005). Fragmentation and flow regulation of the world's large river systems. *Science* 308: 405–408.

Nitsche, M., Rickenmann, D., Kirchner, J.W. et al. (2012). Macroroughness and variations in reach-averaged flow resistance in steep mountain streams. *Water Resources Research* 48, W12518.

Nittrouer, J.A., Mohrig, D., Allison, M.A., and Peyret, A.B. (2011). The lowermost Mississippi River: a mixed bedrock-alluvial channel. *Sedimentology* 58: 1914–1934.

Nittrouer, J.A., Best, J.L., Brantley, C. et al. (2012). Mitigating land loss in coastal Louisiana by controlled diversion of Mississippi River sand. *Nature Geoscience* 5: 534–537.

Noetzli, K., Boll, A., Graf, F. et al. (2008). Influence of decay fungi, construction characteristics, and environmental conditions on the quality of wooden check-dams. *Forest Products Journal* 58: 72–79.

Nonaka, E. and Spies, T.A. (2005). Historical range of variability in landscape structure: a simulation study in Oregon, USA. *Ecological Applications* 15: 1727–1746.

Norris, R.H. and Thoms, M.C. (1999). What is river health? *Freshwater Biology* 41: 197–209.

Nott, J.F. and Price, D.M. (1994). Plunge pools and paleoprecipitation. *Geology* 22: 1047–1050.

O'Connor, J.E. 1993. Hydrology, hydraulics, and geomorphology of the bonneville flood. Geological Society of America Special Paper 274.

O'Connor, J.E. and Webb, R.H. (1988). Hydraulic modeling for paleoflood analysis. In: *Flood Geomorphology* (eds. V.R. Baker, R.C. Kochel and P.C. Patton), 393–402. New York: Wiley.

O'Connor, J.E., Webb, R.H., and Baker, V.R. (1986). Paleohydrology of pool-and-riffle pattern development: Boulder Creek, Utah. *Geological Society of America Bulletin* 97: 410–420.

O'Connor, J.E., Ely, L.L., Wohl, E.E. et al. (1994). A 4500-year record of large floods on the Colorado River in the Grand Canyon, Arizona. *Journal of Geology* 102: 1–9.

O'Connor, J.E., J.H. Hardison, J.E. Costa. 2001. Debris flows from failures of Neoglacial-age moraine dams in the Three Sisters and Mount Jefferson Wilderness Areas, Oregon. US Geological Survey Professional Paper 1606.

O'Connor, J.E., Jones, M.A., and Haluska, T.L. (2003). Flood plain and channel dynamics of the Quinault and Queets Rivers, Washington, USA. *Geomorphology* 51: 31–59.

O'Connor, J.E., Major, J., and Grant, G. (2008). The dams come down: unchaining US rivers. *Geotimes* 53: 22–28.

O'Connor, J.E., Clague, J.J., Walder, J.S. et al. (2013). Outburst floods. In: *Treatise on Fluvial Geomorphology* (ed. E. Wohl), 475–509. Amsterdam: Elsevier.

O'Connor, J.E., Mangano, J.F., Anderson, S.W. et al. (2014). Geologic and physiographic controls on bed-material yield, transport, and channel morphology for alluvial and bedrock rivers, western Oregon. *Geological Society of America Bulletin* 126: 377–397.

O'Connor, J.E., Duda, J.J., and Grant, G.E. (2015). 1000 dams down and counting. *Science* 348: 496–497.

Oerlemans, J. and Klok, E.J.L. (2004). Effect of summer snowfall on glacier mass balance. *Annals of Glaciology* 38: 97–100.

Oguchi, T. and Oguchi, C.T. (2004). Late Quaternary rapid talus dissection and debris flow deposition on an alluvial fan in Syria. *Catena* 55: 125–140.

Oguchi, T. and Ohmori, H. (1994). Analysis of relationships among alluvial fan area, source basin area, basin slope, and sediment yield. *Zeitschrift für Geomorphologie* 38: 405–420.

Oguchi, T., Wasklewicz, T., and Hayakawa, Y.S. (2013). Remote data in fluvial geomorphology: characteristics and applications. In: *Treatise on Fluvial Geomorphology* (ed. E. Wohl), 711–729. Amsterdam: Elsevier.

Ohmori, H. (1991). Change in the mathematical function type describing the longitudinal profile of a river through an evolutionary process. *Journal of Geology* 99: 97–110.

Okazaki, H., Kwak, Y., and Tamura, T. (2015). Depositional and erosional architecture of gravelly braid bar formed by a flood in the Abe River, central Japan, inferred from a three-dimensional ground-penetrating radar analysis. *Sedimentary Geology* 324: 32–46.

Olden, J., Konrad, C.P., Melis, T.S. et al. (2014). Are large-scale flow experiments informing the science and management of freshwater ecosystems? *Frontiers in Ecology and the Environment* 12: 176–185.

Oldknow, C.J. and Hooke, J.M. (2017). Alluvial terrace development and changing landscape connectivity in the Great Karoo, South Africa: insights from Wilgerbosch River catchment, Sneeuberg. *Geomorphology* 288: 12–38.

Olinde, L. and Johnson, J.P.L. (2015). Using RFID and accelerometer-embedded tracers to measure probabilities of bed load transport, step lengths, and rest times in a mountain stream. *Water Resources Research* 51: 7572–7589.

Oliveira, A.G., Baumgartner, M.T., Gomes, L.C. et al. (2018). Long-term effects of flow regulation by dams simplify fish functional diversity. *Freshwater Biology* 63: 293–305.

Ollesch, G., Kistner, I., Meissner, R., and Lindenschmidt, K.-E. (2006). Modelling of snowmelt erosion and sediment yield in a small low-mountain catchment in Germany. *Catena* 68: 161–176.

O'Loughlin, E.M. (1986). Prediction of surface saturation zones in natural catchments by topographic analysis. *Water Resources Research* 22: 794–804.

Onda, Y., Komatsu, Y., Tsujimura, M., and Fujihara, J. (2001). The role of subsurface runoff through bedrock on storm flow generation. *Hydrological Processes* 15: 1693–1706.

Opperman, J.J., Meleason, M., Francis, R.A., and Davies-Colley, R. (2008). "Livewood": geomorphic and ecological functions of living trees in river channels. *BioScience* 58: 1069–1078.

Ortega, J.A., Wohl, E., and Livers, B. (2013). Waterfalls on the eastern side of Rocky Mountain National Park, Colorado, USA. *Geomorphology* 198: 37–44.

Ortega-Becerril, J.A., Gomez-Herez, M., Fort, R., and Wohl, E. (2017a). How does anisotropy in bedrock river granitic outcrops influence pothole genesis and development? *Earth Surface Processes and Landforms* 42: 952–968.

Ortega-Becerril, J.A., Jorge-Coronado, A., Garzon, G., and Wohl, E. (2017b). Sobrarbe Geopark: an example of highly diverse bedrock rivers. *Geoheritage* 9: 533–548.

Osborne, L.L., Bayley, P.B., Higler, L.W.G. et al. (1993). Restoration of lowland streams: an introduction. *Freshwater Biology* 29: 187–194.

Osborne, J.M., Lambert, F.H., Groenendijk, M. et al. (2015). Reconciling precipitation with runoff: observed hydrological change in the midlatitudes. *Journal of Hydrometeorology* 16: 2403–2420.

Osei, N.A., Gurnell, A.M., and Harvey, G.L. (2015). The role of large wood in retaining fine sediment, organic matter and plant propagules in a small, single-thread forest river. *Geomorphology* 235: 77–87.

Osman, A.M. and Thorne, C.R. (1988). Riverbank stability analysis I: Theory. *Journal of Hydraulic Engineering* 114: 134–150.

Osterkamp, T.E., Jorgenson, M.T., Schuur, E.A.G. et al. (2009). Physical and ecological changes associated with warming permafrost and thermokarst in interior Alaska. *Permafrost and Periglacial Processes* 20: 235–256.

O'Sullivan, J.J., Ahilan, S., and Bruen, M. (2012). A modified Muskingum routing approach for floodplain flows: theory and practice. *Journal of Hydrology* 470–471: 239–254.

Oswald, E.B. and Wohl, E. (2008). Wood-mediate geomorphic effects of a jökulhlaup in the Wind River Mountains, Wyoming. *Geomorphology* 100: 549–562.

Oswood, M.W., Milner, A.M., and Irons, J.G. (1992). Climate change and Alaskan rivers and streams. In: *Global Climate Change and Freshwater Ecosystems* (eds. P. Firth and S.G. Fisher), 192–210. New York: Springer.

Overeem, I. and Syvitski, J.P.M. (2010). Shifting discharge peaks in Arctic rivers, 1977–2007. *Geografiska Annaler* 92A: 285–296.

Owen, L.A., Finkel, R.C., Haizhou, M., and Barnard, P.L. (2006). Late Quaternary landscape evolution in the Kunlun Mountains and Qaidam Basin, northern Tibet: a framework for examining the links between glaciation, lake level changes and alluvial fan formation. *Quaternary International* 154–155: 73–86.

Packman, A.I. and Brooks, N.H. (2001). Hyporheic exchange of solutes and colloids with moving bed forms. *Water Resources Research* 37: 2591–2605.

Packman, A.I. and MacKay, J.S. (2003). Interplay of stream-subsurface exchange, clay particle deposition, and streambed evolution. *Water Resources Research* 39 https://doi.org/10.1029/2002WR001432.

Paerl, H.W., Hall, N.S., Peieris, B.L., and Rossignol, K.L. (2014). Evolving paradigms and challenges in estuarine and coastal eutrophication dynamics in a culturally and climatically stressed world. *Estuaries and Coasts* 37: 243–258.

Page, K.J., Nanson, G.C., and Frazier, P.S. (2003). Floodplain formation and sediment stratigraphy resulting from oblique accretion on the Murrumbidgee River, Australia. *Journal of Sedimentary Research* 73: 5–14.

Page, K., Read, A., Frazier, P., and Mount, N. (2005). The effect of altered flow regime on the frequency and duration of bankfull discharge: Murrumbidgee River, Australia. *River Research and Applications* 21: 567–578.

Page, M., Marden, M., Kasai, M. et al. (2008). Changes in basin-scale sediment supply and transfer in a rapidly transformed New Zealand landscape. In: *Gravel-bed Rivers VI: From Process Understanding to River Restoration* (eds. H. Habersack, H. Piégay and M. Rinaldi), 337–358. Amsterdam: Elsevier.

Painter, T.H., Deems, J., Belnap, J. et al. (2010). Response of Colorado River runoff to dust radiative forcing in snow. *Proceedings of the National Academy of Sciences of the United States of America* 107: 17125–17130.

Palmer, M.A., Ambrose, R.F., and Poff, N.L. (1997). Ecological theory and community restoration ecology. *Restoration Ecology* 5: 291–300.

Palmer, M.A., Bernhardt, E.S., Schlesinger, W.H. et al. (2010a). Mountaintop mining consequences. *Science* 327: 148–149.

Palmer, M.A., Menninger, H.L., and Bernhardt, E. (2010b). River restoration, habitat heterogeneity and biodiversity: a failure of theory or practice? *Freshwater Biology* 55: 205–222.

Palmer, M.A., Filoso, S., and Fanelli, R.M. (2014). From ecosystems to ecosystem services: stream restoration as ecological engineering. *Ecological Engineering* 65: 62–70.

Pan, B., Burbank, D., Wang, Y. et al. (2003). A 900 k.y. record of strath terrace formation during glacial-interglacial transitions in northwest China. *Geology* 31: 957–960.

Pandey, A., Himanshu, S.K., Mishra, S.K., and Singh, V.P. (2016). Physically based soil erosion and sediment yield models revisited. *Catena* 147: 595–620.

Paola, C. (2001). Modelling stream braiding over a range of scales. In: *Gravel-bed Rivers V* (ed. M.P. Mosley), 11–46. Wellington: New Zealand Hydrological Society.

Paola, C. and Mohrig, D. (1996). Paleohydraulics revisited: paleoslope estimation in coarse-grained braided rivers. *Basin Research* 8: 243–254.

Paola, C., Mullin, J., Ellis, C. et al. (2001). Experimental stratigraphy. *GSA Today* 11: 4–9.

Papanicolaou, A.N., Strom, K., Schuyler, A., and Talebbeydokhti, N. (2003). The role of sediment specific gravity and availability on cluster evolution. *Earth Surface Processes and Landforms* 28: 69–86.

Papanicolaou, A.N.T., Wilson, C.G., Tsakiris, A.G. et al. (2017). Understanding mass fluvial erosion along a bank profile: using PEEP technology for quantifying retreat lengths and identifying event timing. *Earth Surface Processes and Landforms* 42: 1717–1732.

Park, C.C. (1977). World-wide variations in hydraulic geometry exponents of stream channels: an analysis and some observations. *Journal of Hydrology* 33: 133–146.

Park, H., Yoshikawa, Y., Oshima, K. et al. (2016). Quantification of warming climate-induced changes in terrestrial Arctic river ice thickness and phenology. *Journal of Climate* 29: 1733–1754.

Parker, G. (1976). On the cause and characteristic scales of meandering and braiding in rivers. *Journal of Fluid Mechanics* 76: 457–480.

Parker, G. (1979). Hydraulic geometry of active gravel rivers. *ASCE Journal of the Hydraulics Division* 105: 1185–1201.

Parker, G. (1990). Surface-based bedload transport relation for gravel rivers. *Journal of Hydraulic Research* 28: 417–436.

Parker, G. (1991). Downstream variation of grain size in gravel rivers: abrasion versus selective sorting. In: *Fluvial Hydraulics of Mountain Regions* (eds. A. Armanini and G. DiSilvio), 347–360. Berlin: Springer.

Parker, G. (2008). Transport of gravel and sediment mixtures. In: *Sedimentation Engineering: Theory, Measurements, Modeling and Practice* (ASCE Manuals and Reports on Engineering Practice No. 110) (ed. M. Garcia), 165–251. Reston, VA: American Society of Civil Engineers.

Parker, G. and Peterson, A.W. (1980). Bar resistance of gravel-bed streams. *Journal of the Hydraulics Division* 106: 1559–1575.

Parker, G. and Toro-Escobar, C.M. (2002). Equal mobility of gravel in streams: the remains of the day. *Water Resources Research* 38: 1264.

Parker, G., Klingeman, P.C., and McLean, D.G. (1982). Bedload and size distribution in paved gravel-bed streams. *Journal of the Hydraulics Division* 108: 544–571.

Parker, G., Y. Cui, J. Imran, W.E. Dietrich. 1997. Flooding in the lower Ok Tedi, Papua New Guinea due to the disposal of mine tailings and its amelioration. In International Seminar on Recent Trends of Floods and their Preventive Measures, Sapporo, Japan, 20–21 June 1996, 21–48.

Parker, G., Paola, C., Whipple, K.X., and Mohrig, D. (1998a). Alluvial fans formed by channelized fluvial and sheet flow I. Theory. *Journal of Hydraulic Engineering* 124: 985–995.

Parker, G., Paola, C., Whipple, K.X. et al. (1998b). Alluvial fans formed by channelized fluvial and sheet flow II. Application. *Journal of Hydraulic Engineering* 124: 996–1004.

Parker, G., Paola, C., and Leclair, S. (2000). Probabilistic Exner sediment continuity equation for mixtures with no active layer. *Journal of Hydraulic Engineering* 126: 818–826.

Parker, G., Toro-Escobar, C.M., Ramey, M., and Beck, S. (2003). Effect of floodwater extraction on mountain stream morphology. *Journal of Hydraulic Engineering* 129: 885–895.

Parker, G., Muto, T., Akamatsu, Y. et al. (2008). Unravelling the conundrum of river response to rising sea-level from laboratory to field: Part II, the Fly-Strickland River system, Papua New Guinea. *Sedimentology* 55: 1657–1686.

Parker, G., Shimizu, Y., Wilkerson, G.V. et al. (2011). A new framework for modeling the migration of meandering rivers. *Earth Surface Processes and Landforms* 36: 70–86.

Parrish, J.D., Braun, D.P., and Unnasch, R. (2003). "Are we conserving what we say we are?" Measuring ecological integrity within protected areas. *BioScience* 53: 851–860.

Parsapour-Moghaddam, P. and Rennie, C.D. (2018). Calibration of a 3D hydrodynamic meandering river model using fully spatially distributed 3D ADCP velocity data. *Journal of Hydraulic Engineering* 144.

Parsons, D.R., Jackson, P.R., Czuba, J.A. et al. (2013). Velocity Mapping Toolbox (VMT): a processing and visualization suite for moving-vessel ADCP measurements. *Earth Surface Processes and Landforms* 38: 1244–1260.

Parvis, M. (1950). Drainage pattern significance in air-photo identification of soils and bedrock. *US National Academy of Sciences National Research Council, Highway Research Board Bulletin* 28: 36–62.

Passalacqua, P. (2017). The Delta connectome: a network-based framework for studying connectivity in river deltas. *Geomorphology* 277: 50–62.

Passalacqua, P., Lanzoni, S., Paola, C., and Rinaldo, A. (2013). Geomorphic signatures of deltaic processes and vegetation: the Ganges-Brahmaputra-Jamuna case study. *Journal of Geophysical Research: Earth Surface* 118: 1838–1849.

Pasternack, G.B. (2013). Geomorphologist's guide to participating in river rehabilitation. In: *Treatise on Fluvial Geomorphology* (ed. E. Wohl), 843–860. Amsterdam: Elsevier.

Pasternack, G.B., Baig, D., Weber, M.D., and Brown, R.A. (2018a). Hierarchically nested river landform sequences. Part 1: theory. *Earth Surface Processes and Landforms* 43: 2510–2518.

Pasternack, G.B., Baig, D., Weber, M.D., and Brown, R.A. (2018b). Hierarchically nested river landform sequences. Part 2: bankfull channel morphodynamics governed by valley nesting structure. *Earth Surface Processes and Landforms* 43: 2519–2532.

Pastor, A., Babault, J., Owen, L.A. et al. (2015). Extracting dynamic topography from river profiles and cosmogenic nuclide geochronology in the Middle Atlas and the High Plateaus of Morocco. *Tectonophysics* 663: 95–109.

Patrick, R. (1995). Chemicals in riverine water. In: *Rivers of the United States*, vol. 2, Chemical and Physical Characteristics., 195–228. New York: Wiley.

Patton, P.C. (1988). Geomorphic response of streams to floods in the glaciated terrain of southern New England. In: *Flood Geomorphology* (eds. V.R. Baker, R.C. Kochel and P.C. Patton), 261–277. New York: Wiley.

Paul, M.J. and Meyer, J.L. (2001). Streams in the urban landscape. *Annual Review of Ecology and Systematics* 32: 333–365.

Pavelsky, T.M. and Zarnetske, J.P. (2017). Rapid decline in river icings detected in Arctic Alaska: implications for a changing hydrologic cycle and river ecosystems. *Geophysical Research Letters* 44: 3228–3235.

Payn, R.A., Gooseff, M.N., McGlynn, B.L. et al. (2009). Channel water balance and exchange with subsurface flow along a mountain headwater streams in Montana, United States. *Water Resources Research* 45, W11427.

Pazzaglia, F.J. (2013). Fluvial terraces. In: *Treatise on Fluvial Geomorphology* (ed. E. Wohl), 379–412. Elsevier.

Pazzaglia, F.J. and Brandon, M.T. (2001). A fluvial record of long-term steady-state uplift and erosion across the Cascadia forearc high, western Washington State. *American Journal of Science* 301: 385–431.

Pazzaglia, F.J. and Gardner, T.W. (1993). Fluvial terraces of the lower Susquehanna River. *Geomorphology* 8: 83–113.

Pederson, J.L., Anders, M.D., Rittenour, T.M. et al. (2006). Using fill terraces to understand incision rates and evolution of the Colorado River in eastern Grand Canyon, Arizona. *Journal of Geophysical Research* 111, F02003.

Peipoch, M., Brauns, M., Hauer, F.R. et al. (2015). Ecological simplification: human influences on riverscape complexity. *BioScience* 65: 1057–1065.

Peirce, S., Ashmore, P., and Leduc, P. (2018). The variability in the morphological active width: resultsf rom physical models of gravel-bed braided rivers. *Earth Surface Processes and Landforms* 43: 2371–2383.

Pelletier, J.D. (2012). A spatially distributed model for the long-term suspended sediment discharge and delivery ratio of drainage basins. *Journal of Geophysical Research: Earth Surface* 117 https://doi.org/10.1029/2011JF002129.

Pelletier, J.D. (2013). A robust, two-parameter method for the extraction of drainage networks from high-resolution digital elevation models (DEMs): evaluation using synthetic and real-world DEMs. *Water Resources Research* 49: 75–89.

Pelletier, J.D. and Orem, C.A. (2014). How do sediment yields from post-wildfire debris-laden flows depend on terrain slope, soil burn severity class, and drainage basin area? Insights from airborne-LiDAR change detection. *Earth Surface Processes and Landforms* 39: 1822–1832.

Pelletier, J.D. and Rasmussen, C. (2009). Geomorphically based predictive mapping of soil thickness in upland watersheds. *Water Resources Research* 45, W09417.

Pelletier, J.D., Mayer, L., Pearthree, P.A. et al. (2005). An integrated approach to flood hazard assessment on alluvial fans using numerical modeling, field mapping, and remote sensing. *Geological Society of America Bulletin* 117: 1167–1180.

Peñas, F.J., Barquin, J., and Alvarez, C. (2016). Assessing hydrologic alteration: evaluation of different alternatives according to data availability. *Ecological Indicators* 60: 470–482.

Penkman, K.E.H., Preece, R.C., Keen, D.H. et al. (2007). Testing the aminostratigraphy of fluvial archives: the evidence from intra-crystalline proteins within freshwater shells. *Quaternary Science Reviews* 26: 2958–2969.

Pérez-Peña, J.V., Azañón, J.M., Booth-Rea, G. et al. (2009). Differentiating geology and tectonics using a spatial autocorrelation technique for the hypsometric integral. *Journal of Geophysical Research* 114, F02018.

Perucca, E., Camporeale, C., and Ridolfi, L. (2007). Significance of the riparian vegetation dynamics on meandering river morphodynamics. *Water Resources Research* 43, W03430.

Petroski, H. (2006). Levees and other raised ground. *American Scientist* 94: 7–11.

Pettijohn, F.J. (1957). *Sedimentary Rocks*, 3e. New York: Harper and Row.

Pettit, N.E. and Naiman, R.J. (2006). Flood-deposited wood creates regeneration niches for riparian vegetation on a semi-arid south African river. *Journal of Vegetation Science* 17: 615–624.

Petts, G.E. (1979). Complex response of river channel morphology subsequent to reservoir construction. *Progress in Physical Geography* 3: 329–362.

Petts, G.E. (1984). *Impounded Rivers*. Chichester: Wiley.

Petts, G.E. (1989). Historical analysis of fluvial hydrosystems. In: *Historical Change of Large Alluvial Rivers: Western Europe* (ed. G.E. Petts), 1–18. Chichester: Wiley.

Petts, G.E. and Gurnell, A.M. (2005). Dams and geomorphology: research progress and future directions. *Geomorphology* 71: 27–47.

Pfeiffer, A. and Wohl, E. (2018). Where does wood most effectively enhance storage? Network-scale distribution of sediment and organic matter stored by instream wood. *Geophysical Research Letters* 45: 194–200.

Phanikumar, M.S., Aslam, I., Shen, C. et al. (2007). Separating surface storage from hyporheic retention in natural streams using wavelet decomposition of acoustic Doppler current profiles. *Water Resources Research* 43, W05460.

Phillips, J.D. (2003). Sources of nonlinearity and complexity in geomorphic systems. *Progress in Physical Geography* 27: 1–23.

Phillips, J.D. and Park, L. (2009). Forest blowdown impacts of Hurricane Rita on fluvial systems. *Earth Surface Processes and Landforms* 34: 1069–1081.

Phillips, J.D. and Van Dyke, C. (2016). Principles of geomorphic disturbance and recovery in response to storms. *Earth Surface Processes and Landforms* 41: 971–979.

Pickup, G. and Warner, R.F. (1976). Effects of hydrologic regime on magnitude and frequency of dominant discharge. *Journal of Hydrology* 29: 51–75.

Piégay, H. and Gurnell, A.M. (1997). Large woody debris and river geomorphological pattern: examples from S.E. France and S. England. *Geomorphology* 19: 99–116.

Piégay, H. and Salvador, P.-G. (1997). Contemporary floodplain forest evolution along the middle Ubaye River, Southern Alps, France. *Global Ecology and Biogeography Letters* 6: 397–406.

Piégay, H., Moulin, B., and Hupp, C.R. (2017). Assessment of transfer patterns and origins of in-channel wood in large rivers using repeated field surveys and wood characterisation (the Isère River upstream of Pontcharra, France). *Geomorphology* 279: 27–43.

Pierce, J. and Meyer, G. (2008). Long-term fire history from alluvial fan sediments: the role of drought and climate variability, and implications for management of Rocky Mountain forests. *International Journal of Wildland Fire* 17: 84–95.

Pierson, T.C. (2005). Hyperconcentrated flow – transitional process between water flow and debris flow. In: *Debris-Flow Hazards and Related Phenomena* (eds. M. Jakob and O. Hungr), 159–202. Berlin: Springer.

Pierson, T.C. (2007). Dating young geomorphic surfaces using age of colonizing Douglas fir in southwestern Washington and northwestern Oregon, USA. *Earth Surface Processes and Landforms* 32: 811–831.

Pike, A.S. and Scatena, F.N. (2010). Riparian indicators of flow frequency in a tropical montane stream network. *Journal of Hydrology* 382: 72–87.

Pike, A.S., Scatena, F.N., and Wohl, E.E. (2010). Lithological and fluvial controls on the geomorphology of tropical montane stream channels in Puerto Rico. *Earth Surface Processes and Landforms* 35: 1402–1417.

Pilotto, F., Bertoncin, A., Harvey, G.L. et al. (2014). Diversification of stream invertebrate communities by large wood. *Freshwater Biology* 59: 2571–2583.

Pinter, N. (2005). One step forward, two steps back on US floodplains. *Science* 308: 207–208.

Pišút, P. (2002). Channel evolution of the pre-channelised Danube River in Bratislava, Slovakia (1712–1886). *Earth Surface Processes and Landforms* 27: 369–390.

Pitlick, J. (1994). Relation between peak flows, precipitation, and physiography for five mountainous regions in the western USA. *Journal of Hydrology* 158: 219–240.

Pitlick, J. and Wilcock, P.R. (2001). Relations between streamflow, sediment transport, and aquatic habitat in regulated rivers. In: *Geomorphic Processes and Riverine Habitat* (eds. J.M. Dorava, D.R. Montgomery, B.B. Palcsak and F.A. Fitzpatrick), 185–198. Washington, DC: American Geophysical Union Press.

Pizzuto, J. (2002). Effects of dam removal on river form and process. *BioScience* 52: 683–691.

Pizzuto, J. (2003). Numerical modeling of alluvial landforms. In: *Tools in Fluvial Geomorphology* (eds. G.M. Kondolf and H. Piégay), 577–595. Chichester: Wiley.

Pizzuto, J.E., Hession, W.C., and McBride, M. (2000). Comparing gravel-bed rivers in paired urban and rural catchments of southeastern Pennsylvania. *Geology* 28: 79–82.

Pizzuto, J., O'Neal, M., and Stotts, S. (2010). On the retreat of forested, cohesive riverbanks. *Geomorphology* 116: 341–352.

Planchon, O., Silvera, N., Gimenez, R. et al. (2005). An automated salt-tracing gauge for flow-velocity measurement. *Earth Surface Processes and Landforms* 30: 833–844.

Poff, N.L. (2018). Beyond the natural flow regime? Broadening the hydro-ecological foundation to meet environemental flows challenges in a non-stationary world. *Freshwater Biology* 63: 1011–1021.

Poff, N.L., Tokar, S., and Johnson, P. (1996). Stream hydrological and ecological responses to climate change assessed with an artificial neural network. *Limnology and Oceanography* 41: 857–863.

Poff, N.L., Allan, J.D., Bain, M.B. et al. (1997). The natural flow regime: a paradigm for river conservation and restoration. *BioScience* 47: 769–784.

Poff, N.L., Olden, J.D., Merritt, D.M., and Pepin, D.M. (2007). Homogenization of regional river dynamics by dams and global biodiversity implications. *Proceedings of the National Academy of Sciences of the United States of America* 104: 5732–5737.

Poff, N.L., Richter, B.D., Arthington, A.H. et al. (2010). The ecological limits of hydrologic alteration (ELOHA): a new framework for developing regional environmental flow standards. *Freshwater Biology* 51: 141–170.

Pollen, N. and Simon, A. (2005). Estimating the mechanical effects of riparian vegetation on stream bank stability using a fiber bundle model. *Water Resources Research* 41, W07025.

Pollen-Bankhead, N. and Simon, A. (2010). Hydrologic and hydraulic effects of riparian root networks on streambank stability: is mechanical root reinforcement the whole story? *Geomorphology* 116: 353–362.

Pollock, M.M., Heim, M., and Werner, D. (2003). Hydrologic and geomorphic effects of beaver dams and their influence on fishes. In: *The Ecology and Management of Wood in World Rivers* (eds. S.V. Gregory, K. Boyer and A. Gurnell), 213–233. American Fisheries Society Symposium 37.

Pollock, M.M., Beechie, T.J., and Jordan, C.E. (2007). Geomorphic changes upstream of beaver dams in Bridge Creek, an incised stream channel in the interior Columbia River basin, eastern Oregon. *Earth Surface Processes and Landforms* 32: 1174–1185.

Pollock, M.M., Beechie, T.J., Wheaton, J.M. et al. (2014). Using beaver dams to restore incised stream ecosystems. *BioScience* 64: 279–290.

Pollock, M.M., Lewallen, G., Woodruff, K. et al. (eds.) (2015). *The Beaver Restoration Guidebook: Working with Beaver to Restore Streams, Wetlands, and Floodplains.* Version 1.0. Portland, OR: United States Fish and Wildlife Service.

Polvi, L.E. and Wohl, E. (2012). The beaver meadow complex revisited – the role of beavers in post-glacial floodplain development. *Earth Surface Processes and Landforms* 37: 332–346.

Polvi, L.E. and Wohl, E. (2013). Biotic drivers of river planform – implications for understanding the past and restoring the future. *BioScience* 63: 439–452.

Polvi, L.E., Wohl, E.E., and Merritt, D.M. (2011). Geomorphic and process domain controls on riparian zones in the Colorado Front Range. *Geomorphology* 125: 504–516.

Polvi, L.E., Wohl, E., and Merritt, D.M. (2014). Modeling the functional influence of vegetation type on streambank cohesion. *Earth Surface Processes and Landforms* 39: 1245–1258.

Poole, G.C. (2002). Fluvial landscape ecology: addressing uniqueness within the river discontinuum. *Freshwater Biology* 47: 641–660.

Post, D.A., Littlewood, I.G., and Croke, B.F. (2005). New directions for top-down modeling: introducing the PUB top-down modeling working group. In: *Predictions in Ungauged Basins: International Perspectives on the State of the Art and Pathways Forward* (eds. S. Franks, M. Sivapalan, K. Takeuchi and Y. Tachikawa), 125–133. Wallingford: IAHS Publications no. 301.

Postel, S. (1999). *Pillar of Sand: Can the Irrigation Miracle Last?* New York: Norton and Co.

Potter, P.E. (1978). Significance and origin of big rivers. *Journal of Geology* 86: 13–33.

Poulos, S.E., Ghionis, G., and Maroukian, H. (2009). The consequences of a future eustatic sea-level rise on the deltaic coasts of Inner Thermaikos Gulf (Aegean Sea) and Kyparissiakos Gulf (Ionian Sea), Greece. *Geomorphology* 107: 18–24.

Powell, J.W. (1875). *Exploration of the Colorado River of the West (1869–72).* Washington, DC: US Government Printing Office.

Powell, J.W. (1876). *Report on the Geology of the Eastern Portion of the Uinta Mountains.* Washington, DC: US Government Printing Office.

Powell, D.M. (1998). Patterns and processes of sediment sorting in gravel-bed rivers. *Progress in Physical Geography* 22: 1–32.

Powell, D.M. (2014). Flow resistance in gravel-bed rivers: progress in research. *Earth-Science Reviews* 136: 301–338.

Powell, D.M., Ockelford, A., Rice, S.P. et al. (2016). Structural properties of mobile armors formed at different flow strengths in gravel-bed rivers. *Journal of Geophysical Research: Earth Surface* 121: 1494–1515.

Powers, P.D., Helstab, M., and Niezgoda, S.L. (2019). A process-based approach to restoring depositional river valleys to stage 0, an anastomosing channel network. *River Research and Applications* 35: 3–13.

Prat, N., Gallart, F., Von Schiller, D. et al. (2014). The MIRAGE toolbox: an integrated assessment tool for temporary streams. *River Research and Applications* 30: 1318–1334.

Prestegaard, K.L. (1983). Variables influencing water-surface slopes in gravel-bed streams at bankfull stage. *Geological Society of America Bulletin* 94: 673–678.

Price, W.E.J. (1974). Simulation of alluvial fan deposition by a random walk model. *Water Resources Research* 10: 263–274.

Priesnitz, K. and Schunke, E. (2002). The fluvial morphodynamics of two small permafrost drainage basins, Richardson Mountains, northwestern Canada. *Permafrost and Periglacial Processes* 13: 207–217.

Prince, P.S., Spotila, J.A., and Henika, W.S. (2011). Stream capture as a driver of transient landscape evolution in a tectonically quiescent setting. *Geology* 39: 823–826.

Pringle, C.M. (2001). Hydrologic connectivity and the management of biological reserves: a global perspective. *Ecological Applications* 11: 981–998.

Pringle, C.M., Naiman, R.J., Bretschko, G. et al. (1988). Patch dynamics in lotic systems: the stream as a mosaic. *Journal of the North American Benthological Society* 7: 503–524.

Pringle, C., Vellidis, G., Heliotis, F. et al. (1995). Environmental problems of the Danube Delta. *Ekistics* 62: 370–372.

Procter, J., Cronin, S.J., Fuller, I.C. et al. (2010). Quantifying the geomorphic impacts of a lake-breakout lahar, Mount Ruapehu, New Zealand. *Geology* 38: 67–70.

Prosdocimi, M., Calligaro, S., Sofia, G. et al. (2015). Bank erosion in agricultural drainage networks: new challenges from structure-from-motion photogrammetry for post-event analysis. *Earth Surface Processes and Landforms* 40: 1891–1906.

Prospero, J.M. (1999). Long-range transport of mineral dust in the global atmosphere: impact of African dust on the environment of the southeastern United States. *Proceedings of the National Academy of Sciences of the United States of America* 96: 3396–3403.

Prosser, I.P. and Abernethy, B. (1996). Predicting the topographic limits to a gully network using a digital terrain model and process thresholds. *Water Resources Research* 32: 2289–2298.

Prosser, I.P. and Slade, C.J. (1994). Gully formation and the role of valley-floor vegetation, southeastern Australia. *Geology* 22: 1127–1130.

Prosser, I.P., Dietrich, W.E., and Stevenson, J. (1995). Flow resistance and sediment transport by concentrated overland flow in a grassland valley. *Geomorphology* 13: 71–86.

Prowse, T.D. (2001). River-ice ecology. I: Hydrologic, geomorphic, and water-quality aspects. *Journal of Cold Regions Engineering* 15: 1–16.

Prowse, T.D. and Beltaos, S. (2002). Climatic control of river-ice hydrology: a review. *Hydrological Processes* 16: 805–822.

Prowse, T.D. and Carter, T. (2002). Significance of ice-induced storage to spring runoff: a case study of the Mackenzie River. *Hydrological Processes* 16: 779–788.

Prowse, T.D. and Ferrick, M.G. (2002). Hydrology of ice-covered rivers and lakes: scoping the subject. *Hydrological Processes* 16: 759–762.

Prowse, T., Alfredsen, K., Beltaos, S. et al. (2011). Effects of changes in Arctic lake and river ice. *Ambio* 40: 63–74.

Pulley, S., Foster, I., and Antunes, P. (2015). The uncertainties associated with sediment fingerprinting suspended and recently deposited fluvial sediment in the Nene River basin. *Geomorphology* 228: 303–319.

Pyrce, R.S. and Ashmore, P.E. (2003). The relation between particle path length distributions and channel morphology in gravel-bed streams: a synthesis. *Geomorphology* 56: 167–187.

Radecki-Pawlik, A. (2002). Bankfull discharge in mountain streams: theory and practice. *Earth Surface Processes and Landforms* 27: 115–123.

Raff, D.A., Ramírez, J.A., and Smith, J.L. (2004). Hillslope drainage development with time: a physical experiment. *Geomorphology* 62: 169–180.

Rahel, F.J. (2007). Biogeographic barriers, connectivity and homogenization of freshwater faunas: it's a small world after all. *Freshwater Biology* 52: 696–710.

Rahman, M.M., Khan, M.N.I., Hoque, A.K.F., and Ahmed, I. (2015). Carbon stock in the Sundarbans mangrove forest: spatial variations in vegetation types and salinity zones. *Wetlands Ecology and Management* 23: 269–283.

Rajib, M.A., Merwade, V., and Yu, Z. (2016). Multi-objective calibration of a hydrologic model using spatially distributed remotely sensed/in-situ soil moisture. *Journal of Hydrology* 536: 192–207.

Rämä, T., Norden, J., Davey, M.L. et al. (2014). Fungi ahoy! Diversity on marine wooden substrata in the high North. *Fungal Ecology* 8: 46–58.

Ramanathan, R., Guin, A., Ritzi, R.W. et al. (2010). Simulating the heterogeneity in braided channel belt deposits: 1. A geometric-based methodology and code. *Water Resources Research* 46, W04515.

Rao, P.G. (1995). Effect of climate change on streamflows in the Mahanadi River basin, India. *Water International* 20: 205–212.

Rathburn, S.L., Merritt, D.M., Wohl, E.E. et al. (2009). Characterizing environmental flows for maintenance of river ecosystems: North Fork Cache la Poudre River, Colorado. In: *Management and Restoration of Fluvial Systems with Broad Historical Changes and Human Impacts* (eds. L.A. James, S.L. Rathburn and G.R. Whittecar), 143–157. Boulder, CO: Geological Society of America Special Paper 451.

Rathburn, S.L., Bennett, G.L., Wohl, E.E. et al. (2017). The fate of sediment, wood, and organic carbon eroded during an extreme flood, Colorado Front Range, USA. *Geology* 45: 499–502.

Rathburn, S.L., Shahverdian, S.M., and Ryan, S.E. (2018). Post-disturbance sediment recovery: implications for watershed resilience. *Geomorphology* 305: 61–75.

Ravazzolo, D., Mao, L., Picco, L., and Lenzi, M. (2015). Tracking log displacement during floods in the Tagliamento River using RFID and GPS tracker devices. *Geomorphology* 228: 226–233.

Raymo, M.E., Ruddiman, W.F., and Froelich, P.N. (1988). Influence of late Cenozoic mountain building on geochemical cycles. *Geology* 16: 649–653.

Raymond, P.A., Hartmann, J., Lauerwald, R. et al. (2013). Global carbon dioxide emissions from inland waters. *Nature* 503: 355–359.

Recking, A. (2010). A comparison between flume and field bed load transport data and consequences for surface-based bed load transport prediction. *Water Resources Research* 46, W03518.

Records, R.M., Wohl, E., and Arabi, M. (2016). Phosphorus in the river corridor. *Earth-Science Reviews* 158: 65–88.

Redmond, K.T., Enzel, Y., House, P.K., and Biondi, F. (2002). Climate variability and flood frequency at decadal to millennial time scales. In: *Ancient Floods, Modern Hazards: Principles and Applications of*

Paleoflood Hydrology (eds. P.K. House, R.H. Webb, V.R. Baker and D.R. Levish), 21–45. Washington, DC: American Geophysical Union Press.

Redolfi, M., Tubino, M., Bertoldi, W., and Brasington, J. (2016). Analysis of reach-scale elevation distribution in braided rivers: definition of a new morphologic indicator and estimation of mean quantitites. *Water Resources Research* 52: 5951–5970.

Redolfi, M., Bertoldi, W., Tubino, M., and Welber, M. (2018). Bed load variability and morphology of gravel bed rivers subject to unsteady flow: a laboratory investigation. *Water Resources Research* 54: 842–862.

Regmi, N.R., McDonald, E.V., and Bacon, S.N. (2014). Mapping Quaternary alluvial fans in the southwestern United States based on multiparameter surface roughness of lidar topographic data. *Journal of Geophysical Research: Earth Surface* 119: 12–27.

Reid, L.M. and Dunne, T. (1996). Rapid evaluation of sediment budgets. *Catena.*

Reid, L.M. and Dunne, T. (2016). Sediment budgets as an organizing framework in fluvial geomorphology. In: *Tools in Fluvial Geomorphology* (eds. G.M. Kondolf and H. Piegay), 357–380. Chichester: Wiley-Blackwell.

Reid, D.E. and Hickin, E.J. (2008). Flow resistance in steep mountain streams. *Earth Surface Processes and Landforms* 33: 2211–2240.

Reid, I., Layman, J.T., and Frostick, L.E. (1980). The continuous measurements of bedload discharge. *Journal of Hydraulic Research* 18: 243–249.

Reid, I., Frostick, L.E., and Layman, J.T. (1985). The incidence and nature of bedload transport during flood flows in coarse-grained alluvial channels. *Earth Surface Processes and Landforms* 10: 33–44.

Reid, I., Laronne, J.B., and Powell, D.M. (1998). Flash-flood and bedload dynamics of desert gravel-bed streams. *Hydrological Processes* 12: 543–557.

Reid, S.C., Lane, S.N., Montgomery, D.R., and Brookes, C.J. (2007). Does hydrological connectivity improve modelling of coarse sediment delivery in upland environments? *Geomorphology* 90: 263–282.

Reid, D.E., Hickin, E.J., and Babakaiff, S.C. (2010). Low-flow hydraulic geometry of small, steep mountain streams in southwest British Columbia. *Geomorphology* 122: 39–55.

Reineck, H.E. and Singh, I.B. (1980). *Depositional Sedimentary Environments.* New York: Springer-Verlag.

Rengers, F.J. and Wohl, E. (2007). Trends of grain sizes on gravel bars in the Rio Chagres, Panama. *Geomorphology* 83: 282–293.

Restrepo, J.D. and Kettner, A. (2012). Human-induced discharge diversions in a tropical delta and its environmental implications: the Patia River, Colombia. *Journal of Hydrology* 424–425: 124–142.

Reusser, L.J. and Bierman, P.R. (2010). Using meteoric ^{10}Be to track fluvial sand through the Waipaoa River basin, New Zealand. *Geology* 38: 47–50.

Reusser, L., Bierman, P., and Rood, D. (2015). Quantifying human impacts on rates of erosion and sediment transport at a landscape scale. *Geology* 43: 171–174.

Reynolds, L.V., Cooper, D.J., and Hobbs, N.T. (2012). Drivers of riparian tree invasion on a desert stream. *River Research and Applications* https://doi.org/10.1002/rra.2619.

Rhoads, B.L., Wilson, D., Urban, M., and Herricks, E.E. (1999). Interaction between scientists and nonscientists in community-based watershed management: emergence of the concept of stream naturalization. *Environmental Management* 24: 297–308.

Rhodes, E.J. (2011). Optically stimulated luminescence dating of sediments over the past 200 000 years. *Annual Review of Earth and Planetary Sciences* 39: 461–488.

Ricciardi, A. and Rasmussen, J.B. (1999). Extinction rates of North American freshwater fauna. *Conservation Biology* 13: 1220–1222.

Rice, S. (1994). Towards a model of changes in bed material texture at the drainage basin scale. In: *Process Models and Theoretical Geomorphology* (ed. M.J. Kirkby), 159–172. Chichester: Wiley.
Rice, S.P. (1998). Which tributaries disrupt downstream fining along gravel-bed rivers? *Geomorphology* 22: 39–56.
Rice, S.P. (1999). The nature and controls of downstream fining within sedimentary links. *Journal of Sedimentary Research* 69: 32–39.
Rice, S.P. (2017). Tributary connectivity, confluence aggradation and network biodiversity. *Geomorphology* 277: 6–16.
Rice, S.P. and Church, M. (1996). Sampling surficial fluvial gravels: the precision of size distribution percentile estimates. *Journal of Sedimentary Research* 66A: 654–665.
Rice, S.P., Greenwood, M.T., and Joyce, C.B. (2001). Tributaries, sediment sources, and the longitudinal organisation of macroinvertebrate fauna along river systems. *Canadian Journal of Fisheries and Aquatic Sciences* 58: 824–840.
Rice, S.P., Roy, A., and Rhoads, B.L. (eds.) (2008). *River Confluences, Tributaries, and the Fluvial Network*. Chichester: Wiley.
Rice, S.P., Lancaster, J., and Kemp, P. (2010). Experimentation at the interface of fluvial geomorphology, stream ecology and hydraulic engineering and the development of an effective, interdisciplinary river science: interdisciplinary experiments and integrated river science. *Earth Surface Processes and Landforms* 35: 64–77.
Richards, K.S. (1982). *Rivers: Form and Process in Alluvial Channels*. London: Methuen.
Richardson, K. and Carling, P.A. (2005). *A Typology of Sculpted Forms in Open Bedrock Channels*. Boulder, CO: Geological Society of America Special Paper 392.
Richardson, K., Benson, I., and Carling, P.A. (2003). An instrument to record sediment movement in bedrock channels. In: *Erosion and Sediment Transport Measurement in Rivers: Technological and Methodological Advances* (eds. J. Bogen, T. Fergus and D.E. Walling), 228–235. Wallingford: IAHS Publications no. 283.
Richey, J.E., Mertes, L.A.K., Dunne, T. et al. (1989). Sources and routing of the Amazon River flood wave. *Global Biogeochemical Cycles* 3: 191–204.
Richmond, A.D. and Fausch, K.D. (1995). Characteristics and function of large woody debris in subalpine Rocky Mountain streams in northern Colorado. *Canadian Journal of Fisheries and Aquatic Sciences* 52: 1789–1802.
Richter, B.D. (2010). Re-thinking environmental flows: from allocations and reserves to sustainability boundaries. *River Research and Applications* 26: 1052–1063.
Richter, B.D., Baumgartner, J., Powell, J., and Braun, D. (1996). A method for assessing hydrologic alteration within ecosystems. *Conservation Biology* 10: 1163–1174.
Richter, B.D., Baumgartner, J., Robert, W., and Braun, D. (1997). How much water does a river need? *Freshwater Biology* 37: 231–249.
Richter, B.D., Baumgartner, J., Wigington, R. et al. (1998). A spatial assessment of hydrologic alteration within a river network. *Regulated Rivers* 14: 329–340.
Rickenmann, D. (1991). Bed load transport and hyperconcentrated flow at steep slopes. In: *Fluvial Hydraulics of Mountain Regions* (eds. A. Armanini and G. DiSilvio), 429–441. Berlin: Springer.
Rickenmann, D. (1994). Bedload transport and discharge in the Erlenbach stream. In: *Dynamics and Geomorphology of Mountain Rivers* (eds. P. Ergenzinger and K.-H. Schmidt), 53–66. Berlin: Springer.
Rickenmann, D. (2001). Comparison of bed load transport in torrents and gravel bed streams. *Water Resources Research* 37: 3295–3305.

Rickenmann, D. (2017). Bed-load transport measurements with geophones and other passive acoustic methods. *Journal of Hydraulic Engineering* 143 https://doi.org/10.1061/(ASCE)HY.1943-7900.0001300.

Rickenmann, D., Badoux, A., and Hunzinger, L. (2016). Significance of sediment transport processes during piedmont floods: the 2005 flood events in Switzerland. *Earth Surface Processes and Landforms* 41: 224–230.

Riebe, C.S., Sklar, L.S., Lukens, C.E., and Shuster, D.L. (2015). Climate and topography control the size and flux of sediment produced on steep mountain slopes. *Proceedings of the National Academy of Sciences of the United States of America* 112: 15574–15579.

Riggsbee, J.A., Doyle, M.W., Julian, J.P. et al. (2013). Influence of aquatic and semi-aquatic organisms on channel forms and processes. In: *Treatise on Fluvial Geomorphology* (ed. E. Wohl), 189–201. Amsterdam: Elsevier.

Righini, M., Surian, N., Wohl, E. et al. (2017). Geomorphic response to an extreme flood in two Mediterranean rivers (northeast Sardinia, Italy): analysis of controlling factors. *Geomorphology* 290: 184–199.

Rigon, E., Comiti, F., Mao, L., and Lenzi, M.A. (2008). Relationships among basin area, sediment transport mechanisms and wood storage in mountain basins of the Dolomites (Italian Alps). *WIT Transactions on Engineering Sciences* 60: 163–172.

Riihimaki, C.A., Anderson, R.S., Safran, E.B. et al. (2006). Longevity and progressive abandonment of the Rocky Flats surface, Front Range, Colorado. *Geomorphology* 78: 265–278.

Rinaldi, M. and Darby, S.E. (2008). Modelling river-bank erosion processes and mass failure mechanisms: progress towards fully coupled simulations. In: *Gravel-bed Rivers VI: From Process Understanding to River Restoration* (eds. H. Habersack, H. Piégay and M. Rinaldi), 213–239. Amsterdam: Elsevier.

Rinaldi, M. and Nardi, L. (2013). Modeling interactions between riverbank hydrology and mass failures. *Journal of Hydrologic Engineering* 18: 1231–1240.

Rinaldo, A., Rigon, R., Banavar, J.R. et al. (2014). Evolution and selection of river networks: statics, dynamics, and complexity. *Proceedings of the National Academy of Sciences of the United States of America* 111: 2417–2424.

Ritchie, A.C., Warrick, J.A., East, A.E. et al. (2018). Morphodynamic evolution following sediment release from the world's largest dam removal. *Scientific Reports* 8: 13279.

Ritsema, C.J., Kuipers, H., Kleiboer, L. et al. (2009). A new wireless underground network system for continuous monitoring of soil water contents. *Water Resources Research* 45, W00D36.

Ritter, D.F., Kochel, R.C., and Miller, J.R. (2011). *Process Geomorphology*, 5e. Long Grove, IL: Waveland Press.

Robert, A. (2014). *River Processes: An Introduction to Fluvial Dynamics*. New York: Routledge.

Robert, A., Roy, A.G., and De Serres, B. (1996). Turbulence at a roughness transition in a depth limited flow over a gravel bed. *Geomorphology* 16: 175–187.

Roberts, H.H. (1998). Delta switching: early responses to the Atchafalaya River diversion. *Journal of Coastal Research* 14: 882–899.

Roberts, M.C. and Klingeman, P.C. (1970). The influence of landform and precipitation parameters on flood hydrographs. *Journal of Hydrology* 11: 393–411.

Roberts, G.G. and White, N. (2010). Estimating uplift rate histories from river profiles using African examples. *Journal of Geophysical Research* 115 B02406.

Robichaud, P.R., Pierson, F.B., Brown, R.E., and Wagenbrenner, J.W. (2008). Measuring effectiveness of three postfire hillslope erosion barrier treatments, western Montana, USA. *Hydrological Processes* 22: 159–170.

Robinson, R.A.J., Spencer, J.Q.G., Strecker, M.R. et al. (2005). Luminescence dating of alluvial fans in intramontane basins of NW Argentina. In: *Alluvial Fans: Geomorphology, Sedimentology, Dynamics* (eds. A.M. Harvey, A.E. Mather and M. Stokes), 153–168. Geological Society Special Publication 251.

Robinson, D.A., Campbell, C.S., Hopmans, J.W. et al. (2008). Soil moisture measurement for ecological and hydrological watershed-scale observatories: a review. *Vadose Zone Journal* 7: 358–389.

Rockwell, T.K., Masana, E., Sharp, W.D. et al. (2019). Late Quaternary slip rates for the southern Elsinore fault in the Coyote Mountains, southern California from analysis of alluvial fan landforms and clast provenance, soils, and U-series ages of pedogenic carbonate. *Geomorphology* 326: 68–89.

Rodnight, H., Duller, G.A.T., Tooth, S., and Wintle, A.G. (2005). Optical dating of a scroll-bar sequence on the Klip River, South Africa, to derive the lateral migration rate of a meander bend. *The Holocene* 15: 802–811.

Rodríguez-Iturbe, I. and Rinaldo, A. (1997). *Fractal River Basins: Chance and Self-Organization.* Cambridge: Cambridge University Press.

Rodriguez-Iturbe, I. and Valdes, J.B. (1979). The geomorphologic structure of hydrologic response. *Water Resources Research* 15: 1409–1420.

Rodríguez-Iturbe, I., Rinaldo, A., Rigon, R. et al. (1992). Energy dissipation, runoff production, and the three-dimensional structure of river basins. *Water Resources Research* 28: 1095–1103.

Rodríguez-Rodríguez, M., Benavente, J., Alcalá, F.J., and Paracuellos, M. (2011). Long-term water monitoring in two Mediterranean lagoons as an indicator of land-use changes and intense precipitation events (Adra, southeastern Spain). *Estuarine, Coastal and Shelf Science* 91: 400–410.

Roe, G.H., Montgomery, D.R., and Hallett, B. (2002). Effects of orographic precipitation variations on the concavity of steady-state river profiles. *Geology* 50: 143–146.

Roering, J.J. (2004). Soil creep and convex-upward velocity profiles: theoretical and experimental investigation of disturbance-driven sediment transport on hillslopes. *Earth Surface Processes and Landforms* 29: 1597–1612.

Roering, J.J. (2008). How well can hillslope evolution models explain topography? Simulating soil transport and production with high-resolution topographic data. *Geological Society of America Bulletin* 120: 1248–1262.

Roering, J.J., Almond, P., Tonkin, P., and McKean, J. (2002). Soil transport driven by biological processes over millennial time scales. *Geology* 30: 1115–1118.

Roering, J.J., Marshall, J., Booth, A.M. et al. (2010). Evidence for biotic controls on topography and soil production. *Earth and Planetary Science Letters* 298: 183–190.

Rolls, R.J., Growns, I.O., Khan, T.A. et al. (2013). Fish recruitment in rivers with modified discharge depends on the interacting effects of flow and thermal regimes. *Freshwater Biology* 58: 1804–1819.

Roni, P., Beechie, T., Pess, G., and Hanson, K. (2015). Wood placement in river restoration: fact, fiction, and future direction. *Canadian Journal of Fisheries and Aquatic Sciences* 72: 466–478.

Rood, S.B., Pan, J., Gill, K.M. et al. (2008). Declining summer flows of Rocky Mountain rivers: changing seasonal hydrology and probable impacts on floodplain forests. *Journal of Hydrology* 349: 397–410.

Rosgen, D.L. and Silvey, H.L. (1996). *Applied River Morphology.* Pagosa Springs, CO: Wildland Hydrology.

Roth, D.L., Finnegan, N.J., Brodsky, E.E. et al. (2017). Bed load transport and boundary roughness changes as competing causes of hysteresis in the relationship between river discharge and seismic amplitude recorded near a steep mountain stream. *Journal of Geophysical Research: Earth Surface* 122: 1182–1200.

Rowland, J.C., Jones, C.E., Altmann, G. et al. (2010). Arctic landscapes in transition: responses to thawing permafrost. *Eos, Transactions of the American Geophysical Union* 91: 229–230.

Roy, A.G. and Abrahams, A.D. (1980). Rhythmic spacing and origin of pools and riffles: discussion and reply. *Geological Society of America Bulletin* 91: 248–250.

Roy, N.G. and Sinha, R. (2014). Effective discharge for suspended sediment transport of the Ganga River and its geomorphic implication. *Geomorphology* 227: 18–30.

Roy, A.G., Biron, P., Buffin-Bélanger, T., and Levasseur, M. (1999). Combined visual and quantitative techniques in the study of natural turbulent flows. *Water Resources Research* 35: 871–877.

Roy, S.G., Koons, P.O., Upton, P., and Tucker, G.E. (2016). Dynamic links among rock damage, erosion, and strain during orogensis. *Geology* 44: 583–586.

Rubin, D.M., Nelson, J.M., and Topping, D.J. (1998). Relation of inversely graded deposits to suspended-sediment grain-size evolution during the 1996 flood experiment in Grand Canyon. *Geology* 26: 99–102.

Rubin, Z., Rathburn, S.L., Wohl, E., and Harry, D.L. (2012). Historic range of variability in geomorphic processes as a context for restoration: Rocky Mountain National Park, Colorado, USA. *Earth Surface Processes and Landforms* 37: 209–222.

Rubin, Z., Kondolf, G.M., and Carling, P.A. (2015). Anticipated geomorphic impacts from Mekong basin dam construction. *International Journal of River Basin Management* 13: 105–121.

Rudorff, C.M., Melack, J.M., and Bates, P.D. (2014a). Flooding dynamics on the lower Amazon floodplain: 1. Hydraulic controls on water elevation, inundation extent, and river-floodplain discharge. *Water Resources Research* 50: 619–634.

Rudorff, C.M., Melack, J.M., and Bates, P.D. (2014b). Flooding dynamics on the lower Amazon floodplain: 2. Seasonal and interannual hydrological variability. *Water Resources Research* 50: 635–649.

Rudra, K. (2014). Changing river courses in the western part of the Ganga-Brahmaputra delta. *Geomorphology* 227: 87–100.

Ruffing, C.M., Daniels, M.D., and Dwire, K.A. (2015). Disturbance legacies of historic tie-drives persistently alter geomorphology and large wood characteristics in headwater streams, southeast Wyoming. *Geomorphology* 231: 1–14.

Ruiz-Villanueva, V., Díez-Herrero, A., Stoffel, M. et al. (2010). Dendrogeomorphic analysis of flash floods in a small ungauged mountain catchment (central Spain). *Geomorphology* 118: 383–392.

Ruiz-Villanueva, V., Bodoque, J.M., Diez-Herrero, A. et al. (2013). Reconstruction of a flash flood with large wood transport and its influence on hazard patterns in an ungauged mountain basin. *Hydrological Processes* 27: 3424–3437.

Ruiz-Villanueva, V., Wyzga, B., Hajdukiewicz, H., and Stoffel, M. (2016a). Exploring large wood retention and deposition in contrasting river morphologies: linking numerical modeling and field observations. *Earth Surface Processes and Landforms* 41: 446–459.

Ruiz-Villanueva, V., Piegay, H., Gaertner, V. et al. (2016b). Wood density and moisture sorption and its influence on large wood mobility in rivers. *Catena* 140: 182–194.

Ruiz-Villanueva, V., Piégay, H., Gurnell, A.M. et al. (2016c). Recent advances quantifying the large wood dynamics in river basins: new methods and remaining challenges. *Reviews of Geophysics* 54: 611–652.

Runkel, R.L. (2015). On the use of rhodamine WT for the characterization of stream hydrodynamics and transient storage. *Water Resources Research* 51: 6125–6142.

Runkel, R.L. and Chapra, S.C. (1993). An efficient numerical solution of the transient storage equations for solute transport in small streams. *Water Resources Research* 29: 211–215.

Runkel, R.L., McKnight, D.M., and Rajaram, H. (2003). Modeling hyporheic zone processes. *Advances in Water Resources* 26: 901–905.

Rustomji, R. and Prosser, I. (2001). Spatial patterns of sediment delivery to valley floors: sensitivity to sediment transport capacity and hillslope hydrology relations. *Hydrological Processes* 15: 1003–1018.

Ruther, N. and Olsen, N.R.B. (2007). Modelling free-forming meander evolution in a laboratory channel using three-dimensional computational fluid dynamics. *Geomorphology* 89: 308–319.

Ryan, S.E. (1997). Morphologic response of subalpine streams to transbasin flow diversion. *Journal of the American Water Resources Association* 33: 839–854.

Ryan, S.E., Porth, L.S., and Troendle, C.A. (2002). Defining phases of bedload transport using piecewise regression. *Earth Surface Processes and Landforms* 27: 971–990.

Ryder, J.M. (1971). The stratigraphy and morphology of paraglacial alluvial fans in south-central British Columbia. *Canadian Journal of Earth Sciences* 8: 279–298.

Ryder, J.M. and Church, M. (1986). The Lillooet terraces of Fraser River. *Canadian Journal of Earth Sciences* 23: 869–884.

Safford, H.D., Betancourt, J.L., Hayward, G.D. et al. (2008). Land management in the Anthropocene: is history still relevant? *Eos, Transactions of the American Geophysical Union* 89 (37): 343.

Salant, N.L., Renshaw, C.E., and Magilligan, F.J. (2006). Short and long-term changes to bed mobility and bed composition under altered sediment regimes. *Geomorphology* 76: 43–53.

Salcher, B.C., Faber, R., and Wagreich, M. (2010). Climate as main factor controlling the sequence development of two Pleistocene alluvial fans in the Vienna Basin (eastern Austria) – a numerical modelling approach. *Geomorphology* 115: 215–227.

Samadi, A., Amiri-Tokaldany, E., and Darby, S.E. (2009). Identifying the effects of parameter uncertainty on the reliability of riverbank stability modeling. *Geomorphology* 106: 219–230.

Sambrook Smith, G.H. and Nicholas, A.P. (2005). Effect on flow structure of sand deposition on a gravel bed: results from a two-dimensional flume experiment. *Water Resources Research* 41, W10405.

Sambrook Smith, G.H., Best, J.L., Bristow, C.S., and Petts, G.E. (2006). Braided rivers: where have we come in 10 years? Progress and future needs. In: *Braided Rivers: Process, Deposits, Ecology and Management*, vol. 36 (eds. G.H. Sambrook Smith, J.L. Best, C.S. Bristow and G.E. Petts) Special Publications, 1–10. International Association of Sedimentology.

Šamonil, P., Kral, K., and Hort, L. (2010). The role of tree uprooting in soil formation: a critical literature review. *Geoderma* 157: 65–79.

Sanborn, S.C. and Bledsoe, B.P. (2006). Predicting streamflow regime metrics for ungauged streams in Colorado, Washington, and Oregon. *Journal of Hydrology* 325: 241–261.

Sanborn, P., Geertsema, M., Jull, A.J.T., and Hawkes, B. (2006). Soil and sedimentary charcoal evidence from Holocene forest fires in an inland temperate rainforest, east-central British Columbia, Canada. *Holocene* 16: 415–427.

Sancho, C., Calle, M., Pena-Monne, J.L. et al. (2016). Dating the earliest Pleistocene alluvial terrace of the Alcanadre River (Ebro Basin, NE Spain): insights into the landscape evolution and involved processes. *Quaternary International* 407: 86–95.

Sanders, L.M., Taffs, K.H., Stokes, D.J. et al. (2017). Carbon accumulation in Amazonian floodplain lakes: a significant component of Amazon budgets? *Limnology and Oceanography* 2: 29–35.

Sanderson, J.S., Rowan, N., Wilding, T. et al. (2011). Getting to scale with environmental flow assessment: the watershed flow evaluation tool. *River Research and Applications* https://doi.org/10.1002/rra.1542.

Sandiford, G. 2009. Transforming an exotic species: nineteenth-century narratives about introduction of carp in America. Unpublished PhD dissertation, University of Illinois at Urbana-Champaign.
Savenije, H. (2005). *Salinity and Tides in Alluvial Estuaries*. Amsterdam: Elsevier.
Sawyer, A.H. and Cardenas, M.B. (2009). Hyporheic flow and residence time distributions in heterogeneous cross-bedded sediment. *Water Resources Research* 45, W08406.
Sawyer, A.M., Pasternack, G.B., Moir, H.M., and Fulton, A.A. (2010). Riffle-pool maintenance and flow convergence routing observed on a large gravel-bed river. *Geomorphology* 114: 143–160.
Sawyer, A.H., Cardenas, M.B., and Buttles, J. (2011). Hyporheic exchange due to channel-spanning logs. *Water Resources Research* 47, W08502.
Sayama, T. and McDonnell, J.J. (2009). A new time-space accounting scheme to predict stream water residence time and hydrograph source components at the watershed scale. *Water Resources Research* 45, W07401.
Saynor, M.J., Loughran, R.J., Erskine, W.D., and Scott, P.F. (1994). Sediment movement on hillslopes measured by caesium-137 and erosion pins. In: *Variability in Stream Erosion and Sediment Transport* (eds. L.J. Olive, R.J. Loughran and J.A. Kesby), 87–93. Wallingford: IAHS Publications no. 224.
Scanlon, B.R., Jolly, I., Sophocleous, M., and Zhang, L. (2007). Global impacts of conversions from natural to agricultural ecosystems on water resources: quantity versus quality. *Water Resources Research* 43, W03437.
Scatena, F.N. and Gupta, A. (2013). Streams of the montane humid tropics. In: *Treatise on Fluvial Geomorphology* (ed. E. Wohl), 595–610. Amsterdam: Elsevier.
Scatena, F.N. and Lugo, A.E. (1995). Geomorphology, disturbance, and the soil and vegetation of two subtropical wet steepland watersheds of Puerto Rico. *Geomorphology* 13: 199–213.
Schama, S. (1995). *Landscape and Memory*. New York: A.A. Knopf.
Schanz, S.A. and Montgomery, D.R. (2016). Lithologic controls on valley width and strath terrace formation. *Geomorphology* 258: 58–68.
Scheffer, M., Carpenter, S.R., Foley, J.A. et al. (2001). Catastrophic shifts in ecosystems. *Nature* 413: 591–596.
Scheingross, J.S. and Lamb, M.P. (2016). Sediment transport through self-adjusting, bedrock-walled plunge pools. *Journal of Geophysical Research: Earth Surface* 121: 939–963.
Scheingross, J.S. and Lamb, M.P. (2017). A mechanistic model of waterfall plunge pool erosion into bedrock. *Journal of Geophysical Research: Earth Surface* 122: 2079–2104.
Scheingross, J.S., Winchell, E.W., Lamb, M.P., and Dietrich, W.E. (2013). Influence of bed patchiness, slope, grain hiding, and form drag on gravel mobilization in very steep streams. *Journal of Geophysical Research: Earth Surface* 118: 982–1001.
Scheingross, J.S., Brun, F., Lo, D.Y. et al. (2014). Experimental evidence for fluvial bedrock incision by suspended and bedload sediment. *Geology* 42: 523–526.
Scheingross, J.S., Lo, D.Y., and Lamb, M.P. (2017). Self-formed waterfall plunge pools in homogeneous rock. *Geophysical Research Letters* 44: 200–208.
Schenk, E.R. and Hupp, C.R. (2009). Legacy effects of colonial millponds on floodplain sedimentation, bank erosion, and channel morphology, Mid-Atlantic, USA. *Journal of the American Water Resources Association* 45: 597–606.
Schenk, E.R., McCargo, J.W., Moulin, B. et al. (2015). The influence of logjams on largemouth bass (*Micropterus salmoides*) concentrations on the lower Roanoke River, a large sand-bed river. *River Research and Applications* 31: 704–711.

Scherler, D., DiBiase, R.A., Fisher, G.B., and Avouac, J.P. (2017). Testing monsoonal controls on bedrock river incision in the Himalaya and Eastern Tibet with a stochastic-threshold stream power model. *Journal of Geophysical Research: Earth Surface* 122: 1389–1429.

Schick, A.P. (1974). Function and obliteration of desert stream terraces – a conceptual analysis. *Zeitschrift für Geomorphologie* 21: 88–105.

Schick, A.P., Lekach, J., and Hassan, M.A. (1987). Vertical exchange of coarse bedload in desert streams. In: *Desert Sediments: Ancient and Modern* (eds. L. Frostick and I. Reid), 7–16. Geological Society Special Publication 35.

Schiefer, E., Menounos, B., and Slaymaker, O. (2006). Extreme sediment delivery events recorded in the contemporary sediment record of a montane lake, southern Coast Mountains, British Columbia. *Canadian Journal of Earth Sciences* 43: 1777–1790.

Schiemer, F., Baumgartner, C., and Tockner, K. (1999). Restoration of floodplain rivers: the "Danube restoration project.". *River Research and Applications* 15: 231–244.

Schmadel, N.M., Ward, A.S., and Wondzell, S.M. (2017). Hydrologic controls on hyporheic exchange in a headwater mountain stream. *Water Resources Research* 53: 6260–6278.

Schmeeckle, M.W. (2014). Numerical simulation of turbulence and sediment transport of medium sand. *Journal of Geophysical Research: Earth Surface* 119: 1240–1262.

Schmeeckle, M.W., Nelson, J.M., and Shreve, R.L. (2007). Forces on stationary particles in near-bed turbulent flows. *Journal of Geophysical Research* 112, F02003.

Schmidt, K.-H. and Ergenzinger, P. (1992). Bedload entrainment, travel lengths, step lengths, rest periods – studied with passive (iron, magnetic) and active (radio) tracer techniques. *Earth Surface Processes and Landforms* 17: 147–165.

Schmidt, K.-H. and Gintz, D. (1995). Results of bedload tracer experiments in a mountain river. In: *River Geomorphology* (ed. E.J. Hickin), 37–54. New York: Wiley.

Schmidt, J.C., J.B. Graf. 1990. Aggradation and degradation of alluvial sand deposits, 1965 to 1986, Colorado River, Grand Canyon National Park, Arizona. US Geological Survey Professional Paper 1493.

Schmidt, K.-H. and Morche, D. (2006). Sediment output and effective discharge in two small high mountain catchments in the Bavarian Alps, Germany. *Geomorphology* 80: 131–145.

Schmidt, J.C. and Wilcock, P.R. (2008). Metrics for assessing the downstream effects of dams. *Water Resources Research* 44, W04404.

Schmidt, J.C., Rubin, D.M., and Ikeda, H. (1993). Flume simulation of recirculating flow and sedimentation. *Water Resources Research* 29: 2925–2939.

Schmocker-Fackel, P. and Naef, F. (2010). More frequent flooding? Changes in flood frequency in Switzerland since 1850. *Journal of Hydrology* 381: 1–8.

Schneider, J.M., Turowski, J.M., Rickenmann, D. et al. (2014). Scaling relationships between bed load volumes, transport distances, and stream power in steep mountain channels. *Journal of Geophysical Research: Earth Surface* 119: 533–549.

Schneider, J.M., Rickenmann, D., Turowski, J.M., and Kirchner, J.W. (2015). Self-adjustment of stream bed roughness and flow velocity in a steep mountain channel. *Water Resources Research* 51: 7838–7859.

Schneider, J.M., Rickenmann, D., Turowski, J.M. et al. (2016). Bed load transport in a very steep mountain stream (Riedbach, Switzerland): measurement and prediction. *Water Resources Research* 52: 9522–9541.

Schoklitsch, A. (1962). *Handbuch des Wasserbaus*, 3e. Vienna: Springer-Verlag.

Schoof, R. (1980). Environmental impact of channel modification. *Water Resources Bulletin* 16: 697–701.

Schott, A.M. (2017). Site formation processes and depositional environment of a fine-grained alluvial floodplain at La Playa archaeology site, Sonora, Mexico. *Geoarchaeology* 32: 283–301.

Schrott, L., Hufschmidt, G., Hankammer, M. et al. (2003). Spatial distribution of sediment storage types and quantification of valley fill deposits in an alpine basin, Reintal, Bavarian Alps, Germany. *Geomorphology* 55: 45–63.

Schumm, S.A. (1956). Evolution of drainage systems and slopes in badlands at Perth Amboy, New Jersey. *Bulletin of the Geological Society of America* 67: 597–646.

Schumm, S.A. (1960). The shape of alluvial channels in relation to sediment type. In: *US Geological Survey Professional Paper* 352B, 17–30.

Schumm, S.A. (1963). Sinuosity of alluvial rivers on the Great Plains. *Geological Society of America Bulletin* 74: 1089–1100.

Schumm, S.A. (1967). Meander wavelength of alluvial rivers. *Science* 157: 1549–1550.

Schumm, S.A. (1969). River metamorphosis. *Journal of Hydraulics Division, ASCE* 95: 255–273.

Schumm, S.A. (1973). Geomorphic thresholds and complex response of drainage systems. In: *Fluvial Geomorphology* (ed. M. Morisawa), 299–310. SUNY Binghamton Publications in Geomorphology.

Schumm, S.A. (1977). *The Fluvial System*. New York: Wiley.

Schumm, S.A. (1979). Geomorphic thresholds: the concept and its applications. *Transactions of the Institute of British Geographers* 4: 485–515.

Schumm, S.A. (1981). *Evolution and Response of the Fluvial System: Sedimentologic Implications*, 19–29. Society of Economic Paleontologists and Mineralogists Special Publication 31.

Schumm, S.A. (1985). Patterns of alluvial rivers. *Annual Review of Earth and Planetary Sciences* 13: 5–27.

Schumm, S.A. (1991). *To Interpret the Earth: Ten Ways to Be Wrong*. Cambridge: Cambridge University Press.

Schumm, S.A. (1993). River responses to baselevel change: implications for sequence stratigraphy. *Journal of Geology* 101: 279–294.

Schumm, S.A. and Galay, V.J. (1994). The River Nile in Egypt. In: *The Variability of Large Alluvial Rivers* (eds. S.A. Schumm and B.R. Winkley), 75–102. New York: American Society of Civil Engineers.

Schumm, S.A. and Hadley, R.F. (1957). Arroyos and the semiarid cycle of erosion. *American Journal of Science* 25: 161–174.

Schumm, S.A., R.F. Hadley. 1961. Progress in the application of landform analysis in studies of semiarid erosion. USG.S Circular 437.

Schumm, S.A. and Khan, H.R. (1972). Experimental study of channel patterns. *Geological Society of America Bulletin* 83: 1755–1770.

Schumm, S.A. and Parker, R.S. (1973). Implications of complex response of drainage systems for Quaternary alluvial stratigraphy. *Science* 243: 99–100.

Schumm, S.A., Harvey, M.D., and Watson, C.C. (1984). *Incised Channels: Morphology, Dynamics and Control*. Littleton, CO: Water Resources Publications.

Schumm, S.A., Mosley, M.P., and Weaver, W.E. (1987). *Experimental Fluvial Geomorphology*. Chichester: Wiley.

Schumm, S.A., Rutherfurd, I.D., and Brooks, J. (1994). Pre-cutoff morphology of the lower Mississippi River. In: *The Variability of Large Alluvial Rivers* (eds. S.A. Schumm and B.R. Winkley), 13–44. New York: ASCE Press.

Schumm, S.A., Dumont, J.F., and Holbrook, J.M. (2000). *Active Tectonics and Alluvial Rivers*. Cambridge: Cambridge University Press.

Schürch, P., Densmore, A.L., Ivy-Ochs, S. et al. (2016). Quantitative reconstruction of late Holocene surface evolution on an alpine debris-flow fan. *Geomorphology* 275: 46–57.

Schuster, P.F., Krabbenhoft, D.P., Naftz, D.L. et al. (2002). A 270-year ice core record of atmospheric mercury deposition to western North America. *Environmental Science and Technology* 36: 2303–2310.

Schuur, E.A.G., McGuire, A.D., Schadel, C. et al. (2015). Climate change and the permafrost carbon feedback. *Nature* 520: 171–179.

Schwabe, E., Bartsch, I., Blazewicz-Paszkowycz, M. et al. (2015). Wood-associated fauna collected during the KuramBio-expedition in the North West Pacific. *Deep Sea Research, Part II* 111: 376–388.

Schwendel, A.C., Nicholas, A.P., Aalto, R.E. et al. (2015). Interaction between meander dynamics and floodplain heterogeneity in a large tropical sandbed river: the Rio Beni, Bolivian Amazon. *Earth Surface Processes and Landforms* 40: 2026–2040.

Scott, K.M., G.C. Gravlee. 1968. Flood surge on the Rubicon River, California – hydrology, hydraulics, and boulder transport. US Geological Survey Professional Paper, 422-M.

Scott, D.N. and Wohl, E.E. (2018a). Geomorphic regulation of floodplain organic soil carbon concentration in watersheds of the Rocky and Cascade Mountains, USA. *Earth Surface Dynamics* 6: 1101–1114.

Scott, D.N. and Wohl, E.E. (2018b). Natural and anthropogenic controls on wood loads in river corridors of the Rocky, Cascade, and Olympic Mountains, USA. *Water Resources Research* 54: 7893–7909.

Scott, D.N. and Wohl, E.E. (2019). Bedrock fracture influences on geomorphic process and form across process domains and scales. *Earth Surface Processes and Landforms* https://doi.org/10.1002/esp.4473.

Scott, D.N., Wohl, E., and Yochum, S.E. (2019). Wood Jam Dynamics Database and Assessment Model (WooDDAM): a framework to measure and understand wood jam characteristics and dynamics. *River Research and Applications* https://doi.org/10.1002/rra.3481.

Sear, D.A., Lee, M.W.E., Carling, P.A. et al. (2003). An assessment of the accuracy of the Spatial Integration Method (SIM) for estimating coarse bedload transport in gravel-bedded streams using tracers. In: *Erosion and Sediment Transport Measurement in Rivers: Technological and Methodological Advances* (eds. J. Bogen, T. Fergus and D.E. Walling), 164–171. Wallingford: IAHS Publications no. 283.

Sear, D.A., Wheaton, J.M., and Darby, S.E. (2008). Uncertain restoration of gravel-bed rivers and the role of geomorphology. In: *Gravel-Bed Rivers VI: From Process Understanding to River Restoration* (eds. H. Habersack, H. Piégay and M. Rinaldi), 739–761. Amsterdam: Elsevier.

Sear, D.A., Millington, C.E., Kitts, D.R., and Jeffries, R. (2010). Logjam controls on channel: floodplain interactions in wooded catchments and their role in the formation of multi-channel patterns. *Geomorphology* 116: 305–319.

Segura, C. and Pitlick, J. (2010). Scaling frequency of channel-forming flows in snowmelt-dominated streams. *Water Resources Research* 46, W06524.

Segura-Beltran, F., Sanchis-Ibor, C., Morales-Hernandez, M. et al. (2016). Using post-flood surveys and geomorphologic mapping to evaluate hydrological and hydraulic models: the flash flood of the Girona River (Spain) in 2007. *Journal of Hydrology* 541: 319–329.

Seibert, J., Bishop, K., Rodhe, A., and McDonnell, J.J. (2003). Groundwater dynamics along a hillslope: a test of the steady-state hypothesis. *Water Resources Research* 39: 1014.

Seidl, M.A. and Dietrich, W.E. (1992). The problem of channel erosion into bedrock. In: *Functional Geomorphology: Landform Analysis and Models*, vol. 23 (eds. K.-H. Schmidt and J. de Ploey) Catena Supplement, 101–124.

Seidl, M.A., Dietrich, W.E., and Kirchner, J.W. (1994). Longitudinal profile development into bedrock: an analysis of Hawaiian channels. *Journal of Geology* 102: 457–474.

Seidl, M.A., Finkel, R.C., Caffee, M.W. et al. (1997). Cosmogenic isotope analyses applied to river longitudinal profile evolution: problems and interpretations. *Earth Surface Processes and Landforms* 22: 195–209.

Seker, D.Z., Kaya, S., Musaoglu, N. et al. (2005). Investigation of meandering in Filyos River by means of satellite sensor data. *Hydrological Processes* 19: 1497–1508.

Selby, M.J. (1982). *Hillslope Materials and Processes*. Oxford: Oxford University Press.

Seminara, G. (2006). Meanders. *Journal of Fluid Mechanics* 554: 271–297.

Seminara, G. and Tubino, M. (1989). Alternate bars and meandering: free, forced and mixed interactions. In: *River Meandering* (eds. S. Ikeda and G. Parker), 267–320. Washington, DC: American Geophysical Union Press.

Sendrowski, A. and Passalacqua, P. (2017). Process connectivity in a naturally prograding river delta. *Water Resources Research* 53: 1841–1863.

Senter, A.E., Pasternack, G.B., Piegay, H. et al. (2017). Wood export varies among decadal, annual, seasonal, and daily hydrologic regimes in a large, Mediterranean climate, mountain river watershed. *Geomorphology* 276: 164–179.

Seo, J.I., Nakamura, F., Nakano, D. et al. (2008). Factors controlling the fluvial export of large woody debris and its contribution to organic carbon budgets at watershed scales. *Water Resources Research* 44, W04428.

SER (Society for Ecological Restoration). 2002. The SER Primer on Ecological Restoration. www.ser.org.

Shafroth, P.B., Stromberg, J.C., and Patten, D.T. (2002). Riparian vegetation response to altered disturbance and stress regimes. *Ecological Applications* 12: 107–123.

Shafroth, P.B., Wilcox, A.C., Lytle, D.A. et al. (2010). Ecosystem effects of environmental flows: modeling and experimental floods in a dryland river. *Freshwater Biology* 55: 68–85.

Shafroth, P.B., Schlatter, K.J., Gomez-Sapiens, M. et al. (2017). A large-scale environmental flow experiment for riparian restoration in the Colorado River Delta. *Ecological Engineering* 106B: 645–660.

Shanafield, M. and Cook, P.G. (2014). Transmission losses, infiltration, and groundwater recharge through ephemeral and intermittent streambeds: a review of applied methods. *Journal of Hydrology* 511: 518–529.

Shankman, D. and Pugh, T.B. (1992). Discharge response to channelization of a coastal plain stream. *Wetlands* 12: 157–162.

Sharma, A., Marshall, L., and Nott, D. (2005). A Bayesian view of rainfall-runoff modeling: alternatives for parameter estimate, model comparison and hierarchical model development. In: *Predictions in Ungauged Basins: International Perspectives on the State of the Art and Pathways Forward* (eds. S. Franks, M. Sivapalan, K. Takeuchi and Y. Tachikawa), 299–311. Wallingford: IAHS Publications no. 301.

Sheets, B.A., Hickson, T.A., and Paola, C. (2002). Assembling the stratigraphic record: depositional patterns and time-scales in an experimental alluvial basin. *Basin Research* 14: 287–301.

Shen, H.T. (2016). River ice processes. In: *Advances in Water Resources Management*, Handbook of Environmental Engineering, vol. 16 (eds. L. Wang, C. Yang and M.H. Wang), 483–530. Dordrecht: Springer.

Shieh, C.-L., Guh, Y.-R., and Wang, S.-Q. (2007). The application of range of variability approach to the assessment of a check dam on riverine habitat alteration. *Environmental Geology* 52: 427–435.

Shields, A. (1936). *Arwendung der Aenlich-keits-mechanik and der Turbulenz-forschung auf die Geshienbebewegung. Versuch-sanstalt fur Wasserbau and Schiffsbau*. Berlin, Heft: Mitteilungen der Preussischen.

Shields, F.D. and Smith, R.H. (1992). Effects of large woody debris removal on physical characteristics of a sand-bed river. *Aquatic Conservation: Marine and Freshwater Ecosystems* 2: 145–163.

Shih, W.R. and Diplas, P. (2018). A unified approach to bed load transport description over a wide range of flow conditions via the use of conditional data treatment. *Water Resources Research* 54: 3490–3509.

Shiklomanov, A.I. and Lammers, R.B. (2014). River ice responses to a warming Arctic – recent evidence from Russian rivers. *Environmental Research Letters* 9, 035008.

Shiono, K., Muto, Y., Knight, D.W., and Hyde, A.F.L. (1999). Energy losses due to secondary flow and turbulence in meandering channels with overbank flows. *Journal of Hydraulic Research* 37: 641–664.

Shobe, C.M., Hancock, G.S., Eppes, M.C., and Small, E.E. (2016). Field evidence for the influence of weathering on rock erodibility and channel form in bedrock rivers. *Earth Surface Processes and Landforms* 42: 1997–2012.

Sholtes, J.S. and Doyle, M.W. (2011). Effect of channel restoration on flood wave attenuation. *Journal of Hydraulic Engineering* 137: 196–208.

Shreve, R.L. (1966). Statistical law of stream numbers. *Journal of Geology* 74: 17–37.

Shroba, R.L., P.W. Schmidt, E.J. Crosby, W.R. Hansen, J.M. Soule. 1979. Geologic and geomorphologic effects in the Big Thompson Canyon area, Larimer County. Part B, of Storm and Flood of July 31–August 1, 1976, in the Big Thompson River and Cache la Poudre River Basins, Larimer and Weld Counties, Colorado. US Geological Survey Professional Paper 1115.

Sidle, R.C., Noguchi, S., Tsuboyama, Y., and Laursen, K. (2001). A conceptual model of preferential flow systems in forested hillslopes: evidence of self-organization. *Hydrological Processes* 15: 1675–1692.

Sidle, R.C., Gomi, T., and Tsukamoto, Y. (2018). Discovery of zero-order basins as an important link for progress in hydrogeomorphology. *Hydrological Processes* 32: 3059–3065.

Sieben, J. 1997. Modeling of hydraulics and morphology in mountain rivers. PhD dissertation, Technical University of Delft.

Siewert, M.B., Hugelius, G., Heim, B., and Faucherre, S. (2016). Landscape controls and vertical variability of soil organic carbon storage in permafrost-affected soils of the Lena River Delta. *Catena* 147: 725–741.

Sigafoos, R.S. 1961. Vegetation in relation to flood frequency near Washington, DC. US Geological Survey Professional Paper 424-C.

Sigafoos, R.S. 1964. Botanical evidence of floods and flood-plain deposition. US Geological Survey Professional Paper 485A.

Simco, A.H., Stephens, D.B., Calhoun, K., and Stephens, D.A. (2010). Historic irrigation and drainage at Priestley Farm by Joseph Elkington and William Smith. *Vadose Zone Journal* 9: 4–13.

Simenstad, C.A., Wick, A., Van De Wetering, S., and Bottom, D.L. (2003). Dynamics and ecological functions of wood in estuarine and coastal marine ecosystems. In: *The Ecology and Management of Wood in World Rivers* (eds. S.V. Gregory, K.L. Boyer and A.M. Gurnell), 265–277. Bethesda, MD: American Fisheries Society.

Simon, A. 1994. Gradation processes and channel evolution in modified West Tennessee streams: process, response, and form. US Geological Survey Professional Paper 1470.

Simon, A. and Collison, A.J.C. (2001). Pore-water pressure effects on the detachment of cohesive streambeds: seepage forces and matric suction. *Earth Surface Processes and Landforms* 26: 1421–1442.

Simon, A. and Darby, S.E. (2002). Effectiveness of grade-control structures in reducing erosion along incised river channels: the case of Hotophia Creek, Mississippi. *Geomorphology* 42: 229–254.

Simon, A. and Rinaldi, M. (2006). Disturbance, stream incision, and channel evolution: the roles of excess transport capacity and boundary materials in controlling channel response. *Geomorphology* 79: 361–383.

Simon, A. and Rinaldi, M. (2013). Incised channels: disturbance, evolution and the roles of excess transport capacity and boundary materials in controlling channel response. In: *Treatise on Fluvial Geomorphology* (ed. E. Wohl), 574–594. Amsterdam: Elsevier.

Simon, A., Curini, A., Darby, S.E., and Langendoen, E. (1999). Streambank mechanics and the role of bank and near-bank processes in incised channels. In: *Incised River Channels: Processes, Forms, Engineering and Management* (eds. S.E. Darby and A. Simon), 123–152. Chichester: Wiley.

Simon, A., Curini, A., Darby, S.E., and Langendoen, E. (2000). Bank and near bank processes in an incised channel. *Geomorphology* 35: 193–217.

Simon, A., Dickerson, W., and Heins, A. (2004). Suspended-sediment transport rates at the 1.5-year recurrence interval for ecoregions of the United States: transport conditions at the bankfull and effective discharge? *Geomorphology* 58: 243–262.

Simon, A., Castro, J., and Rinaldi, M. (2016). Channel form and adjustment: characterization, measurement, interpretation and analysis. In: *Tools in Fluvial Geomorphology* (eds. G.M. Kondolf and H. Piégay), 237–259. Chichester: Wiley.

Simons, D.B., E.V. Richardson. 1966. Resistance to flow in alluvial channels. US Geological Survey Professional Paper 422J.

Simons, D.B., E.V. Richardson, W.H. Haushild. 1963. Some effects of fine sediment on flow phenomena. US Geological Survey Water-Supply Paper 1498G.

Simpson, G.H.D. and Schlunegger, F. (2003). Topographic evolution and morphology of surfaces evolving in response to coupled fluvial and hillslope sediment transport. *Journal of Geophysical Research* 108 https://doi.org/10.1029/2002JB002162.

Šimůnek, J., Jarvis, N.J., van Genuchten, M.T., and Gärdenäs, A. (2003). Review and comparison of models for describing non-equilibrium and preferential flow and transport in the vadose zone. *Journal of Hydrology* 272: 14–35.

Singer, M.B. and Dunne, T. (2006). Modeling the influence of river rehabilitation scenarios on bed material sediment flux in a large river over decadal timescales. *Water Resources Research* 42, W12415.

Singh, P., Spitzbart, G., Hübl, H., and Weinmeister, H.W. (1998). The role of snowpack in producing floods under heavy rainfall. In: *Hydrology, Water Resources and Ecology in Headwaters* (eds. K. Kovar, U. Tappeiner, N.E. Peters and R.G. Craig), 89–95. Wallingford: IAHS Publications no. 248.

Singh, P., Arora, M., and Goel, N.K. (2006). Effect of climate change on runoff of a glacierized Himalayan basin. *Hydrological Processes* 20: 1979–1992.

Singh, A.K., Jaiswal, M.K., Pattanaik, J.K., and Dev, M. (2016). Luminescence chronology of alluvial fan in north Bengal, India: implications to tectonics and climate. *Geochronometria* 43: 102–112.

Singh, U., Crosato, A., Giri, S., and Hicks, M. (2017). Sediment heterogeneity and mobility in the morphodynamic modelling of gravel-bed braided rivers. *Advances in Water Resources* 104: 127–144.

Sivapalan, M., Wagener, T., Uhlenbrook, S. et al. (2006). *Predictions in Ungauged Basins: Promise and Progress.* Wallingford: IAHS Publications no. 303.

Skalak, K. and Pizzuto, J. (2010). The distribution and residence time of suspended sediment stored within the channel margins of a gravel-bed bedrock river. *Earth Surface Processes and Landforms* 35: 435–446.

Skalak, K.J., Benthem, A.J., Schenk, E.R. et al. (2013). Large dams and alluvial rivers in the Anthropocene: the impacts of the Garrison and Oahe Dams on the Upper Missouri River. *Anthropocene* 2: 51–64.

Škarpich, V., Galia, T., Ruman, S., and Macka, Z. (2019). Variations in bar material grain-size and hydraulic conditions of managed and re-naturalized reaches of the gravel-bed Bečva River (Czech Republic). *Science of the Total Environment* 649: 672–685.

Sklar, L.S. and Dietrich, W.E. (1998). River longitudinal profiles and bedrock incision models: stream power and the influence of sediment supply. In: *Rivers Over Rock: Fluvial Processes in Bedrock Channels* (eds. K.J. Tinkler and E.E. Wohl), 237–260. Washington, DC: American Geophysical Union Press.

Sklar, L.S. and Dietrich, W.E. (2001). Sediment and rock strength controls on river incision into bedrock. *Geology* 29: 1087–1090.

Sklar, L.S. and Dietrich, W.E. (2004). A mechanistic model for river incision into bedrock by saltating bed load. *Water Resources Research* 40, W06301.

Sklar, L.S., Dietrich, W.E., Foufoula-Georgiou, E. et al. (2006). Do gravel bed river size distributions record channel network structure? *Water Resources Research* 42, W06D18.

Sklar, L.S., Fadde, J., Venditti, J.G. et al. (2009). Translation and dispersion of sediment pulses in flume experiments simulating gravel augmentation below dams. *Water Resources Research* 45, W08439.

Sklar, L.S., Riebe, C.S., Marshall, J.A. et al. (2017). The problem of predicting the size distribution of sediment supplied by hillslopes to rivers. *Geomorphology* 277: 31–49.

Skoulikidis, N.T., Sabater, S., Datry, T. et al. (2017). Non-perennial Mediterranean rivers in Europe: status, pressures, and challenges for research and management. *Science of the Total Environment* 577: 1–18.

Skublics, D., Blöschl, G., and Rutschmann, P. (2016). Effect of river training on flood retention of the Bavarian Danube. *Journal of Hydrology and Hydromechanics* 64: 349–356.

Slaney, P.A. and Martin, A.D. (1997). The watershed restoration program of British Columbia: accelerating natural recovery processes. *Water Quality Research Journal of Canada* 32: 325–346.

Slattery, M.C. and Bryan, R.B. (1994). Surface seal development under simulated rainfall on an actively eroding surface. *Catena* 22: 17–34.

Slingerland, R. and Smith, N.D. (1998). Necessary conditions for a meandering-river avulsion. *Geology* 26: 435–438.

Small, E.E. and Anderson, R.S. (1998). Pleistocene relief production in Laramide mountain ranges, western United States. *Geology* 26: 123–126.

Smart, J.S. (1968). Statistical properties of stream lengths. *Water Resources Research* 4: 1001–1014.

Smart, G.M., M.N.R. Jaeggi. 1983. Sediment transport on steep slopes. Versuchsanstalt für Wasserbau, Hydrologie und Glaziologie, Mitteilungen 64, Eidgenössische Technische Hochschule Zürich, Switzerland.

Smith, N. (1971). *A History of Dams*. London: Peter Davies.

Smith, C.R. (1996). Coherent flow structures in smooth-wall turbulent boundary layers: facts, mechanisms and speculation. In: *Coherent Flow Structures in Open Channels* (eds. P. Ashworth, S. Bennett, J.L. Best and S. McLelland), 1–39. Chichester: Wiley.

Smith, M.W. (2014). Roughness in the Earth Sciences. *Earth-Science Reviews* 136: 202–225.

Smith, H.G. and Dragovich, D. (2008). Post-fire hillslope erosion response in a sub-alpine environment, south-eastern Australia. *Catena* 73: 274–285.

Smith, N.D. and Perez-Arlucea, M. (2008). Natural levee deposition during the 2005 flood of the Saskatchewan River. *Geomorphology* 101: 583–594.

Smith, M.W. and Vericat, D. (2015). From experimental plots to experimental landscapes: topography, erosion and deposition in sub-humid badlands from Structure-from-Motion photogrammetry. *Earth Surface Processes and Landforms* 40: 1656–1671.

Smith, S.M.C. and Wilcock, P.R. (2015). Upland sediment supply and its relation to watershed sediment delivery in the contemporary mid-Atlantic Piedmont (USA). *Geomorphology* 232: 33–46.

Smith, J.A., Villarini, G., and Baeck, M.L. (2011). Mixture distributions and the hydroclimatology of extreme rainfall and flooding in the eastern United States. *Journal of Hydrometeorology* 12: 294–309.

Smock, L.A., Gladden, J.E., Riekenberg, J.L. et al. (1992). Lotic macroinvertebrate production in three dimensions: channel surface, hyporheic, and floodplain environments. *Ecology* 73: 876–886.

Snow, R.S. and Slingerland, R.L. (1987). Mathematical modeling of graded river profiles. *Journal of Geology* 95: 15–33.

Snyder, N.P., Whipple, K.X., Tucker, G.E., and Merritts, D.J. (2000). Landscape response to tectonic forcing: digital elevation model analysis of stream profiles in the Mendocino triple junction region, northern California. *Geological Society of America Bulletin* 112: 1250–1263.

Sobey, I.J. (1982). Oscillatory flows at intermediate Strouhal number in asymmetry channels. *Journal of Fluid Mechanics* 125: 359–373.

Sofia, G., Tarolli, P., Cazorzi, F., and Dalla Fontana, G. (2015). Downstream hydraulic geometry relationships: gathering reference reach-scale width values from LIDAR. *Geomorphology* 250: 236–248.

Soldner, M., Stephen, I., Ramos, L. et al. (2004). Relationships between macroinvertebrates fauna and environmental variables in small streams of the Dominican Republic. *Water Research* 38: 863–874.

Somoza, L., Barnolas, A., Arasa, A. et al. (1998). Architectural stacking patterns of the Ebro delta controlled by Holocene high-frequency eustatic fluctuations, delta-lobe switching and subsidence processes. *Sedimentary Geology* 117: 11–32.

Sousa de Morais, E., dos Santos, M.L., Cremon, E.H., and Stevaux, J.C. (2016). Floodplain evolution in a confluence zone: Paraná and Ivaí Rivers, Brazil. *Geomorphology* 257: 1–9.

Southard, J.B. (1991). Experimental determination of bed-form stability. *Annual Review of Earth and Planetary Sciences* 19: 423–455.

Southard, J.B. and Boguchwal, L.A. (1990). Bed configuration in steady unidirectional water flows; Part 2, Synthesis of flume data. *Journal of Sedimentary Research* 60: 658–679.

Southerland, W.B., F. Reckendorf. 2010. Performance of engineered log jams in Washington State – post project appraisal. 2nd Joint Federal Interagency Conference on Sedimentation and Hydrologic Modeling, Las Vegas, NV, June 27–July 1, 2010.

Spaliviero, M. (2003). Historic fluvial development of the Alpine-foreland Tagliamento River, Italy, and consequences for floodplain management. *Geomorphology* 52: 317–333.

Spence, C. and Woo, M.K. (2003). Hydrology of subarctic Canadian Shield: bedrock upland. *Journal of Hydrology* 262: 111–127.

Spence, C. and Woo, M.K. (2006). Hydrology of subarctic Canadian Shield: soil-filled valleys. *Journal of Hydrology* 279: 151–166.

Spencer, T., Douglas, I., Greer, T., and Sinun, W. (1990). Vegetation and fluvial geomorphic processes in South-East Asian Tropical rainforests. In: *Vegetation and Erosion: Processes and Environments* (ed. J.B. Tornes), 451–469. Chichester: Wiley.

Spencer, T., Schuerch, M., Nicholls, R.J. et al. (2016). Global coastal wetland change under sea-level rise and related stresses: the DIVA Wetland Change Model. *Global and Planetary Change* 139: 15–30.

Spooner, N.A., Olley, J.M., Questiaux, D.G., and Chen, X.Y. (2001). Optical dating of an Aeolian deposit on the Murrumbidgee floodplain. *Quaternary Science Reviews* 20: 835–840.

Spotila, J.A., Moskey, K.A., and Prince, P.S. (2015). Geologic controls on bedrock channel width in large, slowly-eroding catchments: case study of the New River in eastern North America. *Geomorphology* 230: 51–63.

Springer, G.S. (2002). Caves and their potential use in paleoflood studies. In: *Ancient Floods, Modern Hazards: Principles and Applications of Paleoflood Hydrology* (eds. P.K. House, R.H. Webb, W.R. Baker and D.R. Levish), 329–343. Washington, DC: American Geophysical Union Press.

Springer, G.S. (2004). A pipe-based, first approach to modeling closed conduit flow in caves. *Journal of Hydrology* 284: 178–189.

Springer, G.S. and Wohl, E.E. (2002). Empirical and theoretical investigations of sculpted forms in Buckeye Creek Cave, West Virginia. *Journal of Geology* 110: 469–481.

Springer, G.S., Wohl, E.E., Foster, J.A., and Boyer, D.G. (2003). Testing for reach-scale adjustments of hydraulic variables to soluble and insoluble strata: Buckeye Creek and Greenbrier River, West Virginia. *Geomorphology* 56: 201–217.

Springer, G.A., Tooth, S., and Wohl, E.E. (2006). Theoretical modeling of stream potholes based upon empirical observations from the Orange River, Republic of South Africa. *Geomorphology* 82: 160–176.

Springer, G.S., Poston, H.A., Hardt, B., and Rowe, H.D. (2015). Groundwater lowering and stream incision rates in the Central Appalachian Mountains of West Virginia, USA. *International Journal of Speleology* 44: 99–105.

Sridhar, V. and Nayak, A. (2010). Implications of climate-driven variability and trends for the hydrologic assessment of the Reynolds Creek Experimental Watershed, Idaho. *Journal of Hydrology* 385: 183–202.

Staentzel, C., Arnaud, F., Combroux, I. et al. (2018). How do instream flow increase and gravel augmentation impact biological communities in large rivers: a case study on the Upper Rhine River. *River Research and Applications* 34: 153–164.

Staley, D. and Wasklewicz, T.A. (2006). Surficial patterns of debris flow deposition on alluvial fans in Death Valley, CA using airborne laser swath mapping data. *Geomorphology* 74: 152–163.

Stalnaker, C., B.L. Lamb, J. Henriksen, K. Bovee, J. Bartholow. 1995. The instream flow incremental methodology: a primer for IFIM. National Biological Service, US Department of the Interior, Biological Report no 29, Fort Collins, Colorado.

Stanford, J.A. and Ward, J.V. (1988). The hyporheic habitat of river ecosystems. *Nature* 335: 64–66.

Stanley, J.-D. and Clemente, P.L. (2017). Increased land subsidence and sea-level rise are submerging Egypt's Nile Delta coastal margin. *GSA Today* 27: 4–11.

Stanley, E.H. and Doyle, M.W. (2003). Trading off: the ecological effects of dam removal. *Frontiers in Ecology and the Environment* 1: 15–22.

Stanley, E.H., Luebke, M.A., Doyle, M.W., and Marshall, D.W. (2002). Short-term changes in channel form and macroinvertebrate communities following low-head dam removal. *Journal of the North American Benthological Society* 21: 172–187.

Stark, C.P. (2006). A self-regulating model of bedrock river channel geometry. *Geophysical Research Letters* 33, L04402.

Statzner, B. and Peltret, O. (2006). Assessing potential abiotic and biotic complications of crayfish-induced travel transport in experimental streams. *Geomorphology* 74: 245–256.

Statzner, B. and Sagnes, P. (2008). Crayfish and fish as bioturbators of streambed sediments: assessing joint effects of species with different mechanistic abilities. *Geomorphology* 93: 267–287.

Statzner, B., Fuchs, U., and Higler, L.W.G. (1996). Sand erosion by mobile predaceous stream insects: implications for ecology and hydrology. *Water Resources Research* 32: 2279–2287.

Stedinger, J.R. and Baker, V.R. (1987). Surface water hydrology: historical and paleoflood information. *Reviews of Geophysics* 25: 119–124.

Steffen, W., Richardson, K., Rockstrom, J. et al. (2015). Planetary boundaries: guiding human development on a changing planet. *Science* 347: 1259855.

Stella, J.C., Hayden, M.K., Battles, J.L. et al. (2011). The role of abandoned channels as refugia for sustaining pioneer riparian forest ecosystems. *Ecosystems* 14: 776–790.

Stephens, D.B. and Stephens, D.A. (2006). British land drainers: their place among pre-Darcy forefathers of applied hydrogeology. *Hydrogeology Journal* 14: 1367–1376.

Sterling, S.M. and Church, M. (2002). Sediment trapping characteristics of a pit trap and the Helley-Smith sampler in a cobble gravel bed river. *Water Resources Research* 38: 1144.

Sternai, P., Herman, F., Fox, M.R., and Catelltort, S. (2011). Hypsometric analysis to identify spatially variable glacial erosion. *Journal of Geophysical Research* 116, F03001.

Sternberg, H. (1875). Untersuchungen über Längen- und Querprofile geschiebführender Flüsse. *Zeitschrift für Bauwesen* 25: 483–506.

Stewart, I.T., Cayan, D.R., and Dettinger, M.D. (2004). Changes in snowmelt runoff timing in western North America under a "business as usual" climate change scenario. *Climatic Change* 62: 217–232.

Stewart, I.T., Cayan, D.R., and Dettinger, M.D. (2005). Changes toward earlier streamflow timing across western North America. *Journal of Climate* 18: 1136–1155.

Stock, J.D. (2013). Waters divided: a history of alluvial fan research and a view of its future. In: *Treatise on Fluvial Geomorphology* (ed. E. Wohl), 414–457. Amsterdam: Elsevier.

Stock, J.D. and Dietrich, W.E. (2003). Valley incision by debris flows: evidence of a topographic signature. *Water Resources Research* 39: 1089.

Stock, J.D. and Dietrich, W.E. (2006). Erosion of steepland valleys by debris flows. *Geological Society of America Bulletin* 118: 1125–1148.

Stock, J.D. and Montgomery, D.R. (1999). Geologic constraints on bedrock river incision using the stream power law. *Journal of Geophysical Research* 104: 4983–4993.

Stock, J.D., Montgomery, D.R., Collins, B.D. et al. (2005). Field measurements of incision rates following bedrock exposure: implications for process controls on the long profiles of valleys cut by rivers and debris flows. *Geological Society of America Bulletin* 117: 174–194.

Stock, J.D., Schmidt, K.M., and Miller, D.M. (2007). Controls on alluvial fan long-profiles. *Geological Society of America Bulletin* 120: 619–640.

Stoddard, J.L., Larsen, D.P., Hawkins, C.P. et al. (2006). Setting expectations for the ecological condition of streams: the concept of reference condition. *Ecological Applications* 16: 1267–1276.

Stoffel, M. and Corona, C. (2014). Dendroecological dating of geomorphic disturbance in trees. *Tree-Ring Research* 70: 3–20.

Stokes, M. and Mather, A.E. (2015). Controls on modern tributary junction-alluvial fan occurrence and morphology: high Atlas Mountains, Morocco. *Geomorphology* 248: 344–362.

Stokes, S. and Walling, D. (2003). Radiogenic and isotopic methods for the direct dating of fluvial sediments. In: *Tools in Fluvial Geomorphology* (eds. G.M. Kondolf and H. Piégay), 233–267. Chichester: Wiley.

Storey, R.G., Howard, K.W.F., and Williams, D.D. (2003). Factors affecting riffle-scale hyporheic exchange flows and their seasonal changes in a gaining stream: a three-dimensional groundwater flow model. *Water Resources Research* 39: 1034.

Storz-Peretz, Y. and Laronne, J.B. (2018). The morpho-textural signature of large bedforms in ephemeral gravel-bed channels of various planforms. *Hydrological Processes* 32: 617–635.

Stoughton, J.A. and Marcus, W.A. (2000). Persistent impacts of trace metals from mining on floodplain grass communities along Soda Butte Creek, Yellowstone National Park. *Environmental Management* 25: 305–320.

Stout, J.C., Rutherfurd, I., Grove, J. et al. (2017). Using the Weibull distribution to improve the description of riverine wood loads. *Earth Surface Processes and Landforms* 42: 647–656.

Stout, J.C., Rutherfurd, I.D., Grove, J. et al. (2018). Passive recovery of wood loads in rivers. *Water Resources Research* 54: 8828–8846.

Strahler, A.N. (1952). Hypsometric (area–altitude) analysis of erosional topography. *Geological Society of America Bulletin* 63: 1117–1142.

Strahler, A.N. (1957). Quantitative analysis of watershed geomorphology. *Transactions of the American Geophysical Union* 38: 913–920.

Strahler, A.N. (1964). Quantitative geomorphology of drainage basins and channel networks. In: *Handbook of Applied Hydrology* (ed. V.T. Chow), 4-40–4-74. New York: McGraw Hill.

Straumann, R.K. and Korup, O. (2009). Quantifying postglacial sediment storage at the mountain-belt scale. *Geology* 37: 1079–1082.

Strecker, M.R., Hilley, G.E., Arrowsmith, J.R., and Cout, I. (2003). Differential structural and geomorphic mountain-front evolution in an active continental collision zone: the Northwest Pamir, southern Kyrgyzstan. *Geological Society of America Bulletin* 115: 166–181.

Stromberg, J.C., Beauchamp, V.B., Dixon, M.D. et al. (2007). Importance of low-flow and high-flow characteristics to restoration of riparian vegetation along rivers in arid south-western United States. *Freshwater Biology* 52: 651–679.

Sullivan, A.B. and Drever, J.I. (2001). Spatiotemporal variability in stream chemistry at a high-elevation catchment affected by mine drainage. *Journal of Hydrology* 252: 237–250.

Surian, N. (1999). Channel changes due to river regulation: the case of the Piave River, Italy. *Earth Surface Processes and Landforms* 24: 1135–1151.

Surian, N. (2002). Downstream variation in grain size along an Alpine river: analysis of controls and processes. *Geomorphology* 43: 137–149.

Surian, N., Righini, M., Lucia, A. et al. (2016). Channel response to extreme floods: insights on controlling factors from six mountain rivers in northern Apennines, Italy. *Geomorphology* 272: 78–91.

Sutfin, N.A. and Wohl, E. (2017). Substantial soil organic carbon retention along floodplains of mountain streams. *Journal of Geophysical Research: Earth Surface* 122: 1325–1338.

Sutfin, N.A., Shaw, J., Wohl, E.E., and Cooper, D. (2014). A geomorphic classification of ephemeral channels in a mountainous, arid region, southwestern Arizona, USA. *Geomorphology* 221: 164–175.

Sutfin, N.A., Wohl, E.E., and Dwire, K.A. (2016). Banking carbon: a review of organic carbon storage and physical factors influencing retention in floodplains and riparian ecosystems. *Earth Surface Processes and Landforms* 41: 38–60.

Sutherland, R.A. (1991). Caesium-137 and sediment budgeting within a partially closed drainage basin. *Zeitschrift für Geomorphologie* 35: 47–63.

Swanson, F.J., Benda, L.E., Duncan, S.H. et al. (1987). Mass failures and other processes of sediment production in Pacific Northwest forest landscapes. In: *Streamside Management: Forestry and Fishery Implications* (eds. E.O. Salo and T.W. Cundy), 9–38. Seattle, WA: University of Washington, Institute of Forest Resources.

Swanson, K.M., Watson, E., Aalto, R. et al. (2008). Sediment load and floodplain deposition rates: comparison of the Fly and Strickland rivers, Papua New Guinea. *Journal of Geophysical Research* 11, F01S03.

Symons, W.O., Sumner, E.J., Paull, C.K. et al. (2017). A new model for turbidity current behavior based on integration of flow monitoring and precision coring in a submarine canyon. *Geology* 45: 367–370.

Syvitski, J.P.M. and Saito, Y. (2007). Morphodynamics of deltas under the influence of humans. *Global and Planetary Change* 57: 261–282.

Syvitski, J.P.M., Vörösmarty, C.J., Kettner, A.J., and Green, P. (2005). Impact of humans on the flux of terrestrial sediment to the global coastal ocean. *Science* 308: 376–380.

Syvitski, J.P.M., Kettner, A.J., Overeem, I. et al. (2009). Sinking deltas due to human activities. *Nature Geoscience* 2: 681–686.

Syvitski, J.P.M., Kettner, A.J., Overeem, I. et al. (2013). Anthropocene metamorphosis of the Indus Delta and lower floodplain. *Anthropocene* 3: 24–35.

Tabbachi, E., Lambs, L., Guilloy, H. et al. (2000). Impacts of riparian vegetation on hydrological processes. *Hydrological Processes* 14: 2959–2976.

Tadich, T.A., Novaro, A.J., Kunzle, P. et al. (2018). Agonistic behavior between introduced beaver (*Castor canadensis*) and endemic culpeo fox (*Pseudalopex culpaeus lycoides*) in Tierra del Fuego Island and implications. *Acta Ethologica* 21: 29–34.

Tague, C. (2009). Assessing climate change impacts on alpine stream-flow and vegetation water use: mining the linkages with subsurface hydrologic processes. *Hydrological Processes* 23: 1815–1819.

Tague, C., Farrell, M., Grant, G. et al. (2008). Deep groundwater mediates streamflow response to climate warming in the Oregon Cascades. *Climate Change* 86: 1–2.

Tal, M. and Paola, C. (2007). Dynamic single-thread channels maintained by the interaction of flow and vegetation. *Geology* 35: 347–350.

Tal, M., Gran, K., Murray, A.B. et al. (2004). Riparian vegetation as a primary control on channel characteristics in multi-thread rivers. In: *Riparian Vegetation and Fluvial Geomorphology* (eds. S.J. Bennett and A. Simon), 43–58. Washington, DC: American Geophysical Union Press.

Tamura, T., Horaguchi, K., Saito, Y. et al. (2010). Monsoon-influenced variations in morphology and sediment of a mesotidal beach on the Mekong River delta coast. *Geomorphology* 116: 11–23.

Tamura, T., Saito, Y., Nguyen, V.L. et al. (2012). Origin and evolution of interdistributary delta plains; insights from Mekong River delta. *Geology* 40: 303–306.

Taormina, R. and Chau, K.-W. (2015). Data-driven input variable selection for rainfall-runoff modeling using binary-coded particle swarm optimization and Extreme Learning Machines. *Journal of Hydrology* 529: 1617–1632.

Taquet, M., Sancho, G., Dagorn, L. et al. (2007). Characterizing fish communities associated with drifting fish aggregating devices (FADs) in the Western Indian Ocean using underwater visual surveys. *Aquatic Living Resources* 20: 331–341.

Tarnocai, C., Canadell, J.G., Schuur, E.A.G. et al. (2009). Soil organic carbon pools in the northern circumpolar permafrost region. *Global Biogeochemical Cycles* 23, GB2023.

Tarolli, P. and Dalla Fontana, G. (2009). Hillslope-to-valley transition morphology: new opportunities from high resolution DTMs. *Geomorphology* 113: 47–56.

Taylor, M.P. and Kesterton, R.G.H. (2002). Heavy metal contamination of an arid river environment: Gruben River, Namibia. *Geomorphology* 42: 311–327.

Taylor, S., Feng, X., Kirchner, J.W. et al. (2001). Isotopic evolution of a seasonal snowpack and its melt. *Water Resources Research* 37: 759–769.

Tejedor, A., Longjas, A., Zaliapin, I., and Foufoula-Georgiou, E. (2015). Delta channel networks: 1. A graph-theoretic approach for studying connectivity and steady state transport on deltaic surfaces. *Water Resources Research* 51: 3998–4018.

Tesfa, T.K., Tarboton, D.G., Chandler, D.G., and McNamara, J.P. (2009). Modeling soil depth from topographic and land cover attributes. *Water Resources Research* 45, W10438.

Tessler, Z.D., Vorosmarty, C.J., Grossberg, M. et al. (2016). A global empirical typology of anthropogenic drivers of environmental change in deltas. *Sustainability Science* 11: 525–537.

Tharme, R.E. (2003). A global perspective on environmental flow assessment: emerging trends in the development and application of environmental flow methodologies for rivers. *River Research and Applications* 19: 397–441.

Thayer, J.B. and Ashmore, P. (2016). Floodplain morphology, sedimentology, and development processes of a partially alluvial channel. *Geomorphology* 269: 160–174.

Thayer, J.B., Phillips, R.T.J., and Desloges, J.R. (2016). Downstream channel adjustment in a low-relief, glacially conditioned watershed. *Geomorphology* 262: 101–111.

Theule, J., G. Bertoldi, F. Comiti, P. Macconi, B. Mazzorana. 2015. Exploring topographic methods for monitoring morphological changes in mountain channels of different size and slope. Abstract, EGU General Assembly 2015, Vienna, Austria, id 8893.

Thom, R. (1975). *Structural Stability and Morphogenesis: An Outline of a General Theory of Models.* Reading, PA: W.A. Benjamin.

Thomas, M.F., Nott, J., Murray, A.S., and Price, D.M. (2007). Fluvial response to late Quaternary climate change in NE Queensland, Australia. *Palaeogeography, Palaeoclimatology, Palaeoecology* 251: 119–136.

Thompson, A. (1986). Secondary flows and the pool-riffle unit: a case study of the processes of meander development. *Earth Surface Processes and Landforms* 11: 631–641.

Thompson, D.M. (2001). Random controls on semi-rhythmic spacing of pools and riffle in constriction-dominated rivers. *Earth Surface Processes and Landforms* 26: 1195–1212.

Thompson, D.M. (2002). Long-term effect of instream habitat-improvement structures on channel morphology along the Blackledge and Salmon Rivers, Connecticut, USA. *Environmental Management* 29: 250–265.

Thompson, D.M. (2004). The influence of pool length on local turbulence production and energy slope: a flume experiment. *Earth Surface Processes and Landforms* 29: 1341–1358.

Thompson, D.M. (2006). Did the pre-1980 use of in-stream structures improve streams? A reanalysis of historical data. *Ecological Applications* 16: 784–796.

Thompson, D.M. (2008). The influence of lee sediment behind large bed elements on bedload transport rates in supply-limited channels. *Geomorphology* 99: 420–432.

Thompson, D.M. (2012). The challenge of modeling pool-riffle morphologies in channels with different densities of large woody debris and boulders. *Earth Surface Processes and Landforms* 37: 223–239.

Thompson, D.M. (2013). Pool-riffle. In: *Treatise on Fluvial Geomorphology* (ed. E. Wohl), 364–378. Amsterdam: Elsevier.

Thompson, D.M. and Fixler, S.A. (2017). Formation and maintenance of a forced pool-riffle couplet following loading of large wood. *Geomorphology* 296: 74–90.

Thompson, D.M. and Stull, G.N. (2002). The development and historic use of habitat structures in channel restoration in the United States: the grand experiment in fisheries management. *Géographie Physique et Quaternaire* 56: 45–60.

Thompson, D.M., Wohl, E.E., and Jarrett, R.D. (1996). A revised velocity-reversal and sediment-sorting model for a high-gradient, pool-riffle stream. *Physical Geography* 17: 142–156.

Thompson, D.M., Nelson, J.M., and Wohl, E.E. (1998). Interactions between pool geometry and hydraulics. *Water Resources Research* 34: 3673–3681.

Thompson, D.M., Wohl, E.E., and Jarrett, R.D. (1999). Velocity reversals and sediment sorting in pools and riffles controlled by channel constrictions. *Geomorphology* 27: 229–241.

Thorne, C.R. and Lewin, J. (1979). Bank processes, bed material movement and planform development in a meandering river. In: *Adjustments of the Fluvial System* (eds. D.D. Rhodes and G.P. Williams), 117–137. London: George Allen and Unwin.

Thurman, E.M. (1985). *Organic Geochemistry of Natural Waters*. Dordrecht: Martinus Nijhoff/Dr. W Junk.

Tinkler, K.J. and Wohl, E.W. (1998). A primer on bedrock channels. In: *Rivers Over Rock: Fluvial Processes in Bedrock Channels* (eds. K.J. Tinkler and E.E. Wohl), 1–18. Washington, DC: American Geophysical Union Press.

Tobin, G.A. (1995). The levee love affair: a stormy relationship? *Journal of American Water Resources Association* 31: 359–367.

Tockner, K., Schiemer, F., and Ward, J.V. (1998). Conservation by restoration: the management concept for a river-floodplain system on the Danube River in Austria. *Aquatic Conservation: Marine and Freshwater Ecosystems* 8: 71–86.

Tockner, K., Schiemer, F., Baumgartner, C. et al. (1999). The Danube restoration project: species diversity patterns across connectivity gradients in the floodplain system. *Regulated Rivers: Research and Management* 15: 245–258.

Tockner, K., Malard, F., and Ward, J.V. (2000). An extension of the flood-pulse concept. *Hydrological Processes* 14: 2861–2883.

Tolkinnen, M., Mykra, H., Annala, M. et al. (2015). Multi-stressor impacts on fungal diversity and ecosystem functions in streams: natural vs. anthropogenic stress. *Ecology* 96: 672–683.

Tonina, D. and Buffington, J.M. (2009). Hyporheic exchange in mountain rivers I: Mechanics and environmental effects. *Geography Compass* 3 https://doi.org/10.1111/j.1749-8198.2009.00226.x.

Tonina, D. and Jorde, K. (2013). Hydraulic modelling approaches for ecohydraulic studies: 3D, 2D, 1D, and non-numerical models. In: *Ecohydraulics: An Integrated Approach* (eds. I. Maddock, A. Harby, P. Kemp and P. Wood), 31–74. Chichester: Wiley-Blackwell.

Tonina, D., McKean, J.A., Benjankar, R.M. et al. (2019). Mapping river bathymetries: evaluating topobathymetric LiDAR survey. *Earth Surface Processes and Landforms*.

Toone, J., Rice, S.P., and Piégay, H. (2014). Spatial discontinuity and temporal evolution of channel morphology along a mixed bedrock-alluvial river, upper Drôme River, southeast France: contingent responses to external and internal controls. *Geomorphology* 205: 5–16.

Tooth, S. (1999). Floodouts in central Australia. In: *Varieties of Fluvial Form* (eds. A.J. Miller and A. Gupta), 219–247. Chichester: Wiley.

Tooth, S. (2000). Process, form, and change in dryland rivers: a review of recent research. *Earth-Science Reviews* 51: 67–107.

Tooth, S. (2013). Dryland fluvial environments: assessing distinctiveness and diversity from a global perspective. In: *Treatise on Fluvial Geomorphology* (ed. E. Wohl), 613–643. Amsterdam: Elsevier.

Tooth, S. and McCarthy, T.S. (2004). Anabranching in mixed bedrock-alluvial rivers: the example of the Orange River above Augrabies Falls, Northern Cape Province, South Africa. *Geomorphology* 57: 235–262.

Tooth, S. and Nanson, G.C. (2000). Equilibrium and nonequilibrium conditions in dryland rivers. *Physical Geography* 21: 183–211.

Torres, R., Mouginis-Mark, P., Self, S. et al. (2004). Monitoring the evolution of the Pasig-Potrero alluvial fan, Pinatubo Volcano, using a decade of remote sensing data. *Journal of Volcanology and Geothermal Research* 138: 371–392.

Toth, L.A., Obeysekera, J.T.B., Perkins, W.A., and Loftin, M.K. (1993). Flow regulation and restoration of Florida's Kissimmee River. *Regulated Rivers: Research and Management* 8: 155–166.

Tranmer, A.W., Tonina, D., Benjankar, R. et al. (2015). Floodplain persistence and dynamic-equilibrium conditions in a canyon environment. *Geomorphology* 250: 147–158.

Trayler, C.R. and Wohl, E.E. (2000). Seasonal changes in bed elevation in a step-pool channel, Rocky Mountains, Colorado, USA. *Arctic, Antarctic, and Alpine Research* 32: 95–103.

Trimble, S.W. (1983). A sediment budget for Coon Creek basin in the Driftless Area, Wisconsin, 1853–1977. *American Journal of Science* 283: 454–474.

Trimble, S.W. (2013). *Historical Agriculture and Soil Erosion in the Upper Mississippi Valley Hill Country*. Boca Raton, FL: CRC Press.

Trimble, S.W. and Mendel, A.C. (1995). The cow as a geomorphic agent – a critical review. *Geomorphology* 13: 233–253.

Triska, F.J. (1984). Role of wood debris in modifying channel geomorphology and riparian areas of a large lowland river under pristine conditions: a historical case study. *Verhandlungen des Internationalen Verein Limnologie* 22: 1876–1892.

Troiani, F., Galve, J.P., Piacentini, D. et al. (2014). Spatial analysis of stream length-gradient (SL) index for detecting hillslope processes: a case of the Gallego River headwaters (Central Pyrenees, Spain). *Geomorphology* 214: 183–197.

Trumbore, S.E. and Czimczik, C.I. (2008). An uncertain future for soil carbon. *Science* 321: 1455–1456.

Tsai, H., Hseu, Z.-Y., Huang, W.-S., and Chen, Z.-S. (2007). Pedogenic approach to resolving the geomorphic evolution of the Pakua River terraces in central Taiwan. *Geomorphology* 83: 14–28.

Tucker, G.E. and Hancock, G.R. (2010). Modelling landscape evolution. *Earth Surface Processes and Landforms* 35: 28–50.

Tucker, G.E. and Slingerland, R. (1997). Drainage basin responses to climate change. *Water Resources Research* 33: 2031–2047.

Tucker, G.E., Lancaster, S., Gasparini, N., and Bras, R. (2001). The Channel-Hillslope Integrated Landscape Development model (CHILD). In: *Landscape Erosion and Evolution Modeling* (eds. R.S. Harmon and W.W. Doe). New York: Kluwer Academic/Plenum Publishers.

Tucker, G.E., Arnold, L., Bras, R.L. et al. (2006). Headwater channel dynamics in semiarid rangelands, Colorado high plains, USA. *Geological Society of America Bulletin* 118: 959–974.

Turk, J.K., Goforth, B.R., Graham, R.C., and Kendrick, J.K. (2008). Soil morphology of a debris flow chronosequence in a coniferous forest, southern California, USA. *Geoderma* 146: 157–165.

Turley, M.D., Bilotta, G.S., Arbociute, G. et al. (2017). Quantifying submerged deposited fine sediments in rivers and streams using digital image analysis. *River Research and Applications* 33: 1585–1595.

Turowski, J.M. and Cook, K.L. (2017). Field techniques for measuring bedrock erosion and denudation. *Earth Surface Processes and Landforms* 42: 109–127.

Turowski, J.M. and Rickenmann, D. (2009). Tools and cover effects in bedload transport observations in the Pitzbach, Austria. *Earth Surface Processes and Landforms* 34: 26–37.

Turowski, J.M., Hovius, N., Wilson, A., and Horng, M.J. (2008). Hydraulic geometry, river sediment and the definition of bedrock channels. *Geomorphology* 99: 26–38.

Turowski, J.M., Rickenmann, D., and Dadson, S.J. (2010). The partitioning of the total sediment load of a river into suspended load and bedload: a review of empirical data. *Sedimentology* 57: 1126–1146.

Turowski, J.M., Badoux, A., Bunte, K. et al. (2013). The mass distribution of coarse particulate organic matter exported from an Alpine headwater stream. *Earth Surface Dynamics* 1: 1–11.

Turowski, J.M., Hilton, R.G., and Sparkes, R. (2016). Decadal carbon discharge by a mountain stream is dominated by coarse organic matter. *Geology* 44: 27–30.

Tuttle, A.K., McMillan, S.K., Gardner, A., and Jennings, G.D. (2014). Channel complexity and nitrate concentrations drive denitrification rates in urban restored and unrestored streams. *Ecological Engineering* 73: 770–777.

Uchida, T., Ohte, N., Kimoto, A. et al. (2000). Sediment yield on a devastated hill in southern China: effects of microbiotic crust on surface erosion process. *Geomorphology* 32: 129–145.

Uchida, T., Kosugi, K., and Mizuyama, T. (2001). Effects of pipeflow on hydrological processes and its relation to landslide: a review of pipeflow studies in forested headwater catchments. *Hydrological Processes* 15: 2151–2174.

Umar, M., Rhoads, B.L., and Greenberg, J.A. (2018). Use of multispectral satellite remote sensing to assess mixing of suspended sediment downstream of large river confluences. *Journal of Hydrology* 556: 325–338.

Valla, P.G., van der Beek, P.A., and Lague, D. (2010). Fluvial incision into bedrock: insights from morphometric analysis and numerical modeling of gorges incising glacial hanging valleys (Western Alps, France). *Journal of Geophysical Research* 115, F02010.

Valyrakis, M., Diplas, P., and Dancey, C.L. (2011). Entrainment of coarse grains in turbulent flows: an extreme value theory approach. *Water Resources Research* 47 https://doi.org/10.1029/2010WR010236.

Van Breemen, N., Boyer, E.W., Goodale, C.L. et al. (2002). Where did all the nitrogen go? Fate of nitrogen inputs to large watersheds in the northeastern USA. *Biogeochemistry* 57/58: 267–293.

Van Cleve, K., Dyrness, C.T., Marion, G.M., and Erickson, R. (1993). Control of soil development on the Tanana River floodplain, interior Alaska. *Canadian Journal of Forest Research* 23: 941–955.

Van de Lageweg, W.I., Schuurman, F., Cohen, K.M. et al. (2016). Preservation of meandering river channels in uniformly aggrading channel belts. *Sedimentology* 63: 586–608.

Van De Wiel, M.J. and Darby, S.E. (2007). A new model to analyse the impact of woody riparian vegetation on the geotechnical stability of riverbanks. *Earth Surface Processes and Landforms* 32: 2185–2198.

Van de Wiel, M.J., Coulthard, T.J., Macklin, M.G., and Lewin, J. (2007). Embedding reach-scale fluvial dynamics within the CAESAR cellular automaton landscape evolution model. *Geomorphology* 90: 283–301.

Van De Wiel, M.J., Rousseau, Y.Y., and Darby, S.E. (2016). Models in fluvial geomorphology. In: *Tools in Fluvial Geomorphology* (eds. G.M. Kondolf and H. Piegay), 383–411. Chichester: Wiley-Blackwell.

Van den Berg, J.H. (1995). Prediction of alluvial channel pattern of perennial rivers. *Geomorphology* 12: 259–279.

Van den Berg, M.W. and van Hoof, T. (2001). The Maas Terrace sequence at Maastricht, SE Netherlands; evidence for 200 m of late Neogene and Quaternary surface uplift. In: *River Basin Sediment Systems: Archives of Environmental Change* (eds. D. Maddy, M.G. Macklin and J.C. Woodward), 45–86. Lisse: A.A. Balkema.

Van der Beek, P. and Bishop, P. (2003). Cenozoic river profile development in the Upper Lachlan catchment (SE Australia) as a test of quantitative fluvial incision models. *Journal of Geophysical Research* 108: 2309.

Van der Nat, D., Tockner, K., Edwards, P.J., and Ward, J.V. (2003). Large wood dynamics of complex Alpine river floodplains. *Journal of the North American Benthological Society* 22: 35–50.

Van Der Schrier, G., Efthymiadis, D., Briffa, K.R., and Jones, P.D. (2007). European Alpine moisture variability for 1800-2003. *International Journal of Climatology* 27: 415–427.

Van Der Wal, D. and Pye, K. (2003). The use of historical bathymetric charts in a GIS to assess morphological change in estuaries. *The Hydrographic Journal* 110: 3–9.

Van Dijk, W.M., van de Lageweg, W.I., and Kleinhans, M.G. (2012). Experimental meandering river with chute cutoffs. *Journal of Geophysical Research: Earth Surface* 117, F03023.

Van Nieuwenhuyse, E.E. and LaPerriere, J.D. (1986). Effects of placer gold mining on primary production in subarctic streams of Alaska. *Water Resources Bulletin* 22: 91–99.

Vanacker, V., von Blanckenburg, F., Govers, G. et al. (2015). Transient river response, captured by channel steepness and its concavity. *Geomorphology* 228: 234–243.

Vandenberghe, J. and Maddy, D. (2000). The significance of fluvial archives in geomorphology. *Geomorphology* 33: 127–130.

Vandenberghe, J. and Woo, M.-K. (2002). Modern and ancient periglacial river types. *Progress in Physical Geography* 26: 479–506.

Vannote, R.L., Minshall, G.W., Cummins, K.W. et al. (1980). The river continuum concept. *Canadian Journal of Fisheries and Aquatic Sciences* 37: 130–137.

Van Rijn, L.C. (1984). Sediment pick-up functions. *Journal of Hydraulic Engineering* 110: 1494–1502.

Vaux, W.G. 1968. Interchange of stream and intragravel water in a salmon spawning riffle. Spec. Sci. Rep. 405, US Fish and Wildlife Service, Washington, DC.

Vazquez-Tarrio, D. (2017). Using UAS optical imagery and SfM photogrammetry to characterize the surface grain size of gravel bards in a braided river (Veneon River, French Alps). *Geomorphology* 285: 94–105.

Veblen, T.T., J.A. Donnegan. 2005. Historical range of variability for forest vegetation of the national forests of the Colorado Front Range. Final report, USDA Forest Service Agreement 1102-0001-99-033, Golden, CO.

Veldkamp, A. and Tebbens, L.A. (2001). Registration of abrupt climate changes within fluvial systems: insights from numerical modelling experiments. *Global and Planetary Change* 28: 129–144.

Venarsky, M.P., Walters, D.M., Hall, R.O. et al. (2018). Shifting stream planform state decreases stream productivity yet increases riparian animal production. *Oecologia* 187: 167–180.

Venditti, J.G. (2013). Bedforms in sand-bedded rivers. In: *Treatise on Fluvial Geomorphology* (ed. E. Wohl), 138–162. Amsterdam: Elsevier.

Venditti, J.G., Church, M., and Bennett, S.J. (2005). Bed form initiation from a flat sand bed. *Journal of Geophysical Research* 110, F01009.

Venditti, J.G., Dietrich, W.E., Nelson, P.A. et al. (2010). Effect of sediment pulse grain size on sediment transport rates and bed mobility in gravel bed rivers. *Journal of Geophysical Research* 115, F03039.

Venditti, J.G., Nelson, P.A., Bradley, R.W. et al. (2017). Bedforms, patches, structures, and sediment supply in gravel-bed rivers. In: *Gravel-Bed Rivers: Processes and Disasters* (eds. D. Tsutsumi and J.B. Laronne), 439–466. Chichester: Wiley.

Vericat, D., Batalla, R.J., and Garcia, C. (2006). Breakup and reestablishment of the armour layer in a large gravel-bed river below dams: the lower Ebro. *Geomorphology* 76: 122–136.

Vezzoli, G. (2004). Erosion in the western Alps (Dora Baltea Basin); 2. Quantifying sediment yield. *Sedimentary Geology* 171: 247–259.

Vileisis, A. (1997). *Discovering the Unknown Landscape: A History of America's Wetlands.* Washington, DC: Island Press.

Vincent, K.R., Friedman, J.M., and Griffin, E.R. (2009). Erosional consequence of saltcedar control. *Environmental Management* 44: 218–227.

Viseras, C., Calvache, M.L., Soria, J.M., and Fernandez, J. (2003). Differential features of alluvial fans controlled by tectonic or eustatic accommodation space; examples from the Beltic Cordillera, Spain. *Geomorphology* 50: 181–202.

Vivoni, E.R., Bowman, R.S., Wyckoff, R.L. et al. (2006). Analysis of a monsoon flood event in an ephemeral tributary and its downstream hydrologic effects. *Water Resources Research* 46, W02509.

Voller, V.R. and Paola, C. (2010). Can anomalous diffusion describe depositional fluvial profiles? *Journal of Geophysical Research* 115, FA0013.

Vörösmarty, C., Lettenmaier, D., Leveque, C. et al. (2004). Humans transforming the global water system. *Eos, Transactions of the American Geophysical Union* 84 (48): 509–520.

Wagenbrenner, J.W., MacDonald, L.H., and Rough, D. (2006). Effectiveness of three post-fire rehabilitation treatments in the Colorado Front Range. *Hydrological Processes* 20: 2989–3006.

Wagner, T., Themessi, M., Schuppel, A. et al. (2017). Impacts of climate change on stream flow and hydro power generation in the Alpine region. *Environment and Earth Science* 76 https://doi.org/10.1007/s12665-016-6318-6.

Wainwright, J. and Parsons, A.J. (2002). The effect of temporal variations in rainfall on scale dependency in runoff coefficients. *Water Resources Research* 38 (12): 1271.

Wainwright, J., Turnbull, L., Ibrahim, T.G. et al. (2011). Linking environmental régimes, space and time: interpretations of structural and functional connectivity. *Geomorphology* 126: 387–404.

Walcott, R.C. and Summerfield, M.A. (2008). Scale dependence of hypsometric integrals: an analysis of southeast African basins. *Geomorphology* 96: 174–186.

Walden, M.G. (2004). Estimation of average stream velocity. *ASCE Journal of Hydraulic Engineering* 130: 1119–1122.

Walker, J., Arnborg, L., and Peippo, J. (1987). Riverbank erosion in the Colville Delta, Alaska. *Geografiska Annaler* 69A: 61–70.

Wallace, J.B., Webster, J.R., and Meyer, J.L. (1995). Influence of log additions on physical and biotic characteristics of a mountain stream. *Canadian Journal of Fisheries and Aquatic Sciences* 52: 2120–2137.

Wallbrink, P., Blake, W., Doerr, S. et al. (2005). Using tracer based sediment budgets to assess redistribution of soil and organic material after severe bush fires. In: *Sediment Budgets 2* (eds. A.J. Horowitz and D.E. Walling), 223–230. Wallingford: IAHS Publications no. 292.

Walling, D.E. (1983). The sediment delivery problem. *Journal of Hydrology* 65: 209–237.

Walling, D.E. and Foster, I. (2016). Using environmental radionuclides, mineral magnetism and sediment geochemistry for tracing and dating fine fluvial sediments. In: *Tools in Fluvial Geomorphology* (eds. G.M. Kondolf and H. Piegay), 183–209. Chichester: Wiley-Blackwell.

Walling, D.E. and Webb, B.W. (1986). Solutes in river systems. In: *Solute Processes* (ed. S.T. Trudgill), 251–327. Chichester: Wiley.

Walling, D.E., Owens, P.N., and Leeks, G.J.L. (1999). Fingerprinting suspended sediment sources in the catchment of the River Ouse, Yorkshire, UK. *Hydrological Processes* 13: 955–975.

Walling, D.E., Owens, P.N., Foster, I.D.L., and Lees, J.A. (2003). Changes in the fine sediment dynamics of the Ouse and Tweed basins in the UK over the last 100–150 years. *Hydrological Processes* 17: 3245–3269.

Walsh, C.J. and Kunapo, J. (2009). The importance of upland flow paths in determining urban effects on stream ecosystems. *Journal of the North American Benthological Society* 28: 977–990.

Walter, R.C. and Merritts, D.J. (2008). Natural streams and the legacy of water-powered mills. *Science* 319: 299–304.

Walvoord, M.A. and Kurylyk, B.L. (2016). Hydrologic impacts of thawing permafrost – a review. *Vadose Zone Journal* 15 https://doi.org/10.2136/vzj2016.01.0010.

Wang, Z., Tuli, A., and Jury, W.A. (2003). Unstable flow during redistribution in homogeneous soil. *Vadose Zone Journal* 2: 52–60.

Wang, G., Ma, H., Qian, J., and Chang, J. (2004). Impact of land use changes on soil carbon, nitrogen and phosphorus and water pollution in an arid region of Northwest China. *Soil Use and Management* 20: 32–39.

Wang, C., Wang, W., He, S. et al. (2011). Sources and distribution of aliphatic and polycyclic aromatic hydrocarbons in Yellow River Delta Nature Reserve, China. *Applied Geochemistry* 26: 1330–1336.

Wang, H., Wright, T.J., Yu, Y. et al. (2012). InSTAR reveals coastal subsidence in the Pearl River delta, China. *Geophysical Journal International* 191: 1119–1128.

Wang, L., Butcher, A.S., Stuart, M.E. et al. (2013). The nitrate time bomb: a numerical way to investigate nitrate storage and lag time in the unsaturated zone. *Environmental Geochemistry and Health* 35: 667–681.

Wang, T., Hamann, A., Spittlehouse, D., and Carroll, C. (2016). Locally downscaled and spatially customizable climate data for historical and future periods for North America. *PLoS One*: 11, e0156720.

Wang, Y.L., Yeh, T.C.J., Wen, J.C. et al. (2017). Characterizing subsurface hydraulic heterogeneity of alluvial fan using riverstage fluctuations. *Journal of Hydrology* 547: 650–663.

Warburton, J. (1992). Observations of bed load transport and channel bed changes in a proglacial mountain stream. *Arctic and Alpine Research* 24: 195–203.

Warburton, J. and Evans, M. (2011). Geomorphic, sedimentary, and potential palaeoenvironmental significance of peat blocks in alluvial river systems. *Geomorphology* 130: 101–114.

Ward, J.V. (1997). An expansive perspective of riverine landscapes: pattern and process across scales. *River Ecosystems* 6: 52–60.

Ward, J.V. and Stanford, J.A. (1983). The serial discontinuity concept of lotic ecosystems. In: *Dynamics of Lotic Ecosystems* (eds. T.D. Fontaine and S.M. Bartell), 29–42. Ann Arbor, MI: Ann Arbor Science.

Ward, J.V. and Stanford, J.A. (1995). The serial discontinuity concept: extending the model to floodplain rivers. *Regulated Rivers: Research and Management* 10: 159–168.

Ward, J.V., Tockner, K., and Schiemer, F. (1999). Biodiversity of floodplain river ecosystems: ecotones and connectivity. *River Research and Applications* 15: 125–139.

Ward, J.V., Tockner, K., Uehlinger, U., and Malard, F. (2001). Understanding natural patterns and processes in river corridors as the basis for effective river restoration. *Regulated Rivers: Research and Management* 17: 311–323.

Ward, A.S., Gooseff, M.N., and Singha, K. (2010a). Characterizing hyporheic transport processes – interpretation of electrical geophysical data in coupled stream-hyporheic zone systems during solute tracer studies. *Advances in Water Resources* 33: 1320–1330.

Ward, A.S., Gooseff, M.N., and Singha, K. (2010b). Imaging hyporheic zone solute transport using electrical resistivity. *Hydrological Processes* 24: 948–953.

Warne, A.G., Toth, L.A., and White, W.A. (2000). Drainage-basin-scale geomorphic analysis to determine reference conditions for ecologic restoration – Kissimmee River, Florida. *Geological Society of America Bulletin* 112: 884–899.

Warrick, J.A. and Mertes, L.A.K. (2009). Sediment yield from tectonically active semiarid Western Transverse Ranges of California. *Geological Society of America Bulletin* 121: 1054–1070.

Warrick, J.A., Rubin, D.M., Ruggiero, P. et al. (2009). Cobble cam: grain-size measurements of sand to boulder from digital photographs and autocorrelation analyses. *Earth Surface Processes and Landforms* 34: 1811–1821.

Warrick, J.A., Bountry, J.A., East, A.E. et al. (2015). Large-scale dam removal on the Elwha River, Washington, USA: source-to-sink sediment budget and synthesis. *Geomorphology* 246: 729–750.

Waters, J.M., Shirley, M., and Closs, G.P. (2002). Hydroelectric development and translocation of *Galaxias brevipinnis*: a cloud at the end of the tunnel? *Canadian Journal of Fisheries and Aquatic Sciences* 59: 49–56.

Waters, J.V., Jones, S.J., and Armstrong, H.A. (2010). Climatic controls on late Pleistocene alluvial fans, Cyprus. *Geomorphology* 115: 228–251.

Wathen, S.J. and Hoey, T.B. (1998). Morphological controls on the downstream passage of a sediment wave in a gravel-bed stream. *Earth Surface Processes and Landforms* 23: 715–730.

Wathen, S.J., Hoey, T.B., and Werritty, A. (1997). Quantitative determination of the activity of within-reach sediment storage in a small gravel-bed river using transit time and response time. *Geomorphology* 20: 113–134.

Webb, R.H. (1996). *Grand Canyon, a Century of Change: Rephotography of the 1889–1890 Stanton Expedition*. Tucson, AZ: University of Arizona Press.

Webb, R.H. and Baker, V.R. (1987). Changes in hydrologic conditions related to large floods on the Escalante River, south-central Utah. In: *Regional Flood Frequency Analysis* (ed. V.P. Singh), 309–323. Dordrecht: D. Reidel.

Webb, R.H., J.L. Betancourt. 1990. Climatic variability and flood frequency of the Santa Cruz River, Pima County, Arizona. US Geological Survey Open-File Report 90-553.

Webb, R.H. and Jarrett, R.D. (2002). One-dimensional estimation techniques for discharges of paleofloods and historical floods. In: *Ancient Floods, Modern Hazards: Principles and Applications of Paleoflood Hydrology* (eds. P.K. House, R.H. Webb, V.R. Baker and D.R. Levish), 111–125. Washington, DC: American Geophysical Union Press.

Webb, B.W. and Walling, D.E. (1982). The magnitude and frequency characteristics of fluvial transport in a Devon drainage basin and some geomorphological implications. *Catena* 9: 9–23.

Webb, R.H., P.T. Pringle, G.R. Rink. 1989. Debris flows from tributaries of the Colorado River, Grand Canyon National Park, Arizona. US Geological Survey Professional Paper 1492.

Webb, R.H., Belnap, J., and Weisheit, J.S. (2004). *Cataract Canyon: A Human and Environmental History of the Rivers in Canyonlands*. Salt Lake City, UT: University of Utah Press.

Webb, J.A., de Little, S.C., Miller, K.A. et al. (2015). A general approach to predicting ecological responses to environmental flows: making best use of the literature, expert knowledge, and monitoring data. *River Research and Applications* 31: 505–514.

Wegener, P., Covino, T., and Wohl, E. (2017). Beaver-mediated lateral hydrologic connectivity, fluvial carbon and nutrient flux, and aquatic ecosystem metabolism. *Water Resources Research* 53: 4606–4623.

Weissmann, G.S. and Fogg, G.E. (1999). Multi-scale alluvial fan heterogeneity modeled with transition probability geostatistics in a sequence stratigraphic framework. *Journal of Hydrology* 226: 48–65.

Wentworth, C.K. (1922). A scale of grade and class terms for clastic sediments. *Journal of Geology* 30: 377–392.

Wernersson, A.S., Carere, M., Miaggi, C. et al. (2015). The European technical report on aquatic effect-based monitoring tools under the Water Framework Directive. *Environmental Sciences Europe* 27: 7.

Westbrook, C.J., Cooper, D.J., and Baker, B.W. (2006). Beaver dams and overbank floods influence groundwater-surface water interactions of a Rocky Mountain riparian area. *Water Resources Research* 42, W06404.

Westbrook, C.J., Cooper, D.J., and Baker, B.W. (2011). Beaver assisted river valley formation. *River Research and Applications* 27: 247–256.

Westerling, A.L., Hidalgo, H.G., Cayan, D.R., and Swetnam, R.W. (2006). Warmer and earlier spring increase in western US forest wildfire activity. *Science* 313: 940–943.

Western, A.W., Blöschl, G., and R.B. Grayson RB. (2001). Toward capturing hydrologically significant connectivity in spatial patterns. *Water Resources Research* 37: 83–97.

Westoby, M.J., Dunning, S.A., Woodward, J. et al. (2015). Sedimentological characterization of Antarctic moraines using UAVs and Structure-from-motion photogrammetry. *Journal of Glaciology* 61: 1088–1102.

Westra, L., Mioller, P., Karr, J.R. et al. (2000). Ecological integrity: integrating environmental conservation and health. In: *Ecological Integrity and the Aims of the Global Integrity Project* (eds. D. Pimentel, L. Westra and R.F. Noss), 19–44. Washington, DC: Island Press.

Wheaton, J.M., Brasington, J., Darby, S.E., and Sear, D.A. (2010). Accounting for uncertainty in DEMs from repeat topographic surveys: improved sediment budgets. *Earth Surface Processes and Landforms* 35: 136–156.

Wheaton, J.M., Brasington, J., Darby, S.E. et al. (2013). Morphodynamic signatures of braiding mechanisms as expressed through change in sediment storage in a gravel-bed river. *Journal of Geophysical Research: Earth Surface* 118: 759–779.

Whipple, K.X. (2001). Fluvial landscape response time: how plausible is steady-state denudation? *American Journal of Science* 301: 313–325.

Whipple, K.X. (2004). Bedrock rivers and the geomorphology of active orogens. *Annual Review of Earth and Planetary Sciences* 32: 151–185.

Whipple, K.X. and Dunne, T. (1992). The influence of debris-flow rheology on fan morphology, Owens Valley, California. *Geological Society of America Bulletin* 104: 887–900.

Whipple, K.X. and Trayler, C.R. (1996). Tectonic control of fan size: the importance of spatially variable subsidence rates. *Basin Research* 8: 351–366.

Whipple, K.X., Parker, G., Paola, C., and Mohrig, D. (1998). Channel dynamics, sediment transport, and the slope of alluvial fans; experimental study. *Journal of Geology* 106: 677–693.

Whipple, K.X., Hancock, G.S., and Anderson, R.S. (2000a). River incision into bedrock: mechanics and relative efficacy of plucking, abrasion, and cavitation. *Geological Society of America Bulletin* 112: 490–503.

Whipple, K.X., Snyder, N.P., and Dollenmayer, K. (2000b). Rates and processes of bedrock incision by the Upper Ukak River since the 1912 Novarupta ash flow in the Valley of Ten Thousand Smokes, Alaska. *Geology* 28: 835–838.

Whipple, K.X., DiBiase, R.A., and Crosby, B.T. (2013). Bedrock rivers. In: *Treatise on Fluvial Geomorphology* (ed. E. Wohl), 550–572. Amsterdam: Elsevier.

Whitaker, A., Alila, Y., Beckers, J., and Toews, D. (2002). Evaluating peak flow sensitivity to clear-cutting in different elevation bands of a snowmelt-dominated mountainous catchment. *Water Resources Research* 38 https://doi.org/10.1029/WR000514.

Whitbread, K., Jansen, J., Bishop, P., and Attal, M. (2015). Substrate, sediment, and slope controls on bedrock channel geometry in postglacial streams. *Journal of Geophysical Research: Earth Surface* 120: 779–798.

White, W.B. (1999). Karst hydrology: recent developments and open questions. In: *Hydrogeology and Engineering Geology of Sinkholes and Karst* (eds. B.F. Beck, A.J. Pettit and J.G. Herring), 3–20. Rotterdam: A.A. Balkema.

White, W.B. and White, E.L. (2018). Karst geomorphology. In: *Caves and Karst of the Greenbrier Valley in West Virginia* (ed. W.B. White), 45–62. Cham: Springer.

White, W.R., Bettess, R., and Paris, E. (1982). Analytical approach to river regime. *Journal of Hydraulics Division, ASCE* 108: 1179–1193.

White, A.F., Blum, A.E., Schulz, M.S. et al. (1998). Chemical weathering in a tropical watershed, Luquillo Mountains, Puerto Rico: I. Long-term versus short-term weathering fluxes. *Geochimica et Cosmochimica Acta* 62: 209–226.

White, J.Q., Pasternack, G.B., and Moir, H.J. (2010). Valley width variation influences riffle-pool location and persistence on a rapidly incising gravel-bed river. *Geomorphology* 121: 206–221.

Whiting, P.J., Dietrich, W.E., Leopold, L.B. et al. (1988). Bedload sheets in heterogeneous sediment. *Geology* 16: 105–108.

Whittaker, A.C., Cowie, P.A., Attal, M. et al. (2007a). Bedrock channel adjustment to tectonic forcing: implications for predicting river incision rates. *Geology* 35: 103–106.

Whittaker, A.C., Cowie, P.A., Attal, M. et al. (2007b). Contrasting transient and steady-state rivers crossing active normal faults: new field observations from the Central Apennines, Italy. *Basin Research* 19: 529–556.

Whittaker, A.C., Attal, M., Cowie, P.A. et al. (2008). Decoding temporal and spatial patterns of fault uplift using transient river long profiles. *Geomorphology* 100: 506–526.

Wiberg, P.L. and Smith, J.D. (1987). Calculations of the critical shear-stress for motion of uniform and heterogeneous sediments. *Water Resources Research* 23 (8): 1471–1480.

Wiberg, P.L. and Smith, J.D. (1989). Model for calculating bed load transport of sediment. *Journal of Hydraulic Engineering* 115: 101–123.

Wiberg, P.L. and Smith, J.D. (1991). Velocity distribution and bed roughness in high-gradient streams. *Water Resources Research* 27: 825–838.

Wicherski, W., Dethier, D.P., and Ouimet, W.B. (2017). Erosion and channel changes due to extreme flooding in the Fourmile Creek catchment, Colorado. *Geomorphology* 294: 87–98.

Wichmann, V., Heckmann, T., Haas, F., and Becht, M. (2009). A new modelling approach to delineate the spatial extent of alpine sediment cascades. *Geomorphology* 111: 70–78.

Wicks, J.M. and Bathurst, J.C. (1996). SHESED: a physically based, distributed erosion and sediment yield component for the SHE hydrological modeling system. *Journal of Hydrology* 175: 213–238.

Wiele, S.M., Graf, J.B., and Smith, J.D. (1996). Sand deposition in the Colorado River in the Grand Canyon from flooding of the Little Colorado River. *Water Resources Research* 32: 3579–3596.

Wiele, S.M., Wilcock, P.R., and Grams, P.E. (2007). Reach-averaged sediment routing model of a canyon river. *Water Resources Research* 43, W02425.

Wilby, R.L. (1995). Greenhouse hydrology. *Progress in Physical Geography* 19: 351–369.

Wilcock, D.N. (1971). Investigation into the relations between bedload transport and channel shape. *Geological Society of America Bulletin* 82: 2159–2176.

Wilcock, P.R. (1992). Flow competence: a criticism of a classic concept. *Earth Surface Processes and Landforms* 17: 289–298.

Wilcock, P.R. (1997). Entrainment, displacement, and transport of tracer gravels. *Water Resources Research* 22: 1125–1138.

Wilcock, P.R. (2012). Stream restoration in gravel-bed rivers. In: *Gravel Bed Rivers: Processes, Tools, Environments* (eds. M. Church, B.M. Piron and A.G. Roy), 137–149. Chichester: Wiley-Blackwell.

Wilcock, P.R. and Crowe, J.C. (2003). Surface-based transport model for mixed-size sediment. *Journal of Hydraulic Engineering* 129: 120–128.

Wilcock, P.R. and Kenworthy, S.T. (2002). A two-fraction model for the transport of sand/gravel mixtures. *Water Resources Research* 38: 1194.

Wilcock, P.R. and McArdell, B.W. (1997). Partial transport of a sand/gravel sediment. *Water Resources Research* 33: 235–245.

Wilcox, A.C. and Wohl, E.E. (2006). Flow resistance dynamics in step-pool stream channels: 1. Large woody debris and controls on total resistance. *Water Resources Research* 42, W05418.

Wilcox, A.C. and Wohl, E.E. (2007). Field measurements of three-dimensional hydraulics in a step-pool channel. *Geomorphology* 83: 215–231.

Wilcox, A.C., Nelson, J.M., and Wohl, E.E. (2006). Flow resistance dynamics in step-pool channels: 2. Partitioning between grain, spill, and woody debris resistance. *Water Resources Research* 42, W05419.

Wilcox, A.C., Wohl, E.E., Comiti, F., and Mao, L. (2011). Hydraulics, morphology, and energy dissipation in an alpine step-pool channel. *Water Resources Research* 47, W07514.

Wilford, D.J., Sakals, M.E., Innes, J.L., and Sidle, R.C. (2005). Fans with forests: contemporary hydrogeomorphic processes on fans with forest in west central British Columbia, Canada. In: *Alluvial Fans: Geomorphology, Sedimentology, Dynamics* (eds. A.M. Harvey, A.E. Mather and M. Stokes), 25–40. Geological Society Special Publication 251.

Wilkinson, S.N., Prosser, I.P., and Hughes, A.O. (2006). Predicting the distribution of bed material accumulation using river network sediment budgets. *Water Resources Research* 42, W10419.

Willenbring, J.K., Codilean, A.T., and McElroy, B. (2013). Earth is (mostly) flat: apportionment of the flux of continental sediment over millennial time scales. *Geology* 41: 343–346.

Willgoose, G. and Hancock, G. (1998). Revisiting the hypsometric curve as an indicator of form and process in transport-limited catchment. *Earth Surface Processes and Landforms* 23: 611–623.

Willgoose, G., Bras, R.L., and Rodriguez-Iturbe, I. (1990). A model of river basin evolution. *Eos, Transactions of the American Geophysical Union* 71: 47.

Willgoose, G., Bras, R.L., and Rodriguez-Iturbe, I. (1991). Results from a new model of river basin evolution. *Earth Surface Processes and Landforms* 16: 237–254.

Williams, G.P. (1978a). Bank-full discharge of rivers. *Water Resources Research* 14: 1141–1154.

Williams, G.P. 1978b. The case of the shrinking channels – the North Platte and the Platte Rivers in Nebraska. US Geological Survey Circular 781.

Williams, G.P. (1989). Sediment concentration versus water discharge during single hydrologic events in rivers. *Journal of Hydrology* 111: 89–106.

Williams, M.W. and Melack, J.M. (1991). Solute chemistry of snowmelt and runoff in an alpine basin, Sierra Nevada. *Water Resources Research* 27: 1575–1588.

Williams, G.P., D.L. Rosgen. 1989. Measured total sediment loads (suspended loads and bedloads) for 93 United States streams. US Geological Survey Open-File Report 89-67.

Williams, G.P., M.G. Wolman. 1984. Effects of dams and reservoirs on surface-water hydrology; changes in rivers downstream from dams. US Geological Survey Professional Paper 1286.

Williams, M.W., Seibold, C., and Chowanski, K. (2009). Storage and release of solutes from a subalpine seasonal snowpack: soil and stream water response, Niwot Ridge, Colorado. *Biogeochemistry* 95: 77–94.

Williams, R.D., Rennie, C.D., Brasington, J. et al. (2015). Linking the spatial distribution of bed load transport to morphological change during high-flow events in a shallow braided river. *Journal of Geophysical Research: Earth Surface* 120: 604–622.

Williams, R.D., Measures, R., Hicks, D.M., and Brasington, J. (2016). Assessment of a numerical model to reproduce event-scale erosion and deposition distributions in a braided river. *Water Resources Research* 52: 6621–6642.

Willis, I.C., Arnold, N.S., and Brock, B.W. (2002). Effect of snowpack removal on energy balance, melt and runoff in a small supraglacial catchment. *Hydrological Processes* 16: 2721–2749.

Winkley, B.R. (1994). Response of the lower Mississippi River to flood control and navigation improvements. In: *The Variability of Large Alluvial Rivers* (eds. S.A. Schumm and B.R. Winkley), 45–74. New York: ASCE Press.

Winsemann, J., Brandes, C., and Polom, U. (2011). Response of a proglacial delta to rapid high-amplitude lake-level change: an integration of outcrop data and high-resolution shear wave seismics. *Basin Research* 23: 22–52.

Wischmeier, W.H., D.D. Smith. 1978. Predicting rainfall-erosion losses – a guide to conservation planning. Agricultural Handbook 537.

Wishart, D., Warburton, J., and Bracken, L. (2008). Gravel extraction and planform change in a wandering gravel-bed river: The River Wear, northern England. *Geomorphology* 94: 131–152.

Wittmann, H. and von Blanckenburg, F. (2009). Cosmogenic nuclide budgeting of floodplain sediment transfer. *Geomorphology* 109: 246–256.

Wittmann, H., von Blanckenburg, F., Maurice, L. et al. (2011). Recycling of Amazon floodplain sediment quantified by cosmogenic ^{26}Al and ^{10}Be. *Geology* 39: 467–470.

Wobus, C.W., Crosby, B.T., and Whipple, K.X. (2006). Hanging valleys in fluvial systems: controls on occurrence and implications for landscape evolution. *Journal of Geophysical Research: Earth Surface* 111, F02017.

Wohl, E.E. (1992). Bedrock benches and boulder bars: floods in the Burdekin Gorge of Australia. *Geological Society of America Bulletin* 104: 770–778.

Wohl, E.E. (1993). Bedrock channel incision along Piccaninny Creek, Australia. *Journal of Geology* 101: 749–761.

Wohl, E.E. (1998). Bedrock channel morphology in relation to erosional processes. In: *Rivers Over Rock: Fluvial Processes in Bedrock Channels* (eds. K.J. Tinkler and E.E. Wohl), 133–151. Washington, DC: American Geophysical Union Press.

Wohl, E.E. (2000a). Anthropogenic impacts on flood hazards. In: *Inland Flood Hazards: Human, Riparian, and Aquatic Communities* (ed. E. Wohl), 104–141. Cambridge University Press.

Wohl, E. (2000b). Substrate influence on step-pool sequences in the Christopher Creek drainage, Arizona. *Journal of Geology* 108: 121–129.

Wohl, E. (2001). *Virtual Rivers: Lessons from the Colorado Front Range*. New Haven, CT: Yale University Press.

Wohl, E. (2004a). *Disconnected Rivers: Linking Rivers to Landscapes*. New Haven, CT: Yale University Press.

Wohl, E. (2004b). Limits of downstream hydraulic geometry. *Geology* 32: 897–900.

Wohl, E. (2006). Human impacts to mountain streams. *Geomorphology* 76: 217–248.

Wohl, E. (2008). The effect of bedrock jointing on the formation of straths in the Cache la Poudre River drainage, Colorado Front Range. *Journal of Geophysical Research* 113, F01007.

Wohl, E. (2010a). A brief review of the process domain concept and its application to quantifying sediment dynamics in bedrock canyons. *Terra Nova* 22: 411–416.
Wohl, E. (2010b). *Mountain Rivers Revisited*. Washington, DC: American Geophysical Union Press.
Wohl, E. (2011a). *A World of Rivers: Environmental Change on Ten of the World's Great Rivers*. Chicago, IL: University of Chicago Press.
Wohl, E. (2011b). Threshold-induced complex behavior of wood in mountain streams. *Geology* 39: 587–590.
Wohl, E. (2011c). What should these rivers look like? Historical range of variability and human impacts in the Colorado Front Range, USA. *Earth Surface Processes and Landforms* 36: 1378–1390.
Wohl, E. (2013a). Migration of channel heads following wildfire in the Colorado Front Range, USA. *Earth Surface Processes and Landforms* 38: 1049–1053.
Wohl, E. (2013b). Floodplains and wood. *Earth-Science Reviews* 123: 194–212.
Wohl, E. (2014). A legacy of absence: wood removal in US rivers. *Progress in Physical Geography* 38: 637–663.
Wohl, E. (2015a). Legacy effects on sediment in river corridors. *Earth-Science Reviews* 147: 30–53.
Wohl, E. (2015b). Particle dynamics: the continuum of bedrock to alluvial river segments. *Geomorphology* 241: 192–208.
Wohl, E. (2016). Spatial heterogeneity as a component of river geomorphic complexity. *Progress in Physical Geography* 40: 598–615.
Wohl, E. (2017a). Bridging the gaps: an overview of wood across time and space in diverse rivers. *Geomorphology* 279: 3–26.
Wohl, E. (2017b). Connectivity in rivers. *Progress in Physical Geography* 41: 345–362.
Wohl, E. (2018a). Geomorphic context in rivers. *Progress in Physical Geography* 42: 841–857.
Wohl, E. (2018b). The challenges of channel heads. *Earth-Science Reviews* 186: 649–664.
Wohl, E. and Beckman, N.D. (2014). Leaky rivers: implications of the loss of longitudinal fluvial disconnectivity in headwater streams. *Geomorphology* 205: 27–35.
Wohl, E. and Cadol, D. (2011). Neighborhood matters: patterns and controls on wood distribution in old-growth forest streams of the Colorado Front Range, USA. *Geomorphology* 125: 132–146.
Wohl, E.E. and Cenderelli, D.A. (1998). Flooding in the Himalaya Mountains. In: *Flood Studies in India*, vol. 41 (ed. V.S. Kale), 77–99. Memoir Geological Society of India.
Wohl, E.E. and Cenderelli, D.A. (2000). Sediment deposition and transport patterns following a reservoir sediment release. *Water Resources Research* 36: 319–333.
Wohl, E. and David, G.C.L. (2008). Consistency of scaling relations among bedrock and alluvial channels. *Journal of Geophysical Research* 113, F04013.
Wohl, E. and Dust, D. (2012). Geomorphic response of a headwater channel to augmented flow. *Geomorphology* 138: 329–338.
Wohl, E.E. and Enzel, Y. (1995). Data for palaeohydrology. In: *Global Continental Palaeohydrology* (eds. K.J. Gregory, L. Starkel and V.R. Baker), 23–59. Chichester: Wiley.
Wohl, E. and Goode, J.R. (2008). Wood dynamics in headwater streams of the Colorado Rocky Mountains. *Water Resources Research* 44, W09429.
Wohl, E.E. and Ikeda, H. (1997). Experimental simulation of channel incision into a cohesive substrate at varying gradient. *Geology* 25: 295–298.
Wohl, E.E. and Ikeda, H. (1998). Patterns of bedrock channel incision on the Boso Peninsula, Japan. *Journal of Geology* 106: 331–345.

Wohl, E. and Jaeger, K. (2009). A conceptual model for the longitudinal distribution of wood in mountain streams. *Earth Surface Processes and Landforms* 34: 329–344.

Wohl, E.E. and Legleiter, C.J. (2003). Controls on pool characteristics along a resistant-boundary channel. *Journal of Geology* 111: 103–114.

Wohl, E.E. and Merritt, D.M. (2001). Bedrock channel morphology. *Geological Society of America Bulletin* 113: 1205–1212.

Wohl, E. and Merritt, D.M. (2008). Reach-scale channel geometry of mountain streams. *Geomorphology* 93: 168–185.

Wohl, E. and Ogden, F.L. (2013). Wood and carbon export during an extreme tropical storm, Upper Rio Chagres, Panama. *Earth Surface Processes and Landforms* 38: 1407–1416.

Wohl, E.E. and Pearthree, P.A. (1991). Debris flows as geomorphic agents in the Huachuca Mountains of southeastern Arizona. *Geomorphology* 4: 273–292.

Wohl, E. and Scott, D. (2017a). Transience of channel head locations following disturbance. *Earth Surface Processes and Landforms* 42: 1132–1139.

Wohl, E. and Scott, D. (2017b). Wood and sediment storage and dynamics in river corridors. *Earth Surface Processes and Landforms* 42: 5–23.

Wohl, E.E. and Springer, G. (2005). Bedrock channel incision along the Upper Rio Chagres basin, Panama. In: *The Rio Chagres, Panama: A Multidisciplinary Profile of a Tropical Watershed* (ed. R.S. Harmon), 189–209. Dordrecht: Springer.

Wohl, E.E. and Wilcox, A. (2005). Channel geometry of mountain streams in New Zealand. *Journal of Hydrology* 300: 252–266.

Wohl, E.E., Greenbaum, N., Schick, A.P., and Baker, V.R. (1994). Controls on bedrock channel incision along Nahal Paran, Israel. *Earth Surface Processes and Landforms* 19: 1–13.

Wohl, E.E., Anthony, D.J., Madsen, S.W., and Thompson, D.M. (1996). A comparison of surface sampling methods for coarse fluvial sediments. *Water Resources Research* 32: 3219–3226.

Wohl, E.E., Thompson, D.M., and Miller, A.J. (1999). Canyons with undulating walls. *Geological Society of America Bulletin* 111: 949–959.

Wohl, E.E., Cenderelli, D.A., and Mejia-Navarro, M. (2001). Channel change from extreme floods in canyon rivers. In: *Applying Geomorphology to Environmental Management* (ed. D.J. Anthony), 149–174. Littleton, CO: Water Resources Publications.

Wohl, E., Angermeier, P.L., Bledsoe, B. et al. (2005). River restoration. *Water Resources Research* 41, W10301.

Wohl, E., Cooper, D., Poff, L. et al. (2007). Assessment of stream ecosystem function and sensitivity in the Bighorn National Forest, Wyoming. *Environmental Management* 40: 284–302.

Wohl, E., Palmer, M., and Kondolf, G.M. (2008). River management in the United States. In: *River Futures: An Integrative Scientific Approach to River Repair* (eds. G.J. Brierley and K.A. Fryirs), 174–200. Washington, DC: Island Press.

Wohl, E., Ogden, F., and Goode, J. (2009). Episodic wood loading in a mountainous neotropical watershed. *Geomorphology* 111: 149–159.

Wohl, E., Cenderelli, D.A., Dwire, K.A. et al. (2010). Large in-stream wood studies: a call for common metrics. *Earth Surface Processes and Landforms* 35: 618–625.

Wohl, E., Barros, A., Brunsell, N. et al. (2012a). The hydrology of the humid tropics. *Nature Climate Change* 2: 655–662.

Wohl, E., Bolton, S., Cadol, D. et al. (2012b). A two end-member model of wood dynamics in headwater neotropical rivers. *Journal of Hydrology* 462–463: 67–76.

Wohl, E., Dwire, K., Sutfin, N. et al. (2012c). Mechanisms of carbon storage in mountainous headwater rivers. *Nature Communications* 3: 1263.

Wohl, E., Bledsoe, B.P., Fausch, K.D. et al. (2015a). Management of large wood in streams: an overview and proposed framework for hazard evaluation. *Journal of the American Water Resources Association* 52: 315–335.

Wohl, E., Bledsoe, B.P., Jacobson, R.B. et al. (2015b). The natural sediment regime in rivers: broadening the foundation for ecosystem management. *BioScience* 65: 358–371.

Wohl, E., Lane, S.N., and Wilcox, A.C. (2015c). The science and practice of river restoration. *Water Resources Research* 51: 5974–5997.

Wohl, E., Hall, R.O., Lininger, K.B. et al. (2017a). Carbon dynamics of river corridors and the effects of human alterations. *Ecological Monographs* 87: 379–409.

Wohl, E., Lininger, K.B., Fox, M. et al. (2017b). Instream large wood loads across bioclimatic regions. *Forest Ecology and Management* 404: 370–380.

Wohl, E., Lininger, K.B., and Baron, J. (2017c). Land before water: the relative temporal sequence of human alteration of freshwater ecosystems in the conterminous United States. *Anthropocene* 18: 27–46.

Wohl, E., M.K. Mersel, A.O. Allen, K.M. Fritz, S.L. Kichefski, et al. 2017d. Synthesizing the scientific foundation for ordinary high water mark delineation in fluvial systems. US Army Corps of Engineers Cold Regions Research and Engineering Laboratory ERD/CRREL SR-16-5.

Wohl, E., Cadol, D., Pfeiffer, A. et al. (2018a). Distribution of large wood within river corridors in relation to flow regime in the semiarid western US. *Water Resources Research* 54: 1890–1904.

Wohl, E., Lininger, K.B., and Scott, D.N. (2018b). River beads as a conceptual framework for building carbon storage and resilience to extreme climate events into river management. *Biogeochemistry* 141: 365–383.

Wohl, E., Scott, D.N., and Lininger, K.B. (2018c). Spatial distribution of channel and floodplain large wood in forested river corridors of the Northern Rockies. *Water Resources Research* https://doi.org/10.1029/2018WR022750.

Wohl, E., Brierley, G., Cadol, D. et al. (2019a). Connectivity as an emergent property of geomorphic systems. *Earth Surface Processes and Landforms* 44: 4–26.

Wohl, E., Kramer, N., Ruiz-Villanueva, V. et al. (2019b). The natural wood regime in rivers. *BioScience* 69: 259–273.

Wolcott, J. and Church, M. (1991). Strategies for sampling spatially heterogeneous phenomena: the example of river gravels. *Journal of Sedimentary Petrology* 61: 534–543.

Wolf, E.C., Cooper, D.J., and Hobbs, N.T. (2007). Hydrologic regime and herbivory stabilize an alternative state in Yellowstone National Park. *Ecological Applications* 17: 1572–1587.

Wollheim, W.M., Harms, T.K., Peterson, B.J. et al. (2014). Nitrate uptake dynamics of surface transient storage in stream channels and fluvial wetlands. *Biogeochemistry* 120: 239–257.

Wolman, M.G. (1954). A method of sampling coarse river-bed material. *Transactions of the American Geophysical Union* 35: 951–956.

Wolman, M.G. (1959). Factors influencing erosion of a cohesive river bank. *American Journal of Science* 257: 204–216.

Wolman, M.G. (1967a). A cycle of sedimentation and erosion in urban river channels. *Geografiska Annaler* 49A: 385–395.

Wolman, M.G. (1967b). Two problems involving river channels and their background observations. *Northwestern University Studies in Geography* 14: 67–107.

Wolman, M.G. and Gerson, R. (1978). Relative scales of time and effectiveness of climate in watershed geomorphology. *Earth Surface Processes* 3: 189–208.
Wolman, M.G., L.B. Leopold. 1957. River flood plains: some observations on their formation. US Geological Survey Professional Paper 282-C.
Wolman, M.G. and Miller, J.P. (1960). Magnitude and frequency of forces in geomorphic processes. *Journal of Geology* 68: 54–74.
Woltemade, C.J. and Potter, K.W. (1994). A watershed modeling analysis of fluvial geomorphologic influences on flood peak attenuation. *Water Resources Research* 30: 1933–1942.
Womack, W.R. and Schumm, S.A. (1977). Terraces of Douglas Creek, northwestern Colorado: an example of episodic erosion. *Geology* 5: 72–76.
Wondzell, S.M. and Gooseff, M.N. (2013). Geomorphic controls on hyporheic exchange across scales – watersheds to particles. In: *Treatise on Fluvial Geomorphology* (ed. E. Wohl), 203–218. Amsterdam: Elsevier.
Wondzell, S.M. and Swanson, F.J. (1996). Seasonal and storm dynamics of the hyporheic zone of a fourth-order mountain stream. I. Hydrologic processes. *Journal of the North American Benthological Society* 15: 3–19.
Wondzell, S.M. and Swanson, F.J. (1999). Floods, channel change, and the hyporheic zone. *Water Resources Research* 35: 555–567.
Wong, M. and Parker, G. (2006). Reanalysis and correction of bed-load relation of Meyer-Peter and Müller using their own database. *Journal of Hydraulic Engineering* 132: 1159–1168.
Wong, M., Parker, G., DeVries, P. et al. (2007). Experiments on dispersion of tracer stones under lower-regime plane-bed equilibrium bed load transport. *Water Resources Research* 43, W03440.
Woo, M.-K. (2010). Cold ocean seas and northern hydrology: an exploratory overview. *Hydrology Research* 41: 439–453.
Woo, M.-K. and Thorne, R. (2003). Streamflow in the Mackenzie Basin, Canada. *Arctic* 56: 328–340.
Woods, S.W., MacDonald, L.H., and Westbrook, C.J. (2006). Hydrologic interactions between an alluvial fan and a slope wetland in the central Rocky Mountains, USA. *Wetlands* 26: 230–243.
Woolderink, H.A.G., Kasse, C., Cohen, K.M., and Hoek, W.Z. (2018). Spatial and temporal variations in river terrace formation, preservation, and morphology in the Lower Meuse Valley, The Netherlands. *Quaternary Research* 91: 548–569.
Worrall, F., Burt, T.P., and Howden, N.J.K. (2014a). The fluvial flux of particulate organic matter from the UK: quantifying in-stream losses and carbon sinks. *Journal of Hydrology* 519: 611–625.
Worrall, F., Burt, T.P., Howden, N.J.K., and Hancock, G.R. (2014b). Variation in suspended sediment yield across the UK – a failure of the concept and interpretation of the sediment delivery ratio. *Journal of Hydrology* 519: 1985–1996.
Wright, S.A. and Kaplinski, M. (2011). Flow structures and sandbar dynamics in a canyon river during a controlled flood, Colorado River, Arizona. *Journal of Geophysical Research: Earth Surface* 116, F01019.
Wright, S.A., Schmidt, J.C., Melis, T.S. et al. (2008). Is there enough sand? Evaluating the fate of Grand Canyon sandbars. *GSA Today* 18 (8): 4–10.
Wright, N., Hayashi, M., and Quinton, W.L. (2009). Spatial and temporal variations in active layer thawing and their implications on runoff generation in peat-covered permafrost terrain. *Water Resources Research* 45 https://doi.org/10.1029/2008WR006880.
Wu, F.C. and Jiang, M.R. (2007). Numerical investigation of the role of turbulent bursting in sediment entrainment. *Journal of Hydraulic Engineering* 133: 329–334.

Wu, F.C., Chang, C.F., and Shiau, J.T. (2015). Assessment of flow regime alterations over a spectrum of temporal scales using wavelet-based approaches. *Water Resources Research* 51: 3317–3338.

Wyatt, A.M. and Franks, S.W. (2006). The multi-model approach to rainfall-runoff modeling. In: *Predictions in Ungauged Basins: Promise and Progress* (eds. M. Sivapalan, T. Wagner, S. Uhlenbrook, et al.), 134–144. Wallingford: IAHS Publications no. 303.

Wynn, T.M., Henderson, M.B., and Vaughan, D.H. (2008). Changes in streambank erodibility and critical shear stress due to subaerial processes along a headwater stream, southwestern Virginia, USA. *Geomorphology* 97: 260–273.

Wyżga, B. (1991). Present-day downcutting of the Raba River channel (Western Carpathians, Poland) and its environmental effects. *Catena* 18: 551–556.

Wyżga, B. (2001). Impact of the channelization-induced incision of the Skawa and Wisłoka Rivers, southern Poland, on the conditions of overbank deposition. *Regulated Rivers: Research and Management* 17: 85–100.

Xu, J. (1996). Underlying gravel layers in a large sand bed river and their influence on downstream-dam channel adjustment. *Geomorphology* 17: 351–359.

Xu, J. (2004). A study of anthropogenic seasonal rivers in China. *Catena* 55: 17–32.

Yager, E.M. and Schmeeckle, M.W. (2013). The influence of vegetation on turbulence and bed load transport. *Journal of Geophysical Research: Earth Surface* 118: 1585–1601.

Yager, E.M. and Schott, H.E. (2013). The initiation of motion and formation of armor layers. In: *Treatise on Fluvial Geomorphology* (ed. E. Wohl), 87–102. Amsterdam: Elsevier.

Yager, E.M., Kirchner, J.W., and Dietrich, W.E. (2007). Calculating bedload transport in steep, boulder-bed channels. *Water Resources Research* 43, W07418.

Yager, E.M., Dietrich, W.E., Kirchner, J.W., and McArdell, B.W. (2012). Prediction of sediment transport in step-pool channels. *Water Resources Research* 48, W01541.

Yager, E.M., Kenworthy, M., and Monsalve, A. (2015). Taking the river inside: fundamental advances from laboratory experiments in measuring and understanding bedload transport processes. *Geomorphology* 244: 21–32.

Yang, C.T. (1976). Minimum unit stream power and fluvial hydraulics. *Journal of Hydraulics Division, ASCE* 102: 919–934.

Yang, S.L., Milliman, J.D., Li, P., and Xu, K. (2011). 50,000 dams later: erosion of the Yangtze River and its delta. *Global and Planetary Change* 75: 14–20.

Yang, R., Fan, A., Van Loon, A.J. et al. (2018). The influence of hyperpycnal flows on the salinity of deep-marine environments, and implications for the interpretation of marine facies. *Marine and Petroleum Geology* 98: 1–11.

Yanites, B.J. (2018). The dynamics of channel slope, width, and sediment in actively eroding bedrock systems. *Journal of Geophysical Research: Earth Surface* 123: 1504–1527.

Yanites, B.J., Tucker, G.E., Mueller, K.J. et al. (2010). Incision and channel morphology across active structures along the Peikang River, central Taiwan: implications for the importance of channel widths. *Geological Society of America Bulletin* 122: 1192–1208.

Yanites, B.J., Becker, J.K., Madritsch, H. et al. (2017). Lithologic effects on landscape response to basin level changes: a modeling study in the context of the eastern Jura Mountains, Switzerland. *Journal of Geophysical Research: Earth Surface* 122: 2196–2222.

Yanosky, T.M. and Jarrett, R.D. (2002). Dendrochronologic evidence for the frequency and magnitude of paleofloods. In: *Ancient Floods, Modern Hazards: Principles and Applications of Paleoflood Hydrology* (eds. P.K. House, R.H. Webb, V.R. Baker and D.R. Levish), 77–89. Washington, DC: American Geophysical Union Press.

Yao, S. (1943). The geographic distribution of floods and droughts in Chinese history 206 BC–1911 AD. *Far Eastern Quarterly* 2: 357–378.

Yeats, R.S. and Thakur, V.C. (2008). Active faulting south of the Himalayan front; establishing a new plate boundary. *Tectonophysics* 453: 63–73.

Yetemen, O., Istanbulluoglu, E., and Vivoni, E.R. (2010). The implications of geology, soils, and vegetation on landscape morphology: influences from semi-arid basins with complex vegetation patterns in central New Mexico, USA. *Geomorphology* 116: 246–263.

Yevjevich, V. (2001). Water diversions and interbasin transfers. *Water International* 26: 342–348.

Yochum, S.E., Bledsoe, B.P., David, G.C.L., and Wohl, E. (2012). Velocity prediction in high-gradient channels. *Journal of Hydrology* 424–425: 84–98.

Yochum, S.E., Bledsoe, B.P., Wohl, E., and David, G.C.L. (2014a). Spatial characterization of roughness elements in high-gradient channels of the Fraser Experimental Forest, Colorado, USA. *Water Resources Research* 50: 6015–6029.

Yochum, S.E., F. Comiti, E. Wohl, G.C.L. David, L. Mao. 2014b. Photographic guidance for selecting flow resistance coefficients in high-gradient channels. USDA Forest Service General Technical Report RMRS-GTR-323.

Young, M.K., Haire, D., and Bozek, M.A. (1994). The effect and extent of railroad tie drives in streams of southeastern Wyoming. *Western Journal of Applied Forestry* 9: 125–130.

Yue, T.X., Xu, B., and Liu, J.Y. (2004). Apatchconnectivity indexandits changein relation to new wetland at the Yellow River Delta. *International Journal of Remote Sensing* 25: 4617–4628.

Zalamea, M., Gonzalez, G., Ping, C.L., and Michaelson, G. (2007). Soil organic matter dynamics under decaying wood in a subtropical wet forest: effect of tree species and decay stage. *Plant and Soil* 296: 173–185.

Zalewski, M. (2000). Ecohydrology – the scientific background to use ecosystem properties as management tools toward sustainability of water resources. *Ecological Engineering* 16: 1–18.

Zehe, E., Lee, H., and Sivapalan, M. (2005). Derivation of closure relations and commensurate state variables for meso-scale models using the REW approach. In: *Predictions in Ungauged Basins: International Perspectives on the State of the Art and Pathways Forward* (eds. S. Franks, M. Sivapalan, K. Takeuchi and Y. Tachikawa), 134–158. Wallingford: IAHS Publications no. 301.

Zeitler, P.K., Meltzer, A.S., Koons, P.O. et al. (2001). Erosion, Himalayan geodynamics, and the geomorphology of metamorphism. *GSA Today* 11 (1): 4–8.

Zen, E. and Prestegaard, K.L. (1994). Possible hydraulic significance of two kinds of potholes: examples from the paleo-Potomac River. *Geology* 22: 47–50.

Zen, S., Gurnell, A.M., Zolezzi, G., and Surian, N. (2017). Exploring the role of trees in the evolution of meander bends: the Tagliamento River, Italy. *Water Resources Research* 53: 5943–5962.

Zernitz, E.R. (1932). Drainage patterns and their significance. *Journal of Geology* 40: 498–521.

Zha, X., Hang, C., Pang, J. et al. (2015). Reconstructing the palaeoflood events from slackwater deposits in the upper reaches of the Hanjiang River, China. *Quaternary International* 380-381: 358–367.

Zhan, A., Zhang, L., Xia, Z. et al. (2015). Water diversions facilitate spread of non-native species. *Biological Invasions* 11: 3073–3080.

Zhang, T., Barry, R.G., Knowles, K. et al. (2003). Distribution of seasonally and perennially frozen ground in the Northern Hemisphere. In: *Permafrost: Proceedings of the 8th International Conference on*

Permafrost (eds. M. Phillips, S.M. Springman and L.U. Arenson), 1289–1294. Brookfield, VT: A.A. Balkema Publishers.

Zhang, W., Ruan, X., Zheng, J. et al. (2010). Long term change in tidal dynamics and its cause in the Pearl River delta, China. *Geomorphology* 120: 209–223.

Zhang, H., Kirby, E., Pitlick, J. et al. (2017). Characterizing the transient geomorphic response to base-level fall in the northeastern Tibetan Plateau. *Journal of Geophysical Research Earth Surface* 122: 546–572.

Zheng, S., Han, S., Tan, G. et al. (2018). Morphological adjustment of the Qingshuigou channel on the Yellow River Delta and factors controlling its avulsion. *Catena* 166: 44–55.

Ziegler, A.D., Sutherland, R.A., and Giambelluca, T.W. (2000). Runoff generation and sediment production on unpaved roads, footpaths and agricultural land surfaces in northern Thailand. *Earth Surface Processes and Landforms* 25: 519–534.

Zierl, B. and Bugmann, H. (2005). Global change impacts on hydrological processes in Alpine catchments. *Water Resources Research* 41, W02028.

Zimmerman, R.C., Goodlett, J.C., and Comer, G.H. (1967). The influence of vegetation on channel form of small streams. *Symposium on River Morphology*, International Association of Scientific Hydrology 75: 255–275.

Zimmermann, A.E. (2013). Step-pool channel features. In: *Treatise on Fluvial Geomorphology* (ed. E. Wohl), 347–363. Amsterdam: Elsevier.

Zimmermann, A. and Church, M. (2001). Channel morphology, gradient profiles and bed stresses during flood in a step-pool channel. *Geomorphology* 40: 311–327.

Zimmermann, A., Church, M., and Hassan, M.A. (2010). Step-pool stability: testing the jammed state hypothesis. *Journal of Geophysical Research* 115, F02008.

Zolezzi, G., Bellin, A., Bruno, M.C. et al. (2009). Assessing hydrological alterations at multiple temporal scales: Adige River, Italy. *Water Resources Research* 45, W12421.

Zorn, M., Natek, K., and Komac, B. (2006). Mass movements and flash-floods in the Slovene Alps and surrounding mountains. *Studia Geomorphologica Carpatho-Balcanica* 40: 127–145.

Zunka, J.P.P., Tullos, D.D., and Lancaster, S.T. (2015). Effects of sediment pulses on bed relief in bar-pool channels. *Earth Surface Processes and Landforms* 40: 1017–1028.

Index

h

i

t

The manufacturer's authorised representative in the EU for product safety is Oxford University Press España S.A. of El Parque Empresarial San Fernando de Henares, Avenida de Castilla, 2 – 28830 Madrid (www.oup.es/en or product.safety@oup.com). OUP España S.A. also acts as importer into Spain of products made by the manufacturer.

Printed in the USA/Agawam, MA
January 14, 2025

881027.005